Shock Waves @ Marseille I

Springer
Berlin
Heidelberg
New York
Barcelona
Budapest
Hong Kong
London
Milan
Paris
Tokyo

R. Brun   L. Z. Dumitrescu (Eds.)

Shock Waves @ Marseille I

Hypersonics, Shock Tube & Shock Tunnel Flow

Proceedings
of the 19th International Symposium
on Shock Waves
Held at Marseille, France, 26-30 July 1993

With 390 Figures

Springer

Professor Dr. Raymond Brun

Professor Dr. Lucien Z. Dumitrescu

Université de Provence, Centre Saint-Jérôme
IUSTI-MHEQ, Case 321, F-13397 Marseille Cedex 20, France

ISBN 3-540-57710-6 Springer-Verlag Berlin Heidelberg New York
ISBN 0-387-57710-6 Springer-Verlag New York Berlin Heidelberg

Set (Volumes I-IV):
ISBN 3-540-57713-0 Springer-Verlag Berlin Heidelberg New York

CIP data applied for

Typesetting: Camera ready by editors
SPIN: 10132524   55/3144 - 5 4 3 2 1 0 - Printed on acid-free paper

Preface

From the Prefaces written by the chairmen of the previous International Symposia on Shock Waves, it appears that records are broken at each meeting: of participants, of papers, etc. The 19th ISSW, held in Marseille in 1993 is not an exception to this rule: about 450 Abstracts were received by the Papers Committee, which firmly establishes this series of Symposia as the main international forum for discussing topics related to shock waves, and guarantees employment to future organizers.

The diversity of the topics treated is proof of the large number of scientific fields in which shock waves are involved. The same is true for applications, as they cover such traditional fields as aeronautics and astronautics, combustion, lasers, blast waves and detonations, but also a lot of other interesting industrial areas such as high-speed trains, car engines, surface cleaning or underwater explosions. Concern for the environment is also present with forest fires, volcanic eruptions and meteorite fall, not to forget the biological effects of shock waves on cancerous tissues, gallstones or the pulmonary system.

Among the fundamental topics, nonequilibrium hypersonic flow, as well as shock tube and shock tunnel flow continue to be extensively researched, and new diagnostic methods, especially optical, are being developed. In numerous papers, various CFD methods are also applied to these flows in complex configurations. Like in the past, physico-chemical processes and relaxation phenomena, shock wave kinematics: diffraction and refraction, shock-vortex and shock-boundary layer interactions, are well represented, as is the propagation of shock waves in condensed matter and various heterogeneous media.

Finally, after the examination of each paper by two experts from Western Europe, about 300 papers were retained and presented during the five days of the Symposium, 175 in oral sessions and 125 in poster sessions. There were 355 participants from 22 countries and, among them, many young scientists and graduate students, who benefited from the preferential fares reserved for them. Ten invited papers covered the major topics of the Symposium and two Memorial Lectures – the Paul Vieille and the Ernst Mach Lectures – respectively opened and closed the meeting: the Paul Vieille Lecture was given by Prof. K. Takayama and constituted a superb review of optical visualisation methods in shock wave research; the Ernst Mach Lecture was presented by Prof. L.Z. Dumitrescu, who described the successive and often difficult steps of his own involvement with shock waves during 40 years.

In view of the large number of papers and of the diversity of the topics, the present Proceedings of the 19th ISSW have been divided into four volumes, each of them devoted to a group of related topics, as follows:
- **Vol. I:** Hypersonics, Shock Tube and Shock Tunnel Flow
- **Vol. II:** Physico-Chemical Processes and Nonequilibrium Flow
- **Vol. III:** Shock Waves in Condensed Matter and Heterogeneous Media
- **Vol. IV:** Shock Structure and Kinematics, Blast Waves and Detonations.

Each volume includes about 75 papers – contributed, invited and memorial – and an Introductory Survey. The presence of the latter was suggested by the Publisher; well-known personalities have been asked, after the completion of the Symposium, to write the Surveys; they are Profs. J.L. Stollery (Vol. I), J. Kiefer (Vol. II), M. van Dongen (Vol. III) and L.F. Henderson (Vol. IV), to whom warm thanks are due.

As for the previous Symposia, no permanent body was in charge of the 19th ISSW. Therefore an organization had to be entirely set up and financial support to be found; this would not have been possible without the enthusiastic support and generous contribution from many. I am especially endebted to my colleagues and co-workers of the Executive Committee, as well to all the graduate students of our Department; I wish to express special thanks to Prof. D. Zeitoun, the Papers Committee Co-Chairman, who had to "master" the flow of the papers, to Dr. L. Houas for having spent much effort and time on many points of the material organization, and to Mrs. M. Leboisne, the Symposium Secretary, for her efficiency and almost unlimited availability. Thanks are also due Prof. W. Beiglböck (Springer Verlag, Heidelberg) and his team, for their help in publishing, in excellent graphical conditions, these Proceedings

External cooperation has not been lacking either. The members of the International Advisory Committee have given numerous suggestions and advice for improvement. The members of the Papers Committee have examined in a short time the many abstracts submitted. The members of the Sponsoring Committee and the heads of the supporting organisations are at the origin of the decisive financial support for the Symposium; in particular, the participation of colleagues from Eastern Europe is due mainly to their generosity. It must also be underlined that that support represented about 65% of the total expense, only 35% coming from the participants' fees. The ladies of the Companions' Program Committee, helped also by Prof. A. Chauvin, must be especially thanked for the preparation and achievement of a pleasant program, which was followed by more than 80 companions.

Locally, I have received sympathetic and generous help from the Municipalities of Marseille and Cassis, as well as from the Regional Council of Provence-Alpes-Côte d'Azur and, of course, from the Université de Provence. I am pleased to express my gratitude to M. P. Vigouroux, Mayor of Marseille, M. G. Rastoin, Mayor of Cassis, M. J-C. Gaudin, President of the Region and to Prof. V. Kaftandjian, President of the University, as well as to their numerous collaborators, who have offered their precious cooperation that has facilitated the task of the Executive Committee.

The 19th ISSW was held from July 25 to July 30, 1993 on the Saint Charles Campus of the Université de Provence, Marseille. It followed the 18th, held in Sendai, Japan (Chairman, Prof. K. Takayama) and preceeds the 20th, which will be organized in Pasadena, USA (Chairman, Prof. H. Hornung).

In the end, the success of such a meeting is essentially due to the contribution of its participants. In the present case, that contribution was important and multiple, and thus warmly felt by the Chairman and his team. I would like to wish all these participants further fruitful research in the field of shock waves, and new meetings at future Symposia.

Marseille, June 1994   Raymond BRUN
                                                           Chairman, 19th ISSW

19th ISSW - Committees and Supporting Organisations

Host Organisation

Département Milieux Hors d'Equilibre
Institut Universitaire des Sciences Thermiques Industrielles, URA CNRS 1168
Université de Provence, Marseille, France

International Advisory Committee

T. Akamatsu	(Japan)	F. Higashino	(Japan)	F. Obermeier	(Germany)
T. Bazhenova	(Russia)	Z. Han	(China)	C. Park	(USA)
G. Ben-Dor	(Israel)	R. Hanson	(USA)	N. Reddy	(India)
D. Bershader	(USA)	L. Henderson	(Australia)	H. Reichenbach	(Germany)
A. Borisov	(Russia)	A. Hertzberg	(USA)	P. Roth	(Germany)
R. Brun	(France)	R. Hillier	(UK)	D. Russell	(USA)
F. Demmig	(Germany)	H. Honma	(Japan)	J. Sandeman	(Australia)
J. Dewey	(Canada)	H. Hornung	(USA)	S. Sharma	(USA)
L.Z. Dumitrescu	(Romania)	O. Igra	(Israel)	B. Skews	(South Africa)
R. East	(UK)	V. Kedrinskii	(Russia)	R. Stalker	(Australia)
R. Emrich	(USA)	Y. Kim	(USA)	J. Stollery	(UK)
N. Fomin	(Bielorussia)	A. Lifshitz	(Israel)	B. Sturtevant	(USA)
B. Forestier	(France)	H. Matsui	(Japan)	K. Takayama	(Japan)
W. Gardiner	(USA)	B. Milton	(Australia)	M. VanDongen	(Netherlands)
I.I. Glass	(Canada)	H. Mirels	(USA)	Z. Walenta	(Poland)
J. Gottlieb	(Canada)	M. Miyajima	(Japan)	R. Yu	(China)
H. Grönig	(Germany)	R. Nicholls	(Canada)		

Executive Committee

R. Brun (Chairman)	M. Imbert	K. Koffi
M. Autric	M. Leboisne (Secretary)	L. Labracherie
M. Billiotte	G. Leboisne	M. Llorca
Y. Burtschell	G. Méolans	F. Lordet
A. Canova	J. Moutouh	F. Mazoué
A. Chauvin	D. Zeitoun	M. Marti
L.Z. Dumitrescu	D. Benghrib	E. Schall
Y. Fienga	P. Delmer	A. Touat
B. Forestier	M.-P. Dumitrescu	J. Vuillon
L. Houas	S. Granjeaud	

Papers Committee

R. Brun (Chairman)

D.Zeitoun(Co-Chairman)	Marseille (France)	M. Imbert	Marseille (France)
R. Abgrall	Nice (France)	J.C. Lengrand	Meudon (France)
M. Autric	Marseille (France)	A. Lerat	Paris (France)
R. Borghi	Rouen (France)	J.-C. Loraud	Marseille (France)
E. Brocher	Marseille (France)	J.P. Martin	Paris (France)
J. Brossard	Bourges (France)	J.G. Méolans	Marseille (France)
M. Champion	Poitiers (France)	A. Merlen	(Lille (France)
P. Chapron	Courtry France)	R. Monaco	Genoa (Italy)
R. Cheret	Paris (France)	K.W. Naumann	Saint Louis (France,Germany)
F. Demmig	Hannover Germany)	F. Obermeier	Göttingen (Germany)
D. Desbordes	Poitiers (France)	C. Paillard	Orléans (France)
A. Désidéri	Nice France)	R. Perrin	Paris (France)
L. Devéseaux	Chatillon (France)	D. Poll	Manchester (UK)
L.Z. Dumitrescu	Bucharest (Romania)	F. Prat	Lyon (France)
G. Dupré	Orléans (France)	P. Roth	Stuttgart (Germany)
J.-P. Dussauge	Marseille (France)	F. Seiler	Saint Louis (France,Germany)
A. Dyment	Lille (France)	M. Sentis	Marseille (France)
R.A. East	Southhampton (UK)	G. Smeets	Saint Louis (France,Germany)
G. Eitelber	Göttingen Germany)	M. Sommerfeld	Erlangen (Germany)
B. Fontaine	Marseille (France)	J. Stollery	Cranfield (UK)
B. Forestier	Marseille (France)	B. Stoufflet	Paris (France)
R. Gatignol	Paris (France)	T. Srulijes	Saint Louis (France,Germany)
H. Grönig	Aachen (Germany)	M. VanDongen	Eindhoven (Netherlands)
J.-F. Haas	Vaujours (France)	P. Vervisch	Rouen (France)
R. Hillier	London (UK)	J. Warnatz	Stuttgart (Germany)
L. Houas	Marseille (France)	J. Wendt	Brussels (Belgium)

Companions' Program Committee

J. Brun	M. Houas
M.-N. Canova	D. Musso
A. Chauvin	M. Zeitoun
M. Chauvin	

Supporting Organisations

- Aérospatiale
- American Physical Society
- Banque Populaire Provençale et Corse
- Centre National de la Recherche Scientifique
- Centre d'Etudes de Limeil-Valenton
- Centre d'Etudes de Vaujours
- Centre d'Etudes Scientifiques et Techniques d'Aquitaine
- Commissariat à l'Energie Atomique
- Commission of European Communities
- European Office of Aerospace Research & Development
- European Space Agency
- Institut Français du Pétrole
- International Science Foundation
- Ministère des Affaires Etrangères
- Ministère de la Défense
- Ministère de l'Education Nationale
- Ministère de la Recherche et de l'Espace
- Société de Mathématiques Appliquées et Industrielles

- Conseil Général des Bouches du Rhône
- Région Provence, Alpes, Côte d'Azur
- Université de Provence
- Ville de Marseille
- Ville de Cassis

Endorsing Orgnisations

- Institut National de Recherche en Informatique et Automatique
- Institut National de la Santé et de la Recherche Médicale
- Société Française de Physique

Sponsoring Committee

D. Besnard (CELV)	G. Duffa (CESTA)	J. Muylaert (ESA)
R. Cheret (CEA-DAM)	H. Hollanders (Aérospatiale)	J. Périaux (SMAI)
B. Sitt (CEV)		

Contents - Volume I*

* The Contents of the other volumes are given at the end of the book.

Part 2: Combustion and Ram Accelerators

Part 3: Shock Tube Technology and Diagnostic Techniques

Part 4: Numerical Computations

Survey Paper

Plenary Lectures

Volume I: Hypersonics, Shock-Tube and Shock Tunnel Flow - An Introductory Survey

John L. Stollery
Cranfield University College of Aeronautics, Cranfield, Bedford MK43 0AL, UK

Introduction

This article attempts to review four groups of papers, all of which fall in the general category of hypersonic flow or shock tube and shock tunnel flows. No attempt is made to discuss individual papers. Instead the aim is to give a personal view of the most important problem areas, to comment on the way they are being tackled and to describe any novel techniques that have emerged.

I.1. Hypersonic flow and aerospace studies

A slightly cynical view of research activity describes it as 'at best cyclical but often circular'. Hypersonic flow is a good example of this phenomenon. In the 1960's the 'cold war' and the race to the Moon prompted an immense amount of research leading to a vast range of hypersonic missiles and more peaceably, to the design of the Moon lander and the Space Shuttle. With the Moon landing accomplished and the Shuttle fully operational, the interest in hypersonic flow rapidly diminished, as did the funding. A number of experimental facilities were literally destroyed and many of the measuring techniques that had been so painstakingly developed were forgotten.

Then in the 1980's the Star Wars programme and the mounting costs of satellite launching, either by rocket or shuttle, re-kindled enthusiasm. Hypersonics once again attracted funding and the cycle began all over again. Ambitious programmes for anti-missile missiles and single-stage-to-orbit launchers were started. Then came 'Perestroika' and 'Glasnost' plus a growing realisation of what the development of new and novel space launchers would cost.

Today hypersonic flows and aerospace studies are again at the cross-roads. The US NASP programme has given way to a more modest research study. In Europe the work on Hermes, Sänger and Hotol has been drastically reduced. In Japan the development of Hope has been slowed and in Russia turmoil clouds the picture. Everywhere the emphasis is now on 'technology acquisition' ie a more complete understanding of the extremely difficult problems still facing the design of efficient re-entry vehicles.

Many of these problems are long-standing but still with us, for example transition, separation, re-attachment, real gas effects and surface catalysis. Some papers look very similar to those published in the 1950's and 1960's and yet we still cannot fully understand the effects of bluntness, or roughness, or separated flow, on transition from the initially laminar flow over a configuration as simple as a compression corner. One important development is the number of quite large, very high enthalpy facilities now available such as T4 in Australia, T5 in the USA and the HEG in Germany. This means that some familiar geometric configurations can now be tested at much higher total enthalpies where much larger real gas effects may be present. For example the effect of nitrogen dissociation on transition, on shock/shock interaction and on shock/boundary layer interaction are now being measured.

As well as the acquisition of fundamental knowledge there are some investigations tackling the complex flowfields associated with real three-dimensional configurations such as a rocket cluster. It is very important that both categories are pursued. The simple shapes often yield understanding whilst the complex ones sometimes throw up new and difficult areas of interaction that are not easy to predict.

The Symposium Proceedings also reveal the growing interest in duplicating a given experimental environment numerically. Thus the nozzle starting process and flow development over

Shock Waves @ Marseille I
Editors: R. Brun, L. Z. Dumitrescu   © Springer-Verlag Berlin Heidelberg 1995

a given test model can be predicted. The estimates of pressures, shear stresses, heat transfer rates and aerodynamic forces can then be compared with the actual signatures measured. This kind of collaboration between mathematical modelling and experimental measurement is likely to enhance our understanding of both techniques and move us away from the sterile arguments surrounding confrontation.

I.2. Combustion and Ram accelerators

These topics have been linked because they both involve studies of supersonic combustion. Although of enormous interest to the possible development of more economic space vehicles, this section attracted only 8 papers, a number perhaps related to the difficulties associated with the subject. The Symposium reports shock tube and shock tunnel tests in which gaseous or liquid fuels are introduced and under certain conditions "stable combustion is maintained". However, with running times limited to a few milliseconds it is difficult to convince the objective observer that efficient mixing and repeatable combustion geometries are now to hand. Much still needs to be done but the careful experiments now underway are helping to understand and raise the credibility of supersonic combustion.

The conventional approach to engine testing is to place a stationary model in a moving stream. An alternative is to have a moving model in a stationary stream. These alternatives have resulted in the design of hypersonic wind tunnels on the one hand and ballistic ranges on the other. The Ram Accelerator is a most ingenious example of the second approach, in which a ram-jet-shaped-projectile is fired down a close fitting tube filled with a pre-mixed air/fuel mixture. Thus the principle of operation is similar to that of a scramjet in which the fuel burn is only limited by the length of the fuel-filled tube.

The original idea came from Hertzberg and his co-workers at the University of Washington in 1986. Initially it was unclear how well the device would work, or indeed if it would work at all. Since then a number of different groups have taken up the idea and some have built similar facilities. It has now been convincingly demonstrated that supersonic combustion occurs, a net thrust is developed and the ram-jet projectile does accelerate. Further research is aimed at discovering what is the maximum steady value of projectile velocity that can be achieved. Once again understanding and measurement are hampered by the short duration of the test, but already attempts are being made to visualise the flowfield using a transparent section in the launch tube.

The ram-accelerator has added excitement to the study of supersonic combustion because it has confirmed that practical success is achievable, albeit with a pre-mixed fuel. (It is also an excellent way of comparing the efficiency of different fuels and different air/fuel ratios). This stimulus should encourage all the wind tunnel workers to continue and to expand their vitally important mixing and combustion experiments.

I.3. Shock tube technology and diagnostic techniques

The 22 papers in this section form an excellent example of human ingenuity, enthusiasm and determination. Basically the shock tube is a very simple device but by modification, adaption and development it has produced a great spectrum of facilities for studying aerodynamics, physics and chemistry as well as the effects of shock waves on subjects as diverse as buildings and the human body. The range of aerodynamic test conditions now covers everything from subsonic to super-orbital with corresponding total temperatures rising from hundreds to thousands, to tens of thousands. Usually the test times reduce as the temperature rises and instrumentation has had to be developed to cope with running times dropping from seconds to milliseconds to microseconds. The papers contained here give a good state-of-the art picture of how this has all been achieved.

A particulary interesting development is the re-emergence of the expansion tube. First proposed as long ago as 1962, early experiments were unsuccessful. It proved difficult to design a secondary diaphragm that was strong enough to hold the required pressure difference and yet opened quickly enough to give the required rapid expansion. In practice poor diaphragm opening resulted in contaminated, noisy test conditions and the idea was dropped. More recently experiments have shown that pre-deforming the diaphragm and the careful choice of both diaphragm material and mass could significantly improve performance. Moreover it has been found that the addition of a secondary driver section can boost the performance of the facility to even greater enthalpies. Tests are underway in a number of countries but the small pilot Superorbital Expansion Tube facility at the University of Queensland is reportedly producing shock velocities in excess of 13 kms per sec. and total enthalpies in excess of 106 MJ.

As shock tunnel performance has risen so has the problem of heat transfer at the nozzle throat. For the highest enthalpy tunnels now operating it is thermal damage which limits their test conditions and running times. Thermal damage is unacceptable because, even if nozzle throats can be easily and cheaply replaced, the test gas becomes contaminated by metal vapours etc. Moreover the test models can be damaged and their surface condition changed. In order to prevent nozzle throat erosion, a novel form of film cooling has been proposed. A thin film of cool, low molecular weight gas is introduced just upstream of the throat. The weight and temperature of the coolant are chosen to match the acoustic impedance of the test gas, in order to minimise mixing and to avoid disturbing the rapid expansion of the working gas through the nozzle. Analytic and numerical studies suggest this technique has great potential but more experimental data are now needed to determine the practical utility of the method.

Tunnel performance (plus the growing amount of detailed data from CFD calculations) puts increasing pressure on the development and improvement of wind tunnel instrumentation. Measuring forces has always been difficult in short running time facilities. Very fast response is essential and conventional strain gauge balances are often difficult to design for these conditions. Three papers describe novel forms of force 'balances' for millisecond running times.

The first paper describes the extension to three components of a most unusual one- component balance, first demonstrated in 1990. In the one-component version (measuring drag only) the model is rigidly connected to a long, thin, cylindrical, elastic bar. The horizontal bar is strain gauged and freely suspended in the wind tunnel working section, by fine threads. The time history of the drag can be inferred from the output of strain gauges that respond to the passage of stress waves along the bar. The rise time of the balance is so quick that sensible drag measurements have been made with a running time of one millisecond. The extension to three components (lift, drag and pitching moment) involves placing a number of strain gauged struts between the model and the elastic bar. The bar is now suspended at any desired angle of incidence. Considerable development will probably be needed but the method looks promising.

The second novel method allows the model to fly free for the few milliseconds of genuine test time. An ingenious clutch system releases the model just prior to the test gas arriving and catches it again before the afterflow arrives and accelerates the model too much. Small piezo-electric accelerometers measure the accelerations and these together with the pitot pressure history, allow the force coefficients to be calculated directly. The third balance is a clever six component strain gauge design suitable for running times of 5 milliseconds or more. By suitable design and calibration, measurements are possible without acceleration compensation.

There have also been further advances in the development of planar laser-induced fluorescence (PLIF) diagnostics. In the new technique two lasers and cameras are used so that temperature (or alternatively two species) can be measured. The method is being used at Stanford to examine the mixing and combustion of hydrogen injected into a supersonic airflow. At Aachen the PLIF apparatus destined for the high enthalpy shock tunnel in Göttingen (HEG), is being tested. Measurements include NO concentration and NO rotational temperature in the freestream of the Aachen shock tunnel (TH2).

The overall impression is that (i) the older measuring techniques for pressure, heat transfer and skin friction are under continual development to match increased shock tunnel performance, (ii) novel forms of force balances will be capable of coping with millisecond test times and (iii) the search is on for ways of measuring all the chemistry associated with the realities of real gas effects, mixing and combustion.

I.4. Numerical computation

CFD is increasingly powerful and increasingly popular. 22 papers are included here providing a rich diet for the avid reader.

The first choice any mathematical modeller has to make is which equations to solve, e.g. Boltzmann, Navier-Stokes, Euler etc. Often the physics of the flow will dictate the choice but inevitably there are regions of overlap. Hence it is interesting to see some authors using more than one model to see what the similarities and differences in the predictions are. Similarly it is good to see computational efficiency being improved by using multiblock methods in which the Navier Stokes equations are solved when viscous effects are important and the Euler equations when they are not.

Numerical flow visualisation can be very impressive. A further development is a new algorithm to obtain the shock positions from a numerical solution of the flowfield around complete vehicles. The shock envelopes are then pictured isometrically as semi-transparent surfaces surrounding the vehicle.

Many papers deal with some form of shock interaction, reflection or refraction. One important example is the shock/shock interaction explored experimentally by Edney in 1968. A number of authors compute this flow in which an oblique shock meets the bow shock ahead of a circular cylinder. The resultant deformation of the bow shock can result in a supersonic laminar jet which impinges on the cylinder, giving a very localised region of high pressure and intense heat transfer. If a coarse grid is used for the calculations, the peak heat transfer rate is grossly underestimated. A much finer mesh gives the correct result (care is needed in doing the experiments too). The example is a good one in which to show the power of adaptive grids, a point made in a number of the papers.

It is noticeable that most of the calculations reported in this volume are either for inviscid or laminar flow. This is a little disappointing but not surprising. The onset of transition and the modelling of turbulence remain outstanding problems. One welcome trend is to see some authors validating their Navier Stokes solvers against the analytic solution of supersonic compressible laminar boundary layer flow over a flat plate.

The last group of papers confirm a trend noted in §1.1. They report CFD calculations made to help understand the behaviour of the experimental facilities they are using. One study models the shock reflection process at the end of a shock tube. Another paper shows how diaphragm curvature introduces variations in the end wall pressure signature. Two papers model the flow in Stalker tubes. In yet another paper the viscous flow in a super-orbital expansion tube is being modelled for comparison with measurements in the pilot tunnel. In every case such parallel studies are bringing about a much more rapid understanding of the complicated flows involved.

The progress, development and novel ideas described in this volume are most encouraging. Valuable knowledge is being acquired for future design purposes, whenever these may be required. Nevertheless the volume also serves to remind us of the achievements of those involved in the design of past and present re-entry vehicles. With only limited knowledge they produced some outstanding hypersonic spacecraft.

A Life with Shock Waves
(Ernst Mach Memorial Lecture)

Lucien Z. Dumitrescu
Université de Provence, 13397 Marseille, France; formerly, Institute of Aeronautics, Bucharest, Romania

Abstract. In 1957, the author obtained his first flow pictures in a shock tube; paper will retrace the history of a life-long involvement with the field. The first motivation was to use the shock tube as a short-duration aerodynamic test facility, a goal to be belately achieved, with results on e.g. hypersonic flow around flared bodies, or transonic motion of permeable airfoils. In the meantime, two shock tubes and a large facility, combining the shock- and Ludwieg-tube concepts, were built, as well as the necessary measuring equipment. In the same area, a development of the Ludwieg-tube principle, offering a five-fold gain in test time, has been proposed. Our interest centered then on the problem of shock diffraction around obstacles, and in ducts and cavities; for weak shocks, our theory is in excellent agreement with experiments. For the diffraction around a convex corner, a configuration was discovered, which delays BL separation. Shock focusing, amplification, and stability were investigated; a "shock nozzle" was devised, and also an improved version of the spiral contraction. Recently, the author joined Prof. Raymond Brun's Marseille team working on the development of a new Stalker-tube facility; already, very strong shocks ($M_S = 18.5$) were obtained in air, and a new shape for the piston was devised, to minimize pressure disturbances caused by accoustic resonance inside the compression tube.

Key words: Shock waves, Shock tubes

Fig. 1. The first shock tube (1956)

Fig. 2. The first shock tube flow picture (1957)

1. A leap into the unknown

This is the story of a long involvement with Shock Tubes & Waves; it all started in the early fifties, when our Professor, the late Elie Carafoli, was called upon to organize and lead basic scientific research in fluid mechanics, within the Romanian Academy of Sciences. He assembled a small team of his pupils, whose main field was aerodynamics; the subject "in" at the time was supersonic wing theory. He insisted that we should gain first-hand knowledge from our own experiments, since, at the time, there was an almost total lack of communications (a recurring disease). By chance, we discovered an early paper on shock tubes (Bleakney et al. 1949), which prompted the seemingly crazy idea of building such a device, to serve as a poor man's short-duration supersonic facility. With light heart (and budget) we embarked on a course which was to occupy our scientific life for the next forty years (Dumitrescu 1956).

Shock Waves @ Marseille I
Editors: R. Brun, L. Z. Dumitrescu   © Springer-Verlag Berlin Heidelberg 1995

Fig. 3. The hypersonic shock tunnel (1965)   **Fig. 4.** The shock-Ludwieg tube (1976)

2. Struggling with hardware

It took us two years to design and build our first shock tube (Dumitrescu 1959), an 11 m, 200 by 300 mm straight-through machine (Fig.1). The equipment was incredibly modest, mainly "borrowed"; as membranes, we used pieces of Mylar, taken from a cartoon film studio (with pictures still on); for triggering, we set an "aerodynamic delay line" (a piece of hose, closed by a razor-blade contact). Developing the photo plates was the occasion of great suspense, but soon, the very first pictures of supersonic flow were produced, in 1957 (Fig.2). We discovered then that the delay was accidentally set so as to get us into the cold flow regime, behind the contact surface, where we reached Mach 3. Nevertheless, these pictures were the much-awaited proof that the contraption actually worked, and cleared the way to a more permanent job.

A few years later, after gaining knowledge of Hertzberg's pioneering paper (1951), we undertook building a hypersonic shock tunnel (Dumitrescu et al. 1963), which went into operation in 1965 (Fig.3). It produced many beautiful pictures of flow around the customary simple bodies (spheres, cones, &c.) at Mach numbers up to 10; but it was also used as a straight shock tube: with hydrogen drive, we reached $M_S = 8$, recovering many known results, but also finding some new effects (Dumitrescu et al. 1969a, 1969b, 1970).

In 1970, the regime in Romania decided to curtail drastically all basic research, and our team was absorbed into a large aeronautical design and development organisation, while foreign scientific contacts were severely cut. However, we succeeded in luring the authorities into making funds available for a really large-scale project, a combined shock-Ludwieg-tube facility (Dumitrescu 1979): 900 mm ID, 170 m length, 20 atm. max charging pressure, 300 by 800 mm transonic test section with porous walls (Fig.4). Since its commissioning, in 1976, and until 1990, its workload has been evenly divided between aerodynamic tests, in the Ludwieg tube configuration, mainly for airfoil design, high-lift devices, &c., and blast simulation, e.g. for the development of protection devices in shelters (Dumitrescu et al. 1981): with hydrogen drive (not a mean feat in logistics) we reached 22 atm. reflected shock overpressure in the main tube, an impressive hammer to witness.

3. Milliseconds and microseconds

From the beginning, we were aware that instrumentation was the key problem, and it absorbed most of our time for a while (Dumitrescu et al. 1960-1968). We put up the necessary triggers, amplifiers, even a house-made oscilloscope, argon spark gaps, thin-film gages (with gold paint smuggled from a china-ware factory), &c. The main effort was, however, devoted to developing a reliable technique for measuring pressure distributions on models. This proved to be much more

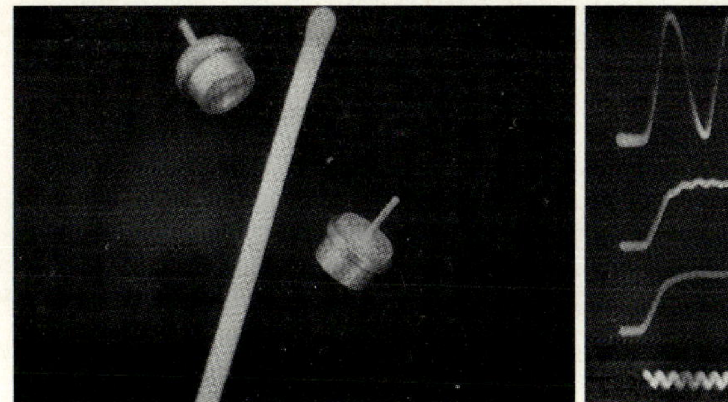

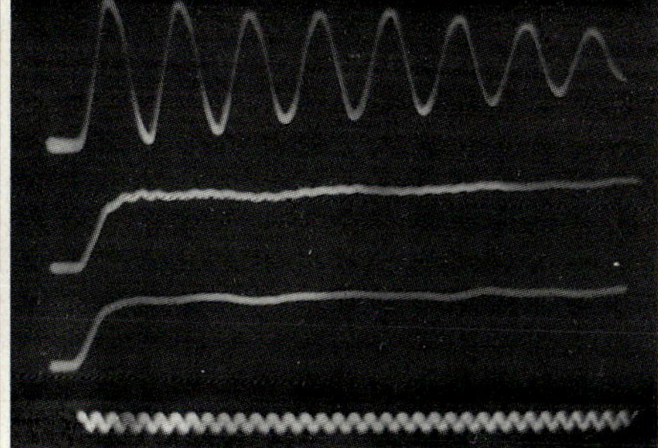

Fig. 5. Miniature pressure pick-ups (1961)

Fig. 6. Pick-up response to shock, with natural frequency compensation. Lower trace, time marking 5 microsecs. (1962)

difficult than expected but, by the early sixties, we achieved a good capability, even by today's standards (Dumitrescu 1967). Very reliable capacitive pick-ups were produced in-house (Fig.5), specially designed to minimize spurious signals, mainly due to mechanical vibrations. Also, a method was devised for getting rid of "ringing" (due to the low inherent damping of the membrane's natural frequency), by suitably compensating (Jakab 1962) the transfer function of the transducer (Fig.6). Later, the concept was extended to accelerometers, strain-gage wind-tunnel balances (Dumitrescu 1975), as well as to the filtering of pressure oscillations in the pressure-tapping tubes of wind-tunnel models (Dumitrescu et al. 1971). In posession of a high-quality pressure-recording capability, we were able to discover many aerodynamic phenomena whose fine details were previously obscured by noise. In particular, we found that great care must be taken to mount the pick-ups flush with the walls: a cavity in front of the transducer, as shallow as 0.2 mm, produced overpressure peaks of 10 percent. A theory explaining these effects was later put up (Dumitrescu 1968). All in all, it was proved to many skeptics that milliseconds, and even microseconds, are domptable beasts, with ingenuity and patience.

4. With the shock tube about the shock tube

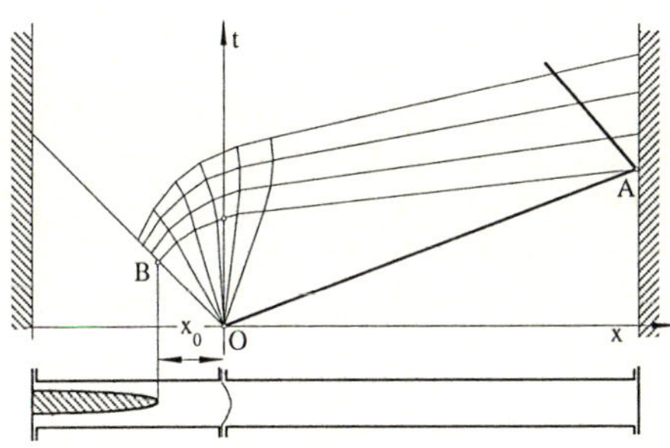

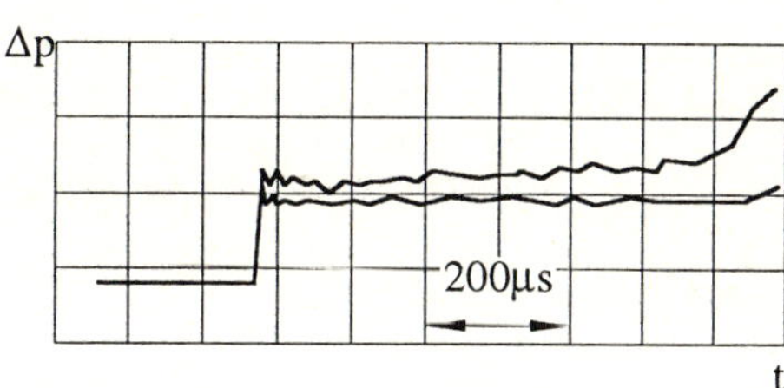

Fig. 7. The attenuation-free shock tube (1972)

Fig. 8. Compensation of end-wall pressure rise

All research tends to become obsessively involved with itself, and we spent a lot of time investigating shock tube flow as such. We put up the obvious (but, for us, unavailable) equations, developed our own version of the optimum tube length (Dumitrescu 1957) and discovered some errors in computing theoretical shock strength limits (Dumitrescu 1960).

Our most notable contribution is, however, the development of a scheme for compensating the attenuation effects (Dumitrescu 1972); the matter became the subject of a fruitful cooperation with the Marseille team (Dumitrescu, Brun et al. 1973, 1976).

Fig.7 shows how, by simply reshaping the driver chamber, the reflected expansion wave can be modified, so that, when it catches up with the primary shock, the ensuing interaction would lead to constant temperature (or pressure) behind the shock.

In Fig.8 an example of the compensation method, applied to keeping constant the reflected-shock pressure, is illustrated. Recalling the importance of even small temperature differences in chemical kinetics studies, we believe that our proposal deserves renewed attention.

5. Bodies and shocks

By the early sixties, all our tools were ready, waiting for a problem to be applied to: the interaction of shocks with obstacles was the obvious choice. Weak shocks ($M_S = 1.1 \ldots 1.25$) still amount to some hard beating and are of considerable practical interest. They have, in addition, the advantage of allowing a significant simplification of the theory. We remarked that, in such cases, the flow velocities are about one order of magnitude smaller than the wave speeds (roughly equal to the sound speed); therefore, the two processes involved, i.e. those of wave diffraction and of flow development, may be decoupled (Figs.9&10).

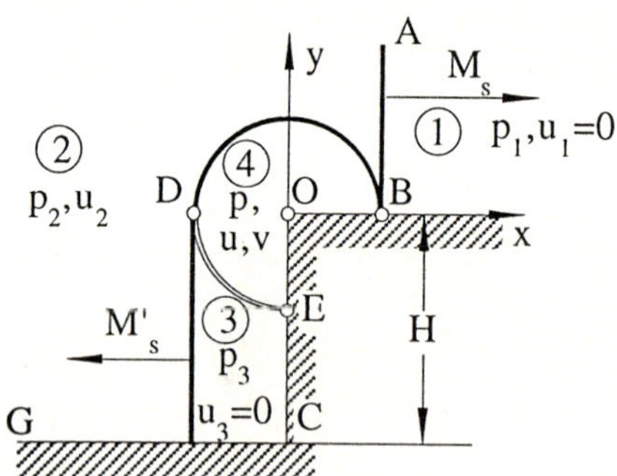

Fig. 9. Shock diffraction ($M_S = 1.2$) at a forward-facing step (1966)

Fig. 10. Simplified wave pattern

Under these assumptions, the equations of motion, following elimination of the time variable by an affine transformation, are linearized and combined into an equation for the pressure; this, in turn, is reduced to a Laplacian by a Busemann transform. Now, the powerful tools of complex functions and conformal mapping can be put to work, and closed-form solutions are obtained for the pressure distribution in the perturbed domain (Dumitrescu 1966,1968), especially in the case of bodies with plane boundaries. In the latter stages of the diffraction process, disturbances begin to overlap, but linearization allows adding their effects. As an example, the computed and experimental time evolution of the pressure, at a point on the face of a forward step hit by a Mach 1.2 shock, are compared in Fig.11.

Many other shapes were investigated; also, ducts with sudden area changes, sharp bends (Fig.12) and closed cavities (Fig.13). In the latter case, we confirmed that even a very shallow recess ($l/d = 0.02$) will produce overpressures of up to 50%; deeper cavities exhibit oscillatory response to a shock strike (Fig.14), while in a conical cavity the pressure peaks can become

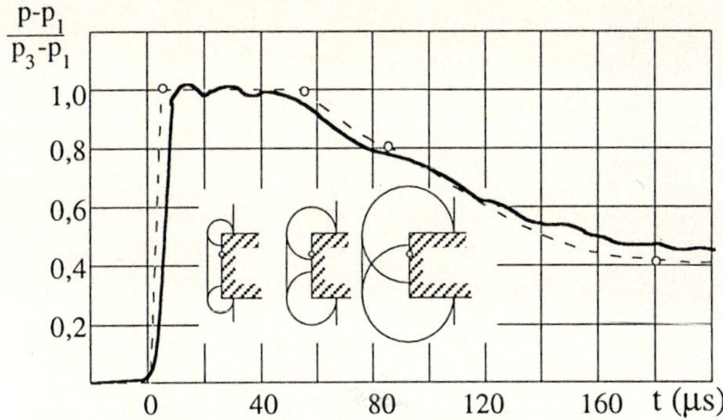

Fig. 11. Pressure evolution in time at a front-facing step: theory and experiment

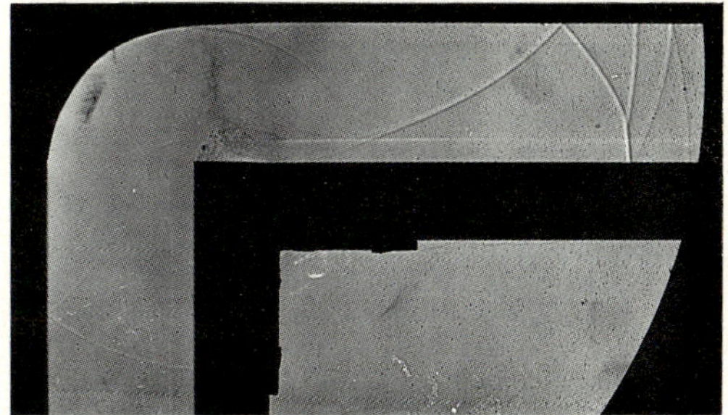

Fig. 12. Successive shock reflections in a sharp bend (1966)

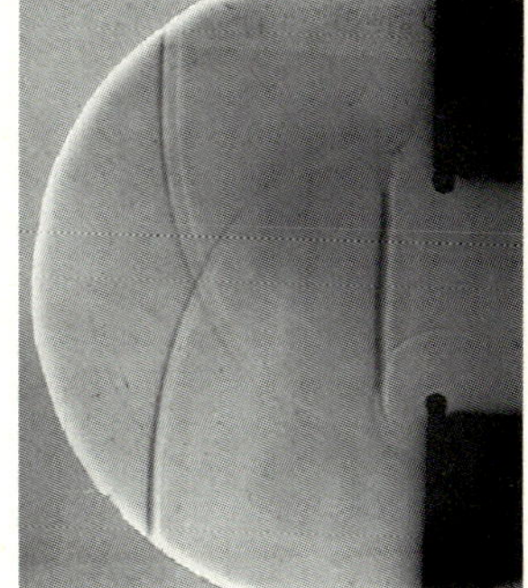

Fig. 13. Shock reflection off a closed cavity

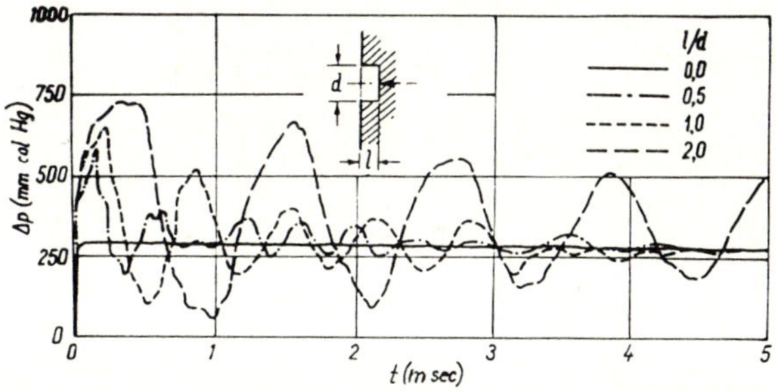

Fig. 14. Pressure oscillations in a cavity

very sharp, sometimes destroying the pick-ups (Dumitrescu 1969). Recently, such an instance of resonant response was found occuring at the end of the compression tube in the Marseille Stalker-tube hypersonic facility, and a simple means to remove it was discovered (M.P.Dumitrescu et al. 1993a, 1993b, see these Proceedings).

Another important finding was that the unsteady flow behind a weak shock will sustain a much higher rate of turning than a steady flow (Fig.15), abstraction made of an attached vortex, which is to be eventually (but much later) swept downstream. At high speeds, however, the pattern is complicated by diffracted shocks, which induce boundary-layer separation, starting at the very edge of a convex corner (the lip shock, Fig.16). However, we found that, if a second expansion corner is added downstream, the flow will become attached at the first (Fig.17); this suggests that

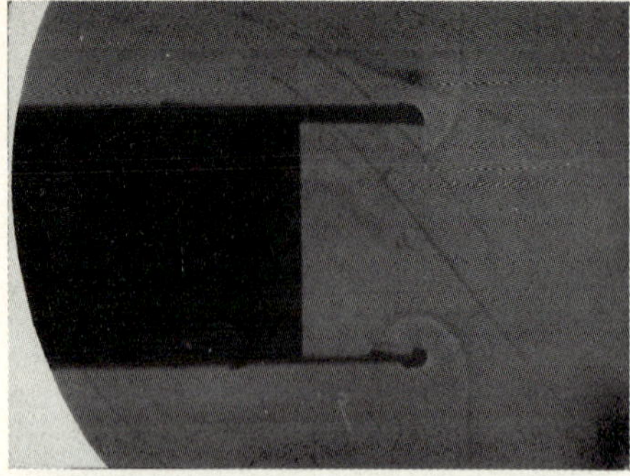

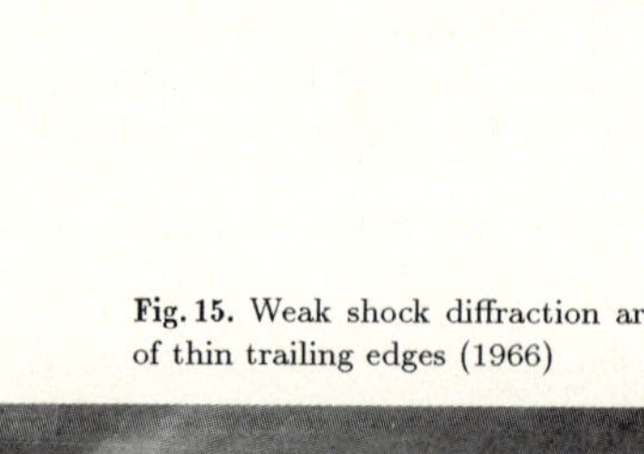

Fig. 15. Weak shock diffraction around a pair of thin trailing edges (1966)

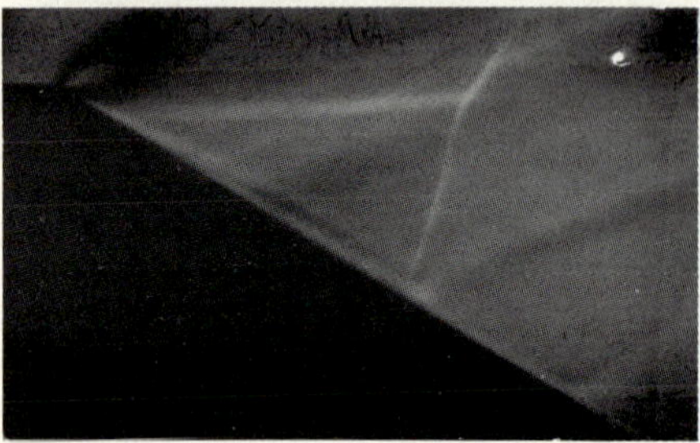

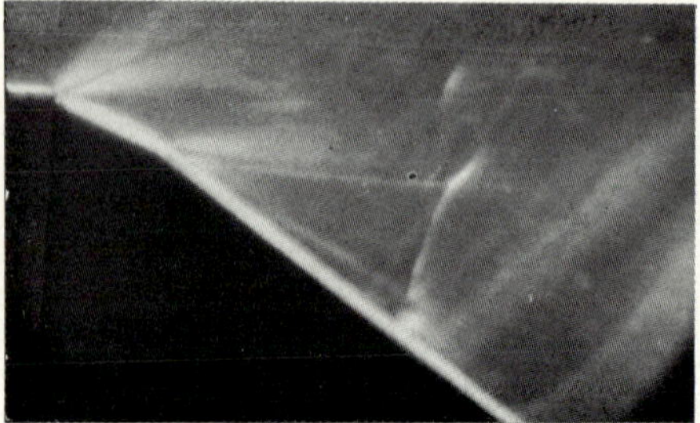

Fig. 16. Shock diffraction at a convex corner ($M_S = 2.2$) (1979)

Fig. 17. Second expansion delays BL separation

higher rates of expansion might be achieved (e.g. for gas laser applications), if one could exploit such unsteady motions in a Laval nozzle.

6. Tests of strength

As a natural extension to our involvement with shock-diffraction patterns, we took on the subject of shock focusing. We devised a "Shock Nozzle" (Dumitrescu 1983), the equivalent of the steady-flow Laval nozzle: by analogy with the Mach characteristics, the Whitham shock-shock lines can be used to shape the walls of a contracting duct, so as to intensify a plane incident shock (Figs.18&19). Another case, already extensively studied by our Australian colleagues, is that of the spiral contraction. We pointed out that what such a device achieves is, actually, a strengthening of a plane shock, which remains however plane, and we developed a slightly modified version (Figs.20&21) which proves to be much more efficient in avoiding unwanted disturbances. It might be said that our Creator was very profficient in gas-dynamics, since one natural example of such a configuration is the ear's cochlea, which might be thought as *the* device producing the most amplification of sound, with the least disturbance.

We also tried (unsuccessfully) to deal with the problem of producing uniform converging shocks, and these thoughts led us to formulate a conjecture, namely that it is impossible to turn a straight, uniform shock, into an *uniform* cylindrical implosion by a 2D configuration, instead of the commonly-used axisymmetric Kantrowitz device (Dumitrescu 1983). At the next Symposium, a paper (Saillard et al. 1985) was presented, purporting to produce a counterexample (but not a definite proof); after much thought, we succeeded (sadly) in rigorously refuting our proposition (Dumitrescu 1992), but found that even if theoretically possible, such a configuration would be impractical. A version of the standard focusing device was then proposed (Fig.22) which should improve the stability of the converging front.

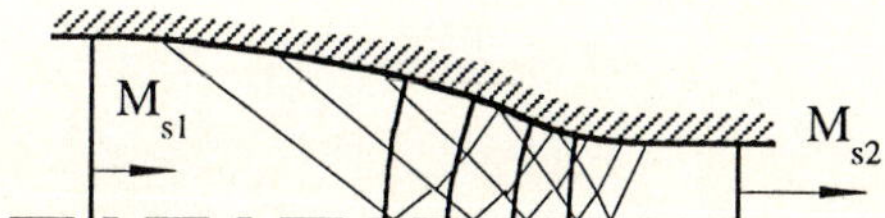

Fig. 18. The "Shock Nozzle" (1983)

Fig. 19. Shock propagation in a contoured nozzle

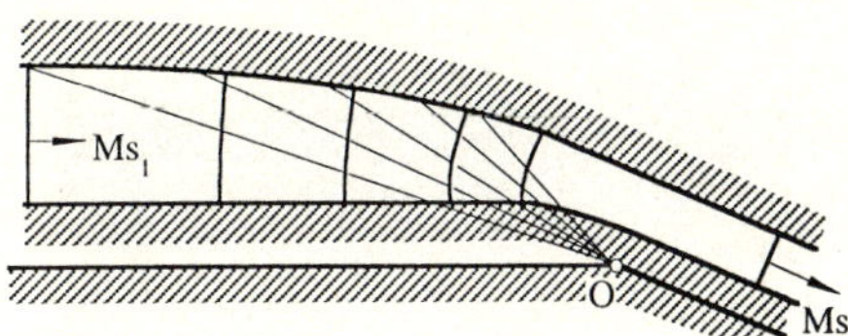

Fig. 20. Spiral contraction with double contoured walls (1983)

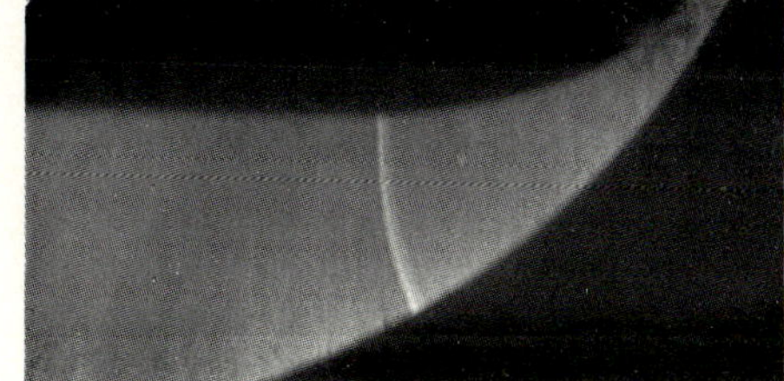

Fig. 21. Plane shock amplification in a spiral contraction

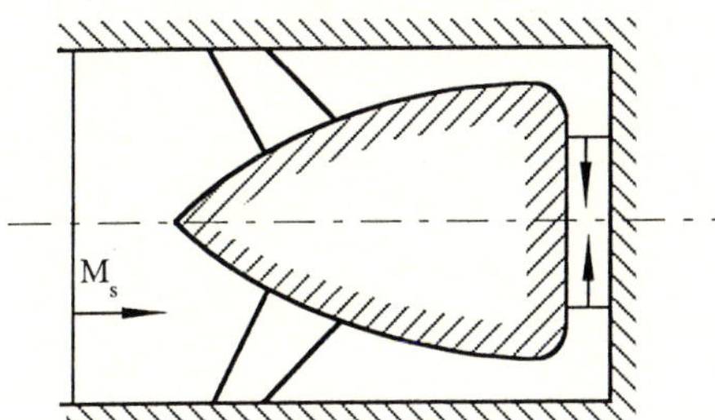

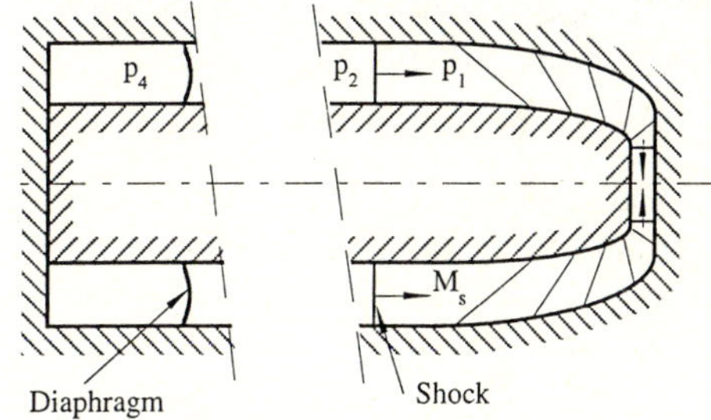

Fig. 22. Standard and modified Kantrowitz device for producing more uniform cylindrical implosions (1992)

7. Bubbles, stars and whirls

From these subjects we went further and ventured to consider the stability of implosions, putting forward the following *Proposition* (Dumitrescu 1983): That, whereas a fully cylindrical (or spherical) implosion is unstable, one that spans only a limited angle might be stable. The reason for that can be better understood if one recalls the essential role played by the walls in stabilizing a shock which travels along a tube; incidentally, this is what made possible the invention of the shock tube (and the eventual birth of our Symposia!). The progressive distortion of converging shocks has been beautifully demonstrated in many researches, and various applications come to mind, e.g. the collapse of cavitation bubbles, extensively studied by our Japanese colleagues; we venture now to suggest some other, far-fetched occurences. One may thus infer that the collapse of a supernova is such a phenomenon, since after the 1987 event it was found that what eventually emerged was a non-symmetrical celestial body. Also, one may conjecture that the gravitational collapse of a black body might not proceed indefinitely, but could end in instability, perhaps with the formation of a pair of new celestial objects (quasars?). Even more hypothetically, the probable implosion of the Universe, before the Big Bang, ought not have condensed the matter to a mathematical point: the remaining non-uniformity could be the origin of all the diversity in our world!

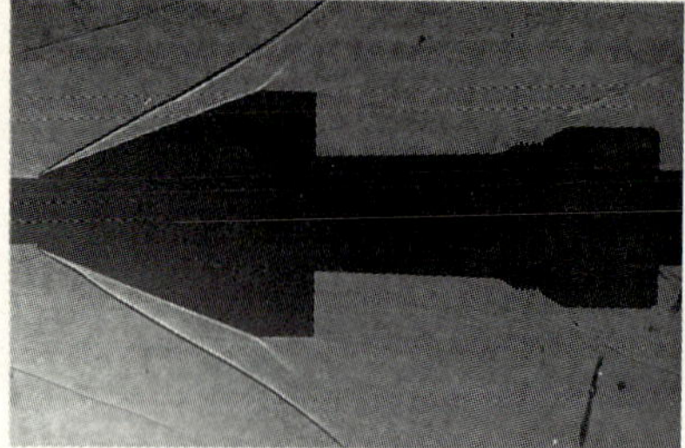

Fig. 23. Separation avoidance by BL bleed on a flared body, $M = 7.0$ (1985)

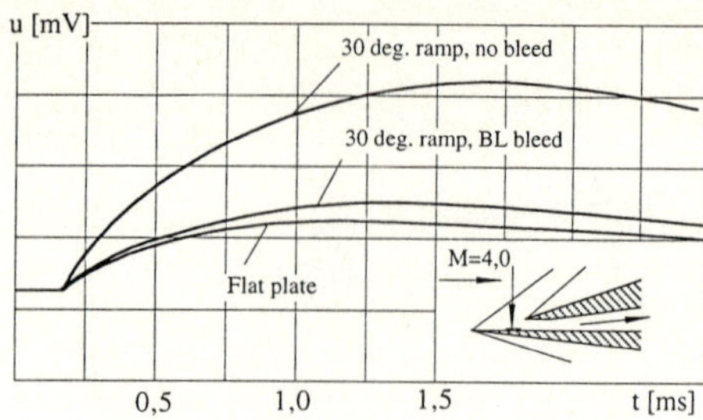

Fig. 24. Thin-film gage records for a 2D flap with (turbulent) BL bleed

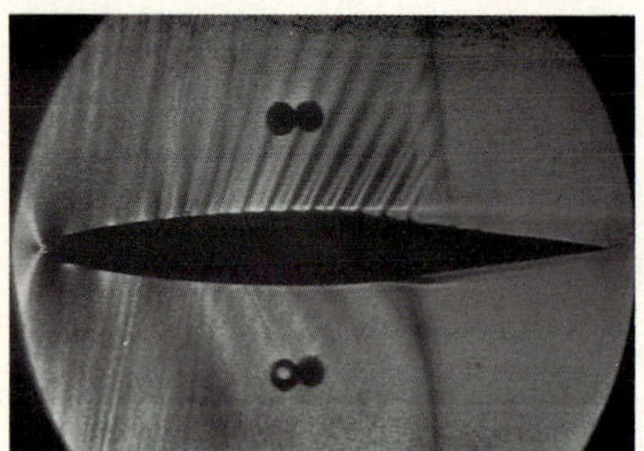

Fig. 25. Shock suppression in transonic flow around an airfoil with porous upper surface (1983)

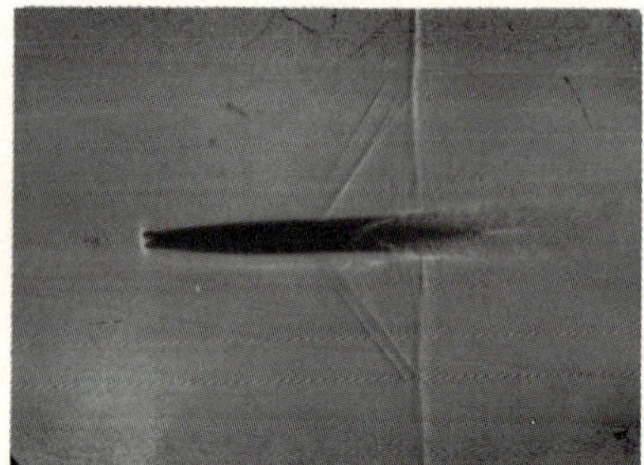

Fig. 26. Transient transonic flow in a shock tube

8. Back to square one

The initial motivation of our endeavour seemed long forgotten, but aerodynamics still remained in the background of our thought. And so, when the necessary tools were finally acquired, we were able to undertake a number of studies in this field. At Prof. Carafoli's suggestion, we investigated various hypersonic problems, among them the flow around flared bodies. We found that boundary-layer separation severely distorted the wave pattern; however, after devising a means for bleeding the boundary-layer at the (concave) corner, the separation was removed (Fig.23) and the shock shape obeyed nicely the Professor's equations. A 2D configuration, modelling the underside flap of the space Shuttle, was also tested. Boundary-layer bleed reduced the heat-transfer at the corner (for a turbulent BL) almost to the flat-plate value (Fig.24).

The bulk of our activity, for some years was, however, dedicated to airfoil aerodynamics, using the large Ludwieg tube built for that purpose. Much of the work was routine, such as the development of customized airfoils, high-lift devices, &c., putting to advantage the high-Reynolds numbers attainable there; however, two pieces of basic research are worth mentioning:

(i) Our pupil and colleague, Dr. George Savu, was the first to put forward a concept for supressing shocks on an airfoil in supercritical flow, by using a permeable upper surface, whose effect is analogous to that of perforated walls in transonic wind-tunnels. In a series of tests, we were able to demonstrate the concept (Fig.25) (Savu et al. 1983); at the Sydney meeting, a high-speed movie was presented, showing many details of the phenomena. Later, the subject has attracted the attention of many researchers world-wide.

(ii) Another finding of general interest is the sharply increasing time, with M, necessary for stabilizing a transonic flow; this was demonstrated by a series of experiments made in the shock tube (Dumitrescu 1972, Fig.26) and in the Ludwieg tube. These results, and more theoretical

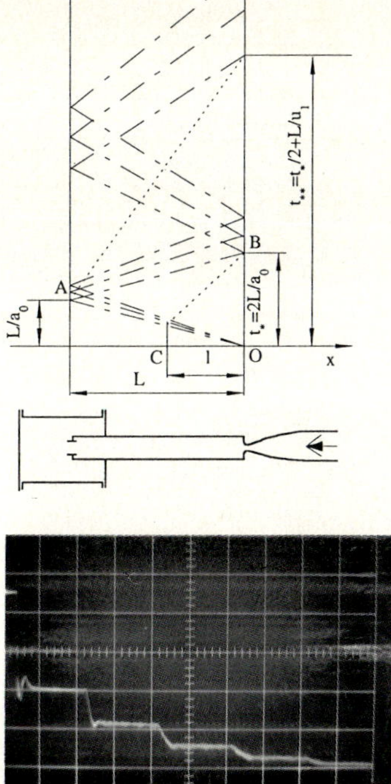

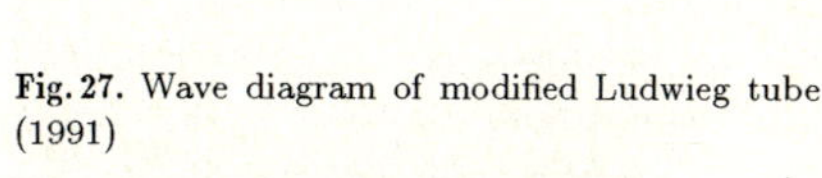

Fig. 27. Wave diagram of modified Ludwieg tube (1991)

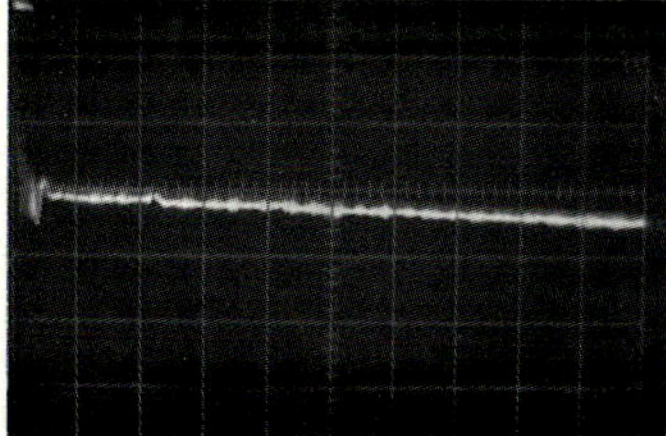

Fig. 28. Succesive pressure plateaux, simple Ludwieg tube

Fig. 29. Coalescence of pressure plateaux in modified Ludwieg tube

arguments, have led us to formulate the following *Conjecture:* that, as the Mach number of a transonic flow is raised towards unity, the duration of the transient regime increases indefinitely: there is no such thing as a steady sonic flow. (Mathematically, when $M \to 1$, the limit taken in the equations of motion, by setting to zero the terms containing time derivatives, is singular).

9. There is always room for improvement

Recently (but, sadly, only after having built our facility), we came across a new idea for increasing the useful run-time of a Ludwieg tube (Dumitrescu 1991). Basically, it consists in adding, at the upstream end of the tube, a larger-diameter tank, connected through a constricted passage. If properly sized, this partial opening will cancel the reflection of the primary expansion wave, which travels upstream the tube; the useful run-time is increased almost five-fold. Fig.27 shows the wave diagram, and Figs.28&29 prove the concept; the opportunities for extending the scope of Ludwieg-tube applications look very promising, as a very simple device for producing a constant flow-rate gas supply.

10. Still going strong

It was only after the collapse of a notoriously infamous regime in Romania that it became possible to renew our scientific contacts with the outside world; and, as, at the same time, one witnessed a renewal of interest in shock tubes (mainly in connection with hypersonic research) the author found, in 1991, a new scientific home at the Université de Provence, in Professor Raymond Brun's team; work is here now in progress for developing a Stalker-tube hypersonic facility of advanced performances. Currently, a less common operating mode of the machine is

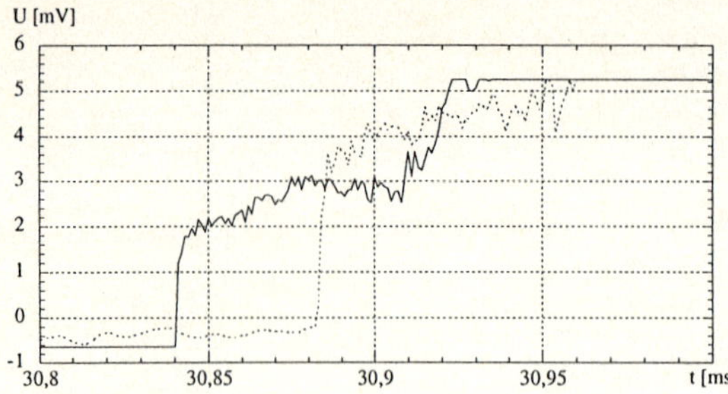

Fig. 30. Pressure traces in the Marseille free-piston shock tube: $M_S = 18.5$, $p_1 = 1$ mb (1993)

being set up, as a shock-tube: already, in air at 1 mbar initial pressure, shocks at $M_S = 18.5$ have been produced (Fig.30). Applications are under way to non-equilibrium chemical kinetics studies of some exotic gas mixtures, such as the atmosphere of Saturn's sattelite, Titan (within the Cassini-Huyghens project). When in full swing, we hope that this machine will prove to be a valuable addition to the range of existing high-enthalpy facilities, as one of its features is the special attention paid to cleanliness, thus, hopefully, improving the accuracy of spectroscopic measurements to be carried out. So, we are looking forward towards another period of exciting involvement with Shock Tubes & Waves @ Marseille!

11. Q.E.D.?

All said, what have we tried to achieve with the above presentation? First, it is hoped that some interesting findings which, due to particular circumstances, were not properly made public, will raise the attention of our colleagues, for possible further development and/or application; but, especially, to illustrate the versatility of that old faithful shock tube, as a device to discover new knowledge. To the author, however, the main benefit provided by his involvement with the shock tube was the opportunity for meeting and gaining the friendship of some of the most distinguished scientists worldwide.

References

Details on the subjects touched upon here are to be found in the reports issued over the years by our team. From the international litterature we cite only three early papers, which were determinant in shaping our endeavour (and future).

Bleakney W, Fletcher CH, Weimer DK (1949) The shock tube: a facility for investigation in fluid dynamics. Rev. Sci. Instrum. 20: 817-815

Hertzberg A (1951) A shock tube method of generating hypersonic flows. Journ. Aeron. Sci. 18: 803

Ludwieg H (1955) Der Rohrwindkanal. Z. f. Flugwiss. 3: 206-216

Dumitrescu LZ (1969) (book) Cercetări în tuburile de şoc (Studies in shock tubes, in Romanian). Editura Academici, Bucharest

Dumitrescu LZ (1956) The shock tube and its applications (in Romanian). Studii şi Cercetări de Mecanică Aplicată (a publication of the Romanian Academy of Sciences, Bucharest) 7:283-291

Dumitrescu LZ (1957) Minimum length of a shock tube, for given working times. Revue de Mécanique Appliquée (a publication of the Romanian Academy of Sciences, Bucharest) 2, 1:61-74

Dumitrescu LZ (1959) A shock tube for aerodynamic research (in Romanian). Studii şi Cercetări de Mecanică Aplicată (a publication of the Romanian Academy of Sciences, Bucharest) 10, 1:271-285

Dumitrescu LZ (1960) On the limiting Mach number of the flow in a shock tube (in Romanian). Studii şi Cercetări de Mecanică Aplicată (a publication of the Romanian Academy of Sciences, Bucharest) 11, 2:523-524

Dumitrescu LZ, Jakab I, Procopovici E, Zaharescu A (1960) Some problems concerning experimental research in high-speed aerodynamics, with the aid of shock tubes (in Romanian). Studii şi Cercetări de Mecanică Aplicată (a publication of the Romanian Academy of Sciences, Bucharest) 11, 6:1599-1605

Procopovici E, Dumitrescu LZ (1961) Measurement of aerodynamic pressures in a shock tube (in Romanian). Studii şi Cercetări de Mecanică Aplicată (a publication of the Romanian Academy of Sciences, Bucharest) 11, 4:185-194

Jakab I, Zaharescu A, Dumitrescu LZ (1961) Measurement of aerodynamic parameters in very short-duration supersonic flows. Proc. III Intl. Congress on Measurement Techniques (IMEKO), Budapest, pp 131-134

Jakab I, Zaharescu A, Dumitrescu LZ (1962) A method for the measurement of the propagation velocity of shock waves. Revue de Mécanique Appliquée (a publication of the Romanian Academy of Sciences, Bucharest) 7, 1:173-183

Jakab I (1962) Ein elektrisches Verfahren zur optimalen Dämpfung von Gebern mechanischer Grössen. Revue de Mécanique Appliquée (a publication of the Romanian Academy of Sciences, Bucharest) 7, 6:1195-1202

Dumitrescu LZ, Jakab I, Zaharescu A (1963) Design and instrumentation of a supersonic shock tube. Report IMF (Inst. Fluid Mechs. Romanian Academy of Sciences, Bucharest)

Dumitrescu LZ, Poeniţă I (1963) Design of a hypersonic shock tunnel. Report IMF (Inst. Fluid Mechs. Romanian Academy of Sciences, Bucharest)

Dumitrescu LZ (1964) The development and use of capacitive pressure transducers for shock tube measurements. Report IMF (Inst. Fluid Mechs. Romanian Academy of Sciences, Bucharest)

Dumitrescu LZ (1964) A method for the dynamic calibration of pressure transducers. Revue de Mécanique Appliquée (a publication of the Romanian Academy of Sciences, Bucharest) 9, 2:467-471

Dumitrescu LZ, Welther L (1964) A method for producing very weak shock waves. Revue de Mécanique Appliquée (a publication of the Romanian Academy of Sciences, Bucharest) 9,4:827-833

Dumitrescu LZ (1966) The propagation of shock waves around obsacles and in bent ducts. In: Broadbrooke J, Bruce J, Dexter RR (eds) Aerospace Proceedings 1966 (Proc. RAeS Cent. Congress, in conj. with Vth ICAS Congress, London). McMillan, London pp 187-205

Dumitrescu LZ (1967) Techniques and results of model pressure measurements in shock tubes.In: Reichenbach H (ed) Proc. VI Intl. Shock Tube Symp. Freiburg iB, Ernst-Mach Inst. Bericht 4/68:578-588

Dumitrescu LZ (1968) The interaction of weak shock waves and two-dimensional obstacles. Revue de Mécanique Appliquée (a publication of the Romanian Academy of Sciences, Bucharest) 13, 3:453-477

Dumitrescu LZ, Brun R (1969a) Sur les conditions réservoir d'une soufflerie à choc. Comptes Rendus de l'Acad. Sci. Paris, Série A, T.268:827-830

Dumitrescu LZ, Popescu C, Brun R (1969b) Experimental study of the shock reflection and interaction in a shock tube. In: Glass II (ed) Proc. VII Intl. Shock Tube Symp. Toronto, Canada, pp 751-770

Dumitrescu LZ, Popescu C (1970) Production of high-intensity shock waves, with application to the generation of high-enthalpy hypersonic flows (in Romanian). Studii şi Cercetări de Mecanică Aplicată (a publication of the Romanian Academy of Sciences, Bucharest) 29, 1:185-201

Dumitrescu LZ, Constantinescu I (1971) Pressure measurements in short-duration aerodynamic facilities. Compensation of errors introduced by connecting ducts. In: Stollery RL, Gaydon AG, Owen PR (eds) Proc. VIII Intl. Shock Tube Symp. London, pp. 16/1-16/3

Dumitrescu LZ (1972) An attenuation-free shock tube. The Physics of Fluids 15, 1:207-209

Dumitrescu LZ, Moisin T (1972) A new concept for correcting the attenuation effects in a shock tube. Revue de Mécanique Appliquée (a publication of the Romanian Academy of Sciences, Bucharest) 17, 3:491-503

Dumitrescu LZ (1972) Experimental techniques for transonic testing in shock tubes. Lecture Notes, AGARD Symp. on Transonic Aerodynamic Testing, von Karman Institute, Bruxelles, Belgium

Dumitrescu LZ, Brun R, Sides J (1973) Computation of gas parameters and compensation of attenuation effects in real shock tube flows. In: Bershader D, Griffith WC (eds) Proc. IX Intl. Shock Tube Symposium, Stanford, pp 439-448

Dumitrescu LZ (1975) Minimum-length axisymmetric Laval nozzles. AIAA J. 13, 4:520-521

Dumitrescu LZ (1975) Compensation des vibrations de l'ensemble maquette- balance-dard dans les essais aérodynamiques en souffleries à rafales. Joint ONERA-IMF Conference, Paris

Dumitrescu LZ, Preda S (1975) Some new results concerning the diffraction of a shock wave around a convex corner. In: Kamimoto G (ed) Proc. X Intl. Shock Tube Symposium, Kyoto, Japan, pp 369-377

Brun R, Emery M, Tchiftchibachian A, Dumitrescu LZ (1976) Détermination de la géométrie de la chambre haute-pression d'un tube à choc destinée à compenser les effets de la couche limite. Entropie, Paris 67:4-10

Dumitrescu LZ (1979) A large-scale aerodynamic test facility combining the shock- and Ludwieg-tube concepts. In: Lifshitz A, Rom J (eds) Proc. XII Intl. Symp. on Shock Tubes and Waves, Jerusalem, pp 137-146

Dumitrescu LZ, Mitrofan A, Preda S (1981) Studies on mechanical and aerodynamical blast-attenuation devices. In: Treanor C, Hall JG (eds) Proc. XIII Intl. Symposium on Shock Tubes and Waves, Niagara Falls,NY, SUNY Press, pp 753-761

Dumitrescu LZ (1983) Studies on shock wave amplifiation, focusing, and stability. In: Archer RD, Milton BE (eds) Proc. XIV Intl. Symp. Shock Tubes and Waves, Sydney, pp 196-204

Savu G, Trifu O, Dumitrescu LZ (1983) Supression of shocks on transonic airfoils. In: Archer RD, Milton BE (eds) Proc. XIV Intl. Symp. Shock Tubes and Waves, Sydney, pp 92-101

Dumitrescu LZ, Preda S (1985) Shock tube studies of separation avoidance by boundary-layer bleeding on flared bodies in hypersonic flow. In: Bershader D, Hanson RK (eds) Proc. XV Intl. Symp. Shock Waves and Shock Tubes, Berkeley, Calif. pp 503-507

Dumitrescu Z (1991) A novel development of the Ludwieg tube concept, for extended test duration. In: Takayama K (ed) Proc. XVIII Intl. Symp. Shock Waves, Sendai, Japan, pp 991-996

Dumitrescu LZ (1992) An efficient shock focusing configuration. In: Davis MR, Walker GJ (eds) Proc. XI Australasian Fluid Mechs. Conference, Hobart, Tasmania, Vol. 2 pp 723-725

Dumitrescu M-P, Brun R, Billiotte M, Bertoni J-M, Burtschell Y, Canova A, Dumitrescu LZ, Houas L, Labracherie L, Zeitoun D (1993) Dealing with pressure oscillations in Stalker tubes. These Proceedings

Labracherie L, Dumitrescu M-P, Burtschell Y, Houas L (1993) On the compression process in a free-piston shock-tunnel. Shock Waves (Springer) 3:19-23

Saillard Y, Barbry H, Mounier C (1985) Transformation of a plane uniform shock into a cylindrical or spherical uniform shock by wall shaping. In: Hanson RK (ed) Proc. XV Intl. Symposium on Shock Waves and Shock Tubes, Berkeley, Calif. pp 147-153

Computational Analysis on Generic Forms in European Hypersonic Facilities: Standard Model Electre and Hyperboloid-Flare

J. Muylaert[*], L.M.G. Walpot[†] and G. Durand[‡]

[*]Aerothermodynamics Section YPA, ESA-ESTEC, P.O. Box 299, 2200 AG Noordwijk, The Netherlands
[†]Department of Aerospace Engineering, Delft University of Technology, 2600 GB Delft, The Netherlands
[‡]Aerothermodynamics and Guidance, Navigation & Control Section, ESA-CNES, 31055 Toulouse, France

Abstract.   Extensive experiments and associated numerical computations have been carried out on two generic forms. The first generic form is a blunt cone called Electre whose purpose is to serve as a standard model in hypersonic wind tunnel testing. The second form is a hyperboloid flare for the study of the separation and reattachment of the boundary layer in front of a deflected control flap. The Electre standard model does not only serve to improve the flow quality of hypersonic tunnels but also to improve the measurement techniques and data reduction procedures. Sensitivity computations with a finite rate catalytic model were performed in high-enthalpy wind tunnel conditions, in particular the influence of the wall temperature has been addressed. The hyperboloid flare turned out to be an excellent test case for CFD validation especially for the high Reynolds laminar perfect gas regime. This axisymmetric generic form provided also the possibility to study the influence of real gas effects on flap efficiency and heating. An increase in flap efficiency resulting from the shrinking of the separated boundary layer region in front of the flap combined with an increase of the pressure recovery on the flap has been computed when extrapolating from high-enthalpy wind tunnel to flight conditions according to the binary scaling law. In order to separate dissociation effects from Mach and Re effects perfect gas computations with variable γ were performed.

1. Introduction

During reentry, real gas- , viscous interaction- and rarefaction effects are influencing the aerothermodynamics of space planes. One of the prime concerns is the influence of real gas effects on aerodynamic stability and aerothermodynamic parameters.

In Europe two high-enthalpy facilities, which are capable to study these real gas effects, have recently been commissioned: the F4 at ONERA Toulouse France and the HEG at DLR Gottingen Germany. The F4 is a 160 MW arc- heated impulse facility with a run time of some 30 msec and the HEG is a free-piston shock tunnel with a run time of 1 msec.

The development of these new high-enthalpy facilities in Europe went along not only with the development of the associated measurement techniques: classical as well as non-intrusive, but also with the development and validation of nonequilibrium Navier-Stokes codes. These highly sophisticated and cpu time-consuming numerical tools are required for the understanding and interpretation of the wind tunnel measurements and for the extrapolation to flight. Because of the fact that at present we are concerned with the understanding as to how these facilities work, what their flow quality really is and how all this influences the data accuracy requirements, simple configurations are being tested which can be computed with present-day tools. The so-called "rebuilding" of these generic shapes in wind tunnel conditions is part of the validation process to understand the physical phenomenon involved i.e. chemical/vibrational modelization and gas surface interactions and to learn what the required level is in terms of time and space convergence. For this reason, simple axisymmetric configuration have been selected so that the cpu time effort is still acceptable.

Shock Waves @ Marseille I
Editors: R. Brun, L. Z. Dumitrescu

© Springer-Verlag Berlin Heidelberg 1995

In the following, first the numerical tools used for the rebuilding will be explained, then we will elaborate mainly on the numerical results obtained in the new European high-enthalpy facilities for the standard model Electre and Hyperboloid-Flare, followed by some statements on the use of variable γ facilities.

2. Numerical simulations

The procedure followed here is a rebuilding from the reservoir conditions, using a 1D chemical and vibrational multi-temperature nonequilibrium Euler code called LORE1D, developed by Walpot (1991). The Edenfield boundary layer correction is included in this Euler code. This 1D code provides the required input parameters to start subsequently a 2D axisymmetric nonequilibrium Navier-Stokes code.

The non-equilibrium Navier-Stokes code used here is the TINA code developed by Netterfield (1991). The formulation of the inviscid fluxes is based on the approximate Riemann solver of Roe, with second order TVD corrections as given by Yee. Central differencing for the viscous fluxes is used. The solution is advanced in time towards a steady state using a line Gauss-Seidel implicit relaxation algorithm which includes a linearization of the chemical and vibrational source terms. A two-temperature model is used to describe the partition of energy between the translational/rotational and vibrational/electronic modes. The governing equations for chemically reacting, viscous 3D flow, are solved comprising a 5 species mixture of gases (N_2, O_2, NO, N, and O) in thermal and chemical non-equilibrium. The reaction rates from Dunn and Kang have been used. The species conductivities are calculated from species viscosities using the Euken formula. The viscosity of each species is found using the Blottner viscosity model, which is a curve fit of experimental data as function of the temperature.

3. Standard model

3.1. Electre as standard model

Standard model testing in different hypersonic facilities with the aim of obtaining redundant measurements for verification of assumptions made during expansion in the nozzle i.e. to consolidate the relations between reservoir conditions, total or stagnation and static test section conditions, as well as to have some indirect means to assess flow quality, in particular influence of possible solid- or gaseous pollutants, was addressed as to be one of the important activities to reduce uncertainties for hypersonic testing.

A blunt cone, called Electre, was selected as standard model because flight data existed, that due to its bluntness it was considered not to be sensitive to boundary layer transition and because enough instrumentation could be mounted on the model, especially for high-quality static pressure measurements. In addition, due to the nose bluntness, the required sensitivity was obtained to detect nonisentropic effects and therefore flow quality could be better understood. Indeed Boudreau and Adams (1988) found that the so-called "melting" effect was measurable on blunt cones. The melting is the process whereby the vibrational energy locked due to the freezing caused by the expansion in the nozzle is relaxed because of interaction with water vapor or other contaminants. This process causes the static temperature in the test section to rise and therefore the Mach number to drop. It was only when testing a blunt cone and not a sharp cone as done in the past that this phenomenon was discovered. Indeed sharp cone static pressure measurements are insensitive to nonisentropic nozzle effects as compared to blunt cone static pressure measurements.

3.2. Electre in F4 and HEG

Computations were done in the past on the Electre model in the F4 hot shot conditions and in flight both with the same speed and binary scaling (density$\times$length) (Muylaert et al. 1992). Here

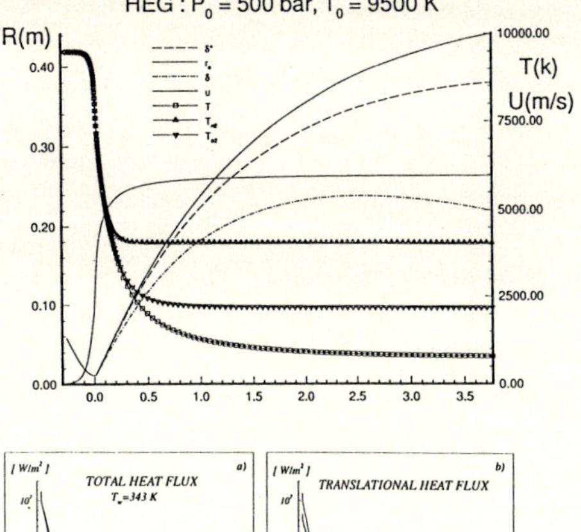

Fig. 1. HEG wall boundary layer- and centerline temperatures as computed with LORE1D

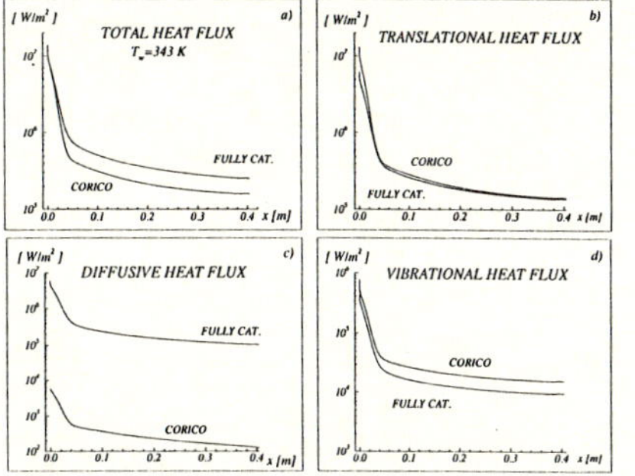

Fig. 2. Total heat flux and its contributions along the Electre model in HEG with $T_w = 343$ K

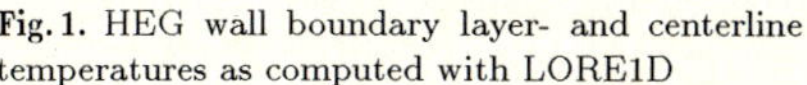

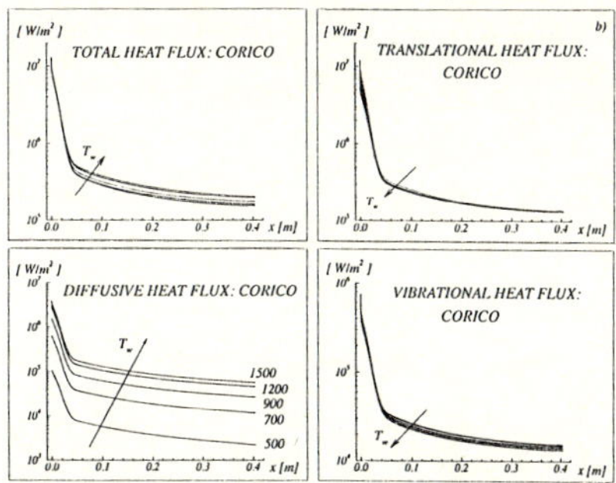

Fig. 3. Total heat flux and its contributions along the Electre model in HEG: influence of silica-coated wall temperature

some sensitivity computations will be shown on Electre in the HEG wind tunnel to study finite rate catalysis and the influence of the wall temperature on some of the important recombination mechanisms. Two reaction mechanisms are possible. The first one is the recombination between an atom from the gas and an adsorbed one: the free gas atom strikes the adsorbed atom, called adatom, and reacts with it. The result is a recombined molecule. This is called the Eley-Rideal mechanism (E-R). After the recombination the molecule is still adsorbed. The second mechanism is the recombination between two adatoms: This is called the Langmuir-Hinshelwood mechanism (L-H).

A finite rate catalytic model CORICO was proposed by Nasuti et al. (1993) for a silica coating and was incorporated in the TINA code by Barbato (1993). Silica was used as a TPS coating on the shuttle. The reservoir conditions for the HEG are shown in Table 2 column 1 together with the test section conditions as computed with LORE1D.

Fig.1 shows for this condition the centerline evolution of the static and vibrational temperatures and the growth of the boundary layer displacement and boundary layer thickness along the nozzle wall. Note the rapid freezing of the vibrational temperatures, the quite thick boundary layers according to the Edenfield correlation and the velocity which quickly reaches constant values. All the important chemical kinetic effects occur at and just downstream of the throat.

Fig.2 shows the Electre wall heat flux distribution with a wall temperature of 343 K. A comparison is shown between full catalytic and CORICO which at that low wall temperature is very close to a non-catalytic wall. A 60 % increase in heat flux is noted for the full catalytic wall relative to the non-catalytic wall. The figure shows also the translational, vibrational and diffusive contributions to the total heat flux. It is clear that the fully catalytic condition is a diffusive dominated process. For the vibrational contribution we see that the fully catalytic wall condition is lower than CORICO. This is because the large number of molecules that leave the wall in local thermal equilibrium cool the gas close to the wall and therefore reduce the vibrational temperature referred to as "quenching".

Fig.3 shows again the heat flux distribution and its contributions on Electre but now with wall temperatures which vary from 500 K to 1500 K as computed with CORICO. The translational heat flux decreases when the wall temperature increases as well as the vibrational contribution where in addition the "quenching" action of the molecules become more intense. The total heat flux increases when increasing the wall temperature due to the diffusive contribution which increases rapidly from 500 K to 900 K and less rapidly from 900 K to 1500 K. The reason for this rapid rise is explained in Figs.4 and 5 which show the contribution of the above-explained E-R and L-H mechanisms to respectively the O and N particle flux. The E-R is the dominating mechanism for the O particle flux over the entire temperature range whereas around 1000 K the L-H mechanism takes over from the E-R for the N particule flux.

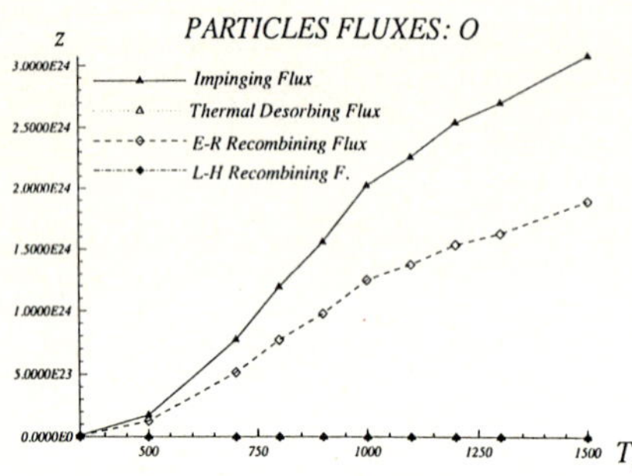

Fig. 4. Oxygen particles flux contributions as a function of the wall temperature

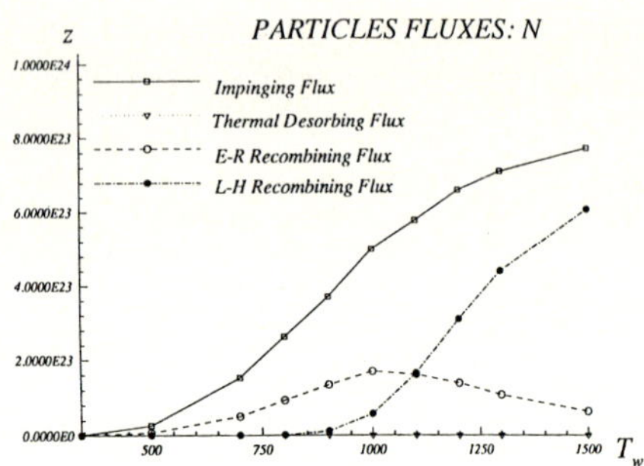

Fig. 5. Nitrogen particles flux contributions as a function of the wall temperature

The above Electre computations show that for this coating the non-catalytic assumption was good enough if the wall remains cold, but that it is necessary to use a finite-rate catalytic model if the wall is hotter than say 700 K in wind tunnel conditions. In addition and perhaps more importantly, the computations have shown that when increasing the wall temperature to flight values as 1500 K the recombination mechanism change! So care should be taken in the analysis of high-enthalpy wind tunnel wall heat fluxes on so called non cat materials and when extrapolating them to flight. More analysis will be done in the future on materials used for wind tunnel and flight testing.

4. Hyperboloid-Flare

4.1. Design of the hyperboloid flare

The hyperboloid flare is an axisymmetric configuration whose contour corresponds to the Hermes windward-centerline. This concept of an "equivalent axisymmetric body" was first proposed by Adams et al. (1977) who showed that an appropriate axisymmetric body at zero degree angle of attack could model the windward-centerline flowfield over the shuttle at a given angle of attack. The results of their investigation demonstrated good agreement between experimental heat transfer data on the windward side of the shuttle at 30° angle of attack and computed heating rates on a 31° half angle hyperboloid. This concept was further extended to include an axisymmetric flare to study the physical phenomenon associated with flap efficiency and heating

(Durand 1992). For the present Hyperboloid-Flare the forward part is not a hyperboloid but the Hermes windward centerline at 30° angle of attack, nevertheless the terminology "Hyperboloid Flare" was kept.

A series of numerical computations were carried out by Schwane and Muylaert (1992) to design the scale of this hyperboloid flare as well as the details of the flare deflection angle and flare length. On one side the separated region must be large enough to be able to obtain the required measurement resolution and on the other side care must be taken to avoid wind tunnel blockage. In the following we will elaborate first on some of the hypersonic perfect gas wind tunnel results and on comparison with computations followed by a computational analysis of this configuration in a high-enthalpy wind tunnel environment.

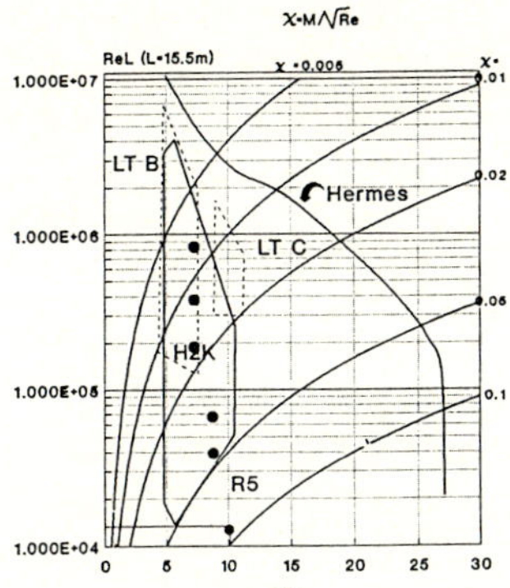

Fig. 6. Mach/ Re performance envelopes for the LTB, H2K and R5 facilities

Table 1. Perfect gas hypersonic wind tunnel conditions for the hyperboloid flare campaigns

	RWG			H2K		R5Ch
M	6.83			8.7		9.95
T_{stag} (K)	700			1155	915	1070
p_{stag} (bar)	9	19	35	7	9	2.7
$Re \times 10^6$/m	3.2	7	15	0.62	1.16	0.19215
$M/\sqrt{Re_L} \times 10^{-2}$	1.57	1.06	0.72	4.53	3.31	9.3

4.2. Cold hypersonic testing

To study the influence of the viscous interaction parameter $M/\sqrt{Re}$ on the shock wave boundary layer interaction- separation and reattachment in the laminar perfect gas regime, three hypersonic research facilities were selected. These facilities are selected so that a large variation of Re nunber can be obtained. These three facilities are the Ludwieg tube at DLR Gottingen, the blowdown facilities H2K at DLR Koln and R5 at ONERA Chalais Meudon. Table 1 shows the test conditions for the three facilities used for the present analysis. Fig.6 shows the details of the performance envelopes of these facilities in terms of Mach / Reynolds together with the testing points corresponding to Table 1, shown as full symbols.

A typical computational result is shown in Fig.7 depicted from Schwane (1993) where the iso- C_p lines are shown. The interactions of the bow shock with the separation and reattachment shocks are clearly visible and well captured. This particular case turned out to be quite difficult to converge as an extreme fine grid was required to approach the measured separation length as seen on Fig.8.

Fig.8 shows the separation length, nondimensionalized with the model length, as a function of the viscous interaction parameter. The separation length was measured from the point of separation up to the hinge line of the flare. The separation length in the R5 facility was depicted from the oil flow visualizations, in the H2K from the pressure measurements and in the Ludwieg tube from liquid crystal and oil flow measurements. A series of perfect gas Navier-Stokes computations are shown with grid refinements from ONERA (Bousquet et al. 1994), DLR (Brenner et al. 1993) and from ESTEC (Schwane 1993). The agreement is not so good except when extremely fine grids are used to capture all the details of this viscous /inviscid interac-

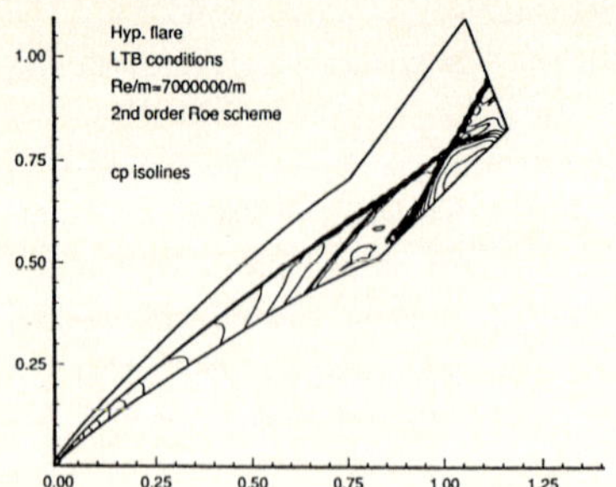

Fig. 7. Iso- C_p lines on hyperboloid flare for LTB conditions: 7×10^6 Re/m, Mach $= 6.83$

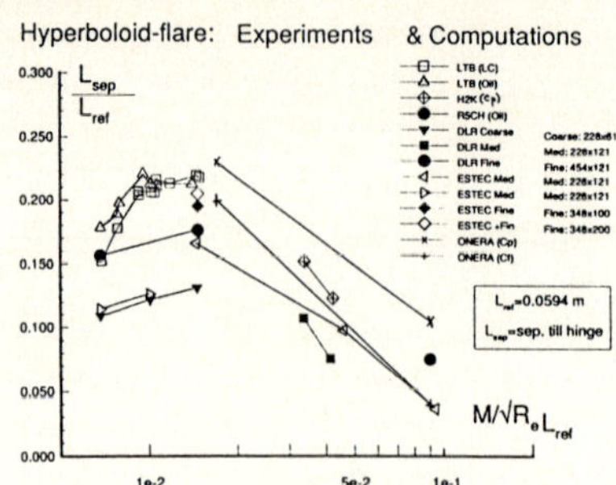

Fig. 8. Comparison between measured and computed hyperboloid flare separation length

tion. A special effort was made to rebuild the Ludwieg Tube 7 million unit Reynolds number case: An extremely fine grid was required to approach the measured separation length. There is some evidence that the remaining discrepancies are due to the misalignment of the separated shear layer with respect to the grid. Apparently the obliqueness of the shear layer with respect to the structured computational grid is not well resolved by the standard 2nd order schemes. The remaining discrepancies between experiment and computations are being analyzed and drive the experimentalist to recheck nozzle flow quality and measurement chain calibration as well as the numericist to make sure that his computations are fully grid and time converged.

Table 2. Real gas wind tunnel and flight conditions for the Hyperboloid-Flare computations

HEG exit conditions: $p_0 = 500$ bar, $T_0 = 9500$ K

	chem/thermal non-equil.	chemical non-equil.	perfect gas	flight extrapolation
Re_L	38000.		38000.	121000.
M	9.75	8.34	10.56	23.32
$M/\sqrt{\mathrm{Re}_L}$	0.05		0.054	0.067
U (m/s)	5953.0	5996.0	5953.0	5953.0
T_∞ (K)	789.29	1051.0	789.29	162.0
$T_{v\infty}$ (K)	3930.0	1051.0		
ρ (kg/m^3)	2.017e-3	2.122e-3	2.017e-3	$\rho/120.$
L (m)	0.1114	0.1114	0.1114	$L \times 120$
C_{N_2}	0.75446	0.7548		0.77
C_N	0.00006	0.0		0.0
C_{O_2}	0.03649	0.0346		0.23
C_O	0.18197	0.1843		0.0
C_{NO}	0.02709	0.0263		0.0
T_w (K)	300	300	300	300

Fig. 9. Influence of nozzle vibrational nonequilibrium on the Hyperboloid-Flare C_p distribution

4.3. Hot hypersonic testing

This section will cover a computational analysis on the hyperboloid flare in high-enthalpy conditions. These conditions were again chosen to be those corresponding to the HEG reservoir conditions of 500 bar and 9500 K. The nozzle exit conditions are computed with the LORE1D code, as already mentioned, and are shown in Table 2. The first column in Table 2 shows the exit conditions when chemical and vibrational nonequilibrium is assumed, the second column shows the nozzle exit conditions when the vibrational excitation is assumed to be in equilibrium and the other columns will be discussed below. When comparing the two columns we can see that the translational temperature is increased by 262 K for the vibrational equilibrium case because of the release of the vibrational energy which is locked into the vibrational modes. This increase in temperature causes the Mach number to drop, similarly to the "melting effect" discussed above in the chapter on "Electre as standard model".

Fig.9 shows the results for both inlet conditions on the hyperboloid flare computed with TINA. It has to be noted that both computations were carried out with chemical and vibrational nonequilibrium TINA. It can be seen that the nonequilibrium vibrational contribution in the nozzle has no measurable effect on the C_p and St distribution.

At present, a more systematic experimental and associated numerical study is being done in the two European high-enthalpy facilities F4 and HEG to assess the influence of different chemical models and vibrational couplings on nozzle flow, their influence on shock standoff distance, shock layer characteristics and wall pressure and heat flux measurements. These sensitivity studies are being performed on the standard model Electre.

5. Flight extrapolation

5.1. Wind tunnel and flight computations

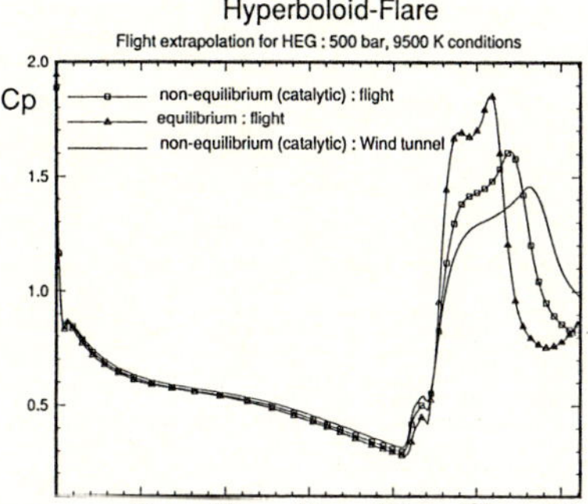

Fig. 10. C_p distribution on hyperboloid flare: comparison between wind tunnel and flight

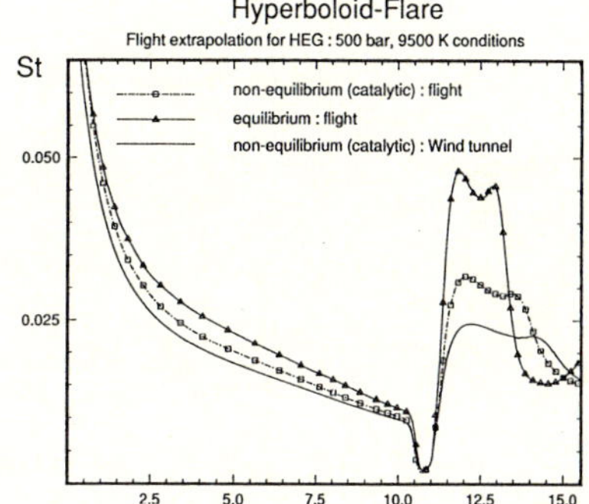

Fig. 11. St distribution on hyperboloid flare: comparison between wind tunnel and flight

Obviously, the way to study the extrapolation to flight is to perform computations in wind tunnel and in flight conditions following some scaling law. The scaling law usually advocated in hypersonics is the binary scaling law which simulates the dissociation reactions and the speed to simulate the kinetic energy. The binary scaling parameter is the product of the density times the length. Table 2, last column, shows the flight conditions where a scale factor of 120 is used. The latter corresponds to the Hermes full scale body length. Note that in Table 2 columns 1 and 4 have the same speed and the same $\rho \times L$, but not the same Reynolds number.

Figs.10,11 show respectively the C_p and the St distribution as computed by TINA for the non-equilibrium HEG wind tunnel condition, for the non-equilibrium flight and the equilibrium flight condition. Some important conclusions can be drawn: an increase in pressure recovery on the flare for flight of approximately 15% relative to the wind tunnel case is noted and another increase for flight in equilibrium of approximately 25% relative to the flight nonequilibrium case can be seen. Similarly, for Fig.11, one can see that an increase in Stanton number on the flare for flight of 30% relative to the wind tunnel and another increase of approximately 50% for equilibrium flight relative to nonequilibrium flight is obtained. Clearly the assumption of chemical and vibrational equilibrium leads to a significant overestimation of the peak pressure and peak heat flux on the flare. Going from wind tunnel to flight, a reduction is seen in separation length combined with a forward shift of the pressure peak giving rise to an increase in flap efficiency.

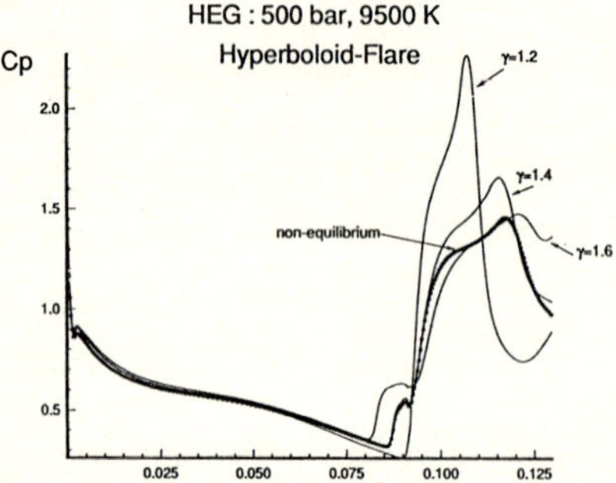

Fig. 12. Influence of γ on hyperboloid flare C_p distribution

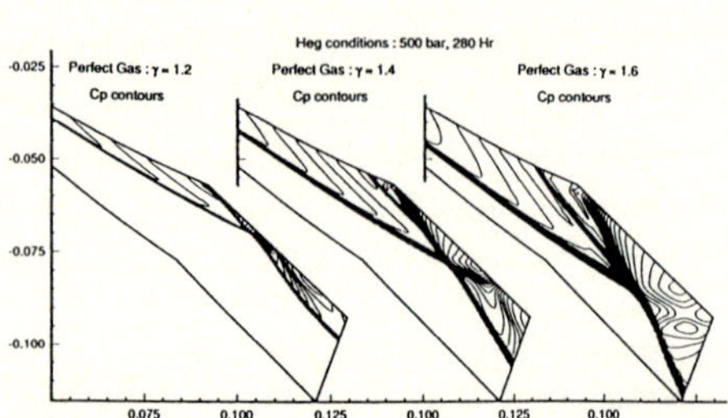

Fig. 13. Influence of variable γ on shock-boundary layer interaction

5.2. Variable γ perfect gas computations

What is still not clear is whether the above increase in flap efficiency is a pure real gas nonequilibrium effect or whether there is also a Re or perhaps a Mach effect. Indeed Table 2 shows that going from wind tunnel (column 1) to flight (column 4), we have an increase in Re primarily due to the much lower static temperature in flight as compared to the wind tunnel, notwithstanding the fact that the density×length and the speed is the same. To answer this question perfect gas computations were performed in wind tunnel and in flight conditions with input conditions as shown in columns 3 and 4 of Table 2. Fig.12 shows a comparison between the nonequilibrium wind tunnel computation and three sets of perfect gas computations with different γ: 1.2, 1.4 and 1.6. The perfect gas computations were started with the exit conditions as Table 2 column 1 but with $\gamma = 1.4$ in order to compare perfect gas with nonequilibrium real gas. Note that there is a slight difference in Mach due to the small increase in γ for the nonequilibrium nozzle case. ($\gamma = 1.44$). Fig.12 shows a remarkable influence of γ. This is also shown in Fig.13 by the iso C_p plots to highlight the change in structure of the shock wave boundary layer interaction. Decreasing γ, with constant Re and only a slight increase in Mach (less than 10% relative to $\gamma = 1.4$), produces a reduction in size of the separation region combined with an increase and forward shift of the pressure peak. Decreasing γ produces also therefore an increase in flap efficiency as obtained above when discussing the extrapolation from wind tunnel to flight. Comparing the $\gamma = 1.4$ perfect gas results with the nonequilibrium wind tunnel results one finds that the separation length is not much altered but that the pressure distribution on the flare follows first the $\gamma = 1.4$ results and more downstream drops to the $\gamma = 1.6$ results and then back to the $\gamma = 1.4$. What is happening?

To try to understand these complex nonequilibrium interactions Fig.14 is plotted. It shows the O_2 and N_2 mass fractions plotted perpendicular to the wall along the hyperboloid flare in wind tunnel and in flight corresponding to Table 2 column 1 and 4; but let's focus here to the wind tunnel N_2 mass fractions. If one follows a line perpendicular to say the $x = 0.05$ m position, then the N_2 mass fraction starts with the nozzle freestream value of 0.754 in accordance with Table 2 column 1; following this line perpendicular to the wall one crosses the bow shock which dissociates the N_2 molecules up to a value in mass fraction of 0.715, further in the shock layer and in the boundary layer due to the cold wall and fully catalytic wall boundary conditions, the N_2 molecules are forced to recombine to reach there equilibrium conditions at the wall i.e. a mass fraction of 0.77. Comparing Fig.12 with Fig.14 in the region of the shock interaction one can see

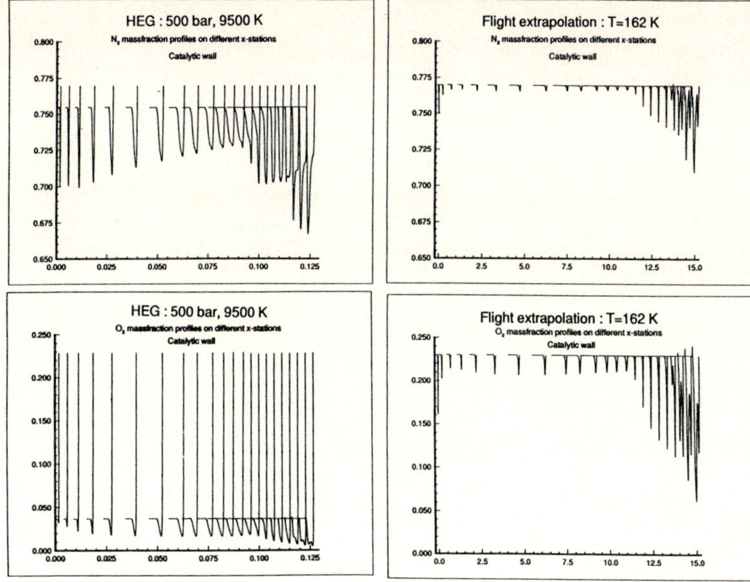

Fig. 14. Comparison between wind tunnel and flight O_2 and N_2 mass fraction profiles perpendicular to wall

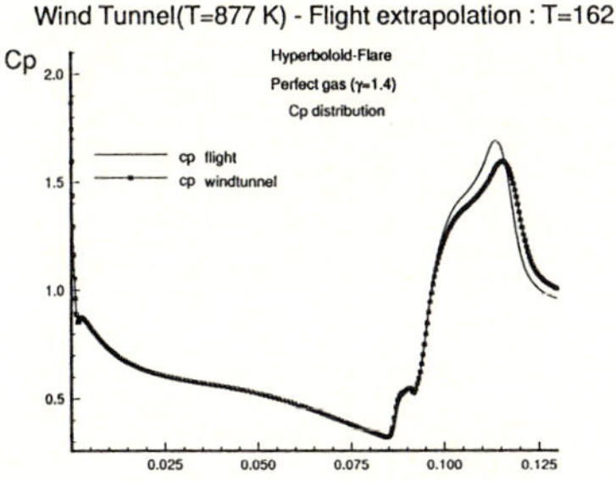

Fig. 15. Comparison between equivalent wind tunnel HEG $\gamma = 1.4$ and perfect gas flight

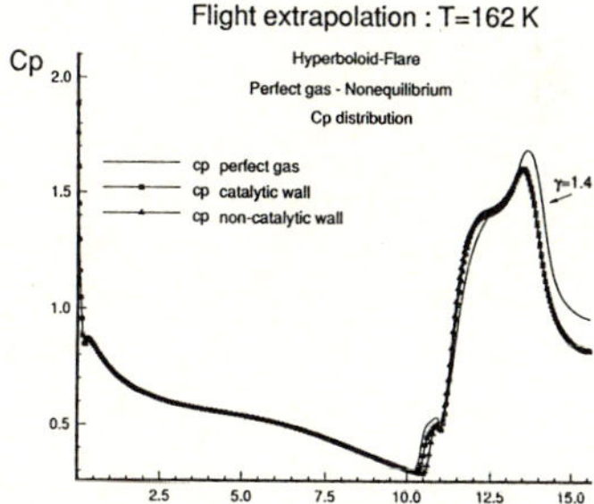

Fig. 16. Comparison between wind tunnel nonequilibrium and flight

that at $x = 0.09$ m the N_2 mass fraction is reduced again because of the existence of the separation and reattachment shock; at $x = 0.1$ m the expansion waves coming from the interaction point with the bow shock balance or freeze the composition at that level until at $x = 0.115$ where the stronger flare shock induces the N_2 molecules to dissociate until up to a level of 0.670. The C_p distribution for the nonequilibrium wind tunnel in Fig.12 follows quite closely the perfect gas $\gamma = 1.4$ distribution until $x = 0.1$ m downstream of which it follows the $\gamma = 1.6$ curve up to the $x = 0.115$ point to follow again the $\gamma = 1.4$ curve.

Let us check now what happens in flight. Fig.15 compares perfect gas wind tunnel with perfect gas flight both with $\gamma = 1.4$ according to Table 2. Only a slight reduction of the separation is seen combined with an increase of the pressure peak and a shift upstream. This increase of pressure peak seems Mach number driven as with a constant γ and same shock inclination angle, the pressure ratio across an oblique shock is only a function of the local normal shock Mach number. In flight this local normal shock Mach number is approximately 3.8 and in the wind tunnel this is 3.2 as taken from the present computations.

Fig.16 compares perfect gas flight with nonequilibrium flight C_p distribution. A reduction of the separation bubble is seen combined with a shift forward in pressure peak. Here the change from perfect gas to nonequilibrium in C_p is not so dramatic as in wind tunnel (Fig.12). A similar behaviour was obtained by Brenner et al. (1993) but for a higher Mach and altitude.

6. Conclusion

The present paper has shown the importance of performing sensitivity studies on simple shapes in the high-enthalpy wind tunnels F4 and HEG. The two shapes considered were the Electre standard model and the hyperboloid flare. The computations on the Electre model revealed the importance of using a finite rate catalytic model when extrapolating from wind tunnel cold wall to flight hot wall conditions. The hyperboloid flare allowed us to study the detail shock interaction effects in cold as well as in hot hypersonic conditions. In addition the influence of nonequilibrium real gas effects and the extrapolation to flight on flap efficiency was examined.

Acknowledgment

The authors would like to thank Pr. Brun and Pr. Dumitrescu for the invitation to present this paper. Special thanks goes to M. Barbato for his contributions related to CORICO. Finally the authors would like to thank their colleagues, especially R. Schwane, M. Netterfield, and G. Simeonides for the fruitful and stimulating discussions.

References

Adams JC, Martindale WR, Mayne AW, Marchand EO (1977) Real gas scale effects on shuttle orbiter laminar boundary layer parameters. J.Spacecraft 14(5)

Barbato M (1993) Heterogeneneous catalysis model for hypersonic flight simulations. Technical Report ESTEC EWP-1731, ESA-ESTEC

Boudreau AH, Adams JC (1988) Characterization of hypersonic wind tunnel fields. In: AIAA 15th Aerodynamic Testing Conference, San Diego, AIAA Paper 88-2006

Bousquet J, Faubert A, Oswald J (1994) Computations of laminar hypersonic flows around the hyperboloid flare.Technical Report H-NT-O-2078-ONERA

Brenner G, Kordulla W, Bruck S (1993) Further simulations of flows past hyperboloid-flare configurations. Technical Report DLR-H-NT-O-2058, DLR-Braunschweig

Durand G (1992) Hyperboloid flare combination: a hypersonic test case for the qualification of compressible Navier-Stokes codes including high temperature effects. Technical Report DLA/ED/3A no 032.09.92, CNES

Muylaert J, Walpot L, Haeuser J, Sagnier P, Devezeaux D, Papirnyk O, Lourme D (1992) Standard model testing in the European high entalpy facility F4 and extrapolation to flight. In: AIAA 17th Aerospace Ground testing Conference, Nashville,TN. AIAA Paper 92-3905

Nasuti F, Barbato M, Bruno C (1993) Material-dependent catalytic recombination modeling for hypersonic flows. In: AIAA 28th Thermophysics Conf., Orlando, FL. AIAA Paper 93-2840

Netterfield MP (1991) Computation of the aerodynamics of spinning bodies using a point implicit method. In: Proceeding of the First European Symposium on Aerothermodynamics for Space Vehicles. AIAA Paper 91-0339

Schwane R (1993) Numerische Simulation von Stoss-Grenzschischt Wechselwirkungen und deren Auswirkung auf den Wirkungsgrad von Aerodynamischen Kontrollelementen. In: 6th STAB Jahresbericht

Schwane R, Muylaert J (1992) Design of the validation experiment: Hyperboloid-flare. Technical Report Document: YPA/1256/RS, ESA-ESTEC

Walpot LMG (1991) Quasi one-dimensional inviscid nozzle flow in vibrational and chemical nonequilibrium. Diploma Thesis, Tech. Rep. ESTEC EWP-1664, TU Delft, Dept. Aerosp. Eng.

Shock Tube Investigations of Combustion Phenomena in Supersonic Flows

G. Smeets
French-German Res. Inst. of Saint Louis (ISL) F-68301 Saint-Louis Cedex, B.P. 34, France

Abstract. A survey is given of recent combustion dynamics research using shock tubes. There are three different fields where phenomena in supersonic gas flows resulting from chemical heat release are of particular interest:
- Combustion in an external flowfield for lift and thrust generation on hypersonic vehicles, for drag reduction or for piloting hypervelocity projectiles.
- Scramjet combustor: detailed information on the supersonic combustion can be obtained with "direct connect" shock tube experiments.
- Ram acceleration of projectiles in a tube by self-synchronized ignition of a combustible gas mixture. The ignition and combustion processes in the flow field around the high-speed projectile can be studied in an expansion tube.

Key words: External combustion, Scramjet, Ram-accelerator

1. Introduction

Combustion research in shock tubes has a long tradition. A review of the preceding meetings - originally termed as Shock Tube Symposia - shows that combustion chemistry covers a nearly constant fraction of about 20% of the contributions: argon gas instantly heated behind the incident or reflected shock is a well defined heat bath for precise investigations in the field of chemical kinetics including pyrolysis, induction, ignition, soot formation and particle growth. Another classical research object has been gas detonations with all its different aspects like initiation and stability of detonations, overdriven detonations, two-phase detonations, etc. Furthermore, combustion physics concerns the operation of some of the shock tube facilities: combustion heated driver gases and combustion phenomena in hydrogen-driven shock tunnels where driver and driven gases get mixed in the contact region.

Table 1. Characteristics of the three types of experiments

	External combustion	Scramjet	Ram accelerator
Facility	Shock tunnel	Shock tube	Expansion tube
U m/s	1500-3000	2000-5000	1500-2300
T K	300	1000-2000	300
Fuel	Gas/Liquid	H_2	Gas
Premixed	No	No	Yes

Except for a very few pioneering experiments, the potential of the shock tube for combustion dynamics (study of phenomena in compressible gas flows resulting from chemical heat release) has been discovered only during the last years. The shock tube competes well as a low-cost alternative with continuous or blowdown high-enthalpy tunnels. As most of the combustion dynamics phenomena occur at near atmospheric pressure, the flows to be generated possess both high enthalpy and high density. A shock tube or shock tunnel as an impulse facility only generating such a flow for some milliseconds is not faced with two problems existing for long duration tunnels: supply of the necessary energy and power (which can be in the range of gigawatts for large test sections) and severe heating of the tunnel and test chamber structures. After a number of powerful non-intrusive and highly time-resolving diagnostic techniques have been developed during

Shock Waves @ Marseille I
Editors: R. Brun, L. Z. Dumitrescu

the last years, there is now quite a choice of diagnostics for the study of combusting shock tube flows.

Presently, shock tube experiments focus on the following three different objectives:
- External combustion: initiating and sustaining a flame at the outer surface of a body in a supersonic or hypersonic air flow. The practical applications under consideration range from drag reduction and lift generation for hypersonic vehicles to piloting hypervelocity projectiles.
- Experiments in relation to the scramjet: combustion in a hot supersonic air stream passing through a duct. Quite a lot of research work on the scramjet has been started only recently, originating from the US initiative to build a National Aerospace Plane and from similar projects in other countries. Shock tubes are particularly well suited for detailed studies of the supersonic combustion with so-called direct connect experiments.
- Ram acceleration: accelerating projectiles in a tube by self-synchronized ignition of a combustible gas mixture. This concept has generated considerable interest in different countries over the past five years. Several laboratories have started activities in this new field of research. For detailed studies of the ignition and combustion phenomena occurring in the flow field around the high-speed projectile, shock tube studies have been undertaken.

The common characteristic of the three types of shock tube experiments is that the combustion takes place at near atmospheric pressure in a supersonic flow. Similar diagnostic techniques can be applied, capable of measuring within a test time of the order of a millisecond. In Table 1, some typical characteristics of the three types of experiments are compared.

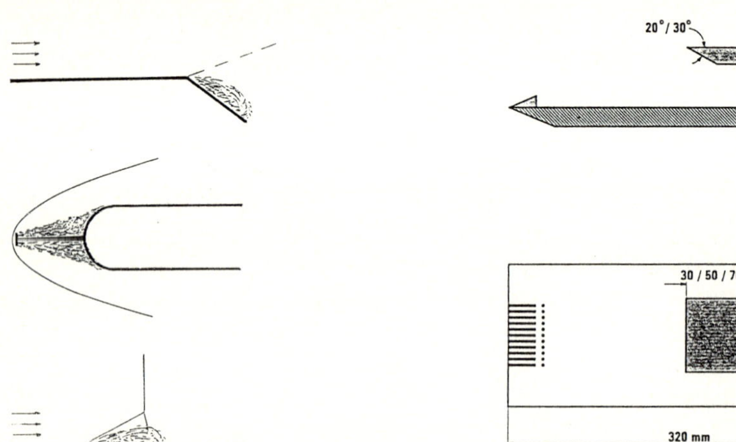

Fig. 1. Combustion in front of a blunt body, on a tangential surface and behind a corner

Fig. 2. Geometry of one of the model configurations

2. External combustion

As explained in Fig.1, a flame can be stabilized in front of a blunted body in the detached region created by a protruding spike, at the tangential wall in the detached flow upstream of a jet spoiler or a strong shock wave and behind a corner at the rear end of a body.

Shock wave-stabilized combustion on tangential surfaces was studied in a shock tunnel at ISL (Smeets 1992). The calculated parameters of the tunnel flow were: flow velocity: 2130 m/s, static pressure: 0.5 bar and static temperature: 530 K. The experiments were performed on a (12 cm wide and 32 cm long) flat plate with a sharp leading edge as basic model. The plate was equipped for injecting gaseous or liquid fuels from various orifices and at different locations on its surface into the tangential flow. In most experiments, a fence of sawteeth was arranged at the front of the plate. By this artificial means, a turbulent boundary layer having well-defined dimensions (8 mm high, 50 mm wide) can be generated.

In order to generate a strong shock wave, different types of structures were fixed to the surface of the plate. The geometry of one of the model configurations is described in detail in Fig.2: a 15 mm thick plate is arranged at 40 mm distance parallel to the ground plate. At its front, it begins with a wedge of a variable angle. The plate is supported by side plates arranged at variable distances behind the leading edge of the parallel plate.

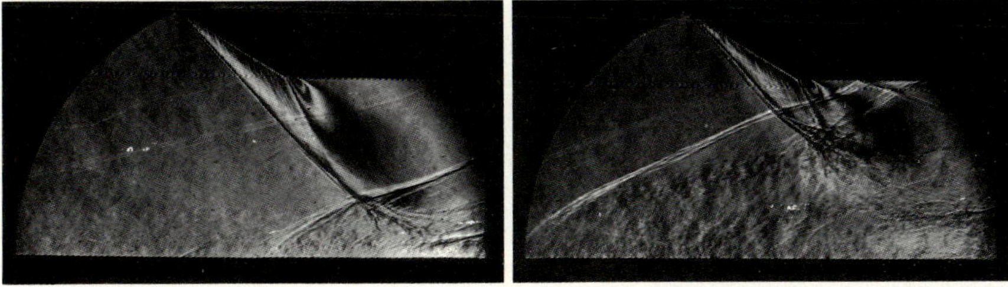

Fig. 3. Schlieren interferograms from experiments without and with H_2 injection

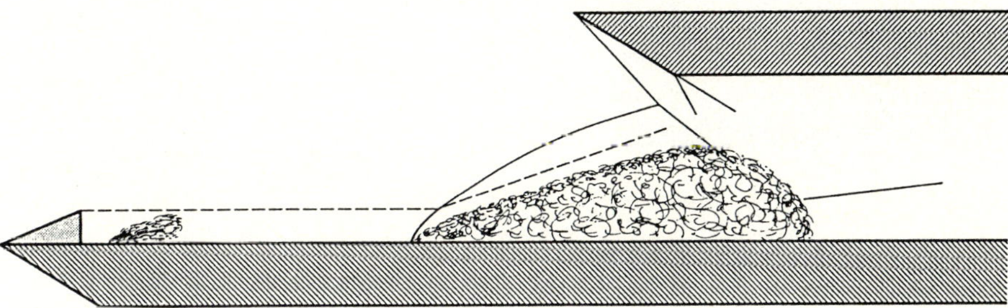

Fig. 4. General lines of the external combustion phenomenon

The Schlieren interferograms in Fig.3 result from experiments made with a 30° wedge angle and with side plates arranged 70 mm downstream. The strong oblique shock wave leads to flow detachment on its interaction with the ground plate boundary layer (upper picture). When hydrogen was injected into the upstream boundary layer, the detached zone expanded widely in the upstream direction and became a stable flame (right-side picture in Fig.3).

Although many details of this rather complicated phenomenon are not quite clear yet, the general lines can at least be described qualitatively. The essential features taken from the experiments supplemented by obvious assumptions are assembled in Fig.4 which is a reproduction of the right-side picture in Fig.3: An artificially turbulent layer (8 mm high and 50 mm wide) is generated by the sawtooth array on top of the plate (11 lamellas, 2 mm thick). In the example under discussion, hydrogen is injected into the artificially enhanced turbulent boundary layer and is premixed with the air by the turbulence. It issues immediately behind the sawtooth array at a rate of 12 g/s from 11 holes having 2 mm in diameter and arranged in a linear array normal to the flow direction. The equivalence ratio of the mixture in the turbulent layer can be estimated to be approximately 2.0 (by relating the above-mentioned fuel rate to the mass flux rate of air, $\rho_\infty \cdot u_\infty$ in the 8 mm by 50 mm cross-section).

As the boundary layer detaches from the wall, the fuel/air mixture going with the detached flow passes over the recirculating flow region. The contact surface dividing the detached supersonic flow from the subsonic detached flow bubble cannot be seen directly on the flow visualization pictures in Fig.3. It shows itself at the point where the oblique shock wave coming down from the wedge suddenly disappears. There is obviously such a strong turbulent exchange across the slip surface that an important amount of the fuel crosses over into the detached zone to support combustion.

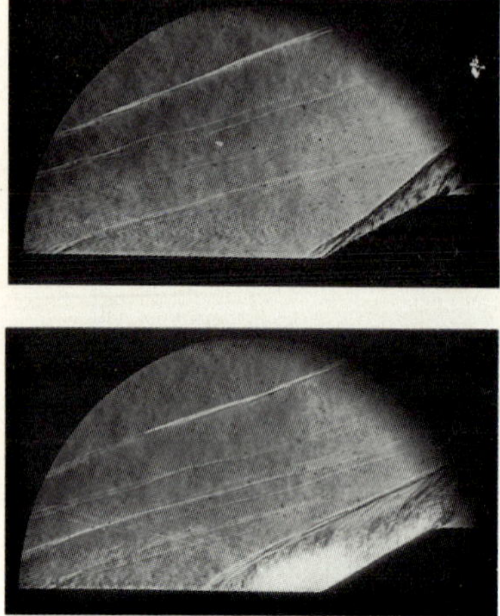
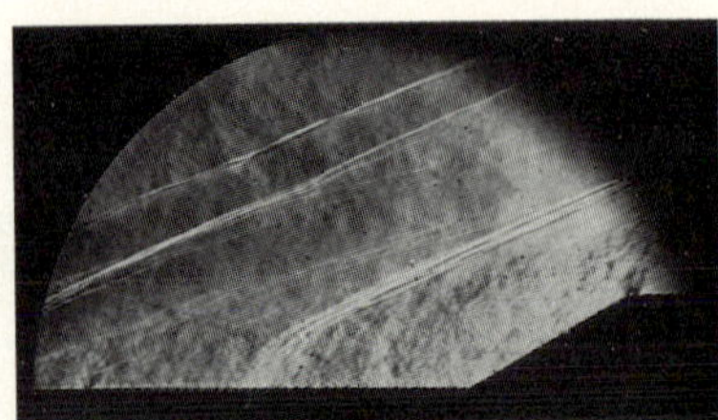

Fig. 5. Detachment in front of a forward-facing corner by fuel injection

In all experiments, auto-ignition occurred. By means of a dual flow experiment (filling nitrogen into the last part of the shock tunnel) it was verified on streak records that when the tunnel flow switches from nitrogen to air, stable combustion appears after a delay of only a fraction of a millisecond. In Fig.5, a detached region in front of a forward-facing corner was created by fuel injection. In the high Mach number flow, the shock wave is very close (and almost still attached) to the 30° angle wedge. By comparison with the no-injection case, the effects of injecting 12 g/s hydrogen and 22 g/s acetylene, respectively can be seen. Acetylene injection results in a luminous combustion zone in a detached flow region in front of the corner.

Although among all tested fuels, hydrogen proved by far to have the greatest effect, a stable external combustion could also be obtained with other combustible gases (acetylene) and even with liquid fuels (heptane).

Base burning behind a corner by injecting hydrogen at the rear end of a body was studied at RAE also in a shock tunnel (Holbeche 1980).

3. Scramjet

A generic aerospace plane with integrated scramjet motor is outlined in Fig.6. Shock tubes are particularly well suited for so-called direct connect experiments. Detailed investigations can be made of the isolated processes of mixing and combustion of hydrogen fuel in the rapidly passing compression-heated supersonic air flow. It is hoped that, by this, general guidelines for optimizing the net thrust generation in future scramjet engines can be obtained. Fig.7 outlines direct connect shock tube experiments of scramjet combustion being performed at ISL (Quenett 1994). The shock heated air is expanded in a divergent in order to increase the flow Mach number to about 3.5.

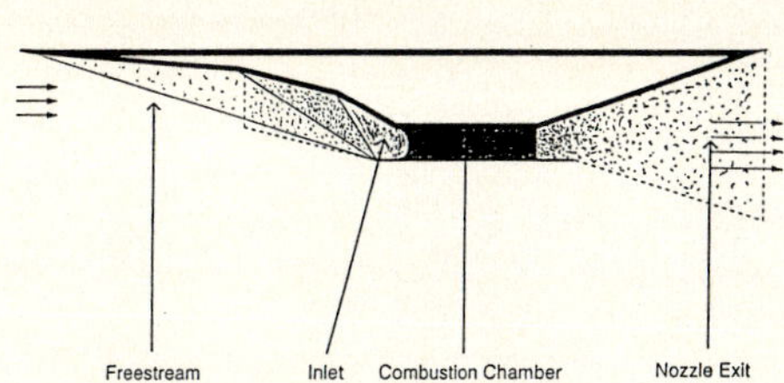

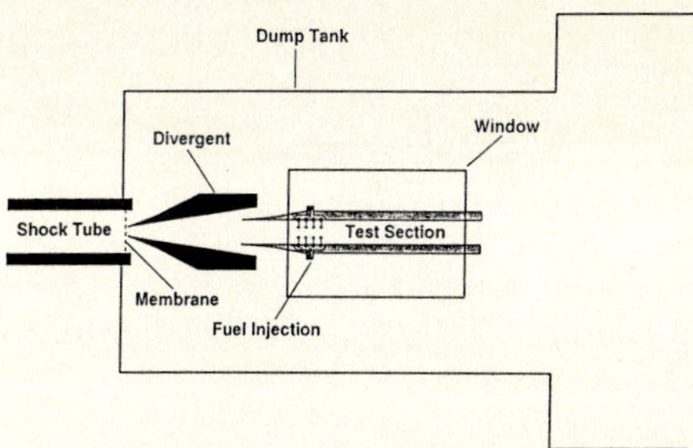

Fig. 6. Generic aerospace plane with integrated scramjet motor

Fig. 7. Direct connect shock tube experiment of scramjet combustion

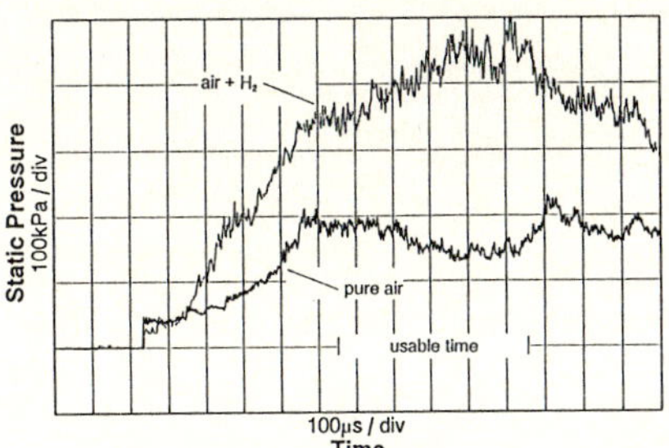

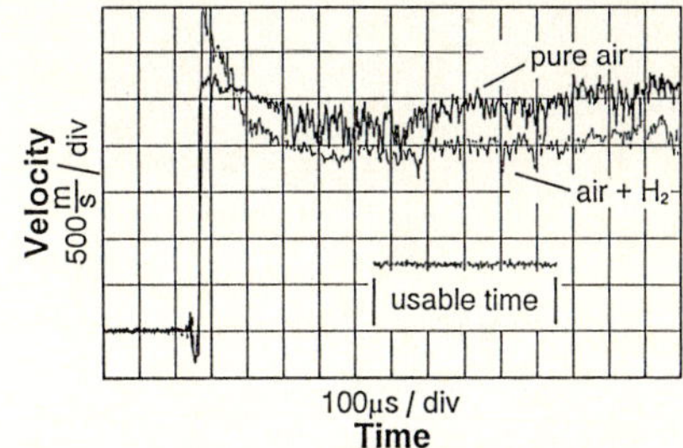

Fig. 8. Pressure records from experiments with and without hydrogen injection

Fig. 9. Recordings of the axial flow velocity obtained by laser Doppler in the center of the duct

The flow parameters at the exit of the divergent are: u of the order 2500 m/s, p between 1 and 2 bar and T between 1200 and 1600 K. The stream of air entering the constant area duct of 60 mm by 60 mm cross section then corresponds to the expected flow after inlet compression when flying at about Mach number 9 at about 40 km flight altitude. Hydrogen is injected from linear arrays of holes in two opposite side walls normally to the high-temperature air stream.

For the optimization of combustion efficiency and thrust generation, the fluxes of both momentum and energy have to be monitored along the duct. For this, the increase of the wall pressure as well as the decrease of the flow velocity have to be recorded simultaneously. A pair of records obtained with wall pressure gages in similar experiments with and without hydrogen injection are compared in Fig.8.

Fig.9 shows two corresponding recordings of the axial flow velocity in the center of the duct obtained by a laser Doppler technique. Scattered light was generated by polydisperse TiO_2 particles having an average size of 0.32 μm. After being added to the air in a special disperser, the particles were filled into the driven tube together with the air. During the delay of a few minutes necessary for preparing the shock tube experiment, they remained suspended because of their small size. For velocity measurements of the scattering particles embedded in the flow, the arrangement shown in Fig.10 is applied. Monochromatic laser light is transmitted by a light fiber and is concentrated in a small measuring volume within the flow field. A bundle of Doppler-shifted light scattered by the tracer particles passing through this volume is collected and transmitted by a second light fiber to a special interference spectrometer. By this technique (see Smeets 1981, 1987, 1993), real-time velocity recordings having a μs time resolution and 1% accuracy are obtained. Table 2 gives a survey of the institutes where shock tubes have been used for scramjet combustion research. The cited references show that most of the research is quite recent and that there is at present important worldwide activity in this field.

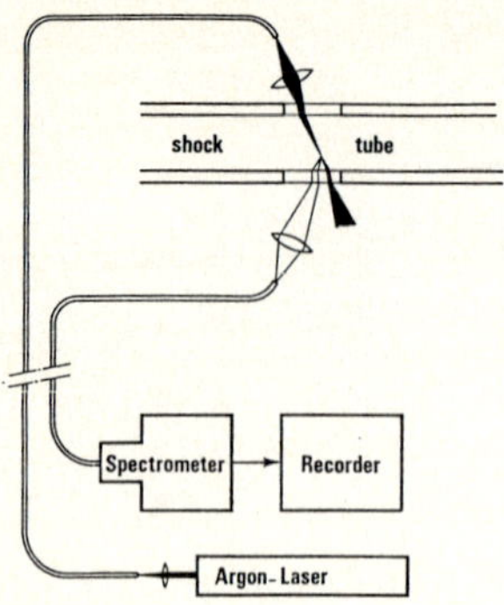

Fig. 10. Optical set-up for Doppler velocity measurement

Table 2. Scramjet experiments

Institute	Reference	Facility	Subject
Tokyo Inst. of Techn.	Takahashi 1981	Shock tube + Det. tube	Comb. in mixing layer of coflowing H_2 and O_2+Ar
Calspan	Dunn 1989	Shock tunnel	Wedge type inlet: pressure and heat flux recording
Univ. of Sheffield	Swithenbank 1989	Shock tunnel	Strut inj. H_2: turb. mixing & comb.
Univ. of Queensland	Stalker 1990, Morgan 1991 Brescianini 1992	Free piston driver shock tunnels (T3, T4)	Strut injection $H2$ in air flow: pressure on thrust surface
GASL	Bakos 1990	Exp. tube (Hypulse)	Annular slot inj. H_2: pressure and heat flux recording
Univ. of Stanford	Lee 1991, Hanson 1993	Shock tube	Temper. image for a comb. H_2 jet in supersonic crossflow
NASA-Ames	Loomis 1992	Shock tunnel	Wedge model, 30 deg. inj. H_2: pressure, heat flux recording
Phys. Sci. Inc., Andover	Parker 1992	Shock tunnel	Inj. H_2 after step, diagnostics of NO and OH distributions
GASL + Univ. of Queensland	Bakos 1992	T4 + Hypulse	Effects of oxygen diss. on combustion experiments in T4
ISL	Quenett 1994	Shock tube	Wall inj. H_2: downstream pressure and velocity recording
Caltech	Hornung	Free piston driver shock tunnel (T5)	Experiments in preparation

4. Ram-accelerator

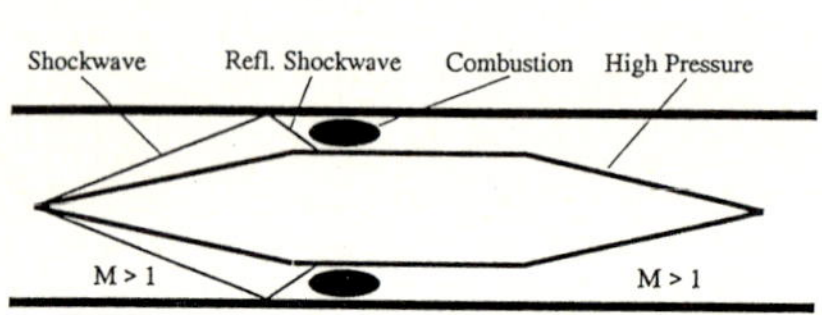

Fig. 11. Flowfield around a projectile flying at superdetonative speed

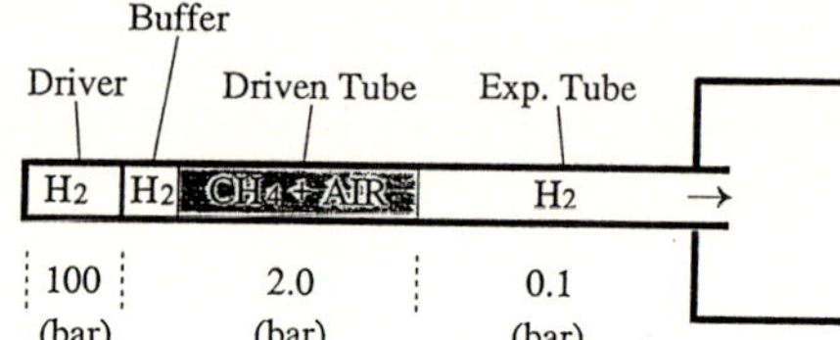

Fig. 12. Expansion tube

Fig.11 explains the flowfield around a projectile flying at super-detonative speed (velocity of the projectile higher than the Chapman-Jouguet detonation velocity of the explosive gas mixture). On the conical front part of the projectile, an attached shock is formed in the high Mach number upstream flow that reflects from the tube wall and compresses and heats the flow. The combustible mixture is expected to ignite and burn in the circular section between the cylindrical body and the tube. There are still debates as to whether ignition occurs in an oblique detonation wave or is initiated by hot spots in the stagnation regions in front of positioning fins or rails between the body and the tube. The combustion-generated high pressure produces thrust on the conical back part of the projectile.

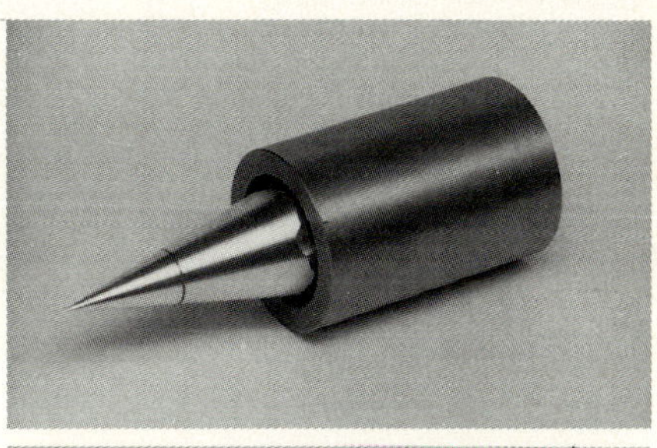

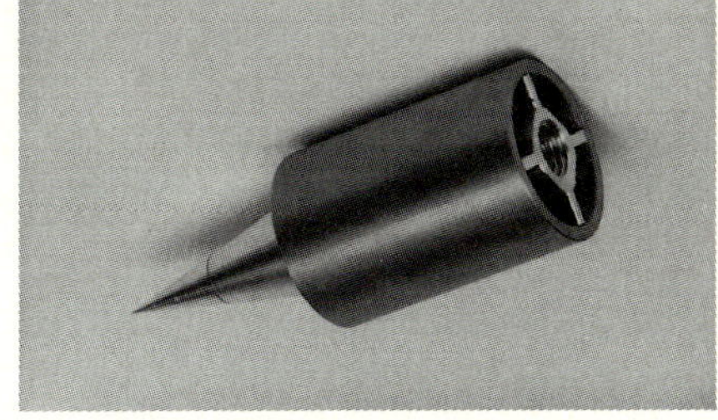

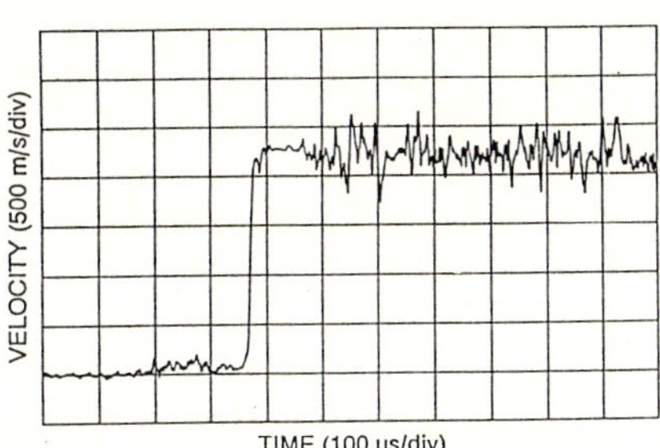

Fig. 13. Velocity of the combustible gas flow issuing from the expansion tube

Fig. 14. Model for studies of ignition and combustion in the external flow

Shock tube studies permit detailed diagnostics of the flow pattern and the combustion processes that are not possible on the travelling projectile in the steel tube. The question was whether explosive gases can be accelerated in shock tubes to high enough velocities without ignition. By means of a two-step process in a so-called expansion tube (see Fig. 12) it has been possible to reach superdetonative speeds. The gas mixture filled into the driven tube is first accelerated in a shock wave, heating it to just below the ignition temperature, and is then expanded by a rarefaction wave within a low-pressure tube branched to the end of the shock tube, leading to an important further increase of stagnation enthalpy.

Fig.13 shows a velocity recording taken at the exit of the open-ended shock tube. It serves as a monitor of the generated flow. The predicted flow velocity of about 2300 m/s was attained in a stoichiometric methane/air mixture (having a C-J-detonation velocity of about 1800 m/s). The fluctuations indicate the turbulence degree. The same tracer particles as in the scramjet experiments can be used, but they could not be introduced in the same way and kept in suspension in the explosive gas mixture. It turned out that by loading the tube walls with TiO_2 powder, a sufficient amount of particles could be entrained into the shock tube flow by turbulent mixing. The calculated temperature of the flow was 290 K and its pressure 0.5 bar.

One of the models exposed to the flow can be seen in Fig.14: a centerbody having a conical front part surrounded by a hollow cylinder of variable wall thickness. The questions to be answered were: what is the limiting dimension of the blunt cylindrical ring for the ignition of the mixture and how does combustion spread into the external flow field?

As an example, a pair of Schlieren interferograms is shown in Fig.15. The cylinder was of 3 mm wall thickness and 36 mm outer diameter. The flow of the combustible mixture is compared to a corresponding inert gas flow field by replacing O_2 in the stoichiometric methane/air by N_2. Quite obviously ignition takes place in the stagnation regions behind the overdriven normal shock

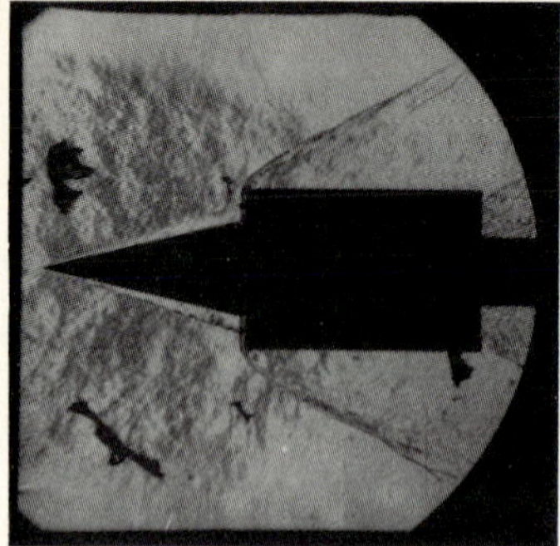 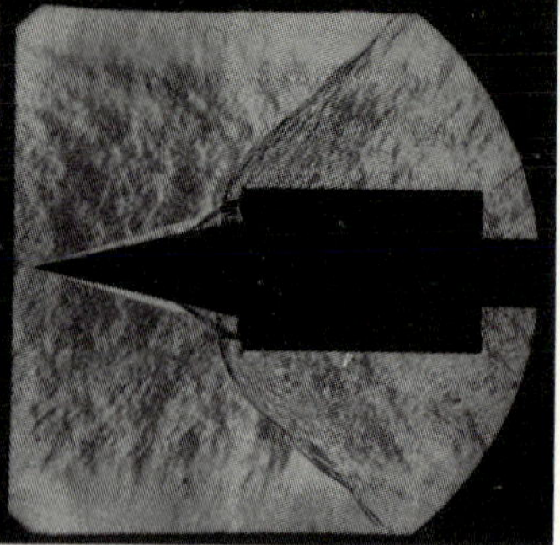

Fig. 15. Schlieren interferograms of flowfield with and without combustion

Fig. 16. Experiment: Ignition in a duct

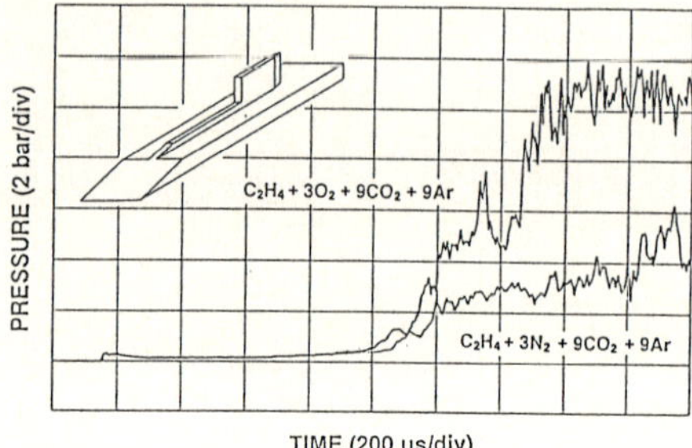

Fig. 17. Pressure records from experiments with combustible and inert mixtures

front and substantial modification of the flow field around the model occurs caused by intense combustion. The oblique shock wave in the outer region steepens up but it seems that it does not yet result in an oblique Chapman-Jouguet detonation wave (Srulijes 1993).

Another combustible mixture of stoichiometric ethylene/oxygen ($C_2H_4+3O_2+9CO_2+9Ar$) could be accelerated to 1800 m/s, being also super-detonative for this relatively heavy mixture. An experiment was performed in order to find out whether a ram projectile preaccelerated by a conventional powder gun to 1800 m/s and passing through this mixture in a ram tube will ignite the gas by the hot spots in front of the fin. Fig.16 shows a two-dimensional model facing the end of the 10 cm diameter shock tube. The flow compression on the front cone of the projectile is reproduced by a compression on a corresponding wedge. The rectangular duct simulates a 90° sector of the slit between the cylindrical body of the projectile and the tube wall with one of the fins in the center (Srulijes 1992). Details of ignition and combustion were investigated by flow visualization and the global effect was quantified by monitoring the downstream wall pressure increase. Ignition was found to sensibly depend on the fin geometry. The two plots in Fig.17 comparing the combustion and quenched combustion cases clearly show the downstream pressure increase resulting from combustion.

Table 3 gives a survey of the institutes where experiments in shock tubes have been done with premixed combustible gas mixtures. Two early experiments are described by Soloukhin (1961) and Kawada (1973). The gas mixtures are accelerated to subdetonative speeds in conventional shock tubes. It is only very recently, that explosive mixtures could be accelerated to superdetonative velocities using the expansion tube principle.

Table 3.Experiments with premixed combustible gas mixtures

Institute	Reference	Facility	Subject
Univ. of Novosibirsk	Soloukhin 1961	Shock tube	Pulsating comb. behind bow shock in subdet. flow speed
Tokyo Inst. of Techn.	Kawada 1973	Shock tube	Stabilization of comb. behind bow shock in subdet. flow speed
ISL	Srulijes 1992,1993	Expansion tube	Ignition by hot stagnation regions in superdet. flow
McGill University	Johnston 1993	Expansion tube	Delaying onset of auto-ignition in exp. tube by inhibitors
GASL	Tamagno	Expansion tube	Attempts to accelerate H_2/air mixtures to superdet. speeds
Univ. of Stanford	Hanson	Expansion tube	Construction of expansion tube in preparation

5. Diagnostics

During the last years very powerful new spectrometric techniques have been developed, leading to quantitative data within the flowfields.

- Laser Doppler velocimetry (see Smeets 1981/87/93): as described above, real-time velocity recordings at selected points within the flow field can be achieved by this technique. In principle, three components can be recorded simultaneously.
- Laser induced fluorescence (LIF and PLIF): this technique gives access to local species concentrations as well as temperatures and - in principle - also to velocities. These quantities can be recorded as a function of time at selected points in the flow (LIF) or as an instantaneous distribution within a laser light sheet (PLIF) (see Lee 1991, Parker 1992, Hanson 1993).
- Rayleigh and Raman scattering and CARS: The same quantities as above can be measured at selected points. During the short test time typical of shock tubes, only instantaneous measurements during single shots of pulse lasers can normally be made.
- Emission and absorption techniques in IR, visible and UV: these are line-of-sight techniques. Access to local quantities is possible only for special geometries (two-dimensional or axisymmetric).

A good survey of modern instrumentation, in great part applicable to combustion diagnostics in shock tubes, was given at two recent meetings and will be documented in their proceedings:(see Boutier 1993 and Kuo 1994).

6. Outlook

It has been shown that the shock tube is quite an effective instrument for experimental research in supersonic combustion and that it competes well as a low-cost alternative with continuous or blowdown high-enthalpy tunnels. It is only recently - maybe as a result of the modern diagnostic techniques - that the shock tube is beginning to attract the attention of researchers in combustion dynamics. Therefore, a wide field for future shock tube research has now opened.

References

External combustion

Holbeche TA, Townend LH, Pratt NH, Cox SG (1980) Analysis of external combustion on lifting-propulsive bodies. RAE Tech. Rep. 80072

Smeets G, Patz G (1992) Flame stabilization at a tangential surface in hypersonic flow by means of shock waves. AIAA Paper 92-3966

Scramjet

Bakos R Tamagno J, Rizkalla O, Pulsonetti MV, Chinitz W (1990) Hypersonic mixing and combustion studies in the Hypulse facility. AIAA Paper 90-2095

Bakos RJ, Morgan RG (1992) Effects of Oxygen dissociation on hypervelocity combustion experiments. AIAA Paper 92-3964

Brescianini C, Morgan RG (1992) An iInvestigation of a wall-injected Scramjet using a shock tunnel. AIAA Paper 92-3965

Dunn MG, Lordi JA, Wittliff CE, Holden MS (1989) Facility requirements for hypersonic propulsion system testing. AIAA Paper 89-0184

Hanson RK (1993) Quantitative absorption and fluorescence diagnostics in combustion systems. Proc. 3rd Intl; Symp. on Spec; Topics in Chem. Prop.: Non-Intrusive Comb. Diagn.

Lee MP, McMillin BK, Palmer JL, Hanson RK (1991) Two-dimensional imaging of shock tube flows using Planar Laser Induced Fluorescence. AIAA Paper 91-0460

Loomis MP, Zambrana HA, Bogdanoff DW (1992) 30 degree injectors at hypervelocity conditions. AIAA Paper 92-3288

Morgan RG, Stalker RJ, Bakos RJ, Tamagno J, Erdos JI (1991) Scramjet testing - ground facility comparisons. 10th ISOABE

Parker TE, Allen MG, Reinecke WG, Legner HH, Foutter RR, Rawlins WT (1992) Supersonic combustor testing using optical diagnostics. AIAA Paper 92-0761

Quenett C (1994) Shock tube experiments on scramjet combustion. Thesis to be completed in 1994

Stalker RJ, Morgan RG, Paull A, Bresciani CP (1990) Scramjet experiments in free piston shock tunnels. NASP Contractor Report 1100, printed by NASA Langley

Swithenbank J, Eames I, Chin S, Ewan B, Yang Z, Cao J, Zhao X (1989) Turbulent mixing in supersonic combustion systems. AIAA Paper 89-0260

Takahashi S, Yoshizewa T, Minegishi T, Kawada H (1981) A study of the Hydrogen-Oxygen diffusion flame in high speed flow. In: Treanor CE, Hall JG (eds) Proc. 13th ISSTW, Niagara Falls

Ram-accelerator

Johnston MH, Zhang F, Lee JHS (1993) Feasibility of using inhibitor to increase the stagnation enthalpy. These Proceedings

Kawada H, Takahashi S (1973) The formation and stabilization of deflagration wave.In: Bershader D, Griffith W (eds) Proc. 9th ISTS, Stanford

Soloukhin RI (1961) Pulsating combustion of gas behind a shock wWave in supersonic flow. Z. Mech u. Techn Phys , 5: 57

Srulijes J, Smeets G, Seiler F (1992) Expansion tube experiments for the investigation of Ram-accelerator-related combustion and gasdynamic problems. AIAA Paper 92-3246

Srulijes J, Smeets G, Seiler F (1993) Expansion tube experiments at ISL for the investigation of Ramac-related phenomena. 1st Intl Workshop on Ram Accelerator, ISL

Diagnostics

Boutier A (1993) New trends in instrumentation for hypersonic research. Nato ASI Series E: Appl. Sciences - Vol. 224

Kuo KK, Pasman HJ (1994) Non-intrusive combustion diagnostics. Proc 3rd Intl Symp on Special Topics in Chemical Propulsion

Smeets G, George A (1981) Michelson spectrometer for instantaneous Doppler velocity measurements. J. Phys. E: Sci; Instrum. 14

Smeets G, Mathieu G (1987) Investigation of turbulent boundary layers and turbulence in shock tubes by means of lLaser velocimetry. In: Grönig H (ed) Proc. 16th ISSTW, Aachen

Smeets G (1993) Doppler velocity measurements using a phase- stabilized Michelson spectrometer. Proc. 3rd Intl. Symp. on Spec. Topics in Chem. Prop.: Non-Intrusive Comb. Diagn.

Part 1: Hypersonic Flow and Aerospace Studies

Interference and Transient Effects on Compression Ramp Flows at Hypersonic Mach Numbers

A.J.D. Smith and R.A. East
Department of Aeronautics and Astronautics, University of Southampton, Southampton, UK

Abstract. The compression corner separated hypersonic flow, and its transient response, have been investigated experimentally. Mechanisms of unsteady separated flow adjustment have been examined, and the development of criteria to predict separated flow establishment times discussed. During the initial experiments, nozzle exit compression waves were found to be a corrupting influence. Methods of detecting this interference, and its effects, have been assessed.

Key words: Hypersonic, Separated flow, Unsteady, Compression ramp

1. Introduction

The compression corner, formed by a sharp-edged flat plate fitted with a trailing edge ramp (wedge), has been tested extensively in the hypersonic flow regime over the last thirty years. A necessary consideration in the design of these experiments when the available flow duration is very limited, or when dynamic testing is to be used as a technique to efficiently generate multiple points of presumed quasi-steady data, is to ensure that the flowfield induced by each test model configuration has sufficient time to achieve steady conditions. This is particularly important when an interaction is sufficiently strong to induce a separated flow. When this occurs, the time required for the attainment of a steady flowfield has been found to take substantially longer than for an interaction in which the boundary layer remains attached.

The present study has investigated both the steady separated flowfield formed at a compression corner, and its transient response induced by a rapidly deflected flap. The separated flows examined in these experiments were transitional in the reattachment region, although their length scales remained more typical of the fully laminar regime (Smith 1993). During the fixed flap experiments it was observed that nozzle exit compression waves, formed in the open jet freestream flow, could be a corrupting influence on the scale, and properties, of a steady separated flow, although these may still appear typical of the expected flowfield type. A particular concern is that the undetected presence of similar interference effects may have influenced some of the large database of experimental results obtained from the many previous investigations of compression corner separated flows. A description of these interference effects, and the development of criteria to determine their occurrence, are presented.

The principal objective of the present study has been to examine the transient response of this particular compression corner separated flow. This has been achieved by comparing measurements made for the rapidly deflected flap surface with those obtained at corresponding fixed flap angles. These experimental results have been compared with analytic models of the separated flow response based on distinct fundamental flow adjustment mechanisms. The conclusions from this have formed a basis for discussing the development of criteria for predicting separated flow establishment times.

2. Experimental set-up and instrumentation

The model configuration is illustrated schematically in Fig.1. For the dynamic tests the flap was deflected from 0 to 35 degrees in approximately 20 ms, generating angular velocities approaching 3000 degs./s (Fig.2). Tests were made both with, and without, side plates attached. These experiments were performed in the University of Southampton isentropic light piston compression tube hypersonic wind tunnel, using a contoured nozzle with a measured exit freestream

Shock Waves @ Marseille I
Editors: R. Brun, L. Z. Dumitrescu   © Springer-Verlag Berlin Heidelberg 1995

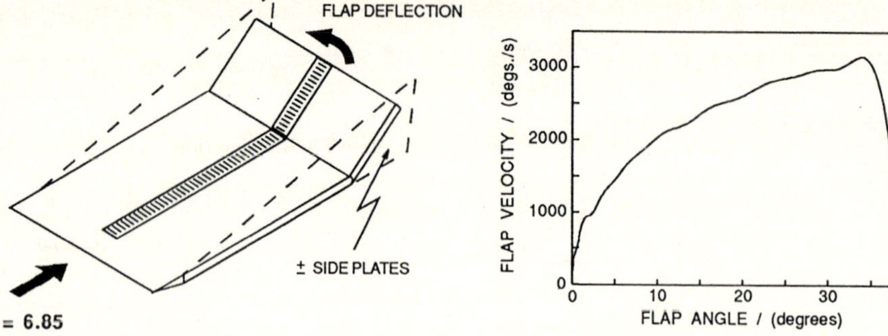

Fig. 1. Test model configuration   Fig. 2. Dynamic flap motion

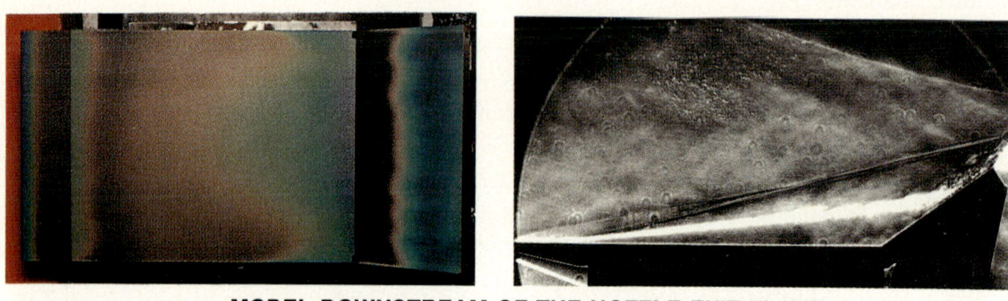

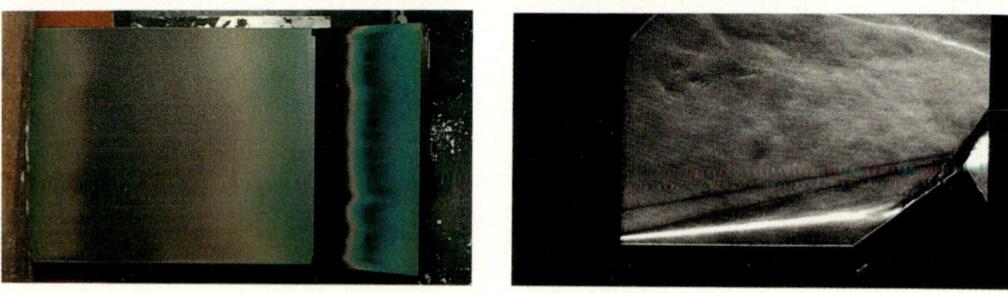

Fig. 3. Liquid crystal surface thermographs, and Schlieren flow visualisation, obtained at two model locations for the highest flap angle of 35 degrees

Mach number of 6.85. The facility was operated at its current lowest unit freestream Reynolds number capability of 2.45×10^6/m.

The principal measurements made for both the fixed, and dynamic, flap conditions were of the model centre line chord heat transfer distributions using thin film resistance gauges. Flow visualisation of both the steady, and unsteady, separated flow structures was obtained using Schlieren photography. Liquid crystal surface thermography was applied in these tests to assess the significance of any 3-dimensional influences in the nominally 2-dimensional fixed flap tests. A separate model was constructed for this incorporating single, and multi-layer, model construction methods (Smith and Baxter 1989).

3. Results

3.1. Nozzle exit compression wave interference effects

The fixed flap tests were initially made with the model located in a position with its leading edge downstream of the nozzle exit plane. The liquid crystal thermographs indicated the introduction, at large flap deflection angles, of very high heating near the flap trailing edge corners, evident from the colour transition to blue, and adjacent clear regions where the liquid crystal had passed above its colour - temperature bandwidth (Fig.3). Simultaneously, there also appeared significant 3-dimensional features on the plate thermograph within the separated flow region. These effects have been attributed to interference from the nozzle exit compression waves. The presence of this had not been initially suspected from either the centre line heat transfer measurements, or the Schlieren photographs, in which the separated flows appeared typical of a transitional wedge type. The impingement of the nozzle exit compression waves on the flap trailing edge corners is not evident in the Schlieren flow visualisation (Fig.3). The normal density gradient of the nozzle shock wave surface at these locations is small, and this is obscured by the reattachment process of the separated shear layer.

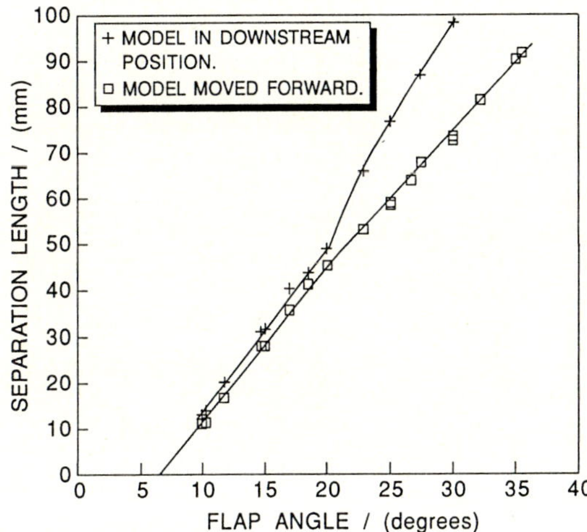

Fig. 4. Effect of nozzle exit wave interference on the separated flow lengths (side plates not attached)

To remove this interference, the model was moved forward, and the open jet length reduced. This minimised flow spillage to the surrounding test section, thereby reducing the strength, and inclination, of the nozzle exit waves. The model was finally located in a forward position where no evidence of interference at the trailing edge corners was observed from the thermographs with the flap at its highest angle of 35 degrees (Fig.3). The heating pattern on the plate surface then appeared much more 2-dimensional (Fig.3).

A comparison of the Schlieren photographs obtained for the two model locations shows that, when the nozzle waves interact with the model flowfield, the scale of the separated flow is increased, and the reattachment process is less clearly defined (Fig.3). The shear layer also then appears to have more prominent irregular edges, suggesting a more transitional / turbulent flow. An initial unawareness of these nozzle wave interference effects led to the determination of a response time of the dynamic flap separated flows (Smith and East 1991) which is now known to be excessive.

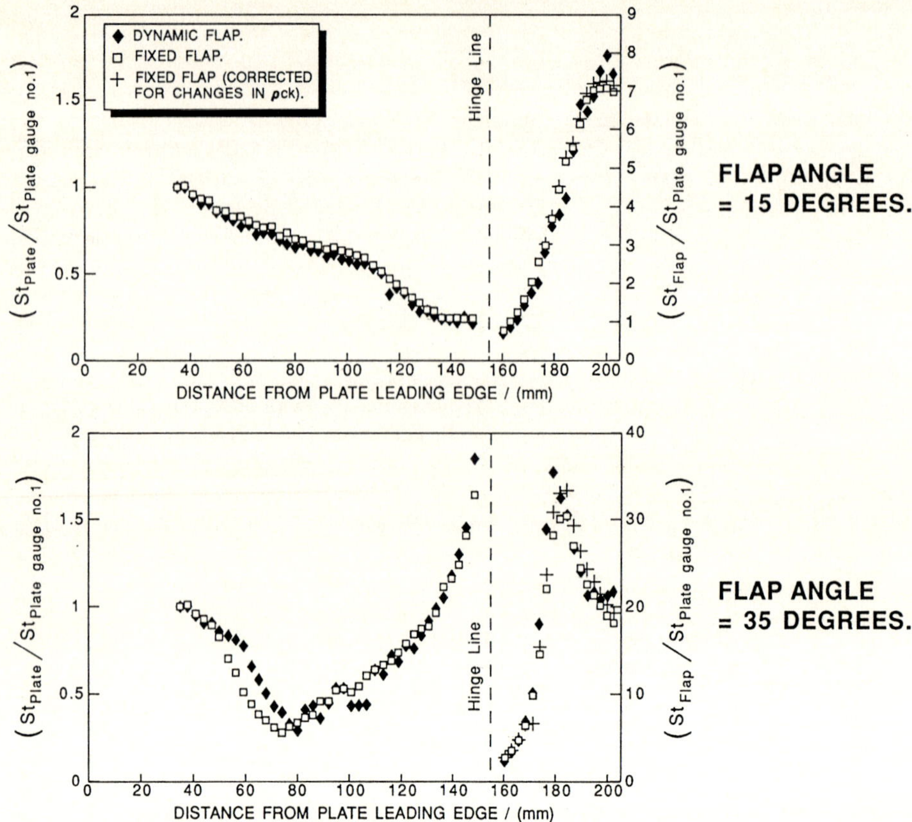

Fig. 5. Measured surface heat transfer distributions (side plates not attached)

The separation lengths with the model in the two positions are shown in Fig.4, defined as the distance from the hinge line to the separation point. With the model moved forward the separation lengths remain nearly linear with flap angle throughout the deflection range. With the model in its downstream position, there is an abrupt increase in the rate of separation length increase for flap deflection angles larger than 20 degrees. This region corresponds to the observation of interference effects from the liquid crystal thermographs. The characteristic of a nearly linear dependence of this separation length on the flap deflection angle is strong evidence of interference-free experiments.

The results of these tests provide yet another endorsement for using liquid crystal thermography as an excellent diagnostic tool in hypersonic testing. The use of this technique, and a check of separation length dependence on flap angle, are recommended procedures to ensure interference free compression corner experiments in open jet facilities.

3.2. Transient response of the dynamic flap separated flow

The Schlieren photographs of the dynamic flap separated flow structures indicated that at any instant its structure was the same as that of the steady wedge type. Examples of the comparisons of fixed flap, and instantaneous dynamic flap, induced separated flow heat transfer distributions are shown in Fig.5. At low to moderate flap angles the differences in separation lengths, and heat transfer distributions, between the dynamic, and fixed, flap conditions are negligible. However, at high flap angles the Schlieren photographs, and heat transfer distributions, are consistent in indicating the development of a lag in the unsteady separated flow growth with respect to the quasi-steady separation lengths. The magnitude of these lags is slightly reduced when the side plates are attached. This has been attributed primarily to the effect of side plates on transition location in the steady separated flows, which influences the rate of mass entrainment into a growing separated flow (sect. 4), and not significantly due to a dynamic lateral venting phenomenon. Throughout the flap deflection range there is no consistent evidence from the heat transfer distributions of a significant unsteady separated flow effect on the location, or process, of transition in the shear layer. Unsteady effects on the plate heat transfer distribution also appear negligible. It has been concluded that the lag which develops in the unsteady separated flow growth is due principally to the finite rate of fundamental flow adjustment mechanisms within this region.

4. Discussion

The unsteady separated flow measurements have been compared with predictions obtained from two analytic models of this flow adjustment (Smith 1993), based respectively on the constraints of pressure wave communication, and mass entrainment requirements. These fundamental flow adjustment mechanisms have been previously described by Ihrig and Korst (1963). The pressure wave adjustment time is that required for pressure changes occurring in the reattachment region to be communicated throughout the separated flow. As the separated flow grows it requires additional mass, and this is entrained at a finite rate from the shear layer. The analytic nature of both these models requires a number of simplifying assumptions to be incorporated, and they are therefore only expected to give approximate results.

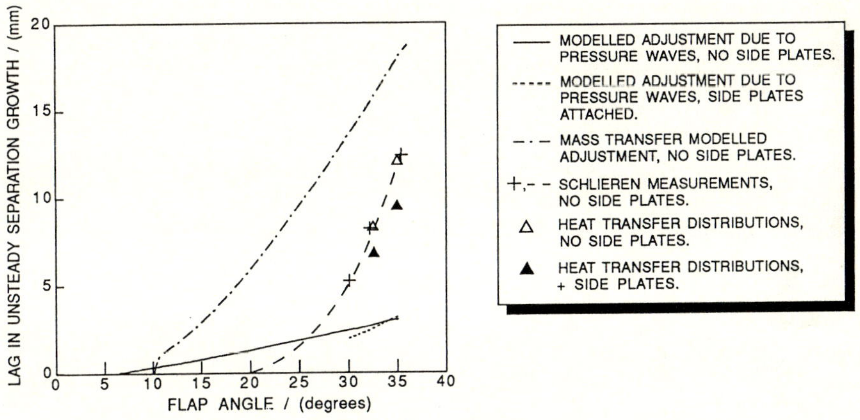

Fig. 6. Comparison of the measured and predicted unsteady separated flow response

Fig.6 illustrates the predicted separated flow responses, together with the experimental results. The measured lags in the unsteady separated flow growth were determined from a comparison of the dynamic tests, and fixed flap tests, with the model in slightly different positions. Although all these tests were free of nozzle wave interference, the steady separation lengths in these two model positions were slightly different. The measured lags shown in Fig.6 have been corrected for this small model position effect. The largest corrected difference in separated flow lengths between the dynamic, and fixed, flap conditions in these experiments was equivalent to a phase lag with respect to the flap angle of 3.8 degrees (equal to 1.3 ms of dynamic flap motion).

At low flap angles the pressure wave adjustment model appears the most accurate, and the mass adjustment model is inadequate. However, at the highest flap angles the mass adjustment model compares better with the experimental measurements, where the lags are clearly much larger than can be attributed only to pressure wave adjustment times. Over a substantial part of the flap deflection range it appears that the response of this separated flow is likely to be characterised by a strong interaction between both pressure wave, and mass entrainment mechanisms.

The present results appear to contradict the correlations of Holden (1971) for laminar wedge separated flow establishment times, in which a pressure wave adjustment model provided an accurate prediction at all conditions. However, Holden's measurements were made at much higher freestream Mach numbers than the present tests, where shear layer deflection angles are reduced, and the thickness of viscous layers are larger. These effects are expected to reduce the significance of mass entrainment adjustment times compared to the present tests.

5. Conclusions

The results of this investigation indicate that it is inappropriate to generalise which adjustment mechanism is important for a particular type of model configuration at all freestream conditions, and for all scales of separated flow. Consequently, establishment times may not always be accurately predicted using criteria developed from simple analytic models. For the purposes of designing, or checking the feasibility, of proposed fully laminar interaction experiments (for which establishment times will be greatest) in short duration facilities, the accurate prediction of separated flow establishment times using computational fluid dynamics appears most promising, as in the recent example of Lees and Lewis (1993). Of course, during an experiment measurements should always be checked to ensure that steady conditions have been achieved.

Acknowledgements

The financial support of Dassault Aviation for funding this research project as part of the Hermes research and development program is gratefully acknowledged.

References

Holden MS (1971) Establishment Time of Laminar Separated Flow. AIAA J. 9:2296-2298

Ihrig HK, Korst HH (1963) Quasi-Steady Aspects of the Adjustment of Separated Flow Regions to Transient External Flows. AIAA J. 1:934-937

Lee JY, Lewis MJ (1993) Numerical Study of the Flow Establishment Time in Hypersonic Shock Tunnels. J. Spacecraft and Rockets 30:152-163

Smith AJD, Baxter DRJ (1989) Liquid Crystal Thermography for Aerodynamic Heating Measurements in Short Duration Facilities. In: Proc. Intl. Congress on Instrumentation in Aerospace Simulation Facilities, pp 104-112

Smith AJD, East RA (1991) The Dynamic Response of Separated Hypersonic Flows. In: Aerothermodynamics for Space Vehicles, Proc. First European Symp., Noordwijk

Smith AJD (1993) The Dynamic Response of a Wedge Separated Hypersonic Flow and It's Effects on Heat Transfer. PhD Thesis, Univ. of Southampton

Effects of Leading-Edge Bluntness on Control Flap Effectiveness at Hypersonic Speeds

D. Kumar and J.L. Stollery
College of Aeronautics, Cranfield Institute of Technology, Cranfield, Bedford, MK43 0AL, UK

Abstract. The effects of leading edge bluntness on boundary layer transition and separation have been studied experimentally. The studies have been carried out in a hypersonic gun tunnel facility at $M_\infty = 8.2$ and $Re_\infty/\mathrm{cm} = 9.0 \times 10^4$. The flow structure over a low aspect ratio flat plate with a full span trailing edge flap has been investigated using high speed Schlieren photography as well as surface pressure and heat transfer measurements.

Key words: Hypersonic control, Leading-edge bluntness, Boundary-layer separation

Nomenclature

C	Chapman-Rubesin constant	$\bar{\chi}$	Viscous interaction parameter, $M^3\sqrt{(C/Re_x)}$
C_D	Leading edge drag coefficient	χ_ϵ	$\epsilon[0.664 + 1.73(T_w/T_0)]\bar{\chi}$
C_H	Heat transfer coefficient	ϵ	$(\gamma - 1)/(\gamma + 1)$
	$q/\rho_\infty U_\infty c_p(T_0 - T_w)$		
c_p	Specific heat (isobaric)	γ	Ratio of specific heats
d	Leading edge diameter	κ_ϵ	$M^3\,C_D\,\epsilon\,(d/dx)$
L	Hingeline length	β	Bluntness-viscous interaction parameter
M	Mach number		$\chi_\epsilon/\kappa_\epsilon^{2/3}$
Re	Reynolds number	δ	Flap deflection angle ($^\circ$)
T	Temperature	δx	Change in distance along the plate
x	Distance from the leading edge	α	Angle of incidence ($^\circ$)

Subscripts

w	wall	∞	freestream
f	flap	s	separation (in figures 1-6)
0	reservoir		

1. Introduction

The aerodynamic requirements of hypersonic flows dictate the intentional blunting of leading edges in order to reduce the heat transfer rate in this region as well as to allow for the installation of active cooling systems if required. Bluntness results in a considerable change in the aerodynamic flow structure from that of a sharp leading edge body. The primary effect is the creation of a bow shock standing off from the leading edge. The increase in the strength of the shock near the leading edge as well as its highly curved structure gives rise to gradients in flow properties, such as Mach and unit Reynolds number along as well as normal to the streamlines over the surface of the body. The shock also creates a region of high temperature, low density flow, known as the entropy layer, near the surface of the body. As a result of the losses associated with the shock, the local Mach number and unit Reynolds number in this layer are much lower than the freestream conditions.

The orthogonal gradients in the entropy layer produce a vortical outer flow which envelops the surface generated vorticity, the latter being confined to the boundary layer. The interaction of the externally generated vortical layer with the boundary layer gives rise to a highly complex transition behaviour associated with bluntness. This involves an initial stabilisation of boundary

Shock Waves @ Marseille I
Editors: R. Brun, L. Z. Dumitrescu   © Springer-Verlag Berlin Heidelberg 1995

layer transition by the introduction of small leading edge bluntness with respect to the equivalent sharp body. Further increases in the leading edge diameter result in a destablisation of the boundary layer with respect to the equivalent sharp leading edge body.

A transition reversal behaviour has been observed at supersonic Mach numbers for flat plates by Jillie and Hopkins (1961). Reshotko and Khan (1979) have modelled the flow over a blunted flat plate by the method of multiple scales. Stability analysis of the boundary layer flow revealed that the vortical layer associated with the outer inviscid flow resulted in a decrease in the critical Reynolds number.

The effects of leading edge bluntness on boundary layer separation upstream of a compression corner, studied by Sanator et al. (1968), Edwards et al. (1968) and by Vermeulen et al. (1992) showed that bluntness promoted the separation of a laminar boundary layer.

However, studies conducted by Townsend (1966) and by Holden (1975) showed that the extent of the laminar separated region was related to the viscous bluntness interaction parameter, β. This effectively accounted for the relative effects of changes in Mach number and Reynolds number on the extent of the interaction. Holden's results indicate that for $\beta < 0.1$, the introduction of leading edge bluntness results in a dramatic reduction in the extent of boundary layer separation while for $\beta > 0.5$, the introduction of leading edge bluntness promotes separation.

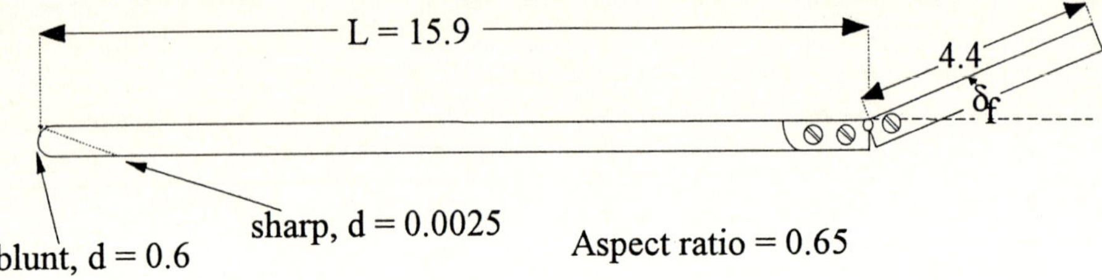

Fig. 1. Geometries of the flat plate configurations used for the present tests (all dimensions in cm)

2. Wind tunnel facility

The Cranfield hypersonic gun-tunnel facility was used for the investigations. The tests were carried out with air as the working gas. An axisymmetric contoured nozzle provided a 15.0 cm diameter test jet at a freestream Mach number of 8.2. The tunnel is equipped with a single pass Schlieren system. This uses a high intensity microsecond duration argon spark source to illuminate density gradients in the flow. Pressure measurements were carried out using Kulite XCS-190 series pressure transducers. Heat transfer measurements were made using surface mounted thin film platinum gauges on a Macor substrate. The temperature history of the gauges was integrated using a low noise, high bandwidth analogue integrator circuit. The models used for the present tests are shown in Fig.1.

3. Results and discussion

3.1. Flat plate flows

The effects of leading edge bluntness on the flat plate pressure distribution are shown in Fig.2(a). This shows that the presence of the bow shock increases in the pressure ratio near the leading edge in comparison to the sharp leading edge pressure distribution. A strong favourable pressure gradient is also established in this region of the blunted body. This diminishes towards the trailing edge. The sharp leading edge pressure distribution in comparison, displays negligible change over the entire length of the flat plate. The effects of viscous-inviscid flow interaction on the pressure distribution over the sharp plate were found to be small under the present test conditions. The

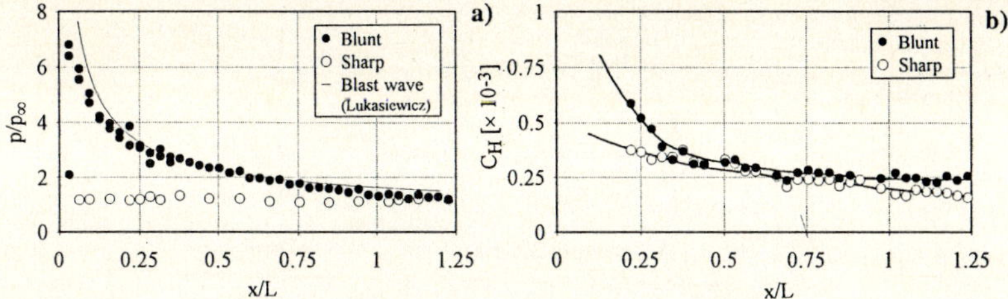

Fig. 2. The effect of leading-edge bluntness on the flat plate pressure and heat transfer distribution ($M_\infty = 8.2$, $Re_\infty/\text{cm} = 9.0\ 10^4$, $Re_d = 5.4\ 10^4$); (a): presure (b): heat transfer

blunt plate pressure ratio tends towards the sharp value near the hingeline. It is well predicted by the second order blast wave theory proposed by Lukasiewicz (1961). The heat transfer rates over the flat plate are compared in Fig.2(b). The blunt plate values are everywhere higher and level off towards the rear of the plate. The sharp values vary approximately as $Re_x^{-1/2}$ but are somewhat overpredicted by the reference enthalpy method.

3.2. Unseparated compression corner flows

Schlieren photographs for the sharp and blunt leading edge configurations with $\delta_f = 5°$ show the presence of a single shock near the flap, indicating an unseparated flow structure. The photographs show the flap shock to be well defined near the hingeline for the blunted configuration while for the sharp configuration, it was seen to be forming over the entire flap region.

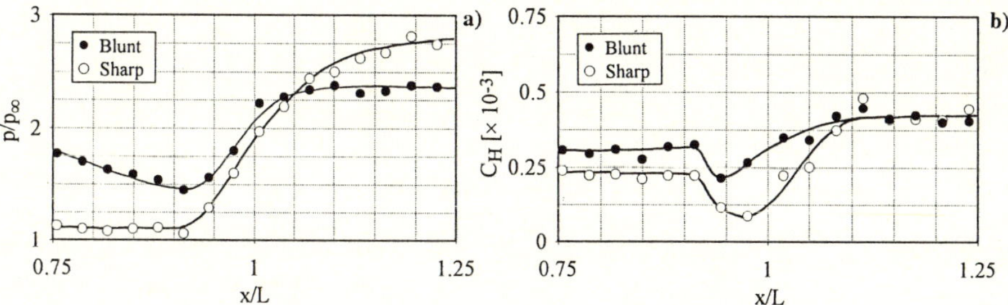

Fig. 3. The effect of leading-edge bluntness on the pressure and heat transfer distribution over a compression corner ($M_\infty = 8.2, Re_\infty/\text{cm} = 9.0\ 10^4, \alpha = 0°, \delta_f = 5°$); (a): pressure (b): heat transfer

The pressure distribution near the hingeline for these configurations is shown in Fig.3(a). At $x/L = 0.9$, both pressure distributions begin to increase as a result of the pressure rise associated with the flap shock wave feeding upstream and causing a local thickening of the boundary layer. The blunt leading edge distribution rapidly reaches a constant pressure level just downstream of the hingeline (at $x/L = 1.05$) whilst the sharp leading edge distribution continues to increase until $x/L = 1.15$.

The heat transfer distributions in the vicinity of the hingeline for $\delta_f = 5°$ are shown in Fig.3(b). The sharp leading edge distribution decreases below the unseparated flat plate level in the interaction region reaching a minimum at the hingeline. This is characteristic of laminar boundary layers (Needham 1966). The decrease is due to a reduction of local skin friction level in this region. The start of this decrease coincides with the start of the increase in pressure distribution associated with this configuration at $x/L = 0.9$. Downstream of the hingeline, the heat transfer distribution is seen to attain a peak before dropping to a near constant level on the remainder of

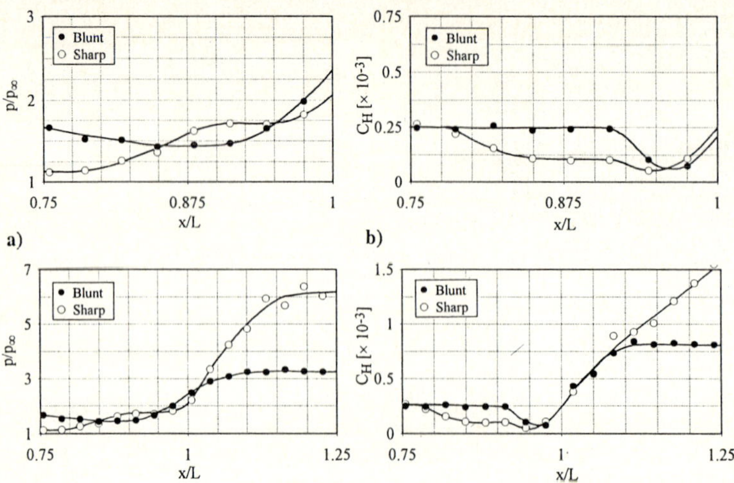

Fig. 4. The effect of leading-edge bluntness on the pressure and heat transfer distribution over a compression corner ($M_\infty = 8.2, Re_\infty/\mathrm{cm} = 9.0\ 10^4, \alpha = 0°, \delta_f = 10°$); (a): pressure (b): heat transfer

the flap. The peak occurs as the compressible boundary layer thins due to the increased pressure region over the flap. The heat transfer distribution for the blunted configuration also starts to decrease at $x/L = 0.9$. However, in this case, a minimum is attained rapidly after the beginning of the interaction and then the heat transfer rate begins to increase well ahead of the hingeline and finally reaches a peak downstream of the hingeline. This increase in the heat transfer level well ahead of the hingeline may indicate transition in this region for the blunted configuration.

3.3. Separated compression corner flows

The Schlieren photograph for the sharp configuration with $\alpha = 0°, \delta_f = 10°$ displays the presence of a dual shock system with a separation and a reattachment shock located upstream and downstream of the hingeline respectively. The photograph for the equivalent blunted configuration shows a highly localised flap shock near the hingeline. There was no evidence to suggest the presence of a dual shock structure for the blunt configuration. The sharp and blunted pressure distributions near the hingeline for $\alpha = 0°, \delta_f = 10°$ are shown in Fig.4(a). The pressure distribution for the sharp leading edge configuration indicates that separation occurs at $x/L = 0.85$. It displays the characteristic plateau associated with well separated flows. The distribution increases steadily in the reattachment region downstream of the hingeline. The blunted configuration pressure distribution shows an increase in pressure from $x/L = 0.9$. The distribution is characteristic of a fully attached flow. The pressure rises steadily to a constant level on the flap, well below the flap pressure level for the sharp leading edge. The loss in the pressure recovered on the flap is due to losses associated with the strong leading edge shock.

The sharp leading edge heat transfer distribution for $\alpha = 0°, \delta_f = 10°$ is shown in Fig.4(b). This displays the characteristic decrease associated with laminar separation from $x/L = 0.8$. The decrease starts from upstream of the separation point due to a slight upstream influence through the subsonic region of the boundary layer. The distribution then exhibits an increase in heat transfer from upstream of the hingeline attaining a plateau downstream of the reattachment point. A second increase in the heat transfer distribution, associated with the transition to turbulence is seen on the flap downstream of reattachment. This also contributes to the disturbed pressure distribution observed in this region. The blunted configuration heat transfer distribution is also shown in Fig.4(b). This is similar to the distribution observed for the unseparated flow with $\alpha = 0°, \delta_f = 5°$, shown in Fig.3(b).

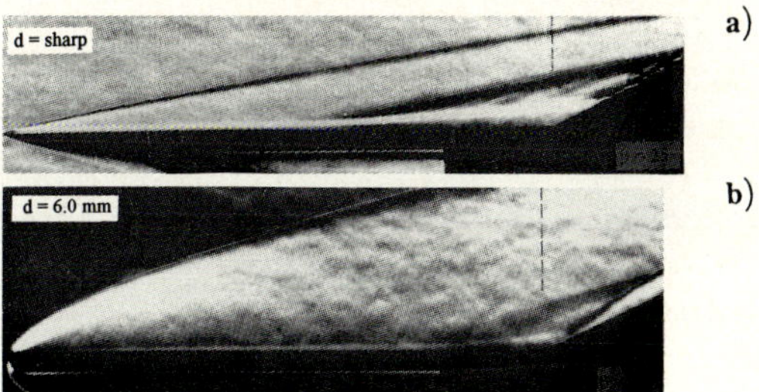

Fig. 5. Schlieren photographs showing the effect of leading-edge geometry on the flow structure for a compression corner ($M_\infty = 8.2, Re_\infty/\text{cm} = 9.0\ 10^4, \alpha = 0°, \delta_f = 25°$); (a): sharp leading edge ($d = 0.0025$ cm);(b): blunt leading-edge ($d = 0.6$ cm)

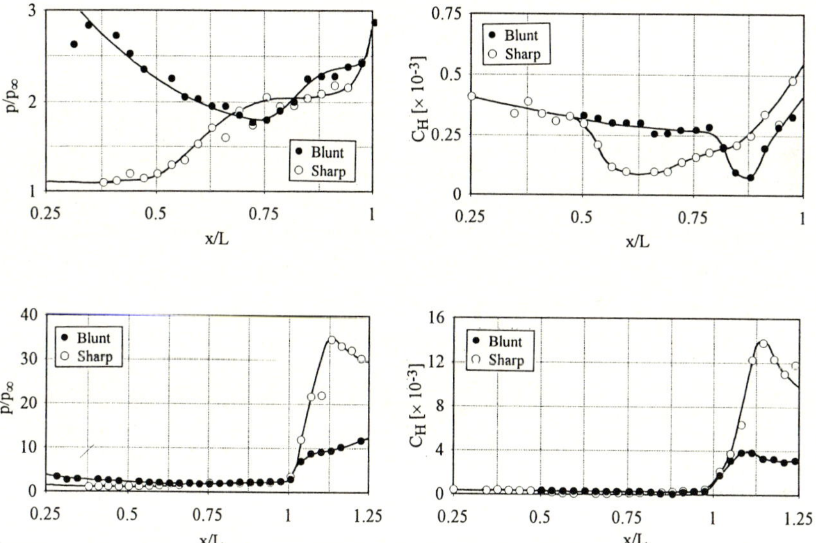

Fig. 6. The effect of leading-edge geometry on pressure and heat transfer distribution over a compression corner ($M_\infty = 8.2, Re_\infty/\text{cm} = 9.0\ 10^4, \alpha = 0°, \delta_f = 25°$)

The Schlieren photograph for the $\alpha = 0°, \delta_f = 25°$ configuration with a sharp leading edge is shown in Fig.5(a). This shows separation occuring at $x/L \approx 0.5$. Downstream of this point, a multiplicity of shock waves are seen at the edge of the shear layer. A strong reattachment shock is present downstream of the hingeline. This turns the shear layer parallel to the flap. The compression associated with this shock causes a decrease in the thickness of the compressible boundary layer in this region. The pressure distribution for this configuration is shown in Fig.6(a). An increase in pressure is seen from upstream of the separation point. A plateau pressure of $P/P_\infty = 2.6$ is attained from $x/L = 0.6$. This point correlates with the location of the trailing edge of the compression fan subsequently forming the separation shock, shown in Fig.5(a). The pressure distribution suggests that reattachment occurs around $x/L = 1.05$. This agrees well with the location of the reattachment shock in the Schlieren photograph. The heat transfer distribution for this configuration is shown in Fig.6(b). This shows a reduction, associated with the free-interaction region, from $x/L = 0.4$. The reduction starts from upstream of the separation point. The extent of the free-interaction region, upstream of the separation point, has increased due to an increase in flap angle. The heat transfer distribution in the separated region increases gradually to attain levels above the initial flat plate level upstream of the hingeline.

The Schlieren photographs of Figs.5(a) and (b) show the delay in separation due to leading edge bluntness. The pressure distribution for the blunt configuration, shown in Fig.6(a) indicates that separation occurs at $x/L = 0.75$. This agrees well with the location of the separation shock in Fig.5(b). The increase in pressure due to reattachment is an order of magnitude lower for the blunted configuration. The distribution also shows a secondary compression downstream of the reattachment region. This is due to an interaction of the inviscid shear layer associated with bluntness with the flap interaction shock structure. The blunt heat transfer distribution, shown in Fig.5(b) shows an initial decrease from upstream of separation. This is followed by a rapid increase in heat transfer from upstream of the hingeline until a reattachment plateau is attained at $x/L = 1.1$. The flap heat transfer levels show an order of magnitude reduction due to bluntness. This is associated with losses sustained due to the strong leading edge bow shock.

4. Conclusions

The study shows that the introduction of a blunted leading edge, under the present conditions, substantially delays separation of the boundary layer upstream of a compression corner. Bluntness also results in a reduction in the pressure recovered downstream of the hingeline and hence a significant loss of control effectiveness.

Acknowledgements

The work presented in this paper was performed as part of the technical section of a Total Technology PhD programme. It was sponsored by the Science and Engineering Research Council with British Aerospace, Stevenage as the industrial sponsor.

References

Edwards CLW, Anders JB (1968) Low density, leading edge bluntness and ablation effects on wedge induced laminar boundary layer separation at moderate enthalpies in hypersonic flows. NASA TN D-4829

Jillie DW, Hopkins EJ (1961) Effect of Mach number, leading edge bluntness and sweep on boundary layer transition on a flat plate. NASA TN-D 1071

Holden M (1975) Experimental studies in shock wave - boundary layer interactions. AGARD AG-203

Lukasiewicz J (1961) Hypersonic flow blast analogy. AEDC TN-61-158

Needham DA (1965) Laminar separation in hypersonic flows. PhD thesis, University of London. See also AIAA Paper 66-455

Reshotko E, Khan MMS (1979) Stability of a laminar boundary layer on a blunted flat plate in supersonic flow - Laminar-Turbulent transition. IUTAM Symposium, Stuttgart, pp 189-200

Sanator RJ, Boccio JL, Shamshins D (1968) Effects of bluntness on hypersonic two-dimensional inlet type flows. NASA CR-1145

Townsend JC (1966) Effects of leading edge bluntness and ramp deflection angle on laminar boundary layer separation in hypersonic flow. NASA TN-D 3290

Vermeulen JP, Simeonides G (1992) Parametric study of shock wave / boundary layer interactions over 2-D compression corners at Mach 6. VKI TN-181.

Measuring the Effect of Nose Bluntness on Drag of a Cone in a Hypervelocity Shock Tunnel Facility

L.M. Porter, A. Paull and D.J. Mee
University of Queensland, St. Lucia, Queensland 4072, Australia

Abstract. Presented in this paper are results obtained from an investigation into the effects of nose bluntness on slender cone drag in the hypersonic ($M > 5$) and hypervelocity (flight speed > 5 km/s) regime. Experiments were performed in a free-piston driver shock tunnel facility. Drag measurements were made using the deconvolution force balance (Sanderson and Simmons 1991) for measuring drag in hypervelocity impulse facilities where test times may be of the order of 1 ms. Experiments were performed on a 5° semi-vertex angle cone with varying degrees of nose bluntness. Results are presented here for drag measurements made in air as test gas at stagnation enthalpies of 3.3 and 14.4 MJ/kg.

Key words: Hypersonic flow, Blunt cone, Drag

1. Introduction

Presented here are results obtained from an investigation into the effects of nose bluntness on slender cone drag in the hypersonic (Mach number greater than 5) and hypervelocity (flight speed greater than 5 km.s^{-1}) flight regime. Experiments were performed in a free piston driver shock tunnel (Stalker and Morgan 1988) using the technique developed by Sanderson and Simmons (1991) for force measurement in hypersonic impulse facilities.

A renewed interest in the development of hypersonic vehicles has motivated research into the aerodynamic characteristics of a blunted cone in the hypersonic and hypervelocity flight regime. The blunted cone simulates the nose geometry common to a number of the different concepts proposed for the design and application of hypersonic vehicles and their propulsion systems. Ideally, a sharp nose tip would be preferred to reduce the drag. However, a pointed slender cone is difficult to cool and does not offer the capability to carry large payloads. Research into blunted nose cones in subsonic and low supersonic flight regimes is quite extensive but at the hypersonic and hypervelocity conditions this is not the case.

Hypervelocity impulse facilities, such as free piston driver shock tunnels, have been developed to enable investigation into the hypersonic and hypervelocity regime. The test time in these facilities is of the order of 1 ms. This short test time inhibits force component measurement using conventional force or accelerometer balances. A novel technique for drag measurement in impulse facilities has been developed by Sanderson and Simmons (1991). The technique relies on interpretation of the transient stress waves propagating within the model and its supporting structure. This approach accounts for the distributed mass effects of the model and force balance structure which become significant in short test times.

2. Experiments

2.1. Variable nose-bluntness slender cone model

Experiments were performed on a 5° semi-vertex angle cone of length 571.5 mm and base radius of 50 mm. A total of 11 variable nose tips was used ranging in nose radius from 0.2 mm to 18.0 mm in steps of 1.8 mm. These correspond to bluntness ratios of 0.004 to 0.36 where the bluntness ratio is defined as the ratio of the nose radius to the cone base radius.

Shock Waves @ Marseille I
Editors: R. Brun, L. Z. Dumitrescu

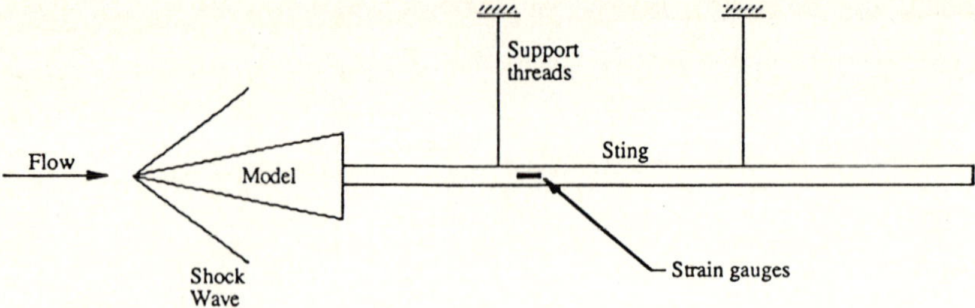

Fig. 1. Drag balance configuration

2.2. Drag measurement technique

The model is attached to a sting which takes the form of a slender elastic bar (Fig.1). The sting is suspended by vertical threads to allow free movement in the axial direction. Strain gauges located on the sting record the passage of stress waves resulting from the impulsively applied drag force as they are transmitted from the model into the sting.

Following Sanderson and Simmons (1991) the dynamic behaviour of the model/sting combination may be modelled as a time-invariant, causal, linear system described by the convolution integral:

$$y(t) \; = \; \int\limits_0^t g(t - \tau)u(\tau) \, d\tau \tag{1}$$

where $u(t)$ is the single input to the system, $y(t)$ is the resulting output and $g(t)$ is the unit impulse response function. Knowing the response of the system to a unit impulsive force it is possible to determine the response of the system to excitation by any arbitrary force via Eq.1. Alternatively, and what is done here, $y(t)$ is obtained from the strain gauge output and a numerical deconvolution process is performed to obtain $u(t)$, the time-history of drag applied to the model. The unit impulse response function is obtained numerically from a dynamic finite element program.

Theoretical analysis of a distributed mass model of the model/sting system shows that the mechanical time constant of the drag force balance is proportional to the mass of the model. In order that the mechanical time constant of the system be kept small, the base area of the cone model was hollowed out to reduce the mass of the cone by almost 50%.

Choice of model and sting materials was also dictated by the system time constant. The model was made of aluminium with a brass sting. The sting was 2.5 m long allowing sufficient time for the strain gauge measurements to be made before interference from the stress wave reflected from the end of the sting. The sting was constructed from brass tubing of 34.9 mm outside diameter and 1.6 mm wall thickness, making its bending stiffness high.

2.3. Test flow conditions

The experiments were performed in the T4 free piston driver shock tunnel facility (Stalker and Morgan 1988). A contoured axisymmetric Mach 5 nozzle was used to expand the test gas from the stagnation region to the appropriate test conditions. The nozzle exit plane was 265 mm in diameter and the nozzle throat diameter was 25 mm. The tunnel was operated in a tailored mode so that the static pressure and enthalpy would be constant throughout the flow test time.

The conditions in the test section were numerically determined using ESTC (McIntosh 1968) and NENZF (Lordi et al. 1966). The shock speed in the shock tube and the stagnation pressure were measured and used as inputs to ESTC to determine the temperature of the test gas in the stagnation region after shock reflection. The test gas undergoes a steady expansion from the stagnation region to the test flow properties at the exit plane of the nozzle. NENZF is a one-dimensional non-equilibrium code which predicts the properties of the test gas at the exit plane of the nozzle given the stagnation pressure and temperature. The test flow properties thus calculated are shown in Table 1. Experiments were performed at two different flow enthalpies in a test gas of air. These two conditions were chosen so as to bring out any differences between high and low enthalpy flows.

Table 1. Flow conditions using air as test gas

Stagnation Enthalpy MJ/kg	Mach No	Static Pressure kPa	Pitot Pressure kPa	Static Temperature K	Flow Velocity km/s	Density kg/m^3
14.4	5.2	16	555	1860	4.5	0.028
3.3	6.4	10	570	355	2.5	0.094

3. Results and discussion

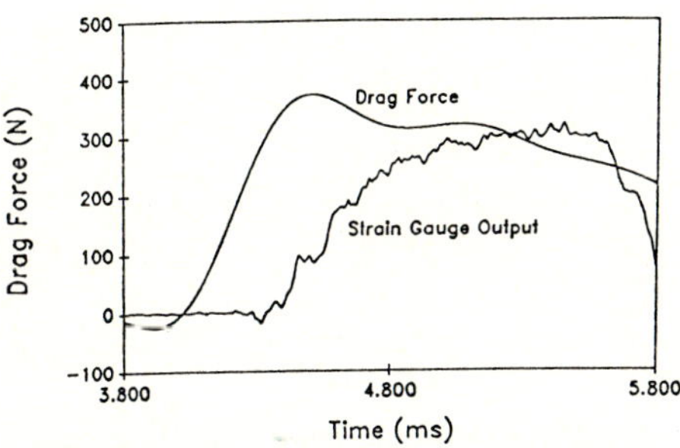
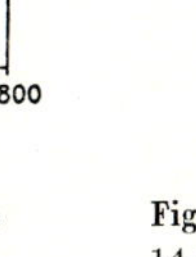
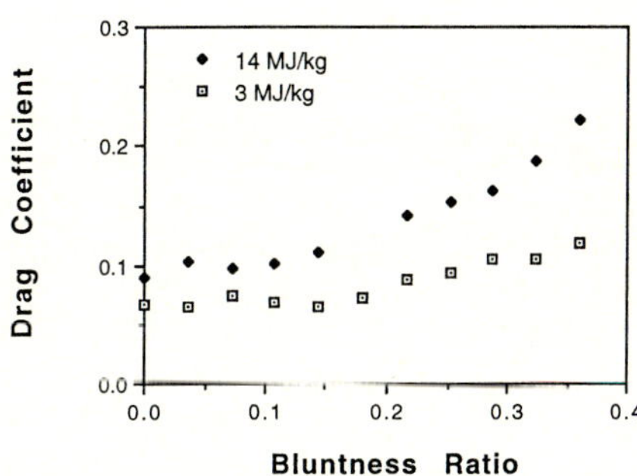

Fig. 2. Comparison between signal from strain gauge bridge before (raw signal) and after (drag signal) deconvolution for a nose bluntness ratio of 0.216

Fig. 3. Drag coefficient vesus nose bluntness ratio at 14 MJ/kg and 3 MJ/kg

The strain gauge signal was deconvoluted numerically as described in Section 2.2 to obtain the time history of the drag on the model. This drag measurement technique is inherently noisy as the deconvolution process tends to amplify any noise present in the original output signal, $y(t)$. It was thus necessary to pass all the drag measurements resulting from the numerical deconvolution process through a 2 kHz, 6 pole Butterworth low-pass digital filter. Fig.2 shows the measured drag in its filtered form for a bluntness ratio of 0.216 (nose radius of 10.8 mm) in comparison with its corresponding strain gauge output signal before deconvolution. An estimate of the accuracy of the technique would suggest that these values are correct to ±10%.

The results from this preliminary investigation into the effects of the nose bluntness has cone drag are summarised in Fig.3. A theoretical prediction of the total drag on a sharp cone at the high enthalpy condition has been found to show good agreement with the measured drag. The pressure drag was predicted using Taylor-Maccoll theory (Taylor and Maccoll 1932) and found to be 159 N for the case being studied here. An estimate for the skin friction drag on the cone was made based on laminar boundary layer theory (White 1974) and found to be 36 N. This gives a total drag of 195 N for a sharp, 5° semi-vertex angle and 571.5 mm long cone travelling in air at the high enthalpy condition of nominally 14 MJ/kg. The value of drag measured for the sharp cone was 200 N.

It would appear that at the smaller nose bluntnesses the effect on the total drag is minimal. For the high enthalpy condition, the drag shows an increase from the sharp nose value of about 20% at a bluntness ratio of 0.144 (nose radius of 7.2 mm). However, beyond this bluntness ratio the drag increases more rapidly so that at a bluntness ratio of 0.36 (nose radius of 18.0 mm) the value of drag is about 145% greater than the drag on the sharp cone.

It is evident from Fig.3 that the total drag increases more rapidly with increasing nose bluntness at the high enthalpy condition than at the low enthalpy condition. At this condition the value of drag at a bluntness ratio of 0.36 (nose radius of 18.0 mm) is only about 70% greater than the sharp cone case. It is postulated that this can be attributed to a real gas effect occuring at the high flow temperatures. The nature of this effect is not fully understood at this stage.

4. Conclusion

The preliminary drag measurement results reveal a steadily increasing effect of nose bluntness on the drag of a cone in hypersonic and hypervelocity flows. The effect at the smaller nose bluntnesses is relatively small, with about a 20% increase in drag at a nose bluntness ratio of 0.144. This is encouraging for the design of a hypersonic space plane or a centrebody for an axisymmetric scramjet where a slightly blunted nose is required to reduce stagnation point heating.

Beyond a nose bluntness ratio of 0.144 the drag increases more rapidly with bluntness. The results presented here suggest the existence of a real gas effect on the total drag of a blunted slender cone with increased flow enthalpy.

Acknowledgements

The authors are grateful for the support received from the Australian Research Council under grant AE9032029 and the Queen Elizabeth II Fellowship Scheme (for Dr. D. Mee).

References

Lordi JA, Mates RE, Moselle JR (1966) Computer program for numerical solution of nonequilibrium expansion of reacting gas mixtures. NASA CR-472

McIntosh MK (1968) Computer program for the numerical calculation of frozen and equilibrium conditions in shock tunnels. Dept. of Phys. ANU, Canberra

Sanderson SR, Simmons JM (1991) Drag balance for hypervelocity impulse facilities. AIAA J. 29: 2185-2191.

Stalker RJ, Morgan RG (1988) The University of Queensland free piston shock tunnel T4 - iInitial operation and preliminary calibration. NASA CR-181721

Taylor GI, Maccoll JW (1932) The air pressure on a cone moving at high speed - I Proc. Roy. Soc. (London) A 139: 278-297.

White FM (1974) Viscous fluid flow. McGraw-Hill Inc, New York

Navier-Stokes Simulation and Measurement of Cone Drag at $M_\infty = 7.9$

C. Jessen*, H. Grönig*, M. Watanabe[†] and K. Takayama[†]

*Stoßwellenlabor, RWTH Aachen, 52056 Aachen, Germany
[†]Institute of Fluid Science, Tohoku University, Sendai, Japan

Abstract. The calculation of a $M_\infty = 7.9$ flow around a 30° cone in a shock tunnel is presented. It starts from the initial reservoir conditions obtained in the experiment in front of the nozzle. The nozzle flow is calculated using axisymmetric Euler equations, while the model flow is simulated by solving the Navier-Stokes equations. The results of this unsteady calculation are presented as plots of the Pitot pressure and velocity within the test section, as pressure contour plots around the cone at different instants and as plots of the friction drag, form drag and the drag coefficient. The latter is compared with an experimental result. The numerical procedure is described and the results are discussed in detail.

Key Words: Cone flow, Numerical simulation, Nozzle starting process, Cone drag

1. Introduction

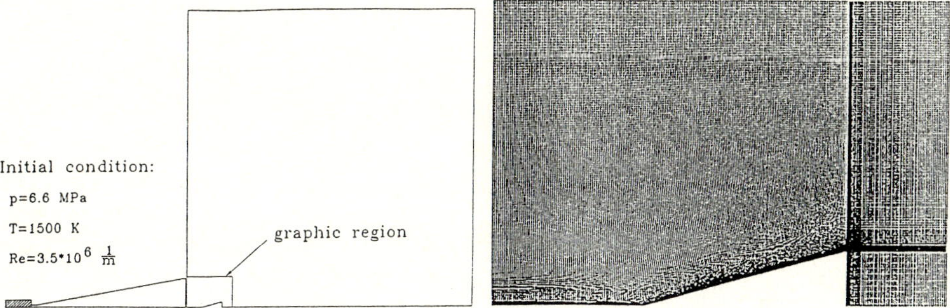

Fig. 1. a) Computational region including the reservoir, the nozzle and the model within the graphic region
b) Grid of the graphic region around the cone (half model)

The dynamic performance of a newly designed strain gauge balance for force measurement in short duration hypersonic facilities (Jessen and Grönig 1993 a) is cross-checked by a full numerical simulation of the nozzle starting process and model flow. The static accuracy of the balance was found to be better than 0.5% for all load cases from the calibration. The construction of the balance, its calibration and the data reduction procedure together with experimental results obtained on a cone in the Aachen Shock Tunnel TH 2 are described in Jessen and Grönig (1993 b).

A pertinent comparison between calculation and measurement is obtained, as the geometries of the nozzle and the model as well as the positions of the model and the Pitot probe with respect to the nozzle exit in the numerical simulation comply exactly with the experimental situation.

The calculations were started with the reservoir conditions in front of the nozzle entrance as they were measured in the experiment. Fig.1 shows the computational regime and the grid of the graphic region in which the presented results were obtained. The flow over a 30° cone at zero angle of attack was investigated. The fully axisymmetric configuration was treated with the Euler

Shock Waves @ Marseille I
Editors: R. Brun, L. Z. Dumitrescu

equations for the nozzle flow and with the Navier-Stokes equations for the model flow, so that the contribution of friction to the drag can be quantified.

2. Numerical simulation

The numerical simulation was conducted using the TVD finite difference scheme (Yee 1987) applied to the Navier-Stokes equations for the cone and to the Euler equations for the nozzle flow.

The conservation form of the axisymmetric Navier-Stokes equations for viscid compressible flow in cartesian coordinates (x, y) can be expressed in generalised curvilinear coordinates (ξ, η) as (Fletcher 1988):

$$U_t + (F - R)_\xi + (G - S)_\eta - (W_1 + W_2) = 0 \,, \tag{1}$$

$$\xi = \xi(x, y) \quad \text{and} \quad \eta = \eta(x, y) \,,$$

where the vector U represents the conserved quantities, F and G represent the ξ- and η-components of the flux vectors, R and S are the ξ- and η-components of the vectors which represent the viscous term and W_1 and W_2 represent inhomogeneous terms in the axisymmetric flow:

$$U = \bar{U}/J \quad, \quad F = \left(\xi_x \bar{F} + \xi_y \bar{G}\right)/J \quad, \quad G = \left(\eta_x \bar{F} + \eta_y \bar{G}\right)/J \quad,$$

$$R = \left(\xi_x \bar{R} + \xi_y \bar{S}\right)/J \quad, \quad S = \left(\eta_x \bar{R} + \eta_y \bar{S}\right)/J \quad, \quad W_1 = \bar{W}_1/J \quad, \quad W_2 = \bar{W}_2/J$$

where J is the Jacobian given as

$$J = \xi_x \eta_y - \xi_y \eta_x \,.$$

The vectors are given by:

$$\bar{U} = \begin{bmatrix} \rho \\ \rho u \\ \rho v \\ e \end{bmatrix} \quad, \quad \bar{F} = \begin{bmatrix} \rho u \\ \rho u^2 + p \\ \rho u v \\ (e + p)u \end{bmatrix} \quad, \quad \bar{G} = \begin{bmatrix} \rho v \\ \rho u v \\ \rho v^2 + p \\ (e + p)v \end{bmatrix} \quad,$$

$$\bar{R} = \frac{1}{\mathrm{Re}} \begin{bmatrix} 0 \\ \tau_{xx} \\ \tau_{xy} \\ u\tau_{xx} + v\tau_{xy} + \frac{\mu}{\mathrm{Pr}(\gamma-1)} \frac{\partial a^2}{\partial x} \end{bmatrix} \quad, \quad \bar{S} = \frac{1}{\mathrm{Re}} \begin{bmatrix} c0 \\ \tau_{xy} \\ \tau_{yy} \\ u\tau_{xy} + v\tau_{yy} + \frac{\mu}{\mathrm{Pr}(\gamma-1)} \frac{\partial a^2}{\partial y} \end{bmatrix} \quad,$$

$$\bar{W}_1 = \frac{1}{y} \begin{bmatrix} \rho v \\ \rho u v \\ \rho v^2 \\ (e + p)v \end{bmatrix} \quad, \quad \bar{W}_2 = \frac{1}{y\,\mathrm{Re}} \begin{bmatrix} 0 \\ \tau_{xy} \\ \tau_{yy} \\ u\tau_{xy} + v\tau_{yy} + \frac{\mu}{\mathrm{Pr}(\gamma-1)} \frac{\partial a^2}{\partial y} \end{bmatrix}$$

with $\quad \tau_{xx} = \frac{4}{3}u_x - \frac{2}{3}v_y \quad, \quad \tau_{xy} = u_y + u_x \quad \text{and} \quad \tau_{yy} = \frac{4}{3}v_y - \frac{2}{3}u_x \,,$

where a, p, ρ, $\underline{\underline{\tau}}$, u, v and e are sound velocity, pressure, density, shearing stress, x-component velocity, y-component velocity and total energy per unit volume, respectively; μ is the kinematic viscosity. In the viscous terms, Pr and Re are Prandtl and Reynolds numbers, respectively. The Reynolds number is defined as

$$\mathrm{Re} = \rho_0 c_0 L/\mu \,,$$

where ρ_0, c_0, L and μ are the initial density ahead of the shock front, the initial sound speed, the characteristic length and the viscosity, respectively. The Prandtl number is defined as

$$\mathrm{Pr} = \mu c_p / k \,,$$

where c_p and k are the specific heat at constant pressure and the thermal conductivity, respectively. For a perfect gas the pressure is given by

$$p = (\gamma - 1)\left[e - \frac{1}{2}\rho(u^2 + v^2)\right] , \tag{2}$$

where γ is the ratio of specific heats and is constant, since real gas effects are neglected.

For the simulation of the nozzle flow, the inhomogeneous terms $\boldsymbol{W}_1$ and $\boldsymbol{W}_2$ are excluded from the governing equations.

The computational flowfield is divided into the nozzle- and the cone flowfield. At each computational time step, the physical variables are transfered between the exit of the nozzle and the entrance of the cone flowfield by using a Lagrange second order interpolation.

The computational grids were generated numerically (David and Robert 1989) and consist of 401×101 knots for the nozzle flowfield and 501×401 knots for the cone flowfield. In the boundary layer of the cone surface, the minimum grid interval is equivalent to 10^{-5} mm for the characteristic length $L = 199$ mm. The stagnation conditions obtained in the experiment (6.6 MPa and 1500 K) are taken as initial conditions upstream of the nozzle throat.

First a complete calculation of the nozzle flow is performed within a computational time step Δt_1. Then the values obtained at the nozzle exit are used as input for the calculation of the flowfield around the model. The coupling between the Euler equations and the Navier-Stokes equations can be described as an in-flow boundary where the results of the left-hand calculations upstream serve as initial conditions for the right-hand calculations (downstream).

The numerical simulation was carried out on the CRAY Y-MP8/4128 of the Supercomputer Center of the Institute of Fluid Science, Tohoku University.

3. Numerical results and comparison with the experiment

Fig.2 shows the calculated and the measured Pitot pressures. The time $t = 0$ corresponds to the rupturing of the diaphragm at the nozzle entrance. It is obvious, that the signature of the two pressure plots is quite similar, whereas the quantitative result of the calculation is too low by a factor of three compared to the experimental result. The reason for this disagreement is not clear up to now, but is has only little influence on the resultant drag coefficient, as the latter is obtained by dividing the drag force (which is too low by roughly the same factor) through the Pitot pressure. Therefore the error is nearly compensated and is not discussed in this context (the argument does not hold for the friction drag).

A slow rise of the Pitot pressure is obtained in the calculation as well as in the experiment and can therewith clearly be attributed to physical reasons and not to a lag caused by the pressure gauge or its installation. The simulation is terminated after 2 ms, when the nozzle starting process is over and a steady flow situation is attained.

Fig.3 shows the velocity in the test section and the resultant forces on a 30° cone at $\alpha = 0°$ angle of attack. These results can be regarded as important for two reasons:

First, it is not possible up to now to measure the velocity profile of the highly unsteady nozzle starting process; and secondly the splitting of the resultant drag into form drag and friction drag allows for a very detailed discussion of the drag establishment by means of analysing pictures of the flow configuration obtained at certain instants. Such pictures are generated within the graphic region (see Fig.1) as pressure contour plots and are presented in Fig.4.

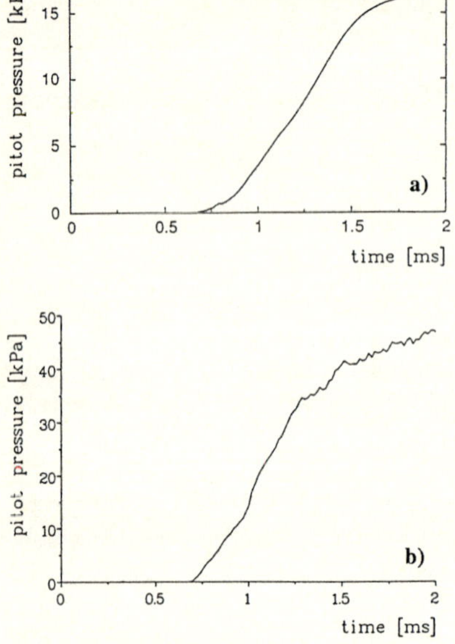

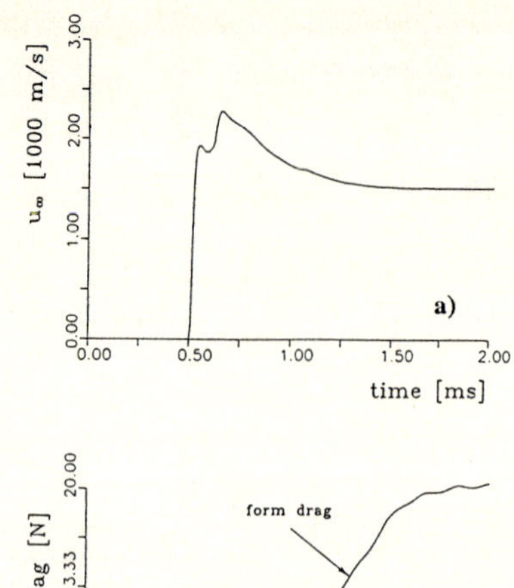

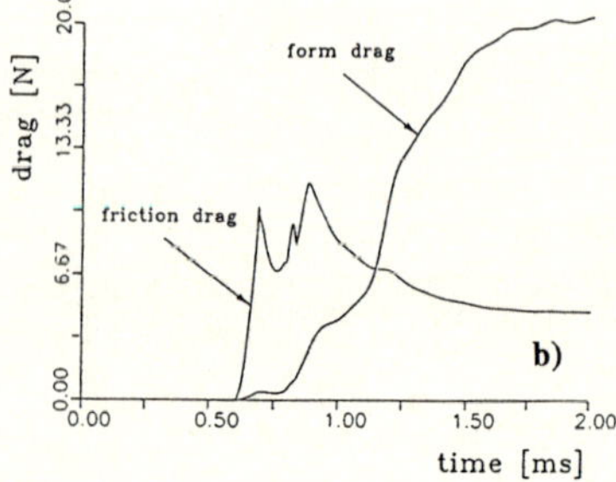

Fig. 2. a) Calculated Pitot pressure in the test section
b) Measured Pitot pressure

Fig. 3. a) Calculated velocity in the nozzle exit
b) Calculated drag force on a 30°-cone at $\alpha = 0°$

Fig.3 clearly shows, that the friction drag roughly follows the signature of the velocity, that rapidly rises over the primary shock. The form drag, which is due to the pressure acting on the surface of the cone, follows the signature of the Pitot pressure in Fig.2 and gains a considerable influence only after several hundred microseconds. In the steady situation, about 2 ms after flow onset at the nozzle entrance, the form drag contributes about 82% to the overall drag, while the friction drag reduces to about 18%.

But let us consider the drag establishment in comparison with the pressure contours of Fig.4. (The increment Δp for the contour lines in Fig.4 is $\Delta p = (p_{max} - p_{min})/100$ for each instant and not the same for all times.) The primary shock produces a steep rise in the velocity, but only a very moderate increase in the pressure. On its way over the model the high velocity gradient in the boundary layer produces strong shear resulting in friction drag. As the shock proceeds (a), its working surface on the cone increases and accordingly the friction drag rises until the shock has passed the model at $t \approx 0.7$ ms. At $t = 0.75$ ms (b) the secondary shock passes over the model, creating a further rise in the friction drag and also increasing even more the pressure, so that the form drag gains influence. In the wake of the secondary shock the conical shock around the model is created and a shock/shock interaction near the boundary layer is generated. This interaction can best be seen at $t = 0.8$ ms (c), where it creates an irregularity in the rise of the friction drag. When the secondary shock has passed the model at $t = 0.88$ ms (d) the friction drag continuously decreases with the decreasing velocity. Presumably the increasing distance of the shock from the body surface results in a lower velocity gradient in the boundary layer, so that the wall shear stress is diminished. At $t = 2.02$ ms (f) a steady situation with regard to the flow around the model is reached. Fig.5 shows the drag coefficient of the cone, defined as

$$C_D = \frac{F_D}{\frac{\varrho}{2}\,u_\infty^2\,A}\,,$$

(3)

(a- computation; b- experiment) where F_D is the drag force, ρ the density and u_∞ the velocity of the free stream; A is a characteristic area, the base area of the cone in our case.

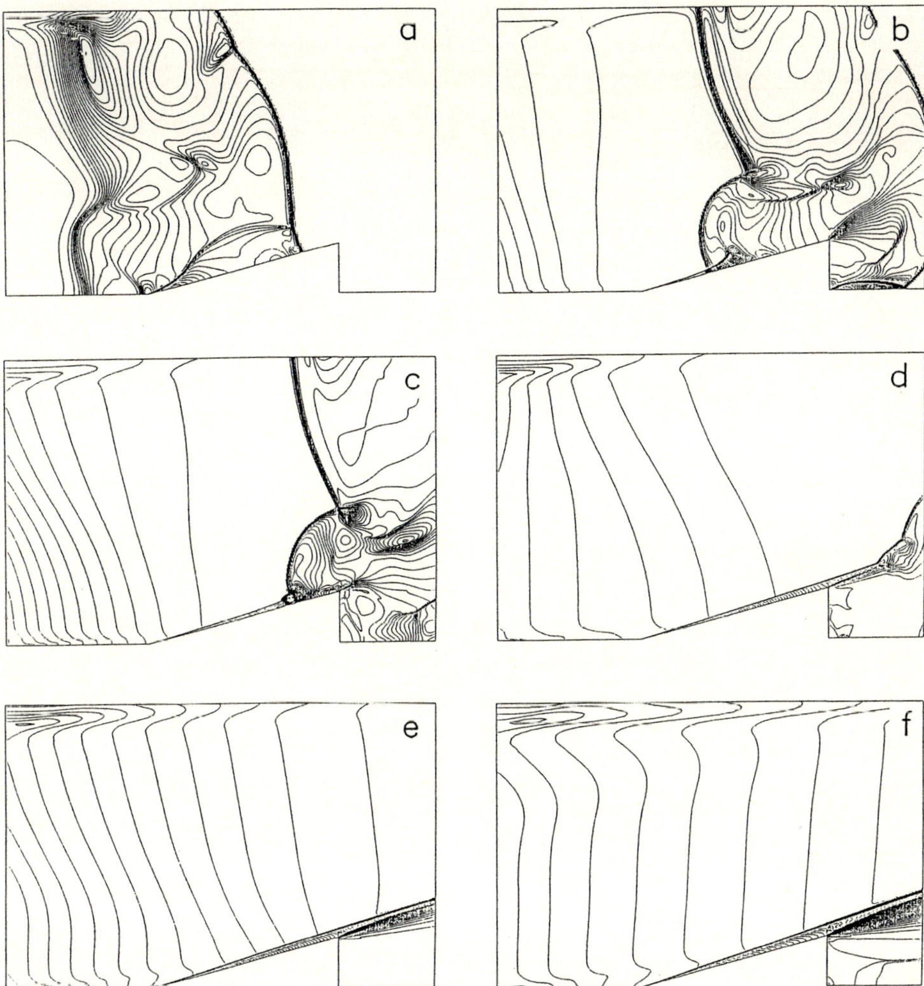

Fig. 4. Pressure contours of the $M_\infty = 7.9$ flow establishment around a cone at zero incidence
a) $t= 0.67$ ms   b) $t= 0.75$ ms   c) $t= 0.8$ ms   d) $t= 0.88$ ms   e) $t= 1.22$ ms   f) $t= 2.02$ ms

The calculated drag coefficient can be regarded as one obtained with "zero response time". The balance used for the measurement has, of course, a certain response time, so that a full agreement with the calculated curve cannot be expected. The first steep rise in the signal is artificial: at zero velocity the drag coefficient is infinity, so that only after a certain period following flow onset a finite drag coefficient can be specified; the earlier values are zero by definition. After the steep artificial rise an oscillating, decreasing signal can be seen. The oscillation is due to the initial transient oscillation of the balance, the decreasing signal agrees with the calculated signature. Not every detail of the calculated signal is reproduced by the measurement, but a good overall agreement can be stated. This is surprising in so far as the balance was not expected to have such a short response time, but was designed for "quasisteady" measurements in the shock tunnel. The steady value at $t = 2.02$ ms is $C_D = 0.166$ for the calculation and $C_D = 0.152$ for the measurement, which corresponds to a deviation of 8.4%. It is attributed to the fact, that the calculation is inviscid in the nozzle, neglecting losses there and leading to a higher drag on the model. Numerical errors and of course errors in the measurement have also to be taken into account, the latter are estimated to be $\pm 5\%$.

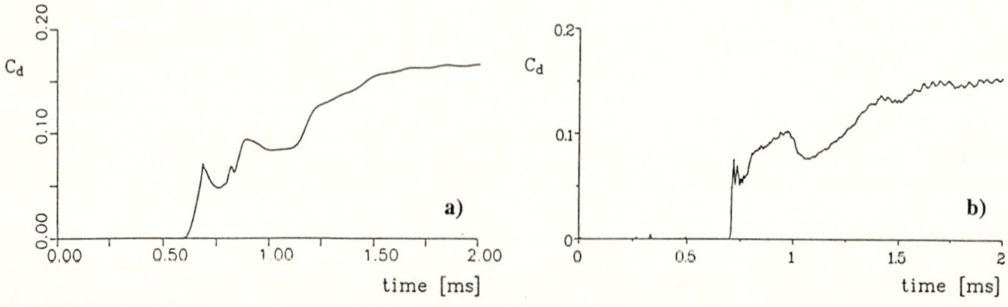

Fig. 5. Drag coefficient of a 30° cone at $M_\infty = 7.9$.   a) Numerical simulation,   b) Measurement

4. Conclusions

An unsteady combined Euler and Navier-Stokes simulation of the $M_\infty = 7.9$ flow around a cone was presented. The nozzle flow was treated with the Euler equations, the model flow with the Navier-Stokes equations, the numerical procedure was briefly described.

From the pressure contours plotted at different instants and from the Pitot pressure and velocity calculated for the test section the drag establishment on the cone could be discussed in detail. It became obvious, that friction dominates the scene during the flow establishment. In the steady situation, the friction contributes about 18% to the overall drag. A good agreement between calculation and measurement of the drag coefficient is obtained.

In conclusion it is stated, that the calculations confirm the balance measurement for the case investigated. Especially it is obviously possible to measure forces in the millisecond time range with a strain gauge balance.

References

David CI, Robert MZ (1989) Conformal mapping and orthogonal grid generation. J. Propulsion 5:327-333

Fletcher CAJ (1988) Computational techniques for fluid dynamics. Vol. II, Springer Verlag, Berlin Heidelberg, pp 431-432

Jessen C, Grönig H (1993 a) A six component balance for short-duration hypersonic facilities. In: Boutier A (ed) New trends in instrumentation for hypersonic research, NATO ASI Series E, Vol. 224, Kluwer Academic Publishers, Dordrecht

Jessen C, Grönig H (1993 b) Six component force measurement in the Aachen shock tunnel. These Proceedings

Yee HC (1987) Upwind and symmetric shock-capturing schemes. NASA TM 89464

The Boundary Layer on a Sharp Cone in High-Enthalpy Flow

P. Germain and H. Hornung
Graduate Aeronautical Laboratories, California Institute of Technology, USA

Abstract. The boundary layer on a 5 deg. half-angle cone is studied by measurement of the heat flux distribution and by qualitative flow visualisation. In the laminar layer, heat flux levels are higher in air than in nitrogen because of a larger heat release from oxygen recombination at the wall. By varying the specific reservoir enthalpy in air and N_2, and from measurements in CO_2, it is found that real-gas effects stabilize the boundary layer.

Key words: Hypervelocity, Boundary layer transition, Experiment

1. Introduction

The well-known linear stability analysis of compressible laminar boundary layers introduced by L. M. Mack in 1969 showed that higher instability modes (Mack modes) dominate the Tollmien-Schlichting (TS) mode at high Mach number. Experimental results e.g. by Kendall (1975), Demetriades (1974) and Stetson et al. (1983) obtained in continuous flow facilities confirmed the higher modes. According to the theory, the TS (or first) mode is stabilized, but the higher modes are destabilized by wall cooling. Most previous experiments have been carried out in cold hypersonic facilities, so that T_w was rarely smaller than T_e. Stetson's experiments (1989), in which he cooled the wall with liquid nitrogen, were an exception, and he observes that wall cooling destabilizes. In our experiments extreme wall cooling is always applied, since the recovery temperature is 2000 K or more, and T_w is always 300 K.

Effects of high-enthalpy real-gas behavior on stability have been computed by Stuckert and Reed (1990) for flat plate flows. The growth rate of the first mode was found to be reduced, while the second mode was found to be destabilized. Since the real-gas effects are endothermic, this effect has some similarity with wall cooling, and is intuitively appealing. So far, no experimental results confirm this trend however.

In the investigation presented here, the transition on a sharp cone is studied experimentally in the high-enthalpy free-piston shock tunnel T5, where dissociative real-gas effects of oxygen and nitrogen can easily be simulated.

2. Free-stream conditions

Table 1 gives the typical reservoir and test section conditions chosen. The reservoir specific enthalpy is determined from the measured shock speed and the measured reservoir pressure, using an equilibrium calculation. The nozzle flow is then computed by using an axisymmetric inviscid non-equilibrium computation. To compute the flow conditions at the outer edge of the boundary layer on the 5 deg. half-angle cone, the Taylor-Maccoll equations were solved for each of the nozzle exit conditions. The conditions thus obtained at the wall of the cone (shown in Table 1) were used to form the Stanton and Reynolds numbers defined below.

Shock Waves @ Marseille I
Editors: R. Brun, L. Z. Dumitrescu   © Springer-Verlag Berlin Heidelberg 1995

3. Instrumentation

Fifteen thermocouples of type E (chromel-constantan) were mounted flush with the surface of the cone to measure the heat transfer distribution. Following Schultz and Jones (1973), the one-dimensional semi-infinite slab theory was used to reduce the temperature traces to heat-transfer rate traces. The heat-transfer rate levels were normalized as follows:

$$\mathrm{St}(x) = \frac{\dot{q}(x)}{\rho_e u_e \left[h_0 - (1-r)\frac{u_e^2}{2} - c_{pw} T_w\right]}.$$

The distance x from the tip along the centerline was normalized by

$$\mathrm{Re}(x) = \frac{\rho_e u_e x}{\mu_e},$$

where the subscripts e and w denote gas properties taken at the outer edge of the boundary layer and at the wall respectively. The recovery factor r is taken to be $\sqrt{\mathrm{Pr}}$ where $\mathrm{Pr} = 0.7$, for both air and pure nitrogen (for a turbulent boundary layer, $r = (\mathrm{Pr})^{1/3}$).

Table 1. Reservoir, free-stream and boundary layer outer edge conditions

p_0, MPa	h_0, MJ/kg	T_0, K	u_e, km/sec	ρ_e, g/m^3	T_e, K	M_e,	Re/m, (10^{-6})
Nitrogen:							
10	4.13	3402	2.54	35.9	360	6.5	4.25
15	5.81	4662	2.97	39.4	494	6.5	4.37
24	12.30	7650	4.51	231.6	1977	5.9	1.96
40	2.89	2452	2.15	206.4	259	6.5	26.86
57	10.65	7359	4.24	70.8	1564	5.4	5.51
60	23.20	9927	5.82	36.8	3686	4.5	2.46
85	13.20	8294	4.70	76.8	1670	5.1	6.39
Air:							
10	3.76	3044	2.36	41.6	323	6.5	4.25
15	5.80	4134	2.75	44.3	437	6.5	4.97
23	11.50	6471	4.20	26.1	1750	5.1	2.06
40	3.41	2826	2.23	187.4	300	6.5	22.96
55	10.13	6150	4.02	68.2	1668	5.0	5.39
59	20.17	9112	5.43	41.0	3156	4.9	3.02
85	12.80	7329	4.47	85.0	2208	4.9	6.35
CO_2:							
55	3.64	2994	2.00	367.4	572	5.3	28.42
55	5.88	3611	2.77	156.2	1492	4.4	8.76
55	8.32	4163	3.01	126.6	1872	4.5	7.25

4. Flow visualization

Except in the case of the lowest enthalpy conditions of the present tests, the sensitivity of classical optical flow visualization techniques is insufficient. To overcome this difficulty, the technique known as resonant enhancement of the refractive index was employed. The refractive index of a gas and the opacity have high extrema in a spectral line of the gas so that the sensitivity of optical techniques can be increased very significantly by working at a suitable wavelength. In our case, this method was implemented at the wavelength of one of the sodium D-lines, using a tunable dye laser as the light source and painting a very thin layer of a saturated saline solution on the portion of the cone surface of interest. This aqueous film was then left to dry, so that a thin layer of salt crystals was left to be ablated during the test. The technique functions satisfactorily for qualitative flow visualization. It turned out to be particularly successful at high enthalpies (lower densities) where it was most needed.

5. Results and discussion

Fig.1 shows examples of the normalized experimental heat-transfer results for nitrogen and air. As may be seen, transition starts at Re=2 million and ends roughly at Re=4 million. Though this is considerably lower than the transition Reynolds number observed by King (1992) in a 'quiet' wind tunnel, it is in the same range as that observed in continuous-flow wind tunnels at low specific reservoir enthalpy, for which real-gas effects are absent. It does not, of course, necessarily mean that the transition mechanism is the same in both cases. However, it supports the view that the boundary layer flow reaches a steady-state in the relatively short test time of the T5 facility.

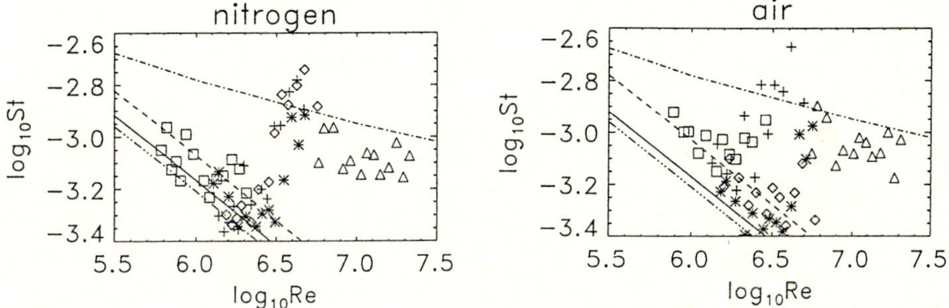

Fig. 1. Five different shots with air and N_2; $\square$: $p_0 = 60$ MPa, $h_0 = 22$ MJ/kg; $*$: 19/6; $+$: 60/11; $\diamond$: 85/13; $\triangle$: 47/3.5. ———: $h_0 = 22$ MJ/kg, frozen gas composition; –···– and –·–·: $h_0 = 3$ MJ/kg; – – –: 22 MJ/kg, chemical equilibrium

The predictions shown in these plots were made from the similarity theory and the Reynolds analogy. The Chapman-Rubesin constant is evaluated with the Eckert reference-temperature concept, with Sutherland's law for the viscosity. The axial symmetry of the flow was taken into account by multiplying the result by $\sqrt{3}$ in accordance with the Mangler transformation for laminar flow. For nitrogen and air at high enthalpy, complete recombination of the atomic species near or at the wall is assumed, and the following correction was made to the corresponding relation for a frozen laminar boundary layer to account for chemical equilibrium. First we evaluate the maximum static enthalpy from a $h = h(u)$ relation. This value and the known static pressure p_e, together with the equilibrium equations of state, determine the chemical composition at the maximum-enthalpy point, denoted by the subscript m. The correction factor is $1 + \alpha_m D/h_m$ where α is the atomic mass fraction and D the dissociation energy per unit mass.

For the laminar case at low enthalpy, the experimental data points fall on the perfect gas behavior line for both gases (see Fig.2). For the laminar case at high enthalpy, the data points fall roughly on the line corresponding to a fully-catalytic wall behavior (see Fig.1). This means that some recombination did occur at the wall, the level of which is difficult to evaluate because of the scatter of the data. For air, the heat flux is higher, suggesting that oxygen atoms are recombining at the wall.

At $p_0 = 19$ MPa and $h_0 = 6$ MJ/kg, the transition Reynolds number is higher. This is a Mach number effect, since $M_e = 6.5$ in this case compared with $M_e < 5$ in the higher enthalpy cases. Since the wall temperature is small compared to the recovery temperature in both cases, the difference is not likely to be connected with wall cooling. The transition Reynolds number increases with Mach number as observed for instance by DiCristina (1970). This is also evident in Fig.2. Further experiments were performed to examine real-gas effects on transition by varying the stagnation enthalpy between 12 MJ/kg and 16 MJ/kg, while M_e stays constant. The results are shown in Fig.4 corresponding to stagnation pressures of 55 MPa and 85 MPa. By decreasing the stagnation enthalpy from around 15 to 11 MJ/kg in steps of 1 or 2 MJ/kg, we see that in both cases, air flows become transitional at lower Reynolds number than nitrogen flows.

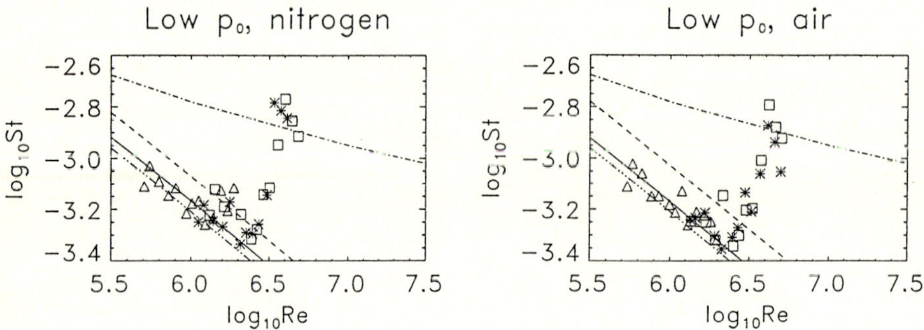

Fig. 2. Low reservoir pressure shots; □: $p_0 = 10.5$ MPa, $h_0 = 3.5$ MJ/kg; ∗: 16/5.8; △: 24/11.6

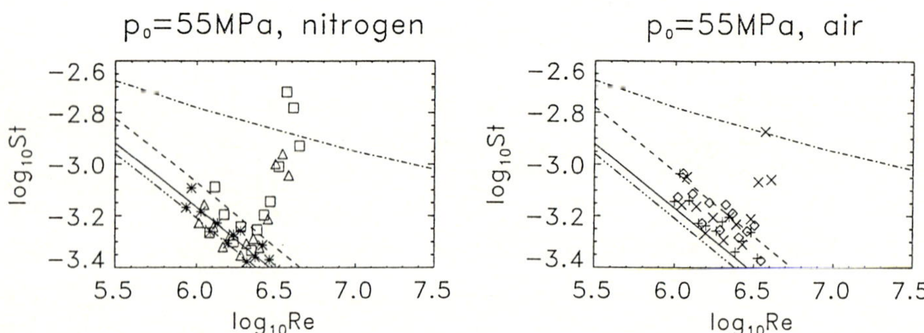

Fig. 3. Six different shots; □: $h_0 = 11$ MJ/kg; ∗: 15; +: 14; ◇: 12.2; △: 14; ×: 11.9

Fig.4 shows the results of carbon dioxide runs. Although the Mach number is slightly lower here than in air or nitrogen, the transition Reynolds number is much higher (onset around 5 million). There are more active chemical reactions taking place in CO_2 than in air or pure nitrogen. These two sets of results suggest that non-equilibrium chemistry stabilizes the flow. If the second mode of instability is predominant, despite M_e being as low as 5, this seems to contradict the earlier suggestion that it is destabilized by real-gas effects whereas the TS mode is stabilized.

The pictures presented in Figs.6 and 7 show that the nature of the instability is that of the TS mode. The second mode, undetectable with our method at these free-stream conditions, may also be present but cannot be seen since it would appear in the form of periodic density fluctuations,

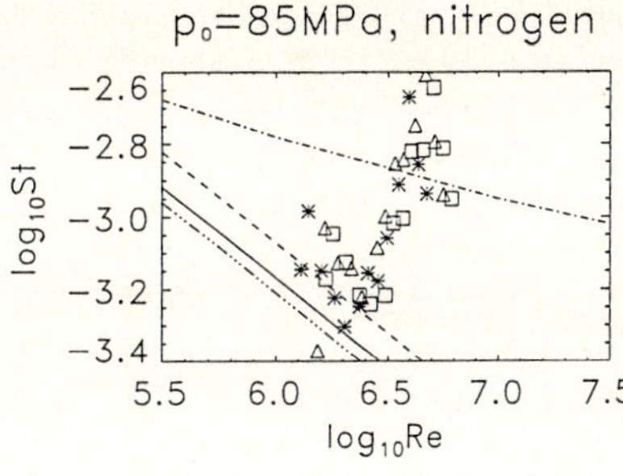

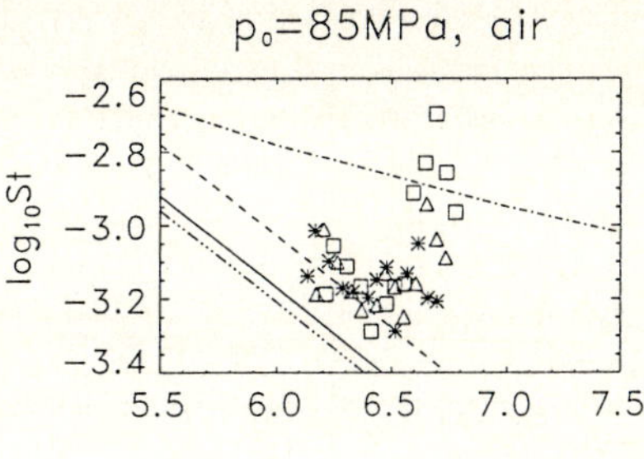

Fig. 4. Three different shots with air and N_2; $\square$: $h_0 = 12.2$ MJ/kg; $*$: 14.4; $\triangle$: 13.1

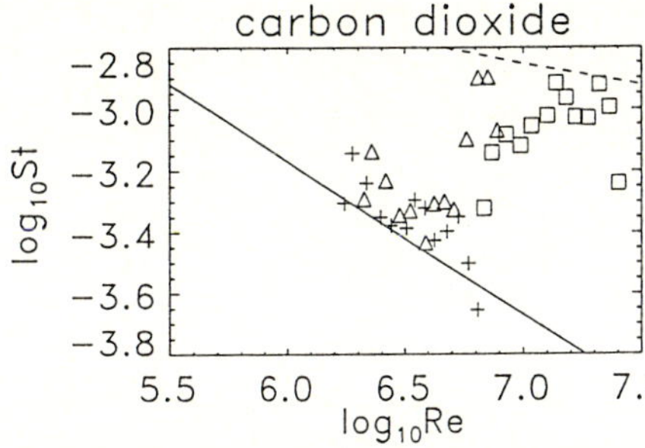

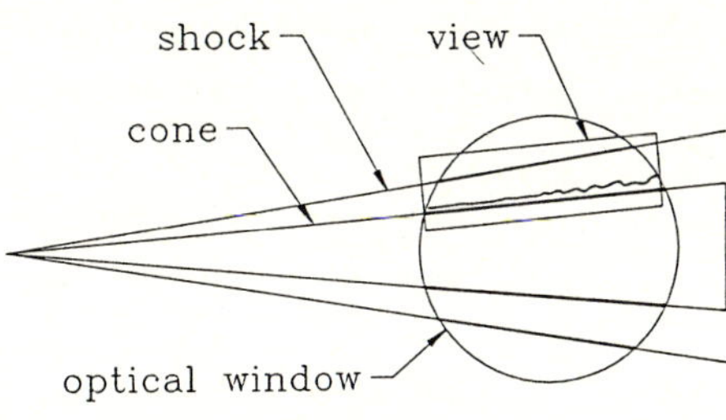

Fig. 5. Left: Three different shots with CO_2; $\square$: $p_0 = 55$ MPa, $h_0 = 3.6$ MJ/kg; $+$: 55/8.3; $\triangle$: 55/5.9. Right: Schematic explaining the photographs shown below

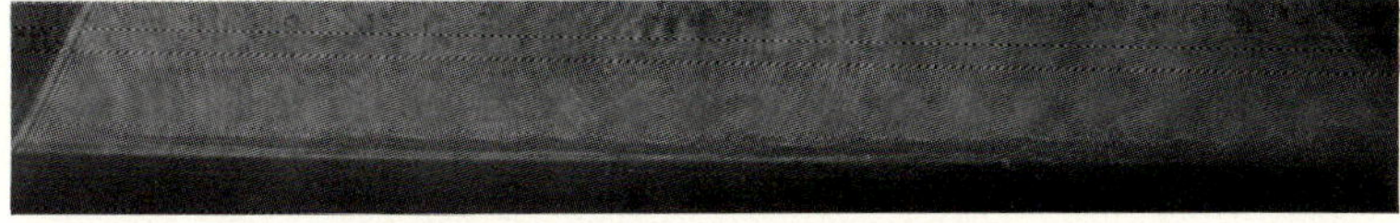

Fig. 6. Flow of N_2, $p_0 = 60$ MPa, $h_0 = 11$ MJ/kg. See Fig.5 for an explanation of the picture. Re = 1.1 million on the left and Re = 1.5 million on the right

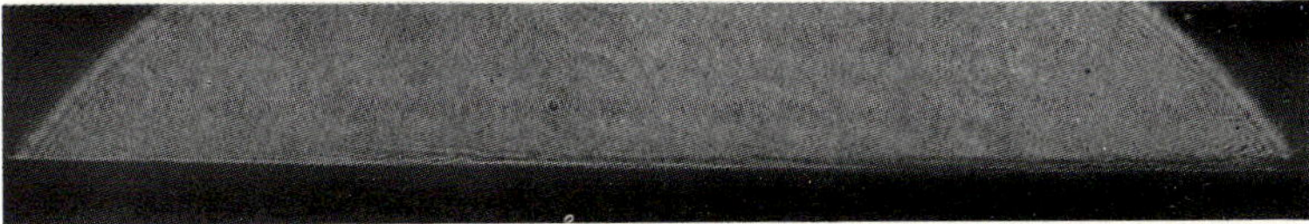

Fig. 7. Flow of N_2, $p_0 = 60$ MPa, $h_0 = 14$ MJ/kg. Re = 0.88 million on the left and Re = 1.2 million on the right

Fig. 8. Flow of air, $p_0 = 55$ MPa, $h_0 = 10.4$ MJ/kg. Re = 1.1 million on the left and Re = 1.5 million on the right. See Germain et al. (1993) for other pictures

see e.g. Demetriades (1974). The previous experimental results mentioned above indicate that the frequency of the most unstable mode (the second mode), at Mach number above 5, is of order 1 MHz at T5 conditions. No mechanical vibrations of such a high frequency are to be expected in the facility. The only source of important noise is likely to be the nozzle-wall boundary layer, and this also may not be in the right frequency range, since the nozzle-wall boundary layer is much thicker than the laminar boundary layer on the cone. These observations indicate that if one linear mode of instability dominates the transition process at the conditions studied, it is the TS mode in the shock tunnel experiments. Fig.8 shows the picture of a turbulent boundary layer.

6. Conclusions

(1) Comparison of air and N_2 flows show that oxygen recombination at the wall increases the heat flux in the laminar boundary layer.
(2) At low enthalpy, increasing the Mach number stabilizes the flow.
(3) Small changes in the specific reservoir enthalpy in air and N_2, and carbon dioxide runs reveal that real-gas effects stabilize the boundary layer.
(4) At the conditions tested, the most strongly amplified linear mode appears to be the Tollmien-Schlichting mode.

Acknowledgement

This work was supported by NASA Grant NAG-1-1209 (Dr. Griffin Anderson), and by AFOSR Grant F49610-92-J-0110 (Dr. L. Sakell).

References

Demetriades A (1974) Hypersonic viscous flow over a slender cone, Part III: Laminar instability and transition. AIAA Paper 74-535, Palo Alto, California

DiCristina V (1970) Three-dimensional laminar boundary-layer transition on a sharp 8° cone at Mach 10. AIAA J. 8, 5: 852-856

Germain P, Cummings E, Hornung H (1993) Transition on a sharp cone at high enthalpy; new measurements in the shock tunnel T5 at GALCIT. AIAA Paper 93-0343, Reno, Nevada

Kendall JM (1975) Wind tunnel experiments relating to supersonic and hypersonic boundary-layer transition. AIAA J 13, 3: 290-299

King RA (1992) Three-dimensional boundary-layer transition on a cone at Mach 3.5. Experiments in Fluids 13: 305-314

Schultz DL, Jones TV (1973) Heat transfer measurements in short duration facilities. AGARD Report 165

Stetson KF, Thompson ER, Donaldson JC (1983) Laminar boundary layer stability experiments on a cone at Mach 8, Part 1: Sharp cone. AIAA Paper 83-1761, Danvers, Massachusetts

Stetson KF, Thompson ER, Donaldson JC, Siler LG (1989) Laminar boundary layer stability experiments on a cone at Mach 8, Part 5: Tests with a cooled model. AIAA Paper 89-1895, Buffalo, New York

Stuckert GK, Reed HL (1990) Stability of hypersonic, chemically reacting viscous flows. AIAA Paper 90-1529, Seattle, Washington.

Shock Wave Interactions in Hypervelocity Flow

S. R. Sanderson and B. Sturtevant
California Institute of Technology, Pasadena, CA 91125, USA

Abstract.   The impingement of shock waves on blunt bodies in steady supersonic flow is known to cause extremely high local heat transfer rates and surface pressures. Although these problems have been studied in cold hypersonic flow, the effects of dissociative relaxation processes are unknown. In this paper we report a model aimed at determining the boundaries of the possible interaction regimes for an ideal dissociating gas. Local analysis about shock wave intersection points in the pressure–flow deflection angle plane with continuation of singular solutions is the fundamental tool employed. Further, we discuss an experimental investigation of the nominally two-dimensional mean flow that results from the impingement of an oblique shock wave on the leading edge of a cylinder. The effects of variations in shock impingement geometry were visualized using differential interferometry. Generally, real gas effects are seen to increase the range of shock impingement points for which enhanced heating occurs. They also reduce the type IV (Edney 1968 a,b) interaction supersonic jet width and influence the type II–III transition process.

Key words: Shock-on-shock interaction, Shock impingement, Hypervelocity flow

1. Introduction

Shock impingement phenomena, which inhibit the further development of hypervelocity vehicles, are discussed in the literature by many authors, notably by Edney (1968 a,b). For the inviscid, compressible flow of a perfect gas a sufficient set ϕ of dimensionless parameters to describe any quantity in the flow is, $\phi = \phi[M, \gamma, \beta_1, \Lambda, \Gamma]$, where M is the freestream Mach number, γ is the ratio of specific heats, β_1 is the impinging shock angle, Λ describes the position of the impinging shock and Γ is a set of parameters defining the body geometry. Consider the case of a given gas and fixed freestream condition, impinging shock strength and body geometry. The only remaining dependence is then the location of the impingement point relative to the body, i.e., $\phi = \phi[\Lambda]$. As the incident shock wave is translated relative to the body it potentially intersects with all possible shock strengths of both positive and negative slope. On the basis of an experiment such as this, with spherical and modified spherical bodies, Edney (1968 a,b) observed and categorized six interaction regimes; known as types I–VI. Edney rationalized the observed flowfields through local analysis about shock wave intersection points in the pressure–flow deflection angle plane (p-δ plane). The key conclusion drawn from such an analysis is the role of the three-shock solutions (or λ-shocks) and this is discussed further in §2. Based on the assumption of straight shocks in the vicinity of the interaction, the flow field for the global type IV interaction, whereby a supersonic jet penetrates a region of low subsonic flow (see Edney 1968 a,b or Fig.4), can be solved approximately as a freestreamline flow up to some unknown length scale. Typically this length scale is specified in terms of the width of the jet.

According to Edney's model, the heat transfer is determined by the attachment at the body surface of the shear layers generated at the shock impingement points. A correlation was obtained between local pressure and local heat transfer rate at the surface. A further observation made by Edney is that the jet curvature increases as the intersection point moves up. Some recent contributions to the literature are noted in the references of this paper.

Shock Waves @ Marseille I
Editors: R. Brun, L. Z. Dumitrescu

2. Local analysis for an ideal dissociating gas

Local analysis about shock-surface intersection points in the p-δ plane is well known in the Mach reflection literature (e.g. Hornung 1986). Our purpose here is to generalize these tools to the case of oblique translational shock waves where both the upstream and downstream conditions are non-equilibrium states with respect to the internal degrees of freedom of the gas. This analysis is performed for a one-dimensional shock and subsequently extended to oblique waves.

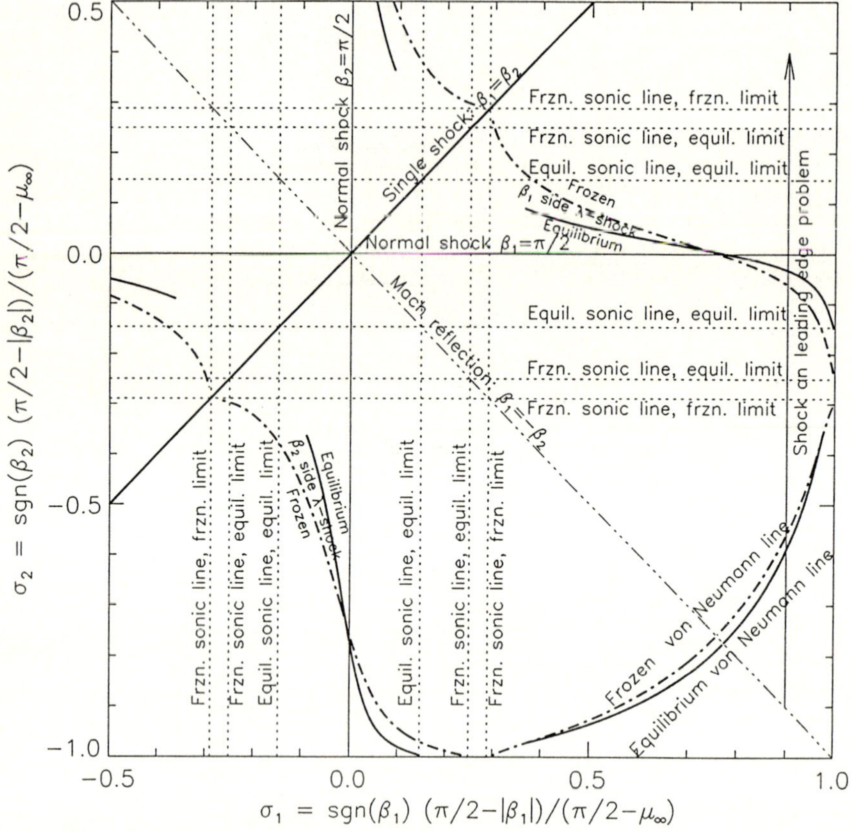

Fig. 1. Map of interaction regimes for two intersecting oblique shock waves of angles β_1 and β_2. Shock angles are normalized so that $\sigma = 0$ corresponds to a normal shock wave. Mach waves correspond to $\sigma = \pm 1$. The freestream conditions are taken as $H_0 = 0.5$, $P = 0.03$ (or $M_\infty = 5$ with $\gamma = 4/3$), $\alpha_1 = 0$ and $\hat{\rho}_d = 10^6$.

Consider the dimensionless form of the continuity and momentum equations that apply across a one-dimensional translational discontinuity and throughout the downstream relaxation zone,

$$\hat{\rho}\hat{u} = 1 , \qquad \hat{p} = 1 + \frac{1}{P}\frac{\hat{\rho}-1}{\hat{\rho}} , \tag{1,2}$$

where $P = p_1/\rho_1 u_1{}^2$. The symbols have the usual meanings and generally $\hat{\phi} = \phi_2/\phi_1$.

We use Lighthill's (1957) model for the thermodynamics of an ideal dissociating gas (IDG):

$$p = \frac{k}{2m}(1+\alpha)\rho T , \qquad h = \frac{k}{2m}[(4+\alpha)T + \alpha\theta_d] , \tag{3,4}$$

where k is Boltzmann's constant, m is the mass of one atom of the gas, θ_d is a temperature characterizing the dissociation energy and α is the dissociated mass fraction. Using Eq.4 the dimensionless energy conservation equation becomes,

$$(4+\alpha_2)\frac{\hat{T}}{D} + \alpha_2 + \frac{K}{\hat{\rho}^2} = H_0 ,$$ (5)

where the dimensionless parameters are $D = \theta_d/T_1$, $K = mu_1^2/k\theta_d$ and $H_0 = 2mh_0/k\theta_d$. H_0 is the conserved stagnation enthalpy.

For a nonequilibrium binary mixture, it is sufficient to specify P, H_0 and α_1. The remaining parameters are then given by the following identities, obtained from H_0 written in terms of conditions upstream of the discontinuity,

$$K = \frac{H_0 - \alpha_1}{1 + 2P\frac{4+\alpha_1}{1+\alpha_1}} \qquad D = \frac{4 + \alpha_1}{H_0 - K - \alpha_1} .$$ (6,7)

From Eq.3 it follows that $\hat{T} = \frac{\hat{p}(1+\alpha_1)}{\hat{\rho}(1+\alpha_2)}$ and, eliminating $\hat{T}$ and D from Eq.5, with the pressure given by Eq.2, leads to the following quadratic equation for the density,

$$(H_0 - \alpha_2)\hat{\rho}^2 - \frac{\lambda(P+1)}{P}\hat{\rho} + \frac{\lambda - KP}{P} = 0 , \qquad \lambda = 2KP\frac{4+\alpha_2}{1+\alpha_2} .$$ (8)

One solution describes the variation of density downstream of a translational shock. The remaining solution describes relaxation of the (possibly) nonequilibrium upstream state without any discontinuity. Both solutions are parameterized in terms of the dissociation mass fraction α_2.

Eq.8 may be combined with an equilibrium expression (Lighthill 1959) to obtain the asymptotic state far downstream from the translational shock.

The extension to oblique shock waves is obtained in the usual way. The normal components of P and K are given in terms of β, the angle of the oblique shock wave, by

$$P_{1_N} = \frac{P_1}{\sin^2 \beta} , \qquad K_{1_N} = K_1 \sin^2 \beta .$$ (9,10)

These are used in Eqs.6, 8, and 2, and the flow deflection δ is given by $\hat{\rho} = \tan\beta/\tan(\beta - \delta)$.

To facilitate the computation of multiple shock jumps for the interaction problem we must find a convenient means of determining P and H_0 downstream of the oblique shock. Trivially $H_{0_2} = H_{0_1}$. It can be shown that $K_2 = K_1(\cos^2\beta + \sin^2\beta/\hat{\rho}^2)$. P_2 then follows from Eq.6.

3. Continuation of singular solutions

One technique for reducing the amount of information that arises as the number of parameters is increased is to consider only the singular cases that delineate regions of solutions with similar forms. Various special cases arise that are significant in terms of transition to Mach reflection (e.g. Hornung 1986). Now generalize these cases to the asymmetrical interaction of oblique shock waves:

i. Interaction of shock waves that produce a Mach stem with zero curvature. This is a generalization of the von Neumann condition in Mach reflection.

ii. Sonic flow downstream of the transmitted shock waves.

iii. Coalescence of the strong and weak regular solutions such that the shock loci intersect only tangentially. This is a generalization of the maximum deflection condition in Mach reflection.

iv. Sonic flow downstream of the incident shock waves.

v. One transmitted shock wave—λ-shocks.

Given the solution for one such set of these singular points, solutions for neighboring values of the parameters may be determined using path-following techniques (Keller 1987). The frozen solution applies in regions near the shock wave intersection point and the equilibrium solution is valid far from the shock waves. The regular 4-shock intersections map into the 2nd and 4th quadrants of Fig.1 and lie on the surface bounded by the axes and the von Neumann curves. Outside this region the only solutions are the λ-shocks which exist only along lines. If the impingement point on a blunt body maps to a region of Fig.1 where no regular solution exists then the flow must deform in such a way that the λ-shock condition is maintained at the intersection point. This highlights the physical mechanism determining the gross nature of the flowfield. For weak impinging shock waves, shown by the vertical line in Fig.1, the separation of the frozen and equilibrium λ-shock solutions is relatively large. This suggests some effect on the transition from type II to type III. This occurs in the vicinity of the sonic line which is itself sensitive to non-equilibrium processes.

4. Experimental results

We have conducted an experimental investigation of the nominally two-dimensional mean flow that results from the impingement of an oblique shock wave on the leading edge of a cylinder. These initial experiments, comprising 35 shots, were conducted in the GALCIT T5 free piston reflected shock tunnel with nitrogen test gas at nozzle reservoir enthalpies of 3 MJ/kg and 12.5 MJ/kg and reservoir pressures of 12 MPa and 25 MPa, respectively. The reservoir gas was expanded through a contoured axisymmetric nozzle with area ratio 109.5 to yield a nominal Mach number of 5.5 in the test section. The cylinder was 37.5 mm in diameter with aspect ratio 4.5 and the flow was deflected 7 degrees by the incident planar shock wave. The effects of variations in shock impingement geometry were visualized using infinite fringe differential interferometry with vertical beam shear. Figs.2–4 illustrate type II, III and IV interactions at stagnation enthalpies of 3MJ/kg and 12.5 MJ/kg. The incident shock is concave down at the edge of test section and some fringe shift is observed below this wave. However, the main centerline disturbance is generally at the top of the line-of-sight-integrated image.

a
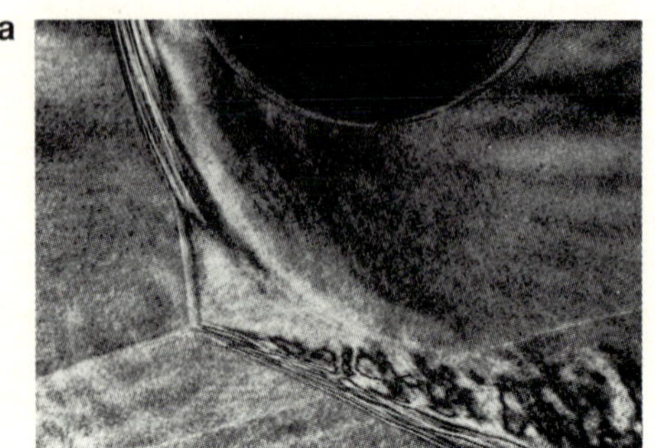
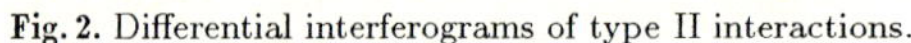
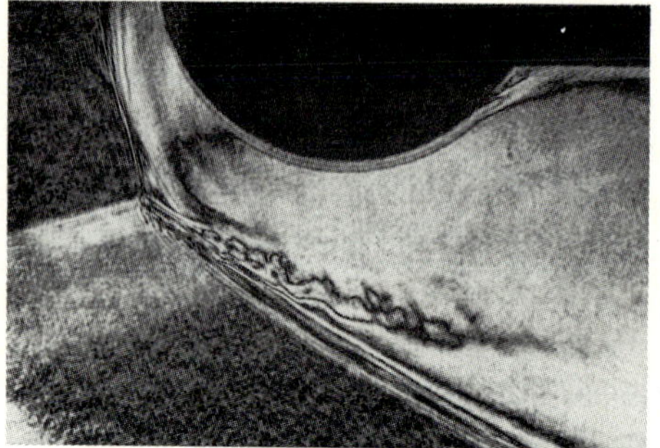
b

Fig. 2. Differential interferograms of type II interactions.   (a) 3 MJ/kg   (b)12.5 MJ/kg

At the lowest impingement point reported in this paper the type I–II transition has already occurred with the incident shock wave impinging somewhat below the lower sonic line (Fig.2). In the low enthalpy case a Mach stem connects the λ-shock pattern at the impingement point to the bow shock of the cylinder. At the intersection point a strong vortex sheet is generated separating the upper region of subsonic flow from the lower region of supersonic flow. No transmitted wave is resolvable at the upper end of the Mach stem which appears to join smoothly into the bow shock in the vicinity of the sonic line. The upper portion of the bow shock remains symmetrical.

In the high enthalpy case the bow shock remains undisturbed. However, a point of inflection is observed immediately above the intersection point.

a
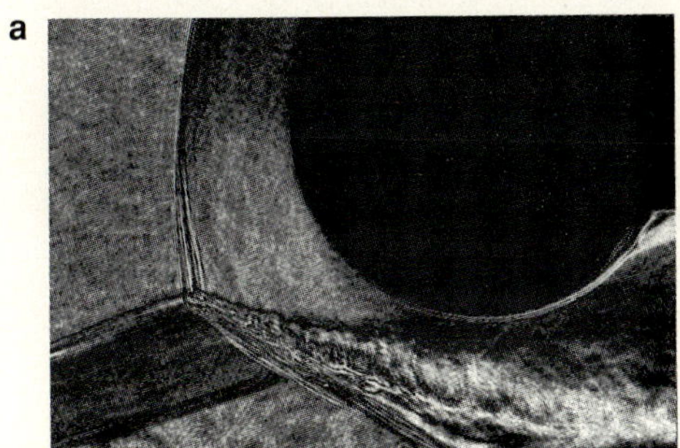
b
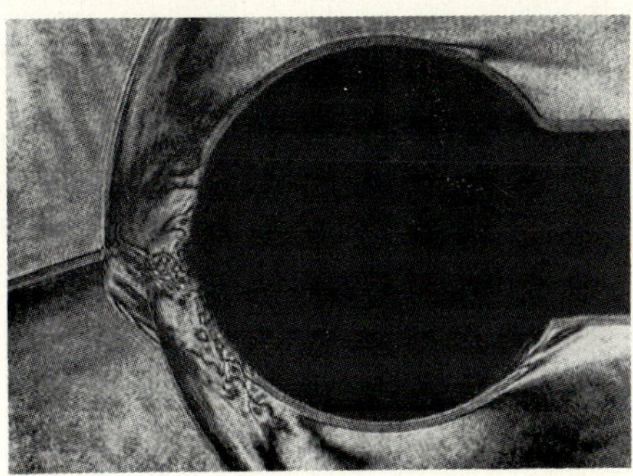

Fig. 3. Differential interferograms of type III interactions.   (a) 3 MJ/kg   (b)12.5 MJ/kg

a

b
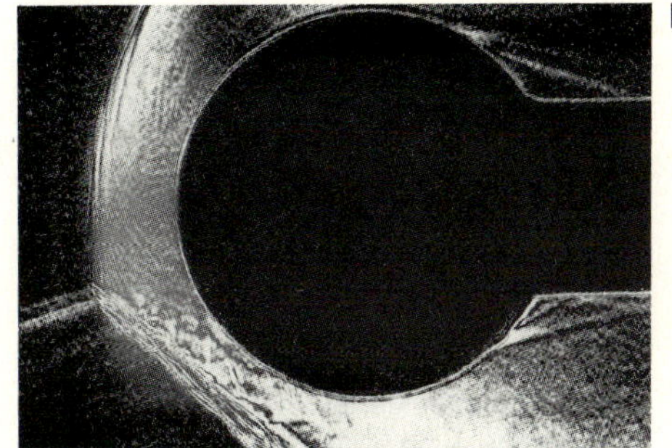

Fig. 4. Differential interferograms of type IV interactions   (a) 3 MJ/kg   (b)12.5 MJ/kg

In the type III experiments (Fig.3) the incident shock wave intersects the bow shock wave in the subsonic region, somewhat below the geometric stagnation point. For both enthalpies the Mach stem has merged with the bow shock wave and the λ-shock pattern is preserved at the intersection point. The influence of the impinging shock wave is now global and both the radius of curvature and the standoff distance of the asymmetrical upper portion of the bow shock has increased to match the λ-shock pattern at the impingement point. Dominated by the inertia of the supersonic stream, the shear layer is deflected only in the immediate vicinity of the cylinder. Generally, the standoff distance decreases with increasing enthalpy and this influences the impingement of the shear layer on the body. In the high enthalpy case, the resulting compression waves, that are observed as a series of nearly vertical white lines below the impingement point, steepen the lower shock. Kinks are observed where these waves meet the supersonic continuation of the bow shock. Weak reflected expansions are observed as horizontal disturbances emanating from the intersection points. Note the analogy between this phenomenon and the mechanism of complex Mach reflection.

When the incident shock wave impinges in the vicinity of the geometrical stagnation point a type IV flow results (Fig.4). The λ-shock pattern is preserved and the supersonic transmitted portion of the bow shock forms a second inverted λ-shock with the subsonic continuation of the bow shock below the interaction. The flow behind the oblique shock wave connecting the two λ-shocks is supersonic and forms a jet that turns upwards under the action of the pressure gradient produced by the second λ-shock. The strongly curved shear layer generated at the upper λ-shock forms large plumes as it passes above the cylinder. The standoff distance and jet width are reduced in the high enthalpy case but this also depends strongly on the location of the impinging shock. A portion of the supersonic flow is now turned so that it passes above the cylinder and hence the stagnation streamline must pass through the supersonic jet. The stagnation density is higher for streamlines that pass through the supersonic jet than for streamlines that cross the adjacent strong bow shock. The higher density and large unsteady velocity gradients produced by the impingement of the supersonic jet provide the mechanism for local increase of the heat transfer rate.

5. Conclusions

Local analysis about shock wave intersection points in the p-δ plane has been extended to the ideal dissociating gas model. The application of continuation methods to singular solutions illustrated the role of λ-shocks and determined the influence of real gas effects on local analysis. Experiments conducted in a free piston shock tunnel provided insight into real gas effects on the global flow. Generally, real gas effects are seen to increase the range of shock impingement points for which enhanced heating occurs. They further influence the Mach stem form in type II flows, the deflection of the shear layer in type III flows and the length scales associated with the type IV jet.

Acknowledgement

This work was supported by the Air Force Office of Scientific Research under Grant No. F 49620-92-J-0110.

References

Edney BE (1968a) Effects of shock impingement on the heat transfer around blunt bBodies. AIAA J. 6:15–21

Edney BE (1968b) Anomalous heat transfer and pressure distributions on blunt bodies at hypersonic speeds in the presence of an impinging shock. FFA-115, The Aeronautical Research Institute of Sweden

Hornung HG (1986) Regular and Mach reflection of shock waves. Ann. Rev; Fl. Mech. 18:33–58

Keller HB (1987) Numerical methods in bifurcation problems. Springer-Verlag

Klopfer GH, Yee HC (1988) Hypersonic shock-on-shock interaction on blunt cowl lips. AIAA Paper 88-0233

Lighthill MJ (1957) Dynamics of a dissociating gas. Part I. Equilibrium flow. J. Fluid Mechs. 2:1-32

Wieting AR, Holden MS (1989) Experimental shock wave interference heating on a cylinder at Mach 6 and 8. AIAA J. 27:1557-1565

Experimental Investigation of Shock-on-Shock Interactions in the High-Enthalpy Shock Tunnel Göttingen (HEG)

S. Kortz[*], **T.J. McIntyre**[†] **and G. Eitelberg**[*]
[*]Institute for Fluid Mechanics, DLR, Göttingen, Germany
[†]Dept. Physics and Theoretical Physics, ANU, Canberra, Australia

Abstract.   The outcome of an experimental investigation of shock-on-shock interactions, created when an oblique shock impinges on a bow shock formed in front of a circular cylinder in a high enthalpy flow of 20 MJ/kg is presented. This interaction is known as the Edney type of interaction. The current study concentrated on Edney types III and IV, where the real gas dissociation effect results in the bow shock being much closer to the cylinder than in case of a perfect gas at the same incident flow Mach number. Flowfield pictures as well as pressure and heat transfer distributions on the cylinder are shown. The jet flow behind the bow shock, typical for the type IV interaction, was not observed and the peak pressure and heat loads are much less severe than was expected from the previous ideal gas investigations. Under the investigated flow conditions, the type III interaction can lead to heat transfer and pressure loads as high as those observed for the type IV interaction.

Key words: Pressure, Heat transfer, Interferogram, Dissociation, Non-equilibrium

1. Introduction

Shock-on-shock interactions occur whenever supersonic flowfields of different directions interfere with each other, as is usually the case for the flow around any aerodynamically stabilized vehicle, flying at high supersonic or hypersonic velocities. The severe heat and pressure loads associated with these shock interferences may reach several times the stagnation point values without interaction and lead to severe structural damages, as was visible on the NASA X-15 A2 experimental aircraft after a flight at Mach 6 (1968). Reviews of the many experimental investigations, which have been performed in the past, are given by Korgeki (1971) and Holden et al. (1992). The systematic investigation of the flow phenomena as well as of the heat transfer and pressure loads, encountered when a weak shock, produced by a wedge-type shock generator, impinges on the bow shock around a blunt body, lead to the fundamental classification by Edney (1968). He distinguished six types of interactions, depending on the location, where the impinging shock intersects with the bow shock, and noticed that the type IV interaction leads to the highest peak heating and pressure rates. Holden et al. found peak pressures and peak heating rates of up to 24 or respectively 32 times the stagnation point value without interaction. But none of the experimental investigations have been performed at conditions leading to a significant dissociation of oxygen or nitrogen behind a bow shock.

To complete the experimental data base and to gain insight on the influence of the finite rate chemistry of the nitrogen dissociation on shock-on-shock interactions a set of 20 experiments has been performed in the High Enthalpy Shock Tunnel Göttingen (HEG) (Eitelberg et al. 1992).

2. Test conditions and experimental procedure

The experimental investigation became possible with the recent completion of the commissioning and initial calibration phase of the HEG, showing that the facility provides an inviscid core flow of 0.5 m in diameter, large enough to mount a circular cylinder, having a length of 0.45 m and an outer diameter of 0.09 m transversely in the test section, together with a 10° wedge-type shock

Shock Waves @ Marseille I
Editors: R. Brun, L. Z. Dumitrescu   © Springer-Verlag Berlin Heidelberg 1995

generator of 0.515 m in length (see Fig.1). To alter the location of the interaction, the cylinder can be displaced vertically and horizontally.

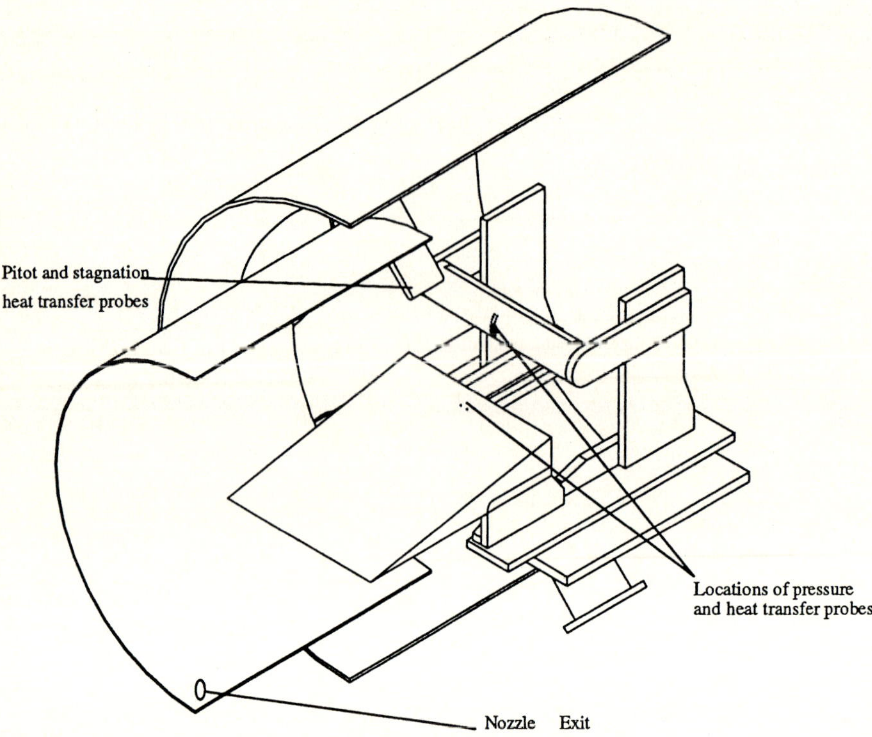

Fig. 1. Cylinder with wedge-type shock generator mounted in the test section of the HEG

In order to get the flow conditions required to obtain nitrogen dissociation behind the bow shock around the cylinder, and to keep the chemical reactions in a nonequilibrium state for at least a length in the order of magnitude of the shock stand-off distance, the test runs were performed at total enthalpies of 20 MJ/kg and reservoir pressures of 40 MPa with N_2 as test gas, leading to a reaction rate parameter Ω of 3. The reaction rate parameter, as introduced by Hornung (1972), is the ratio of the reaction rate times a characteristic length to the flow velocity. Reservoir conditions are calculated from the measured reservoir pressure and shock speed using an equilibrium model. Test section flow conditions are calculated by means of a one-dimensional nonequilibrium nozzle flow calculation. The free-stream conditions were then used to calculate the stagnation point heat transfer to a sphere of 20 mm diameter via the relation introduced by Fay and Riddell, and then compared with the measured stagnation-point heat transfer. Also the Pitot pressure was calculated. The agreement between the calculated and measured values lay within the error bars of the measured values. Free-stream conditions are summarized in Table 1.

Table 1. Calculated mean free-stream conditions for the 20 test runs

V_∞ [km/s]	ρ_∞ [kg/m^3]	p_∞ [Pa]	T_∞ [K]	α_∞	Ma	Re
5.78	0.00168	421	773	0.075	9.9	1.274×10^4

Steady flow conditions have been obtained for at least 1.0 ms, operating the shock tunnel in the tailored mode. This turned out to be crucial for the pressure and heat transfer measurements, because of the relatively long time needed for the shock-on shock interaction phenomenon to come to a steady state.

The test time was identified by the temporal evolution of the ratio of the Pitot pressure to the reservoir pressure, showing the time of steady flow conditions in the test section, and by the temporal pressure evolution in the interaction region. The useful test time was found to be of the order of 0.4 ms.

Pressures on the cylinder surface were measured at 16 different angular positions in the middle section of the cylinder, giving a circumferential pressure resolution of 5°. It is possible to vary the circumferential location of the sensors from one test run to the other by turning the cylinder around its axis. The two-dimensionality of the flow was checked by two pressure sensors having the same angular positions but in different planes. Fast responding Kistler transducers of type 603 B, having a natural frequency of more than 400 kHz were used.

For the heat transfer measurements, calorimetric heat transfer gauges, developed in-house (Grauer-Carstensen 1991), having linear time response, were used. The 16 heat transfer gauges were mounted at the same angular position as the pressure probes.

The data were recorded on a KRENZ transient recorder at a sample frequency of 1 MHz, and later reduced to 100 samples/ms by taking the average over 10 samples. Mean pressure values were determined by averaging the reduced data over the test time with the method of the least squares.

Flow visualisation was realized using the holographic interferometer of the HEG (Eitelberg et al. 1992).

3. Results

3.1. Determination of interaction types III and IV

To distinguish the type III from the type IV interaction, the flow features visible on the interferograms were compared with the typical flow features as determined by Edney. In addition to this straightforward procedure, triple point calculations are carried out for the perfect and ideal dissociating gas in chemical equilibrium, and the shock polar representation is used to find the maximum flow deflection. At the interaction types III and IV, the flow behind the bow shock has to be deflected to fit the cylinder contour. If the maximum flow deflection is less than the flow deflection required by the cylinder surface at the stagnation point the resulting flow phenomena are classified as type IV interactions. If the maximum flow deflection is more than the required flow deflection at the stagnation point, then the flow phenomena are classified as type III interactions (Kortz et al. 1992).

3.2. Type III interaction

Eight of the test runs lead to interactions identified as type III. A typical interferogram is shown in Fig.2. The impinging shock is running from the left into the lower subsonic region of the bow shock. A slightly curved shear layer is emerging at the triple point and hits the cylinder surface at -40°, where a shock wave is reflected, which runs under a slight curvature into the transmitted shock, originating at the triple point, as is shown in the flowfield sketch of Fig.3. It is somewhat surprising, that the stagnation point is almost 10° downstream of the impingement point of the edge of the shear layer at -40°, as can be seen in Fig.2 as well as from the location of the peak pressure as is displayed in Fig.4. Pressure jumps of up to 5 times the reference pressure, which is the rounded calculated Pitot pressure, have been measured. The pressure distribution found by Brück et al. (1993) with the numerical simulation of the shock-on-shock interaction agrees reasonably well with the experimental results, but shows a slight disagreement in the peak

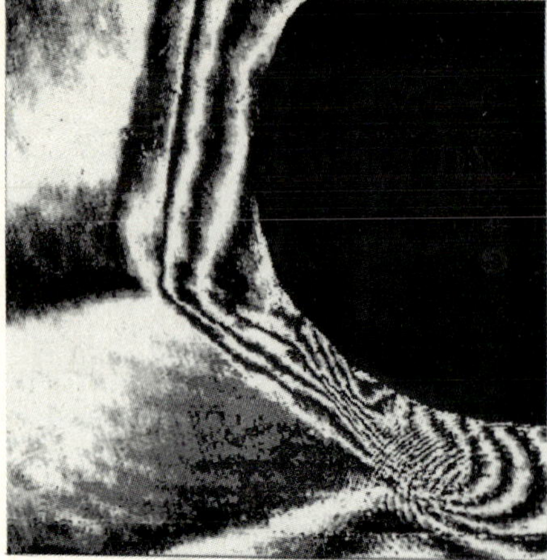

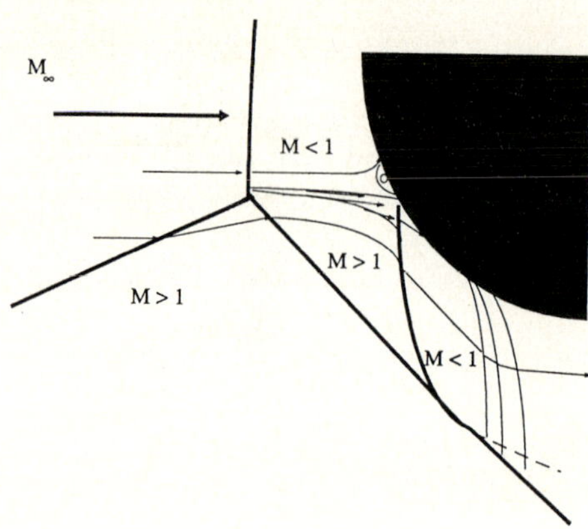

Fig. 2. Interferogram for Test # 49   **Fig. 3.** Flowfield sketch for Test # 49

pressure ratio with the experimental values, as shown in Fig.4. A typical heat transfer distribution in comparison with the numerical results is displayed in Fig.5. The heat transfer ratio shows a similar behaviour as the pressure ratio, and is of the same order of magnitude. The reference heat transfer is the numerically-calculated stagnation value for a flow without interaction (Brück et al. 1993). The location of the peak heating is typically shifted by 5° downstream from the location of the peak pressure. During the tests several heat transfer gauges have been replaced by sensors, which have not been calibrated. The non calibrated sensors (NC) have no error bars and are distinguished by the open symbols.

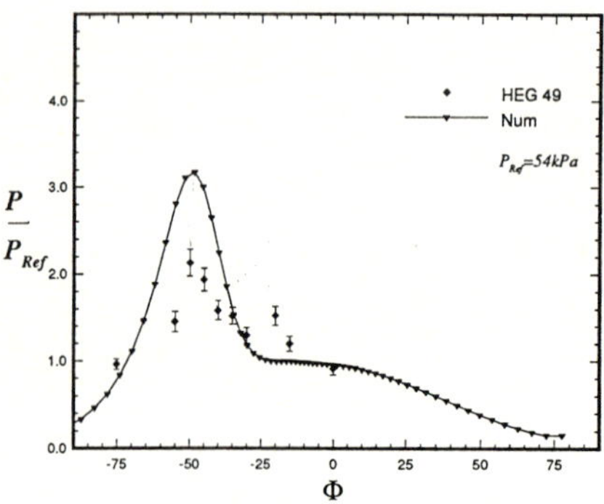

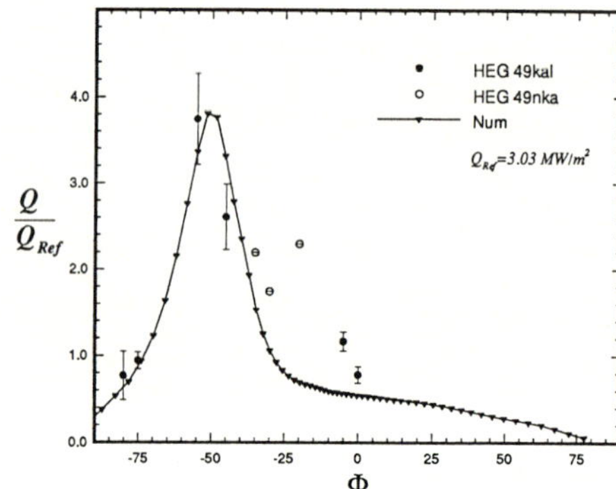

Fig. 4. Pressure distribution compared with numerical data of Brück et al.

Fig. 5. Heat-transfer distribution compared with numerical data

3.3. Type IV interaction

Ten test runs were classified as type IV interactions, although the interferograms do not present the typical flow features specified by Edney. On the interferogram in Fig.6 the impinging shock, running into the upper portion of the subsonic part of the bow shock is clearly visible, but the embedded supersonic region and the typical jet flow are not recognized. According to

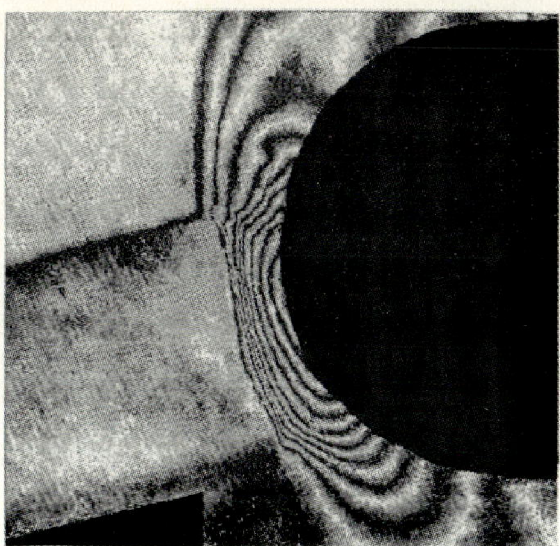

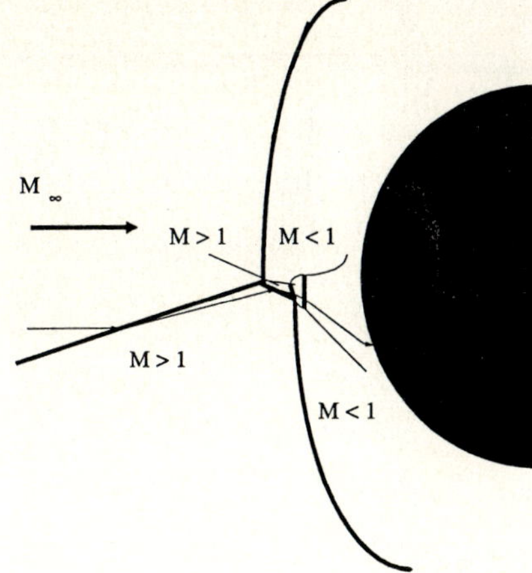

Fig. 6. Interferogram for Test # 46   **Fig. 7.** Flowfield sketch for Test # 46

Henderson (1964) there exist four physically possible triple point solutions, but for the given free-stream conditions only one is possible for this particular test case, consisting of a weak shock of 17° running into a strong shock of almost 90°. And this solution requires a weak transmitted shock followed by a supersonic region. But the establishment of a jet flow as expected from the previous perfect gas cases is not a necessity, and the flowfield sketch shown in Fig.7 has been deduced from the interferograms.

Peak pressure ratios of up to not more than 5.4 times the reference pressure support the observation, that a supersonic jet flow is not established. A comparison with the numerical data (see Fig.8) shows good agreement with the experimental data even though the peak pressure is not visible in the numerical data. The heat transfer profile shows a similar behaviour as the pressure profile, but with the location of the peak heat transfer shifted by 5° downstream of the location of the peak pressure, as was already observed for the type III interaction. The heat transfer ratios are always of the same order of magnitude as the pressure ratios. The slight discrepancies between the numerical and the experimental heat transfer data are probably due to the difficulty to identify precisely the location of the interaction from the interferogram.

In four of the test runs high peak pressure ratios of up to 27 times the reference pressure were observed for two single pressure transducers. Although malfunctioning of the sensors or the experimental set-up as well as a calibration error could not be traced back, these pressures, since they did not occur in a repeatable manner and could not be correlated neither by the interferograms nor by the heat transfer measurements, are not presented in the graphs.

4. Conclusions

In the described experiments, the build-up of a supersonic jet behind the shock intersection point, as expected from previous perfect gas investigations, was not observed. This can be attributed to a drop in total pressure due to the dissociation of nitrogen behind strong shocks, leading to the choking of the supersonic jet flow. As a consequence, peak pressure and heating ratios are strongly reduced. It is observed, that the type III interaction can lead to peak pressure and heat transfer rates as high as those observed for the type IV interaction.

Further measurements are needed to investigate also the other Edney interference types as well as the influence of different reaction rates on the phenomena observed.

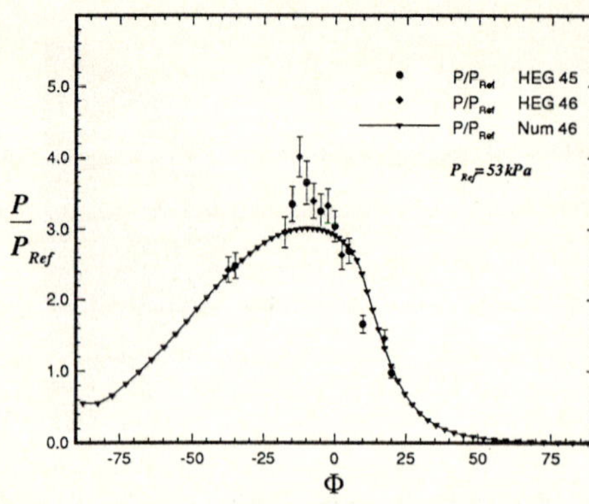

Fig. 8. Pressure distribution compared with numerical data of Brück et al. (1993)

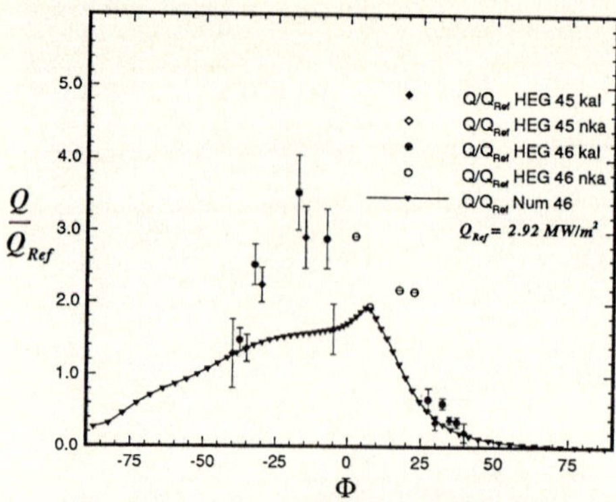

Fig. 9. Heat-transfer distribution compared with numerical data

Acknowledgement

The work of D. Kastell, who operated the holographic interferometer and created a set of high quality interferograms, is highly appreciated.

References

Brück S, Brenner G, Kortz S (1993) Numerical simulations of shock-shock interactions in non-equilibrium flow. DLR IB 221-93 A 09

Edney B (1968) Anomalous heat transfer and pressure distributions on blunt bodies at hypersonic speeds in the presence of an impinging Shock. FFA Rep. 115

Eitelberg G, Fleck B, McIntyre TJ (1992) Holographic interferometry ion the high enthalpy shock tunnel in Göttingen (HEG). Proc. NATO Advanced research Workshop on new trends in instrumentation for hypersonic research, ONERA

Eitelberg G, McIntyre TJ, Beck WH, Lacey J (1992) The high enthalpy shock tunnel in Göttingen. AIAA Paper 92-3942

Grauer-Carstensen H (1991) Messung von Druck und Wärmefluss im HEG, Teil 1: Messung des Wärmeflusses. DLR IB 222-91 A26

Henderson LF (1964) On the cConfluence of three shock waves in a perfect gas. Aeronautical Quarterly, May

Holden MS, Moselle JR, Lee J (1992) Studies of aerothermal loads generated in regions of shock/shock interaction in hypersonic flow. NASA-CR-181893

Hornung H (1972) Non-equilibrium dissociating nitrogen flow over spheres and circular cylinders. J. Fluid Mechs. 53, 1:149-A6

Korgeki RH (1971) Survey of viscous Interactions associated with high Mach number flight. AIAA J. 9 , 5

Kortz S, McIntyre TJ, Eitelberg G (1992) Theoretische Untersuchung des Einflusses der Stickstoffdissoziation auf Stoß/StoßWechselwirkungen zur Vorbereitung von Messungen im HEG. DLR IB 222-92 A25

Watts JD (1968) Flight experience with shock impingement and iInterference heating on the X-15-2 research airplane. NASA TM-X-1669

Oblique Shock Interactions with Mach Number Distributions

D.R. Buttsworth and R.G. Morgan
Department of Mechanical Engineering, The University of Queensland, Brisbane 4072, Australia

Abstract.   An equation governing the interaction of oblique shock waves with steady, continuous, spatial distributions of Mach number is derived from the Rankine-Hugoniot shock jump conditions. This equation essentially describes the shape an oblique shock assumes as it traverses a region of varying Mach number. Because the analysis also provides a flow solution downstream of the shock, it is seen as a particularly useful tool in the study of shock wave - mixing region interactions, a topic of current interest for shock induced ignition in scramjets. As a demonstration of this analysis, a published experiment involving an oblique shock in a mixing region is examined.

Key Words: Oblique shock, Interaction, Non-uniform flow

1. Introduction

Regions of non-uniform flow occur frequently in supersonic flow problems, and since shock waves are an inherent feature of supersonic flows, shocks and non-uniform flow regions often interact. Examples of non-uniform flow regions which regularly interact with shock waves include boundary layers, free shear layers, the inviscid flow field behind a curved bow shock, and the turbulent gas interface region in a shock tube. Interaction between shocks and flow non-uniformities will also occur frequently in proposed scramjet engines. In a scramjet engine (for example, Swithenbank 1967), a series of oblique shocks compress the intake air flow to a state sufficient to support rapid combustion of the fuel which is injected either during or after the compression process. Thrust is generated when the combustion products are accelerated in the exhaust nozzle. Significant transverse property variations are likely to exist within a scramjet due to the mixing, combustion and large boundary layer regions. Shock waves generated by the intake compression or fuel injection process will inevitably interact with these regions. Such interactions or those resulting from specifically designed oblique shocks may be crucial in augmenting supersonic mixing and successfully implementing scramjet propulsion (Kumar et al. 1989). Therefore, understanding and predicting the interaction of oblique shocks and non-uniform flow has relevance in supersonic flows generally, but is likely to be of particular importance in the development of scramjet engines. To further the understanding of the interaction process, a model for the interaction of an oblique shock and non-uniform region is developed. The value of the current model lies in its analytical nature. It adds to the physical understanding of the interaction mechanisms and affords the ability to rapidly predict the steady flow downstream of an oblique shock in a non-uniform region. Following the model development, the present analysis is applied to a recent experiment involving a weak shock wave in a mixing region.

2. Non-uniform region shock wave model

Consider the situation depicted in Fig.1. The incident wave is assumed to be a two-dimensional, steady shock wave, generated by a flat plate. In the region upstream of the shock wave, the flow is assumed to be parallel, free from all pressure gradients and is characterized by a steady, transverse variation in Mach number alone. A perfect gas with a constant value of γ has been assumed throughout the flow field. An additional assumption included in this model is that the

Shock Waves @ Marseille I
Editors: R. Brun, L. Z. Dumitrescu

post-shock Mach number distribution remains supersonic throughout the flow in order to support
the required system of reflected isentropic waves.

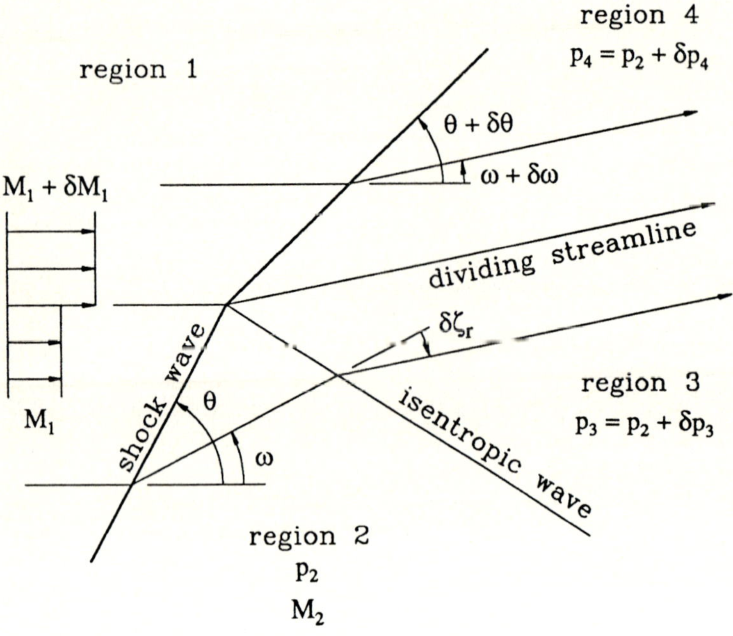

Fig. 1. Shock wave - Mach number distribution interaction model

Transmission and reflection of the incident shock wave will be such that on either side of
the dividing streamline, Fig.1, the flow will have the same pressure and be moving in the same
direction. Conditions of matched deflection and pressure are given by:

$$\delta\omega = -\delta\zeta_r \tag{1}$$

$$\delta p_4 = \delta p_3 \tag{2}$$

The Rankine-Hugoniot oblique shock wave relations (Liepmann and Roshko 1957) necessary for
this analysis are:

$$\frac{p_2}{p_1} = 1 + \frac{2\gamma}{\gamma + 1}\left(M_1^2 \sin^2\theta - 1\right) \tag{3}$$

$$\tan\omega = 2\cot\theta\,\frac{M_1^2 \sin^2\theta - 1}{M_1^2(\gamma + \cos 2\theta) + 2} \tag{4}$$

$$M_2 = \left[\frac{1 + \frac{\gamma-1}{2}M_1^2 \sin^2\theta}{\left(\gamma M_1^2 \sin^2\theta - \frac{\gamma-1}{2}\right)\sin^2(\theta - \omega)}\right]^{\frac{1}{2}} \tag{5}$$

The pressure jump condition, Eq.3, for the transmitted wave, may be written:

$$\frac{\delta p_4}{p_1} = \frac{4\gamma}{\gamma + 1}\left(M_1 \sin^2\theta\ \delta M_1 + M_1^2 \sin\theta\ \cos\theta\ \delta\theta\right) \tag{6}$$

once second order terms such as $\delta M_1^2, \delta\theta^2$, and $\delta M_1 \times \delta\theta$ have been neglected. Similarly, using Eq.4, the flow deflection by the transmitted shock wave is given by

$$\left[4M_1\sin^2\theta - (\gamma + 1 - 2\sin^2\theta)2M_1\tan\theta\tan\omega\right]\ \delta M_1$$
$$+\left\{2M_1^2\tan\theta(2\cos^2\theta - \sin^2\theta) + 2\tan\theta - \left[2 + (\gamma + 1)M_1^2\right]\tan\omega + 6M_1^2\sin^2\theta\tan\omega\right\}\ \delta\theta \quad (7)$$
$$+\left\{2M_1^2\sin^2\theta(\tan\theta - \tan\omega) + 2\tan\omega - \left[2 + (\gamma + 1)M_1^2\right]\tan\theta\right\}\ \delta\omega = 0$$

once the second order terms have been neglected and the approximation equation

$$\tan(\omega + \delta\omega) = \frac{\tan\omega + \delta\omega}{1 - \delta\omega\tan\omega} \quad (8)$$

for infinitesimal values of $\delta\omega$, has been employed.

In the limit as $\delta M_1 \rightarrow 0$, the wave reflected from the dividing streamline will be isentropic. The pressure change caused by such a wave (Liepmann and Roshko 1957) is given by

$$\frac{\delta p_3}{p_2} = \frac{\gamma M_2^2}{\sqrt{M_2^2 - 1}}\ \delta\zeta_r \quad (9)$$

By the chosen convention, a positive value of $\delta\zeta_r$ represents a compression wave and a negative value, an expansion wave. The requirement of matched flow direction on either side of the dividing streamline, Eq.1, allows Eq.9 to be expressed as

$$\frac{\delta p_3}{p_2} = \frac{-\gamma M_2^2}{\sqrt{M_2^2 - 1}}\ \delta\omega \quad (10)$$

and the matched pressure condition, Eq.2, may be written:

$$\frac{\delta p_4}{p_1} = \frac{\delta p_3}{p_2}\frac{p_2}{p_1} \quad (11)$$

Eqs.3, 6 and 10 may be combined with Eq.11 to yield the expression

$$4M_1\sin^2\theta\ \delta M_1 + 4M_1^2\sin\theta\cos\theta\ \delta\theta + \frac{M_2^2}{\sqrt{M_2^2 - 1}}\left(1 - \gamma + 2\gamma M_1^2\sin^2\theta\right)\ \delta\omega = 0 \quad (12)$$

By eliminating $\delta\omega$ between Eqs.7 and 12, the following expression may be obtained:

$$\lim \delta M_1 \rightarrow 0\quad \left(\frac{\delta\theta}{\delta M_1}\right) = \frac{\partial\theta}{\partial M_1} = f(M_1, \theta)$$

$$= -\left[4M_1\sin^2\theta - 2(\gamma + 1)M_1\tan\theta\tan\omega + 4M_1\sin^2\theta\tan\theta\tan\omega - 8M_1 A\sin^2\theta\tan\omega\right.$$
$$+8M_1 A\sin^2\theta\tan\theta + 8M_1^3 A\sin^4\theta\tan\omega + 4(\gamma + 1)M_1^3 A\sin^2\theta\tan\theta - 8M_1^3 A\sin^4\theta\tan\theta\left.\right]$$
$$/\ \left[2\tan\theta - 2\tan\omega + 4M_1^2\sin\theta\cos\theta - 2M_1^2\sin^2\theta\tan\theta - (\gamma + 1)M_1^2\tan\omega\right. \quad (13a)$$
$$+6M_1^2\sin^2\theta\tan\omega + 8M_1^2 A\sin^2\theta - 8M_1^2 A\sin\theta\cos\theta\tan\omega - 8M_1^4 A\sin^4\theta$$
$$\left.+8M_1^4 A\sin^3\theta\cos\theta\tan\omega + 4(\gamma + 1)M_1^4 A\sin^2\theta\right]$$

where

$$A = \frac{\sqrt{M_2^2 - 1}}{M_2^2(1 - \gamma + 2\gamma M_1^2\sin^2\theta)} \quad (13b)$$

This is the differential equation which governs the direction of the shock wave due to the non-uniform Mach number distribution upstream of the shock. In the current problem, the primary source of shock curvature is the Mach number distribution. However, additional changes in shock direction may arise due to the intersection of the shock wave and isentropic waves reflected off

the post-shock Mach number gradients. If such effects can be neglected, then the shock direction
becomes a function of the Mach number alone. That is,

$$\frac{d\theta}{dM_1} = f(M_1, \theta) \tag{14}$$

where f is the same function given in Eq.13. For initial values of θ and M_1, it is therefore
possible to integrate the above equation (for a known value of γ) to find the shock angle at any
particular Mach number (provided $M_2 > 1$). Having obtained the shock wave direction, all other
flow variables may be determined using the Rankine-Hugoniot equations.

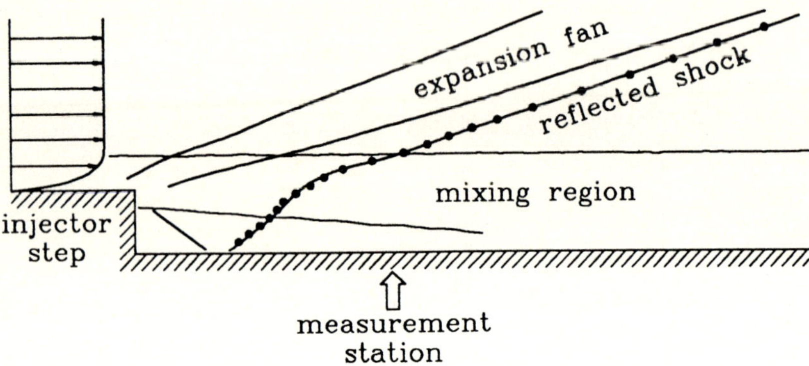

Fig. 2. Wall slot injection configuration (Kwok et al. 1991) with measurements and curve fit for reflected shock
location

3. An application of the analysis

As a demonstration of the current analysis and its application in practical supersonic flows, an
experiment reported by Kwok et al. (1991) has been chosen for consideration. Air was injected
from a wall slot, parallel to a co-flowing Mach 3 stream, producing the shock and expansion
wave system shown in Fig.2. The injector step was of height, $H = 12.1$ mm. By measuring
the reflected shock wave trajectory from the published shadowgraph (Kwok et al. Fig. 4b) and
applying the present wave analysis, a prediction of the Mach number distribution along the shock
path within the layer can be made. Measurements of the reflected shock wave location, taken
from the shadowgraph are given in Fig.2, along with a curve fit for this data.

From the leading and trailing waves of the expansion fan, it was determined that the freestream
flow behind the expansion was heading towards the lower wall at an angle of $3.07°$ and a Mach
number of 3.02. The reflected shock angle in the freestream was found to be $18.2°$ with respect to
the tunnel wall, and was therefore at an angle of $21.3°$ with respect to the flow direction upstream
of the shock. Using the initial conditions $(M_1, \theta) = (3.02, 21.3°)$, Eq.14 was integrated numerically
to obtain the function shown in Fig.3. From this result, the Mach numbers corresponding to the
experimental shock wave angles were obtained, and are plotted in Fig.4 along with a Mach number
distribution (Kwok et al. Fig. 9) determined from probe measurements at the station indicated
in Fig.2. Since the shock passes through the flow ahead of the measuring station for values of

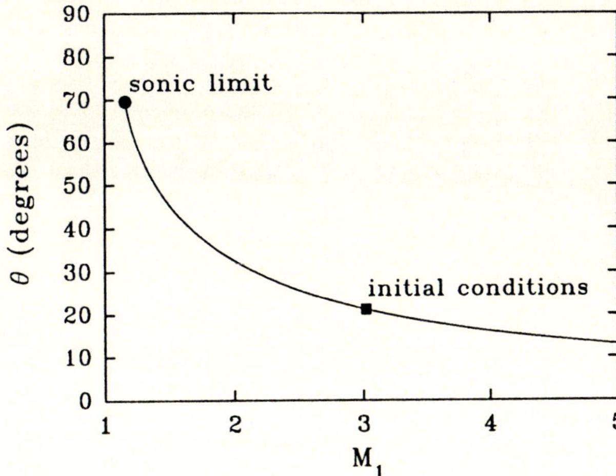

Fig. 3. Integration of Eq.14 for the reflected shock using the initial conditions shown and $\gamma = 1.4$

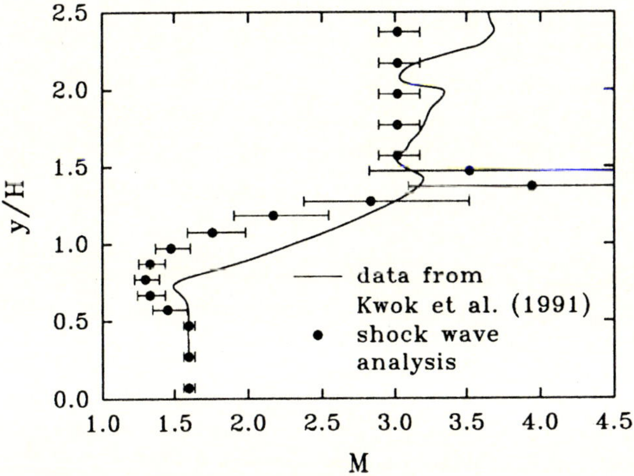

Fig. 4. Comparison of Mach number distributions according to the wave analysis and probe measurements

$y/H \leq 1.5$, M_2 is actually presented in Fig. 4 at these locations. The error bars given in Fig.4 are based on uncertainties of $\pm 2°$ and $\pm 4°$ in the freestream and non-uniform region shock wave angles respectively.

According to the wave analysis, the wake region is more extensive than probe measurements indicate. This is reasonable since the shock passes through the developing layer ahead of the measurement station, and the wake region, clearly visible in the shadowgraph at the shock location, is barely apparent at the probe measurement station. At $y/H \approx 1.4$, the wave analysis predicts a peak Mach number significantly higher than obtained from the probe measurements. The decrease in precision of the wave analysis at large Mach numbers (see Fig.3) and fundamental differences between the two methods may contribute to this apparent discrepancy. The shadowgraph is effectively an instantaneous spatial integral of properties across the duct width, whereas probe measurements are typically time integrated but are approximately point measurements.

4. Conclusion

An analytical model for the steady interaction of an oblique shock and a non-uniform region characterized by its Mach number distribution has been developed to assist in the study of shock interactions with supersonic mixing regions. Provided the flow remains supersonic, the shock trajectory and the post-shock flow throughout the varying Mach number region may be solved using this analysis. Using the inverse approach, the analysis may be used as a tool in experiments to determine the Mach number distribution along an oblique shock wave traversing a non-uniform region, provided the shock direction can be determined with sufficient precision.

Caution must be exercised in applying this analysis to oscillating or turbulent flows since a steady flow has been assumed. However, in such flows the model provides a benchmark against which unsteady effects may be assessed. The essentially analytic nature of the solution permits a rapid assessment of the shock wave influence on steady flow properties. Since the effects of oblique shock - mixing region interactions and mixing augmentation remain important issues in scramjet development, the current analysis may be used in future work to further examine these effects.

Acknowledgments

Financial assistance provided by an Australian Postgraduate Research Award and NASA under grant NAGW-674 are gratefully acknowledged.

References

Kumar A, Bushnell DM, Hussaini MY (1989) Mixing augmentation technique for hypervelocity scramjets. J. Prop. 5:514-522

Kwok KT, Andrew PL, Ng WF, Schetz JA (1991) Experimental investigation of a supersonic shear layer with slot injection of helium. AIAA J. 29:1426-1435

Liepmann HW, Roshko A (1957) Elements of gasdynamics Wiley, New York

Swithenbank J (1967) Hypersonic air-breathing propulsion. In: Küchemann D (ed) Progress in aeronautical sciences Vol. 8. Pergamon, Oxford, pp 229-294

Shock Wave/Boundary Layer Interaction in High-Enthalpy Compression Corner Flow

S.G. Mallinson, S.L. Gai and N.R. Mudford
Department of Aerospace and Mechanical Engineering, University College, UNSW, ADFA Canberra ACT 2601, Australia

Abstract.   Shock wave/boundary layer interaction in laminar, two-dimensional compression corner flow is investigated at high enthalpy. High enthalpy flows exhibit a smaller scale of interaction which is in accordance with calculations based on momentum integral theory and experimental observations behind a rearward facing step.

Key words: Shock wave/boundary layer interaction, Compression corner, Hypervelocity.

Nomenclature

$C^* = $ Chapman-Rubesin constant $ = \mu(T^*)T_\infty/\mu(T_\infty)T^*$

$C_p = $ pressure coefficient $ = 2(p_w - p_\infty)/\rho_\infty U_\infty^2$

$h = $ enthalpy

$h_r = $ recovery enthalpy $ = h_0 + 0.5(\sqrt{\mathrm{Pr}} - 1)U_\infty^2$

$l_i = $ upstream influence length

$L = $ upstream plate fetch

$M = $ Mach number

$p = $ pressure

$\mathrm{Pr} = $ Prandtl number

$q = $ heat flux

$\mathrm{Re}_x = $ Reynolds number $ = \rho_\infty U_\infty x/\mu_\infty$

$\mathrm{St} = $ Stanton number $ = q_w/\rho_\infty U_\infty(h_r - h_w)$

$T = $ temperature

$T^* = $ Eckert reference temperature $ = 0.5(T_w + T_\infty)$
$\qquad +0.11\,\mathrm{Pr}^{1/2}(\gamma_f - 1)M_\infty^2 T_\infty$

$U = $ flow speed

$x = $ distance from leading edge

$\alpha_o = $ oxygen dissoc. mass fraction

$\delta = $ boundary layer thickness

$\gamma_f = $ frozen ratio of specific heats

$\chi_L = M_\infty^3 \sqrt{C^*/\mathrm{Re}_L}$

$\mu = $ viscosity

$\rho = $ density

$\theta_w = $ wedge angle

Subscripts and superscripts:

$f = $ frozen condition

$0 = $ reservoir condition

$w = $ value at the wall

$\infty = $ free-stream condition

1. Introduction

Shock wave/boundary layer interaction has received wide attention and is often studied using a compression corner configuration. Fig.1 presents the main features of compression corner flow. The shock due to the corner interacts with the boundary layer on the flat plate. For a sufficiently strong shock, the boundary layer separates from the surface to form a region of reversed flow. The flow subsequently reattaches on the face of the ramp. The interaction can cause reduced control surface effectiveness for re-entry vehicles and proposed Space Planes. Space Planes will rely on scramjet engines in the cruise phase of their trajectories. The engine inlets experience shock wave/boundary layer interactions which can decrease inlet efficiency. The high enthalpy flows around these vehicles may exhibit real gas behaviour such as vibration, dissociation and ionization. To date these effects have been largely neglected in relation to shock wave/boundary layer interaction. The exceptions are one experimental (Stalker and Rayner 1985), one theoretical (Ikawa 1979) and one computational study (Ballaro and Anderson 1991). Stalker and Rayner observed no real gas effects, whereas Ikawa and Ballaro and Anderson showed that real gas behaviour can affect separation and reattachment. It was therefore decided to investigate real gas effects on laminar boundary layer/shock wave interaction at a compression corner.

Shock Waves @ Marseille I
Editors: R. Brun, L. Z. Dumitrescu

Experiments were conducted using the Australian National University's free-piston shock tunnel, T3 (Stalker 1972), in reflected mode. The static pressure and surface heat transfer distributions were measured in compression corners. The important features of these distributions are discussed and compared with results from previous studies.

2. Experimental procedure

2.1. Free-piston shock tunnel

The free-piston shock tunnel T3 consists of a free-piston driven shock tube and a nozzle separated by a thin Mylar diaphragm. For the present experiments, the 13 mm inlet, 305 mm exit conical nozzle (half-angle 7.5°) was used. The test gas for these experiments was air. The nozzle reservoir conditions were calculated using the ESTC code (MacIntosh 1968). The free-stream conditions were calculated using the NENZF code (Lordi et al. 1966). Three conditions were used, which are representative of highly dissociated (B), moderately dissociated (D), and undissociated (G) flow. Nitrogen dissociation was negligible for the present conditions. Typical reservoir and free-stream conditions are presented in Table 1. Viscosity has been calculated using data from Hirschfelder et al. (1967). Prandtl number was assumed constant, with Pr $=0.72$.

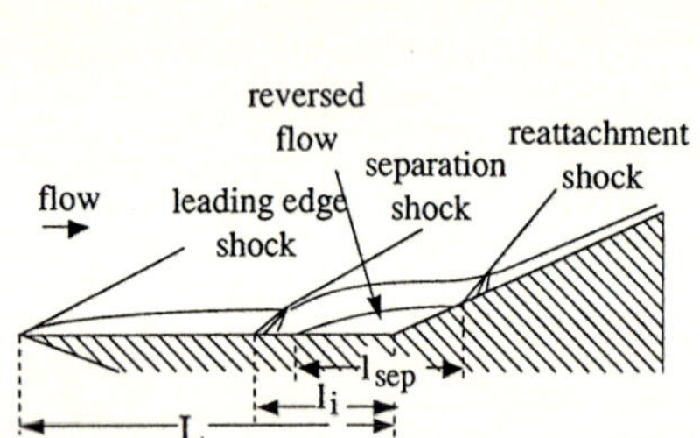

Fig. 1. Shock wave/boundary layer interaction in a compression corner

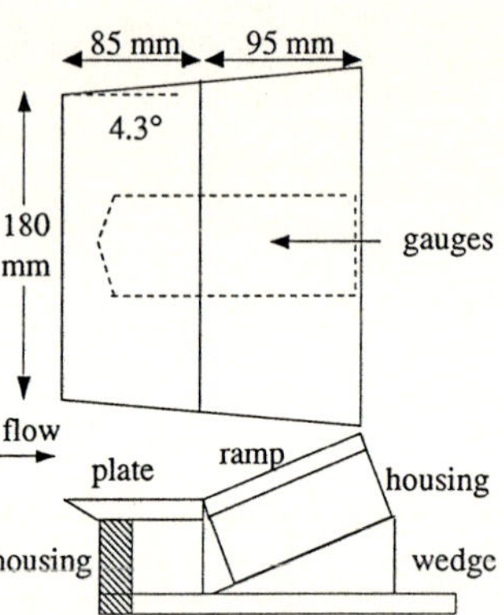

Fig. 2. Details of the model

Free-piston shock tunnels produce short duration flows of order 1 ms or less. The maximum available test time is dictated by the onset of driver gas contamination (Stalker and Crane 1978). The nozzle starting process and the establishment of steady boundary layer flow reduce the uncontaminated test time to approximately 0.1-0.5 ms (East et al. 1980). Mallinson and Gai (1992) demonstrated that even though steady separated flows require a finite time to establish (Holden 1971b), driver gas contamination is not a problem for the present study.

Table 1. Reservoir and free-stream conditions, designated by '0' and '∞', respectively

	h_0 MJ.kg^{-1}	p_0 MPa	T_0 K	p_∞ kPa	T_∞ K	ρ_∞ g.m^{-3}	u_∞ km.s^{-1}	M_∞	Re_∞ $\times 10^{-5}$ m^{-1}	$\gamma_{f\infty}$	$\alpha_{0\infty}$
B	19.0	22.2	8400	0.99	1160	2.60	5.47	7.5	3.10	1.45	0.8
D	13.7	22.2	7200	0.99	940	3.43	4.72	7.5	4.08	1.43	0.4
G	2.83	22.4	2400	0.73	160	16.0	2.28	9.1	32.2	1.40	0.0

2.2. Model and instrumentation

Compression corner models made of mild-steel were placed in the free-stream at the exit of the nozzle (Fig.2). Separate models were used for pressure and heat transfer measurements. Each model consisted of a flat plate and a ramp plate. The aspect ratio was slightly greater than unity which was sufficient to ensure two-dimensional flow along the centre-line, with or without side fences (Lewis et al. 1968). The leading edge was nominally sharp, with the Reynolds number based on leading edge thickness less than 200. The corner was sealed from below to prevent leakage. The model sides were inclined to match the source flow from the nozzle virtual origin. Skirts were used to prevent upwash affecting the main flow. Several wedge angles were tested at each condition. This paper will discuss results of angles of 0°, 15°, 18° and 24°.

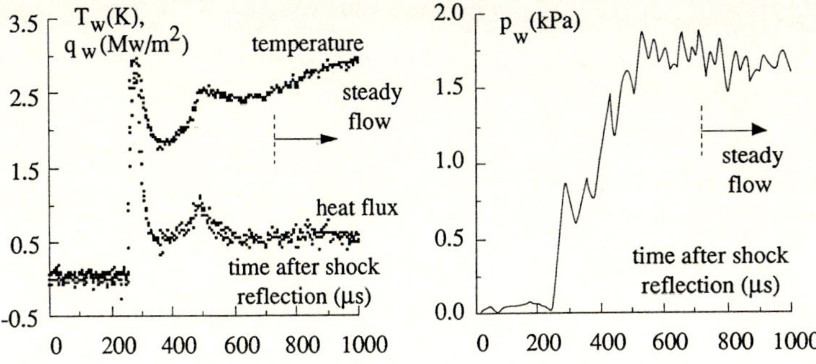

Fig. 3. Typical temperature, heat transfer and pressure signals

Surface temperatures were measured using chromel-alumel coaxial thermocouples. The heat flux was deduced from the variation of temperature (Schultz and Jones 1973). Surface pressures were measured using PCB model 113M165 piezo-electric pressure transducers. Fig.3 shows examples of pressure, temperature and heat transfer signals.

3. Results

3.1. Heat transfer and pressure distributions

The flat plate heat transfer is shown is Fig.4(a). The heat transfer data lie slightly below the Blasius limit of $St\sqrt{Re_x} = 0.332$. The corresponding pressure data are shown in Fig.4(b). The comparison with the correlation of Holden (1971a) is fair.

The heat transfer and pressure distributions for condition B at wedge angles of 15°, 18° and 24° are presented in Figs.5(a&b) respectively. The heat transfer distribution at 15° has a sharp minimum which is a characteristic of attached flow (Délery 1989). The pressure distribution follows the flat plate distribution until just before the corner after which the pressure rises to the final value. At an angle of 18°, the minimum in heat transfer becomes rounded while an inflection in the pressure distribution is noticeable both of which indicate incipient separation. At an angle of 24°, a large region of minimum heat transfer and a pressure plateau are evident near the corner. These features are typical of well separated flows.

The heat transfer and pressure distributions for condition D were similar. Fig.6 presents a comparison between high enthalpy (conditions B and D) and low enthalpy (condition G and Holden and Moselle (1970)) heat transfer results for a wedge angle of 24°. There are two obvious differences. Firstly, the upstream influence, as determined by the point at which the heat transfer (or pressure) distribution deviates from that of the flat plate, is smaller and the reattachment appears to occur earlier for the high enthalpy flows. Secondly, the reattachment process at high enthalpy appears to be more gradual than that for low enthalpy which is consistent with observation of flow reattachment behind a rearward facing step in hypervelocity flow (Gai et al. 1989).

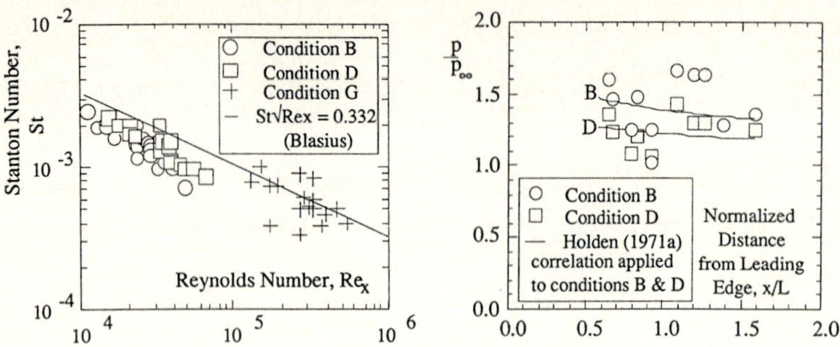

Fig. 4. Flat plate distributions of heat transfer and pressure

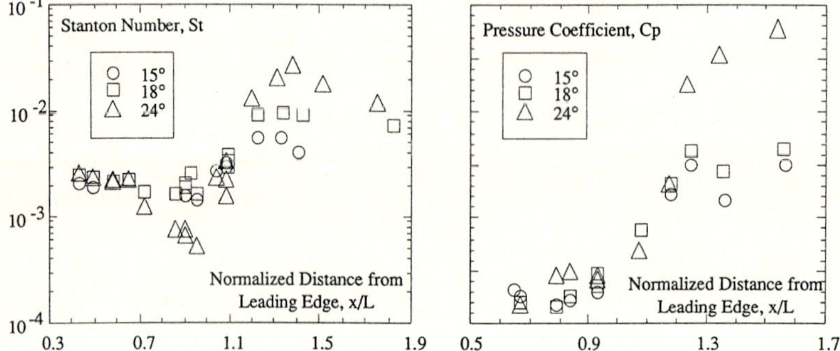

Fig. 5. Heat transfer and pressure distributions for condition B: wedge angles =15°, 18°, and 24°

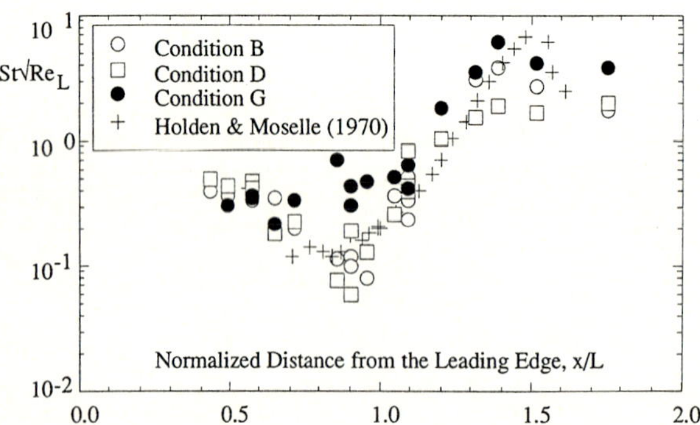

Fig. 6. Heat transfer distribution: comparison between high and low enthalpy results

3.2. Upstream influence

Upstream influence in shock wave/boundary layer interaction increases with increasing boundary layer thickness, Reynolds number, wedge angle and wall-to-total temperature ratio, and decreases with increase in Mach number (Lewis et al. 1968; Délery 1989). That is,

$$l_i/\delta = F(\mathrm{Re}_L, \theta_w, 1/M_\infty, T_w/T_0) \tag{1}$$

Lewis et al. (1968) showed that a large change in T_w/T_0 causes only a relatively small change in upstream influence so that it may be ignored. The boundary layer thickness may be calculated using the correlation of White (1974). The Mach number and Reynolds number may be combined to form the hypersonic viscous interaction parameter, which scales with upstream influence as the inverse square root (Délery 1989). Eq.1 may now be simplified as

$$l_i/\delta^* = F\left(\frac{\theta_w}{\sqrt{\chi_L}}\right) \tag{2}$$

The data from conditions B, D and G are presented in Fig.7. Also shown are the experimental data from Holden (1967), Lewis et al. (1968), Holden and Moselle (1970), Bloy and Georgeff (1974). The dependence expressed by Eq.2 appears to correlate the low enthalpy data reasonably well. The high enthalpy data also seem to correlate well, but separately from the low enthalpy results.

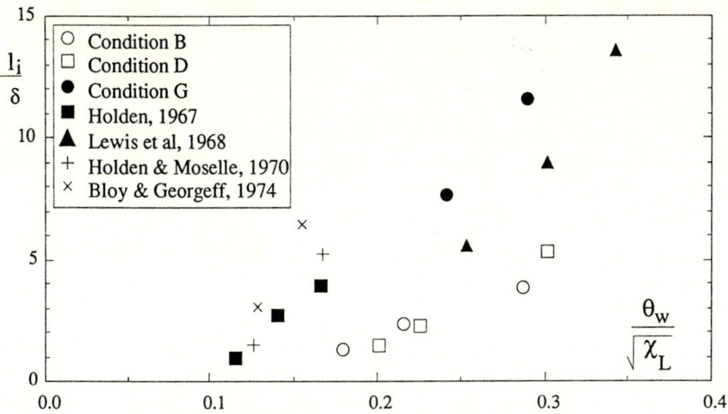

Fig. 7. Correlation of upstream interaction lengths in hypersonic flow

It would appear that upstream influence is less in high enthalpy flows. This might be because high enthalpy gas can have excited internal states such as vibration, dissociation and ionization. Thus, the gas within the boundary layer, which may be vibrationally excited, would have reduced enthalpy and temperature and increased density (Park 1990). The stream-tubes within the boundary layer would then contract so that the boundary layer thickness, and hence upstream influence, decreases. Ikawa (1979) calculated the pressure distributions for the experimental conditions of Lewis et al. (1968) and Bloy and Georgeff (1974), except that the free-stream was assumed to be completely dissociated and to behave as an ideal dissociating gas in chemical equilibrium (Lighthill 1957). Ikawa noted the same trend in the calculated results that has been observed in the present experiments, that is reacting flows have a smaller upstream influence. The reduced scale of interaction has also been observed by Gai et al. (1989) in separated hypervelocity flows over rearward-facing steps.

4. Conclusions

Shock wave-boundary layer interaction has been investigated in two-dimensional laminar high enthalpy compression corner flows. The presence of real gas effects causes a reduction in scale of the interaction. This accords with the previous theoretical and experimental studies.

Acknowledgement

The authors wish to gratefully acknowledge the assistance of Mr. P. Walsh in conducting the experiments. The first author would like to gratefully acknowledge the financial assistance provided by the Royal Aeronautical Society's Letitia Eadon Award.

References

Ballaro CA, Anderson JD Jr (1991) Shock strength effects on separated flows in non-equilibrium chemically reacting air shock wave/boundary layer interaction. AIAA Paper 91-0250

Bloy AW, Georgeff MP (1974) The hypersonic laminar boundary layer near sharp compression and expansion corners. J. Fluid Mech. 63:431-447

Crane KCA, Stalker RJ (1977) Mass-spectrometric analysis of hypersonic flows. J. Phys. D: Appl. Phys. 10:679-695

Délery J (1989) Shock/shock and shock-wave/boundary-layer interactions in hypersonic flows. AGARD Rep. 761, Pt. 9

East RA, Stalker RJ, Baird JP (1980) Measurements of heat transfer to a flat plate in a dissociated high-enthalpy laminar air flow. J. Fluid Mech. 97:673-699

Gai SL, Reynolds NT, Ross C, Baird JP (1989) Measurements of heat transfer in separated high-enthalpy dissociated laminar hypersonic flow behind a step. J. Fluid Mech. 199:541-561

Hirschfelder JO, Curtiss CF, Bird RB (1967) Molecular theory of gases and liquids. Wiley& Sons

Holden MS (1967) Theoretical and experimental studies of laminar flow separation on flat plate-wedge compression surfaces in the hypersonic strong interaction regime. ARL Rep. 67-0112

Holden MS (1971a) Boundary-layer displacement and leading-edge bluntness effects on attached and separated laminar boundary layers in a compression corner. Part II: Experimental study. AIAA J. 9:84-93

Holden MS (1971b) Establishment time of laminar separated flows. AIAA J. 9:2296-2298

Holden MS, Moselle JR (1970) Theoretical and experimental studies of the shock wave-boundary layer interaction on compression surfaces in hypersonic flow. ARL Rep. 70-0002

Ikawa H (1979) Real gas laminar boundary layer separation methodology as applied to Orbiter control surface effectiveness prediction. AIAA Paper 79-0212

Lighthill MJ (1957) Dynamics of a dissociating gas. Part I Equilibrium flow. J. Fluid Mech. 2:1-32

Lordi JA, Mates RE, Moselle JR (1966) Computer program for the numerical solution of nonequilibrium expansions of reacting gas mixtures. NACA CR-472

Mallinson SG, Gai SL (1992) Establishment of laminar separated flows in a free-piston shock tunnel. IUTAM Symp. on Aerothermochemistry of Re-entry Vehicles and Associated Hypersonic Flows, Marseille, France

Park C (1990) Nonequilibrium hypersonic aerothermodynamics. Wiley-Interscience, p 200

Schultz DL, Jones TV (1973) Heat-transfer measurements in short-duration hypersonic facilities. AGARDograph No. 165

Stalker RJ (1972) Development of a hypervelocity wind tunnel. Aeronautical J. 76:374-384

Stalker RJ, Rayner JP (1986) Shock wave-laminar boundary layer interaction at finite span compression corners. In: Bershader D, Hanson R (eds) Shock Waves and Shock Tubes, Proc. 15th ISSWST, Berkeley, CA, Stanford Univ Press, pp 509-515

White FM (1974) Viscous fluid flow. McGraw-Hill. p 592

Strength of Characteristics at a Curved Shock Wave

Sannu Mölder

DLR Göttingen, Germany and Ryerson Polytechnic University Toronto, Canada

Abstract. The strength of characteristic waves is related to the local gradient and streamline curvature. This relationship and the equations giving the pressure gradient and streamline curvature are used to determine the relative strengths (the reflection coefficient) of characteristics just downstream of a two-dimensional curved shock wave. It is shown that the characteristics' strengths are a complex function of the specific heat ratio, the upstream Mach number and shock angle and vary directly with the shock curvature. The reflection coefficient, which is independent of shock curvature, is used to characterise four different types of shock wave whose existence depends on the specific heat ratio of the gas and the upstream flow Mach number. The nature of reflection at the shock's downstream surface may change up to four times and this is posed as the explanation for the inflected shocks that have been observed both experimentally and computationally. It is concluded that such approximate analytical methods as the Tangent-Wedge should not be used when strong curved shocks are present and that the nature of wave reflections behind a weak shock in air is not properly simulated by tests with helium.

Key words: Supersonic flow, Shock waves, Characteristic strength

Nomenclature

$\mu_{\pm}$	Mach angle$(= \pm \sin(1/M)$	$A \cdots C$	coefficients in compatibility equations
$\eta_{\pm}$	distance measured along characteristics	$A' \cdots C'$	coefficients in compatibility equations
s	distance measured along streamline	S_a	shock wave curvature measured in a
n	distance measured normal to streamline		plane which contains both the up- and
$C_{\pm}$	characteristic lines		downstream flow vectors at the shock
p	static pressure	S_b	shock wave curvature measured in a
ρ	static density		plane perpendicular both to the shock
V	flow speed		and to the plane containing S_a
$\pi_{\pm}$	strengths of $C_{\pm}$ - characteristics	θ	angle between upstream flow vector
P	non-dim. press. gradient along streamline		and plane of shock
D	streamline curvature	γ	ratio of specific heats
δ	flow deflection	λ	reflection coefficient $(= \pi_-/\pi_+)$
M	Mach number	R	radius of curvature of the shock
			corresponding to curvature $S = -1/R$

Subscripts and superscripts

1, 2	up- and downstream sides of the shock wave
con	conical flow
$*$	sonic conditions
$+$	refers to the characteristic which is inclined at acute angle $+\mu$ to the streamline
$-$	refers to the characteristic which is inclined at acute angle $-\mu$ to the streamline
a, b	refer to streamwise and transverse shock curvatures

Shock Waves @ Marseille I
Editors: R. Brun, L. Z. Dumitrescu

1. Introduction

For supersonic flow, the notion of characteristic lines is the basis of the most accurate method of numerically calculating such flow. Furthermore, characteristic lines are physically significant because pressure disturbances propagate along characteristics, and characteristic lines and surfaces delineate regions of influence and domains of dependence (Zucrow and Hoffman 1976). Thus a great deal of insight can be had from an examination of the strengths and behaviour of characteristic lines. This becomes particularly useful in designing computer strategies for dealing with the difficulties of embedded shock waves (Rusanov 1976) and mixed flows (Newland and Spee 1973). Less precise methods of calculation than the method of characteristics, such as the shock-expansion method, depend on their usefulness on the rear surface of the shock wave being a weak reflector of characteristic waves. In this regard, the reflective properties of two-dimensional shock waves have been examined in Waldman and Probstein (1977) and Chernyi (1961), where it has been shown that, in general, two-dimensional shocks are weak reflectors of characteristic waves except near the sonic point where the reflections are of strength comparable to the incident waves. But even for two-dimensional flows no sweeping conclusions can be made regarding the reflectivity of shocks since the sign and magnitude of the reflectivity vary in a complicated fashion with shock angle and Mach number (Chernyi 1961). Intuitively it seems that strong compression waves, reflecting from the downstream surface of a shock, would eventually coalesce to create embedded shock waves. At the very least, one would expect packets of strong waves to cause an undulating shock wave. Results of method of characteristic calculations (Rusanov 1976) have demonstrated the existence of a shock wave, with a number of inflection points, enveloping an otherwise monotonically curving axisymmetric solid body surface. These physical and calculational situations provide a need for understanding the nature of characteristics' behaviour behind curved shock waves.

It is the purpose of this paper to derive explicit expressions for the reflectivity of two-dimensional, curved shock waves and to point out the significance to highly curved shock waves and aerodynamic simulation in helium.

1.1. The strength of characteristics

Characteristics are waves or lines in supersonic flow oriented at angles of $\pm\mu$ to the local streamline. Pressure disturbances, travelling along these characteristics, are reflected from bounding walls, sonic surfaces, shear layers and the back sides of shock waves. There exist simplified theories for calculating supersonic flowfields which neglect the reflected waves (the Tangent-Wedge, Tangent-Cone and Shock-Expansion theories), and it thus becomes important to calculate the strength of the reflected waves in order to assess the errors incurred by the above approximate theories.

Let η_+ and η_- denote distances along the C_+ and C_- characteristics, where s and n are distances measured along and normal to the streamline so that for the C_+ characteristic:

$$\partial \bullet /\partial\eta_- = \cos\mu\, \partial \bullet /\partial s + \sin\mu\, \partial \bullet /\partial n \tag{1}$$

and for the C_- characteristic

$$\partial \bullet /\partial\eta_+ = \cos\mu\, \partial \bullet /\partial s - \sin\mu\, \partial \bullet /\partial n \tag{2}$$

The *strength* of the C_+ characteristic is now defined as the pressure gradient across the C_+ characteristic measured in the direction of the C_- characteristic i.e. $\partial p/\partial\eta_-$, and similarly the strength of the C_- characteristic is $\partial p/\partial\eta_+$, so that the expressions for the strengths of the C_+ and C_- characteristics, when normalized with respect to twice the local dynamic pressure (ρV^2), become:

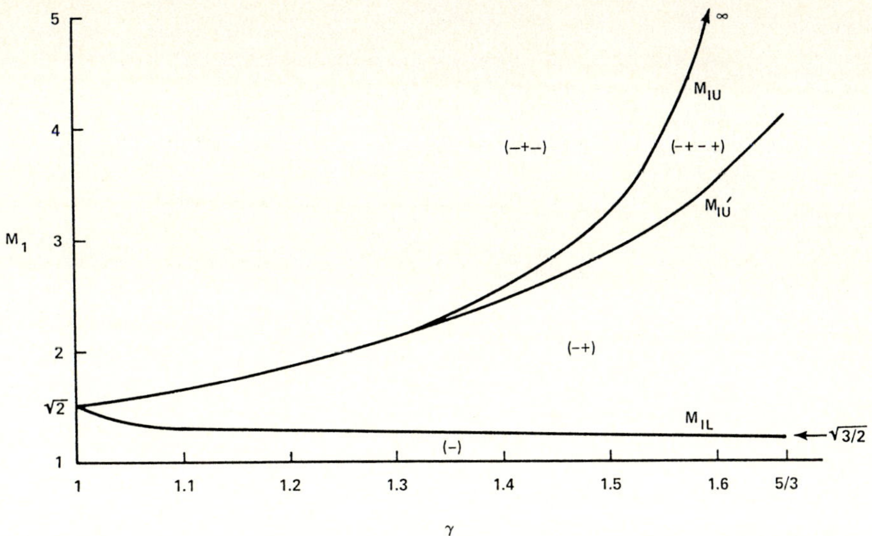

Fig. 1. Variation of reflection coefficient with shock angle and upstream Mach number at the rear surface of a two-dimensional shock wave

$$\pi_+ \equiv \frac{1}{\rho V^2}\frac{\partial p}{\partial \eta_-} = \frac{\cos\mu}{\rho V^2}\frac{\partial p}{\partial s} + \frac{\sin\mu}{\rho V^2}\frac{\partial p}{\partial n} \tag{3}$$

$$\pi_+ \equiv \frac{1}{\rho V^2}\frac{\partial p}{\partial \eta_+} = \frac{\cos\mu}{\rho V^2}\frac{\partial p}{\partial s} - \frac{\sin\mu}{\rho V^2}\frac{\partial p}{\partial n} \tag{4}$$

The C_+ characteristic is a compression wave if $\pi_+ > 0$ and an expansion wave if $\pi_+ < 0$, and similarly for the C_- characteristic. A Mach wave (wave of zero strength) appears when $\pi_\pm = 0$.

1.2. Streamwise pressure gradient and streamline curvature

We now define a normalized pressure gradient along the streamline by:

$$P \equiv \frac{1}{\rho V^2}\frac{\partial p}{\partial s} \tag{5}$$

and further define a streamline curvature:

$$D \equiv \frac{\partial \delta}{\partial s} \tag{6}$$

where δ is the flow inclination angle; and we note from the cross-stream momentum equation (Eq.6), that,

$$\frac{\partial p}{\partial n} = -\rho V^2 \frac{\partial \delta}{\partial s} = -\rho V^2 D. \tag{7}$$

Using these relations in Eqs.3 and 4 and the fact that $\sin\mu = 1/M$ we can write

$$\pi_+ = \frac{1}{M}\left[\sqrt{M^2-1}\,P + D\right] \tag{8}$$

$$\pi_- = \frac{1}{M}\left[\sqrt{M^2-1}\,P - D\right] \tag{9}$$

These are now the strengths of the C_+ and C_- characteristics written in terms of the local Mach number, M, the normalized streamwise pressure gradient and the streamline curvature. The two expressions are quite general in that they apply to supersonic flow which may be rotational, non-adiabatic, and three-dimensional as long as the flow is inviscid and steady.

We will now find the expressions for π_+ and π_- on the downstream surface of a two-dimensional shock wave such as one would find on a blunt unswept leading edge of a wing in supersonic flow. For such a shock, in uniform upstream flow, we can write (from Mölder 1979) for the gas state (2), behind the shock,

$$P_2 = \frac{[BC]}{[AB]} S_a \tag{10}$$

$$D_2 = \frac{[CA]}{[AB]} S_a \tag{11}$$

where (see Mölder 1979)

$$[AB] \equiv A_2 B_2' - A_2' B_2$$

$$[BC] \equiv B_2 C' - B_2' C$$

$$[CA] - CA_2' - C'A_2$$

and

$$A_2 = \sin 2\theta / 2 \sin(\theta - \delta)$$

$$B_2 = -\sin 2\theta / 2 \cos(\theta - \delta)$$

$$C = -2 \sin 2\theta / (\gamma + 1)$$

$$A_2' = \left\{ 1 + \left(M_2^2 - 2 \right) \sin^2(\theta - \delta) \right\} \frac{\sin \theta \cos \theta}{\sin(\theta - \delta) \cos(\theta - \delta)}$$

$$B_2' = -\sin 2\theta$$

$$C' = \sin \theta \tan(\theta - \delta) \sin \delta$$

and S_a is the shock curvature which is defined to be positive when the shock is concave towards the upstream direction. The coefficients A_2, B_2, C, A_2', B_2' and C' are functions of the freestream Mach number M_1, the shock angle θ and the specific heat ratio γ only, which implies that P_2 and D_2 are also functions of these variables, as well as being linearly dependent on the shock curvature S_a. Eqs.8 and 9 can then be written as:

$$\pi_{+_2} = \frac{1}{M_2} \left[\sqrt{M_2^2 - 1} \frac{[BC]}{[AB]} + \frac{[CA]}{[AB]} \right] S_a \tag{12}$$

$$\pi_{-_2} = \frac{1}{M_2} \left[\sqrt{M_2^2 - 1} \frac{[BC]}{[AB]} - \frac{[CA]}{[AB]} \right] S_a \tag{13}$$

Subscript 2 on M denotes conditions behind the shock wave. Note that the characteristics' strengths are both directly proportional to the shock curvature and the proportionality factor is a complex function of γ, M_1, and θ. In (Mölder 1979) it is shown that the two terms $[BC]/[AB]$ and $[CA]/[AB]$ remain positive for all shock angles for which the flow behind the shock is super-sonic. Thus the C_+ characteristics are compression ($\pi_+ > 0$) or rarefaction waves ($\pi_+ < 0$) when the shock is curved positively or negatively respectively. This confirms the well-known fact that compression waves incident on the rear surface of a shock will cause the shock to steepen and expansion waves will cause it to bend in the downstream direction. The behaviour of the reflected C_- characteristic is however not so simple because the coefficient of S_a in Eq.13 can change sign. The above expressions for π_{+_2} and π_{-_2} have been derived for use in computation. However, for the sake of compactness of future discussion and comparison with previous similar results we will henceforth refer to the ratio π_{-_2}/π_{+_2}, and call this ratio λ - the reflection coefficient.

2. The reflection coefficient

We define the ratio of strengths of the C_- and C_+ characteristics, at any point, $\lambda = \pi_-/\pi_+$. Using Eqs.8 and 9 gives (Zucrow and Hoffman 1976):

$$\lambda = \frac{\sqrt{M^2 - 1}P - D}{\sqrt{M^2 - 1}P + D} \tag{14}$$

Eq.14 is quite general in that it applies to any point in steady inviscid sonic or supersonic flow without any particular restrictions regarding rotationality, symmetry or gas properties. For the restricted case of flow behind two-dimensional shocks in a uniform upstream flow, P and D are given by Eqs.10 and 11 giving the reflection coefficient (Rusanov 1976),

$$\lambda_{2D} = \frac{\sqrt{M_2^2 - 1}[BC] - [CA]}{\sqrt{M_2^2 - 1}[BC] + [CA]} \tag{15}$$

The regimes where λ_{2D} is negative and positive, and the lines where it is zero are shown in Fig.1 in terms of the upstream Mach number and the shock angle for $\gamma = 1.4$. The topmost line of this figure is the sonic condition where M_2 (behind the shock) equals 1. As expected from Eqs.14 and 15 the value of $\lambda_{2D} = -1$ at the sonic point on the shock. It is well known that the reflection of characteristics from the sonic line is also equal and opposite i.e. $\lambda = -1$. We may therefore conclude that wave reflections from the sonic line, and its contiguous shock surface, are similar in nature. Next to the shock sonic point the shock wave reflects characteristics in an opposite sense (i.e. $\lambda < 0$). In terms of shock angle this region is narrow at very low and at high Mach numbers, however it does exist at all Mach numbers and it means that approximate (shock-expansion) methods of flow calculation are not very good if a substantial segment of the shock surface has a near-sonic downstream Mach number. An examination of this graph shows that $|\lambda|$ is less than 0.01, i.e. the strength of the shock-reflected wave is less than 1% of the incident wave as long as the shock angle is below 50 degrees for the whole range of supersonic Mach numbers up to 4.

There are points on a shock wave where the character of the reflection process changes. On one side of these points $\lambda_{2D} > 0$, and characteristic waves reflect in a like sense, i.e. compression waves reflect as compression waves and rarefactions reflect a rarefactions. On the other side of these points $\lambda_{2D} < 0$, and waves reflect in the opposite sense; and at the point the reflected wave is of zero strength. From Eq.14 we see that the points for zero-strength reflection are given by:

$$P/D = (M_2^2 - 1)^{-1/2} \tag{16}$$

which for planarly symmetric flow becomes:

$$[BC]/[CA] = (M_2^2 - 1)^{-1/2} \tag{17}$$

The solutions of this equation show that there are either one, two, three or four such points depending on freestream Mach number (Niewland and Spee 1973). Above a freestream Mach number M, given by

$$M_{1U}^2 = \frac{1 - \sqrt{(\gamma + 1)/(\lambda - 1)}}{1 - \frac{1}{2}\sqrt{(\lambda + 1)/(\gamma - 1)}}$$

there are three points where $\lambda_{2D} = 0$. This means that if we follow the downstream side of a curved shock from the sonic point towards the limit where the shock becomes a Mach wave then we would first find a region of shock, near the sonic point, where the characteristics reflect in an opposite sense (λ neg.) then there would be a region of same sense (λ pos.) reflection followed by another opposite sense region (λ neg.). We denote this behaviour by $(- + -)$. Below the Mach number M_{1U}, but above a freestream Mach number given by:

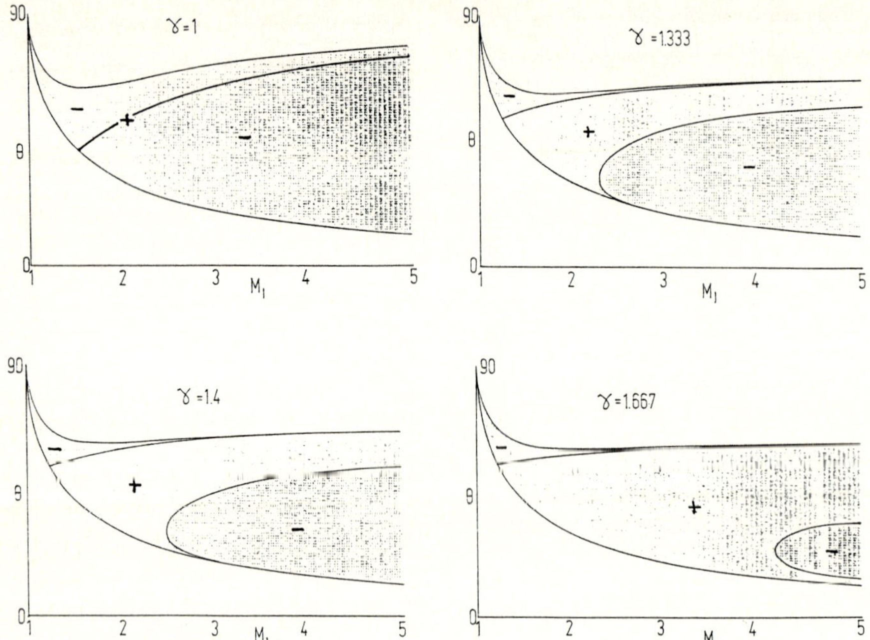

Fig. 2. The polarity of the reflection coefficient for various values of γ as a function of shock angle and freestream Mach number

$$M_{1L}^2 = \frac{1 + \sqrt{(\gamma + 1)/(\gamma - 1)}}{1 + \frac{1}{2}\sqrt{(\gamma + 1)/(\gamma - 1)}}$$

there exist two points where $\lambda_{2D} = 0$. This Mach number range is denoted by $(- +)$. Below M_{1L} there is only one point, this being the degenerate point where $\theta = \mu$. This region is denoted by $(-)$. In the narrow range of Mach numbers between $M_{1U'}$ and M_{1U} there are four values of θ where $\lambda_{2D} = 0$. The existence of this $(- + - +)$ region does not seem to have been noted previously as far as wave reflections are concerned. For $\gamma = 1.4$, $M_{1U} = 2.53958$ and $M_{1L} = 1.24519$. The value of $M_{1U'}$ is difficult to determine algebraically. For $\gamma = 1.4$, the value was found by iteration as 2.462 ± 0.0005.

Based on the reflective property we may now speak of four different types of shock wave: type $(-)$, type $(- +)$, type $(- + -)$ and type $(- + - +)$. In each case the designation refers to unlike $(-)$ or like $(+)$ reflectivity as proceeding from the sonic condition to the weak shock (Mach wave) condition, and for $\gamma = 1.4$ we have seen that all four types are possible. This leads us to ask what happens for other values of γ which are of interest in compressible gas flow analysis.

3. Effect of specific heat ratio γ

Plots similar to Fig.1 are given in Fig.2 for $\gamma = 1$, $4/3$ and $5/3$. For $\gamma = 1$ (Fig.2a) M_{1L} and M_{1U} merge and the positive region between them disappears and $M_{1U'}$ approaches ∞, so that the shock becomes type $(-)$ over all of its length. The reflections are generally weak except near the sonic point where $\lambda = -1$. At $\gamma = 4/3$ $M_{1U'}$ has merged with M_{1U} and three types of shock appear: type $(-)$, $(- +)$ and $(- + -)$. For $\gamma = 5/3$, $M_{1U} \to \infty$, which brings in to being three shock types $(-)$, $(- +)$ and $(- + - +)$. The values of M_{1L}, M_{1U} and $M_{1U'}$ are plotted against

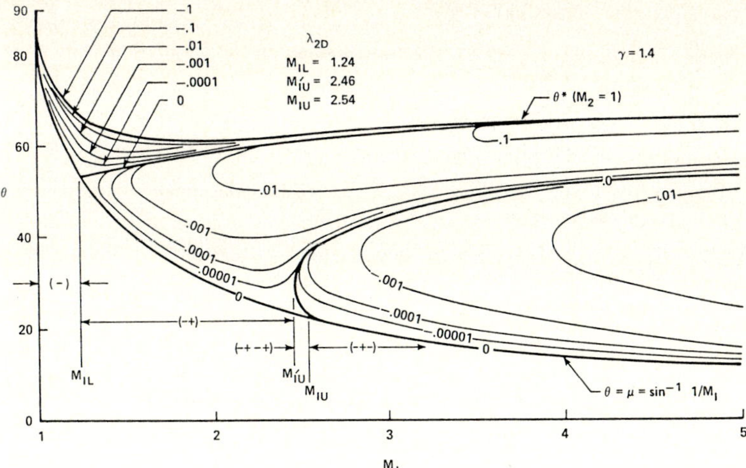

Fig. 3. Influence of the specific heat ratio (γ) on the upstream Mach number (M_1) which delineates various types of shock reflectivity

γ in Fig.3. It is now appropriate to discuss two practical implications of the theoretical results presented so far.

4. Applications

4.1. Planetary entry vehicles, AOTV's

Considering the hypersonic limit where $M_1 \to \infty$ and $\gamma \to 1$, we see that the "like" reflection region disappears and the whole downstream side of the shock reflects in the opposite sense ($-$). These reflections are generally weak (λ between 0 and -0.01) except near the sonic condition behind the shock where $\lambda \to 1$. This means that for the mushroom-shaped re-entry configurations, where, at the shoulder of these bodies, there is a sonic flow with γ close to one and a strong expansion (C_+) wave set, impinging on the back of the shock, there will be strong (C_-) compression waves reflecting off the back of the shock. Detailed flowfield calculations will show whether the compression waves will impinge on the body, or the wake, or coalesce to form a shock which may then impinge on the body.

4.2. Aerodynamic simulation using helium

The simulation of true flight conditions for high Mach numbers in wind tunnels is difficult because air, when accelerated from tunnel stagnation conditions to freestream conditions reaches temperatures low enough to condense. This has led to the development of tunnels where the reservoir/stagnation air is heated or to the use of a low liquefaction point gas such as helium in place of air. In the latter case it is important to examine the effects, on proper simulation, of the usual parameters Mach number and Reynolds number as well as the differences in specific heat ratio γ between air and helium.

From Fig.3 we see that, for helium ($\gamma = 5/3$) at Mach numbers above 4.15 only the type ($- + - +$) shock is possible, and that for air ($\gamma = 1.4$) this type of shock exists only between Mach 2.462 and 2.540. Thus the rear surface of a high Mach number shock wave in helium behaves as a shock in air only in this restricted Mach number range. However the situation is not as bad as it seems since the two shocks differ only by the last + sign in the type ($- + - +$) string for helium. This means that this qualitative difference exists only for weak shock waves. Thus the nature of wave reflections behind a weak shock in air is not simulated by tests in helium and it is conceivable that the formation of embedded shock waves is not properly represented so that both quantitative as well as qualitative flow differences are possible in the simulation.

5. Conclusions

The strengths of characteristics are defined in terms of pressure gradient and flow curvature. The ratio of these strengths for the C_+ and C_- characteristic (called the reflection coefficient) is calculated for conditions at the downstream side of a curved two-dimensional shock wave. Depending on the freestream Mach number, it is possible to have four distinctly different types of shock wave surface when it pertains to the reflection of characteristic waves. These results indicate that shock waves are likely to have inflection points when facing a freestream of Mach number 2.5 in air. Shock waves in monatomic gases are less likely to be inflected and shocks in flow with $\gamma = 1$ are least likely to be inflected. Tangent-wedge methods should not be used when shocks are strong enough to cause subsonic or sonic downstream flow. Highly curved shock waves are accompanied by strong downstream characteristics especially at conditions where the downstream flow is near sonic. The flow of helium downstream of weak shocks does not properly simulate the flow of air.

Acknowledgement

A year of most pleasant sabbatical leave was made possible by Prof. G.E. A. Meier and Dr. G. Eitelberg of DLR and Dr. S. Matar of RPU.

References

Chernyi GG (1961) Introduction to Hypersonic Flow. Academic Press

Crocco L (1937) Singolarita delle corrente gassosa iperacustica nell' intorno di una prora a diedro. L'Aerotechnica 17: 519-536

Henderson LF (1964) On the confluence of three shock waves in a perfect gas. Aeronautical Quarterly 15:181

Mölder S (1979) Flow behind curved shock waves. UTIAS Rep. 217, University of Toronto, Institute for Aerospace Studies

Nieuwland GY, Spee BM (1973) Transonic airfoils: recent developments in theory, experiment and design. Ann. Rev. of Fluid Mechs. Vol. 5

Rusanov VV (1976) A blunt body in a supersonic stream. Ann. Rev. of Fluid Mechs. Vol. 8

Waldman GD, Probstein RF (1960) An analytic extension of the shock-expansion method. J. Aeron. Sci. 28, 2

Zucrow MJ, Hoffman JD (1976) Gas Dynamics, Vol. 1. Wiley & Sons Inc.

Blunt Body Flow - The Transonic Region

Sannu Mölder
DLR Göttingen, Germany and Ryerson Polytechnic University, Toronto, Canada

Abstract.   For supersonic flow of an ideal gas over planar and axisymmetric bodies we consider the transonic region bounded by the sonic line, the body surface, the downstream surface of the shock wave and the limiting characteristics. Using curved shock theory, we establish the conditions under which the shock surface is (type I), or is not (type II), a boundary of the transonic region. It is found that the existence of type I or type II flow is determined solely by the values of free stream Mach number, specific heat ratio of the gas and transverse-to-longitudinal curvature of the shock surface. When the curvature ratio is greater than 1, only type II flow is possible irrespective of the specific heat ratio. When the curvature ratio is less than $-2/(\gamma - 1)$ then only type I is realized. When the curvature ratio is between 1 and $-2/(\gamma - 1)$ the appearance of type I or type II is determined by the free stream Mach number. With proper choice of curvatures the results apply to general, three dimensional, shock waves as well.

Key words: Supersonic flow, Shock waves, Transonic flow

1. Introduction

Steady supersonic flow over a blunt body generates a shock wave ahead of the body. Directly ahead of the body the shock is strong and the flow between the shock and the body is subsonic. Off to the side, the shock is weaker and the flow behind the shock is supersonic. Separating these two flows is a sonic surface. Since the flow behind the sonic surface is supersonic it is there populated by positive and negative characteristics. Some of these impinge on the back of the sonic surface (where they are perpendicular to the streamlines) and propagate from there to intercept either the shock or the body surface. The characteristic(s) of this type which are furthest downstream are called limiting characteristics, and they form the downstream boundary of the transonic region between the shock and the body. It is the characteristics which *impinge* upon the sonic line that form the downstream boundary of the *domain of dependence* for the sonic surface, and it is the last characteristics that reflect from the sonic surface that form the *region of influence* of the sonic surface.

The topography of the shock/body/sonic-line/limiting characteristics is shown in Fig.1. Three distinctly different types of transonic region are possible. For type I, a section of the body forms the lower boundary of the domain of dependence but the shock is not a boundary. For type II both the shock and the body form boundaries of the domain of dependence and for type III the shock forms the upper boundary to the domain of dependence. A small disturbance at any point in the domain of dependence is capable of influencing the flow at the sonic surface and hence also the subsonic flow upstream of the sonic surface. Consequently a calculation of the subsonic flow between the bow shock and the body must include the points on the body and the shock which lie within the domain of dependence downstream of the sonic surface. Also a flow calculation in the supersonic region must start at an initial data line which is downstream of the limiting characteristic(s) *not* just downstream of the sonic surface (Anderson 1989). For this reason it is important to know the location of the limiting characteristics and the conditions which presage the appearance of type I, II or III sonic regions. Hayes and Probstein (1959) described the three types for both axisymmetric and planar flow. For planar flow they proved that type III cannot exist and that transition between types I and II occurs at a Mach number less than 2. Rusanov (1976) proved that for planar flow transition between types I and II occurs at a certain freestream Mach number $M_0(\gamma)$ and he argued, but did not conclusively prove, that type I flow is not possible for axisymmetric flow.

Shock Waves @ Marseille I
Editors: R. Brun, L. Z. Dumitrescu

It is the purpose of this paper to confirm as well as contradict some of the assertions of Hayes and Probstein and Rusanov with respect to the flow behind the shock using the curved shock theory of Mölder (1979). In particular we establish the criteria for the transition between type I and II transonic flow for plane shocks as well as axisymmetric shocks for both external and internal flow.

2. Transonic flow behind curved shocks

The topography of transonic flow behind a curved shock can be discussed in terms of a type I, II and III behaviour.

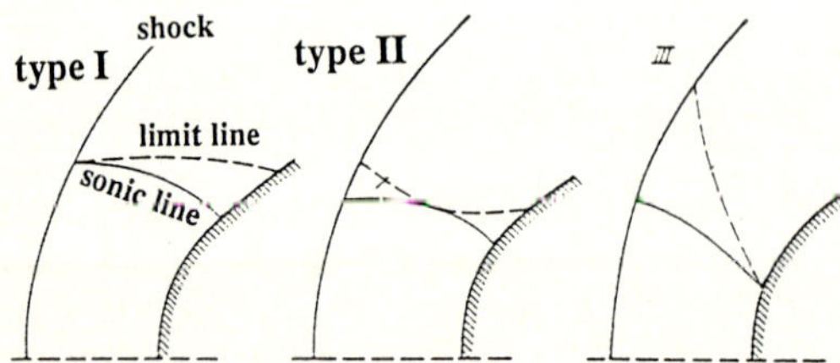

Fig. 1. Three types of transonic flow behind a curved shock as characterized by the position of the sonic and limiting lines

The differences arise from the shape of the transonic region bounded on the upstream by the sonic line and on the downstream by the limiting characteristics and on the side by the body (type I), the shock (type III) or by both (type II). Rusanov points out that types I and II are distinguished by the behaviour of the flow immediately behind the shock, whereas types II and III are differentiated by flow details on the body surface. In both cases the transition point from type I to type II and from type II to type III occurs when the characteristic and the sonic line become coincident. Since at sonic conditions the characteristics are perpendicular to the streamline then at transition it must be that the sonic line is also perpendicular to the streamline. Thus the problem of finding the transition point reduces to finding the conditions when the sonic line is normal to the streamline at the shock or on the surface. Rusanov solves this problem for the body surface computationally by calculating the flow over various specific bodies. No general solution is obtained. For the shock, however, Rusanov solves the problem exactly for a 2D body and gives the transition Mach number as

$$M_0(\gamma) = \sqrt{(\gamma^2 + 4\gamma + 5)/(\gamma + 3)},$$

which is 1.6895 for $\gamma = 1.4$. Thus for planar bodies at $M_1 < M_0(\gamma)$ one would obtain type I flow and at $M_1 > M_0(\gamma)$ type II flow.

It is the purpose of the present report to examine the transition from type I to type II transonic flow behind an oblique shock and to verify and generalize the transition criteria to doubly curved shocks with both positive and negative curvature.

3. Theory

In flow with constant enthalpy, a line of constant velocity must also be a line of constant temperature, hence constant sound speed, and hence a line of constant Mach number. Thus the sonic line is a line of constant velocity. If l is the distance measured along the sonic line then :

$$\frac{dV}{dl} = \frac{\partial V}{\partial s}\frac{\partial s}{\partial l} + \frac{\partial V}{\partial n}\frac{\partial n}{\partial l} = 0 \tag{1}$$

where V is the flow velocity and s and n are streamline coordinates.

With α_v the angle between the constant velocity line and the streamline Eq.1 becomes:

$$\frac{\partial V}{\partial s}\cos\alpha_v + \frac{\partial V}{\partial n}\sin\alpha_v = 0 \qquad \text{or} \qquad \tan\alpha_v = -\frac{\partial V/\partial s}{\partial V/\partial n}. \tag{2}$$

For steady inviscid flow the Euler equations in streamline coodinates can be used to write

$$\tan\alpha_v = \left[\frac{1}{\rho_2 V_2^2}\left(\frac{d\rho}{ds}\right)_2\right]\Big/\left[-\frac{\xi_2}{V_2^2} + \left(\frac{\partial\delta}{\partial s}\right)_2\right]. \tag{3}$$

where $(\partial p/\partial s)_2$ is the pressure gradient, $(\partial\delta/\partial s)_2$ is the flow curvature and ξ_2 is the vorticity, all evaluated just behind the shock. For the sonic line to be perpendicular to the streamline we must then have that $\tan\alpha_v \to \infty$, or

$$\xi_2/V_2^2 = (\partial\delta/\partial s)_2. \tag{4}$$

From Hayes and Probstein (1966) we find that

$$\frac{\xi_2}{V^2} = -\frac{\sin(\theta-\delta)}{\tan\theta}\left[\frac{\tan\theta}{\tan(\theta-\delta)} - 1\right]^2 S_a = K_{\text{vort}}\cdot S_a, \tag{5}$$

and from Mölder (1979)

$$\left(\frac{\partial\delta}{\partial s}\right)_2 \equiv D_2 = \frac{[CA]}{[AB]}S_a - \frac{A_2 G'}{[AB]}S_b \tag{6}$$

In these expressions

$$[CA] \equiv C \times A_2' - C' \times A_2 \qquad\qquad [AB] \equiv A_2 \times B_2' - A_2' \times B_2$$

where

$$A_2 = (q_2/q_1)\cos(\theta-\delta) \qquad\qquad A_2' = \left(1 + (M_2^2 - 2)\right)\sin^2(\theta-\delta) \times q_2/q_1 = \frac{q_2}{q_1}\cos^2(\theta-\delta)$$

$$B_2 = -\sin(2\theta)/\left(2\cos(\theta-\delta)\right) \qquad\qquad B_2' = -\sin(2\theta)$$

$$C = -2/(\gamma+1)\times\sin(2\theta) \quad C' = -\sin(2\delta)/\left(2\cos(\theta-\delta)\right) \quad G' = -\sin\delta\times\sin\theta\times\tan(\theta-\delta)$$

and S_a and S_b are the longitudinal and transverse curvatures of the shock wave.

All quantities are evaluated at the conditions where the flow downstream of the shock is sonic so that it is appropriate to attach the superscript $*$ to both S_a and S_b. Substituting Eqs.5 and 6 into Eq.4 gives

$$\left(\frac{S_b}{S_a}\right)^* = \frac{-\frac{\sin(\theta-\delta)}{\tan\theta}\left[\frac{\tan\theta}{\tan(\theta-\delta)} - 1\right]^2 [AB] + [CA]}{A_2 G'}.$$

This expression gives the ratio of shock curvatures at the transition between type I and type II transonic flow in terms of γ, free stream Mach number and the corresponding sonic values of θ and δ. The equation is plotted in Fig.2 where the line is seen to separate the two types of transonic flow. On the horizontal axis $(S_b/S_a)^* = 0$ and we have two-dimensional flow. Below a freestream Mach number of 1.69 there is type I flow and above 1.69 it is of type II. This transition point is well confirmed by Rusanov's value of $M_0(\gamma)$, and as Rusanov states "... for plane flow the question of the type of transonic region was completely cleared up."

For axisymmetric flow the transition Mach number depends on the local curvatures of the shock at the sonic point. Shocks with a sonic point curvature ratio $(S_b/S_a)^* \geq 1$ are always of type II, and shocks with a sonic point curvature ratio < 5 are always of type I. Above the line $(S_b/S_a)^* = 0$ there can exist a "shell shock" with both negative curvatures or a "cup shock"

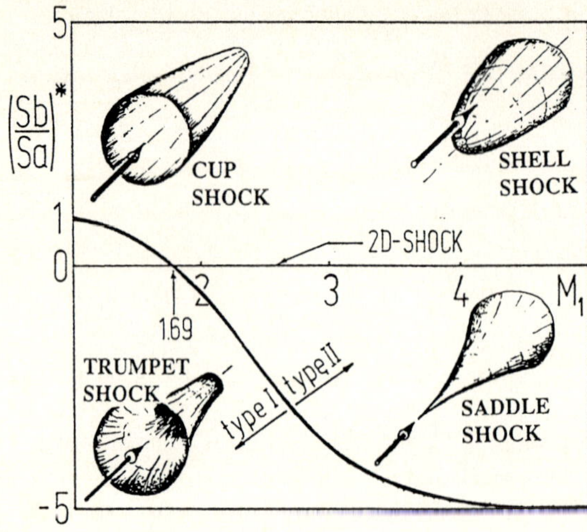

Fig. 2. Existence of type I and type II transonic flow as governed by upstream Mach number M_1 and shock curvatures at the shock-sonic point

with both positive curvatures. Flow behind the first is expansive whereas behind the second it is compressive, so that the limit line for the expansive flow is downstream of the sonic line and for the compressive flow it is in front. Below the $(S_b/S_a)^* = 0$ line there is either a "saddle shock" or a "trumpet shock". For these the curvatures have opposite signs and each may have a type I or II sonic region. From this we see that for all four types of shock wave it is possible to have type I and type II flows and the existence of one or the other is determined exclusively by the ratio of shock curvatures and the Mach number in front of the shock at the place on the shock where the downstream Mach number is sonic. Contrary to the results of Rusanov there is no direct influence of the surrounding shock wave or the underlying body surface. For other values of γ, the transition curve changes position but not overall shape: for all γ's, the curves go through the $(S_b/S_a)^*$-axis at 1; the intercept on the M_1-axis changes according to $M_0(\gamma)$ and the lower asymptote is at $(S_b/S_a)^* = -2/(\gamma - 1)$

4. Conclusion

The problem of the shape of the transonic region at the back of a curved shock is now solved not only for a planarly or axially symmetric flow but also for doubly curved shock waves.

Acknowledgement

A most pleasant year of sabbatical leave was made possible by Prof. G.E.A. Meier and Dr. G. Eitelberg of DLR and Dr. S. Matar of RPU.

References

Anderson JD (1989) Hypersonic and High Temperature Gas Dynamics. McGraw-Hill
Hayes WD, Probstein RF (1966) Hypersonic Flow Theory, 2nd ed. Academic Press
Mölder S (1979) Flow behind curved shock waves. UTIAS Rep. 217, University of Toronto
Rusanov VV (1976) A blunt body in a supersonic stream. Ann. Rev. of Fluid Mechs. Vol. 8

Shock Tube Application to the Study of Compressible Turbulent Boundary Layer with Mass Injection

J.-Y. Liaw and D. Bershader

Department of Aeronautics and Astronautics, Stanford University, Stanford, California, USA

Abstract.   Paper explores the feasibility of using the shock tube in studying the heat transfer in a compressible turbulent boundary layer with mass injection. It aims to gain further understanding of the film cooling in the compressible flow regime. The test model is a flat plate equipped with a 2D 30° injection slot. Air is used for both driver and driven gases. The injection flow is measured by using the slot as a flow measuring device. Heat transfer data are obtained with thin-film heat gauges. The shock-induced flow has Mach number 0.9, Reynolds number 4.62×10^7 per meter, and gas to wall temperature ratio 1.6. The resulting heat transfer measurements are correlated using the scaling parameter suggested by Forth et al. and compared with that of Metzger et al. (1968, 1971)(at Mach 0.03) and Forth et al. (1986) (at Mach 0.55). This comparison revealed a possible Mach number effect on film cooling that has never been addressed before. We use the Prandtl-Glauert formula to model this effect. The agreement is good within measurement uncertainty.

Key words: Compressible turbulent boundary-layer, Mass injection

1. Introduction

Many applications of heat transfer analysis are in the compressible flow regime. One typical example is the heat transfer analysis of turbine blades. Despite its common presence in many applications, the compressible turbulent boundary layer remains a virgin territory. This is due to the lack of experimental data for turbulence model development and code validation. Even the data so fundamental as those on flat plate are few. Heat transfer analyses on turbine blades also involve film cooling as a means to protect the blades from hot gas in addition to impingement and convective cooling in the internal passages of the blades. Most of the correlations for film cooling have been developed based on experiments conducted in a low-speed regime. Gladden et al. (1987) calculated the blade temperatures under real engine conditions with the state-of-the-art computational tools and the correlation derived from existing experimental data. Their results showed significant discrepancy between calculations and measurements. They attributed this discrepancy to the lack of understanding of film cooling in the compressible flow regime.

The shock tube is well-known for its simplicity and capability of generating high enthalpy flows for gasdynamic research. In addition to high enthalpy, high-Reynolds-number flows can be easily obtained if the shock moves into a stationery driven gas initially at atmospheric condition. The shock Mach number is generally less than 3 such that the ideal normal shock relations can be applied. This shock-induced flow is well-behaved in terms of uniformity and repeatability.

The shock-induced boundary layer on the flat plate (see Fig.1) possesses two distinct domains, one near the travelling shock, where the flow is quasi-steady in the shock-fixed frame (Mirels 1955, 1956; Bershader and Allport 1956) and one unsteady flow that approaches the steady state asymptotically (Lam 1958, Lam and Crocco 1969). The mass injection is non-stationary in the shock-fixed frame. This disturbs the quasi-steadiness and excludes the quasi-steady region for the present study. The flow under consideration is then the asymptotically steady region.

Lam and Crocco (1958, 1969) had provided the analysis of the unsteady (relative to the laboratory frame instead of shock-fixed frame) laminar boundary layer induced by a travelling shock wave on the semi-infinite plate. They transformed the coordinate system from x (streamwise

Shock Waves @ Marseille I
Editors: R. Brun, L. Z. Dumitrescu

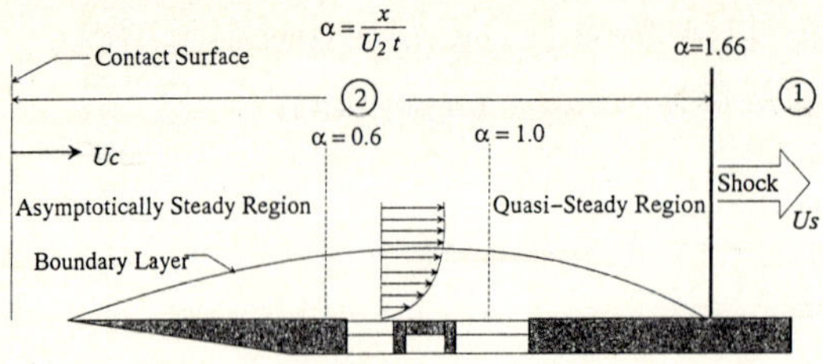

Fig. 1. Shock-induced boundary layer flow on a flat plate

coordinate measured from the leading edge of the plate), y (transverse coordinate), and t (time) into a set of three nondimensional variables α, β, and γ defined as

$$\alpha = \frac{x}{u_2 t}, \qquad \beta = \frac{u}{u_2}, \qquad \gamma = \frac{u_2^2 t}{\nu_2} \tag{1}$$

where u_2 and ν_2 are the velocity and kinematic viscosity in the region 2 (between shock wave and contact surface) of the shock tube flow. The shock path on the x–t diagram is a line of $\alpha = A$. The constant A is a finite value that is always greater than one whose magnitude depends on the shock strength. The quasi-steady region lies between $1 < \alpha < A$. Flow in this region is steady in a shock-fixed frame. It does not know the existence of the plate leading edge until the condition $\alpha = 1$ is reached. Different positions on the plate are affected by the plate leading edge at different times. The line $\alpha = 1$ characterizes the onset of this process. The region bounded by $0 \leq \alpha \leq 1$ is unsteady relative to both shock-fixed and laboratory-fixed frames. The flow in this region approaches to a regular steady state on flat plate asymptotically. We term this region the *asymptotically steady region*. The numerical results of Lam and Crocco (1968) indicate that at $\alpha = 0.3$ the flow reaches a steady state. The *establishment time* for a laminar boundary layer induced by a shock wave is thus $x/0.3u_2$. Davies and Bernstein (1969) confirmed the theory of Lam and Crocco experimentally. They went further to extend the establishment time for turbulent flow too. The establishment time for turbulent flow is found to be $x/0.6u_2$.

The present case differs from that of Davies and Bernstein by the addion of a mass injection into the flow. However, it is reasonable to argue that the establishment time criterion proposed by Davies and Bersnstein is valid for the mass injection case. The argument is as follows: The progress of the flow approaching steady state involves two types of reaction: the external flow reaction and the boundary layer reaction. The external flow reaction is a response at the speed of sound that is much faster than the boundary layer reaction. Thus the boundary layer reaction is the dominant mechanism for this purpose. The boundary layer establishes steady state at the time $x/0.6u_2$ after the shock reaches the leading edge of the plate for a flow without mass injection. Under the same mechanism, the flow with mass injection will approach steady state at the time $(x - x_J)/0.6u_2$ after the shock reaches the slot, where x_J is the distance of the injection slot measured from the leading edge of the plate. The time for the shock to reach the slot is x_J/Au_2 after the shock has reached the leading edge of the plate. Thus the establishment criterion for the time measured from the moment at which the shock reaches the plate leading edge should be the time that satisfies two conditions simultaneously:

$$t > \frac{x}{0.6u_2} \qquad\qquad t > \frac{x_J}{Au_2} + \frac{x - x_J}{0.6u_2} \tag{2}$$

The first condition was originally proposed by Davies and Bernstein (1969) for induced flow without mass injection. The second condition extends the same principle to the case with mass injection that must be satisfied in conjunction with the first condition. The present experimental design indeed is such that the second condition is satisfied whenever the first condition is satisfied.

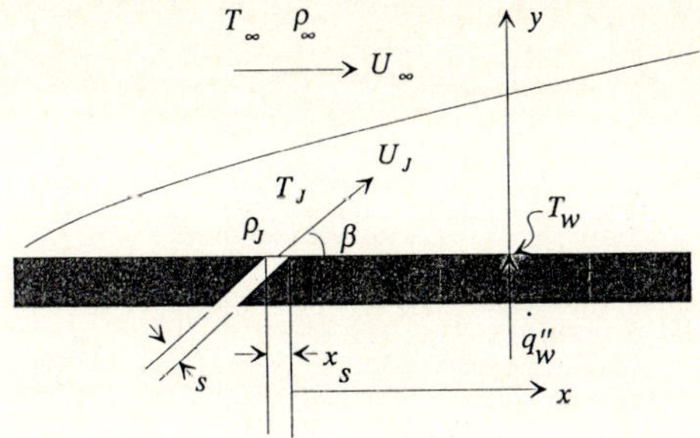

Fig. 2. Ideal film cooling model

The problem under consideration is an idealized film cooling model with a 2D slot injection on an isothermal flat plate (see Fig.2). Fig.2 also shows the relevant flow variables for both the main stream (designated by subscript ∞) and injection flow (designated by subscript J), as well as geometric parameters. For the convenience of film cooling correlation, from this point we will define x as measured from the downstream lip of the slot shown in Fig.2 instead of being measured from the plate leading edge. Note that Lam's treatment used β as a transformed variable. We will not use this transformed variable and thus designate β to be the injection angle as shown in Fig.2.

2. Experimental apparatus

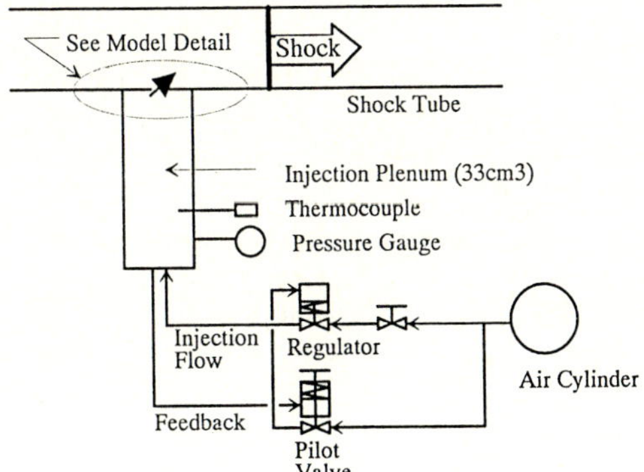

Fig. 3. Injection system schematic

We used the shock tube in the Department of Aeronautics and Astronautics at Stanford University. Two Kulite pressure transducers located 10.2 cm (4") apart on the driven section measure the time of flight of the shock wave for shock speed calculation. This research has achieved excellent repeatability in the time-of-flight measurement by using Mylar sheet as the diaphragm. The standard deviation was 0.54 % over 35 runs.

We added a mass injection system (see Fig.3) to the shock tube for film cooling research. The air cylinders supply the injection flow through a set of pilot-controlled regulators with pneumatic feedback to the injection plenum. The injection slot itself was used as a flow measuring device, by using correlation developed by Fitt et al. (1985) and Forth et al. (1986) relating the momentum flux ratio to the pressure difference ratio as

$$I_J = \frac{(\rho u^2)_J}{(\rho u^2)_2} = \frac{1.25}{\sin^2 \beta} \left[\frac{p_{tJ} - p_2}{p_{t2} - p_2} \right]^3 \tag{3}$$

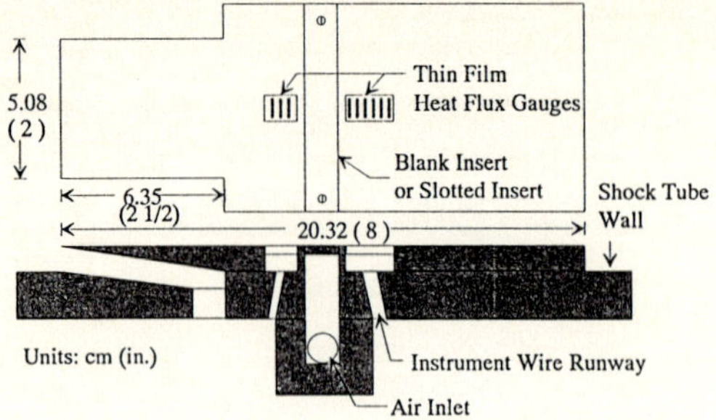

Fig. 4. Flat plate test model

where the subscript J denotes the injection flow, subscript 2 the condition in region 2 of the shock tube, and β is the injection angle.

The test model (see Fig.4) is a flat plate equipped with a 2D 30° injection slot. This model is 20.3 cm (8") long and 5.1 cm (2") wide. The angle of the sharp leading edge is 8.5°. Three gauges measure the heat transfer rates upstream of the injection. Six gauges measure the heat transfer rates downstream of the injection.

Table 1. Test conditions of Metzger, Forth and the present work

	Metzger	Forth	Present
Facility	Wind tunnel	Piston tube	Shock tube
M_∞	0.03-0.04	0.55	0.90
Re (1/m)	1.6×10^6	2.7×10^7	4.6×10^7
x/s	$35 \sim 70$	$2 \sim 98$	$26 \sim 45$
m	$0.25 \sim 1.0$	$0.12 \sim 0.73$	$0.47 \sim 0.89$
β	$20°, 40°, 60°$	$30°$	$30°$
$T_{t\infty}/T_W$	<1.0	0.77-2.0	1.86

The model is then bolted onto the lower surface of the shock tube wall. Detailed descriptions of the experimental apparatus and their dimensions can be found in Liaw (1992).

3. Results and discussion

The resulting heat transfer measurements were analyzed based on the superposition scheme for film cooling. The heat transfer rate equation is written as

$$q_w^{"} = h(T_w - T_{r\infty}) \tag{4}$$

where h is the heat transfer coefficient and T_w and $T_{r\infty}$ are the wall and free-stream recovery temperature respectively. Denote by h_0 the heat transfer coefficient without mass injection and by h' the heat transfer coefficient with mass injection. The film cooling effectiveness Φ in a superposition scheme is defined as

$$\Phi = \frac{h'}{h_0} \tag{5}$$

in which the heat transfer coefficients were evaluated at same position with the same boundary condition. Previous investigation by Metzger et al. (1968, 1971) and by Forth et al. (1986) have shown that the film cooling effectiveness is a linear function of injection flow temperature written in nondimensional form as

$$\Phi = A + B\Theta_J \tag{6}$$

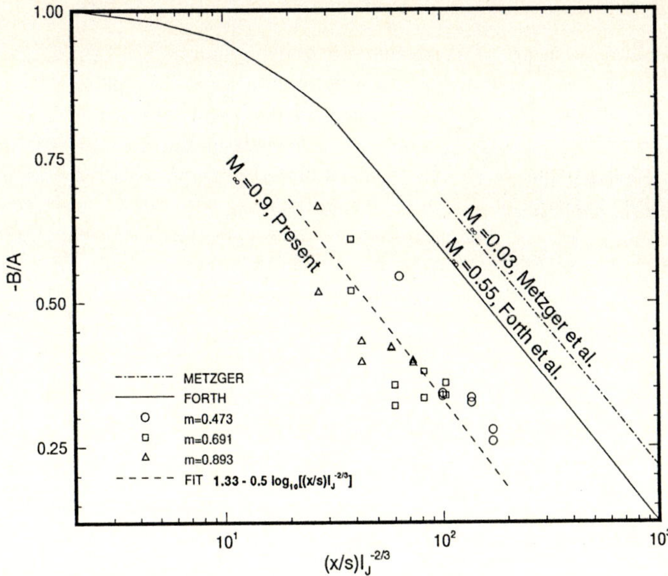

Fig. 5. Measured $-B/A$ vs. $(x/s)I_J^{-2/3}$ showing the Mach number effect

where Θ_J is the injection temperature parameter defined as

$$\Theta_J = \frac{T_J - T_{r\infty}}{T_w - T_{r\infty}} \tag{7}$$

In this scheme, the parameter A characterizes the aerodynamic mixing effect under isoenergetic injection condition. The parameter B characterizes the thermal effect, i.e., the sensitivity of the film cooling effectiveness to the injection temperature. The scaling parameter suggested by Forth et al. (1986) is to correlate the ratio B/A vs. $(x/s)I_J^{-2/3}$. This scaling parameter enables correlation of the film cooling effectiveness covering a range of injection flow rates and locations.

The data analysis (see Liaw (1992) and Liaw and Bershader (1993) for detailed description of this analysis) covers that of Metzger et al. (1968, 1971) (at Mach 0.03) and Forth and al. (1986) (at Mach 0.55) together with the present one (Mach 0.9). Table 1 summarizes the test conditions of these experiments. The symbol m in Table 1 is the injection mass flux ratio, defined as the ratio of the injection mass flux $(\rho u)_J$ to the mainstream mass flux $(\rho u)_\infty$. Fig.5 shows the final result. There exists a region where the parameter $-B/A$ varies linearly with respect to $\log\left[(x/s)I_J^{-2/3}\right]$. A closed examination of the correlation reveals the following approximation for the possible effect of he Mach number:

$$(-B/A)_{M_\infty=0} - (-B/A)_{M_\infty} \approx K\left(1 - \sqrt{1 - M_\infty^2}\right) \tag{8}$$

where $K \approx 0.56$ fits the data reasonably well. This Mach number effect is by no means our final conclusion. We believe that more analysis and experimental study is needed to be done to confirm this finding. However, we believe that the Mach number is the dominant parameter for this variation by the following arguments: The Prandtl numbers in these data are relatively constant. The Reynolds numbers in these data, although differing significantly, are well beyond the critical value to ensure full turbulence of the flow and are considered to be of minor importance compared to the Mach number effect. Thus the variation of the correlation between the different sets of data may possibly be due to Mach number.

4. Conclusion

This paper explores the feasibility of using the shock tube for studying film cooling in the compressible flow regime. The establishment time criteria proposed by Davies and Bernstein (1969) was extended to the case with mass injection and was used to guide the design of the experimental apparatus and test model. The advantage of using a shock tube for film cooling research is that it is capable of generating flow that simulates Mach numbers, gas-to-wall temperature ratios, and Reynolds numbers that are comparable to those in real turbine engine operation. The data analysis covers three Mach numbers obtained in different facilities. The results show a possible Mach number effect on film cooling that have never been addressed before. We used the Prandtl-Glauert formula to model this effect. The agreement is good within measurement uncertainty. Should this be valid, vast data of film cooling obtained in the low-speed regime can be casted into high speed regime for application. More study is needed to confirm this Mach number effect.

References

Bershader D, Allport J (1956) On the laminar boundary layer induced by a traveling shock wave. NR061-020, Department of Physics, Princeton University

Davies DR, Berstein L (1969) Heat transfer and transition to turbulence in the shock-induced boundary layer on a semi-infinite flat plate, Part 1. J. Fluid Mechs. 36: 87-112

Fitt AD, Ockendon JR, Jones TV (1985) Aerodynamics of slot film cooling: theory and experiment. J. Fluid Dynamics 160: 15-27

Forth CJP, Jones TV (1986) Scaling parameters in film-cooling. In: Tien CL, Carey VP, Ferrell JK (eds) Heat Transfer, Vol. 3, pp 1271-1276

Gladden HJ, Yeh FC, Austin PJ Jr.(1987) Computation of full-coverage film-cooled airfoil temperature by two methods and comparison with high heat flux data. ASME-GT-213, Gas Turbine Conference and Exhibition, Anaheim, California, May 31 - June 4

Lam SH (1958) Shock induced unsteady laminar compressible boundary layers on a semi-infinite plate. Ph.D. Thesis, Princeton University

Lam SH, Crocco L (1969) Note on the shock-induced unsteady laminar boundary layer on a semi-infinite flat plate. J. Aero/Space Sci. 26: 54-56

Liaw J-Y (1992) Shock tube application to the study of the compressible turbulent boundary layer with mass injection. PhD Dissertation, Dept. Aeronautics and Astronautics, Stanford University, Stanford, California

Liaw J-Y, Bershader D (1993) A study of film cooling in the compressible turbulent boundary layer induced by a traveling shock wave. AIAA Paper 93-2558 29th AIAA/ASME/SAE/ASEE Joint Propulsion Conference, Monterey, California

Metzger DE, Capper HJ, Swank LR (1968) Heat transfer with film cooling near nontangential injection slots. Journal of Engineering for Power: 157-163

Metzger DE, Fletcher DD (1971) Evaluation of heat transfer for film-cooled turbine components. J. Aircraft 8, 1: 33-38.

Mirels H (1955) Laminar boundary layer behind shock advancing into stationary fluid. NACA TN 3401, Washington, USA

Mirels H (1956) Boundary layer behind shock or thin expansion wave moving into stationary fluid. NACA TN 3712, Washington, USA

Interaction of Thermal Protection Materials with the High-Enthalpy Flow of the Arc-Heated Wind Tunnel LBK

A. Gülhan

DLR Wind Tunnel Division WT-WK, Linder Höhe, 51140 Köln, FRG

Abstract. Gas surface interaction phenomena behind the shock wave in front of space vehicles during the reentry phase can only be investigated in high-enthalpy facilities with a long testing time. Sintered silicon carbide and carbon fibre-reinforced silicon carbide samples have been tested at different test conditions in the LBK. Experimental results concerning the gas parameters have been supported by numerical computation including chemical non-equilibrium processes in the nozzle flow coupled with frozen or equilibrium shock solutions. Spectroscopic analyses of the material composition of the samples tested in the air atmosphere indicate the development of a protective SiO_2 -layer.

Key words: Hypersonic flow, Surface reaction, Protective layer

1. Introduction

During the reentry phase of space vehicles into the atmosphere the nose cap and wing leading edges are exposed to high thermal loads caused by shock waves heating the air up to temperatures at which real gas effects are important. Therefore the simulation of the reentry phase of the flight requires not only the duplication of Mach number and Reynolds number, but also setting real gas similarity parameters (Koppenwallner 1990). While similarity parameters for dissociation processes like the binary scaling factor and the flow velocity can be duplicated in shock tunnels (Olivier et al. 1991) and other short time high enthalpy facilities, longer testing time is necessary to investigate recombination and surface reactions.

Despite restrictions in the simulation of flight environment arising from strong nonequilibrium phenomena in their flow field due to low reservoir pressures, arc heated wind tunnels with their long running times are very useful tools for studying real-gas processes (Wagner et al. 1990). Arc heaters with a rod-shaped cathode have normally a tungsten cathode, which is very sensitive to chemical reaction with oxygen. Only inert gases or nitrogen can be supplied along the cathode and oxygen or other gases are injected into the main flow downstream of the nozzle throat (Auweter-Kurtz 1991). At high powers arc heaters with hollow electrodes find more application.

2. Experimental set-up and measurement technique

The arc heated wind tunnel LBK consists of two test legs called L2K and L3K (Fig.1). All parts of the arc heater and nozzle are water cooled and interchangable (Gülhan 1993). For the experiments presented here a conical nozzle with a nozzle diameter of 29 mm and a half angle of 12° has been used. A vacuum pumping system consisting of three Roots blowers, one rotary plunger vacuum pump and one liquid seal pump allow to produce pressures down to 1 Pascal in the test chamber. Stagnation enthalpies up to 28 MJ/kg, which correspond to gas temperatures up to 6650 K, can be achieved at low mass flow rates. Because of dissociation and ionization processes at these high temperatures noxious gas NO is produced. It is decontaminated in a NO-absorption facility washing it with NaOH-solution.

Shock Waves @ Marseille I

Editors: R. Brun, L. Z. Dumitrescu   © Springer-Verlag Berlin Heidelberg 1995

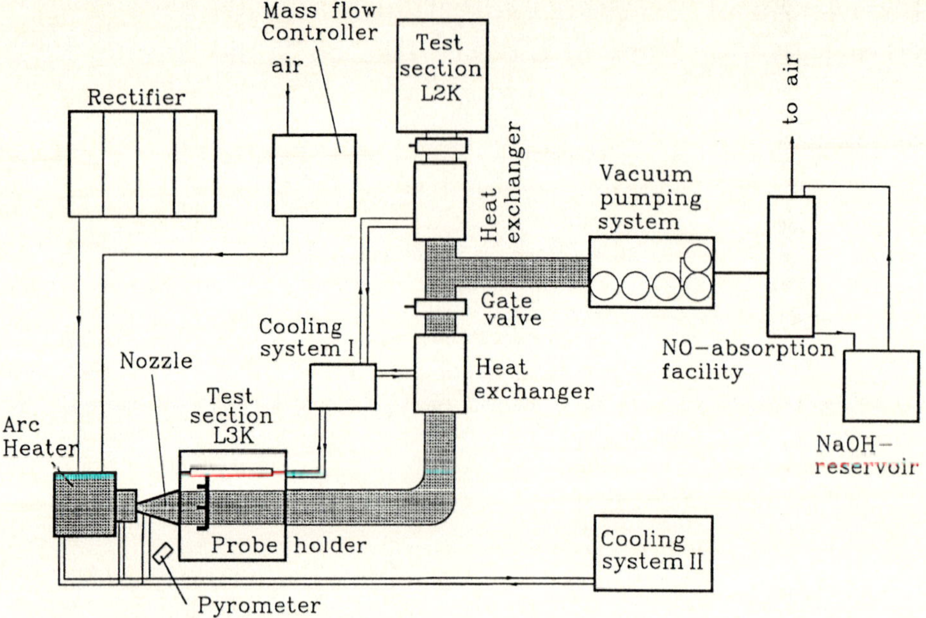

Fig. 1. Experimental set-up of the LBK

A high pressure DC-arc heater with a maximum power of 1 MW is connected to a tyristor-controlled rectifier system. The test gas with a maximum mass flow rate of 75 g/s is injected tangentially into the space between two hollow electrodes to force the rotation of the arc (Fig.2). The rotation of the arc foot points on the inner surface of electrodes reduces the erosion rate of the electrodes, i.e. flow contamination, but produces swirl in the flowfield. This will be reduced by the expansion of the gas inside the settling chamber and the nozzle. The length of the arc and its stability is controlled by magnetic coils placed around the electrodes.

Mass flow rate, reservoir pressure, current and voltage of the arc are continuously measured and controlled during the tests. A Pitot probe, model holder and heat flux probe are mounted on the same shaft and can be moved into the flowfield successively. The surface temperature of models has been determined using two spectral pyrometers. The low temperature pyrometer works in the wavelength range of 2000-4500 nm and measures temperatures between 353 K and 1273 K. The other spectral pyrometer can be used for temperatures between 1373 K and 2273 K and has a spectral range of 830-1030 nm. An additional two-color (920 nm, 1040 nm) pyrometer with a measurement range between 1173 K and 2273 K, which eliminates the influence of emissivity changes on the measurement has been used for control measurements and tests on samples with unknown emissivity. The heat flux probe consists of a copper slug with two embedded thermocouples. This cylindrical slug is mounted adiabatically in a water cooled holder having the same geometry as the sample holder in order to produce the same shock stand-off distance. The heat flux is evaluated from the temperature history of thermocouples under the assumption of one-dimensional heat conduction in the cylinder. Heat flux rates up to 1800 kW/m^2 have been measured in the LBK.

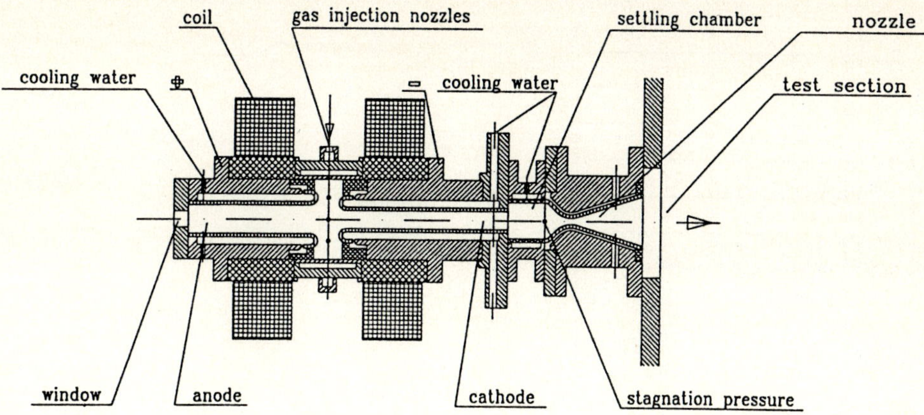

Fig. 2. The arc heater

3. Numerical computation

The determination of vibrational temperature and species concentrations would be a big contribution to understand gas-surface interaction phenomena. This can experimentally be achieved applying sophisticated spectroscopic measurement techniques like CARS, LIF, etc., which can satisfy such requirements partly. A numerical code including options for flow solutions based on chemical equilibrium, frozen chemistry and chemical nonequilibrium has been used to fill the gap (Bade and Los 1975). N_2, O_2, N, O, NO, NO^+ and electrons are the species of argon free air model for temperatures below 6000 K. At higher temperatures four additional ions, i.e. N_2^+, O_2^+, N^+ and O^+, are additional species. The measured mass flow rate, reservoir pressure and the nozzle geometry are the input parameters for the computation. The flow is always assumed to start from a state of thermochemical equilibrium in the reservoir. The nonequilibrium solution is obtained under the basic approximation of a quasi one-dimensionality of the flowfield. A perturbation method is used until the departure from equilibrium is large enough to allow numerical integration without stiffness problems. The inviscid nozzle flow solution is coupled with the boundary layer displacement thickness resulting from an approximate laminar boundary layer calculation.

The calculation of stagnation point conditions begins with an approximate normal shock solution assuming either equilibrium flow behind the shock or a frozen shock with the species mole fractions behind the shock being equal to those in front of it. Gas parameters on the surface are calculated using the incompressible Bernoulli equation between the point directly behind the shock and the stagnation point. This equation is reasonably accurate to compute the actual isentropic flow behind the shock for typical Mach numbers in the LBK.

Computed temperature distribution for the test condition, at which spectroscopic measurements using the LIF system were performed, shows a nearly frozen flow in the divergent part of the nozzle (Fig.3). Good agreement has been achieved between measured and computed free-stream temperatures by our code NATA and other codes of ONERA (Devezeaux et al. 1992).

4. Results and discussion

The test conditions have been defined by the temperature and pressure on the sample surface. The three test points have been chosen in such a way that either the temperature or the pressure is to be varied while keeping the other parameter nearly constant. Silicon carbide (SiC) and carbon fibre reinforced silicon carbide (C-SiC) samples have been tested. Computed and measured test parameters are listed in Table 1.

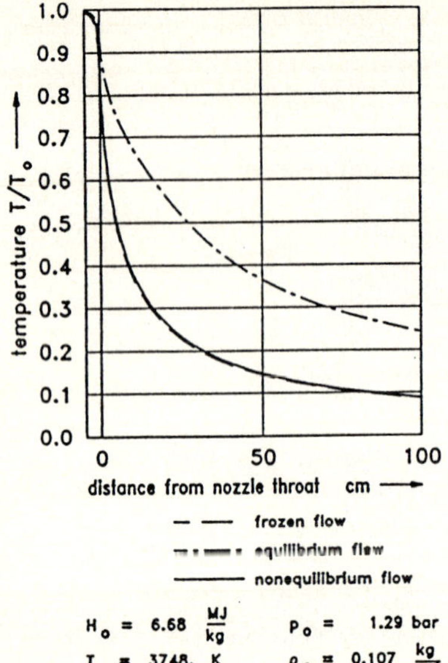

Fig. 3. Computed temperature distribution along the flow axis

Table 1. Computed and measured test parameters

Samples	SiC-1	SiC-2	SiC-3	C-SiC-1	C-SiC-2	C-SiC-3
h_0 (MJ/kg)	6.76	3.96	9.15	6.26	4.05	9.65
T_0 (K)	3631	2960	4540	3508	2995	4751
$^n N_2$	0.6800	0.7527	0.6530	0.6907	0.7504	0.6511
$^n O_2$	0.0562	0.1681	0.0094	0.0734	0.1649	0.0048
$^n N$	0.13×10^{-6}	0.13×10^{-9}	0.88×10^{-6}	0.56×10^{-7}	0.18×10^{-8}	0.89×10^{-4}
$^n O$	0.2235	0.0385	0.3171	0.1931	0.0429	0.3303
$^n NO$	0.0403	0.0406	0.0205	0.0428	0.0418	0.0137
$^n NO^+$	0.29×10^{-6}	0.14×10^{-7}	0.58×10^{-6}	0.0428	0.0418	0.0137
p_s (mbar)	24.3	73.1	75.3	24.2	73.6	74.7
T_s (K)	1563	1608	1873	1668	1688	1953
ΔM (g/m^2·h)	3.75	2.5	63.8	738.	1022.	789.

The enthalpy h_0 and temperature T_0 in the reservoir are calculated from the measured mass flow rate and reservoir pressure and nozzle throat diameter. Species mole fractions n_i behind the bow shock are determined by performing a nonequilibrium flow computation in the free stream and a frozen shock calculation without relaxation phenomena behind it. The surface pressure, i.e. Pitot pressure p_s and surface temperature T_s, are directly measured parameters. The erosion rates ΔM are determined by weighing the samples before and after each test cycle. A test cycle begins with moving the sample holder into the flow field in such a way that the temperature gradient is smaller than 6°C/s. Therefore this transition phase to reach the final surface temperature has a duration of 5 minutes. After a testing time of 20 minutes at constant temperature and pressure, the sample is taken out of the flowfield with the reversed transition procedure.

Comparable heat flux rates of 780 kW/m^2 and 800 kW/m^2 have been measured with the heat flux probe at the test conditions for SiC-1 and SiC-2 samples successively. The effect of low enthalpy levels on the heat flux rate for SiC-2-conditions is compensated by a higher surface pressure of 73.1 mbar. Because of low surface temperatures the influence of different gas composition on the erosion rate of SiC-1 and SiC-2 samples is negligible. The high surface temperature of the SiC-3 sample caused a high erosion rate, which decreased during two further cycles on the same sample. This phenomenon can only be explained by the development of a protective SiO_2-layer on the sample surface. The detection of oxygen by energy dispersive X-ray analysis (EDX) on the surface of SiC-samples indicates the presence of such a layer (Table 2). EDX-analysisses of the material composition inside these samples have not detected any oxygen lines and indicate a layer thickness smaller than 1 μm.

Sample	SiC-1	SiC-2	SiC-3
n_C	0.1151	0.2432	0.1003
n_O	0.5531	0.4884	0.4686
n_{Si}	0.3319	0.2684	0.4311

Table 2. Atomic species concentration on the surface of SiC-samples measured by the EDX-method

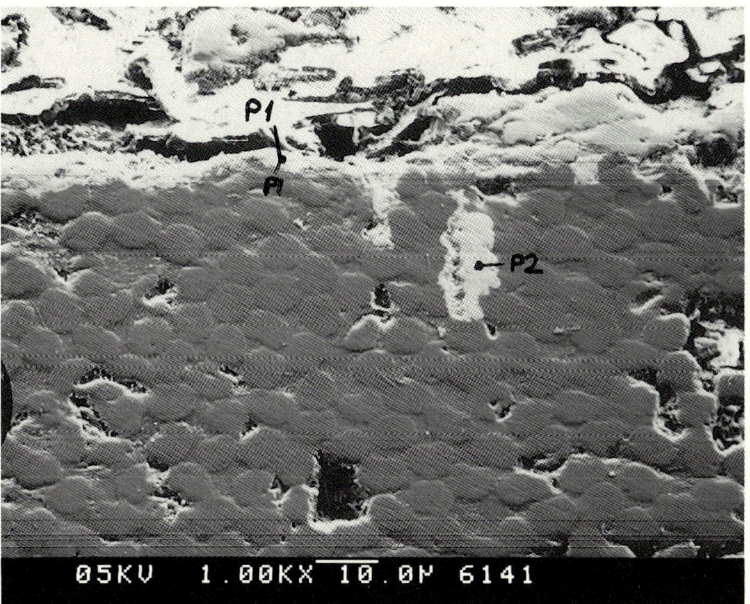

Fig. 4. The structure of the C-SiC-1 sample

C-SiC-samples have much higher erosion rates than the SiC-samples (Table 1). Strong oxidation of the carbon fibres producing carbon oxide gases is responsible for these high mass losses. Compared to the C-SiC-1 and C-SiC-2 samples the C-SiC-3-sample was exposed to an air flow with a higher mole fraction of molecular oxygen and has the highest weight loss. This fact shows that molecular oxygen is more agressive than atomic oxygen for C-SiC-samples. Oxygen, carbon, silicon elements have also been detected by EDX-measurements on the surface of these samples. The C-SiC-samples have been produced by the liquid impregnation technique (Krenkel and Schanz 1992). Fig.4 shows the inner structure of the C-SiC-1 sample and spectroscopically analysed points on it. The corresponding atomic species concentrations are listed in Table 3.

The facts, that a small amount of oxygen is detected on the free surface SiC area (point 1) and that a SiC-island in the sample (point 2) contains no oxygen, are again an indication of active oxidation on the surface due to chemical reaction with the high enthalpy flow.

	Point 1	Point 2
n_C	0.7184	0.6832
n_O	0.0228	0.0000
n_{Si}	0.2588	0.3168

Table 3. Atomic species concentration at two points in the C-SiC-1 sample

Acknowledgement

The author gratefully acknowledges the effort of Dr. von Bradke from the Institut für Technische Thermodynamik of the DLR-Stuttgart for performing EDX-analysisses. The financial support of the ESA through CNES for some parts of this work is also gratefully acknowledged.

References

Auweter-Kurtz M (1991) Qualification of thermal protection systems by laboratory simulation techniques. Space Course on Low Earth Orbit Transportation and Orbital Systems, Aachen

Bade WL, Los TM (1975) The NATA Code. NASA CR-2547

Devezeaux D, Gülhan A, Kindler K, Koch U, Sagnier Ph (1992) High enthalpy experiments in LBK and comparison with numerical simulations. Euromech 296, Göttingen

Gülhan A (1993) The arc heated wind tunnel LBK and test equipment. DLR IB-39113-93A06

Koppenwallner G (1990) Low density facilities. The Third Joint Europe/US Short Course in Hypersonics, Aachen

Krenkel W, Schanz P (1992) Fiber ceramic structures based on liquid impregnation technique. Acta Astronautica 28, 159-169

Olivier H, Vetter M, Grönig H (1991) High-enthalpy testing in the Aachen shock tunnel TH2. Proc. Ist European Symposium on Aerothermodynamics for Space Vehicles, ESTEC, Noordwijk

Wagner DA, Smith RK, Gunn JA, Hasegawa S (1990) Hypersonic test facility requirements for the 1990's. AIAA Paper 90-1389, Seattle

An Attractor-Driven Approximation for Turbulent Burst Dynamics in a Supersonic Free Shear Layer

J.A. Johnson III[*], L.E. Johnson[*], and J. Zhang[†]
[*]CeNNAs, Florida A&M Univ., Tallahassee, FL 32310 USA
[†]City College, CUNY, New York, NY 10031 USA

Abstract.   LIF measurements of density fluctuations in a turbulent supersonic free shear layer produced in a Ludwieg tube's asymmetric nozzle provide evidence of turbulent shocklets and the underlying turbulent burst dynamics. Correlations of shocklet phenomena with changes in chaotic dimension motivate a quasi-one dimensional approximation for an estimator of shocklet properties which makes predictions consistent with the measurements reported and has implications for shocklet observations in other systems.

Key words: Compressible turbulence, Supersonic flow, Chaotic dimension

1. Introduction

For supersonic flows, regions of strong compression can be formed in free shear layers resulting in the formation of shock waves in the flow. The subsequent propagation of such shocks or eddy shocklets will interact with the flow field and other structures (Hussaini 1986, Lele 1989). Turbulent bursts are also believed to exist not only in wall boundary shear flow but also in free shear layers (Lee et al. 1990). In the flow visualization studies of Kim et al. (1971) and of Corrino and Brodkey (1969), it was found that, within the viscous sublayer of the wall boundary, the low velocity fluid tends to accumulate into longitudinal structures known as streaks. This process is associated with a major part of the Reynolds stress and turbulent energy production. We, therefore, are motivated to conduct experimental research aimed at identifying the coherent structures, quantitatively describing the spatial and temporal evolution of the structures and establishing the role of each of the structures in the production of turbulence.

2. Set-up and procedures

Our Ludwieg tube-wind-tunnel is a conventional shock tube modified by inserting a layer-spilling asymmetric nozzle into the test section as described in Johnson et al. (1988). At a time determined by the ratio of the nozzle's throat to its exit area, which is about 4 to 6 ms, the nozzle is choked, the mass flow rate is frozen, and stable, steady supersonic flow is established in the exit region. The high pressure section is filled by an admixture of 98% N_2 and 2% NO_2 which provides a target for the incident light and is the source of the fluorescence.

The free shear layer is generated by the nozzle producing a plane two-dimensional boundary layer which detaches at the sharp corner of the nozzle. The shear layer is thin compared to the step of the nozzle and it is not affected by the presence of the tube's wall. The flow at the outlet of the nozzle and flow in the free shear layer are characterized by the angles. By changing the nozzles in the Ludwieg tube, Mach numbers in the range 1.6 to 2.5 are obtained. For a given nozzle, we achieve a variation in Reynolds number by changing the starting pressure, P_4, in the high pressure section from 10^6 to 10^8 in free-stream unit Reynolds number. The flat free shear layer from the outlet of the nozzle can be produced by controlling the ratio of filling high pressure P_4 to filling low pressure P_1. In fluorescence measurements, with the high pressure section filled by an admixture of 2% NO_2 and 98% N_2 the high pressure and low pressure which can give flat free shear layer in our facility become $P_4 = 51.84$ psi, and $P_1 = 110.4$ torr respectively. Using the

Shock Waves @ Marseille I
Editors: R. Brun, L. Z. Dumitrescu

appropriate downstream and upstream static pressure settings for the nozzle, we know that for our $M = 2$ nozzle, the starting stage is 6 ms long; after 6 ms, a supersonic flow with $M = 2$ will be established. Simultaneously, the boundary layer which is appropriate for $M = 2$ free-stream will be formed after the starting stage.

The optical system used to collect the fluorescence signals from the test region is set as follows: A laser light beam from an argon ion laser at 488.0 nm wave length is split into three beams with the same diameter and the same intensity; they are separated by 1.5 mm. The cross section of the set of beams is a triangular form. The beams are sent through the free shear layer along the y direction. The fluorescence light signals emitted from arranged measuring points are collected by sets of optical systems, then passed on to the photomultiplier tubes. A filter which can only allow light with wavelength longer than 5900 Ångstrom to pass is placed before the photomultiplier tubes to remove scattered laser light.

In general, the NO_2 spectrum extends from 800 nm to near 400 nm and this entire 400 nm region contains a very high density of lines. Within the range 570 nm to 670 nm the spectrum is mainly the vibronic bands (Smally 1975). In our measurement, the NO_2 molecules, which are mixed with N_2, provide a target for the incident light and are pumped to vibrational excitation states by the Argon ion laser at 488 nm. The intensity of the fluorescence emitted by NO_2 molecules, at a given temperature, depends on the density of the NO_2 molecules in the flow and the intensity of the incident laser beam. Therefore, the density or density fluctuation of NO_2 can be calibrated so as to represent the flow density and density fluctuation.

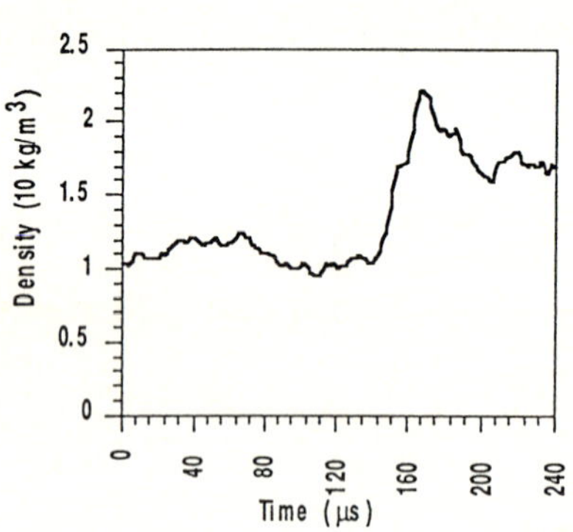

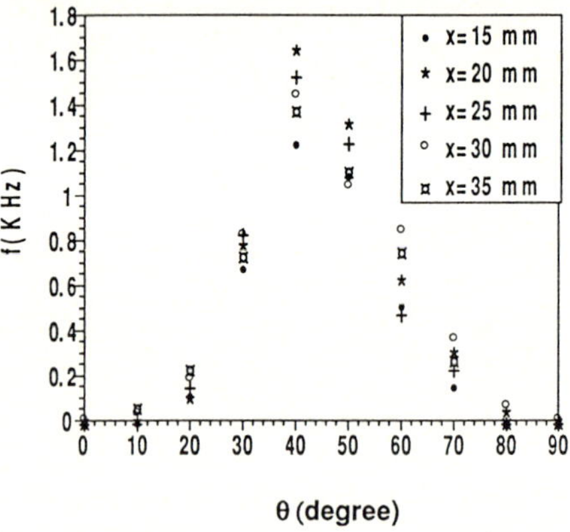

Fig. 1. A sample shocklet profile at $x = 2.0$ cm downstream from the nozzle lip

Fig. 2. The frequency of shocklet production at various shocklet Mach angles

3. Data and analysis

3.1. Shocklet dynamics

A typical shocklet profile is shown in Fig.1. In this figure, the density measurements are taken at the spatial position 2.0 cm downstream from the nozzle's lip. Since point density measurements were taken, the signals give good spatial resolution. When shocklets appear in the outer layer of the shear flow, they seem to show up as a sharp increase in the density. The apparent times of appearance of shocklets are random. This result agrees with the association of shocklets with eddies convecting in shear flow; since eddies appear in a random manner, their interactions with the flowfield are also random.

Our measurement results show that the shocklets have a localized spatial and directional distribution which varies with spatial position and angular direction. The velocity of a shocklet can be represented as $V_s = M_s a$, where a is the local speed of sound in the supersonic free stream, and the Mach number of shocklets is given by $M_s = (\sin \theta)^{-1}$. Fig. 2 shows that the mean frequency of shocklets varies with the angle θ, which is the angle between the streamline of main flow and the shocklet. This angle is related to the velocity of the shocklet, or, more precisely, the velocity difference between the mean flow velocity and the convective velocity. The angular distribution of shocklets suggests that within a certain velocity range the shocklets have relatively higher concentration. From our measurements, the shocklets are mainly distributed in the angle between 20 and 70° and the associated velocities are between 276 m/s to 520 m/s. The maximum value is at about 38°, which is about 422 m/s. Fig. 2 also shows that the angular distribution varies at different streamwise locations. Near 28° the distribution increases sharply and then decreases relatively slowly at angles larger than 40°.

The distribution of shocklets varies in both axial and transverse location. The trend that the shocklets concentrate near the upper edge of the shear layer suggests that only the small size structures can generate shocklets. Near the upper edge, the mean shear flow speed is supersonic and the structures originating from the lower layer with large size are broken during the motion from the low layer to the upper region. The production of shocklets thereby mainly remains in the convective motion of small size structures. Fig. 3 shows the mean shocklet frequency distribution along the streamwise direction (where R_1 is Reynolds number based on the free stream flow at the outlet of the nozzle of 7.6×10^6, $R_2 = 2.3 \times 10^6$ and $R_3 = 7.1 \times 10^5$). The mean apparent frequency sharply increases at about 1.5 cm downstream from the nozzle's lip, and slowly decreases as the distance increases. Beyond 2.0 cm from the nozzle's lip, the mean frequency slowly decreases. This is apparently caused by a slowing down of the mean velocity of the shear flow. Therefore, only the eddies with a lower convective velocity can satisfy the condition to generate the shocklet. The mean frequency distribution increases as the Reynolds number increases. At higher Reynolds number flow, more small size structures are produced in the shear layer. The mean periods associated with the Reynolds numbers are 0.133 ms, 0.152 ms and 0.189 ms.

Experimentally, one expects streaky structures and subsequent burst process to have different velocity and density from the environmental fluid field. This is found to be the case as shown in Figs.4 and 5. In those plots, relative large structures are presented; when they move to the upper layer during the convection, they break up into smaller structures. To distinguish the burst, a normalized parameter is defined as $H = \frac{|\rho'|^2}{(\rho_0)^2}$ where ρ_0 is the density at initial state. When burst measurements at different streamwise locations are studied, we find, near the nozzle lip where the shear flow has higher shear rate, the density measurements show more thin peaks; further from the nozzle's lip the peaks are grouped. This means that in high shear rate regions, the streaky structures more frequently break up, or bursts have higher production frequencies. At other regions further away, the shear rate is lower and the flow is more vorticial; bursts have lower production frequencies and the structure scales associated with the ejection are larger.

3.2. Dynamical analysis of turbulence in free shear flow

The dimension of a chaotic or turbulent system is the first level of knowledge necessary to characterize its properties. The dimension is also a lower bound on the number of essential variables needed to model the dynamics. From a time-delay approach (Ruelle 1971), a single time series $x(t)$ can be converted into a multiple time series $x(t) = [x_0(t), x_1(t), \ldots, x_{n-1}(t)]$ by selecting a delay τ and defining $x_k(t) = x(t + k\tau)$ for purposes of determining complexity. A dimension estimator D_2 can be computed by the correlation integral from $D_2 = \lim_{l \to 0} \dfrac{\log C(l)}{\log l}$ where $C(l)$ is a

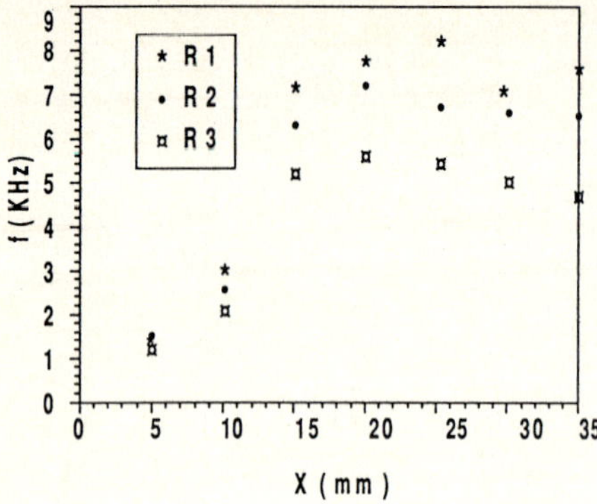

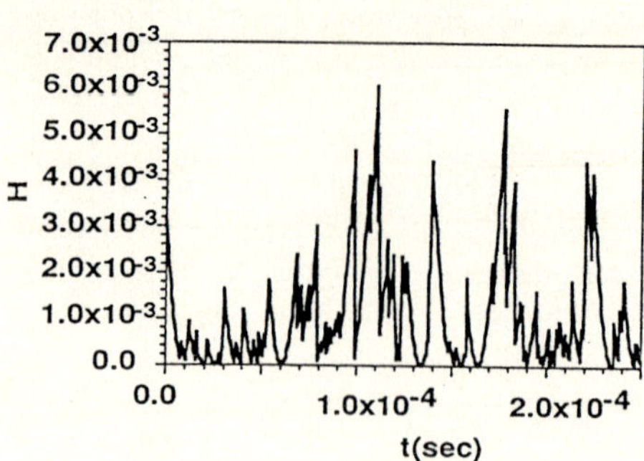

Fig. 3. The mean production frequency distribution of shocklets along the streamwise direction

Fig. 4. Evolution of bursts at $y = 1.5$ mm and 1.0 cm downstream from the nozzle lip

correlation integral function defined by $C(l) = \lim\limits_{N \to \infty} \dfrac{1}{N^2} \sum\limits_{ij} \theta[l - |\boldsymbol{x}_i - \boldsymbol{x}_j|]$ and where $\theta[l - |\boldsymbol{x}_i - \boldsymbol{x}_j|]$ is the Heaviside function. D_2 provides a lower bound on the degree of complexity of the system.

The results of our determination of the axial evolution of the D_2 dimension in the shear layer are plotted in Fig 6. In this plot four sets of data were used; these were taken with transverse location at $y = 4.5$ mm, $y = 3.0$ mm, $y = 1.5$ mm, and $y = 0.0$ mm. The run conditions for these measurements were the same. With regard to variations in the transverse direction, in the lower layer the flow speed is subsonic, structures have relative larger size and the motion is relatively simple. However, at the upper layer region the flow speed is supersonic; a strong intermittent effect is causing fluctuations in density, pressure and velocity, and the shocklet generated in this region also interacts with those structures. All those processes and effects make the flow relatively more complicated.

Generally stated, the dimension has higher value at $x = 3.5$ cm than that at 0.5 cm. More specifically, the results show that at a certain value x_o, which is 1.8 ± 0.4 cm in this free shear layer, downstream from the nozzle lip the dimension increases sharply. This means that the shear layer detaching at the nozzle lip has a more complicated structure than that near the nozzle lip. When the shear layer detaches before it becomes fully developed turbulence, it goes through a transition process forming a transition region. Although the flow in the transition region should not be laminar under our supersonic conditions, it has a relatively simple structure. In this region, if we neglected the structures coming from the boundary layer of the nozzle wall, the instability is still at an early stage and the structures formed due to the instability waves are quite ordered, i.e. the structures have a relatively higher order of coherence. When these structures convected downstream, they break up due to the entrainment motion and interaction with other structures. Therefore, at the region further from the nozzle's lip, the turbulent structures are less coherent and the dimension of the system is higher.

3.3. An analytical model for shocklets

To model the shocklet process, we can consider that the vortex structures or eddies are rigid. The interaction is generally not strong. Therefore, it will be assumed that the effects of gas dissociation and relaxation phenomena can be neglected. We also consider that the Reynolds number is high, the flow is ideal gas flow and that the stagnation temperature does not exceed about 2000 K.

Let us in addition only concern ourselves with the formation of the shock waves and ignore the internal structures of the shocks; we treat only the eddy interactions with the flow field. For

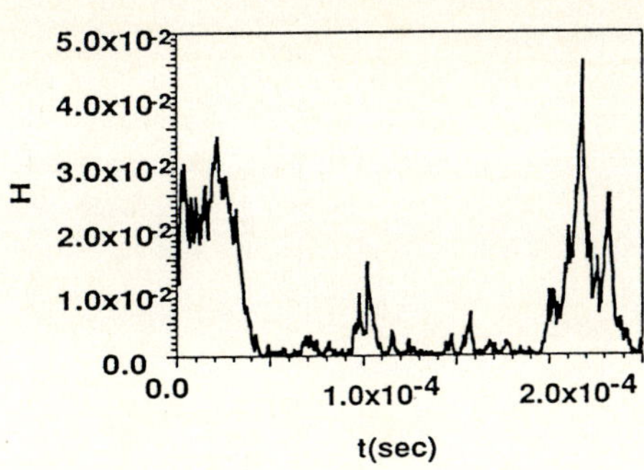

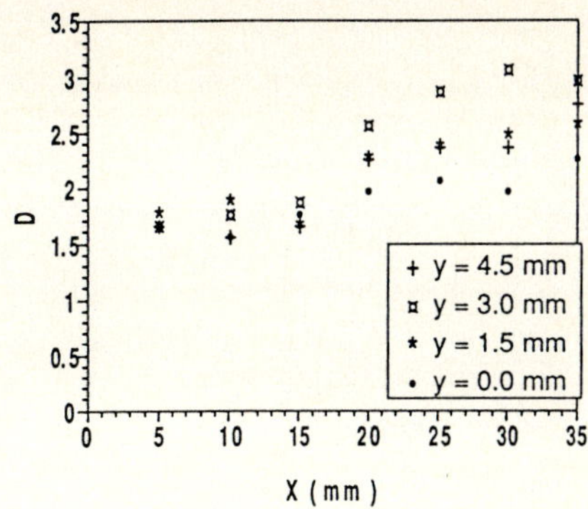

Fig. 5. Evolution of bursts at $y = 1.5$ mm and 2.5 cm downstream from the nozzle lip

Fig. 6. The D_2 chaotic dimension variation at different y along the streamwise direction

a turbulent shear layer there is the possibility that we can consider separately the processes of generation of the pressure waves and the subsequent development of the shocks.

In most cases the boundary approximation can be applied to shear layer problems. We can consider that, in a turbulent shear zone, the mean velocity u_1 is a function of the x_2 position coordinate only. The necessary condition for generation of shocklet in shear layer is that we have U greater than a_0, the speed of sound at infinity. For the sake of analytical convenience, we suppose that all mean point properties of the motion in the shear zone are independent of x_1, x_3 and time t, being a function of x_2 alone, the position coordinate normal to the zone.

To further simplify the analyses we can neglect products of two fluctuating quantities with zero means, on the understanding that these will be small (in mean square) compared with the products of mean and fluctuating quantities. The terms of generation, the processes of convection and refraction of the shock wave by the mean flow, and by variation with x_2 of the local mean speed of shock wave are retained. With these assumptions and the standard dimensionless form, including $A^2(y_2) = \frac{\bar{a}^2}{a_0^2}$, $M = \frac{U}{a_0}$ one achieves

$$\left\{ \left(\frac{\partial}{\partial \tau} + V \frac{\partial}{\partial y_1} \right)^2 - \frac{1}{M^2} \frac{dA^2}{dy_2} \frac{\partial}{\partial y_2} - \frac{A^2}{M^2} \frac{\partial^2}{\partial y_i^2} \right\} \log \left(\frac{p}{p_0} \right) = \gamma \frac{\partial v_i}{\partial y_j} \frac{\partial v_j}{\partial y_i} \tag{1}$$

In free shear layers, the sizes of turbulent eddies are widely distributed from the dissipation range to the flow dimension. The shock waves generated by the interactions of the eddies and the flow should be the function of eddy wave number, k, and Mach number of the shear flow. To study the mechanism of shocklets and the eddy wave number, we now can therefore us a Fourier transformation. By using the standard definition (Lighthill 1958), the equation relating the generalized Fourier transform ϖ and Γ corresponding to Eq.3 is

$$\frac{d^2}{dy^2} \varpi(y, \mathbf{k}, n) + \left\{ \frac{M^2}{A^2} (n + V k_1)^2 - k^2 - \frac{1}{A} \frac{d^2 A}{dy^2} \right\} \varpi(y, \mathbf{k}, n) = -\frac{M^2}{A^2} \Gamma(y, \mathbf{k}, n) \tag{2}$$

where $k^2 = k_1^2 + k_3^2$, A and V are functions of y only.

From our experimental results using both shadowgraph and multi-point fluorescence measurements we know that the shocklets are generated in the upper edge of the shear flow and propagate into main flow. Thus Eq.2 becomes $\frac{d^2 \varpi}{dy^2} + \left\{ M^2 (n + k_1)^2 - k^2 \right\} \varpi = 0$. Since $\varpi(y)$ must be bounded as $y \to \infty$, the solutions to this equation must be either exponentially decreasing or oscillatory according as the coefficient of ϖ is negative or positive. From this equation, it is also clear that the oscillatory solution corresponds to a radiated pattern of shock waves and the Mach number, M. For given Mach number the oscillatory solutions are associated with wave number $\mathbf{k}$ and frequency n such that $M^2(n + k_1)^2 - k^2 > 0$.

If the the flow is supersonic, it is sufficient to neglect (for the moment) the evolution of the eddy pattern as it is carried along; with this, the frequency of the components of wave number $\mathbf{k}$ is just the frequency with which the wave component is swept past the fixed observation point at a rigid convected pattern. In this way we have $n = k_1 V_c$. Let θ be the angle between the vector wave number $\mathbf{k}$ and the direction of the mean velocity. Then $\cos\theta = k_1/k$ and this condition becomes $\cos^2\theta > [M(1 - V_c)]^{-2}$, in good agreement with our results (Fig.2).

4. Implications

From LIF, we find the shocklets observed in a supersonic free shear layer to be strongly concentrated in the upper edge of the layer where the flow has a higher shear rate and the mean flow speed is supersonic. We also find a correlation between the streaks and shocklets; the streaks are formed at the lower regions of the free shear layer and break into small structures near the upper region where the flow speed is supersonic.

There is furthermore evidence for a clear possible connection between dynamical systems theory and fully developed turbulence. Aubry et al. (1988) and Holmes (1989) showed that low dimensional methods might indeed be applied in concert with the proper orthogonal decomposition to the study of a class of turbulent flows possessing coherent structures. The instantaneous field in shear layer can be expanded in so-called empirical eigenfunctions. The results of such an expansion allow us to retain correct dynamical representation of the turbulence production phenomenon and keep as few modes as possible in order to obtain a low-dimensional system. A proper way to do that is find the cutoff modes from the experimental measurement.

Our results now define regions in the free shear layer where such measurements can be attempted, viz, regions with low dimensionality and with low shocklet production frequency.

Acknowledgements

This work was supported in part by grants from the NASA, USA.

References

Aubry N, Holmes P, Lumley JL, Stone E (1988) The dynamics of coherent structures in the wall region of a turbulent boundary layer. J. Fluid Mechs. 192:115

Corino ER, Brodkey RS (1969), A visual study of turbulent shear flow. J. Fluid Mechs. 37:1

Holmes PJ, Marsden JE, Scheurie J (1989) Hamiltonian dynamical system. Cont. Meth; 81: 213

Johnson III JA, Zhang Y, Johnson LE (1988) Evidence of Reynolds number sensitivity in supersonic turbulent shocklets. AIAA J. 26:502

Kim HT, Kline SJ, Reynolds WC (1971) The production of turbulence near a smooth wall in a turbulent boundary layer. J. Fluid Mechs. 50:133

Lee MJ, Kim J, Moin P (1990) Structure of turbulence at high shear rate. J. Fluid Mechs. 216:516-583

Lele SK (1989) Direct numerical simulation of compressible free shear flows., AIAA Paper 89-0374

Lighthill MJ (1958) Fourier analysis and generalized equations. Cambridge University Press

Ruelle D, Takens F (1971) On the nature of turbulence. Comm. Math. Phys. 82:137-151

Smalley RE, Wharton L, Levy DH (1975) The fluorescence excitation spectrum of rotationally cooled NO_2. J. Chem. Phys. 63:4977

Effect of Nozzle Configurations on Unsymmetrical Supersonic Flows

Y. Watanabe, S. Matsuo and F. Higashino
Department of Mechanical Systems Engineering, Tokyo Noko University, Koganei-shi, Tokyo 184, Japan

Abstract.   In the present study, to analyze the effects of the flow Mach number at nozzle entrance and the divergence angle of the nozzle on the flow; six kinds of nozzle with different divergence angles and wall lengths were used. Furthermore, the appearance and oscillation of a pseudo-shock wave was investigated. A Schlieren system, and a Mach-Zehnder interferometer composed of an Argon laser (CW) as the light source with both an acoustic-optic modulator (AOM) and an electro-optic modulator (EOM) were used to visualize the flow.

Key words: Supersonic nozzle flow, Pseudo-shock wave, Flow visualization

1. Introduction

The nozzle configuration of scramjet engines utilized for space vehicles may be unsymmetrical in general (Henry and Anderson 1973). In supersonic flows generated in such unsymmetrical nozzles, the interaction of shock waves and expansion waves with the boundary layer significantly affects nozzle performance. In the internal flow of nozzles and ducts, shock waves which oscillate in the flow direction are generated, and as the degree of interaction with the boundary layer becomes stronger, pseudo-shock waves (Crocco 1958) are generated. Problems of such supersonic flows may be characterized by a supersonic flow containing a shock wave on a nozzle wall.

Lewis et al. (1966) suggested that the supersonic underexpanded jet from an unsymmetrical nozzle may be a useful means of producing lift without the necessity for physically deflecting the jet. Wlezien et al. (1988) have analyzed the noise generation characteristics of supersonic jets from nonaxisymmetric nozzles in a supersonic wind tunnel. Hopkins et al. (1979) have investigated the supersonic nozzle flow generated in a detonation tube to simulate a scramjet engine and measured both pressure and heat transfer over a simulated cowl, and over an afterbody. The study of unsymmetrical jet flow is important not only for basic research in gasdynamics but also for mechanical applications.

In the present study, to analyze the effects of flow Mach number at the nozzle entrance and that of the nozzle configuration on the flow, six kinds of nozzle with different divergence angles and wall lengths were used. Furthermore, the appearance and oscillation of a pseudo-shock wave generated as a result of interaction of the supersonic flow with the boundary layer on the wall was investigated experimentally.

2. Experimental apparatus and method

A shock tube was used to generate supersonic flows in two-dimensional unsymmetrical nozzles. A schematic diagram of the experimental apparatus is shown in Fig.1. The low-pressure channel is 3000 mm long and 50 mm ID. The high-pressure chamber is 1000 mm long and 70 mm ID. The test chamber is 260 mm high, 40 mm wide and 630 mm long. Piezo-electric pressure transducers (PCB 113A24) were flush mounted on the upper and lower walls of the nozzles.

The supersonic flowfield in the nozzle at a certain instant was observed by taking sequential Schlieren photographs. In addition, a Mach-Zehnder interferometer (Sigma Koki) was used for taking infinite-fringe interferograms (Watanabe et al. 1992). The Mach-Zehnder interferometer

Shock Waves @ Marseille I
Editors: R. Brun, L. Z. Dumitrescu

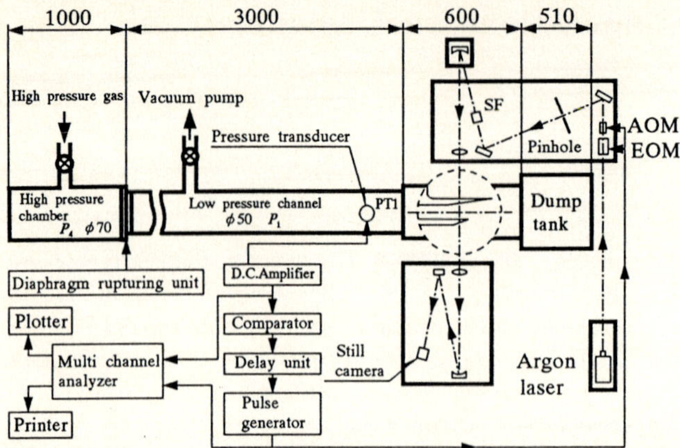

Fig. 1. Schematic diagram of experimental apparatus: p_1 - initial pressure of low pressure section; p_4 - initial pressure of high pressure chamber; AOM - acoustic-optic modulator; EOM - electro-optic modulator; PT1 - pressure transducer; SF - spatial filter. Dimensions in mm

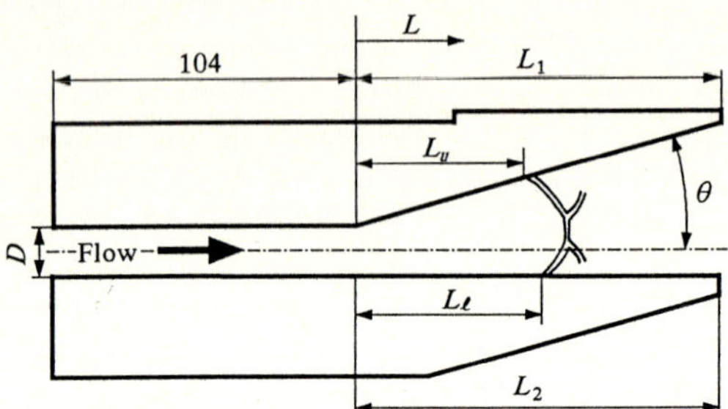

Fig. 2. Detail of nozzle configuration: L - distance from the nozzle entrance; L_1 - length of the upper wall; L_2 - length of lower wall; L_u - distance from nozzle entrance to leading shock wave on the upper wall; L_l - distance from nozzle entrance to leading shock wave on the lower wall; θ - divergence angle of nozzle; D - height of nozzle entrance. Dimensions in mm

consisted of an Argon laser (Spectra Physics 164-08: 600 mW) as the light source, and an acoustic-optic modulator (Japan Laser AOM-80B) and an electro-optic modulator (Conoptics Model 10) as shutters. Exposure time was changed at will by controlling the modulators through an electrical pulse generator. Using an AOM along with an EOM, it was possible to minimize laser power coming though these modulators when the shutters were closed. In Fig.1, the arrangement of the Mach-Zehnder interferometer is shown.

Table 1. Nozzle dimensions

	Upper wall length L_1 (mm)	Lower wall length L_2 (mm)	Divergence angle θ (degs)	Height of entrance D (mm)
Nozzle 1	220	147	15	29
Nozzle 2	125	80	15	16
Nozzle 3	125	125	10	16
Nozzle 4	125	125	15	16
Nozzle 5	125	125	20	16
Nozzle 6	125	80	20	16

In Fig.2, the configuration of the unsymmetrical nozzles is shown. The dimensions of the nozzles used in the present experiments are shown in Table 1. In the case of Nozzles 1 and 2, the flow Mach number could be changed from 1.18 to 1.53. In the case of Nozzles $3\cdots6$, the flow Mach number was 1.42. Air was used as both driver and driven gas.

3. Experimental results and discussion

Fig.3 shows infinite-fringe interferograms of the flow in the case of Nozzles $3\cdots6$. The expansion waves emanating from the corner of the upper wall interact with the boundary layer on the lower wall. Upstream of the shock waves a very stable flowfield is observed. Pseudo-shock waves are observed downstream as a result of interaction of the boundary layer with the shock waves. The flow Mach number immediately in front of the leading shock waves on the upper and lower walls was the same.

As seen from these photographs, the divergence angle of the nozzle affects the direction of the flow behind the shock waves. In the case of Nozzle 3 the flow is along the lower wall, while in the case of Nozzle 5 it is along the upper wall. Even though the Mach numbers immediately in front of the first shock waves on the upper and lower walls are almost the same, the regions of the expansion waves on the upper and lower walls are different due to the unsymmetrical configuration of the nozzle. As a result, the flow direction behind the shock wave may be deflected to the side of the upper or lower wall.

Furthermore, the flow direction behind the shock wave in the case of the short wall (Nozzle 6) is deflected to the side of the upper wall to a greater extent compared with the case of the long wall (Nozzle 5). It is found that the pressure-time history measured on the short lower wall is higher than that on the long lower wall. As a result, the flow direction behind the shock wave is deflected to the side of the upper wall, due to an increase in the shock angle of the leading shock wave on the lower wall.

Oscillation of the first shock waves was investigated by taking sequential Schlieren photographs. Figs.4a,b show the relation between both frequency and amplitude of the oscillation of the shock wave, and the flow Mach number at the nozzle entrance in the case of Nozzles 1 and 2, and the relation between both frequency and amplitude of the oscillation of the shock wave, and the divergence angles in the case of Nozzles $3\cdots6$, respectively. The abscissae in Figs.4a,b are the flow Mach number M_2 at the nozzle entrance and divergence angle θ, respectively. On the ordinates, the normalized amplitude $\Delta L/D$, along with the frequency f, of the oscillation are shown. As seen from Fig.4a, the amplitude and frequency is hardly affected by the flow Mach number for each nozzle. In Fig.4b, the amplitude and frequency of oscillation increase with the increase of the divergence angle. This is considered to be due to a change in the degree of interaction of the expansion wave with the boundary layer on the lower wall. Furthermore, the amplitude for the case of the short wall (Nozzle 6) becomes greater than that for the case of the long wall (Nozzle 5). This is considered to be due to an increase in turbulence caused by separation of the boundary layer behind the shock wave.

4. Conclusion

The present optical system with a Mach-Zehnder interferometer is successful for the visualization of the supersonic flow. In the case of nozzles with the same wall length, the amplitude and frequency of oscillation are hardly affected by the flow Mach number at the nozzle entrance, but increase with the increase of the divergence angle. In the case of nozzles with the same divergence angle, amplitude increases and frequency decreases with the decrease in length of the lower wall. The flow direction behind the first shock wave depends on the relation between the position of the first shock wave and the region of expansion waves on the wall. Furthermore, for nozzles having the same divergence angle the flow direction behind the shock wave is deflected to the upper wall

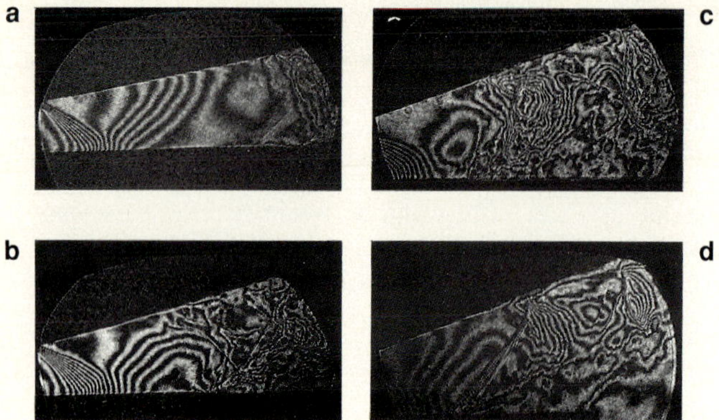

Fig. 3a-d. Infinite-fringe interferograms: P - laser power; d - exposure time; θ - divergence angle of nozzle. a) Infinite-fringe interferogram for Nozzle 3, P=600 mW, $d = 2$ μsec, b) Infinite-fringe interferogram for Nozzle 4, P=600 mW, $d = 2$ μsec, c) Infinite-fringe interferogram for Nozzle 5, P=600 mW, $d = 2$ μsec, d) Infinite-fringe interferogram for Nozzle 6, P=600 mW, $d = 2$ μsec

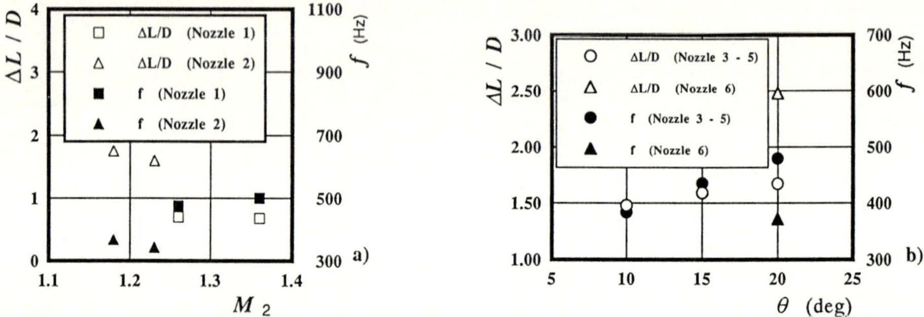

Fig. 4a-b. Frequency and amplitude of the first shock wave. a) Relation between frequency and amplitude of the shock wave, and the flow Mach number at the nozzle entrance. b) Relation between frequency and amplitude of the shock wave, and divergence angle

side to a larger extent in the case of the short lower wall compared to that of the case with the long lower wall.

References

Crocco L (1958) One-dimensional treatment of steady gasdynamics. In: Emmons HW (ed) Fundamentals of Gas Dynamics. B, Princeton Univ. Press, pp110-130

Henry JR, Anderson GY (1973) Design considerations for the airframe - integrated Scramjet. NASA TM X-2895

Hopkins HB, Konopka W, Leng J (1979) Validation of Scramjet exhaust simulation technique at Mach 6. NASA CR-3003

Lewis WGE, Herd RJ, Herbert MV (1966) Lift characteristics of asymmetric exhaust nozzles at high flight speeds. J. Roy. Aeron. Society 70: 1036-1040

Watanabe Y, Matsuo S, Higashino F (1992) Visualization of supersonic nozzle flows. Proc. 20th Symposium on Flow Visualization 12 Suppl: 147-150

Wlezien RW, Kibens V (1988) Influence of nozzle asymmetry on supersonic jets. AIAA Journal 26:27- 33

Radiative Heat Transfer from a Shock Layer Generated around a Projectile Launched in a Ballistic Range

Kimiya Komurasaki, Jiro Kasahara, Shujiro Yano and Toshi Fujiwara
Department of Aeronautical Engineering, Nagoya University, Nagoya 464-01, Japan

Abstract. Radiative heat transfer from a strong shock layer over a hypersonic flight vehicle is experimentally investigated using a ballistic range facility. A plastic projectile of 1.2 cm diameter is accelerated in this facility up to 5.3 km/s ($M = 15$), and the radiation spectra and global power emission from the launched projectile are measured. As a result, the radiant intensity from the shock layer became approximately 300 W/m^{2}sr, and the carbon radiation spectra were observed along with the air spectra. In order to take account of model-dimension effects of the projectile, which is two orders of magnitude smaller than the actual size of the vehicles, a scaling law has been developed by analytical and numerical considerations. The computed results are compared with the measured ones, and the validity of the calculation model including the wall condition is discussed.

Key words: Radiative heat transfer, Shock layer structure, Ballistic range, Hypersonic flow

Nomenclature

L	: shock layer thickness or shock stand-off distance, m
M	: Mach number
$[M]$	: total mole density, mole/m^3
p	: pressure, Pa
R	: nose radius of the blunt body, m
T, T_V	: translational and vibrational temperature, K
T_a	: geometric average temperature, K
u	: local flow velocity, m/sec
δ_T	: thermal boundary thickness, m
λ_V, λ_R	: vibrational-relaxation distance and reaction distance, m
τ_V, τ_R	: vibrational-relaxation time and reaction time, sec
ρ	: density, kg/m^3

1. Introduction

In recent years, hypersonic flight missions such as space planes and aero-assisted orbital transfer vehicles have been studied along with re-entry missions of re-usable space vehicles. In such hypersonic flights, aerodynamic heating in the shock layer generated around a space vehicle will be of the order of 10 W/cm^2, and thermal protection becomes a serious problem in developing these vehicles. Thus, reliable data on heat transfer rate from the shock layer are necessary for the vehicle design. However, the radiative heat transfer has not been well studied compared with the convective one.

In order to investigate the hypersonic flow and the accompanying radiation phenomena, experimental studes using the Nagoya university ballistic range have been sustained for three years (Chang et al. 1993). This apparatus is a two-stage light gas gun of 6.8 m in length, utilizing a driver powered by high pressure air stored in a reservoir. With this apparatus, high enthalpy shocks can be produced in the laboratory. Furthermore, computational work has also been conducted along with the experimental efforts. In the flowfield calculation, thermochemical non-equilibrium effect is taken account by Murayama et al. (1991) by employing the two-temperature model (Park

Shock Waves @ Marseille I
Editors: R. Brun, L. Z. Dumitrescu   © Springer-Verlag Berlin Heidelberg 1995

1993). The radiative intensity and absorption coefficient are computed by the program NEQAIR of Park (1984), and are three-dimensionally integrated (Sasoh et al. 1991).

From these studies, it was found that the radiation from the shock layer is sensitively affected by the thermochemical non-equilibrium phenomena. The thickness of the non-equilibrium region depends on the free-stream conditions: air pressure, temperature, and velocity. On the other hand, the shock layer thickness itself changes with the dimension of the flight model. Therefore, in the experiments, these flow conditions and model dimension should be adjusted to those of the re-entry vehicle. However, it is impossible to test with the actual size of the flight model using a ballistic range, so that the model-dimension effect must be considered. The objective of this study is to develop a scaling law for the shock structure, and to compare the measured radiative properties with the computed ones for an evaluation of the conventional chemical reaction and radiation model, including the wall boundary conditions.

2. Experimental apparatus

The ballistic range consists of a high-pressure reservoir, a pump tube, a launch tube, and a test section (vacuum chamber). A projectile of about 1 gram can be accelerated faster than 5 km/s, although the total length is only 6.8 m. A quick-valve system is utilized in the first stage instead of a diaphragm for the purpose of good reproducibility and operating convenience. Because of the system compactness, it takes only three hours to clean and to set up the facility again. When the quick-valve system acts, a free piston is driven by the high pressure air stored in the reservoir and compresses the helium gas filling up the pump tube. The peak pressure in the downstream end of the pump tube ranges from two to three hundred MPa with several hundred μsec duration. Owing to this high pressure, the diaphragm set at the coupling is ruptured and a projectile is accelerated in the launch tube toward the test section.

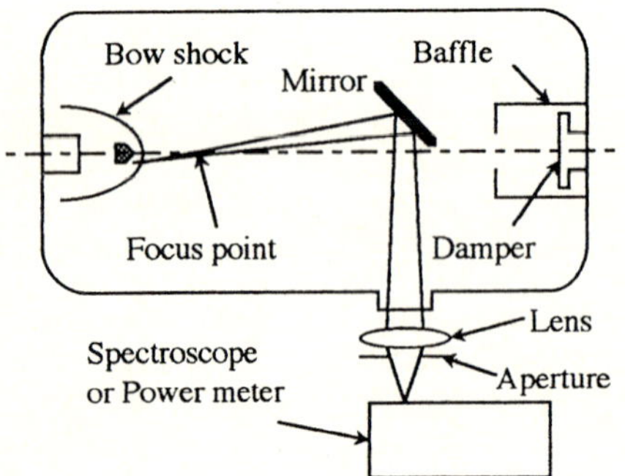

Fig. 1. Collection optics arrangement

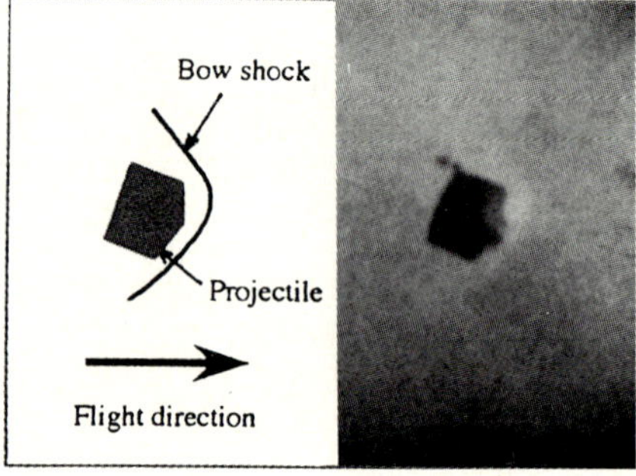

Fig. 2. Shadowgraph of the projectile. Flight speed is 5 km/sec, and exposure time is 100 nsec

The projectile flight velocity is measured by a laser-cut method and checked by the microwave stationary-wave method, which is adopted to obtain the profile of the projectile acceleration in the launch tube. Shadowgraphs are taken using a high-speed camera (NAC 20 Mfps) to watch the flight attitude of the projectile. The spectrogram is taken on an instant film (Polaroid, 57 type; ASA 3000) using a collection optics arrangement shown in Fig. 1. The spectrum in the range of 180 nm wavelength can be photographed in a single exposure using a spectroscope (JAS.Co. CT50N). Because of the film performance, the measurable spectrum is limited from 330 to 650 nm. The global power emission upward the projectile is measured with a Si-pin photo-diode power meter

(YOKOGAWA 3296), whose response time is less than 5 nsec. Recorded power profile has a peak when the projectile passes the focal point shown in the figure.

3. Measured results

The projectile flight velocity is set at 5 km/s for all the experiments. A shadowgraph of the projectile flying in the test section is shown in Fig.2. As seen in the figure, the projectile is hardly damaged, but slightly tilted with respect to the flight direction. This seems to be caused by the interaction with the copper wire which is used to get the trigger signal for the shutter of the high-speed camera. A bow shock can be seen ahead of the projectile.

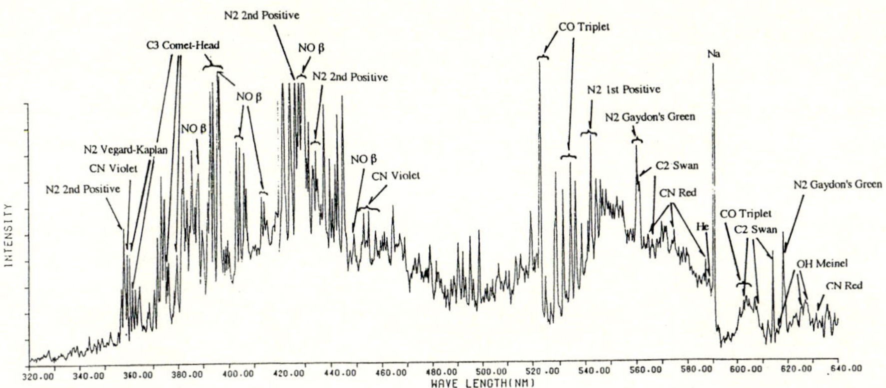

Fig. 3. Measured radiation spectra distribution. Spectral resolution is 0.5 nm at the 500 nm wave length

The measured radiation spectral distribution is shown in Fig.3. The spectra from some carbon compounds are identified along with the air spectra. These carbon compounds are thought to be generated by ablation and several chemical reactions in the shock layer. The measured global power emission profile is shown in Fig.4. At the focal point, the field of view becomes a circle of 3 mm diameter and the solid angle of the collection cone is 8×10^{-6} sr. The measured radiant intensity was estimated as 310 W/m^2.sr under the conditions of $p = 500$ Pa and $T = 300$ K. Besides, this profile indicates that the light collected in this optics system is only that coming from the shock layer around the projectile.

4. Calculation of the shock layer structure

Computational study was conducted to examine the shock layer structure. In the flowfield calculation, 11 chemical species and 17 reactions including the ionization reactions are considered (Murayama 1991). Chemical reaction rate coefficients are determined from the translational and vibrational temperatures. In the radiation calculation, bound-free, bound-bound, and free-free transitions are calculated for atoms and 6 dominant transitions for molecules (Sasoh 1991). The radiative intensity and absorption coefficient are three-dimensionally integrated line by line, and the emission power in front of the blunt body is computed. Blunt bodies with nose radii $R_0 = 1$ m and $R = 0.01$ m $(0.01R_0)$ are calculated. R_0 is the same order of magnitude of the re-entry vehicle and $0.01R_0$ is of the projectile launched in the ballistic range. The air temperature and density are chosen as $T_0 = 218$ K and $\rho_0 = 8.7 \times 10^{-5}$ kg/m^3, respectively, to simulate the atmosphere

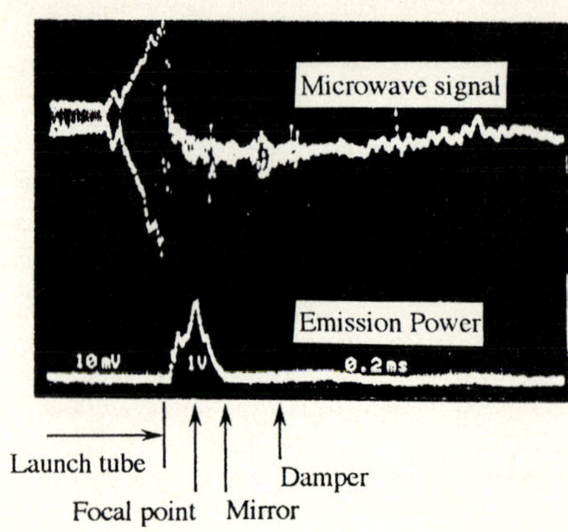

Launch tube
Focal point   Mirror
Damper

Fig. 4. Measured global power emission profile. $p = 5 \times 10^2$ Pa

Table 1. Shock layer thickness

R, m	ρ, kg/m^3	L, m	L/R
1.00	8.7×10^{-5}	7.51×10^{-2}	0.075
0.01	8.7×10^{-5}	9.49×10^{-4}	0.095
0.01	8.7×10^{-3}	8.97×10^{-4}	0.090

$M = 15$, $T_0 = 218$ K

at 70 km altitude. The flow Mach number is set at 15, corresponding to a vehicle flight speed of 5 km/s. The calculated shock layer thickness is shown in Table 1.

As seen in the table, the shock layer thickness is found to change almost proportionally to the nose radius of the blunt body, but not to change with the density. This relationship can be expressed by $L = \alpha R$. In the case of the spherical-conical blunt body, α becomes of the order of 0.1. At $M = 15$, the post-shock translational temperature, before reaction, can be computed as 7700 K from the Rankine-Hugoniot relation. On the other hand, the vibrational temperature is assumed to remain the same as the free-stream temperature of 218 K. Both temperatures are relaxed in the shock layer, and are finally converged to the wall temperature when the wall is isothermal.

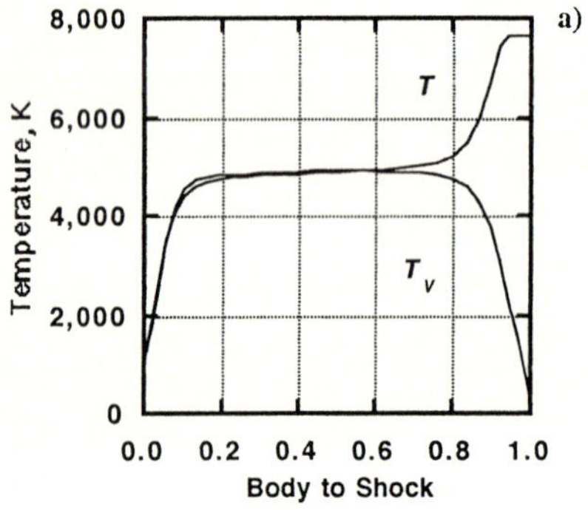

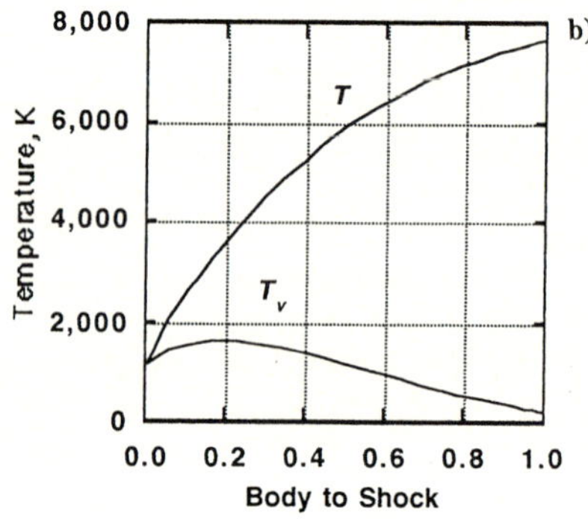

Fig. 5. Calculated distributions of translational and vibrational temperatures on the stagnation streamline. (a) $R = 1$ m, $\rho = 8.7 \times 10^{-5}$ kg/m^3.
(b) $R = 0.01$ m, $\rho = 8.7 \times 10^{-3}$ kg/m^3

In the case of $R = 0.01\, R_0$, the vibrational temperature is not raised at all, because the shock layer is too thin to excite vibrational motions. Since the vibrational relaxation time is inversely proportional to the pressure, the vibrational relaxation distance is also inversely proportional to the density as

$$\lambda_V = \tau_V u \propto 1/p \propto 1/\rho \tag{1}$$

All reaction rate coefficients for the chemical reactions are expressed as functions solely of the temperature, and the reaction time is given by $\tau_R = 1/k_f[M]$. Therefore, neglecting the temperature dependence, the relaxation distance becomes

$$\lambda_R = \tau_R u \propto 1/[M] \tag{2}$$

The reaction distance is also inversely proportional to the density. Since $L = \alpha R$, the fraction of the shock layer which is in thermal or chemical non-equilibrium, becomes inversely proportional to the product ρR:

$$\lambda_{V,R} \propto 1/\rho R \tag{3}$$

Fig.5 shows the calculated distributions of the translational and vibrational temperatures on the stagnation streamline for the cases of (a) $R = R_0$ and (b) $R = 0.01\ R_0$. The characteristic distances and some flow properties at the post-shock point and at the middle point on the stagnation streamline are listed in Table 2. As shown in the table, in spite of the good coincidence in the post-shock conditions, the relaxation and dissociation distances differ at the middle point. It should be noted that the relaxation rate is enhanced with the increase in atomic fraction. For example, the relaxation time due to the collisions between N_2-O is one order of magnitude shorter than that due to N_2-N_2 collisions when the fraction $[O]/[M]$ is 0.3. This means that the vibrational relaxation is accelerated by the dissociation. Since the reaction rates are sensitive to the average temperature, the difference in the relaxation distance can be also attributed to the difference in the average temperature.

5. Thermal boundary layer

When the wall temperature of the blunt body is maintained constant by cooling, sharp temperature gradient appears near the wall surface, as seen in Fig.5(a). In addition to this boundary, another boundary layer is formed on the shock side, owing to the vibrational-mode heat conduction. The thickness of these layers can be estimated as

$$\delta_T \approx \sqrt{\kappa/\rho C_p} \tag{4}$$

Table 2. Characteristic distances and flow properties

Nose radius	R, m	1.00	0.01
Free-stream density	ρ, kg/m^3	8.7×10^{-5}	8.7×10^{-3}
Shock layer thickness	L/R	7.50×10^{-2}	9.00×10^{-2}
Post-shock condition	T_a, K	1295	1295
	λ_V/R	4.95×10^{-2}	4.87×10^{-2}
	λ_R/R	5.42×10^{13}	5.36×10^{13}
Middle-point condition	T_a, K	4915	2661
	$[O]/[M]$	0.30	0.05
	λ_V/R	1.07×10^{-2}	1.00×10^{-1}
	λ_R/R	2.36×10^{-1}	4.64×10^{3}

Since both the thermal conductivity and specific heat are functions of the temperature, the thickness can be represented as a function of the density:

$$\delta_T = \beta/\sqrt{\rho/\rho_0} \quad \text{meter} \tag{5}$$

The constant β can be determined from the temperature distribution shown in Fig.5(a). When the wall temperature is fixed to 1200 K, β is approximately 0.01. The ratio of the thermal boundary layer to the shock layer thickness can be deduced to be

$$\frac{\delta_T}{L} = \frac{\beta}{\alpha R \sqrt{\rho/\rho_0}} = \frac{0.1}{\sqrt{R}} \tag{6}$$

when $\rho R = \rho_0 R_0$. This equation indicates that the effect of the thermal boundary layer becomes larger with the decrease in the model size, and the shock layer is entirely covered with the thermal boundary layers when the nose radius is of the order of 0.01 m or less.

6. Calculated results

Flowfield and radiation calculations were conducted for the conditions $R = 0.01$, $p = 500$ Pa, and $T = 300$ K, to simulate the experimental results. Adiabatic wall condition was assumed. Because of the relatively high number density, ionized species are rare in the shock layer. Calculated results show that atomic radiation does not appear due to the low electron density, whereas NO and N_2 radiation is very strong.

The calculated global power emission is 5×10^4 W/m^2.sr on the front of the projectile. This is almost two orders of magnitude larger than the measured one. This discrepancy is thought to come from the uncertainty of the wall condition, because the thermal boundary layer around the projectile is so thick that the wall condition sensitively affects the temperature distributions.

7. Summary

From the radiation measurement, the global power emission was estimated as 300 W/m^2.sr, and carbon compound spectra were detected along with the air spectra. In the calculations, the thermal boundary layer thickness was found to change as $\delta_T/L = 0.1/\sqrt{R}$, when $\rho R = $ constant. This relationship indicates that, in the case of $R = 0.01$ m, the shock layer is entirely covered with a thermal boundary layer, and the wall condition sensitively affects the temperature distribution in the shock layer. The calculated global emission becomes two orders of magnitude larger than the measured one. In order to accurately estimate the heat transfer rate, future work should be focused on the determination of the temperature in the vicinity of the wall surface by taking into account the ablation layer and the accompanying carbon radiation.

References

Chang H, Hemmi M, Komurasaki K, Fujiwara T (1993), Radiation measurement by using a ballistic range. AIAA Paper 93-0635, Reno, Nevada

Murayama T, Sasoh A, Fujiwara T (1991) Chemical and thermal processes in a hypersonic shock layer. In: Takayama K (ed) Proc 18th ISSW, Sendai, pp 691-696

Park C (1984) Calculation of nonequilibrium radiation in AOTV flight regimes. AIAA Paper 84-0306

Park C (1991) Chemical-kinetic problems of future NASA missions. 1. Earth entries: a review. AIAA Paper 91-0464, Reno, Nevada

Sasoh A, Chang X, Murayama T, Fujiwara T (1991) Radiative heat transfer from non-equilibrium high-enthalpy shock layers. In: Takayama K (ed) Proc 18th ISSW, Sendai, pp. 723-726

Aerodynamic Heating in Three-Dimensional Bow Shock Wave/Turbulent Boundary Layer Interaction Region

Syozo Maekawa[*], Shigeru Aso[*], Shigehide Nakao[*], Kazuo Arashi[†], Kenji Tomioka[‡] and Hiroyuki Yamao[§]

[*]Department of Aeronautics and Astronautics, Kyushu University, Fukuoka 812, Japan
[†]Churyo Engineering, Co. Ltd., Nagoya, Japan
[‡]NASDA, Tokyo, Japan
[§]Mitsubishi Heavy Industry, Nagoya, Japan

Abstract. Aerodynamic heating phenomena in the shock wave/turbulent boundary layer interaction region are one of the most important research tasks and this subject demands further investigation. To explain these phenomena, we investigate the bow shock wave/turbulent boundary layer interaction region induced by a blunt body, that is to say, the model case of the internal flow between a main rocket body and an auxiliary rocket booster. In this study, selecting as the main parameter the distance from the flat plate to the blunt body, the effect of the height is investigated by use of oil flow technique, surface pressure measurements and surface heat flux measurements. The results show that the distance has a major impact on the interaction region. Moreover, this parameter influences the flow behaviour.

Key words: Aerodynamic heating, Shock wave, Turbulent boundary layer, Interaction

1. Introduction

Presently, the development of various kinds of launchers with multi-solid rocket boosters is in progress. In these configurations, a bow shock wave generated by the auxiliary rocket booster interacts with the turbulent boundary layer developed on the main rocket and cause severe aerodynamic heating (Korkegi 1971, Dolling and Bogdonoff 1982, Kaufmann and Johnson 1974). To get reliable data for the design of heat protection, the detailed flow structure of three-dimensional shock wave/turbulent boundary layer interaction regions between the rocket main body and an auxiliary rocket booster should be investigated.

In the present study, the structures of the flowfields are studied in detail by oil flow technique and surface pressure and heat flux measure- ments and the effects of the distance between the rocket main body and the rocket booster on the interacting flowfields are mainly investigated. For the experiments, a flat plate model is installed in the high-speed wind tunnel in place of a main rocket body, and the bow shock wave interacts with the turbulent boundary layer on the plate. A schematic diagram of the flow field is shown in Fig.1. In these experiments, the detailed flow structure of the interaction region and the effects of the displacement, between main rocket body and rocket booster on the interacting flow fields are mainly investigated by changing the distance between the blunt body and the flat plate.

Table 1. Testing conditions

	h [mm]	M_∞	p_0 [MPa]	T_0 [K]	T_w/T_0	Re [$\times 10^7$]
	5	3.86	1.24	421	0.69	1.51
Oil flow	10	3.89	1.26	422	0.69	1.50
	15	3.87	1.23	422	0.68	1.48
	20	3.79	1.23	397	0.73	1.66
	5	3.84	1.26	422	0.69	1.56
Pressure	10	3.83	1.26	414	0.70	1.60
Heat flux	15	3.87	1.27	426	0.70	1.52
	20	3.89	1.26	420	0.69	1.53

Shock Waves @ Marseille I
Editors: R. Brun, L. Z. Dumitrescu

© Springer-Verlag Berlin Heidelberg 1995

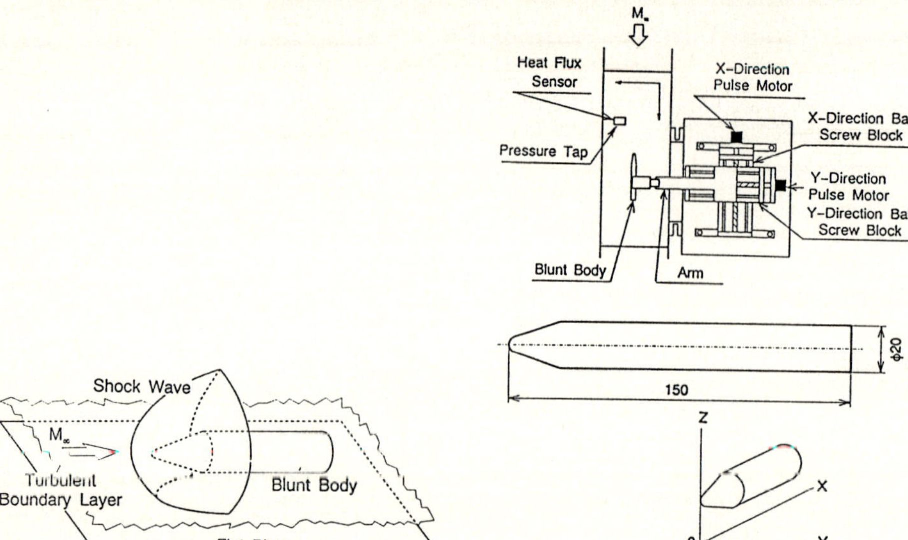

Fig. 1. Schematic diagram of the flowfield

Fig. 2. Schematic diagram of the blunt body model and coordinate system

A blown-down type supersonic wind tunnel with square test section of 150×150 mm and nominal Mach number of 4 is used for the present experiments. A scanning system of the blunt body enables to measure the whole interaction region efficiently. The schematic diagram of the blunt body model and the coordinate system are shown in Fig.2. The diameter of the blunt body is 20 mm and the distance h, between the flat plate and the blunt body rocket is also specified on the figure.

For the measurements of heat flux, a new measuring method is used. This method is based on a new type of thin-film heat transfer gauge with high spatial resolution and fast response (Aso et al. 1990). The principle of this method is based on measuring the temperature gradient across the heat resistant layer with two thin-film resistance thermometers on its upper and lower surfaces. These layers are deposited by the vacuum evaporation technique. On one metal probe, 10 sets of heat flux sensors are formed at every 2 mm. The shape of each sensing portion of the sensor is spiral. The sensing portion is 2.0×1.6 mm.

2. Experimental apparatus and procedure

All experiments are performed under cooled wall conditions, and representative test conditions are tabulated in Table 1.

3. Results and discussion

a. Oil flow technique

The surface flow is visualized by the oil flow technique. Pictures and sketches of oil flow induced by the blunt body at different distances h, are shown in Fig.3. These results show quite interesting features of aerodynamic heating in various shock wave/turbulent boundary layer interactions, and these experimental studies are quite important in order to understand the mechanism of aerodynamic heating. In the figures, S_1, S_2 and A mean the primary separation line, secondary

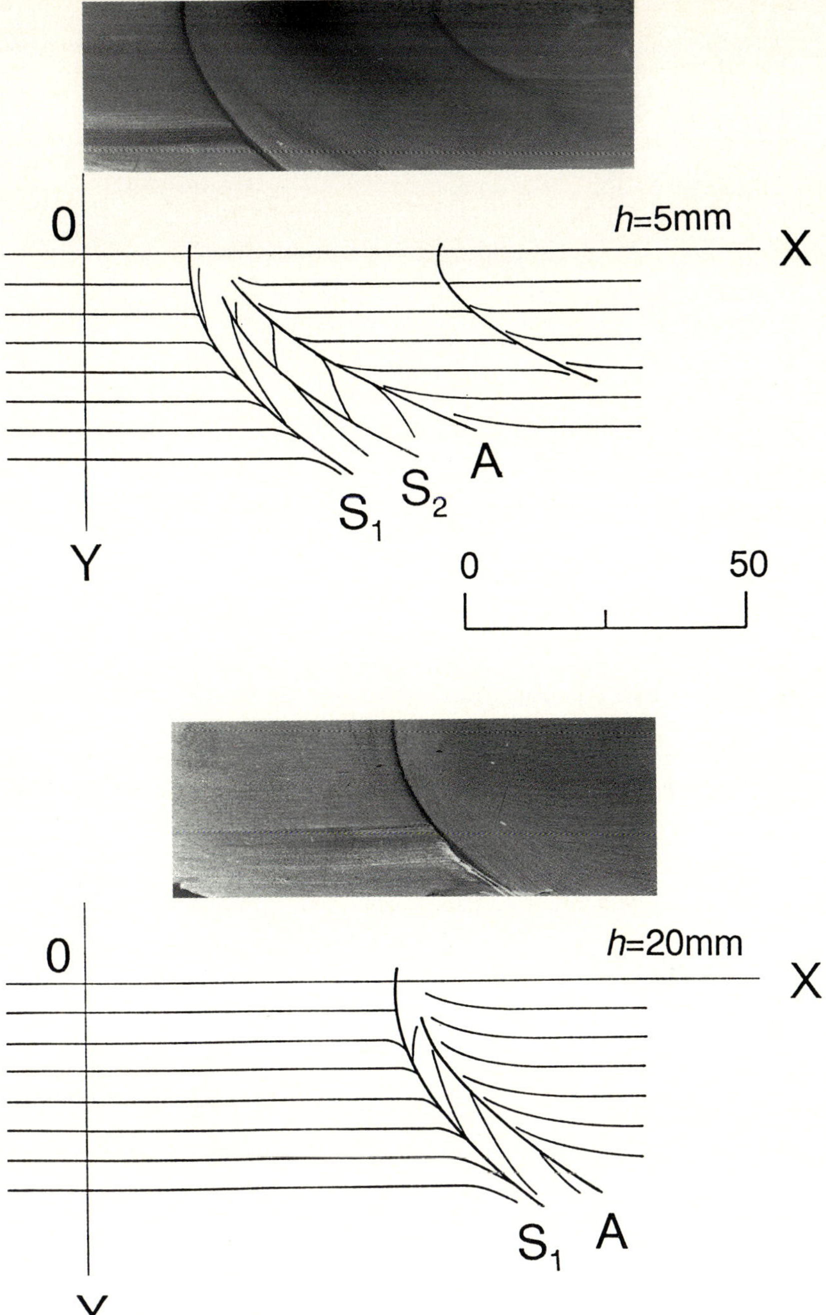

Fig. 3. Representative oil flow pictures (above: $h = 5$ mm, below: $h = 20$ mm)

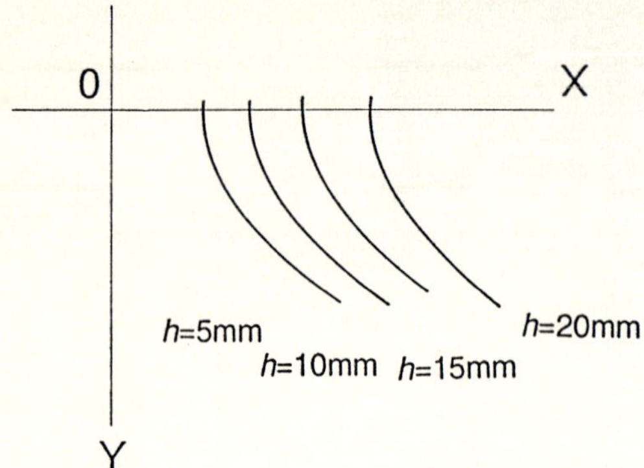

Fig. 4. Shift of primary separation line with h

separation line and attachment line, respectively. The separation line is defined as a convergent line of oil flow lines, and the attachment line is a divergent line. As the left end of each picture indicates the location of the top of the blunt body, the primary separation line moves downstream when h is increased. For $h = 5$ mm, a primary separation and a secondary separation are observed. The primary separated region becomes larger away from the centerline. The result indicates that the separated vortex generated at the centerline becomes larger away from the centerline. The same tendency is observed for $h = 10$ mm. However for $h = 15$ mm and 20 mm, no secondary separation line is observed. The shift of the primary separation line with h is shown in Fig.4. The results show that the extent of the interaction moves downstream linearly as h is increased. The surface oil flow changes drastically with distance h.

b. Pressure and heat flux distribution

Representative streamwise pressure and heat flux distributions at the centerline for various distances h are shown in Fig.5. For $h = 5$ mm, a pressure peak is observed on the centerline and the peak pressure is about 4.1 times the undisturbed value. The peak pressure decreases rapidly away from the centerline. Also a local minimum pressure in the pressure plateau due to reacceleration of the expansion fan generated from the shoulder of the blunt fin is observed. For $h = 10$ and 15 mm, the pressure distributions show almost the same tendency. For $h = 20$ mm only a single peak is observed in the streamwise pressure distribution and the peak pressure decreases rapidly while a stepwise pressure rise is observed away from the centerline.

The streamwise surface heat flux distributions are also shown. Owing to the turbulent boundary layer, the Stanton Number is calculated from the heat flux. For $h = 5$ mm, a heat flux peak is observed on the centerline and this peak is about 3.3 times. Downstream of this maximum peak, a local small peak is observed. This peak is due to reacceleration of the expansion fan. The peak heat flux decreases rapidly away from the centerline. For larger values of Y a local minimum heat flux before the maximum peak due to the separated vortex is observed. For $h = 20$ mm only a single peak is observed in the streamwise heat flux distribution. By comparing both distributions, a strong correlation between pressure and heat flux is observed.

Detailed surface pressure and heat flux contours are shown in Fig.6 for $h = 5$ mm and 20 mm. By comparing both contours, strong correlation between pressure and heat flux is observed. This strong correlation is preserved for various distances h. These findings are quite important for the design of a practical configuration.

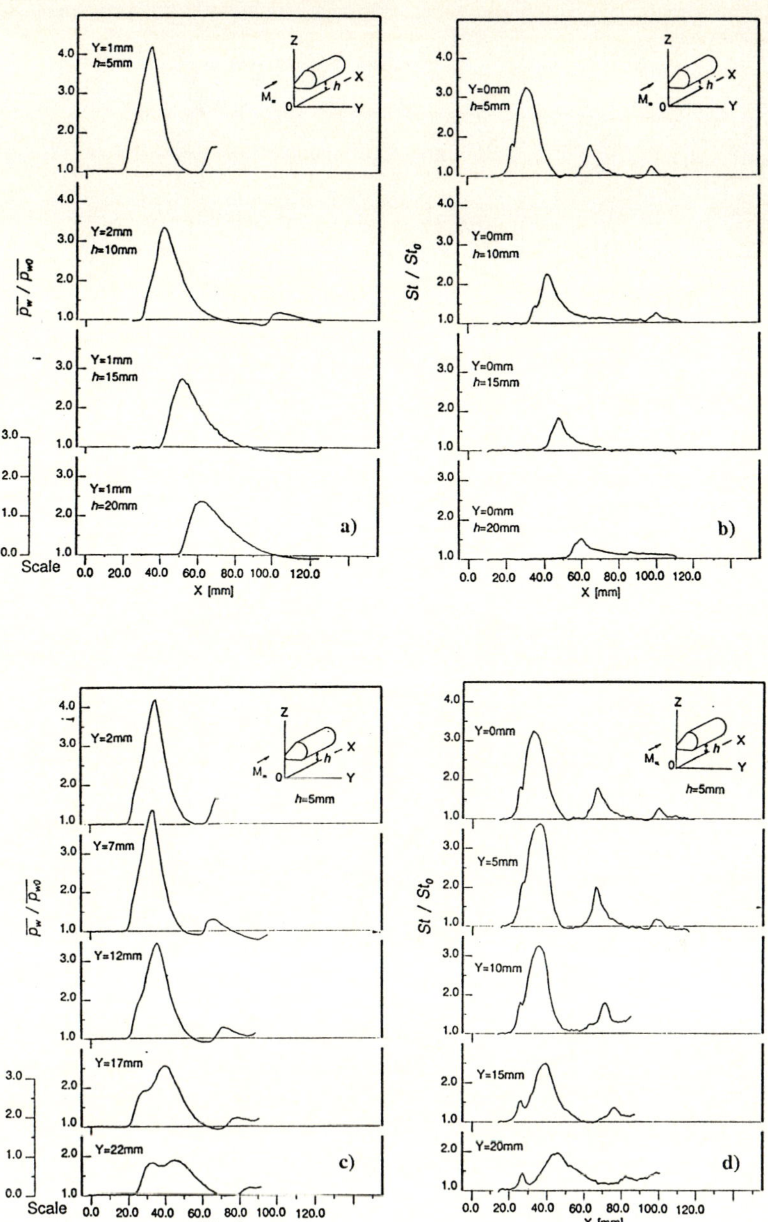

Fig. 5. Representative streamwise pressure and heat flux distributions.
(a) Pressure distributions on centerline; (b) Heat flux distributions on center line;
(c) Streamwise pressure distributions ($h = 5$ mm); (d) Streamwise heat flux distributions ($h = 5$ mm)

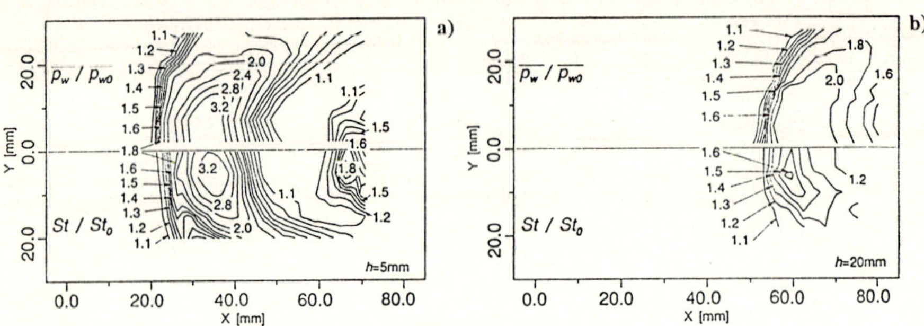

Fig. 6. Surface pressure and heat flux contours: (a) $h = 5$ mm; (b) $h = 20$ mm

4. Conclusions

Bow sock wave/turbulent boundary layer interaction is investigated by oil flow, surface pressure distribution and heat flux measurement. The major conclusions of this study are summarized as follows:

1) In the interaction region primary and secondary separations are observed and a quite complicated flow structure is revealed.

2) The effect of distance h, of the auxiliary rocket booster from the main rocket, on pressure and heat flux distribution is quite significant.

3) The range of the interaction region is bounded by a primary separation line. The range shifts downstream linearly as the disptance h is increased.

4) For small h a secondary separation with primary separation is observed. A sharp pressure peak is observed on the centerline and the pressure peak decreases rapidly away from the centerline. Also a local minimum pressure in the pressure plateau due to the separated vortex is observed.

5) For large h no secondary separation is observed. Only a single peak is observed in the streamwise distribution and this peak decreases rapidly, while a stepwise rise is observed away from the centerline.

References

Aso S, Kuranaga S, Nakao S (1990) Aerodynamic heating phenomena in three-dimensional shock wave/turbulent boundary layer interactions induced by sweptback blunt fins. AIAA Paper 90-0381

Dolling DS, Bogdonoff SM (1982) Blunt fin induced shock wave/turbulent boundary layer interaction. AIAA J. 20: pp.1674-1680

Kaufmann II LG, Johnson CB (1974) Interference heating from interactions of shock waves with turbulent boundary layer at Mach 6. NASA TM D-7469

Korkegi RH (1971) Survey of viscous interaction associated with high Mach number flight. AIAA J. 9, 5:771-784

Parabolic Viscous Shock Layer Theory for 3D Hypersonic Gas Flow

S.V. Peigin
Institute of Applied Mathematics and Mechanics Tomsk University 634050, Lenina 36, Tomsk, GSP-14, Russia

Abstract.   A new gasdynamic model of 3D viscous gas flow over blunt bodies with permeable surfaces is suggested - named the Parabolic Viscous Shock Layer theory (PVSL). The advantage of this theory in comparison with other mathematical flow models (thin viscous shock layer (TVSL), full viscous shock layer (FVSL), parabolized Navier-Stokes equations) consists in the possibility to calculate the flowfield on the front body surface up to its middle section and to use very economical and effective numerical methods for obtaining these solutions. In the framework of the PVSL theory, systematic numerical solutions of some two- and three-dimensional hypersonic viscous gas flow problems past blunt bodies for a wide range of Reynolds numbers and body shape are obtained; the accuracy and the applicability range of the suggested model are estimated on the basis of comparisons with theoretical and experimental data.

Key words: Hypersonic flow, Viscous flow

1. Introduction

As it is known the mathematical simulation of supersonic 3D viscous gas flow over bodies in the framework of full Navier-Stokes equations requires large computer resources. The estimations by Agarwal (1992) demonstrate, that for a direct numerical simulation of these problems for large Reynolds numbers in an acceptable time one has to use computers with a performance of more than 10^{18} Flop. In this connection a number of approximate gasdynamic flow models have found wide use in the scientific literature. One of these pioneering models was the TVSL theory by Cheng (1961). Asymptotic analysis demonstrates that the TVSL model has a correct character and adequately describes the viscous gas flowfield between the shock wave and the body surface. Later the TVSL model was widely applied for investigations of 2D and 3D supersonic viscous gas flow problems over smooth blunt bodies (Peigin and Tirsky 1988). From a mathematical point of view the TVSL system of equations is a composite, since it contains all members of the boundary layer equations and all members of the hypersonic approximation of the inviscid shock layer equations. The attractiveness of this model for practical applications (in particular for heat transfer investigations connected to the "Buran" spacecraft) is explained by at least two reasons. The first of them lies in the fact that, as comparisons with solutions obtained for more complete mathematical flow models and experimental data demonstrate, the TVSL model has a rather high accuracy on the stagnation region of the body surface, where the shock layer is sufficiently thin and heat fluxes are at a maximum. The second reason is that the TVSL equations system is of parabolic type, which permits to apply a fast and economical marching numerical procedure for obtaining the solution and, by this means, to increase greatly the calculations efficiency.

However, the TVSL model has a disadvantage, because in the framework of this theory a zero pressure line (shock layer separation line) appears on the front part of a convex body surface and the solution of the TVSL equations cannot be continued behind this line. The asymptotics of a singularity on this line for a 2D inviscid shock layer was investigated by Gonor and Ostapenko (1974). It was proved that the singular line appearance (Newton separation line) is completely determined by using a simplified momentum equation in projection on the normal to the body surface direction and applying the understated value of boundary condition on the shock wave for the pressure (as a result of the assumption that the shock and body are identical in form).

Shock Waves @ Marseille I
Editors: R. Brun, L. Z. Dumitrescu   © Springer-Verlag Berlin Heidelberg 1995

In this paper a serious effort was made to resolve the singularity and to construct a correct analytical theory (on the basis of the method of double approximation) and in doing so to bring the separation point nearer to the mid-body section.

Another way of solving the separation layer problem was suggested by Tolstyhk (1966). He proposed to use the FVSL equations as the initial mathematical model, which contains all members of the boundary layer equations together with all members of Euler equations. Because the second order derivatives with respect to x^1 in the PVSL equations are absent, this system is of the evolution type relative to marching coordinates. As a consequence this system of equations can be solved as an initial boundary-value problem using marching numerical methods. But the analysis by Lin and Rubin (1973) demonstrates, that this approach will be only correct when the flowfield is supersonic with respect to coordinate x^1, where the inviscid operator is of the hyperbolic type, and will be not correct for subsonic domains, where this operator is elliptic. In this connection, for the regularization problem in the subsonic domain it was suggested to use either a time-asymptotic numerical technique or a global iteration method, by which the shock wave shape and the elliptic members in the equations are determined from the previous global iteration. At the same time it is important to note, that by gaining advantage over the TVSL model in the size of the computation domain we strongly worsen the calculation efficiency. Altough the global iteration method enables one to decrease the computation time more than ten times (with respect to time-asymptotic methods), this method in turn requires for its realization more than one order of computer time in comparison with the marching procedures. For 3D problems this difference in efficiency between these methods takes a great significance.

In this connection, in this paper a new asymptotically exact mathematical model of 3D supersonic viscous gas flow past blunt bodies is suggested - named the parabolic viscous shock-layer theory (PVSL). This model is intermediate between the TVSL and FVSL models. The main idea of the suggested model consists in computing the velocity, temperature and density fields between the shock wave and the body surface using the FVSL equations with the assumption that longitudinal and circumferential gradients of pressure at the shock layer are known. For defining the pressure gradients an additional system of two differential equations is used, which is deduced from the hypersonic approximation of the momentum equation in projection on the normal to the body surface direction. The PVSL model possesses the merits of the TVSL theory (the initial equations are parabolic, which permits to apply for the solution a marching numerical procedure), and at the same time the demerits of the TVSL model (existence of a shock-layer separation line) are absent here.

2. Statement of the problem

For the substantiation of the PVSL model let us consider the FVSL system of equations, in which a number of members, which are immaterial for the hypersonic regime, have been deleted. These equations, written in a curvilinear coordinate system x^i and describing the 3D viscous gas flow between the shock and the body surface have the following dimensionless form:

$$D_i(\rho u^i a^{1/2} a_{(ii)}^{-1/2}) = 0, \quad 2D_3 P = -(1+\epsilon)\rho(\epsilon Du^3 + A_{\alpha\beta}^3 u^\alpha u^\beta), \quad P = \rho T,$$

$$\rho(Du^\alpha + A_{\beta\delta}^\alpha u^\beta u^\delta) = -2\epsilon(1+\epsilon)^{-1} a^{\alpha\beta} a_{(\alpha\alpha)}^{1/2} D_\beta P + D_3(\mu K^{-1} D_3 u^\alpha),$$

$$\rho DT = 2\epsilon(1+\epsilon)^{-1} D^* P + D_3(\mu(\sigma K)^{-1} D_3 T) + \mu K^{-1} B_{\alpha\beta} D_3 u^\alpha D_3 u^\beta, \tag{1}$$

$$D_i = \partial/\partial x^i, \quad D^* = u^\alpha a_{(\alpha\alpha)}^{-1/2} D_\alpha, \quad D = D^* + u^3 D_3, \quad B_{\alpha\beta} = a_{\alpha\beta} a_{(\alpha\alpha)}^{-1/2} a_{(\beta\beta)}^{-1/2},$$

$$K = \epsilon Re, \quad Re = \rho_\infty U_\infty L/\mu(T_0), \quad T_0 = T_\infty(\gamma - 1)M_\infty^2, \quad \epsilon = (\gamma - 1)/(\gamma + 1)$$

The system (1) is solved with boundary conditions on the shock wave and body surfaces. On the shock wave the hypersonic approximation of the generalized Rankine-Hugoniot conditions is used, which take into account the effects of molecular transfer into the shock wave.

$$x^3 = x^3_s(x^1, x^2): \quad \rho(u^3 - D^* x^3_s) = u^3_\infty, \;\; 2P = (1+\epsilon)(u^3_\infty)^2, \;\; u^3_\infty(u^\alpha - u^\alpha_\infty) = \mu K^{-1} D_3 u^\alpha \, (2)$$

$$u^3_\infty(T - 0.5(u^3_\infty)^2 - B_{\alpha\beta}(u^\alpha - u^\alpha_\infty)(u^\beta - u^\beta_\infty)) = \mu(\sigma K)^{-1} D_3 T$$

The boundary conditions on the body surface are the following:

$$x^3 = 0: \quad u^\alpha = 0, \quad T = T_w(x^1, x^2), \quad (\rho u^3)_w = G(x^1, x^2). \tag{3}$$

Here x^3 is taken along the normal to the body surface, and x^1, x^2 are selected on the surface. The summation is made over repeating subscripts, not enclosed in brackets. Latin indices are equal to 1,2,3 and Greek indices are equal to 1,2. $V_\infty u^\alpha$, $\epsilon V_\infty u^3$ are the physical components of the velocity vector; $PP_\infty \epsilon^{-1} T_0/T_\infty, \rho\rho_\infty \epsilon^{-1}, TT_0, \mu\mu_0$ are respectively the pressure, density, temperature and viscosity; $a_{\alpha\beta}$ are covariant components of the first quadratic form of the body surface, $A^i_{\alpha\beta}$ - known functions of the body form. Subscripts w, ∞, s correspond to values on the body surface, at infinity and on the inner boundary of the shock wave surface.

As noted above, the boundary-value problem (Eqs.1-3) is of elliptic type with the result that within the framework of this system upstream propagation of perturbations is possible and the direct application of a marching method to these equations is impossible. As the analysis demonstrates, the reason of these properties of the system (Eqs.1-3) has a double character: on the one hand the shock shape in conditions (2) is unknown beforehand and on the other hand the initial equations (Eqs.1) are of elliptic type in the subsonic flow domain. Therefore, any way of regularizating the problem (Eqs.1-3), which is to render correct the use of marching techniques, has to consist of two (possibly independent) parts: "shock wave" regularity and "equations" regularity. So, for example, for the TVSL model, the first of these lies in the assumption that the shapes of the shock and body surfaces coincide, while the equations regularity lies in the neglect of the term $\epsilon D u^3$ in the momentum equation in normal projection to the body surface.

Alternatively, it is shown (Anderson et al. 1984) that the elliptic character of Eqs.1 is completely determined by terms containing the longitudinal pressure gradients $\mathbf{P}_\alpha \equiv \partial P/\partial x^\alpha$ in the momentum equation projected on axes x^α. The following analysis demonstrates, that if $\mathbf{P}_\alpha$ are known functions, then the system (Eqs.1) is of hyperbolic type. If we recall that terms with longitudinal pressure gradients appear in the system (Eqs.1) with multiplier ϵ and use the asymptotical analysis results by Gershbein (1981) a sufficiently good approximation for defining $\mathbf{P}_\alpha$ can be obtained by using a Cauchy problem solution with boundary conditions on the shock wave for the following equations:

$$2D_3\mathbf{P}_\alpha = -(1+\epsilon)D_\alpha(\rho A^3_{\delta\beta} u^\delta u^\beta) \tag{4}$$

This system can be deduced by differentiation (with respect to the coordinates x^α) of the hypersonic approximation of the momentum equation in projection along the normal to the body surface.

So the main idea of the PVSL theory consists in solving the boundary-value problem (Eqs.1-3) (for determining the velocity field, temperature, pressure and density) with the assumption that the shapes of the shock and body surface coincide, while the longitudinal and circumferential gradients of pressure $\mathbf{P}_\alpha$ are found from the additional system of Eqs.4.

For the numerical solution of the problem the system (Eqs.1-4) is written in variables of Dorodnicyn type (see Borodin and Peigin 1989) and in this way the singularities of the initial equations are resolved:

$$\xi^\alpha = x^\alpha, \quad \zeta = \Delta^{-1}\int_0^{x^3} \rho \, dx^3, \quad \Delta = \int_0^{x^3_s} \rho \, dx^3, \quad \theta = \frac{T}{T_*}, \quad l = \frac{\mu\rho}{K\Delta^2}, \quad u^\alpha = D_3 f_\alpha, \tag{5}$$

$$\mathbf{P}_1 = (\xi^1)^{-1} D_1 P, \quad \mathbf{P}_2 = (\xi^1)^{-2} D_2 P, \quad u^1_* = u^1_\infty, \quad u^2_* = \xi^1, \quad T_* = 0.5(u^3_\infty)^2,$$

$$\rho u^3 a^{1/2} = -D_\alpha\big(\psi^*_{(\alpha)}f_\alpha\big) - \psi^*_{(\alpha)}D_3 f_\alpha D_\alpha \zeta, \quad \psi^*_{(\alpha)} = \Delta a^{1/2}\phi^*_\alpha = \Delta a^{1/2}u^\alpha_*/a^{1/2}_{(\alpha\alpha)}$$

As result one obtains the following system of equations:

$$D_3(lD_3 u^\alpha) = Du^\alpha + C^\alpha_{\beta\delta}u^\beta u^\delta + 2\epsilon(1+\epsilon)^{-1}L^\alpha_\beta \rho^{-1}\mathbf{P}_\beta, \quad D_3 = \partial/\partial\zeta, \tag{6}$$

$$D_3(l\sigma^{-1}D_3\Theta) = D\Theta + F_\alpha u^\alpha\Theta - 2\epsilon(1+\epsilon)^{-1}L^3_{(\beta)}u^\beta\rho^{-1}\mathbf{P}_\beta - lC^3_{\alpha\beta}D_3 u^\alpha D_3 u^\beta, \quad D_\alpha = \partial/\partial\xi^\alpha$$

$$2D_3\mathbf{P}_\alpha = -(1+\epsilon)(K^\alpha_{\beta\delta}u^\delta u^\beta + M^{(\alpha)}_{\beta\delta}u^\beta D_\alpha u^\delta), \quad D = \phi^*_{(\alpha)}u^\alpha D_\alpha - (\phi^*_{(\alpha)}D_\alpha f_\alpha + B_\alpha f_\alpha)D_3.$$

$$\zeta = 0: \qquad u^\alpha = 0, \quad \Delta\big(\phi^*_{(\alpha)}D_\alpha f_\alpha + D_\alpha f_\alpha\big) = -G(\xi^1,\xi^2), \quad \Theta = \Theta_w(\xi^1,\xi^2). \tag{7}$$

$$\zeta = 1: \qquad \Delta\big(\phi^*_{(\alpha)}D_\alpha f_\alpha + D_\alpha f_\alpha\big) = -u^3_\infty, \quad u^3_\infty\big(u^\alpha - u^\alpha_s\big) = l\Delta D_3 u^\alpha \tag{8}$$

$$u^3_\infty\big[\Theta - 1 - 0.5C^3_{\alpha\beta}\big(u^\alpha - u^\alpha_s\big)\big(u^\beta - u^\beta_s\big)\big] = l\Delta\sigma^{-1}D_3\Theta, \quad u^1_s = 1, \quad u^2_s = u^2_\infty/\xi^1$$

$$\mathbf{P}_1 = \mathbf{P}_{1s} = (1+\epsilon)(\xi^1)^{-1}u^3_\infty D_1(u^3_\infty), \quad \mathbf{P}_2 = \mathbf{P}_{2s} = (1+\epsilon)(\xi^1)^{-2}u^3_\infty D_2(u^3_\infty)$$

Here the coefficients $\phi^*_\alpha, B_\alpha, C^i_{\beta\delta}, L^i_\beta, F_\alpha, K^\alpha_{\beta\delta}, M^\alpha_{\beta\delta}$ are known functions and are presented in the paper by Borodin and Peigin (1989).

The system (Eqs.6-8) has to be closed by the mass conservation equation, the equation of state and the momentum equation with respect to the normal direction. After some conversions these equations can be reduced to the following equation for the determination of x^3 as an unknown function of ξ^1, ξ^2, ζ:

$$D_3\big[(\gamma+1)\gamma^{-1}T_*\Delta\Theta(D_3 x^3)^{-1} + \Delta^{-1}\epsilon B^2 D_3 x^3 - \epsilon B\phi^*_{(\alpha)}u^\alpha D_\alpha x^3\big] = \tag{9}$$

$$R_{\beta\delta}u^\delta u^\beta - \epsilon\Delta\phi^*_{(\alpha)}u^\alpha D_\alpha\big[\phi^*_{(\beta)}u^\beta D_\beta x^3 - \Delta^{-1}BD_3 x^3\big] - \epsilon D_3 B\big[\phi^*_{(\beta)}u^\beta D_\beta x^3 - \Delta^{-1}BD_3 x^3\big].$$

$$B = a^{-1/2}D_\alpha\big(\phi^*_{(\alpha)}\Delta a^{1/2}f_\alpha\big), \quad D_3 B = a^{-1/2}D_\alpha\big(\phi^*_{(\alpha)}\Delta a^{1/2}u^\alpha\big).$$

$$\zeta = 0: \quad x^3 = 0; \qquad \zeta = 1: \quad D_3 x^3 = \Theta_s\Delta/(1+\epsilon) \tag{10}$$

After determination of the value x^3 the pressure, the density and the normal component of the velocity vector are obtained by using the following relations:

$$u^3 = \phi^*_{(\beta)}u^\beta D_\beta x^3 - \Delta^{-1}BD_3 x^3, \quad \rho = \Delta(D_3 x^3)^{-1}, \quad P = T_*\Delta\Theta(D_3 x^3)^{-1}$$

3. Analysis of results

For the numerical solution a cylindrical coordinate system on the body surface (x^1, x^2) was used $(x^1$ - marching coordinate, x^2 - circumferential coordinate). Numerical solution of the problem was obtained on the basis of an implicit (relatively to x^3) finite-difference method, having $O(\Delta x^1) + O(\Delta x^2)^2 + O(\Delta x^3)^4$ approximation order and representing the generalization of a scheme by Petuhkov (1964). For convective operator approximation the derivatives with respect to the marching coordinate x^1 were replaced by upwind differences and the derivatives with respect to the coordinate x^2 were replaced by central difference operators on the basis of the solution obtained during the previous global iteration on the current circle $x^2 + \Delta x^2 = $ const. Each of Eqs.6,9 was reduced to a first order system and was solved in the following order: momentum equations, energy equation, equation for x^3 determination and equations for the pressure gradients. The thickness of the shock layer was obtained using the cyclical tridiagonal solver method after finishing the global iteration. The important characteristic property of this numerical algorithm is that it does not imply the existence of symmetry planes within the flow and enables one to investigate the 3D flow around bodies at incidence and yaw angles.

As an example the flow around a three-axis ellipsoid at angle of attack α and yaw angle β was considered (of equation, in a Cartesian coordinate system: $(z^1/a - 1)^2 + (z^2/b)^2 + (z^3/c)^2 = 1$).

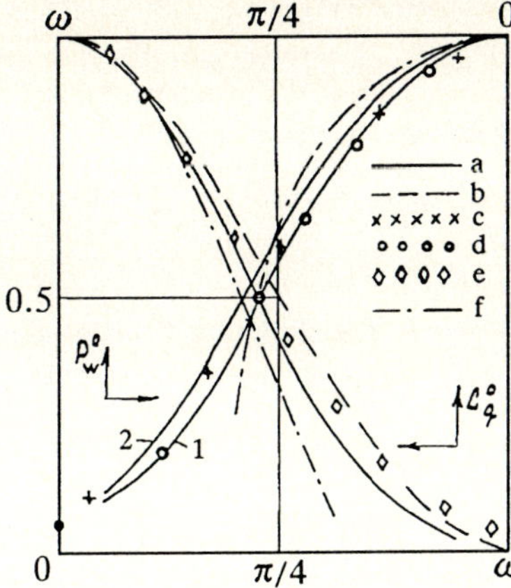

Fig. 1. Relative distributions of pressure P_w^o and heat flux C_q^o for sphere. Lines 1, 2 - $\Theta_w = 0.1; 0.25$. Lines: a is solution of the PVSL equations, b - Newton's formula for pressure, c - global iteration method for TVSL model, d - experimental data, e - solution of the Euler equations, g - solution of the TVSL equations

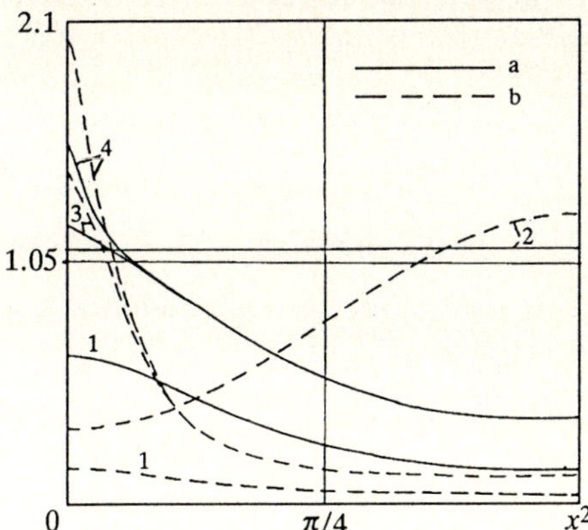

Fig. 2. Relative distributions of the heat flux C_q^o for ellipsoids. $x^1 = 0.4; 0.76$ - lines a, b. $b = 0.7$, $c = 0.3$ - line 1; $b = 2.0, c = 3.0$ - line 2; $b = 4.0, c = 0.5$ - line 3; $b = 8.0, c = 0.5$ - line 4.

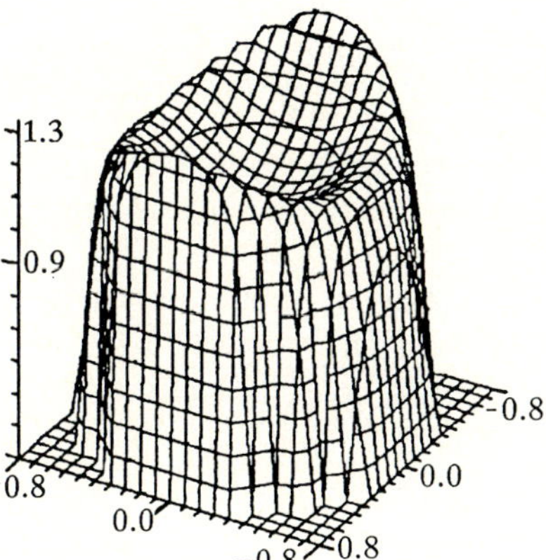

Fig. 3. Distribution of C_q^o for ellipsoids. $b = 2.0, c = 3.0$, $\alpha = \beta = 0°$

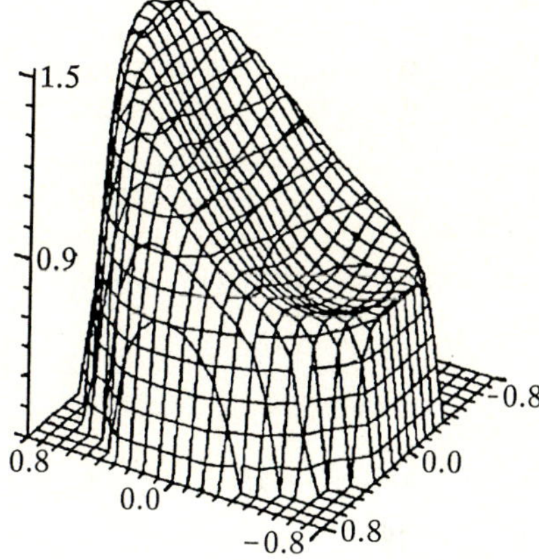

Fig. 4. Distribution of C_q^o for ellipsoids. $b = 0.3, c = 0.7$, $\alpha = 75°, \beta = 0°$

In Fig.1 the results of test calculations for a sphere are presented. Globally the comparison results enable one to draw the conclusion, that the suggested PVSL theory has a good accuracy for the distribution of pressure and heat fluxes along the body surface up to the mid-body section (in comparison with the TVSL model the size of the calculation domain increases more than twice) and it may be successfully employed for 3D viscous gas flow simulation over blunt bodies.

Let us consider some characteristic properties of the ratio of the heat flux C_q^o referred to the value at the stagnation point, along the lateral body surface. In Fig.2 the dependence of the distribution of C_q^o on the coordinate x^2 along the surface for different body shapes is given. It is shown that for a large value of axis b the distinctions between proper distributions take place only in a thin layer near the symmetry plane $x^2 = 0$. At the same time the influence of parameters b and c on the value of C_q and on x^2 dependence character is considerable and can exceed 100%.

In conclusion we present some distributions of C_q^o (Figs.3-4) for the general case, when $\alpha \neq 0, \beta \neq 0$. It is shown that the character of heat flux distribution in a qualitative sense depends on the body shape and the values of α and β. In particular, the location of the high heat flux zone has a good correletion with the location of the domain where the body surface exhibits a high middle curvature.

Acknowledgements

This investigation was accomplished with the support of Russia Foundation of Fundamental Investigations (Project 93-013-17957). The author would like to thank A.Borodin for assistance in computations.

References

Agarwal RK (1992) Parallel computers and large problems in industry. In: Hirsch Ch.(ed) Computational methods in applied sciences. Elsevier. Amsterdam, pp 1-11

Andreson DA, Tannehill JC, Pletcher RH (1984) Computational fluid mechanics and heat transfer. Hemisphere, New York

Borodin AI, Peigin SV (1989) 3D thin viscous shock layer without symmetry planes in the flow. Izvestiya Acad. Nauk, Mehan. Zhidkosti i Gaza 2: 150-158 (in Russian)

Cheng HK (1961) Hypersonic shock-layer theory of the stagnation region at low Reynolds number. Proc. Heat Transfer and Fluid Mechanics Institute, Stanford Press, pp.161-175

Gershbein EA(1981) Asymptotical investigation of 3D viscous gas flow problem over blunt bodies with permeable surface. In: Gershbein EA, Tirsky GA (eds) Hypersonic 3D flows with physical-chemical effects. Moscow Univ. Moscow, pp 29-51 (in Russian)

Gonor AL, Ostapenko NA (1974) Analitical investigation of hypersonic gas flow over blunt bodies. In: Grigoryan SS (ed) Some questions of modern mechanics. Moscow Univ. Moscow, pp 109-121 (in Russian).

Lin TC , Rubin SG (1973) Viscous flow over cone at moderate incidence. J. Comp. and Fluids 1, 1:37-57

Peigin SV, Tirsky GA (1988) Three-dimensional problems super- and hypersonic gas flow past bodies. Advances in Science and Technology VINITI, Ser. Mech. Fluid and Gas, 22:62-177 (in Russian)

Petuhkov IV (1964) Numerical solution of 2D flows in boundary layer. In: Numerical methods of differential and integral equations solutions and quadrature formulas. Nauka, Moscow, pp.305-325 (in Russian)

Tolstyhk AI (1966) Numerical solution of supersonic viscous gas flow problems ober blunt bodies. J. Comp. Math. and Math. Phys. 6,1:113-120 (in Russian).

Viscous Hypersonic Flow over a Body Flying Through a Thermal in the Atmosphere

I.F. Muzafarov, V.U. Nabiev, S.V. Utyuzhnikov and N. K. Yamaleev

Moscow Institute of Physics and Technology, Dept. of Computer Mathematics, Dolgoprudnyi, 141700 Russia

Abstract. The free movement of a blunt slender body of revolution through a rising large-scale cloud of hot gas in the stratified atmosphere is investigated. An effective numerical method to determine the trajectory of the body is proposed. The method is based on the simultaneous solution of the flow and ballistic problems. The flow problem is solved by a highly economical (with respect to computer resources) method, which is based on the expansion in a small parameter (the angle of attack) (Tirskiy et al. 1992, 1994), in conjunction with the method of global iterations (Tirskiy et al. 1992, 1994). The proposed method allows to determine the aerodynamic coefficients, trajectory, and orientation of the body with a high accuracy. In addition, it minimizes the computer time needed to attain this accuracy. By this method it is determined how a raising cloud of hot gas influences on the trajectory, space orientation, and stability of the body. It is shown that the body can overturn when it goes through a low density raising cloud.

Key words: Hypersonic flow, Body, Thermal, Atmosphere

1. Statement of the problem

The cloud of hot gas has the following parameters

$$T(z, l) = T_a(z) + (T_{\max} - T_a(z)) \exp\left[-(l/R_T)^2\right]$$

at an initial instant in the stratified atmosphere. Here $R_T = 4$ km, $T_{\max} = 10^4$ K, $T_a(z)$ is the temperature of the undisturbed atmosphere at height z, l is the distance from the center of a spherical volume. The temperature maximum is situated at the height $H = 20$ km.

Under the action of buoyancy forces, the cloud raises and generates a vortex ring. The vortex formation is accompanied by an intensive turbulent agitation of the cool and hot air layers. In 15 seconds after the cloud begins to raise a body flights into it at a height about 22 km with the velocity of 2000 m/sec. The body is a sphere-blunt cone, which has a nose radius $R_0 = 0.1$ m, a semi-vertex angle of 15°, a length of 2 m, and a mass of 1000 kg. We assume that the center of mass is situated at a distance L from the tip. The time, at which the body is situated at the distance of 8300 m away from the axis of symmetry of the thermal, is taken as the initial time t_0. It is assumed that the body moves in a plane, which goes through the axis of symmetry of the thermal and the point, in which the body is situated at the initial instance.

At the instant when the body flies into the gas, the cloud curls up in a torus-like vortex ring rising in the atmosphere. Thus, the space structure of the free-stream flow changes in time.

2. Numerical method for solving the ballistics problem

The numerical method to calculate the air convection and diffusion in a thermal is described in Muzafarov et al. (1993). There, the flight of a blunt body along a straight line is investigated.

The gas motion is governed by the system of viscous shock layer equations. It is assumed that the gas flow is quasi-stationary between the body surface and the shock wave at every point of the trajectory because the characteristic time, during which a fluid particle goes through the shock wave, is well below the time, during which the problem parameters change.

Shock Waves @ Marseille I

Editors: R. Brun, L. Z. Dumitrescu   © Springer-Verlag Berlin Heidelberg 1995

In order to solve the equations of motion, the small parameter method is applied (Tirskiy et al. 1992, 1994). The essential of the method resides in the expansion of the unknown space solution into an asymptotic series in the angle of attack.

Application of the small parameter method yields the following representation of the aerodynamic coefficients: the drag coefficient $C_x = C_{x_0}$; the lift coefficient $C_y = \alpha C_{y_1}$; and the pitching moment $M_z = \alpha M_{z_1}$. Here C_{x_0}, C_{y_1}, and M_{z_1} are independent of α.

The system of the governing equations is solved in the following way: At an initial instant in time the system of the viscous shock layer equations is solved for given parameters of the free-stream flow. Then the system of the ballistics equations is investigated for a time Δt_g to determine the position of the center of mass and the orientation of the body in space for fixed C_{x_0}, C_{y_1}, and M_{z_1}, which depend solely on the parameters of the free-stream flow. Note that the angle of attack varies during the integration. The time step Δt_g is determined by the characteristic time, for which the parameters of the free-stream flow change. The step of the integration of the ballistic system $\Delta t_b \ll \Delta t_g$ because the characteristic time, for which the angle of attack changes, is well below the characteristic time, for which parameters of the free-stream flow change (the body oscillates around the position corresponding to zero angle of attack). After the time Δt_g the system is solved in the same sequence, however, Δt_b and Δt_g might be now different. The characteristic ratio $\Delta t_b / \Delta t_g$ is about $10^{-2} \cdots 10^{-3}$. The time step Δt_g is $0.2 \cdots 1$ sec. The actual computer time necessary to calculate one step in the physical time t, is several dozens of minutes for an IBM 386/387.

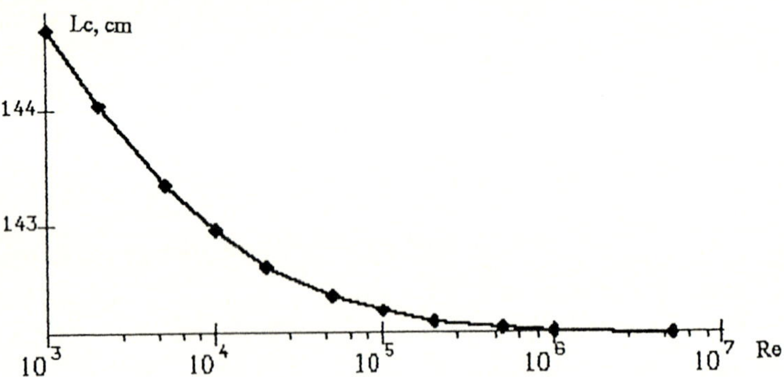

Fig. 1. The location of the center of pressure

3. Results

The gas curls up in a vortex ring under the action of the buoyancy forces. The thermal rises and the velocity field forms a torus-like structure.

In the undisturbed atmosphere, the center of pressure is behind the center of mass and the body motion is stable. When the body moves to the center of the thermal the density of the free-stream flow rapidly decreases and the temperature increases. Thus, the Mach number M and the Reynolds number Re decrease. In the following Table the values of the gas temperature T, the gas velocity W in the thermal, and the angle between the vector W and the axis of symmetry of the thermal are given at several points of the trajectory. The values of M and Re, which are determined by the parameters of the free-stream flow, are also given in the Table.

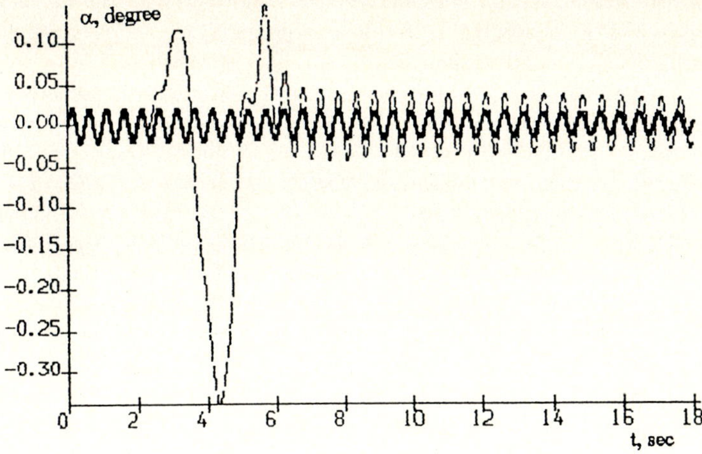

Fig. 2. The amplitude of oscillations of the angle of attack. The distance of the center of mass from the cone tip is 50 cm

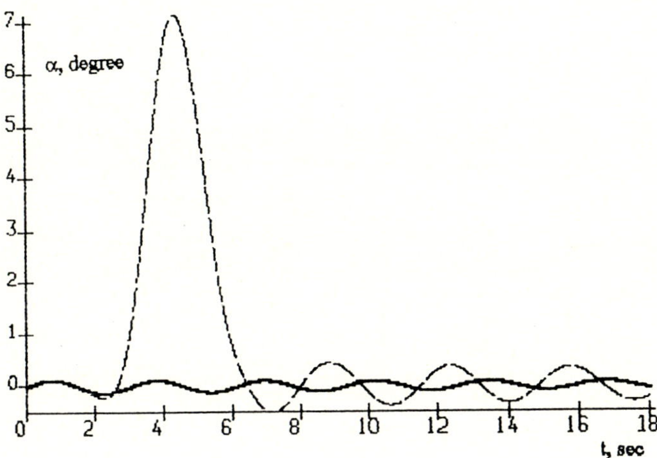

Fig. 3. The amplitude of oscillations of the angle of attack. The distance of the center of mass from the cone tip is 140 cm

Table 1.

t, sec	T, K	W, m/sec	θ	M_∞	$\mathrm{Re}_\infty \times 10^5$
2.0	220.2	15.4	108.1	6.44	8.33
3.2	420.5	60.1	26.7	4.52	2.41
3.8	623.8	184.1	4.4	3.81	1.41
4.2	663.8	235.6	0.0	3.74	1.34
4.7	614.4	199.1	3.2	3.88	1.56
5.3	440.2	79.2	15.3	4.61	2.82
6.6	219.0	17.6	145.6	6.18	8.50

If M_∞ decreases when Re_∞ is fixed then the center of pressure shifts in the directions of the cone apex. On the other side, if Re_∞ decreases when M_∞ is fixed then the center of pressure shifts in the opposite direction, as may be seen in Fig.1. However, the center of pressure shifts to the cone apex under the total influence of the change in M_∞ and Re_∞ along trajectory, i.e., its stability diminishes.

The body oscillates around a position, which corresponds to zero angle of attack, during the motion. When the body flies through the undisturbed atmosphere the amplitude of oscillations is low (Fig.2). We considered two positions of the center of mass: $L = 50$ cm (Fig.2) and $L = 140$ cm (Fig.3). When $L = 50$ cm the body is stable and the angle of attack oscillates with a low amplitude. When $L = 140$ cm the period and amplitude of the angle of attack is several times larger than in the case when $L = 50$ cm. The position, where the angle of attack is maximal, coincides with the position of the maximum of the gas velocity in the neighborhood of the center of the thermal. In this region, as one may see in Fig.2, the center of pressure is situated ahead of the center of mass and, hence, the body is unstable. In the last case the body can turn over.

When $L = 50$ cm the trajectory of the body is virtually not disturbed by the thermal as one may see in the plot. In the second case, i.e. when $L = 140$ cm, the trajectory is essentially disturbed

4. Conclusions

An efficient numerical method for the coupled solution of flow and ballistics problems is proposed. The method allows to determine the motion of a body and its orientation in space. The motion of a blunt body through a rising large-scale cloud of hot gas is considered. The position of the center of mass is investigated in dependence of the Mach and Reynolds numbers. It is shown that the body can be unstable when it flies through a thermal.

References

Tirskiy GA, Utyuzhnikov SV, Vershinin IV, Yamaleev NK (1992) Efficient numerical method for solution of some problems of supersonic viscous 3-D flows over blunt-nosed bodies. In: Proc. IUTAM Symposium on Aerothermochemistry and Associated Hypersonic Flows, Marseille, pp 214-219

Tirskiy GA, Utyuzhnikov SV, Yamaleev NK (1993) An efficient numerical method for simulation of supersonic viscous flow over a blunt body at small angle of attack. Computers and Fluids 23, 1: 103-114

Muzafarov IF, Tirskiy GA, Utyuzhnikov SV, Yamaleev NK (1992) Numerical simulation of viscous flow over body flying through a cloud of hot gas. Matemat. Modelir. (Mathematical modeling) 4, 9:55-68

Part 2: Combustion and Ram Accelerators

Hydrogen Mixing and Combustion in a High-Enthalpy Hypersonic Stream

R.T. Casey* and R.J. Stalker[†]

*School of Mechanical and Manufacturing Engineering, University of New South Wales,
PO Box 1, Kensington 2033, NSW Australia
[†] Department of Mechanical Engineering, University of Queensland, 4072 QLD Australia

Abstract. Experimental profiles of Pitot pressure measured across the wake of a hydrogen jet injected into a high enthalpy hypersonic stream were compared to the results of a semi-empirical model. The agreement between the two was good and the effect of confinement on the flow was demonstrated. Analysis demonstrated the sensitivity of the results to the stagnation enthalpy and velocity of the freestream flow.

Key words: Hypersonic flow, Combustion, Mixing

1. Introduction

Supersonic combustion offers attractive potential for airbreathing engines operating at high flight speeds (Ferri 1964). From a thermochemical point of view, temperatures at the combustor inlet should be below 2000 K or so since above this temperature, dissociation of the combustion products significantly reduces the heat release from the fuel. This then requires that the Mach number at the combustor inlet of an airbreathing engine operating at near orbital velocity, is greater than five. Although hypersonic combustion has been shown to be effective (Casey et al. 1992), there are still difficulties yet to be overcome before such an engine is brought to optimal use. The high velocity and short residence time of the working fluid in the engine accentuates the need for faster mixing. This then requires an optimal configuration for the combustion chamber to be found.

From an experimental point of view, it is necessary to correctly reproduce the stagnation enthalpy of the flow in order to correctly model the thermochemical processes. Experimental data in the literature for hypersonic combustion processes at stagnation enthalpies that are representative of near orbital flight speeds is limited and so this paper is intended to expand on the available data base. The experimental results are interpreted in terms of a semi-empirical/theoretical model which was intended to extrapolate existing theory.

2. Experimental setup

The experiments were conducted in the free piston shock tunnel T4 (Stalker and Morgan 1980) which was run in reflected shock mode. The shock heated test gas was expanded through a contoured nozzle of area ratio 110 designed to expand the flow to nominally Mach 5 flow. Both air and nitrogen were used as the test gas. The experimental test conditions were determined using the ESTC and NENZF codes (McIntosh 1968, Lordi et al. 1966) with the initial shock tunnel filling conditions, shock speed and final static pressure input to the programs. The conditions determined in this manner are set out in Table 1. The Mach number was in excess of 5 for all the experiments and the stagnation enthalpy was approximately 7.5 MJ/kg which corresponds to a flight speed of 3.9 km/s.

Shock Waves @ Marseille I
Editors: R. Brun, L. Z. Dumitrescu

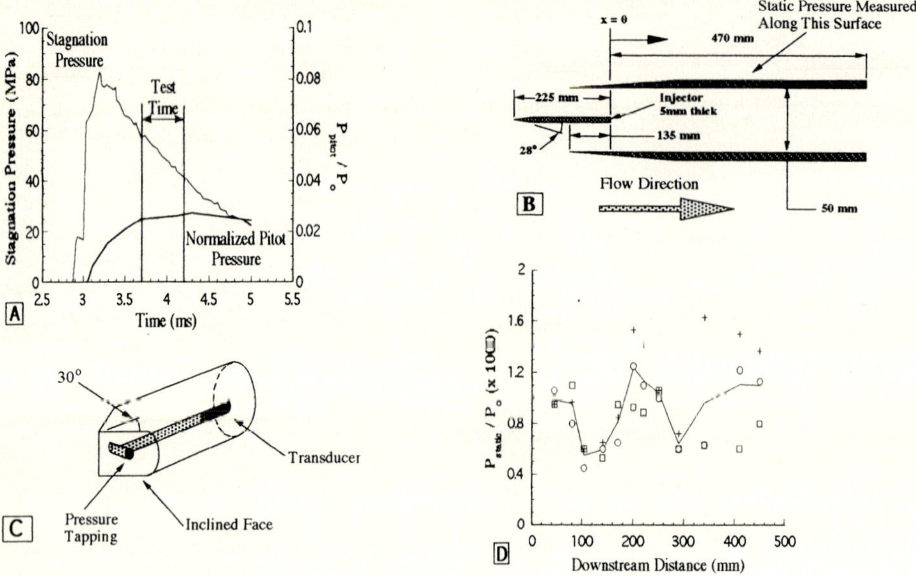

Fig. 1a-d. a. Stagnation pressure and normalized Pitot pressure, **b.** experimental setup, **c.** tips used for the Pitot probes, **d.** static pressure measurements; □ Fuel-Off, o Fuel-N_2, + Fuel-Air

Table 1. Experimental test conditions

Freestream	Conditions		Hydrogen Jet Conditions	
Quantity	**N_2**	**Air**	**Quantity**	
Total Enthalpy (MJ/kg)	7.53	7.48	Reservoir Pressure (kPa)	720
Temperature (K)	1164	1212	Reservoir Temperature (K)	298
Pressure (kPa)	40.2	40.9	Velocity (m/s)	2191
Density (kg/m^3)	0.125	0.117	Density (kg/m^3)	0.081
Velocity (m/s)	3536	3442	Mach Number	2.52
Mach Number	5.2	5.1		
Reu$\times 10^{-6}$ m^{-1}	7.7	7.7		

Reu is the unit Reynolds Number

The shock tunnel was run in undertailored mode resulting in a stagnation pressure that decreased with time. However, the Mach number variation associated with this stagnation pressure decrease was negligibly small. Moreover, the static pressure decrease was of the same order as the stagnation pressure decrease and so by normalizing Pitot pressure with the stagnation pressure, an approximate time invariant quantity was derived over the 'test time' (Fig.1a).

Hydrogen was injected from an ambient temperature reservoir through a slot injector with an aspect ratio of 10:1 (width to height) which was thought to be sufficient to make the flow essentially two dimensional. The state of the hydrogen stream at the injector exit was determined by using isentropic theory to expand the gas from the known reservoir pressure through the converging-diverging throat of the injector. The equivalence ratio for the experiments in which hydrogen was injected into air was 0.74 and the other hydrogen stream conditions are as set out in Table 1. Three types of experiment were performed in which hydrogen was injected into air (now termed Fuel-Air), hydrogen was injected into nitrogen (now termed Fuel-N_2) and no injection took place (now termed Fuel-Off). The injector spanned the mid-plane of a 50 mm × 50 mm duct that was approximately 600 long. The sharp leading edge of the injector was extended past the intake of the duct to ensure that the shock generated at the leading edge was not swallowed by the duct. Static pressure was measured along the upper wall of the duct as shown in Fig.1b.

The transverse Pitot pressure profile was measured at 150 mm, 250 mm and 450 mm downstream of the exit plane of the injector. A Pitot rake containing nine Kulite XCQ-080-250 transducers equispaced at 5.5 mm was positioned along the centerline of the duct. Experiments verified that the flow was sensibly two-dimensional over 70% of the duct width. Experiments were repeated with the Pitot probes repositioned up half the probe spacing which increased the resolution of the measurements, subject to experimental repeatability. Therefore, the experimental Pitot pressure profiles are in fact composite results of two experiments. It was expected that the pressure levels exposed to the transducers would exceed the safe limit and so probe tips were designed as shown in Fig.1c. The front surface of the probe tip had two inclined faces machined symmetrically onto it with the wedge‚angle of each surface being 30° to the flow direction. It was experimentally determined that this resulted in a reduction of 3.76 in the pressure level which is in agreement with the factor of 4 predicted by Newtonian theory.

3. Theoretical model

Weinstein et al. (1956) derived a theoretical model to describe the free isobaric diffusion of longitudinal momentum of an incompressible turbulent jet that issued into a turbulent co-flowing secondary stream in two-dimensional flow. Intrinsic in their derivation was the use of the Reichardt hypothesis which involves making the following assumptions on the mixing process: 1) The diffusion of momentum from a point source relative to a moving secondary stream is described by a Gaussian distribution, 2) No external forces exist on the mixing region and 3) Momentum from individual point sources can be superimposed to give the momentum diffusion from a source of finite area.

Table 2. Sensitivity analysis for variation of Pitot pressure

Variable	Value	Variable	Value
Stagnation Enthalpy	5.02	Freestream Velocity	10.2
Static Pressure	0.70	Spreading Coefficient[*]	0.75
Jet Velocity	1.90		

[*]The value associated with Spreading Coefficient is for $\left(\frac{\Delta Width}{Width}\right) / \left(\frac{\Delta Spread.coeff.}{Spread.Coeff.}\right)$

The value in the table is for $\left(\frac{\Delta P_{Pitot}}{P_{Pitot}}\right) / \left(\frac{\Delta Variable}{Variable}\right)$

Integration of the momentum conservation equation under the above conditions leads to

$$\frac{u - u_\infty}{u_j - u_\infty} = \left\{ \left(\frac{1}{2\epsilon}\right)^2 + \frac{1+\epsilon}{2\epsilon}\left[\mathrm{erf}\left\{\frac{a}{b}\left(\frac{y}{a}+1\right)\right\} - \mathrm{erf}\left\{\frac{a}{b}\left(\frac{y}{a}-1\right)\right\}\right]\right\}^{1/2} - \frac{1}{2\epsilon} \tag{1}$$

where

$$\epsilon = \frac{u_j - u_\infty}{2u_\infty} \tag{2}$$

In Eqs.1 and 2, u_j and u_∞ are the mean jet velocity and freestream velocity respectively, a is half the jet exit width, b is the spreading coefficient, y is the transverse coordinate and erf is the error function. It can be seen from Eq.1 that the flow is characterized by the parameter ϵ which is fixed for a given freestream and jet velocities. Moreover, although Eq.1 was derived for turbulent mixing, it contains no mathematical turbulence model to describe the nature of the turbulence, nor does it contain physical fluid properties to describe the growth rate of the mixing layer. All of this information is contained in the spreading coefficient term b/a which is derived empirically.

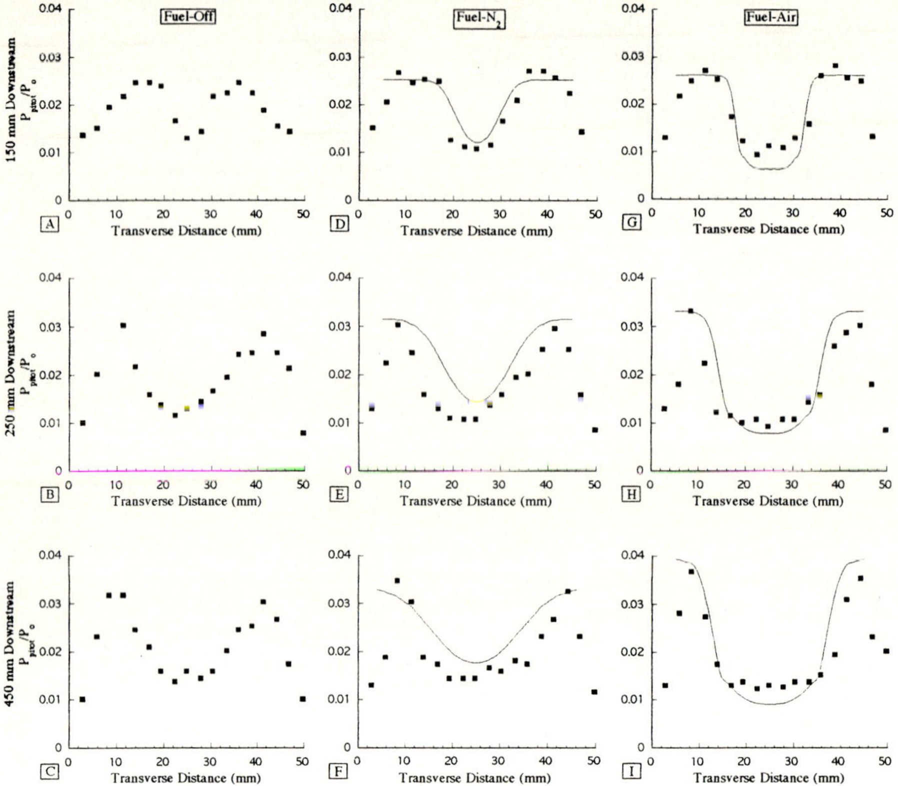

Fig. 2. Experimental and theoretical Pitot pressure profiles

Furthermore, although the initial assumption of incompressibility was used, no density terms appear in the equation.

This model was extended by incorporating the Schwarb-Zeldovich transformation to give a first-order model for the variation of species concentrations. For the case where fuel was injected into air, a chemical model for the reaction scheme

$$H_2 + \tfrac{1}{2}O_2 \rightleftharpoons H_2O \tag{3}$$

was used and consequently, the heat release associated with the combustion process was solved. Inherent in this transformation is the assumption that the Lewis and Prandtl numbers are unity which is restrictive, however it suffices as an approximation. This transformation is well documented in Kanury (1975) where the reader is referred to for greater depth of information. Equilibrium dissociation of the products of combustion was used which 'smeared out' the thin reaction zone and hence reduced the peak temperature. In total, the chemical species that were considered were, N_2, H_2, O_2, OH and NO. Pitot pressure was predicted with the Rayleigh Pitot formula, where the Mach number upstream of the Pitot probe was predicted by the theoretical model and the static pressure upstream of the probe was experimentally determined. The input parameters to the theoretical model were the static pressure, the spreading coefficient, the velocity and the stagnation enthalpy. The temperature was implicitly contained in the stagnation enthalpy. A sensitivity analysis was carried out on the theoretical model at a typical condition.

The results are presented in Table 2 where the values in the table are for the parameter

$$(\Delta P_{Pitot}/P_{Pitot})/(\Delta X/X)$$

where X is either static pressure, spreading coefficient, stagnation enthalpy or velocity. The values presented in Table 2 are representative of the whole range of conditions used for the experiments. From the table it can be seen that a 1% variation in freestream velocity will lead to over 10% variation in Pitot pressure and a 1% variation in stagnation enthalpy will lead to over 5% variation in Pitot pressure.

4. Results and discussion

The normalized wall static pressure traces for the three injection cases are presented in Fig.1d. The line in this figure joins the mean value of the three pressures for each downstream location and is intended to aid in the visual interpretation of the data but is not intended to be an interpolation of the data. There is no significant increase of the Fuel-N_2 data points over the Fuel-Off data and so there is little displacement effect due to the injected fuel. The injection conditions were the same for the Fuel-Air and Fuel-N_2 cases and so the increase in the Fuel-Air data over that for the Fuel-N_2 case is taken as evidence of the presence of combustion since hydrogen is essentially inert in the presence of nitrogen. The static pressure used in the theoretical model was read off this figure. Pitot pressure measurements were taken at 150 mm and 250 mm downstream where it can be seen from Fig.1d that the static pressure is varying sharply with downstream distance at these locations and so represents a potential source of error to the theoretical results. This is mitigated somewhat by the factor of 0.7 indicated in the sensitivity analysis for the static pressure.

The normalized Pitot pressure profiles for the Fuel-Off case are given in Fig.2a-c. The effect of the fuel jet can be seen as a dip in the center of the profile and the effect of the wall can be seen as the decrease in pressure adjacent to the upper and lower walls. The peak pressure in the Pitot pressure profile increases with increasing downstream distance however the pressure in the center of the profile, corresponding to the centerline pressure, remains approximately constant over the distance range considered. The development of the wake downstream of the injector is apparent in the widening of the profiles.

The experimental Fuel-N_2 data, presented in Fig.2d-f, shows a similar trend to the Fuel-Off data, in the increase of the peak Pitot pressure. However, there is a marginal increase with distance in the centerline Pitot pressure level for the Fuel-N_2 data. The theoretical results in these plots were obtained by varying the value of the spreading coefficient input to the model until the width of the mixing layer approximately corresponded to the experimental value. The theoretical results show good agreement for the 150 mm downstream case, in the region away from the walls. It is thought that the use of the wall static pressure in the theoretical model contributed somewhat to the mismatch in the center of the duct for the other two downstream locations in that more reliable results could be obtained if the static pressure profile across the duct was known. Since the theoretical model assumes free mixing, the wall effects cannot be predicted although increases in static pressure levels due to confinement of the flow are empirically accounted for.

The Fuel-Air data is given in Fig.2g-i where it can be seen that the agreement between the theory and experiment is good. The experimental profiles are slightly flatter in the center of the duct in comparison with the theoretical results. Again, the value of the spreading coefficient was varied until the width of the theoretical profile approximately matched that for the experimental results. From the experimental results, the mixing layer can be seen to widen as it develops however the low Pitot pressure levels in the center are maintained over the downstream range considered. The values of the spreading coefficient used for the theoretical plots given in Fig.2 are plotted in Fig.3 against normalized downstream distance. The spreading coefficients for the Fuel-N_2 case are consistently higher than that for the Fuel-Air case. Also plotted in this figure is an

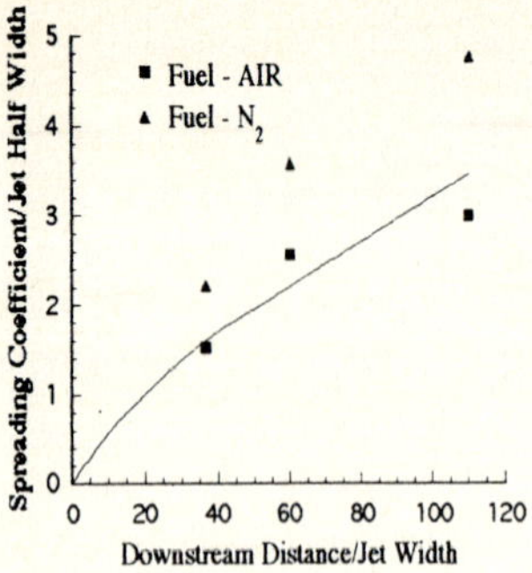

Fig. 3. Spreading coefficient versus downstream distance. The line in this figure is an extrapolation from the incompressible results.

extrapolation of the experimental results of Weinstein et al. (1956) for incompressible flow. Recent work suggests that it would be expected that the compressible experimental profile would develop more slowly than that for incompressible flow, if free mixing took place (Papamoshou and Roshko 1986). The experimental results show comparable growth between the confined compressible flow and the free incompressible flow.

5. Conclusions

The results and analysis presented above have indicated the importance of correctly modelling the stagnation enthalpy and velocity when dealing with Pitot pressure measurements in hypersonic flows. The results suggest that the influence of confinement on mixing over the range of downstream distances considered was to increase the growth rate of the compressible mixing layer to approximately the same levels as for free incompressible mixing. Finally, it may be necessary to model transverse static pressure variation to better the agreement between theory and experiment for the Fuel-N_2 case.

References

Casey RT, Stalker RJ, Brescianini C (1992) Hydrogen combustion in a hypersonic airstream. Aero. J. 96(955):200-202

Ferri A (1964) Problems in application of supersonic combustion. J; Roy. Aeron. Soc; 68:575-597

Kanury A (1975) Introduction to combustion phenomena. Gordon and Breach, NY

Lordi JA, Mates RE, Moselle JR (1966) Computer program for the numerical solution of non-equilibrium expansions of reacting gas mixtures. NASA CR-472

McIntosh MK (1968) Computer program for the numerical calculation of frozen equilibrium conditions in shock tunnels. Rep; Dept. Phys; Aust. Nat. Univ. Canberra, Australia

Papamoschou D, Roshko A (1986) Observation of supersonic free shear layers. AIAA 24th Aerosp. Sci. Meet. Reno, Nevada, USA

Stalker RJ, Morgan RG (1980) The University of Queensland free piston shock tunnel T4-initial operation and preliminary results. Proc 4th Natl. Space Eng. Symp. Adelaide, Australia

Transverse Jet Mixing and Combustion Experiments in the Hypersonic Shock Tunnel T5 at GALCIT

J. Bélanger and H. Hornung
Graduate Aeronautical Laboratories, California Institute of Technology,USA

Abstract. For ground simulation of the flows that occur in SCRAM-Jet propulsion of hyper-velocity vehicles in which the hydrogen fuel is used as a coolant, it is essential that not only the enthalpy and pressure of the air flow, but also that the pressure and speed of the hydrogen be reproduced. Both can be achieved in the free-piston shock tunnel T5 together with the new small, combustion-heated hydrogen injection shock tunnel at GALCIT. Critical elements that need to be examined are the combustion efficiency and the turbulent mixing rate, especially at the unavoidable zero-shear condition, where the fuel and air speeds are the same. Results are presented from tests with a simple configuration consisting of a rectangular duct in which the injection system was mounted flush with one of the walls. The conditions cover a range of pressures from low values chosen to match previous experiments with the same duct from the HYPULSE expansion tube facility at the General Applied Sciences Laboratory, to values corresponding to the conditions closer to those in a real SCRAM-Jet engine.

Key words: SCRAM-Jet engine, Hydrogen combustion, Hypervelocity flow

1. Introduction

The conditions at which air enters the combustor of a SCRAM-Jet engine are in the ranges $6\,\mathrm{MJ/kg} < h_0 < 18\,\mathrm{MJ/kg}$, and $40\,\mathrm{kPa} < p < 120\,\mathrm{kPa}$, where h_0 is the total enthalpy and p is the static pressure. These conditions are achieved in T5 up to a pressure of 80 kPa with a Mach number of 5.2. To reach 120 kPa, the Mach number would have to be reduced to 4.5. In designing the hydrogen injection system, several alternatives were considered, including the Ludwieg tube, previously employed by Rocketdyne in tests in T5, and the gun tunnel, in use at the T4 laboratory. The combustion-heated shock tunnel system was chosen because it can provide hydrogen flows at significantly higher speeds (reservoir temperatures up to 2000 K), and because the total amount of hydrogen needed is minimized. The latter is important for safety reasons.

2. Experimental setup

The free-piston reflected shock tunnel T5 produces test flows at the conditions required for the present experiments for a test duration of approximately 1.5 ms. For details of the facility and its performance the reader is referred to Hornung and Bélanger (1990) or Hornung et al. (1991).

2.1. The combustion-driven hydrogen-injection shock tunnel

The driver section of the combustion-driven shock tunnel is a stainless steel tube with an inside diameter of 50 mm, wall thickness of 19 mm and length of 1.5 m. Gas mixtures commonly used are 14% hydrogen, 7% oxygen and 79% helium, and 16% - 8% - 76%, respectively, by volume. The mixture is ignited by twelve automotive spark plugs uniformly distributed along the length of the tube and triggered simultaneously to avoid detonation. The driven section of the shock tunnel is a 1.92 m long stainless steel tube with an inside diameter of 25 mm and a wall thickness of 12.5 mm. This tube had to be curved because of space constraints. The radius of curvature was nowhere smaller than 11 tube diameters. No effects of the bend could be detected in the performance.

Synchronization of the two devices, T5 and hydrogen-injection tunnel, was achieved by discharging a capacitor into a probe in contact with the diaphragm, after the driver mixture is burned from an earlier signal. Tests showed repeatability to within 50 μs. For more information about the combustion-driven shock tunnel see Bélanger and Hornung (1992) or Bélanger (1993).

2.2. Instrumentation

For the present tests 19 channels of the data acquisition system were used for pressure measurements in the test model, with a sampling rate of 200 kHz. The flow visualization system is a differential interferometer using a Wollaston prism. To increase the sensitivity of the system, resonant enhancement of the refractive index of the medium was used. The technique relies on the fact that the refractive index of a gas reaches very much larger values in a spectral line than the broad-band value. This technique was implemented by seeding the hydrogen jet with sodium in the form of salt and using a light source with a wavelength just on the red side of the higher energy D-line of sodium. The light source was a tunable dye laser built in-house. The laser is pumped by a Nd:YAG laser and can produce a 40 mJ pulse at about 589 nm wavelength with a bandwidth of about 3 GHz.

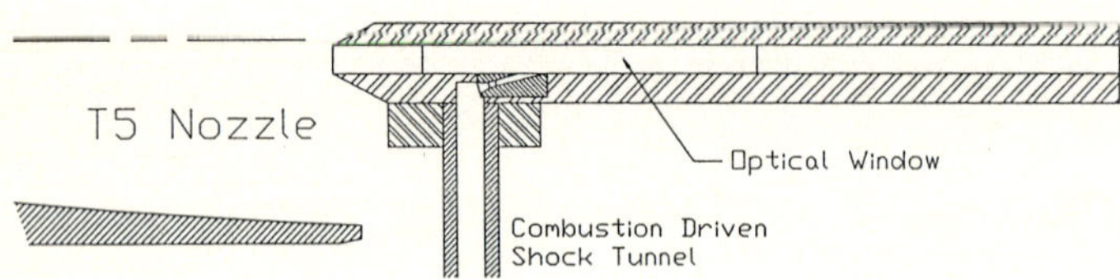

Fig. 1. Sketch of the combustor and end of the combustion-driven shock tunnel with respect to T5 exit nozzle

Fig. 2. Sketch of the opposite wall and the injector wall of the combustor with the injector position, pressure transducers and optical window locations

3. Test model

The combustor has a cross-section of 50.8×25.4 mm and is 711 mm long. Optical windows in the model side walls permit flow visualization extending from 83 mm to 387 mm from the inlet of the combustor. The injector used during these tests has a Mach 1.7, flush-mounted nozzle injecting at an angle of 15° relative to the main flow. The injector nozzle is conical with a half angle of 1.25° and exits into the combustor 177 mm downstream of the inlet from the lower wall, see Figs.1, 2.

4. Test conditions

The combustor was tested at two airflow conditions. One reproduces the conditions of the HYPULSE experiments and the second was at a much higher pressure. For each of these conditions, air and nitrogen were used to compare combustion and no-combustion tests, see Table 1.

Table 1. Main flow conditions for the HYPULSE tests and for the two sets of experiments done in air and nitrogen with the free piston shock tunnel T5

Facility Test Gas	Hypulse air	T5 air	T5 N_2	T5 air	T5 N_2
Reservoir Conditions					
Total Pressure (MPa)	142	37.5	38.0	85.0	85.0
Total Temperature (K)	8350	7860	8476	8100	8875
Total Enthalpy (MJ/kg)	15.2	15.3	15.3	15.4	15.7
Test Section Conditions					
Pressure (kPa)	16.5	18.3	16.8	43.9	38.5
Temperature (K)	2090	2125	2015	2340	2210
Density (kg/m^3)	0.028	0.028	0.027	0.063	0.063
Velocity (m/s)	5060	4785	4885	4805	5005
Mach Number	5.75	5.26	5.54	5.17	5.45
Concentration					
$[O_2]/[O_2]_{air}$	0.973	0.521	-	0.672	-
$[O]/2[O_2]_{air}$	0.003	0.361	-	0.201	-
$[NO]/2[O_2]_{air}$	0.024	0.118	-	0.127	-

For the low-pressure case, three different fuel injection conditions were tested. The first two used cold hydrogen injection, and for these the combustion driven shock tunnel was converted into a Ludwieg tube. These conditions reproduce the experiments done in HYPULSE with fuel equivalence ratios of 1 and 2. The third was done with a fuel equivalence ratio of 2 but with hydrogen reservoir temperature 1250 K. Two injection conditions were used with the high pressure tests, with fuel equivalence ratios of 1 and 2, and with hot hydrogen at reservoir temperature of 1500 K.

5. Experimental results

The set of low-pressure experiments with the fuel equivalence ratio of 2 is compared with the HYPULSE tests in Figs.3 to 6, Figs.3 and 4 showing the results for the two different gases on the injector wall and Figs.5 and 6 on the opposite wall. All the pressure measurements are nondimensionalized by the inlet pressure. For HYPULSE, this inlet pressure is the exit static pressure of the expansion tube and, for the present work, it is the average pressure of the two front pressure transducers.

The results show excellent agreement between the present experiments and those done in the expansion tube HYPULSE. The impingement of the incident bow shock on the opposite wall (Figs.5 and 6) at a distance of 250 mm from the inlet and the first reflection of the shock on the injector wall at a distance of 375 mm from the inlet (Figs.3 and 4) are very well reproduced. The large pressure levels measured on both walls between 500 and 600 mm from the inlet are believed to be caused by a side wall reflection of the bow shock. The overall pressure level was also well reproduced, except in the first part of the combustor, where the HYPULSE tests show deviations caused by a weak oblique shock coming from the HYPULSE diffuser. Very good agreement was also found in the experiments with fuel equivalence ratio of 1.

One of the main goals of our experiments was to determine if it is possible to burn the hydrogen before it exits the combustor. To do so, combustion tests in air and non-combustion

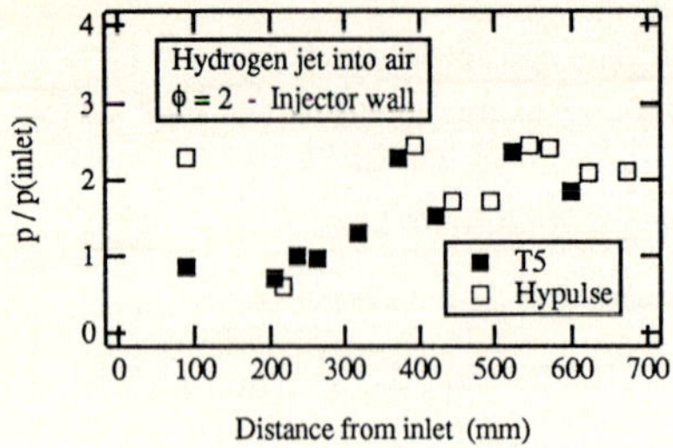

Fig. 3. Pressure measurements on the injector wall for the low-pressure case in air with a fuel equivalence ratio of 2

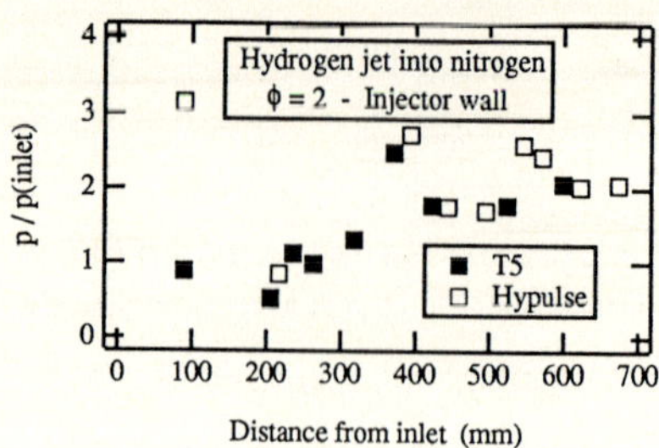

Fig. 4. Pressure measurements on the injector wall for the low-pressure case in nitrogen with a fuel equivalence ratio of 2

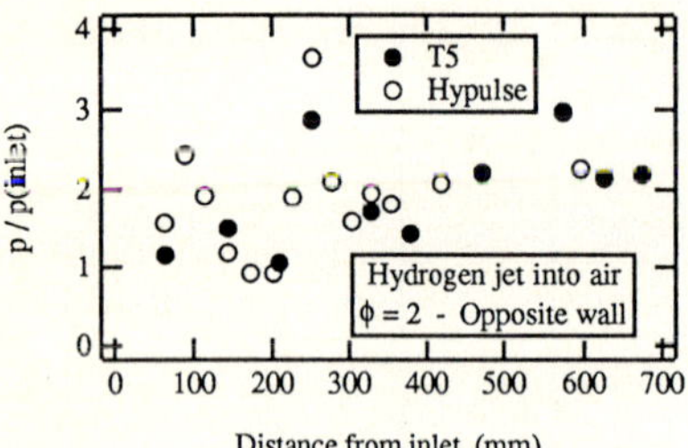

Fig. 5. Pressure measurements on the opposite wall for the low-pressure case in air with a fuel equivalence ratio of 2

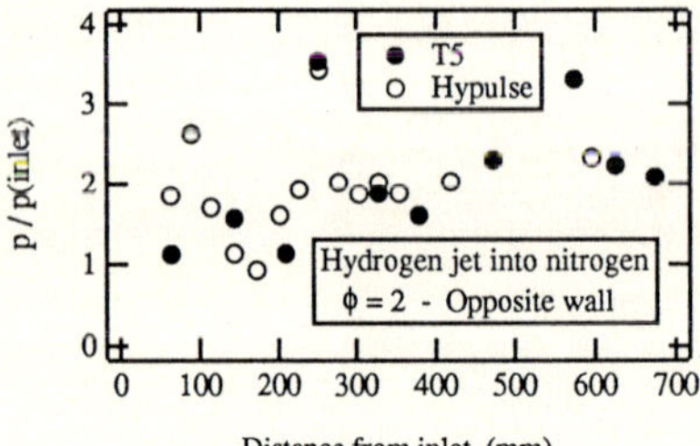

Fig. 6. Pressure measurements on the opposite wall for the low-pressure case in nitrogen with a fuel equivalence ratio of 2

tests in nitrogen are compared. As was shown in Table 1, the static pressure in the air flow is expected to be 10% higher than the one in nitrogen due to the real gas effects in the nozzle of T5. Fig.7 shows the measured pressure ratio between the air and nitrogen flow for the low-pressure case with no injection. As may be seen, the pressure ratio is very close to the expected value of 1.1 and stays at that level along the whole length of the combustor.

Fig.8 shows the pressure ratio for the low-pressure case with cold injection at a fuel equivalence ratio of 2. The very large pressure ratio of 1.6 measured on the injector wall at 525 mm from the inlet indicates some combustion in that region of the flow. This location in the combustor is exactly where the reflected shock impinges on the injector wall. Fig.9 presents the pressure ratio results for the low-pressure case with hot injection at a fuel equivalence ratio of 2. Although some of the pressure ratios are very high in the upstream part of the combustor, the ratios settle rapidly to about 1.1, indicating that no significant combustion occurred in these tests. Fig.9 shows the results for the high-pressure case with hot injection at a fuel equivalence ratio of 2. The pressure ratio is very significantly higher than the no-injection pressure ratio of 1.1. By evaluating the pressure ratio at the exit of the combustor to be about 1.6 and using a 1-D combustion model, the combustion efficiency is estimated to be close to 40% for these tests.

Finally, Figs.11 and 12 show flow visualization photos for the high pressure tests with fuel ratio of 2 in air and in nitrogen, respectively. They were taken at the same delay time, and the stagnation conditions of the combustion driven shock tunnel and T5 were within 2% of each other. As may be seen, the air and nitrogen flows are quite different. The penetration of the jet is much stronger in the air flow, and the large structures on the interface of the jet and the main flow exhibit significantly different length scales. The nitrogen flow in Fig.12 shows large structures

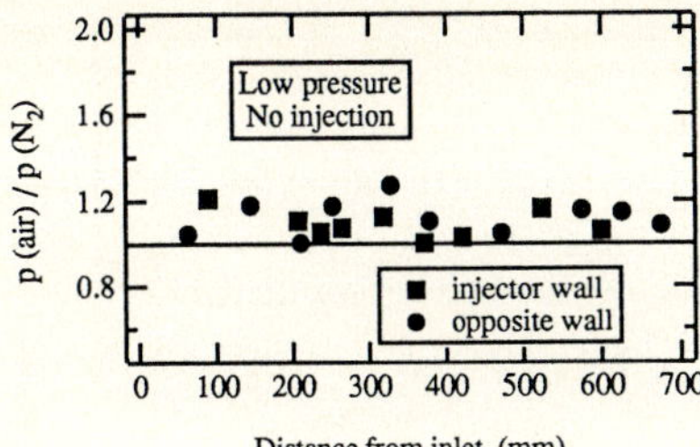

Fig. 7. Pressure ratio between the air and nitrogen flow for the low-pressure case with no injection

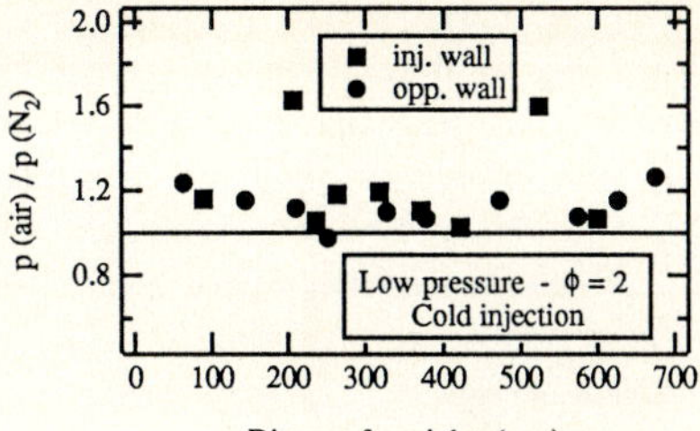

Fig. 8. Pressure ratio between the air and nitrogen flow for the low-pressure case with cold hydrogen injection with fuel equivalence ratio of 2

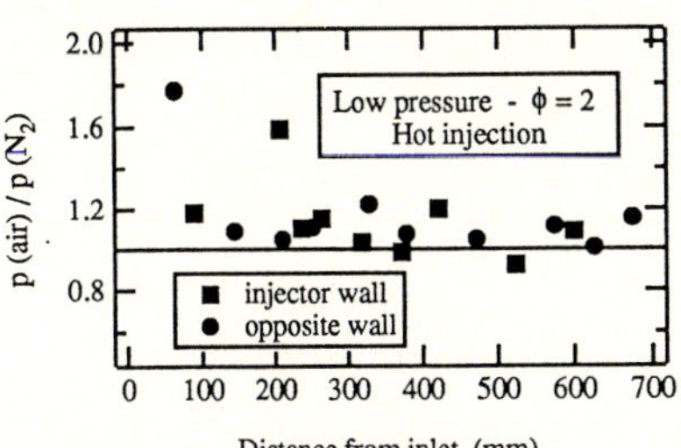

Fig. 9. Pressure ratio between the air and nitrogen flow for the low-pressure case with hot hydrogen injection and fuel equivalence ratio of 2

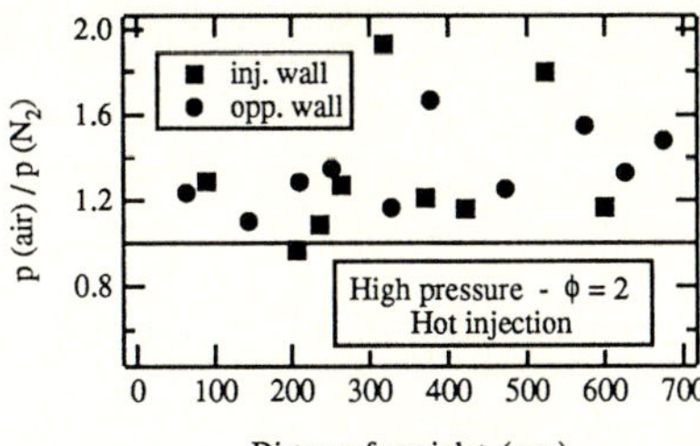

Fig. 10. Pressure ratio between the air and nitrogen flow for the high-pressure case with hot hydrogen injection and fuel equivalence ratio of 2

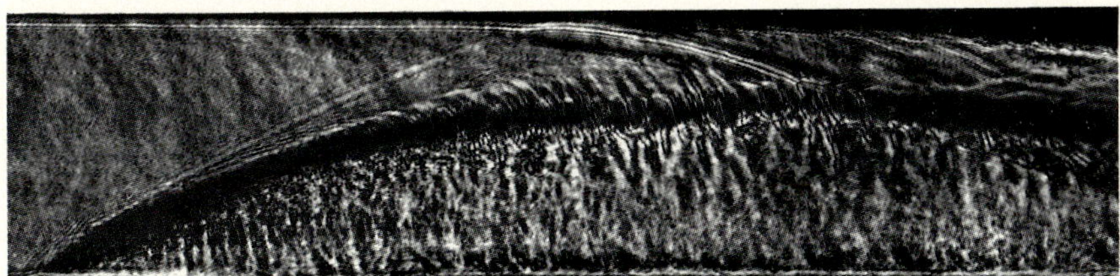

Fig. 11. Resonantly enhanced interferogram of the high-pressure case in air with hot hydrogen injection and fuel equivalence ratio of 2

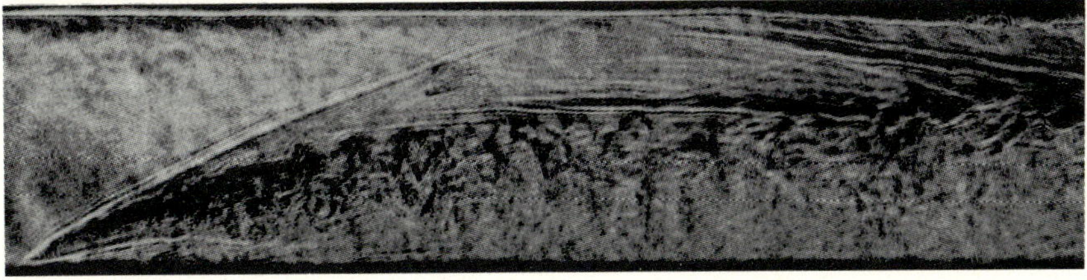

Fig. 12. Resonantly enhanced interferogram of the high-pressure case in nitrogen with hot hydrogen injection and fuel equivalence ratio of 2

with characteristic length of about 4 mm, whereas the structures in the airflow of Fig.11 display much smaller scales.

According to Hermanson and Dimotakis (1989), this difference in the length scale of the structures is due to a significant heat release at the interface of the jet and the main flow. For the case of a shear layer, they have shown experimentally that the growth rate of the shear layer thickness decreases with increasing heat release, and the size of the large structure inside the shear layer decreases with respect to the shear layer thickness as the heat release increases. These two effects add up to significantly reduce the size of the large structures for the cases where combustion occurs.

6. Conclusions

(1) Good agreement is observed between the HYPULSE test results and those obtained in T5.
(2) Some differences between the cold and hot hydrogen injection for the low-pressure tests were observed, but no evidence of combustion could be found in the hot injection case.
(3) The high-pressure case shows significant combustion according to both the pressure measurements and the flow visualization.

Acknowledgement

This work was supported by NASA Grant NAG-1-1209 (Dr. Griffin Anderson).

References

Bélanger J, Hornung HG (1992) A combustion driven shock tunnel to complement the free piston shock tunnel T5 at GALCIT. AIAA Paper 92-3968, Nashville

Bélanger J (1993) Studies of mixing and combustion in hypervelocity flows with hot hydrogen injection. CALTECH PhD thesis

Hermanson JC, Dimotakis PE (1989) Effects of heat release in a turbulent, reacting shear layer. J. Fluid Mech. 19:333-375

Hornung HG, Bélanger J (1990) Role and techniques of ground testing for simulation of flows up to orbital speed. AIAA 16th Aerodynamic Ground Testing Conference

Hornung HG, Sturtevant B, Bélanger J, Sanderson S, Brouillettte M, Jenkins M (1991) Performance data of the new free piston shock tunnel T5 at GALCIT. In: Takayama K (ed) Proc. 18th ISSW, Sendai, pp 603-610. See also Hornung (1992) AIAA Paper 92-3943, Nashville.

Three-Dimensional Mixing Flow Field in Supersonic Flow Induced by Injected Secondary Flow through a Traverse Circular Nozzle

Shigeru Aso*, Shozo Maekawa*, Michiaki Tan-Nou*, Satosh Okuyama*, Yasunori Ando[†] and Yoshiyuki Yamane[†]
*Dept. Aeronautics and Astronautics, Kyushu University, Fukuoka 812, Japan
[†]Ishikawajima-Harima Heavy Industry, Co. Ltd., Yokohama 235, Japan

Abstract. Shock wave/turbulent boundary layer interaction regions induced by gaseous secondary flows injected into supersonic flows through circular nozzles have been experimentally and computationally investigated.

In the experiments the flowfields are visualized by the Schlieren method, oil flow technique, surface pressure; spatial total pressure distributions are measured in the whole interaction region. The total pressure ratio, P_c/P_0, (P_c: total pressure of injected secondary flow, P_0: total pressure of freestream) are varied and the changes of the flowfield are investigated for circular injection. The detailed flow structures in three-dimensional mixing flow structures have been revealed. Especially, the surface oil patterns show a quite complicated flow in the interacting region with primary and secondary separations. Also, quite interesting Pitot pressure fields are revealed. The same flowfields have been simulated by solving Navier-Stokes equations with turbulent modelling. Surface pressure and spatial Pitot pressure distributions show quite good agreement with experiments. The results suggest that the numerical code is quite useful for supersonic mixing flows.

Key words: Mixing flow, SCRAM -jet engine, Shock-boundary layer interaction

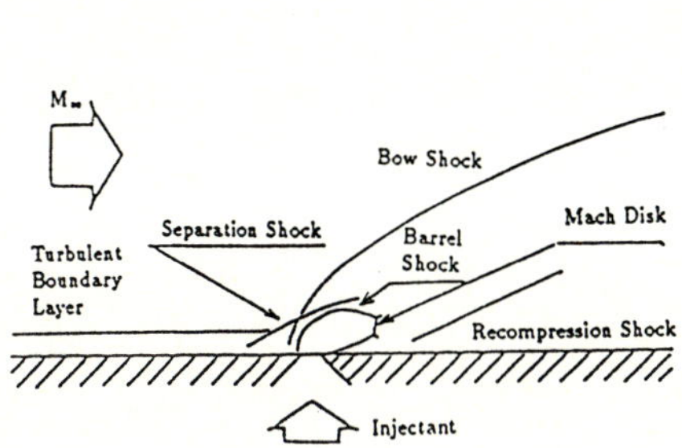

Fig. 1. Schematic diagram of the flowfield

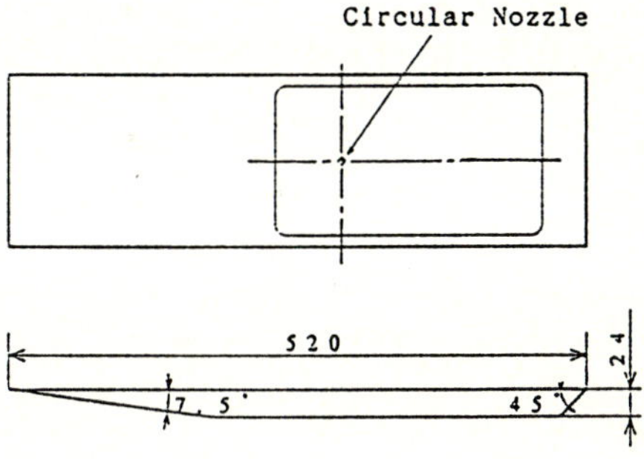

Fig. 2. Schematic diagram of the flat plate model

1. Introduction

Recently the development of SCRAM-jet engines is quite important for propulsion systems of hypersonic transports and space planes (Anderson 1987). However there are many unsolved problems. One of these is the mixing process of fuel with supersonic flow in supersonic combustors. For the design of supersonic combustors the mechanism of mixing processes between free stream supersonic flows and secondary gaseous flow should be investigated completely.

Shock Waves @ Marseille I
Editors: R. Brun, L. Z. Dumitrescu   © Springer-Verlag Berlin Heidelberg 1995

A schematic diagram of the flowfield interacting with a secondary flow through a circular nozzle is shown in Fig.1. When a secondary gaseous flow is injected into a supersonic flow through a nozzle, the flow expands in the freestream and interacts with it, then a shock wave system of Mach disk and barrel shock waves can be generated. As the freestream is blocked partly by the secondary flow, a strong bow shock wave is formed in front of the injection point. Also ahead of the injection point the boundary layer separates due to the interaction between shock waves and boundary layer. Just after the injection point the boundary layer reattaches and a recompression shock wave is generated. Hence the flowfield is quite complicated and various shock wave/boundary layer interactions can be observed in the whole region.

These flowfields have been investigated by many authors (Zukoski and Spaid 1964, Schetz et al. 1967, Hawk and Amick 1967, Spaid and Zukoski 1968, Young and Barfield 1972, King et al. 1989, Thompson 1989, Rhie et al. 1990) experimentally and numerically. These results suggested that the major controlling parameters of the flowfields are Reynolds number, freestream Mach number, boundary layer thickness, size of nozzles, total pressure ratio of the secondary flow to the main flow, and the molecular number of the secondary flow.

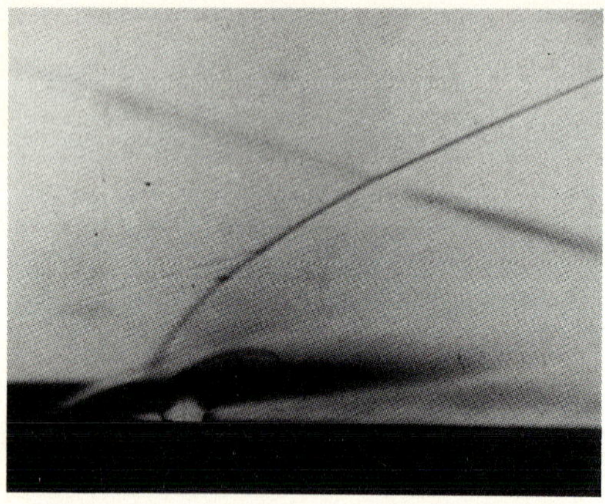

Fig. 3. Representative Schlieren picture due to circular injection ($P_c/P_0 = 0.46$)

In the previous studies fundamental experiments of mixing processes injected in supersonic flows through traverse slot were conducted (Aso et al. 1991a, Aso et al. 1991b). The flowfields were visualized by the Schlieren method. The flow structure and flow properties such as separation length and penetration height were investigated for various total pressure ratios. Also detailed surface pressure distributions have been measured.

As well as experimental studies, numerical simulations calculating the two-dimensional mixing flowfields have been conducted (Rizzetta 1992, Clark and Chan 1992). The results suggest that the whole flow patterns are well predicted qualitatively but for example the range of separated region is not predicted quantitatively.

In the present study fundamental experiments and numerical simulations of mixing processes injected in supersonic flows have been conducted. In the experiments three-dimensional mixing flows with circular injection are investigated for the practical application. The flowfields are visualized by the Schlieren method and surface pressure distributions are measured in the whole interaction region. In the experiments the total pressure ratio of the secondary flow to the stream is varied and the change of the flowfield is investigated carefully. Also surface flow patterns are visualized by the oil flow technique. The present fundamental experiments can be quite important

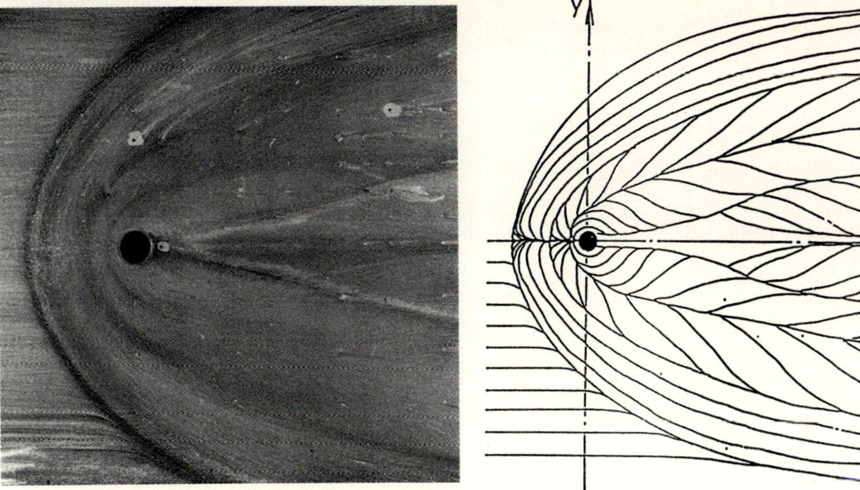

Fig. 4. Representative oil flow picture ($P_c/P_0 = 0.46$)

in order to provide a precise data set for the code validation of numerical simulations. Also as mentioned later a new experiment using Helium injection is in progress and spatial measurements of the fraction of injected Helium is in progress.

Also it is quite important to develop a reliable numerical code for testing various flow conditions of supersonic mixing flow in the design of SCRAM -jet engine. In the present study the same flowfields have been simulated by solving the three-dimensional full Navier-Stokes equations with turbulent modeling. Surface pressure and spatial Pitot pressure distributions are compared with experiments. The results show quite good agreement with experiments and the numerical code is quite useful for supersonic mixing flows.

2. Experimental apparatus and procedures

In the experiments a supersonic wind tunnel of nominal Mach number 4,with a test section of 150×150 mm, is used. A flat plate model, of width is 150 mm, is installed in the wind tunnel. A diagram of the flat plate model is shown in Fig.2. On the plate a fully developed turbulent boundary layer is established. A circular nozzle is prepared normal to the freestream in the flat plate model and sonic secondary jet flow is injected vertically through the circular nozzle into the supersonic flows of Mach number of 4. The diameter of the circular nozzle is 5 mm. Nitrogen gas (N_2) has been used as an injectant.

The experiments are conducted under the conditions of free stream Mach number of 3.75~3.81, total pressure of 1.20 MPa, total temperature, T_0, of 283~299 K and Reynolds number based on the distance between the leading edge of the flat plate and the nozzle of $1\sim 2 \times 10^7$ under the almost adiabatic wall condition. Due to the high Reynolds number, it is considered that a fully developed turbulent boundary layer is established at the injection nozzle.

In the flowfield a primary parameter is the pressure ratio, P_c/P_0, (P_c: total pressure of injected secondary flow, P_0: total pressure of free stream). The interacting flowfield between the supersonic flow and the secondary flow is visualized by the Schlieren method. Streamwise static pressure distributions on the flat plate models and spatial Pitot pressure distributions are measured with multi-tube manometers. Also some flow fields are visualized by the oil flow technique.

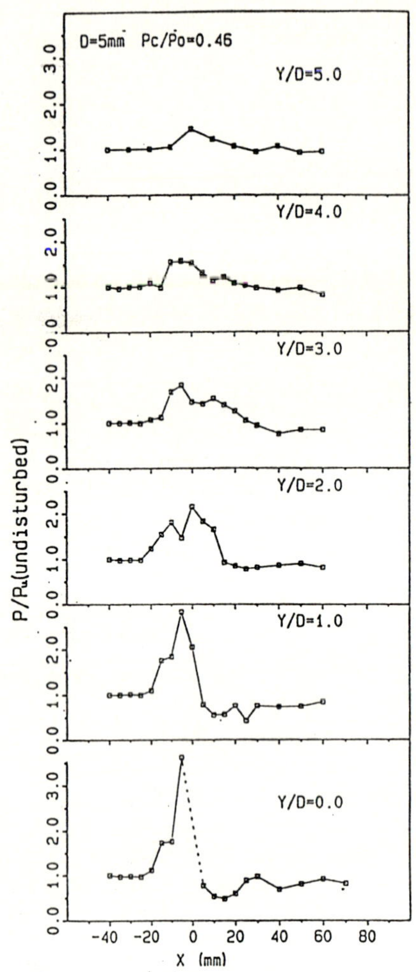

Fig. 5. Representative surface pressure distributions $(P_c/P_0 = 0.46)$

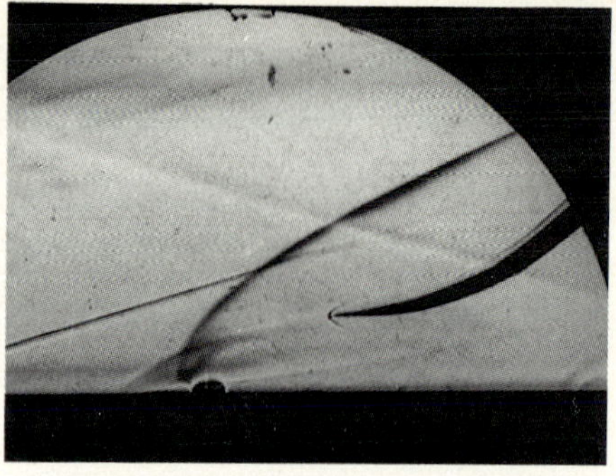

Fig.6. Schlieren picture with Pitot tube $(P_c/P_0 = 0.46)$

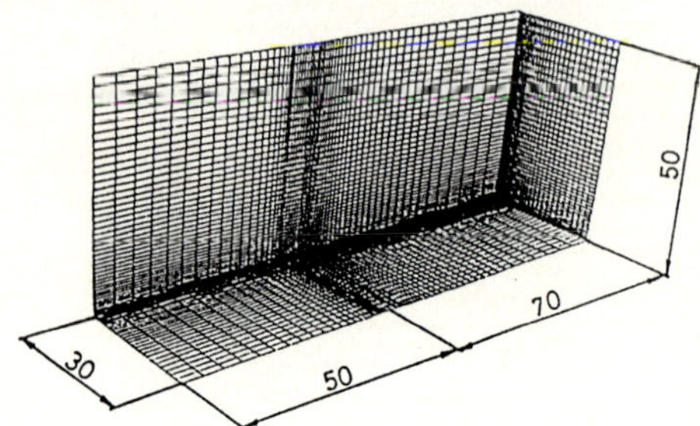

Fig.7. Mesh system(IMAX=62, JMAX=32, KMAX=50)

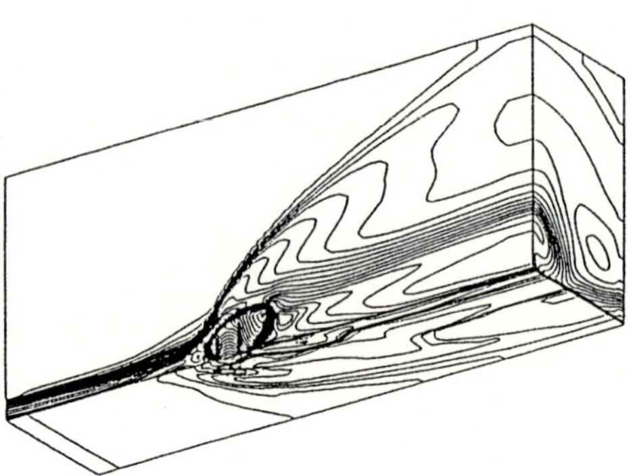

Fig.8. Mach contour $(P_c/P_0 = 0.46)$

3. Experimental results and discussions

Representative Schlieren pictures of the flowfield due to an injected jet through a circular nozzle are shown in Fig.3 for $P_c/P_0 = 0.46$. The pictures show that the barrel shock and Mach disc are bent more downstream compared with those for slotted injection. Also as P_c/P_0 is increased, the bow shock wave and barrel shock become strong and interact with the separation shock wave.

The flowfields are also visualized by the oil flow technique. Representative oil flow pictures and their sketches are shown in Fig.4 for total pressure ratios of $P_c/P_0 = 0.46$. A primary separation line is observed much forward of the circular nozzle and quite complicated surface flow patterns are observed. A secondary separation line is also observed. The results suggest that a horse-shoe vortex is generated in front of the circular nozzle and that the vortex is extended downstream apart from the centerline. Behind the circular nozzle the flow attachment lines expanding downstream and away from the centerline are observed.

Representative surface pressure distributions are shown in Fig.5 for total pressure ratio $P_c/P_0 = 0.46$. At the center line a pressure plateau and a peak pressure are observed in front of the circular nozzle. Behind the nozzle a local minimum pressure followed by local maximum pressure due to flow recompression is observed. Apart from the centerline the pressure peak decreases. However in the pressure plateau region a local minimum is observed. The oil flow picture and local minimum pressure suggest a horse-shoe separated vortex in the separated region.

In the present experiments a Pitot tube is installed in the flowfield and spatial Pitot pressure (P_{02}) distributions are measured. The measuring portion of the Pitot tube is designed small and flat enough to produce small disturbances near the tube, as shown in Fig.6. The measured spatial Pitot pressure distributions, together with calculated results, are shown in Fig.10. Quite significant changes in Pitot pressure distributions for smaller x and quite mild changes for larger x are observed. Also at the cross sections of $x = 0, 5$ and 10 mm Pitot pressure distributions of $y = 5, 10$ and 20 mm the Pitot pressures drop and keep constant when z is increased from the wall. A bow shock wave is formed above the interacting flowfields and the three-dimensional shape of the bow shock wave can be detected by those measurements.

4. Numerical procedures

For the numerical simulations the three-dimensional supersonic, compressible heat and mass transfer analysis code AIKOF3 (Ando et al. 1991) is used. In general curvilinear coordinate system (ξ, η, ζ) the governing equations are as follows:

$$\frac{\partial \tilde{Q}}{\partial t} + \frac{\partial \tilde{E}}{\partial \xi} + \frac{\partial \tilde{F}}{\partial \eta} + \frac{\partial \tilde{G}}{\partial \zeta} = \frac{1}{\mathrm{Re}} \left(\frac{\partial \tilde{E}_v}{\partial \xi} + \frac{\partial \tilde{F}_v}{\partial \eta} + \frac{\partial \tilde{G}_v}{\partial \zeta} \right)$$

For convective terms Yee and Harten's (1987) TVD scheme is used. For the viscous terms a central difference scheme is used. For turbulent modeling the Baldwin-Lomax (1978) model is used.

For the boundary conditions non-slip wall conditions are assumed on the wall and zero-derivative for pressure. The wall is assumed adiabatic. For incoming boundary condition a turbulent boundary layer thickness which is observed in the experiment is used. At exit boundary all values are extrapolated. The mesh size is $62 \times 32 \times 50$ as shown in Fig.7. In the figure the size of the computational domain is expressed by millimeters. A symmetric condition is assumed on the centerline and the half domain is solved numerically. In the mesh system the circular nozzle exit on the wall is expressed by 7×4 points.

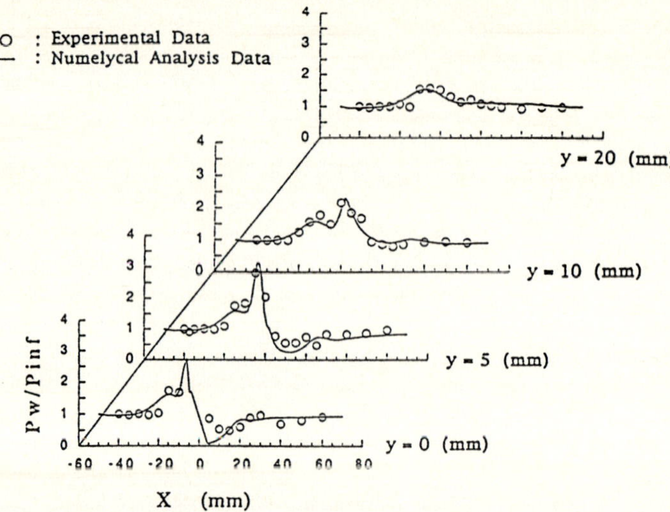

Fig. 9. Surface pressure distributions ($P_c/P_0 = 0.46$)

5. Numerical results and discussions

The Mach contour is shown in Fig.8. In the figure, Mach contours on the axis-symmetric plane, a horizontal plane near the wall ($z=0.5$ mm) and the exit boundary are shown. The shape of shock wave patterns at axis-symmetric plane shows quite good agreement with experiments of Fig.9.

Comparisons of calculated surface pressure distributions with experiments are shown in Fig.9. The calculated pressure plateau and peak pressure in the axis-symmetric plane ($y=0$ mm) and other planes ($y=5,10$ and 20 mm) parallel to the freestream show good agreements with experiments.

Comparison of calculated Pitot pressure distributions (P_{02}) with experiments are also shown in Fig.10. Data are normalized by freestream total pressure, P_0. Considering local flow deflection, the calculated P_{02} is obtained by using the velocity component parallel to the freestream. In the whole domain the calculated P_{02} shows quite good agreement with experiment except near the nozzle exit. The disagreement of calculated results with experiments near the nozzle exit may be generated by flow disturbances due to the Pitot tube, which are not included in the calculation.

Those results show that the newly developed numerical code is quite useful for the prediction of three-dimensional supersonic mixing flow structure. And calculated results show good agreement with experiments quantitatively.

6. Conclusions

The conclusions of the present studies are summarized as follows:

For three-dimensional circular injection, complicated flow structures of the three-dimensional circular injection with primary and secondary separations are revealed. This is quite different from those of slotted injection. Also a pressure plateau and peak pressure is observed at the center line. Oil flow pictures and local minimum pressure in the separated region suggest a horse-shoe separated vortex.

Also the same flowfields have been simulated by solving Navier-Stokes equations with turbulent modeling. Surface pressure and spatial Pitot pressure distributions show quite good agreement

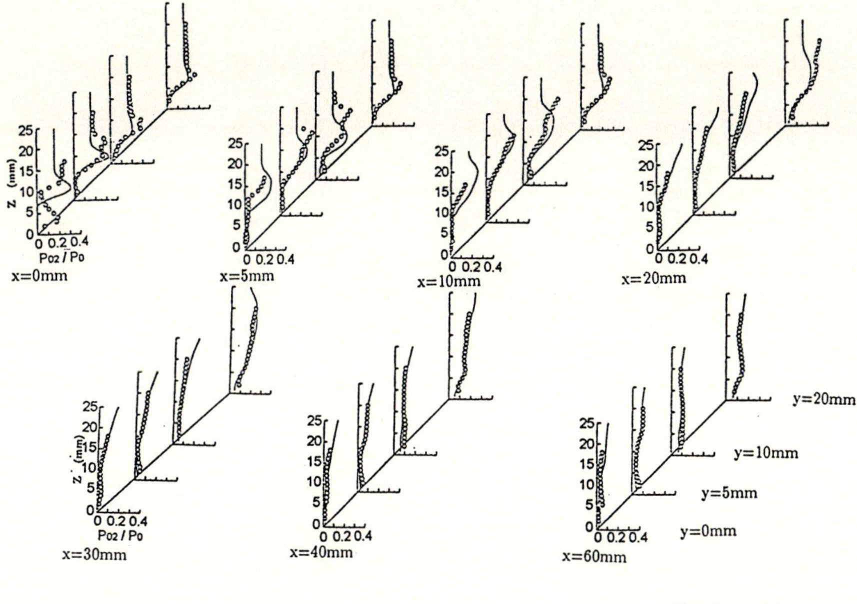

Fig. 10. Spatial Pitot pressure distributions $(P_c/P_0 = 0.46)$

with experiments quantitatively. The results show that the newly developed numerical code is quite useful for the prediction of three-dimensional supersonic mixing flow structure.

References

Anderson GY (1987) An outlook on hypersonic flight. AIAA Paper 87-2074

Ando et. al. (1991) A study of supersonic aerodynamic mixing in the Scramjet combustor. Technology Reports of Ishikawajima-Harima Heavy Industry, 31, 1: 1-7

Aso S, Okuyama S, Kawai M, Ando Y (1991a) Experimental study on mixing phenomena in supersonic flows with slot injection. AIAA Paper 91-0016

Aso S, Okuyama S, Ando Y (1991b) Experimental study on interacting secondary flow through a slot nozzle into supersonic flow. The Memoirs of the Faculty of Engineering, Kyushu University, 51, 1: 53-62

Baldwin BS, Lomax H (1978) Thin layer approximation and algebraic model for separated turbulent flows. AIAA Paper 78-0257

Clark SW, Chan SC (1992) Numerical investigation of a transverse jet for supersonic aerodynamic control. AIAA Paper 92-0639

Hawk NE, Amick JL (1967) Two-dimensional secondary jet interaction with a supersonic stream. AIAA J. 5/ 655-660

King PS, Thomas RH, Schetz JA (1989) Combined tangential-normal injection into a supersonic flow. AIAA Paper 89-0622

Rhie CM, Syed SA (1990) Critical evaluation of three-dimensional supersonic combustor calculations. AIAA Paper 90-0207

Rizzetta DP (1992) Numerical simulation of slot injection into a turbulent supersonic stream. AIAA Paper 92-0827

Schetz JA, Hawkins PF, Lehman H (1967) Structure of highly underexpanded transverse jets in a supersonic stream. AIAA J. 5: 882-884

Spaid FW, Zukoski EE (1968) A study of the interaction of gaseous jets from transverse slots with supersonic external flows. AIAA J. 6: 205-212

Thompson DS (1989) Numerical solution of a two-dimensional jet in a supersonic crossflow using an upwind relaxation scheme. AIAA Paper 89-1869

Yee HC (1987) Upwind and symmetric Shock-Capturing Schemes. NASA-TM-89464

Young CT, Barfield BF (1972) Viscous interaction of sonic transverse jets with supersonic external flows. AIAA J. 10

Zukoski EE, Spaid FW (1964) Secondary injection of gases into a supersonic low. AIAA J. 2:1689-1696

Limitations of the Ram Accelerator

M. Brouillette[*], **D.L. Frost**[†], **F. Zhang**[†], **R.S. Chue**[†], **J.H.S. Lee**[†], **P. Thibault**[‡] and **C. Yee**[‡]

[*]Département de Génie Mécanique, Université de Sherbrooke, Sherbrooke, Québec, Canada
[†]Mechanical Engineering Department, McGill University, Montréal, Québec, Canada
[‡]Combustion Dynamics Ltd, Medicine Hat, Alberta, Canada

Abstract.   We present an investigation of the validity of the pseudo steady-state approximation of the one-dimensional control volume conservation equations for the ram accelerator. It is found that this assumption is justified for most current ram accelerator operating modes. We also expose the limitations of one-dimensional conservation equations to predict the performance of ram accelerators. These shortcomings are exposed with the use of a simple, ideal ram accelerator cycle.

Key words: Ram accelerators

1. Introduction

The ram accelerator concept has generated considerable interest in different countries over the past five years. In particular, small projectiles have been accelerated to speeds greater than the Chapman-Jouguet (CJ) detonation velocity for the gas mixture (Hertzberg et al. 1991). However, from the experimental demonstration of the ram accelerator concept, it is not clear what is the maximum steady value of projectile velocity that can be attained. This is due to the highly transient nature of the acceleration process, the non-uniformity of the combustion mixtures used in multiple staging along the tube, and an insufficient tube length to establish limiting conditions.

Insight into the various operating limits for the ram accelerator concept can be obtained by performing simple one-dimensional control volume analyses. This paper examines first the validity of neglecting the gas unsteady terms in the 1-D control volume analysis of the ram accelerator. We then discuss previous inconsistencies with the Hugoniot analysis of ram accelerator performance when the pseudo-steady approximation is justified. We then present the results of an ideal ram accelerator thermodynamic cycle which exposes the limitations of one-dimensional control volume analyses in predicting the limits of ram accelerator configurations.

2. Control volume analysis

We first consider the conservation of mass, linear momentum and energy for a control volume which contains both gas and projectile (Fig.1). The control volume has a length L, with one end on the tip of the projectile, and a cross-sectional area A equal to that of the tube. The control volume is moving at the projectile velocity u_p with respect to the laboratory reference frame and its dimensions remain constant; the gas in front of the control volume is therefore at rest and the gas velocity behind the control volume is u_f in the laboratory reference frame. Assuming purely one-dimensional inflow and outflow, the conservation of mass, momentum and energy can be written as:

$$\int_{V_g} \frac{\partial \rho}{\partial t} dV - \rho_1 u_p A + \rho_2 (u_p - u_f) A = 0 \, , \tag{1}$$

$$\int_{V_g} \frac{\partial (\rho u)}{\partial t} dV + m_p \frac{du_p}{dt} + \rho_2 u_f (u_p - u_f) A + (p_1 - p_2) A = 0 \, , \tag{2}$$

Shock Waves @ Marseille I
Editors: R. Brun, L. Z. Dumitrescu

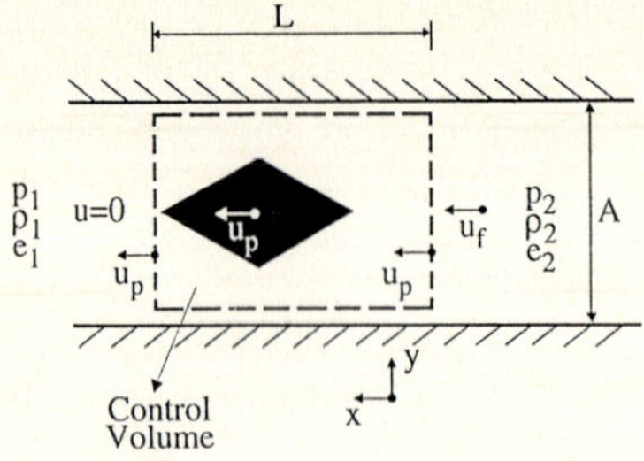
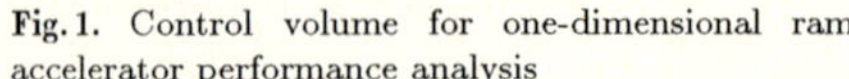

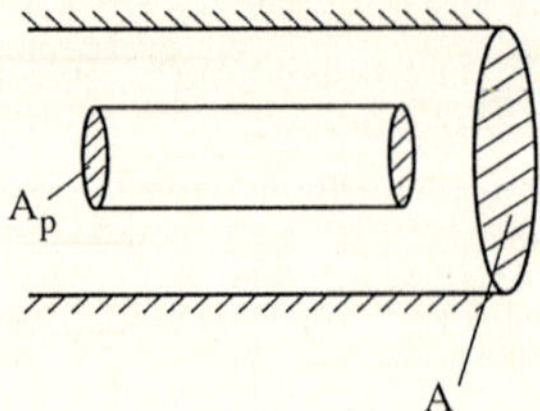

Fig. 1. Control volume for one-dimensional ram accelerator performance analysis

Fig. 2. Simplified ram accelerator geometry for Eq.11

$$\int_{V_g} \frac{\partial}{\partial t}\left[\rho\left(e+\frac{u^2}{2}\right)\right]dV + \frac{1}{2}m_p\frac{du_p^2}{dt} - \rho_1 u_p A h_1 + \rho_2(u_p-u_f)A\left(h_2+\frac{u_f^2}{2}\right)$$

$$+ (p_1-p_2)u_p A - \rho_1 u_p A \Delta q = 0, \tag{3}$$

where ρ, p, e, h and u are respectively the density, pressure, internal energy, enthalpy and velocity of the gas, Δq is the chemical heat release per unit mass of gas and m_p is the mass of the projectile. Since the control volume encloses both gas and projectile, the control volume integrals are split into the gas volume V_g, which yields the gas unsteady integral terms in the three equations, and the volume of the projectile, which yields the projectile unsteady terms in the momentum and energy conservation equations.

To quantify the relative contribution of the unsteady terms in Eq.1 to Eq.3, these equations are non-dimensionalized with respect to the upstream flow conditions ρ_1 and h_1, projectile velocity u_p and projectile characteristic time $t_p \equiv (L/a_p)^{1/2}$, where $a_p \equiv du_p/dt$ is the acceleration of the projectile. Defining $\epsilon \equiv L/(u_p t_p)$ and denoting non-dimensional quantities by $(\hat{\ })$, the conservation equations can be rewritten as:

$$\epsilon\int_{\hat{V}_g}\frac{\partial\hat{\rho}}{\partial\hat{t}}d\hat{V} - 1 + \hat{\rho}_2(1-\hat{u}_f) = 0 \tag{4}$$

$$\epsilon\int_{\hat{V}_g}\frac{\partial(\hat{\rho}\hat{u})}{\partial\hat{t}}d\hat{V} + \epsilon^2\frac{m_p}{m_g} + \hat{\rho}_2\hat{u}_f(1-\hat{u}_f) + (\hat{p}_1-\hat{p}_2) = 0 \tag{5}$$

$$\epsilon\int_{\hat{V}_g}\frac{\partial}{\partial\hat{t}}\left[\hat{\rho}\left(\hat{e}+\hat{u}^2/2\right)\right]d\hat{V} + \epsilon^2\frac{1}{2}\frac{m_p}{m_g} - 1 + \hat{\rho}_2(1-\hat{u}_f)\left(\hat{h}_2+\hat{u}_f^2/2\right) + (\hat{p}_1-\hat{p}_2) - \hat{q} = 0 \tag{6}$$

with m_g the mass of the gas within the control volume.

For typical ram accelerator conditions with $L \approx 0.1$ m, $u_p \approx 2000$ m/s and $a_p \approx 150000$ m/s² (Bruckner et al. 1991), $\hat{\rho}_2$ and $\hat{u}_f$ are $O(1)$ while $\hat{p}_2$ and $\hat{p}_1$ are respectively one and two order of magnitude smaller; therefore $\epsilon \approx 0.1$. To estimate the magnitude of the gas unsteady integral terms, it is more appropriate to use the gas flow time scale t_g rather than the projectile characteristic time scale t_p. In fact, it can be shown that:

$$\frac{\partial\hat{\rho}}{\partial\hat{t}} \approx \frac{t_p}{u_p}a_g = \epsilon\frac{a_g}{a_p} \tag{7}$$

where a_g is the average acceleration of the gas within the control volume, which can differ from that of the projectile.

To neglect the unsteady integral term in the mass conservation equation, the condition $\epsilon^2 a_g/a_p \ll O(1)$ has to be fulfilled. Because the analysis is performed with respect to the laboratory frame, the incoming gas velocity is zero and the average gas acceleration within the control volume can be represented by: $a_g \approx du_f/dt$. Since $a_p \equiv du_p/dt$, then $a_g/a_p \approx du_f/du_p$. For ram accelerator modes with constant $M_2 \equiv (u_p - u_f)/c_2$, where c is the speed of sound, and constant ratio of specific heats γ, it can be shown that:

$$\frac{a_g}{a_p} \approx \frac{du_f}{du_p} = 1 - \frac{M_1 M_2 (\gamma - 1)(1 + Q)^{1/2}}{(2 + (\gamma - 1)M_1^2)^{1/2}(2 + (\gamma - 1)M_2^2)^{1/2}} \, , \tag{8}$$

where $M_1 \equiv u_p/c_1$ and $Q = \Delta q/C_p T_1$. For example, with M_1 large and $M_2 = 1$

$$\frac{du_f}{du_p} \approx 1 - \sqrt{\frac{\gamma - 1}{\gamma + 1}}(1 + Q) \, , \tag{9}$$

which is smaller than $O(1)$ since $Q \approx 3 - 10$ for most experimental ram accelerators. Therefore, a_g/a_p is also smaller than $O(1)$ and the unsteady integral term in the conservation of mass is $O(\epsilon^2) = O(0.01)$; this term can thus be neglected in the mass conservation Eq.1.

Following a similar approach for the momentum and energy conservation equations, it is found that the unsteady integral terms can be neglected with respect to the projectile acceleration term only if $\epsilon^2 a_g/a_p \ll \epsilon^2 m_p/m_g$ or $a_g/a_p \ll m_p/m_g$. For typical ram accelerators, $m_p/m_g \approx 10$, and $a_g/a_p < O(1)$ from above; therefore the gas unsteady terms are at least an order of magnitude smaller than the projectile unsteady terms and can be neglected in the conservation of momentum and energy. This approximation is not as good as in the mass conservation equation, where the gas unsteady integral term was two orders of magnitude smaller than the other terms. Anyway, the gas unsteady terms are neglected in the conservation of momentum and energy but the projectile unsteady terms are kept.

This analysis points out that for most ram accelerator operating conditions encountered so far the pseudo-steady approximation seems justified. This assumption is likely to break down, however, if the projectile is very light or if high pressure gases are used, such that $m_p/m_g \approx 1$. The condition $a_g/a_p \ll m_p/m_g$ will also break down when the ram accelerator operating mode is changing rapidly, e.g., when the combustion mode is evolving from a subsonic combustion mode behind the projectile to supersonic combustion on the body of the projectile.

3. Pseudo-steady conservation relations

With the pseudo steady-state approximation, the unsteady integral terms can be neglected in Eqs.1-3, but the terms containing the projectile acceleration are kept. In the following sections, we examine the consequence of keeping the projectile acceleration terms in the momentum and energy equations on possible operating conditions of the ram accelerator.

3.1. Mass and momentum conservation

By combining mass and momentum conservation equations and by assuming a perfect gas, the following relation is obtained:

$$\mathcal{P} = 1 + \gamma_1 M_1^2 (1 - \mathcal{V}) + I \, , \tag{10}$$

where $\mathcal{P} \equiv p_2/p_1$ is the pressure ratio, $\mathcal{V} \equiv \rho_1/\rho_2$ the specific volume ratio and $I \equiv m_p a_p/p_1 A$ is the projectile specific impulse.

This relation is reminiscent of the Rayleigh line in gas dynamics which describes the path, in $\mathcal{P} - \mathcal{V}$ space, of the thermodynamic process taken between the initial state and the final state. Furthermore, because a Rayleigh line represents a 1-D process, all intermediate states between the end states also satisfy conservation laws.

The relation Eq.10 obtained for the pseudo-steady ram accelerator is not a Rayleigh line, however. First, the curve described by Eq.10 does not pass through the initial state $\mathcal{P} = \mathcal{V} = 1$! Therefore, in the present form, this relation **cannot** represent the thermodynamic path for the ram accelerator process, since this process has to start at $\mathcal{P} = \mathcal{V} = 1$. This important difference is due to the fact that the impulse of the projectile $m_p a_p = p_1 A I$ represents the net force exerted by the gas on the projectile. This force is the surface integral of pressure and frictional shear stresses, which depends on the flowfield around the projectile and therefore on the value of the final state $\mathcal{P}$, $\mathcal{V}$. Thus I is not an independent adjustable parameter but is a function of $\mathcal{P}$ for a given projectile velocity, geometry and heat addition. This interdepedence between the end state and the impulse parameter can be seen with a simple model. We consider a thin cylindical projectile, of cross-sectional area A_p, aligned with the tube (Fig.2). If friction is neglected and the pressure is assumed constant on both faces, with $p = p_1$ on the front face and $p = p_2$ on the back face, then the specific impulse is given by $I = (\mathcal{P} - 1)A_p/A$. Eq.10 then becomes:

$$\mathcal{P} = 1 + \frac{\gamma_1 M_1^2}{1 - \frac{A_p}{A}}(1 - \mathcal{V}) , \tag{11}$$

which does indeed pass through the initial state $\mathcal{P} = \mathcal{V} = 1$.

Second, even if Eq.10 is fixed up to start from the initial state by relating the specific impulse to the final state, the intermediate states along Eq.10 between the initial and end states will not satisfy 1-D conservation laws. This is simply because not all thermodynamic processes taking place in the ram accelerator are one-dimensional. The path in the $\mathcal{P} - \mathcal{V}$ plane between the initial and final states can indeed be quite complicated, even for idealized cases as will be seen below.

Third, unlike the conventional Rayleigh line, the speed of the projectile cannot be represented anymore by the slope of Eq.10. In particular, increasing slopes of the line joining the initial and final ram accelerator states do not necessarily reflect increasing projectile speeds. For example, consider the ram accelerator process from $\mathcal{P} = \mathcal{V} = 1$ to $\mathcal{P} = 11$ and $\mathcal{V} = 0.5$. The line joining these two points has a slope -20, but the speed of the projectile depends on the value of the specific impulse at the final state, which depends on projectile geometry and heat release. For this example, it is easy to see that if $I = 0$, i.e., that the projectile has achieved a steady speed, then $\gamma_1 M_1^2 = 20$ from Eq.10. But if $I = 5$ (projectile accelerating) then $\gamma_1 M_1^2 = 10$ and if $I = -5$ (projectile decelerating) then $\gamma_1 M_1^2 = 30$!

To conclude this point, since Eq.10 does not represent the path taken by any ram accelerator process, it can be used, at best, to compute the projectile acceleration if the end state and projectile velocity are known, or to calculate the projectile speed for a given acceleration and final state.

3.2. Mass and energy conservation

By combining mass and energy conservation equations and assuming a perfect gas behavior, one can obtain the equation:

$$\mathcal{P} = \frac{\dfrac{\gamma_1(\gamma_2 - 1)}{\gamma_2(\gamma_1 - 1)}(Q + 1)\mathcal{V}}{\mathcal{V}^2\left(1 + \dfrac{\gamma_2 - 1}{2}M_2^2\right) - \dfrac{\gamma_2 - 1}{2}M_2^2} , \tag{12}$$

which is reminiscent of the Fanno line in 1-D gas dynamics. It is not a Fanno line however, since ram accelerator processes are not one-dimensional. This curve represents all the possible end states for ram accelerator processes that conserve mass and energy for a given M_2. Even though Eq.12 is independent of the specific impulse I and projectile speed M_1, the relation between the two can be obtained for a given end state through Eq.10.

For example, for the constant volume ram accelerator processes (CVRA), $\mathcal{V} = 1$ and the pressure ratio obtained from Eq.12 becomes

$$\mathcal{P}_{\mathcal{V}=1} = \frac{\gamma_1(\gamma_2 - 1)}{\gamma_2(\gamma_1 - 1)}(Q + 1) \,, \tag{13}$$

which is independent of M_2. It can also be shown from Eq.12 that all ram accelerator processes that reduce the specific volume (i.e., $\mathcal{V} < 1$) produce final pressures larger than $\mathcal{P}_{\mathcal{V}=1}$ from Eq.13. Therefore, the minimum pressure ratio achievable with a ram accelerator process that reduces the specific volume is that attained for a constant volume process.

For the CVRA, it is not necessary to use Eq.10 to calculate the speed because $\mathcal{V} = 1$ implies $\rho_1 = \rho_2$ and from Eq.1 this forces $u_f = 0$. Therefore M_1 and M_2 can be related by:

$$M_1^2 = \frac{\gamma_2}{\gamma_1} M_2^2 \mathcal{P}_{\mathcal{V}=1} = M_2^2 \frac{(\gamma_2 - 1)}{(\gamma_1 - 1)}(Q + 1) \,, \tag{14}$$

which means that, as long that M_2 is finite, M_1 is not infinite for the CVRA, even though the slope of the line joining the initial and the final state **is** indeed infinite. Since $\mathcal{V} = 1$ for the CVRA, then Eq.10 is simply a relation between the end state and the specific impulse, i.e.,

$$I_{\mathcal{V}=1} = \frac{\gamma_1(\gamma_2 - 1)}{\gamma_2(\gamma_1 - 1)}(Q + 1) - 1 \,. \tag{15}$$

Therefore the projectile acceleration for the CVRA is independent of the projectile speed and is only a function of gas properties and heat release.

3.3. Mass, momentum and energy conservation

By combining the equations for the conservation of mass, momentum and energy the following relation is obtained:

$$\mathcal{P} = \frac{\dfrac{2\gamma_1}{(\gamma_1 - 1)}(Q + 1) - (\mathcal{V} + 1)(I + 1)}{\mathcal{V}\left(\dfrac{\gamma_2 + 1}{\gamma_2 - 1}\right) - 1} \,, \tag{16}$$

which is analogous to the conventional Hugoniot. However, since the Hugoniot represents the locus of all possible final states, Eq.16 cannot be thought as a Hugoniot curve since the specific impulse I is not an independent parameter but depends on projectile speed, geometry and heat release. Therefore Eq.16 cannot be used to calculate possible solutions since it depends on the solution!

4. Ideal ram accelerator cycles

To examine the existence of possible ram accelerator operating conditions, one has to consider the entire thermodynamic process from the initial to the final state. This can best be accomplished numerically because of the complexity of the real flowfield around the ram accelerator projectile. Some interesting insight can be obtained, however, by considering idealized ram accelerator cycles. In particular, we consider the frictionless process comprised of: supersonic isentropic compression from state 1 to state a, supersonic constant area heat addition (a–b), supersonic isentropic expansion (b–2) (Fig.3). This simplified process allows closed-form calculation of the entire flowfield around the projectile, in particular if the gas is assumed to have a constant ratio of specific heats.

As an example, a projectile with an area ratio $A_1/A_a = A_2/A_b = 5$ is considered. For a projectile velocity $M_1 = 10$, a gas with $\gamma = 1.4$ and heat release $Q = 6$, it is found that the final state is $\mathcal{P} = 4.84$, $\mathcal{V} = 1.047$. From Eq.10 the impulse is found to be $I = 10.42$ and this solution also satisfies the two other conservation equations. The path of the thermodynamic process in the

$\mathcal{P} - \mathcal{V}$ plane is shown in Fig.4; processes $1-a$ and $b-2$ are isentropes and although it looks like it because of the scale, process $a-b$ is not a constant volume process. Two interesting observations can be made from this solution. First, ram accelerator processes that increase both the pressure and the specific volume are allowable. This is contrary to 1-D conventional gas dynamics, which does not allow solutions that increase pressure and specific volume simultaneously, since they would represent solutions with imaginary initial Mach number M_1. The difference here is that the projectile can produce expansion or compression processes that are not present in conventional 1-D analyses. Second, this example shows that ram accelerator processes can produce a specific impulse I that exceeds the value of the heat release parameter Q. Previous speculations on the maximum value of specific impulse had set the limit at $I_{max} = Q$ for constant γ gases (Knowlen and Bruckner 1993). Furthermore, if it is assumed that there exists a ram accelerator process that can convert all chemical heat release into an increase in projectile kinetic energy, then the ultimate limit $I_{ult} = \gamma Q/(\gamma - 1)$ is obtained. For this example, with $Q = 6$, $I_{ult} = 21$!

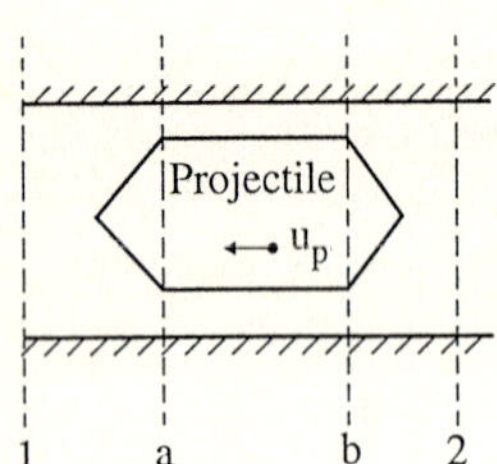

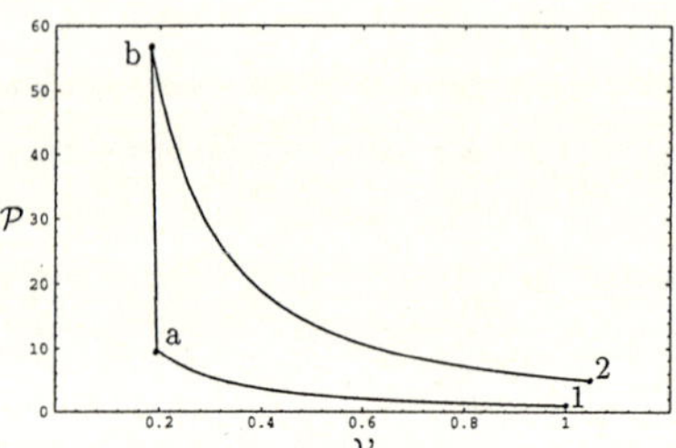

Fig. 3. Projectile geometry for ideal ram accelerator cycle

Fig. 4. Thermodynamic path in the $\mathcal{P} - \mathcal{V}$ plane for the ideal ram accelerator cycle with $A_1/A_a = 5$, $M_1 = 10$, $\gamma = 1.4$ and $Q = 6$

Even more troubling is the case of the ideal ram accelerator cycle with $\gamma = 1.4$, $Q = 0$, $A_1/A_a = A_2/A_b = 1.15$. For $M_1 = 1.46$ the solution represents an isentropic diffuser to $M_2 = 0.63$, with $\mathcal{P} = 2.64$ and $\mathcal{V} = 0.5$. The specific impulse is calculated from Eq.10 as $I = 0.15$ and this solution also satisfies the other two conservation equations. In this case, positive projectile acceleration is obtained even for $Q = 0$. Obviously, such a ram accelerator process would be able to operate isentropically only at the design Mach number and is further handicapped by subsonic outflow conditions. To be realistic, the calculation for the flowfield has to be coupled to the motion of the projectile, even though projectile acceleration terms are indeed small in these equations.

We may conclude that even if a ram accelerator solution satisfies the three 1-D conservation equations and provides a consistent thermodynamic path from the initial to the final state, pseudo-steady calculations are limited in providing the global picture. configuration.

References

Bruckner AP, Knowlen C, Hertzberg A, Bogdanoff DW (1991) Operational characteristics of the thermally choked ram accelerator. J. Propulsion 7:828-836

Hertzberg A, Bruckner AP (1988) The ram accelerator and its applications. In: Grönig H (ed) Shock Tubes and Waves, Proc 17th ISSTW, Aachen. VCH, Wendheim, pp. 117-128

Hertzberg A, Bruckner AP, Knowlen C (1991) Experimental investigation of ram accelerator propulsion modes. Shock Waves 1:17-25

Knowlen C, Bruckner AP (1993) A Hugoniot analysis of the ram accelerator. In: Takayama K (ed) Shock Waves, Proc 18th ISSW, Sendai, pp. 471-476

Stability Studies of Detonation Driven Projectiles

F. Zhang[†], D.L. Frost[*], R.S. Chue[*], J.H.S. Lee[*], P. Thibault[†] and C. Yee[†]
* McGill University, Montreal, Quebec, Canada
† Combustion Dynamics Ltd., Medicine Hat, Alberta, Canada

Abstract. The feasibility of attaining a standing detonation wave configuration around a super-sonic projectile has been studied numerically with a quasi-1D model. Instead of the classical CJ criterion, a generalized sonic choking condition has been obtained which includes the influence of area change. The transient internal flow including one-step Arrhenius kinetics is coupled with the unsteady motion of the projectile. The results indicate that the competing effects of area change and heat release drive a detonation in the supersonic moving nozzle to become more unstable.

Key words: Instability, Detonation, Supersonic flow, Ram accelerator

1. Introduction

The concept of using a stabilized detonation wave for supersonic engines or a ram accelerator has stimulated a number of recent numerical simulations of the reactive flow field around a projectile. However, the stability of a standing wave configuration around a supersonic projectile has not been considered. It forms the basis of whether or not a standing wave pattern attached to the projectile can actually be obtained from the transient development of the flow. This is the objective of the present study which continues earlier work by Lee et al. (1992). To elucidate the basic instability phenomena, a fully transient quasi-one-dimensional model is used where the influence of supersonic flow, area change and heat release on the stability of the detonation wave configuration is analyzed.

2. Physical model

A hollow projectile geometry is used which consists of a short tube with inlet and nozzle sections (Fig.1). This geometry has the essential features of supersonic engines and the ram accelerator (effectively turned inside out). The internal flow for this model is described by quasi-1D theory. In a coordinate system moving with the projectile, the quasi-1D governing equations can be derived from conservation considerations, i.e.,

$$\frac{\partial \rho}{\partial t} + \frac{\partial(\rho u)}{\partial x} = -\frac{1}{A}\frac{dA}{dx}\rho u \tag{1}$$

$$\frac{\partial(\rho u)}{\partial t} + \frac{\partial}{\partial x}\left(\rho u^2 + p\right) = -\frac{1}{A}\frac{dA}{dx}\rho u^2 - \rho a_p \tag{2}$$

$$\frac{\partial(\rho e)}{\partial t} + \frac{\partial}{\partial x}\left[u(\rho e + p)\right] = -\frac{1}{A}\frac{dA}{dx}u(\rho e + p) - \rho u a_p \tag{3}$$

$$\frac{\partial(\rho Y)}{\partial t} + \frac{\partial(\rho Y u)}{\partial x} = -\frac{1}{A}\frac{dA}{dx}\rho Y u - w \tag{4}$$

where ideal gas is considered, and

$$e = \frac{1}{\gamma - 1}\frac{p}{\rho} + QY + \frac{u^2}{2}$$

The variable u is the flow velocity in the projectile-attached frame. The reactant depletion rate w is assumed to obey a simple one-step Arrhenius law.

It is assumed that the area change has the form

Shock Waves @ Marseille I
Editors: R. Brun, L. Z. Dumitrescu   © Springer-Verlag Berlin Heidelberg 1995

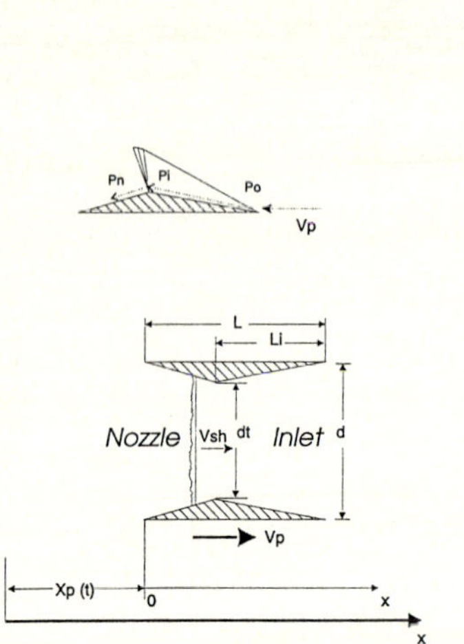

Fig. 1. Hollow projectile and coordinate systems

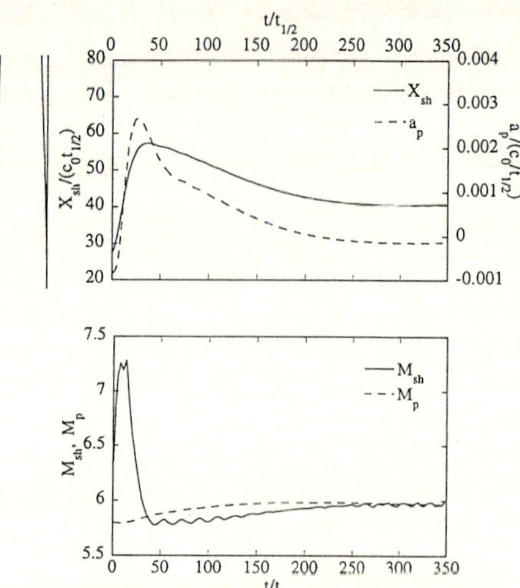

Fig. 2. Time evolution of shock front trajectory X_{sh}, acceleration of projectile a_p, detonation Mach number M_{sh}, and Mach number of projectile M_p. Projectile parameters are: $d/L = 0.84$, $L_i/L = 0.6$, $d_t/d = 0.7$, $M_{pi} = 5.8$. Detonation parameters are: $\gamma = 1.2$, $Q = 50$, $E = 22$. $t_{1/2}$ is the standard half reaction time

$$\frac{1}{A}\frac{dA}{dx} = \frac{\pi}{L - L_i} \ln \frac{d_t}{d} \sin \frac{\pi x}{L - L_i} \qquad \text{for } 0 \le x \le L - L_i \qquad (5.1)$$

$$\frac{1}{A}\frac{dA}{dx} = \frac{\pi}{L_i} \ln \frac{d_t}{d} \sin \frac{\pi(2L_i - L + x)}{L_i} \qquad \text{for } L - L_i \le x \le L \qquad (5.2)$$

which smoothes the area change displayed in Fig.1. This helps alleviate numerical noise. The projectile acceleration is determined in terms of the thrust divided by the projectile mass according to the Newtonian law, i.e.,

$$a_p = \frac{I}{m} = -\frac{1}{m} \left(\int_0^{x_{sh}} p \frac{dA}{dx} dx + D_w \right) \qquad (6)$$

where the variable x_{sh} denotes the shock front position. The thrust I is the horizontal component of the force acting on the internal projectile surface. Although the presence of the oblique shock attached to the inlet cannot be modeled directly in a one-dimensional formulation, a supersonic wave drag D_w is included to effectively account for the contribution of the oblique shock to pressure on the inlet section. For a qualitative analysis, we will assume a Cartesian geometry and neglect reflected waves (see Fig.1), and so the drag D_w is determined by the pressures behind the oblique shock wave and the Prandtl-Meyer expansion at the throat, i.e.,

$$D_w = (p_i - p_n)(A_0 - A_t) \qquad \text{for } x_{sh} < 0 \qquad (7.1)$$

$$D_w = p_i(A_0 - A_t) - p_n[A(x_{sh}) - A_t] \qquad \text{for } 0 \le x_{sh} \le L - L_i \qquad (7.2)$$

$$D_w = p_i[A_0 - A(x_{sh})] \qquad \text{for } L - L_i \le x_{sh} \le L \qquad (7.3)$$

The influence of the unsteady expansion on D_w is negligible because the pressure behind the expansion at the throat (i.e., p_n) is one to two orders less than that behind the oblique shock (i.e., p_i) for the incoming flow velocities considered. Note that D_w is used in the expression for a_p only. The equations are numerically solved using a higher order extension of Godunov's method with Strang's splitting algorithm. In this study the diameter ratio of the throat to the projectile d_t/d and the initial projectile Mach number (i.e., $M_{pi} = V_{pi}/c_0$) are variables, whereas the other projectile parameters are fixed as follows: $L_i/L = 0.6$, $d/L = 0.84$, $L = 120$ mm, $m = 51$ g, $p_0 = 11$ bar. The value of the nondimensionalized mixture parameters are $\gamma = 1.2$, $Q = 50$ (it yields a CJ detonation velocity of $D_{CJ} = 6.2c_0$), and an activation energy $E = 22$ that is much below the detonation stability limit for a planar detonation (i.e., $E_c = 25$). The detonation is initiated at the exit of the nozzle by imposing a local region of high pressure and temperature.

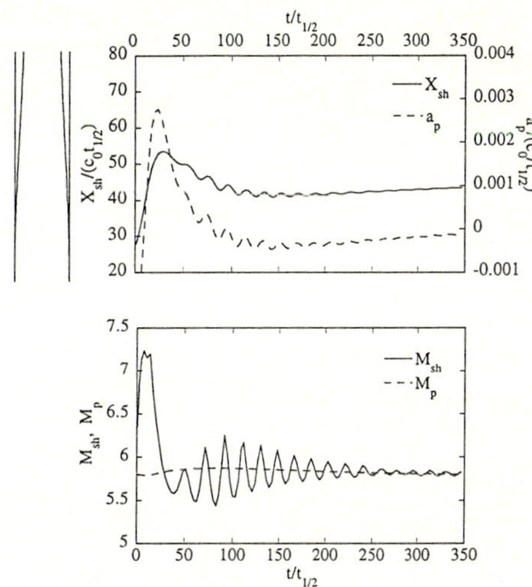

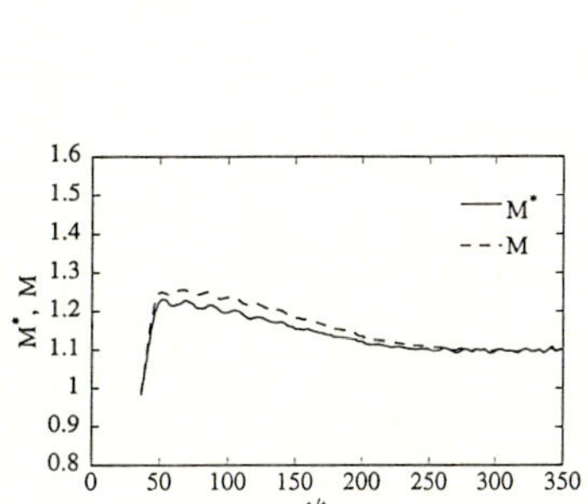

Fig. 3. Time evolution of Mach number of the product flow with respect to shock (i.e., M^*) and to projectile (i.e., M). Parameters are the same as those in Fig. 2

Fig. 4. Time evolution of shock front trajectory X_{sh}, acceleration of projectile a_p, detonation Mach number M_{sh}, and Mach number of projectile M_p. Parameters are the same as those in Fig.2 except for $d_t/d = 0.6$

3. Criteria for a stationary wave configuration

From the classical mechanical point of view, a detonation wave can be stationary over a supersonic projectile only if the projectile and the detonation velocities are equal and the acceleration of the projectile vanishes, i.e.,

$$V_p = V_{sh}, \qquad a_p = 0 \tag{8}$$

Any acceleration of the projectile will result in a movement of the detonation front with respect to the projectile, and vice versa. The question arises as to how the internal flow adjusts itself to satisfy this criterion? To answer this question let us look at the state along the characteristic C^+, because any disturbance from downstream acting on the shock front is transmitted along a right running acoustic wave. Upon manipulating Eqs.(1-4), the characteristic equation along C^+ can be obtained, i.e.,

$$C^+ : \qquad \frac{dx}{dt} = u + c \tag{9}$$

$$\left(\frac{dp}{dt}\right)^{+} + \rho c \left(\frac{du}{dt}\right)^{+} = \Psi - \rho c a_p \tag{10}$$

where

$$\Psi = (\gamma - 1)Qw - \frac{1}{A}\frac{dA}{dx}\rho c^2 u \tag{11}$$

$$\left(\frac{d}{dt}\right)^{+} = \frac{\partial}{\partial t} + (u + c)\frac{\partial}{\partial x}$$

The function Ψ describes a competing effect of the source terms of heat release and area change. If the competition between the source terms which interact with the shock front results in the flow having a right- running characteristic ($dx/dt = 0$) along which a steady state exists, such a characteristic must correspond to the sonic locus x_s (see Eq.9) on which the right-hand side of Eq.10 vanishes. A sufficient condition for disappearance of the right-hand side of Eq.10 is $\Psi = 0$ and $a_p = 0$. Then one can use a control volume analysis between the two steady states, i.e., at the sonic locus and at the projectile front to obtain a steady wave solution. Thus a condition for a steady solution (i.e., $a_p = 0$) is

$$V_p - v(x_s) = -u(x_s) = c(x_s) \tag{12}$$

$$\Psi(x_s) = 0 \tag{13}$$

where the variable v is the flow velocity in the laboratory coordinate system. Upon replacing Eq.8 into Eq.12 one obtains

$$V_{sh} - v(x_s) = c(x_s) \tag{14}$$

The criterion (13-14) gives a sonic choking plane with respect to the shock front. At this plane the competition between heat release and area change attains a balance (Zhang and Lee 1993). Any wave configuration which does not satisfy the criterion (8) or (13-14) is a transient state. For the supersonic projectile studied, the criterion (8) or (13-14) cannot be satisfied for the detonation located in the inlet because the classical supersonic diffuser instability will occur. This fact is illustrated as follows. If a detonation wave is stationary in the inlet (i.e., $V_p = V_{sh}$ and $a_p = 0$), any disturbance which, for instance, moves the shock front upstream by an increment will cause a deceleration of the projectile due to the positive area change (see Eq.6). This reduces the projectile velocity which, in turn, causes the shock front to move further forward with respect to the projectile. On the other hand, the compression waves created by the area change coupled with the reacting flow, which moves in leeward direction with respect to the projectile, amplify the shock to increase the detonation velocity (see the second term of Eq.11). Thus, the condition (8) or (13-14) is not satisfied, and finally the detonation front will be expelled out of the inlet. However, for a detonation wave propagating in the nozzle, the criterion (8) or (13-14) must be satisfied, because the oblique shock wave in the inlet creates a negative thrust which enables a balance to exist with the positive thrust behind the shock front (see also Thibault et al. 1992).

Fig.2 indicates the equivalence of the conditions (8) and (13-14). In this calculation the diameter ratio of throat to projectile is $d_t/d = 0.7$ and the initial projectile Mach number $M_{pi} = 5.8$. After an initial overshoot, the detonation Mach number (i.e., $M_{sh} = V_{sh}/c_0$) decays. Then the high pressure behind the shock front overcomes the wave drag and creates a thrust which accelerates the projectile. As the projectile Mach number (i.e., $M_p = V_p/c_0$) equals M_{sh}, the detonation penetrates furthest into the inlet while a_p reaches a maximum. Because of the projectile acceleration, M_p increases further and the detonation front moves downstream, causing a_p to decrease. Finally the detonation wave is stabilized at the position where $M_p = M_{sh}$ and $a_p = 0$ holds. In the entire transient process, the Mach number of the product flow (corresponding to the position where the reactant is consumed as up to a fraction of 10^{-3}) with respect to the shock front and to the projectile (i.e., M^* and M respectively) are supersonic due to the influence of the area change (see Fig.3). They asymptotically approach the same value as the detonation front becomes stabilized. The slightly supersonic asymptotic value is attributed to the expansion behind the sonic choking plane described by Eqs.13-14.

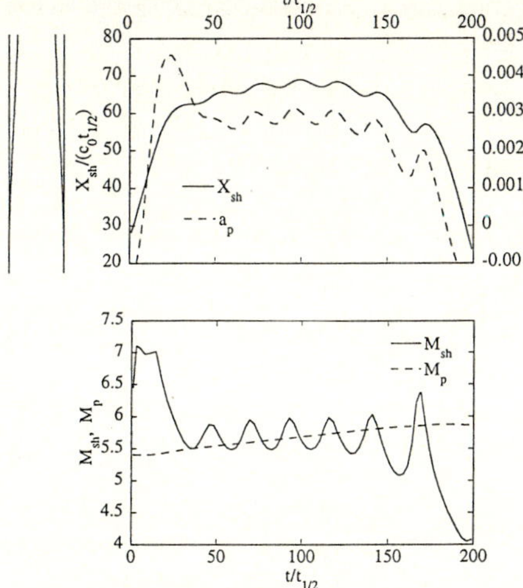

Fig. 5. Time evolution of shock front trajectory X_{sh}, acceleration of projectile a_p, detonation Mach number M_{sh}, and Mach number of projectile M_p. Parameters are the same as those in Fig.2 except for $d_t/d = 0.6$, $M_{pi} = 5.4$

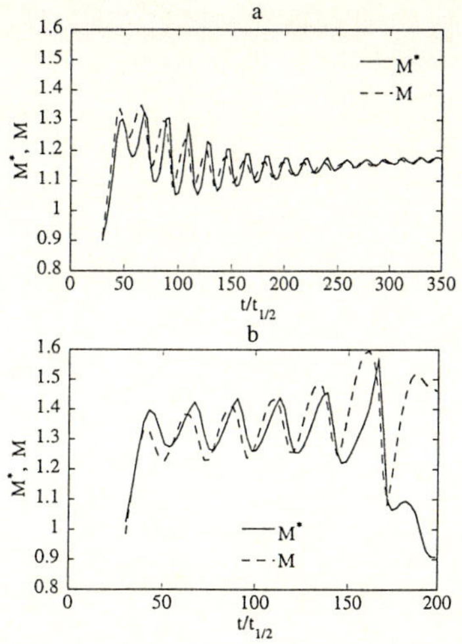

Fig. 6. Time evolution of Mach number of the product flow with respect to shock (i.e., M^*) and to projectile (i.e., M). a): Parameters are the same as those in Fig.4. b): as in Fig.5

4. Instability of the detonation wave in the moving nozzle

Chester (1960) pointed out that amplification or attenuation of a shock wave propagating into a converging channel depends on the flow conditions ahead of the shock front. This principle is also valid for a detonation wave propagating into the nozzle of the projectile studied. For a subsonic projectile, the detonation wave will be strengthened, because the velocity of the reactive flow behind the shock front is in the forward direction with respect to the projectile, whereas the area change is negative. Their coupling provides a positive term in the Ψ function (see Eqs.10-11), thus amplifying the detonation velocity. However, for a supersonic projectile, the reacting flow with respect to the projectile moves in the downstream direction. This results in a negative term in the Ψ function associated with the area change, and hence reduces the post-shock temperature and the detonation velocity.

It is well established in a one-dimensional detonation that an oscillatory wave solution occurs as the activation energy of a mixture is increased above a stability limit (e.g., Fickett and Wood 1966). A drop of the post-shock temperature has the same effect on the reaction rate as that of an increase of the activation energy, because the reaction rate is exponentially proportional to E/RT. Thus, the increase of the area change in the nozzle of a supersonic projectile in general drives a detonation to become more unstable.

For a small area change, the detonation can be stabilized within the projectile (see Fig.2). For a very large area change, the detonation decays after initiation and eventually fails. However, for a range of intermediate values of the area change, the detonation executes an oscillatory behavior as it propagates (see. Fig.4 with $d_t/d = 0.6$). Whether it can be restabilized or not, depends highly on the rate of area change $dA/(Adx)$. In Fig.4 we choose the same projectile Mach number as that used in Fig.2 while d_t/d is decreased. The detonation wave penetrates to the furthest position ($x = 52$), where $M_p = M_{sh}$ and $a_p = 0$ are attained. The oscillation of the wave front is damped and it returns downstream, due to the reduction of the rate of the area change and the increase of the pressure ratio across the front. This causes the velocity of the product flow at the end of reaction to approach its supersonic asymptotic value in an oscillatory fashion (see Fig.6.a). Thus the wave can be restabilized and a standing wave configuration can exist in the cycle-averaged sense. However, upon reducing the initial Mach number to $M_{pi} = 5.4$, the penetration depth of the detonation wave is increased beyond > 62, where the amplitude of the area change $dA/(Adx)$ is sufficiently large to induce rapid growth of the oscillation (see Fig.5). The Mach number of the product flow at the end of the reaction runs away from unity (see Fig.6.b). As a consequence, the detonation velocity cannot recover and the wave is swallowed out of the exit of the nozzle. The above analysis is based on a very low activation energy (i.e., $E = 22$), If the mixture has an activation energy over the detonation stability limit (i.e., $E_c = 25$), an oscillatory solution will occur even for a small area change.

5. Conclusions

The unsteady quasi-1D stability analysis of a detonation-driven projectile demonstrates that the detonation front stability is decreased by the area change in the supersonic moving nozzle. Depending on the kinetic properties of the mixture and the shape of the projectile, this can lead to a complete loss of thrust due to the expulsion of the detonation outside the projectile. For a given mixture and projectile shape, propulsion may be stabilized by increasing the initial velocity of the projectile, or by making the projectile larger while preserving geometric similarity. The stable detonation configuration may be analyzed by replacing the classical CJ criterion with a generalized sonic choking condition that takes into account the competition between area change and heat release.

References

Chester W (1966) The propagation of shock waves along ducts. Adv. Appl. Math. 6: 119-152, New York

Fickett W, Wood WW (1966) Flow calculations for pulsating one-dimensional detonations. Phys. Fluids 9:903-916

Lee JH, Zhang F, Chue RS (1992) Some fundamental problems of detonation instabilities and its relation to engine operation. 2nd ICASE/NASA LaRC Meeting on Combustion, Newport News, VA, Oct. 12-14

Thibault PA, Penrose JD, Sulmistras A, Murray SB, Labbe JLDS (1992) Studies on detonation driven hollow projectiles. 2nd ICASE/NASA LaRC Meeting on Combustion, Newport News, VA, Oct. 12-14

Zhang F, Lee JHS (1993) Friction-induced oscillatory behavior of one-dimensional detonations. Proc. Roy. Soc. Lond. A (in press)

Bow Shock Wave Heating and Ablation of a Sharp-Nosed Projectile Flying inside a Ram Accelerator

F. Seiler and K.W. Naumann
French-German Research Institute of Saint-Louis (ISL), F-68301 Saint -Louis, France

Abstract.   In a ram accelerator, as designed by Hertzberg et al. (1986), a sharp-nosed-body flies at supersonic velocity through a tube initially filled with a highly compressed combustible gas mixture. By shock compression, i.e., the bow wave and its reflections at the tube wall, the gas mixture is heated progressively so that it becomes ignited at the body's back giving a forward thrust to the ram-projectile. Due to the transfer of heat from gas to projectile, the latter's surface temperature increases on the one hand in the nose region (no ignition) and on the other hand at the back contour (ignition). This temperature rise can lead to melting processes which are naturally undesirable, especially at the sharp nose of the ram-projectile. The control of the heating at the nose is necessary for successful ram accelerator operation, therefore a prediction of the heat flux becomes needed. For this reason a boundary layer and an ablation model have been developed by which the projectile's nose heating and its ablation can be estimated for an optimal choice of projectile material in a desired velocity range.

Key words: Ram acceleator, Shock heating, Ablation

1. Introduction

Good experimental hypersonic test facilities are nowadays of utmost importance. In shock tubes, wind tunnels, etc. a gas flow is accelerated to high velocities and then used for developing a hypersonic flow around a stationary fixed body. In all these test facilities, as a result of the hypersonic flow formation, the gas density and temperature are generally very low. Hypersonic flight in a rarefied atmosphere at high altitudes can be studied in an excellent manner herewith. For studying the hypersonic flight at ground level in a dense atmosphere, the application of commonly used test facilities for flow acceleration is not always possible. Several authors have therefore made many proposals, e.g., Naumann (1990) to accelerate a test vehicle at high supersonic speeds for testing in free flight the gasdynamic behavior of the flow around a body.

2. Hypersonic flight in the dense ram-tube atmosphere

For vehicle acceleration to super speeds, several possibilities are known to this date, e.g., electromagnetic acceleration and light gas gun. In the last years a new type of mass accelerator, called ram accelerator, was built and successfully tested by Hertzberg et al. (1986). Following their publications, the ram accelerator can be used as an in-tube mass driver for velocities higher than 10 km/s with the gasdynamic principles well known from Ramjet- and Scramjet-engines. Fig.1, e.g., shows the thermally choked subdetonative combustion mode (in subsonic flow) and the superdetonative one (in supersonic flow) for ram accelerator operation.

Owing to ISL's need for a hypersonic mass accelerator facility, the decision was taken to build two ram accelerators: a 30-mm-tube, called RAMAC 30 and a 90-mm-one, RAMAC 90. As described by Hertzberg et al. (1986) the ram-tube is filled with a high-pressure combustible gas mixture containing, e.g., methane, oxygen and some inert gases. By pre-acceleration a subcaliber test projectile is fired supersonically with 1 km/s up to 2 km/s into the ram-tube, closed initially by membranes. By further acceleration, for example in the ISL 30-mm-caliber ram accelerator with a ram-tube length of 12 m and an initial gas pressure of 20 bars projectiles of, e.g., 60 g -

Shock Waves @ Marseille I
Editors: R. Brun, L. Z. Dumitrescu

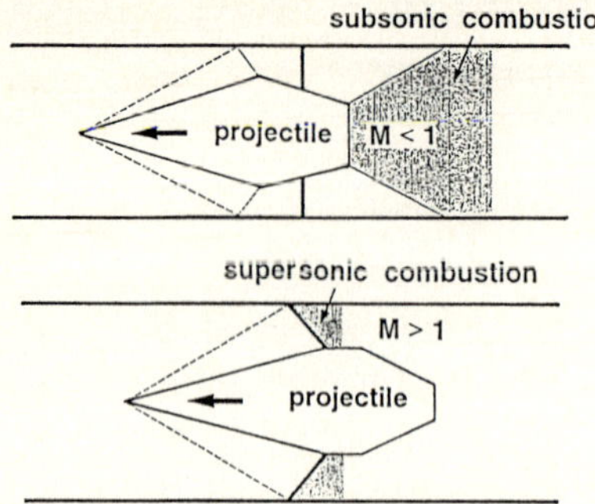

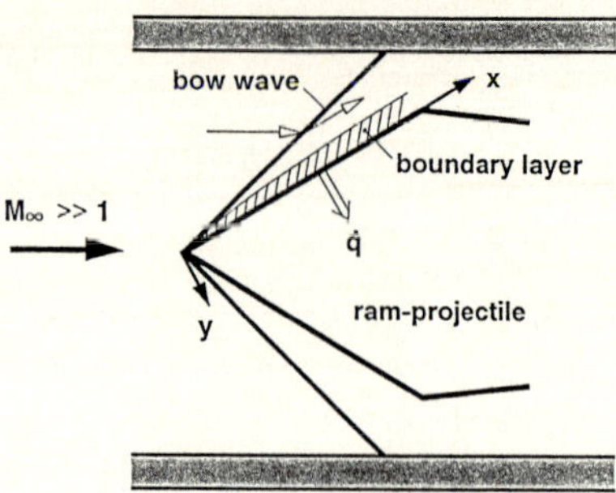

Fig. 1. Sub-and superdetonative operation modes   Fig. 2. Flow against the ram-projectile

200 g may be brought to about 3 km/s up to 4 km/s. To date the RAMAC 30 ram-tube-length is 4.5 m. The no-fin and no-sabot ram-projectiles are guided with rails fixed at the inner diameter of the ram-tube. With pre-acceleration to 1800 m/s the projectiles are fired directly into the superdetonative combustion mode.

At these high gas pressures and high mass velocities, surface heating during the in-tube flight of the projectile will be excessively high. Since no large surface deformation by ablation can be tolerated, the surface temperature during in-bore movement has to be kept lower than melting temperature so that melting processes at the projectile nose can be suppressed (or limited) in order to avoid preignition of the combustible gas mixture followed by a ram unstart.

3. Modeling of heating and ablation

3.1. Heating model

Assumptions

For calculating the temperature distribution at the surface and inside of the sharp-nose of the ram-projectile as shown in Fig.2, a flow model was developed in the reference frame of the projectile. The flow between the conical bow wave and projectile is assumed parallel to its surface. It is also assumed that at the cylindrical cone surface a compressible and turbulent boundary layer develops that will be simulated as two-dimensional in the flow model. This assumption is justified since the boundary layer thickness at the cone surface is much smaller than the radius of the projectile's cone. Although this assumption fails near the cone's tip, this small error is tolerated for obtaining an analytical solution for the description of the ram nose heating.

Solution of the boundary layer equation

Beginning with Prandtl's boundary layer equations, a differential equation was analytically found that satisfies the given boundary conditions. As a solution, for the prediction of surface and in-tube temperature distribution the heat flux $\dot{q}_w$ at the cone surface at $y = 0$ is given by:

$$\dot{q}_w(x) = a_f \left(\frac{n+1}{n+3}\right)^{\frac{2}{n+3}} [B(n)\,\varphi]^{\frac{n+1}{n+3}} \; c_p\left(T_r - T_w\right) P_r^{-2/3} \rho_e \left(\frac{\delta^{**}}{\delta}\right)^{\frac{2}{n+3}} u_e^{\frac{n+1}{n+3}} \left(\frac{\nu_e}{x}\right)^{\frac{2}{n+3}}. \qquad (1)$$

In Eq.1, u_e is the flow velocity behind the bow wave, parallel to the cone surface outside of the boundary layer. The other quantities used, i.e. $n, B(n), \varphi, \delta, \delta^{**}, T_r, T_w, c_p, \rho_e$ and ν_e are explained in Heiser et al. (1992)

Solution of the heat conduction equation

For taking into account the heat flux to the sharp-cone zone of the ram-projectile nose the calculated flat-plate heat flux of Eq.1 is transformed using a factor well known for this transformation.

For the determination of the factor a_f the calculated heat flux $\dot{q}_w$ of Eq.1 is fitted to the experimental heat transfer results of Chien (1974) on a 5° sharp cone. The best agreement, see Fig.3, is found with $a_f = 1.07$ using n = 9 for the exponent of the velocity profile inside the turbulent boundary layer.

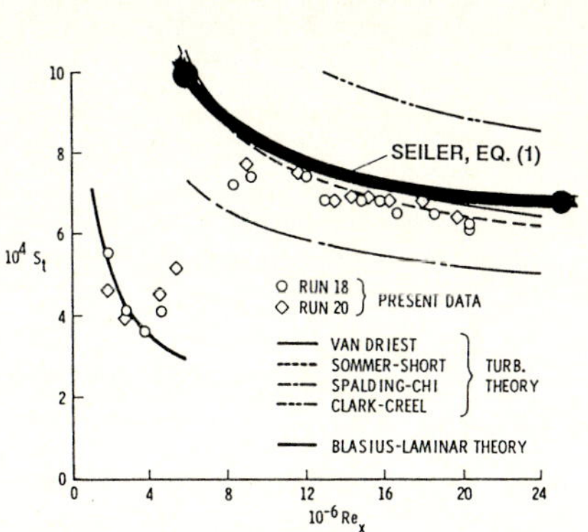

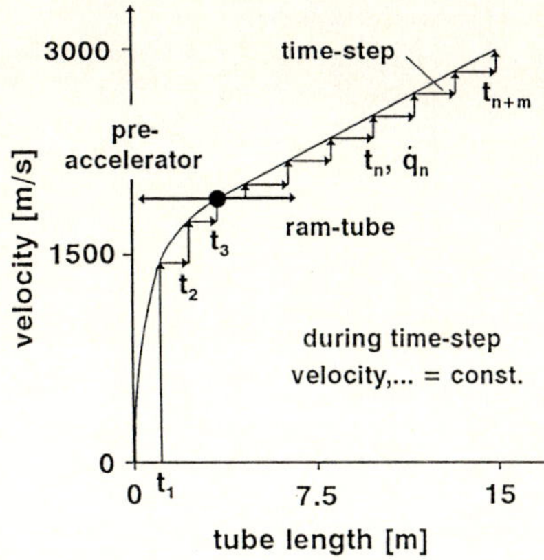

Fig. 3. Comparison of heat flux with Chien (1974)   Fig. 4. Velocity cycle and time-step approach

Owing to small variations in the calculated heat flux tangentially along the x-coordinate at the cone's surface, the heat conduction was considered one-dimensional in the y-direction, i.e., depth. Therefore, the one-dimensional heat-conduction equation could be applied inside the wall region. Thus, an analytical solution can be found for the temperature distribution ΔT_w inside the nose of the ram-projectile as a function of the heat flux at the surface at $y = 0$. The instationary, time-dependent acceleration process was taken into account by a time-step-procedure as shown in Fig.4. The real acceleration cycle expected in our RAMAC 30 is approached by successive time intervals $\Delta t_1, \Delta t_2, \ldots, \Delta t_n, \ldots$ with constant flow quantities during these steps, but changing with ongoing time. In this procedure the heat flux $\dot{q}_w(x)$ becomes additionally a function of time: $\dot{q}_w(x,t)$. Thus, the temperature distribution $T_w(x,y,t)$ is found:

$$\Delta T_w(x,y,t) = \frac{1}{\sqrt{\pi\ \rho_w\ c_w\ \lambda_w}} \int_0^t \frac{\dot{q}_w(x,t)}{\sqrt{t-\tau}}\ \exp\left(\frac{\rho_w\ c_w\ y^2}{4\ \lambda_w\ (t-\tau)}\right)\ d\tau. \tag{2}$$

3.2. Analytical ablation model

Assumptions

It is assumed that ablation occurs only by melting erosion with no evaporation. Melting erosion often takes place when a hot gas flow with a high stagnation temperature is in contact with colder walls. This process is extensively treated by many authors, see Adams (1959). In this paper an analytical ablation model for the sharp-cone ablation of a ram-projectile is described, see Seiler (1992). A similar numerical model was developed by Naumann (1993). In the analytical model it is assumed that the sharp-cone geometry remains approximately unchanged by heating and ablation (i.e. small amount of ablation). Therefore, the boundary layer formation is considered to be uninfluenced. Due to strong shear stress it is supposed that the melting is wiped away from the cone surface immediately as it is produced by heat input in case of the wall temperature exceeding the melting temperature. This means that no liquid layer develops on the solid surface. Heat addition from melt to gas flow is not considered.

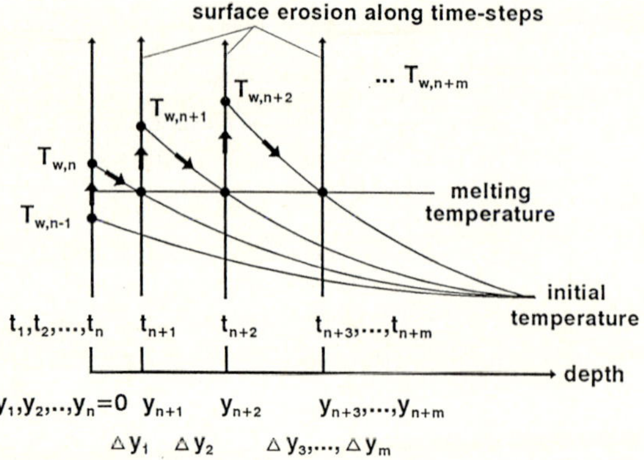

Fig. 5. Temperature $T_{w,1},\ldots T_{w,n+m}$ and ablation $\Delta y_1 \ldots \Delta y_m$ along time-step-procedure

Ablation equations

Heating and melting ablation are decoupled during the time interval Δt_n at each point x along the sharp-cone. For $t < t_n$ the wall temperature $T_{w,n-1}\,(x,y,t<t_n)\;<T_{melt}$ (melting temperature), see Fig.5. At $t = t_n$ the wall temperature $T_{w,n}\,(x,y,t)$ exceeds the melting temperature T_{melt} by the heat input $\dot{q}_{w,n}\,(x,t)$ of Eq.1. Now melting occurs in the layer $\Delta y_j\ (j = 1\ldots m)$ at time interval $\Delta t_n = t_{n+1} - t_n$ and the melting heat h_{melt} of the projectile material has to be taken into account. The total heat flux input $\dot{q}_{w,n}$ must be divided into one part $\dot{q}_{c,n}\,(x,t)$ for heat conduction and one part for heat of melting:

$$\dot{q}_{w,n}(x)\,\Delta t_n = \dot{q}_{c,n}(x)\,\Delta t_n + \rho\,h_{melt}\,\Delta y_j \qquad\qquad j = 1\ldots m. \qquad\qquad (3)$$

For obtaining the molten layer Δy_j Eq.3 can be solved together with Eq.2 and the additional assumption

$$T_{w,n}(x, y = y_{n+1}) = T_{melt} \qquad\qquad\qquad (4)$$

At the end of time interval Δt_n the molten layer Δy_j is wiped away by the shearing forces exerted by the strong wall shear stress τ_w . The described processes of melting and shearing are continued during the whole ram acceleration cycle, as:

$$t < t_n : T_{w,n-1} < T_m$$

$$t = t_n : T_{w,n} \quad > T_m \quad\xrightarrow{\;\;\Delta y_1 = y_{n+1} - y_n\;\;}\quad T_{w,n+1} = T_m,$$

$$t > t_n : T_{w,n+1} > T_m \quad\xrightarrow{\;\;\Delta y_2 = y_{n+2} - y_{n+1}\;\;}\quad T_{w,n+2} = T_m, \qquad (5)$$

$$\vdots$$

$$T_{w,n+m} > T_m \quad\xrightarrow{\;\;\Delta y_m = y_{n+m} - y_{n+m-1}\;\;}\quad T_{w,n+m} = T_m.$$

Then, the total erosion e at position x is:

$$e(x) = \sum_{j=1}^{m} \Delta y_j(x), \qquad\qquad j = 1 \ldots m. \qquad\qquad (6)$$

4. Results calculated with the heating and the ablation model

Results for the heating of the ram-projectile nose are given for the acceleration cycle of Fig.4 expected for the ISL 30-mm-caliber ram accelerator. The calculations are carried out for a projectile of magnesium and for a combustible gas mixture used nowadays in the operation of RAMAC 30 with only one stage: $C_2H_4+3O_2+6CO_2$. The data of the magnesium alloy used are listed by Naumann (1993). Initially the pressure is 20 bars and the nose cone has a length of 10 cm and an angle of $20°$, see Fig.6.

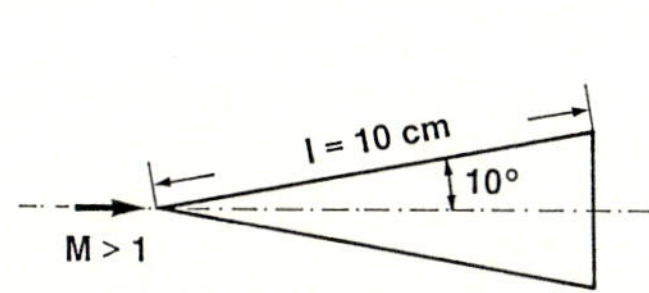

Fig. 6. Geometry of nose cone

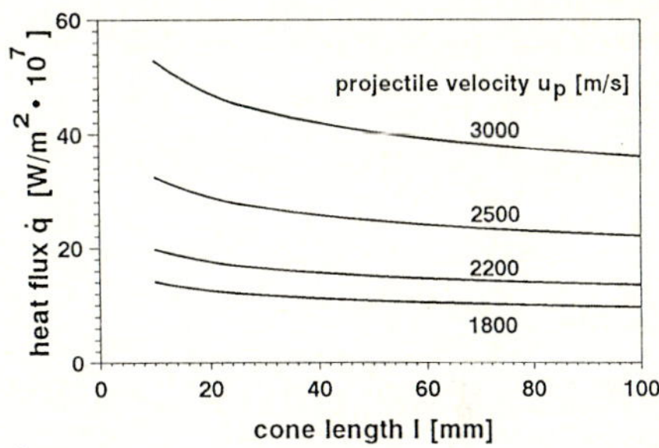

Fig. 7. Surface heat flux in ram accelerator

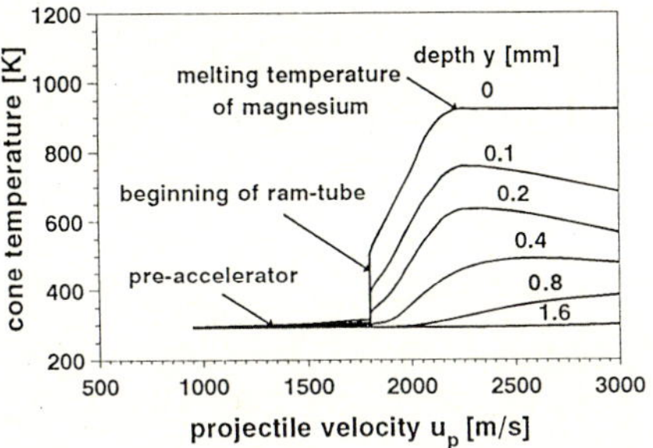

Fig. 8. Temperature profiles vs. projectile speed

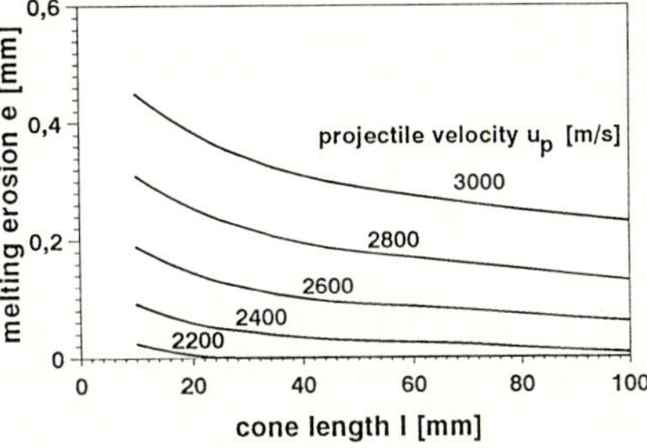

Fig. 9. Melting erosion vs. cone position x

An impression of the distribution of surface heat flux $\dot{q}_w(x)$ along the length x of the magnesium projectile nose cone is given in Fig.7. The heat flux development during flight in the pre-accelerator up to 1800 m/s is very small. Only in the ram accelerator flight the heat flux becomes important for the heating of the cone with $u_p \geq 1800$ m/s.

The heat transfer in depth y is presented in Fig.8 at a distance $x = 10$ mm from the cone tip as a function of the projectile velocity. The melting temperature of magnesium (about 923 K) is reached at a projectile velocity of about 2200 m/s. Ablation will then occur and the cone tip erosion will begin. At $x = 1$ mm the erosion onset is at 2000 m/s and at $x = 0.1$ mm at

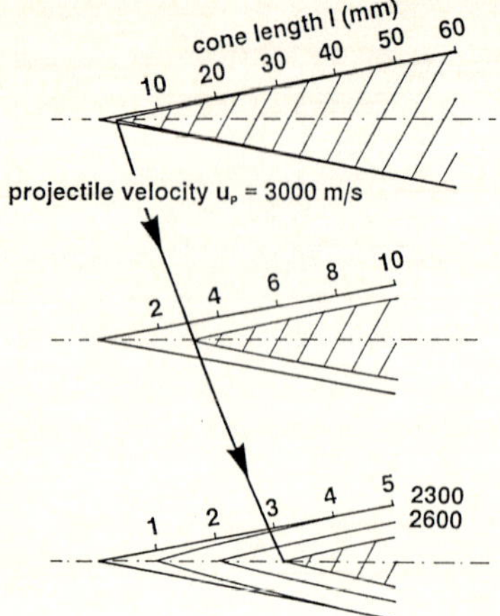

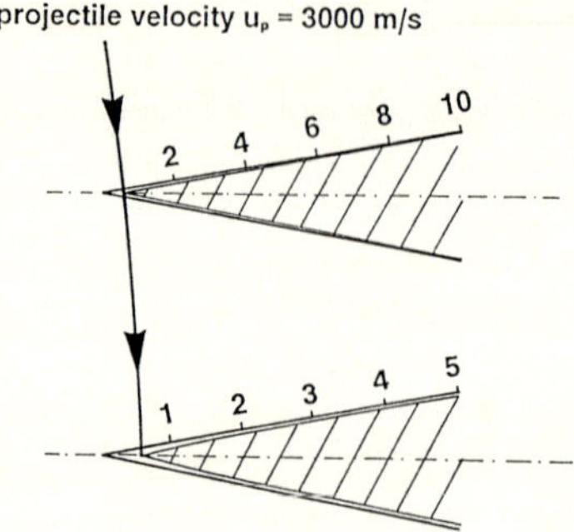

Fig. 10. Magnesium nose geometry variation   Fig. 11. Titanium nose geometry variation

1900 m/s. The melting erosion on magnesium can be calculated with Eq.6 as shown in Fig.9. The variation of the cone tip geometry by ablation with increasing projectile velocity is shown in Fig.10 for magnesium and additionally in Fig.11 for titanium by means of increasing sharp-cone magnification. As mentioned, in this model the cone does not become blunt.

5. Conclusions

Analytical methods for the heating and the ablation of a sharp-nosed body flying at hypersonic speed in a tube filled with high-compressed gas, i.e., in a so-called ram accelerator, was developed. The heat flux from gas to solid is calculated using Prandtl's boundary layer equations. The flat plate boundary-layer result is fitted to the sharp-cone measurements of Chien (1974): $a_f = 1.07$. Erosion calculations show erosion onset for magnesium projectiles at $x = 0.1$ mm from the cone tip at a projectile velocity of 1900 m/s in a single-stage combustible mixture: $C_2H_4 + 3O_2 + 6CO_2$. At 3000 m/s the calculation estimates a cone shortening of about 3 mm (see Fig.10). In contrast to that high magnesium erosion calculations with titanium, Fig.11, show about a tenth of the magnesium erosion rate.

References

Adams Mac C (1959) Recent advances in ablation. ARS Journal 29, 9:625-632

Chien K-Y (1974) Hypersonic, turbulent skin-friction and heat-transfer measurements on a sharp cone. AIAA J. 12, 11: 1522-1526

Heiser R, Seiler F, Zimmermann K (1992) Experimental and theoretical investigation of heat transfer in a gun barrel. 13th Intl. Symp. on Ballistics, Stockholm

Hertzberg A, Bruckner AP, Bogdanoff DW (1986) The Ram accelerator: a new chemical method of achieving ultra-high velocities. 37th ARA-Meeting, Quebec, Canada

Naumann KW (1990) Ram-Beschleuniger als Höchstgeschwindigkeitsversuchsanlage. ISL N602/90

Naumann KW (1993) Heating and ablation of projectiles during acceleration in a ram accelerator tube. AIAA/SAE/ASME/ASEE 29th Joint Propulsion Conference, Monterey, CA USA

Seiler F (1992) Heating and ablation of a sharp-nosed body flying at hypersonic velocity through a tube filled with highly compressed gas mixture. IUTAM Symp. Aerothermochemistry and Associated Hypersonic Flows, Marseille

In-Tube Photography of Ram Accelerator Projectiles

C. Knowlen, A.J. Higgins, A.P. Bruckner and A. Hertzberg
Aerospace and Energetics Research Program, University of Washington, Seattle, Washington, USA

Abstract.   Techniques for photographing ram accelerator projectiles operating in a transparent tube section are currently being developed at the University of Washington. The shock waves and intense pressure pulses generated in these flowfields necessitate relatively thick-walled tubes to allow for the safe observation of accelerating projectiles. Short transparent tube inserts made from polycarbonate and acrylic have sufficient strength for ram accelerator operation in propellant mixtures at reduced fill pressures (< 10 bar) and are transparent enough for direct observation of the projectile. In-tube photographs of strobe-illuminated projectiles in a 38 mm bore facility, taken with image intensifying cameras, have shown that many flow characteristics are discernible.

Key words: Ram accelerator, Flow visualization, Reacting flow, Detonation

1. Introduction

The ram accelerator is a ramjet-in-tube projectile accelerator whose principle of operation is similar to that of a supersonic air-breathing ramjet (Hertzberg et al. 1988). The projectile resembles a ramjet centerbody and is accelerated at supersonic velocity through a stationary tube filled with premixed gaseous propellants. Combustion occurs on and behind the projectile, elevating the pressure at its base and propelling the projectile forward. Acceleration through the Chapman-Jouguet detonation speed (V_{CJ}) of the undisturbed propellant mixture has been experimentally demonstrated. The ram compression of combustible gases in the tube generates a complicated flow field which exhibits many of the phenomena associated with scramjet combustors. More detailed descriptions of the ram accelerator propulsive modes can be found in Bruckner et al. (1991) and Hertzberg et al. (1991).

Techniques for visualizing the flow field phenomena associated with accelerating projectiles are currently being developed to gain a better understanding of ram accelerator operating characteristics. The intense pressure pulses arising in these flow fields necessitate relatively thick-walled tubes to allow photography of projectiles operating in combustible mixtures. Transparent tube inserts made from polycarbonate and acrylic have proven to be strong enough for ram accelerator operation at reduced fill pressures (< 10 bar) and sufficiently transparent to allow direct observation of the projectile. Preliminary experiments have been conducted to determine the durability of the transparent tubes and to examine the self-luminous phenomena observable via photography. This paper presents the first in-tube photographs of operating ram accelerator projectiles and the corresponding tube-wall pressure and luminosity data.

2. Experimental facility

The ram accelerator facility at the University of Washington consists of a light gas gun, ram accelerator test section, final dump tank, and projectile decelerator. The 16 m long test section consists of eight 2 m long steel tubes having a bore of 38 mm, and an outer diameter of 102 mm. The nominal instrument port spacing is 40 cm, though additional resolution can be obtained with short (~ 3 cm) inserts placed between existing tubes, as shown in Fig.1 (Burnham et al. 1992). The primary diagnostic instruments for the test section consist of piezoelectric pressure transducers, electromagnetic (EM) sensors, and fiber-optic luminosity probes.

Polycarbonate and cast acrylic tubes having a 38 mm bore and a wall thickness of 20 mm were found to be sufficiently transparent and robust for the initial in-tube photography attempts at

Shock Waves @ Marseille I
Editors: R. Brun, L. Z. Dumitrescu   © Springer-Verlag Berlin Heidelberg 1995

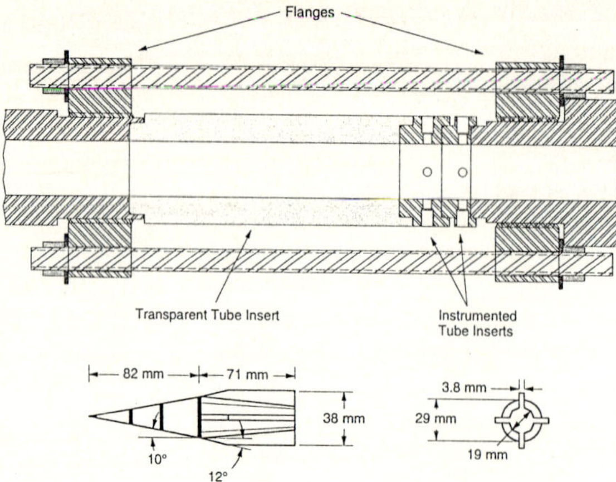

Fig. 1. Transparent tube assembly with instrumented tube inserts. Experimental projectile with dimensions is shown below at same scale

low pressures. Prototypes of these tube designs were hydrostatically burst at pressures of 420 bar and 260 bar, respectively. The 17 cm long transparent tube is clamped between two flanges along with the two four-port inserts (Fig.1). This assembly is installed at the end of the ram accelerator test section and is three-quarters enclosed by an 8 mm thick steel cylinder to shield the laboratory from a potential burst failure. Polaroid photographs are taken with Beckman & Whitley (BW) and Space Technology Laboratories (STL) image intensifying cameras having shutter speeds of 1 μsec and 0.2 μsec, respectively. The BW camera provides a single frame image and the STL camera projects three images at 10 μsec intervals onto the film. Illumination of the projectile is provided by a single high-intensity strobe light. Signals from two EM probes located ~ 70 cm upstream of the transparent tube are input into a timing circuit which determines the appropriate delay before triggering the cameras and the strobe. The nominal projectile configuration and its dimensions are shown to scale in Fig.1. These projectiles have a 10° nose half angle and four fins whose blunt leading edges ramp back at 12.5°. They are fabricated in two hollow pieces, nose cone and body, which thread together at the throat (point of maximum cross-sectional area). Magnets are placed at the projectile throat to enable tracking with the EM sensors.

3. Experimental results

In order to ensure adequate safety margins for the transparent section, preliminary experiments were performed to validate consistent ram accelerator operation at reduced fill pressures. To enable reliable initiation of the ram accelerator process with an entrance velocity of ~ 1150m/sec, the first stage required a fill pressure of at least 13 bar for a propellant mixture consisting of $2.8CH_4 + 2O_2 + 5N_2$. The highest projectile velocities in the transparent tube were attained by using a first stage fill pressure of 20 bar to increase the initial acceleration. It was found that continuous and stable ram acceleration can be maintained when the projectile makes a transition between propellant mixtures having a pressure ratio of approximately 2:1. Thus, the ram accelerator test section was partitioned into the appropriate number of stages to accelerate the projectile to the desired velocity before injecting it into the transparent section. To maintain ram accelerator operation within this low pressure stage, the heat of combustion of the propellant mixture was increased by reducing its diluent concentration to approximately 90% of that in the first stage mixture.

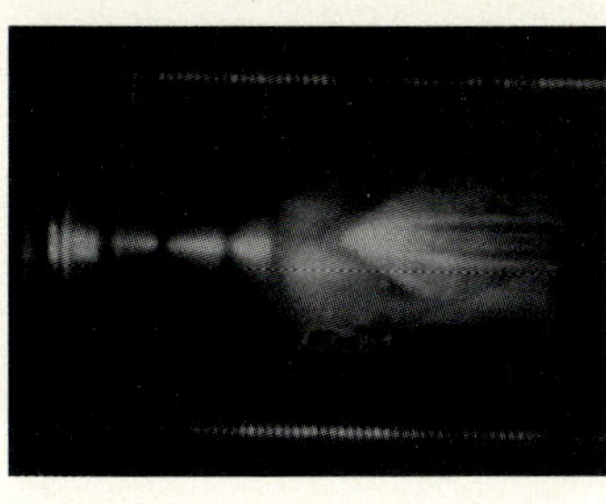
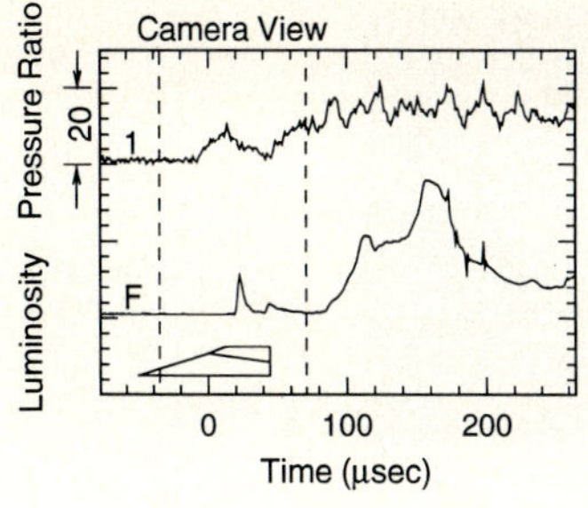

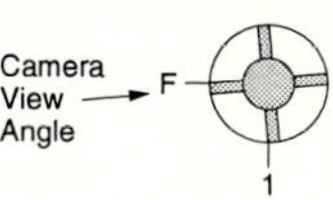

Fig. 2. Projectile operating at 1590 m/sec (Mach 4.4) in $2.8CH_4+2O_2+5.6N_2$ at 6 bar. Pressure ratio data from 20 cm ahead of transparent tube (Lexan) are shown. Dashed lines indicate camera view

In the initial experiments the test section was partitioned into three stages filled to successively lower pressures. The first stage was 2 m long and filled to 20 bar, and the third stage, where the transparent tube was installed, was 10 m long and had a fill pressure of 6 bar. The intermediate stage was used to reduce the pressure differential between the first and third stages; it was 4 m long and filled to 10 bar. For the reasons mentioned above, a slightly more energetic propellant mixture ($2.5CH_4+2O_2+5.6N_2$) was used in the low pressure section than in the prior two stages ($2.7CH_4+2O_2+5.8N_2$).

A photograph through a Lexan tube of a projectile operating at 1590 m/sec (90% V_{CJ}) in the 6 bar propellant mixture is shown in Fig.2. The leading edges of the projectile fins were machined to form a wedge-shaped junction at the projectile throat, and equally-spaced black rings were drawn on the nose cone to provide reference marks. Luminous oblique shock waves emanating from the fin leading edges can be seen intersecting with each other and the tube wall. A thin luminous streak can be seen on each side of the projectile fin, possibly indicating the thickness of either a reacting boundary layer or an entropy layer. Note the dark region behind projectile, where some luminosity would be expected if the recirculating zone at the projectile base were acting as a flame holder.

The corresponding pressure and logarithmically amplified luminosity data from an instrument station located 20 cm upstream of the transparent tube are shown in Fig.2, along with a schematic indicating the orientation of the projectile relative to the camera and sensors. The luminous regions observed behind the projectile throat and at the intersection point of the shock waves near its base are also seen in the data from the fiber-optic probe. In addition, the dark zone behind the projectile is also indicated by these data. A region of intense luminosity was detected several tube diameters farther behind but was not in the field of view at the time of the camera trigger.

As a control experiment, the photograph of the projectile was repeated with a methane-nitrogen mixture that simulated the propellant used in the experiment discussed above, but without the possibility of combustion. The photograph shows no luminosity, confirming the supposition that the light emissions in Fig.2 are associated with combustion.

A two-stage configuration consisting of a 2 m long initial stage at 13 bar and a 14 m long second stage at 7 bar was able to accelerate a projectile to 1620 m/sec (93% V_{CJ}) at the entrance

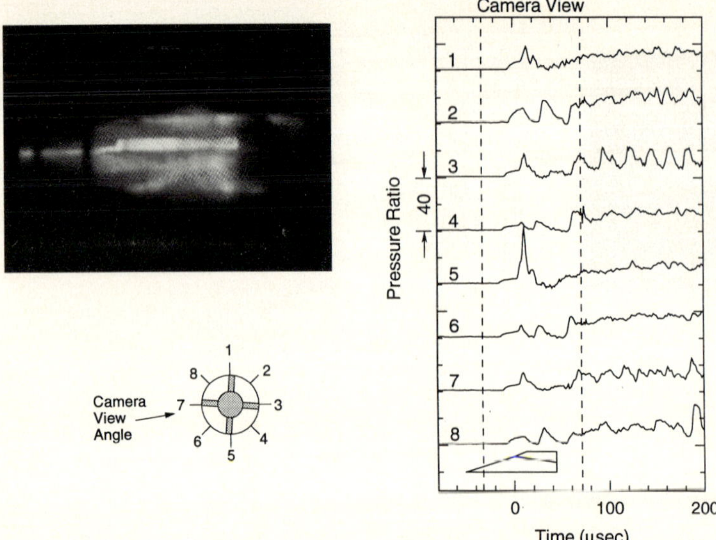

Fig. 3. Projectile operating at 1620 m/sec (Mach 4.5) in $2.8CH_4+2O_2+4.5N_2$ at 7 bar in acrylic tube. Pressure ratio data from instrumented tube inserts. Dashed lines indicate camera view

to the transparent section, which contained a mixture of $2.8CH_4+2O_2+4.5N_2$. A projectile having the nominal blunt fin leading edges (see Fig.1) was photographed through an acrylic tube in this experiment (Fig.3). The wave structure apparent in Fig.2 is not nearly as distinct in this case. A diffuse glow is evident on the body of the projectile and a faint luminous wake behind each fin is apparent. The corresponding pressure data from the instrumented tube inserts are also shown in Fig.3, along with a reference schematic showing the orientation of the projectile relative to the pressure transducers and camera viewing angle. The differences in arrival time and amplitudes of the leading pressure pulse indicate that the projectile was canted as it passed through the transparent section (Hinkey et al. 1993).

In another experiment a three-stage configuration was used to accelerate a projectile up to 1740 m/sec (100% V_{CJ}). In this case the timing circuit was adjusted to photograph only the projectile nose cone when it was midway through the transparent tube (Lexan), as shown in Fig.4. The corresponding pressure data are from the instrumented tube inserts and the luminosity data are from 20 cm farther upstream. The pressure in the transparent section was 7 bar and the propellant mixture was the same as in Fig.3. A faint glow at the nose tip and a slight haziness over the nose cone can be seen. No signs of combustion activity are observed on or behind the projectile in this case. The fiber-optic probe detected the nose tip light and did not observe any luminosity on the body, in agreement with the photograph. No acceleration was measured while the projectile passed through the low pressure section, which is consistent with the observed lack of combustion activity.

Photographs of a projectile coasting through the transparent section at 2030 m/sec (120% V_{CJ}) were taken in the superdetonative velocity regime. In Fig.5 three photographs are shown that were snapped at 10 μsec intervals of a projectile that was injected into a 6-m long, low pressure section after being accelerated by two prior ram accelerator stages. The fill pressure of the transparent section (Lexan) was 6 bar and the propellant mixture was composed of $2.7CH_4+2O_2+5.7N_2$. The projectile image becomes progressively darker as the strobe light intensity wanes. The luminous

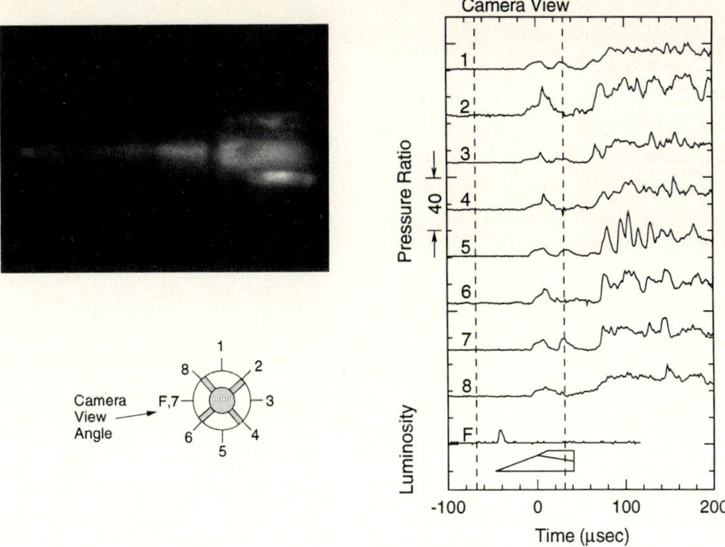

Fig. 4. Projectile operating at 1740 m/sec (Mach 4.8) in $2.8CH_4+2O_2+4.5N_2$ at 7 bar. Pressure ratio data from instrumented tube inserts and luminosity data from 25 cm ahead of transparent tube (Lexan) are shown. Dashed lines indicate camera view

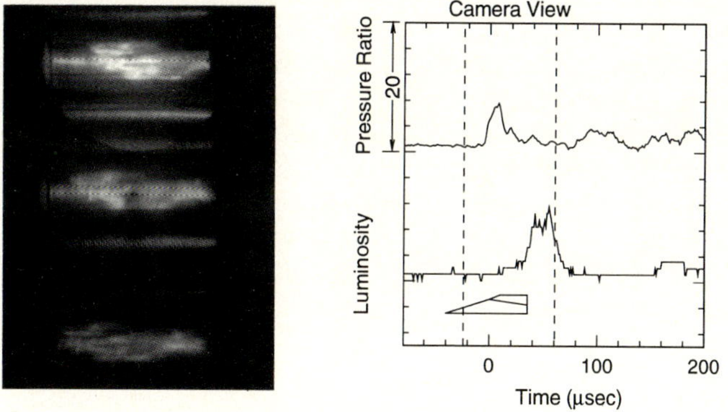

Fig. 5. Projectile operating at 2030 m/sec (Mach 5.6) in $2.7CH_4+2O_2+5.7N_2$ at 6 bar. Pressure and luminosity data from 25 cm ahead of transparent tube (Lexan) are shown. Dashed lines indicate camera view in the first of the three images

region just behind the projectile is asymmetrically distributed and a distinct structure that begins approximately one tube diameter behind the projectile is observed to travel with it. The corresponding data from an instrument station located 25 cm ahead of the transparent tube indicate that there were no pressure pulses present, other than those associated with the oblique shock system generated by a totally supersonic projectile. Luminosity on and behind the projectile body was detected by the fiber-optic probe that was positioned almost in line with the camera view.

4. Discussion

Luminous phenomena in the vicinity of the projectile are apparent in the photographs taken in low-pressure propellant mixtures. The distinctive wave character seen with the wedge-shaped fins at subdetonative velocities was not nearly so apparent in the photographs of projectiles having the standard, blunt leading edge fins. This may be due to the projectile geometry, propellant chemistry, or the orientation of the fin with respect to the camera. The factors which allow such wave phenomena to be discerned are still being investigated.

In general, the luminosity data from the fiber-optic probes in the low pressure test section were much less intense than typically detected in higher pressure propellant mixtures. Consequently, the luminous phenomena in the photographs are not as bright as they would be if higher pressures were used. In addition, normal ram accelerator operation was more difficult to establish at low fill pressures for reasons that are still under investigation. Nevertheless, this preliminary photographic survey of ram accelerator operation in methane-oxygen-nitrogen propellant mixtures confirmed that in-tube visualization of the projectile is feasible, that shock wave structures can be observed, and that combustion luminosity can be asymmetric. Future work will incorporate shadowgraph and smear photography to visualize the flow field around the nose cone and behind the projectile body.

5. Conclusion

Techniques for photographing ram accelerator projectiles have been developed to observe their operation in low pressure (< 10 bar) propellant mixtures contained by transparent tube inserts having a 38 mm bore. Self-luminous phenomena that highlight shock wave and boundary layer phenomena have been observed in the photographs. Near the CJ speed diffuse luminous regions are observed on the projectile body and a faint glow can be detected at the tip of the nose cone. An asymmetric luminous wake is observed behind the projectile when it is traveling at 120% V_{CJ}. The distribution of luminosity in the photographs correlates well with data from fiber-optic probes, and the projectile orientation is readily observed, which facilitates interpretation of pressure measurements. The results of these exploratory experiments indicate that low pressure operation of the ram accelerator is feasible and that projectile photographs can be obtained through transparent tube inserts.

Acknowledgment

This work was supported by ARO Grant No. DAAL03-92-G-0100. The assistance of Ed Burnham, John Hinkey, Gilbert Chew, and Malcom Saynor is much appreciated.

References

Bruckner AP, Knowlen C, Hertzberg A, Bogdanoff DW (1991) Operational characteristics of the thermally choked ram accelerator. J. Propulsion & Power 7:828-836

Burnham EA, Hinkey JB, Bruckner AP (1992) Investigation of starting transients in the thermally choked ram accelerator. 29th JANNAF Combustion Subcommittee Meeting, Hampton, Virginia, USA

Hertzberg A, Bruckner AP, Bogdanoff DW (1988) Ram accelerator: a new chemical method for accelerating projectiles to ultrahigh velocities. AIAA J. 26:195-203

Hertzberg A, Bruckner AP, Knowlen C (1991) Experimental investigation of ram accelerator propulsion modes. Shock Waves 1:17-25

Hinkey JB, Burnham EA, Bruckner AP (1993) Investigation of ram accelerator flow fields induced by canted projectiles. AIAA Paper 93-2186

Part 3: Shock Tube Technology and Diagnostic Techniques

Hypervelocity Aerodynamics in a Superorbital Expansion Tube

A.J. Neely and R.G. Morgan
Department of Mechanical Engineering, The University of Queensland, Brisbane, 4072, AUSTRALIA

Abstract. The viability of the Superorbital Expansion Tube concept as a ground-based testing facility has been demonstrated by the operation of a small scale tube able to produce test flows in air of up to 13 km.s^{-1} for durations of 15 μs. The useability of these test flows for aerodynamic research has been confirmed by initial experiments measuring heat transfer rates to the surface of a model placed in the flow. The flat plate heating rates measured at superorbital velocity are found to be of the same order as Eckert's laminar correlation.

Key words: Superorbital, Laminar, Heat transfer, Hypervelocity

1. Introduction

The Superorbital Expansion Tube is a new hypervelocity facility designed to produce flow conditions in a variety of test gases at velocities exceeding Earth orbital velocity. Such flight speeds are of interest in the design of any proposed vehicle which will be used to enter the atmosphere of another planet or return to the Earth's atmosphere from beyond Earth orbit. Recent studies (Lyne et al. 1993) have proposed Earth entry velocities exceeding 15 km.s^{-1} for future manned Mars missions. Existing ground based test facilities, including free-piston shock tunnels and expansion tubes, are limited in the flow enthalpies they can produce and will not be able to directly reproduce these extreme flight velocities. Superorbital flows have been produced in non-reflected shock tunnels (Sharma and Park 1990), however these flows are unsuitable for aerodynamic testing as complete energy addition across the shock results in a dissociated, ionised plasma. At present a small scale Superorbital Expansion Tube facility is in operation at The University of Queensland and is being used to investigate these hypervelocity flight regimes.

2. Description of the facility

The superorbital expansion tube uses the phenomenon of enthalpy multiplication of the test flow through an unsteady expansion such as occurs in a standard expansion tube. However it differs from the standard expansion tube layout through the addition of a secondary driver section (Fig.1). The device is driven by a helium filled, free-piston driver which is used to shock heat a helium filled secondary driver section which in turn exhausts into the test gas contained in the shock tube. This gas is separated from the secondary driver by a thin cellophane diaphragm (23 μm, 90 kPa burst pressure). Morgan and Stalker (1991) have described the performance boosting effect of the secondary driver. The incident secondary shock wave heats, compresses and accelerates the quiescent test gas as it traverses the intermediate tube and ruptures the light tertiary diaphragm (9 μm cellulite grocery wrap, 20 kPa burst pressure) and then accelerates to a higher velocity as it passes into low pressure helium contained in the acceleration tube. The test gas following the shock is also further accelerated as it is expanded unsteadily into the acceleration tube. The test gas then passes over the model located at the exit of the acceleration tube.

Shock Waves @ Marseille I
Editors: R. Brun, L. Z. Dumitrescu   © Springer-Verlag Berlin Heidelberg 1995

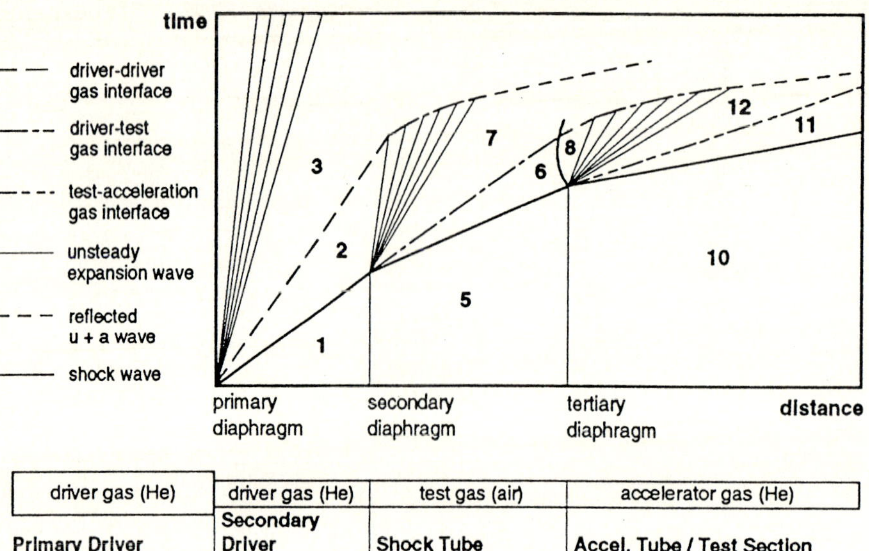

Fig. 1. Wave diagram illustrating Superorbital Expansion Tube operation

3. Superorbital test flow conditions

Preliminary results obtained from the commissioning of the pilot Superorbital Expansion Tube facility were presented at the 18th ISSW (Morgan and Stalker 1991). That paper reported observed tertiary shock velocities of the order of 18 km.s^{-1}. At that time only very limited flow diagnostics were used to determine the state of the test gas, and no period of useful steady flow was defined. To date, quasi-steady flow conditions using air as the test gas have been produced in the Superorbital Expansion Tube with shock velocities exceeding 13 km.s^{-1} for durations approaching 15 μs (Neely and Morgan 1993). Total enthalpies of 108 MJ.kg^{-1} have been achieved at corresponding total pressures of 335 MPa.

3.1. Experimental determination of flow state

Flow diagnostics including measurements of shock speeds, wall static pressures and centreline Pitot pressures were used to confirm the presence of a period of steady flow. To investigate the behaviour of the accelerator-test gas interface, heat transfer rates were measured along the wall of the acceleration tube using flush mounted platinum-quartz thin film resistance thermometers. The gauges have a sensing surface of 2 mm diameter. In Fig.2 a sidewall heat transfer rate history measured at the end of the acceleration tube is compared with an empirical laminar calculation, using Eckert's reference enthalpy method (1955), which assumes that only a test gas boundary layer, growing from the traversing tertiary shock, is present.

The predicted peak that occurs with shock arrival, and the laminar trend that can be seen in the flat plate results discussed in the next section, are not observed in the measured sidewall trace. The level is also greater than predicted. These effects are thought to be due to the presence of the underlying helium boundary layer (Fig.3), which is neglected in the empirical calculation, and its interaction with the growing air boundary layer above, but the mechanism is not fully understood.

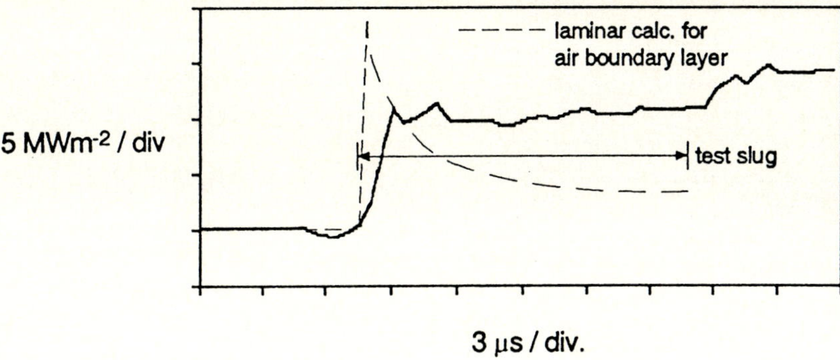

Fig. 2. Heat transfer rate measurements at acceleration tube wall near the exit

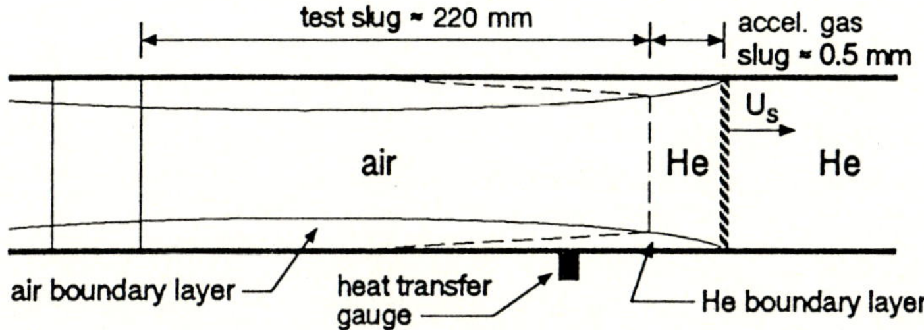

Fig. 3. Schematic of boundary layer development at the end of acceleration tube (after Mirels 1963)

Calculations show that full Mirels (1963) development of the accelerator gas slug occurs by the exit flow, indicating significant entrainment of helium in the boundary layer (Fig.3). On a surface placed away from the wall in the freestream, there will be little effect from helium. The small amounts of quiescent helium and shock heated accelerator gas are quickly flushed from the surface leaving only a boundary layer of test gas. Measured heating rates in the test flow are indicative of this and are reported in the next section. A parallel numerical investigation (Akman and Morgan 1993) into the viscous flow mechanisms occurring in the acceleration tube should provide more detail of the behaviour of the accelerator gas and its effect on the test flow.

3.2 Analytical determination of the flow state

A real gas analytical model has been developed to determine the final flow states produced in the facility which uses the experimentally measured secondary and tertiary shock speeds. Consideration of boundary layer growth and non ideal diaphragm rupture effects have allowed a better understanding of the dominant flow mechanisms and enabled more accurate matching of calculated and observed test flow states. In particular the entropy increase associated with the reflection of the secondary shock at the tertiary diaphragm is accounted for. The model is described in detail elsewhere (Neely and Morgan 1993). The test flow state used in the experiments was calculated using this model and is set out in Table 1.

Table 1. Hypervelocity test flow state

FLOW CONDITION	108 MJ.kg^{-1}
primary shock velocity (m.s^{-1})	6 450
secondary shock velocity (m.s^{-1})	7 280
Test Flow	
velocity† (m.s^{-1})	13 000
pressure (Pa)	6 028
temperature (K)	5 976
density (10^{-2} kg.m^{-3})	0.222
static enthalpy (MJ.kg^{-1})	23.6
Mach number	6.57
gas constant	454
ratio of spec. heats	1.43
Pitot pressure (kPa)	375
total pressure* (MPa)	286.8
total temperature* (K)	29 665
dissociation fractions : N	0.648
O	1.00

$\dagger$ set equal to measured shock speed
$*$ determined by isentropic stagnation

4. Flat plate heat transfer rate experiments

To investigate the heating rates which occur at the high freestream enthalpies associated with atmospheric flight of a vehicle at superorbital velocity, experiments were conducted with an instrumented wedge placed in the flow. The model was instrumented along the centerline of one surface with thin film gauges of the type described above, flush mounted at 8 mm intervals beginning 13.2 mm from the leading edge.

Measurements were taken with the instrumented surface at zero incidence to the flow and at 5° incidence to the flow. Comparisons of the measured heating rates were made with an empirical prediction made using Eckert's reference enthalpy method (Eq.1) to determine the nature of the boundary layer and the validity of the correlation at superorbital velocities:

$$\dot{q}_w = 0.332\rho^* \, u_e \, \mathrm{Pr}^{-2/3} \, (\mathrm{Re}^*)^{-1/2} \, (h_r - h_w) \tag{1}$$

with the starred values evaluated at a reference enthalpy defined by

$$h^* = 0.5 \, (h_e + h_w) + 0.22 \, (h_r + h_e) \tag{2}$$

4.1. Flat plate at zero incidence

For laminar boundary layers with similar profiles, the coefficients in Eq.1 are all constant except for the local Reynolds number. In Fig.4 the Stanton number is normalised by $x^{-1/2}$ which is proportional to $q/\mathrm{Re}^{-1/2}$, and the steady level when plotted against the wetted length is seen to be constant, indicating laminar flow. The level is also seen to be in good agreement with Eckert's empirical correlation. This is an interesting result as the correlations are unproven at these flow velocities and would not be expected to account for the increased importance of mechanisms such as radiative heating at the elevated temperatures present.

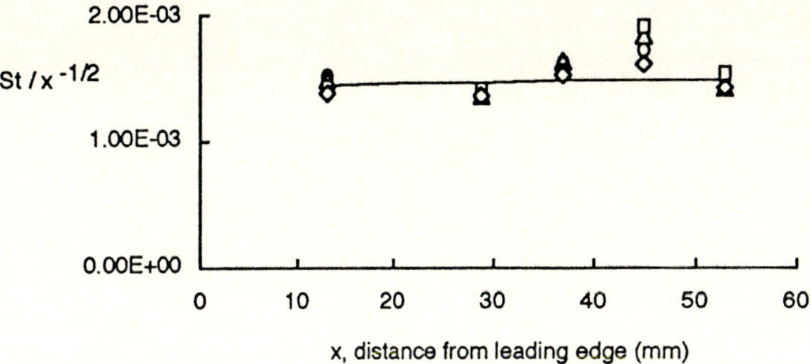

Fig. 4. St normalised by $x^{-1/2}$ versus wetted length for flat plate at zero incidence

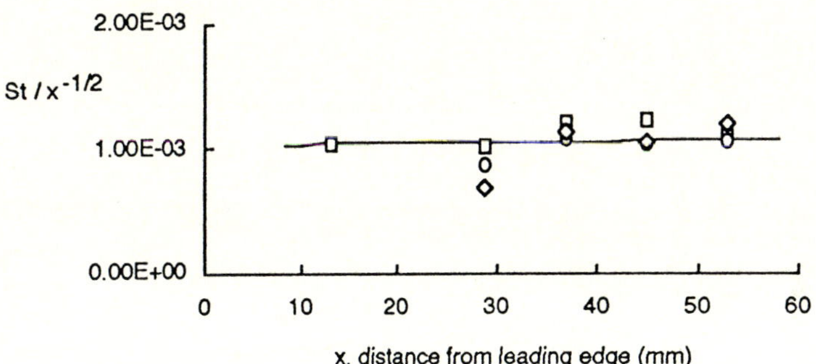

Fig. 5. St normalised by $x^{-1/2}$ versus wetted length for flat plate at 5° incidence

4.2. Flat plate at incidence

For the surface at incidence the inviscid flow state was calculated by assuming the test flow passed through an oblique shock with a flow deflection angle of 5° and an equilibrium gas state.

Once again the Stanton number is normalised by $x^{-1/2}$, and the steady level when plotted against the wetted length (Fig.5) is seen to indicate laminar flow. Again the level is seen to be of the same order as Eckert's empirical correlation.

5. Conclusions

The combination of an added secondary driver section to further shock heat the driver gas and the enthalpy multiplication effect of the unsteady expansion of the test gas allow the Superorbital Expansion Tube to produce flows at superorbital velocities, without the dissociation and ionisation problems experienced by conventional shock tubes at these speeds. The small scale facility used to investigate the concept has been operated to produce quasi-steady test flows with shock velocities in excess of 13 $km.s^{-1}$ and with a duration of useable test flow of approximately 15 μs. No other existing ground based test facility can produce test flows at these velocities, with gas composition suitable for aerodynamic testing.

Measurements of heating rates to the wall of the acceleration tube during the passage of the test flow are not of the form that would be expected for a boundary layer of test gas. They are thought to indicate the entrainment of a significant amount of helium accelerator gas in the boundary layer.

Heat transfer rate measurements to the surface of a model in the hypervelocity flow at zero incidence and at 5° incidence are indicative of a laminar boundary layer. They were also found to correlate well with Eckert's empirical model for laminar flow. These results show that the test flow produced by the facility can be used for meaningful aerothermodynamic research.

These results continue to demonstrate the viability of the Superorbital Expansion Tube concept. It is foreseen that larger facilities using this concept will be invaluable for the basic aerodynamic research required for future interplanetary missions involving atmospheric manoeuvring as well as for the design and testing of vehicles to be used on such missions.

Acknowledgements

The authors gratefully acknowledge the support of the Australian Research Council.

References

Akman N, Morgan RG (1993) Numerical simulation of viscous flow in a superorbital expansion tube. These Proceedings

Eckert ERG (1955) Engineering relations for friction and heat transfer to surfaces in high velocity flow. J. Aeron. Sciences, August:585-587

Lyne JE, Tauber ME, Braun RD (1992) Parametric study of manned aerocapture Part I: Earth return from Mars. J. Spacecraft and Rockets 29, 6:808-813

Mirels H (1963) Test time in low-pressure shock tubes. The Physics of Fluids, 6, 9

Morgan RG, Stalker RJ (1991) Double diaphragm free-piston driven expansion tube. in: Takayama K (ed) Proc. 18th ISSW, Sendai, pp 1031-1038

Neely AJ, Morgan RG (1993) The superorbital expansion tube concept, experiment and analysis. Submitted to the Royal Aeronautical Journal

Sharma SP, Park C (1990) Operating characteristics of a 60- and 10 cm electric arc-driven shock tube - Part II: The driven section. J. Thermophysics 4, 3:266-272

Influence of Secondary Diaphragm on Flow Quality in Expansion Tubes

G.T. Roberts[*], R.G. Morgan[†] and R.J. Stalker[†]
[*]Department of Aeronautics and Astronautics, University of Southampton, UK.
[†]Department of Mechanical Engineering, University of Queensland, Australia

Abstract. Experiments are described in which the influence of secondary diaphragm thickness and pre-deformation on Pitot and sidewall (static) pressures are investigated. The University of Queensland TQ expansion tube facility was operated in shock tube mode with argon as the driver and test gases. The results suggest that shock reflections between the diaphragm and the driver-test gas interface can cause disturbances to be propagated into the test gas flow, causing early termination of the run. This is best avoided by using diaphragms which are as thin as possible and pre-deformed prior to the run.

Key words: Expansion tube

1. Introduction

The expansion tube is an impulse-type wind tunnel facility, capable of producing high enthalpy test gas flows for aerothermodynamic investigations, where the test gas, initially heated and accelerated by a primary shock, is further accelerated by an unsteady expansion centred at the location of a (thin) secondary diaphragm. Flow velocities above 10 km.s^{-1} have been attained in such facilities (e.g. Neely et al. 1991, Morgan and Stalker 1991).

The main advantage of the expansion tube, compared with conventional shock tunnels, is that (in the ideal case) the flow is not processed by a reflected shock; consequently the test gas temperatures remain relatively modest and so radiation losses from the test gas, which would otherwise limit the stagnation enthalpy achieved, are reduced (Stalker et al. 1991). The major disadvantages are the rather short run times available (typically an order of magnitude shorter than those of reflected shock tunnels, or less) and, for any particular facility, the narrow range of operating conditions in which the flow quality is acceptable (e.g. Miller 1975). The latter, in particular, has tended to inhibit their acceptance in the aerodynamic community.

Paull and Stalker (1992) discuss the various factors which limit the run time and have shown that disturbances, which appear to originate in the driver gas and subsequently propagate into the test gas to cause early corruption of the flow, can be eliminated by a judicious choice of driver-test gas sound speed ratio, thus extending the range of useful operating conditions. In particular, if the facility is operated at overtailored conditions, where the shock-heated test gas sound speed is greater than that of the expanded driver gas (i.e. $a_2 > a_3$), the driver-test gas interface acts as a filter for transverse acoustic waves which are probably caused by primary diaphragm rupture. However, the role of the secondary diaphragm in the production, transmission or amplification of such disturbances, and hence its effect on the flow quality and the run duration, has hitherto remained little explored and is the subject of this study.

2. Experimental details

The University of Queensland TQ expansion tube facility was used in this study. This is a small facility, often used for pilot studies, comprising a free-piston driver section (in which the driver gas is compressed by a heavy (approx. 3.4 kg) piston, itself accelerated by high pressure air behind it), a shock tube section in which the test gas initially resides and a section containing the acceleration

Shock Waves @ Marseille I
Editors: R. Brun, L. Z. Dumitrescu   © Springer-Verlag Berlin Heidelberg 1995

gas (usually, but not necessarily, the same as the test gas). Downstream of the acceleration section is the test section/dump tank. The overall length of the facility is approximately 7.75 m and the internal diameter of both the shock and acceleration tubes is 37 mm. A full description of the facility appears elsewhere (Neely et al. 1991).

The primary aim of these experiments was to explore the influence of secondary diaphragm thickness and pre-deformation on the test gas flow duration and quality. Deformation always occurs with the thin plastic diaphragms usually employed, due to the pressure difference initially applied across them when operating in expansion tube mode, and was thought to be a further possible source of transverse acoustic wave disturbances. For this reason, the facility was operated in shock tube mode (i.e. with no applied pressure difference) so that comparisons between runs with either planar or pre-deformed diaphragms could be made. The diaphragms tested were polyethylene (13 μm thick), cellophane (25, 50 μm) and mylar (102, 191 μm) and, in the tests described here, were located 3.74 m downstream of the primary diaphragm (approximately 1.26 m upstream of the test section/dump tank).

A probe was mounted in the test section to monitor the Pitot pressure ($p_{2,t}$) and measurements of sidewall (static) pressure (p_2) were obtained at various locations along the shock and acceleration tubes; the latter were also used to determine the variation in shock speed (U_s). All pressures were measured using PCB piezo-electric transducers, which were pre-calibrated. The data were recorded on a Physical Data Inc. multichannel transient recorder operating at sample rates of either 40 kHz or, on some channels, 2 MHz.

The requirement to operate at overtailored conditions led to the choice of argon driving argon as the test gas (the latter was preferred to air to avoid causing real gas effects due to vibration and dissociation). A summary of the run conditions employed is given in Table 1 (overleaf). The sound speed ratio a_2/a_3 was calculated from ideal shock tube theory assuming argon behaves as a perfect gas, with allowance made for the area reduction at the primary diaphragm and the known non-ideal driver gas compression and expansion processes (Morgan and Stalker 1991).

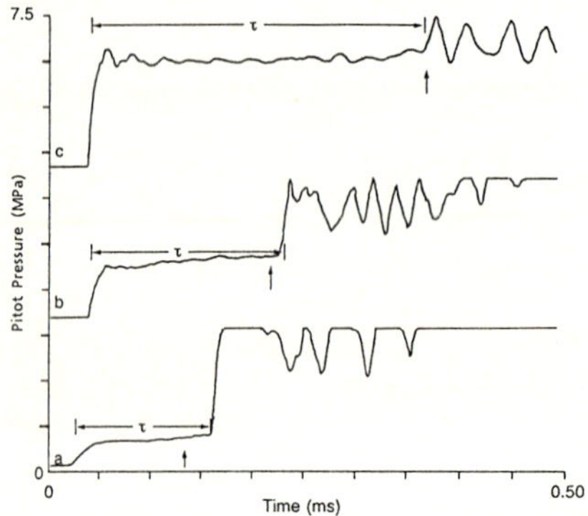

Fig. 1. Variation of Pitot pressure with time
a. $p_1 = 1$ kPa; **b.** $p_1 = 4$ kPa; **c.** $p_1 = 20$ kPa.
τ is the steady run period; ↑ indicates Mirels' run time

Table 1. Typical run conditions

p_1 (kPa)	U_s (km/s)	a_2/a_3
20	1.65	2.0
4	2.25	3.7
1	2.75	7.3

Air reservoir pressure: 1.85 MPa
Primary diaphragm: 0.6 mm mild steel
Burst pressure $p_4 = 19.5$ MPa
Driver reservoir pressure: 75 kPa (Argon)

3. Experimental results

3.1. No secondary diaphragm

Fig.1 shows typical Pitot pressure signals obtained at each run condition without secondary diaphragms. The initial rise in pressure signifies the arrival of the incident shock and commencement of the run; the rise is less rapid at lower pressures because viscous effects become more dominant and alter the response time of the Pitot probe/transducer. Ideally the Pitot pressure should remain constant until the arrival of the driver-test gas interface terminates the run. This is evident as a second abrupt rise in Pitot pressure, which becomes comparitively larger at higher shock speeds (i.e. lower p_1) as both the test gas velocity and the density ratio across the interface increases.

As expected, the run period (τ) decreases with increasing shock speed; also shown is the predicted time of arrival of the interface according to the theory of Mirels (1964), assuming turbulent boundary layer growth behind the shock. The theoretical run time is slightly less than that observed, the discrepancy becoming greater at the higher shock speeds. This is thought to be due to an overestimation of the viscous effects at the high temperatures (>7000 K) predicted by ideal gas shock wave theory for argon.

After the arrival of the interface, the Pitot signals exhibit large, periodic fluctuations (frequency 25 kHz $< f <$ 45 kHz) which were attributed by Paull and Stalker (1992) to disturbances arising from the non-ideal primary diaphragm rupturing process. Similar, but smaller scale, fluctuations are observed during the run period in Fig.1(a) and may be due to the comparitively weaker filtering effect of the driver-test gas interface at this run condition, where the sound speed ratio a_2/a_3 is relatively low, since they are not observed on the signals obtained at the higher shock speeds where the sound speed ratio is greater.

3.2. Effect of secondary diaphragm thickness

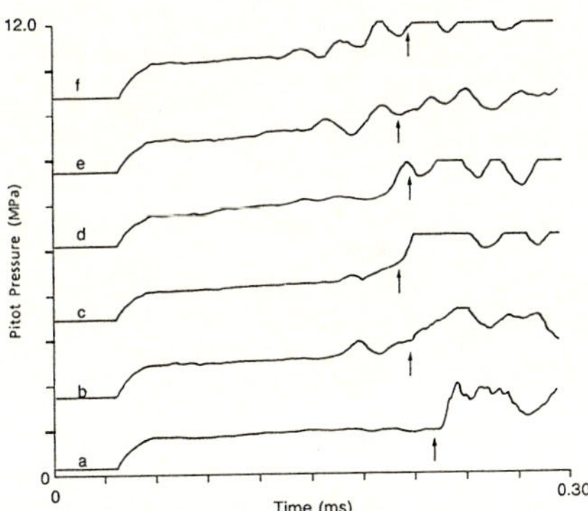

Fig. 2. Effect of diaphragm thickness
on Pitot pressure; p_1=4 kPa
a. no diaphragm; b. 13 μm polyethylene; c. 25 μm
cellophane; d. 50 μm cellophane; e. 102 μm Mylar;
f. 191 μm Mylar; ↑ indicates Mirels' run time

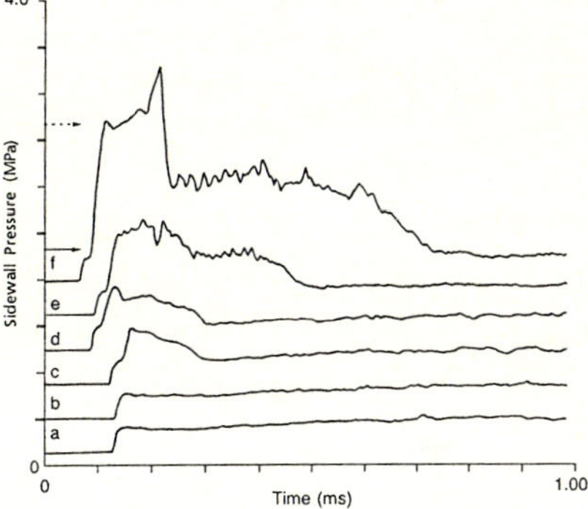

Fig. 3. Effect of diaphragm thickness on
sidewall pressure (measured 15 mm upstream of the
diaphragm); p_1=4 kPa
a. no diaphragm; b. 13 μm polyethylene; c. 25 μm
cellophane; d. 50 μm cellophane; e.102 μm Mylar;
f. 191 μm Mylar
For Fig.3f: ⟶ is predicted incident shock pressure
- - ⟶ is predicted reflected shock pressure

Fig.2 shows the effect on Pitot pressure of varying the secondary diaphragm thickness at an initial pressure p_1 = 4 kPa; for comparison, the Pitot signal obtained with no diaphragm is also included together with an estimate of the run times using Mirels' theory (these differ slightly due to run-to-run variations in shock speed). It is clear that, as the diaphragm thickness is increased, the period of steady pressure is reduced due to the early arrival of large scale periodic fluctuations; the abrupt increase in Pitot pressure signifying the arrival of the driver-test gas interface also becomes progressively smeared. Similar observations were made at the other operating conditions described in Table 1.

It is possible to estimate the time taken for the diaphragm to rupture from the measured shock trajectory; the results are shown in Table 2. As may be expected, the rupture time increases as both the diaphragm thickness increases and test gas pressure reduces. The negative value obtained for the 13 μm polyethylene diaphragm at $p_1 = 1$ kPa indicates the approximate nature of the estimation but does suggest that it ruptures practically instantaneously even at low test gas pressures.

Table 2. Estimation of diaphragm rupture time

Diaphragm	Burst Time μs		
Material	$p_1=20$ kPa	$p_1=4$ kPa	$p_1=1$ kPa
191 μm Mylar	61	112	207
102 μm Mylar	50	53	120
50 μm Cellophane	15	23	44
25 μm Cellophane	13	16	37
13 μm Polyethylene	6	5	-1

Fig.3 shows the effect of diaphragm thickness on the sidewall (static) pressure measured 15 mm upstream of the diaphragm (also for $p_1 = 4$ kPa); also indicated are the theoretically predicted incident and reflected shock pressures (the latter estimated assuming the diaphragm acts as a solid boundary). Clearly, the thicker the diaphragm (or longer the burst time) the greater is the reflected shock overpressure. Generally, however, the burst times are so short that the reflected shock is weakened (by the unsteady expansion resulting from diaphragm rupture) before it passes over the pressure transducer in the upstream direction. The overpressures are consequently less than predicted assuming a solid reflection boundary; indeed, for the thinnest diaphragm (13 μm polyethylene) this weakening process is so rapid that no evidence of shock reflection was detected by the transducer under any of the conditions tested.

These observations are consistent with a simple theory for the diaphragm rupture and acceleration processes described in Morgan and Stalker (1991). As the shock tube was operated at overtailored conditions, even though weakened the reflected shock will itself reflect back off the driver-test gas interface; for the thickest diaphragm at low test gas pressures multiple reflections are likely to occur between the interface and diaphragm before the latter bursts. It is thought that the interaction between the reflected shock and the driver-test gas interface is the probable cause of the early transmission of disturbances into the test gas flow and the smearing of the Pitot pressure increase associated with the arrival of the driver-test gas interface; however, the details of this process have not yet been fully analysed.

3.3. Effect of diaphragm deformation

Experiments were performed with the 25 μm cellophane diaphragm material deliberately deformed before running the shock tube; the deformation was created by maintaining a pressure differential of about 80 kPa (approximately 80% of the burst pressure) between the two sections of shock tube for 20 minutes, after which both sides are pumped down prior to introduction of the test gas. Inspection of the diaphragms subjected to this loading indicated that they had deformed by about 10 mm (i.e. approx. 25% of the tube diameter) at the centre and maintained their shape even after the pressure differential was removed. Tests were conducted at $p_1 = 1$ and 4 kPa only.

Fig.4 shows the effect of the deformation on the sidewall pressure measured 15 mm upstream of the diaphragm for $p_1 = 1$kPa. Apparently, the deformation reduces the pressure after shock reflection considerably; with $p_1 = 4$ kPa (not illustrated) there was no increase in pressure at all when the diaphragm was predeformed. The shape of the signal suggests that the deformation has simply caused the diaphragm to burst early under impulsive loading conditions rather than radically alter the shock reflection process. The diaphragm burst times estimated from the shock trajectories (as described in 3.2) were 29 μs and 12 μs for $p_1 = 1, 4$ kPa, respectively. Both are slightly less than the corresponding values for the planar diaphragm (37, 16 μs - see Table 2), although the degree of uncertainty in this estimation is probably greater than the differences calculated.

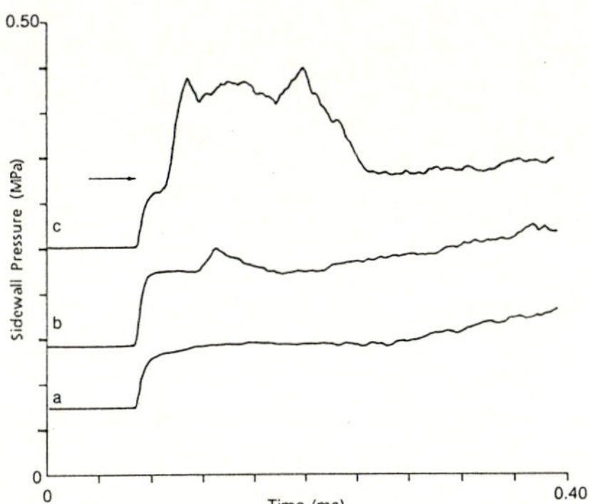

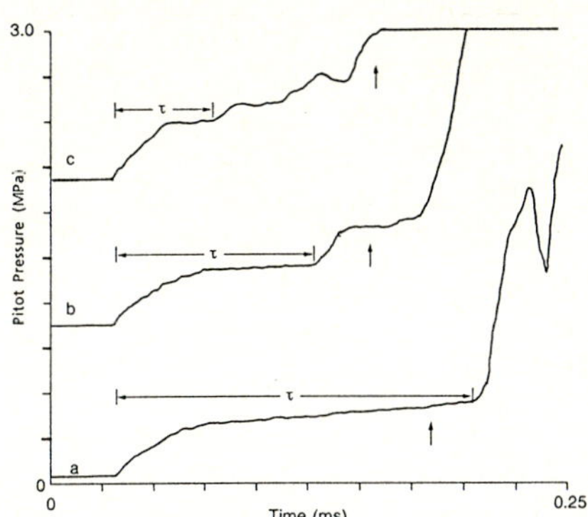

Fig. 4. Effect of diaphragm deformation on sidewall pressure (measured 15 mm upstream the diaphragm); $p_1 = 1$ kPa
a. no diaphragm; b. 25 μm cellophane, predeformed;
c. 25 μm cellophane, planar
For fig.4c: $\longrightarrow$ is predicted incident shock pressure

Fig. 5. Effect of diaphragm deformation on Pitot pressure; $p_1 = 1$ kPa
a. no diaphragm; b. 25 μm cellophane, predeformed;
c. 25 μm cellophane, planar
τ is the steady run period; $\uparrow$ indicates Mirels' run time

Fig.5 shows the effect of deformation on the observed Pitot pressure: it appears that, rather than produce additional disturbances which cause early truncation of the run, deformation delays the arrival of and weakens those disturbances which are thought to be associated with the interaction of the reflected shock with the driver-test gas interface.

4. Conclusions

From the above, it is clear that the secondary diaphragm considerably influences the flow quality in expansion tubes: the thickness of the diaphragm affects its rupture time which, in turn, affects the magnitude of the reflected shock pressure, particularly if it is sufficiently long for multiple shock reflections to occur between the diaphragm and the driver-test gas interface.

It is likely that such interactions cause disturbances to be propagated into the test gas flow, leading to early termination of the run. The strong shock reflection occuring with thick diaphragms would increase the dissociation fraction in a diatomic test gas, which may be undesirable if, say, the facility was to be used for hypersonic airbreathing combustion studies (Tamagno and al (1990). Therefore it is advisable to operate such faclilities with secondary diaphragms which are as thin as practically possible.

Pre-deforming the diaphragms apparently has the beneficial effect of reducing the reflected shock pressure (and hence, temperature and dissociation fraction) and delaying the arrival of disturbances which corrupt the test gas flow.

Acknowledgements

This work was carried out with the support of the Department of Mechanical Engineering, University of Queensland, the Royal Society of Great Britain (under the Anglo-Australian Exchange Scheme) and the UK Defence Research Agency (Aerospace).

References

Miller CG (1975) Shock shapes on blunt bodies in hypersonic-hypervelocity helium, air and CO2 flows, and calibration results in Langley 6-inch expansion tube. NASA TN D-7800

Mirels H (1964) Shock tube test time limitation due to turbulent boundary layer. AIAA J. 2:84-93

Morgan RG, Stalker RJ (1991) Double diaphragm driven free piston expansion tube. In: Takayama K (ed) Proc. 18th Intl. Symp. on Shock Waves, Sendai, pp 1031-1038

Neely AJ, Stalker RJ, Paull A (1991) High enthalpy, hypervelocity flows of air and argon in an expansion tube. Aeronautical Journal 95:175-186

Paull A, Stalker RJ (1992) Test flow disturbances in an expansion tube. J. Fluid Mech. 245:493-521

Stalker RJ, Paull A, Neely AJ (1991) Comparative features of free-piston shock tunnel and expansion tube facilities. Paper 241, presented at 10th Intl. Aerospace Plane Technology Symp., Canberra, Australia

Tamagno J, Bakos R, Pulsonetti M, Erdos J (1990) Hypervelocity real gas capabilities of GASL's expansion tube (HYPULSE) facility. AIAA paper 90-1390, presented at 16th Aerodynamic Ground Testing Conf. Seattle

Noise Reduction in Argon Driven Expansion Tubes

A.Paull
Mechanical Engineering Department, University of Queensland, Australia, 4072

Abstract.   Analytical results are used to show that the presence of noise in the test flow of an expansion tube is dependent on the sound speed of the driver gas. If the flow enthalpy is fixed it is shown that lowering the sound speed of the driver makes the test flow quieter. It is also shown that it is more difficult to obtain quiet test flows at lower enthalpies. The analysis also predicts that a decrease in the driver gas sound speed may be required if the reduction in noise in large diameter expansion tubes is to be the same as that in small tubes.

Key words: Expansion tube, Noise reduction

1. Introduction

In all impulse facilities disturbances in the test gas are cause for concern. In general, these disturbances are relatively small in shock tunnels. However, the existence of these disturbances significantly hindered the use and development of expansion tubes. Paull and Stalker (1992) modelled these disturbances as acoustic waves and showed that test flows could be made acceptably quiet by ensuring that the driver gas sound speed at the driver-test gas interface was sufficiently less than that of the test gas. The implications of this restriction on the choice of driver gases is examined here.

Expansion tubes traditionally use helium as the driver gas. However, with the development of the free-piston driven expansion there is no need to use such a light driver gas to generate hypersonic flows. Mixtures of helium and argon can also drive the expansion tube without a change in test flow Mach number and velocity. If the test flow Mach number and sound speed are specified then this imposes a lower limit on the range of driver gas sound speeds which can be used. In addition to this well understood lower limit, there also exists a less understood upper limit. In this paper, the theory developed by Paull and Stalker have been extended to show that a mixture with an excessive driver gas sound speed will produce a noisy test flow. This extended theory also shows that in practice the lower the test flow enthalpy, the more difficult it is to achieve noiseless flows.

The expansion tube used by Paull and Stalker (1992) had an exit diameter of 37 mm. This only permits the testing of small models. Hence, expansion tubes with diameters up to 300 mm are now being developed. The theoretical results presented in this paper also indicate some of the difficulties which may arise in producing quiet test flows in large diameter expansion tubes.

2. Expansion tube operation

Fig.1 is an $x-t$ diagram for an expansion tube. A shock is produced when the primary diaphragm ruptures. This shock accelerates in the acceleration tube because the acceleration tube has a lower filling pressure than the shock tube. The gas in which test are made has the properties of region 5. The total enthalpy of the test gas is increased between regions 3 and 5 by the unsteady expansion centred at the secondary diaphragm. This means of increasing the total enthalpy of the test gas is the prime advantage of this type of facility. For a more detailed explanation on the theory and limitations of an expansion tube see Paull and Stalker (1992).

Shock Waves @ Marseille I
Editors: R. Brun, L. Z. Dumitrescu   © Springer-Verlag Berlin Heidelberg 1995

3. Acoustic wave theory

Paull and Stalker (1992) proposed that the noise seen in an expansion tube could be modelled as acoustic waves. It was shown that small disturbances in the velocity and pressure can be represented by

$$u = u_0 + \nabla \Phi \tag{1}$$

and

$$p = p_0 - \rho_0 \partial \Phi / \partial t, \tag{2}$$

respectively, where u_0, p_0 and ρ_0 are the unperturbed velocity, static pressure and density, respectively. For a circular tube the potential function, Φ, has the form

$$\Phi = J_0(\lambda r)\exp(i\omega(t \pm \beta x/a)). \tag{3}$$

J_0 is a zero-order Bessel function of the first kind, ω is the fundamental frequency, a is the sound speed, x is the axial co-ordinate with x increasing in the direction of the flow, r is the radial co-ordinate, t is time and the dispersive term is

$$\beta = \left(1 - (\lambda a/\omega)^2\right)^{1/2} \tag{4}$$

The origin of the spatial co-ordinates is fixed in the frame of the gas.

The product λr_0, where r_0 is the radius of the shock tube, is one of the infinite zeros of J_1. The first zero of J_1 is zero. Solutions of this form are called longitudinal waves as they have no radial dependence. Paull and Stalker (1992) have shown that longitudinal waves are not the disturbances observed in an expansion tube, but instead the majority of the noise can be modelled as first order lateral waves. That is to say, solutions for which λr_0 is equal to the first non-zero zero of J_1 ($\lambda r_0 \approx 3.83$).

As the origin of the spatial co-ordinates is fixed in the frame of the gas, relative to the laboratory, the origin is moving down the tube with the velocity of the gas, u. It follows from Eq.3 that the frequency observed from the laboratory, ν, of a disturbance is not ω but is Doppler shifted to

$$\nu = \omega(1 \mp u\beta/a)2\pi \tag{5}$$

Paull and Stalker (1992) also show that the frequency of a disturbance which encounters a centred unsteady expansion is Doppler shifted by both the change in velocity and sound speed across the expansion. It is shown that as the disturbance moves through the expansion the dispersive term changes as

$$\frac{\beta - 1/2(\gamma-1)}{\beta - 1} = \frac{\beta_0 - 1/2(\gamma-1)}{\beta_0 - 1}\left(\frac{a}{a_0}\right)^{\frac{3-\gamma}{\gamma-1}}, \tag{6}$$

where the subscript 0 represents values upstream of the expansion. The Doppler shift in the frequency can be obtained from the substitution of β in Eq.5.

4. Interface filtering

In this section the ability of the driver-test gas interface to filter noise which originates in the driver gas is discussed.

From Eq.4 it can be seen that the value of β is dependent on the sound speed of the medium. Hence, in general the value of β in the test gas and the driver gas will be different for the same value of the fundamental frequency.

It can be seen from Eq.3 that if a lateral wave ($\lambda \neq 0$) is present in the driver gas then the fundamental frequency is such that β is real. However, if the wave is transmitted from the driver gas into the test gas and the sound speed of the test gas is sufficiently greater than that of the driver gas, so that β in the test gas is now imaginary, then a wave originating in the driver gas with this fundamental frequency will decay in the test gas. Hence, disturbances with this fundamental frequency will not be disruptive to the test flow.

It should be seen that not only disturbances with this fundamental frequency, but disturbances with this and lower frequencies decay in the test gas. That is to say, if the sound speed of the test gas is greater than that of the driver gas there is a band of frequencies which will decay when transmitted from the driver gas to the test gas.

It can be shown that in the laboratory frame of reference, if ν_L is the lowest frequency in the driver gas which does not decay in the test gas then

$$\nu_L/\nu_0 = a_3/a_2 \left(1 + M_2(1 - a_2/a_3)^2)\right)^{1/2}, \tag{7}$$

where ν_0 is lowest observable frequency in the driver gas,

$$\nu_0 = \lambda a_2/(2\pi), \tag{8}$$

$M_2 = u_2/a_2$ is the Mach number in region 2 and a_3 is the sound speed of the test gas in region 3. If the sound speed of the driver gas is greater than that of the test gas there is no decay of the noise in the test gas.

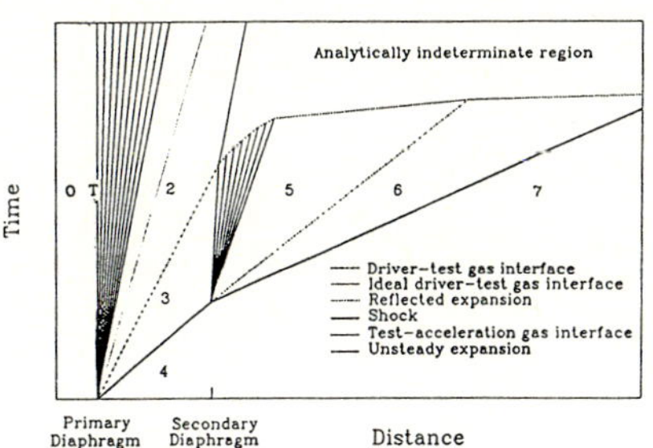

Fig. 1. $x - t$ diagram for an expansion tube

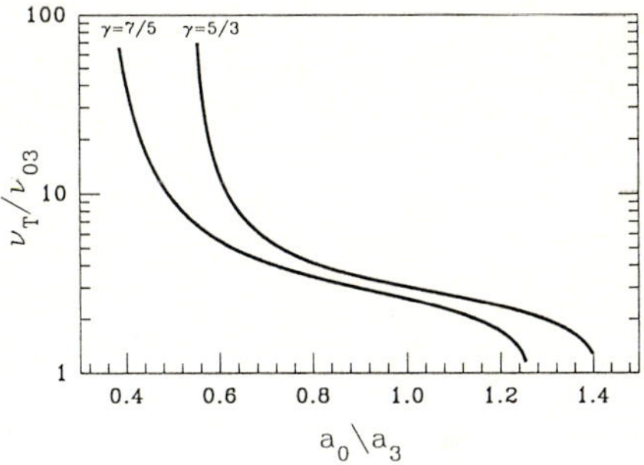

Fig. 2. Lowest frequency as measured from the primary diaphragm which does not decay as a function of sound speed ratio across the interface

5. Postulations

An understanding of an expansion tube's limitations will be obtained, only if the source of the noise which disrupts the test flow is identified. In this section two postulates relating to the source of the noise are made. These postulates are the bases for the theory which follows.

It is postulated that;

(i) The noise which disrupts the test flow originates at the primary diaphragm. It is present in the driver gas and is transmitted through the unsteady expansion centred at the primary diaphragm and then through the driver-test gas interface. This occurs before the test gas is subjected to the unsteady expansion centred at the secondary diaphragm.

(ii) The spectrum of the noise induced into the driver gas at the primary diaphragm is dependent only upon the physical parameters of the driver-shock tube connection and the mechanics involved in rupturing the primary diaphragm. It is not dependent upon the properties of the driver gas.

It could be argued that the second postulation is an oversimplification of the process. Indeed results presented by Jacobs (1992) would support such arguments. However, the object of the analysis here is to obtain some insight into the complex problem at hand and it is argued that this postulation is of sufficient complexity to gain this insight.

It was seen by Paull and Stalker (1992) that the first postulation is also an oversimplification, and there are other sources of noise. However, it was also shown by Paull and Stalker that if the driver-test gas interface eliminates noise generated from the primary diaphragm then it also eliminates noise produced by other sources. This occurs because the noise generated by the primary diaphragm would appear to cause the disturbance with the highest frequency.

6. Driver sound speed variations

In this section postulations (i) and (ii) are assumed and the spectrum of noise which is incident to the driver-test gas interface is determined. In the frame of reference of the driver gas, the spectrum of the noise induced into the driver gas at the primary diaphragm will be different for each gas. This follows from postulation (ii) and the Doppler shift expressed by Eq.5. As the driver gas expands, this spectrum is changed in accordance with Eq.6. It is this spectrum of noise that the interface has to eliminate to produce a quiet test flow. For all practical purposes measurements will be made in the laboratory frame of reference. Hence, another Doppler shift in the spectrum is introduced if measurements are to be made. This Doppler shift is opposite in sign of that at the primary diaphragm, but is of a different magnitude.

It is assumed that

(i) all gases are inviscid and perfect,

(ii) strong shock approximations can be made for all shocks,

(iii) the secondary diaphragm is massless and ruptures instantaneously and

(iv) the driver gas is at rest in the compression tube.

Using assumptions (i) and (ii) it can be shown that the velocity u_5 and sound speed a_5 of the test flow are related to the test gas sound speed before it is expanded by

$$a_3 \left(M_3 + 2/(\gamma_t - 1) \right) = u_5 + 2/(\gamma_t - 1)a_5, \tag{9}$$

where γ_t is the specific heat ratio for the test gas and $M_3 = u_3/a_3$ is the Mach number of the test gas upstream of the expansion. The left hand side of Eq.9 is uniquely determined by the Mach number of the shock in the shock tube which, in turn, is uniquely determined by the test flow velocity and Mach number. However, the Mach number of the shock is not uniquely determined by the driver gas sound speed at the time of rupture of the primary diaphragm because the same shock Mach number can be obtained using different driver gas sound speeds by varying the shock tube filling pressure. It should be noted that as the driver gas sound speed in the compression tube, a_0, is decreased the burst pressure of the primary diaphragm must be increased if the static pressure in the test flow is to be maintained.

If relative to the laboratory frame of reference, a disturbance with frequency ν_T is induced into the driver gas, then since the flow at the diaphragm is sonic, from Eqs.4 and 5, the dispersive term β_T for this disturbance can be obtained from

$$\nu_T/\nu_{03} = a_0/a_3 \left(\frac{2}{\gamma + 1} \right)^{1/2} \left(\frac{1 + \beta_T}{1 - \beta_T} \right)^{1/2}, \tag{10}$$

where

$$\nu_{03} = \lambda a_3/(2\pi). \tag{11}$$

A disturbance induced at the primary diaphragm is then transmitted downstream through the unsteady expansion centred at the primary diaphragm. The unsteady expansion will produce the shift in frequency prescribed by Eq.6. If β_T and β_2 are the dispersive terms upstream and downstream of the expansion, respectively, then from Eq.6 it can be shown that

$$\beta_T = \frac{\beta_2 \left(\frac{\gamma-1}{2} \left(\frac{a_2}{a_T} \right)^{\frac{3-\gamma}{\gamma-1}} - 1 \right) + \frac{\gamma-1}{2} \left(1 - \left(\frac{a_2}{a_T} \right)^{\frac{3-\gamma}{\gamma-1}} \right)}{\beta_2 \left(\left(\frac{a_2}{a_T} \right)^{\frac{3-\gamma}{\gamma-1}} - 1 \right) + \frac{\gamma-1}{2} - \left(\frac{a_2}{a_T} \right)^{\frac{3-\gamma}{\gamma-1}}}, \tag{12}$$

where a_T is the sound speed at the primary diaphragm and γ is the specific heat ratio of the driver gas.

If in the laboratory frame of reference ν_2 is the frequency which corresponds to the dispersive term β_2 then by Eqs.4 and 5 it follows that

$$\beta_2 = \left(\nu_2/\nu_0 \left((\nu_2/\nu_0)^2 + M_2^2 - 1 \right)^{1/2} - M_2 \right) \left(M_2^2 + (\nu_2/\nu_0)^2 \right)^{-1}. \tag{13}$$

If ν_T is now the lowest frequency which does not decay when transmitted through the driver-test gas interface then ν_2 must equal ν_L, where ν_L is given by Eq.7. Alternatively, this will occur if

$$\beta_2 < \left(1 - (a_2/a_3)^2 \right)^{1/2}. \tag{14}$$

Finally, if a_T is the sound speed at the primary diaphragm, then for an inviscid perfect gas,

$$a_T(\gamma + 1)/2 = M_3 a_3(\gamma - 1)/2 + a_2. \tag{1}$$

By assumption (iv) it follows from the conservation of energy that

$$a_0(2/(\gamma + 1))^{1/2} = a_T. \tag{16}$$

From Eqs.10-16 it can be seen that ν_T/ν_{03} is only a function of M_3 and a_0/a_3. For strong shocks

$$M_3 \approx [2/(\gamma_T(\gamma_T - 1))]^{1/2}. \tag{17}$$

and is therefore approximately constant. Hence, ν_T/ν_{03} can be explicitly written in terms of a_0/a_3. This relationship is plotted in Fig.2 for both $\gamma = 5/3$ and $7/5$ with $\gamma_T = 7/5$.

7. Discussion

Fig.2 displays the relationship between the driver gas sound speed in the compression tube, the test flow conditions (which are encompassed in a_3 via Eq.9) and the highest frequency induced at the primary diaphragm which will decay in the test gases for this driver gas and test condition. If the highest frequency induced at the primary diaphragm, ν_{TH}, is known and a_3 is specified by the test flow conditions, then an appropriate choice of the driver gas sound speed would be one for which $\nu_T > \nu_{TH}$. This choice of a_0 will ensure that no noise induced at the primary diaphragm will be transmitted into the test gas. It can be seen from Fig.2 that this places an upper limit on the choice of a_0. A lower limit on a_0 exists because the driver gas has a limit to which it can be expanded. This limit is reflected in the sharp rise in the curve of Fig.2 at the lower values of a_0/a_3. If the total enthalpy of the test flow is decreased by decreasing the value of a_3 then since

$$\nu_{TH}/\nu_{03} = 2\pi\nu_{TH}/(\lambda a_3), \tag{18}$$

ν_{TH}/ν_{03} will increase. Hence, it can be seen from Fig.2 that if the sound speed of the driver gas is unchanged, less frequencies will be filtered at the driver-test gas interface. Thus, in order to maintain a quiet test flow as the enthalpy is decreased the driver gas sound speed must also be decreased.

Eq.17 can also be written so as to show the dependence on the diameter of the tube. Since $\lambda r_0 = Z_n$, the nth zero of $J_1, Z_n > 0$,

$$\nu_{TH}/\nu_{03} = 2\pi r_0 \nu_{TH}/(Z_n a_3). \tag{19}$$

Thus, as the diameter of the expansion tube increases the value of ν_{TH}/ν_{03} also increases. Therefore, as occurred with a decrease in total enthalpy, the driver gas sound speed must be increased if the quality of the test flow is to be maintained. It should be understood that this assumes that the frequencies induced into the driver gas at the primary diaphragm are unchanged as the diameter of the tube is increased. This may not be the case. It may be possible that lower frequencies at higher modes (larger n) may be induced at the primary diaphragm as the tube diameter is increased. This will mean that the restriction on a_0 is not as severe. What actually happens as the diameter is increased is at this stage unknown.

8. Conclusions

Test flow conditions place a lower limit on the driver gas sound speed. The spectrum of noise induced into the driver gas when the primary diaphragm ruptures places an upper limit on the driver gas sound speed. Otherwise, the driver gas sound speed can be chosen independently of test flow conditions.

The degree of freedom associated with the choice of driver gas is one which is readily available on a free-piston driven expansion tube and one which is essential for the operation of an expansion tube. In general, without it, the test flow will not be quiet.

Acknowledgements

The author wishes to acknowledge that this work was completed under grants supplied by NASA LaRC and the Australian Research Council.

References

Jacobs PA (1992) Numerical simulation of transient hypervelocity flow in an expansion tube, NASA CR 189601.

Paull A, Stalker RJ (1992) Test flow disturbances in an expansion tube. J. Fluid Mechs. 245:493-521

On the Principle, Design, and Performance of an Expansion–Shock Tube for Nucleation Studies

K.N.H. Looijmans, J.F.H. Willems and M.E.H. van Dongen
Faculty of Applied Physics, Eindhoven University of Technology, P.O. box 513, NL–5600 MB Eindhoven , The Netherlands

Abstract.   A new expansion–shock tube for homogeneous condensation experiments, using the nucleation pulse technique, is described. The nucleation pulse is created by reflections of a shock wave at a local widening. Droplet sizes and concentrations are measured with a 90° Mie–scattering technique and a light–extinction method. Droplet growth rates and nucleation rates of water and n–nonane have been measured. The nucleation experiments of water cover a temperature range from 197 K up to 260 K. Nucleation rates vary from 10^{13} m^{-3}s^{-1} up to 10^{17} m^{-3}s^{-1} and are in agreement with Classical Nucleation Theory within two orders of magnitude over the whole range of temperatures. Results of nucleation experiments of n–nonane are in accordance with experimental data found in literature.

Key words: Shock tube, Nucleation, Condensation

1. Introduction

Homogeneous condensation, which is the vapour-to-liquid phase transition in the absence of foreign nuclei, is of interest in many fields. It plays an important role in cloud formation, combustion engines, turbines, and gas transport.

The process of homogeneous condensation can be considered as a two stage process. First, microscopic small droplets, called nuclei, are formed by statistical fluctuations; this process is called nucleation. The main parameter that characterizes the nucleation process is the nucleation rate; this is the number of droplets formed per unit time and space. Second, the formed droplets grow to macroscopic sizes by incorporating single vapour molecules.

Several techniques for experimental investigation of the homogeneous condensation process, and in particular for measuring nucleation rates, have been applied since the invention of the Wilson cloud chamber (Wilson 1897). In expansion chambers, nozzles, and expansion–shock tubes, a vapour undergoes a rapid adiabatic expansion such that a supersaturated state is reached in which the condensation takes place in the form of a cloud of droplets.

To be able to measure the nucleation rate in an expansion cloud chamber, Allard and Kassner (1965) developed a nucleation pulse technique. A nucleation pulse is a very short time period in which the nucleation takes place. A programmed piston expands a gas–vapour mixture to induce nucleation, and then slightly recompresses it after a short period of time. Consequently, nucleation is extinguished, but by maintaining a supersaturated state, formed nuclei keep on growing to macroscopic sizes. The number of droplets formed is then counted from a photograph.

The nucleation pulse technique was first applied in an expansion-shock tube by Peters (1983). A constriction in the low pressure section (LPS) of the shock tube partially reflects the shock wave back into the high pressure section (HPS) where it causes a small recompression. This idea was further improved (Peters and Paikert 1989) by positioning the observation section near the endwall of the HPS. Droplet size and concentration were determined from a 90° Mie–scattering method.

For our expansion–shock tube, we modified the principle of Peters to generate the nucleation pulse (Looijmans et al. 1993). Furthermore, the optical detection of droplets has been extended with an extinction method to measure droplet concentrations. The principle and the performance

Shock Waves @ Marseille I
Editors: R. Brun, L. Z. Dumitrescu   © Springer-Verlag Berlin Heidelberg 1995

of the tube and results of condensation experiments with water and n-nonane are discussed in the following sections.

2. Principle of the nucleation pulse method

2.1. Homogenous condensation

For the first stage of the homogeneous condensation process to occur, it is required that the saturation ratio S of the vapour be much larger than unity. For a pure substance, S is defined as the ratio of vapour pressure and equilibrium vapour pressure evaluated at the same temperature T; $S = p(T)/p_s(T)$. The nucleation rate, denoted by J, is extremely dependent on S. A variation in S of a few percents will result in a variation in J of several orders of magnitude. For the droplet growth stage it is sufficient that S exceeds unity.

The extreme dependency of J on S is used in the nucleation pulse technique to separate nucleation from droplet growth. By applying a proper pressure history to the vapour (expansion followed by a small recompression), supersaturation is first raised to the critical value to induce nucleation, and then it is lowered in such a way that new droplets are no longer formed, but S still exceeds unity. Consequently, existing droplets grow to macroscopic sizes and are all of nearly the same size.

2.2. Nucleation pulse in a shock tube with a local widening

The place–time diagram of Fig.1 shows the wave propagation in the new expansion–shock tube with the local widening from which the nucleation pulse arises.

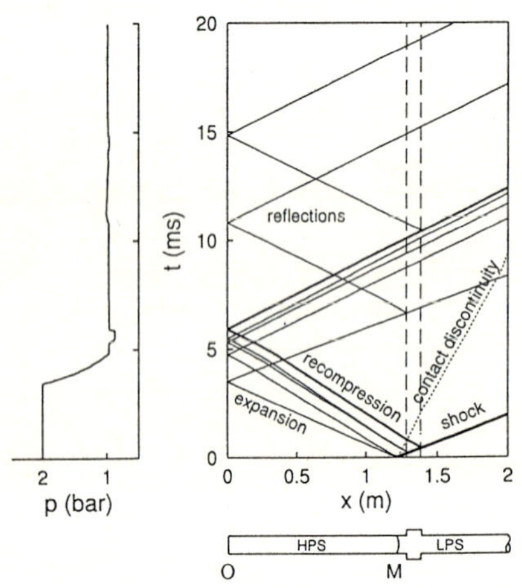

Fig. 1. $x–t$ diagram of wave propagation in an expansion–shock tube for nucleation studies, with a local widening in the low pressure section. At the endwall of the high pressure section, a so–called nucleation pulse is created. The diagram has been calculated with the Random Choice numerical Method, for nitrogen gas

At inital time $t = 0$, the diaphragm that separates the HPS from the LPS is removed. An expansion wave travels into the HPS causing a pressure drop at the observation point O at the endwall. The shock wave partially reflects at the widening, first as a small expansion wave, followed by a small recompression. These reflections run behind the initial expansion wave back into the HPS and cause the desired nucleation pulse at the endwall. An important property of the widening can be seen when the reflected expansion wave passes it. This wave is also partially reflected at the widening as a compression and an expansion, but due to the spatial extent of the expansion fan, which is much larger than the length of the widening, both waves are largely cancelled by interference. Finally, at the observation point, only a positive disturbance is noticed; there will not be a second nucleation period. Thus, only one single well defined nucleation pulse has been obtained.

3. Experimental method

The expansion–shock tube was constructed according to the principle described in Section 2.2. Fig.2 gives a schematic drawing of the tube.

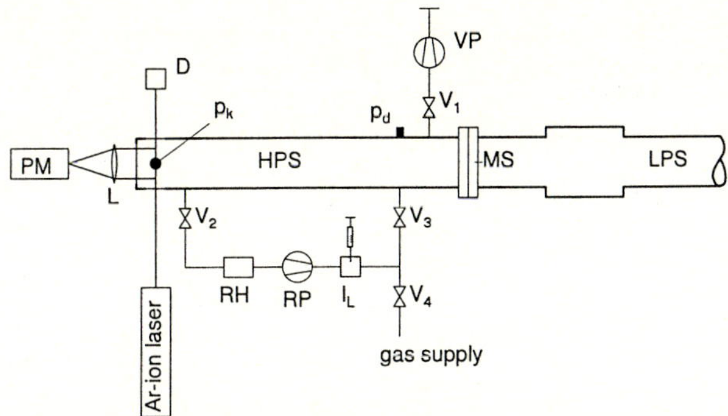

Fig. 2. Experimental set-up. **HPS:** High-Pressure Section; **LPS:** Low-Pressure Section; **MS:** Membrane Section; **PM:** Photo Multiplier; **D:** PhotoDiode; **L:** Lens; **V:** Valve; p_k: Dynamic Pressure Transducer; p_d: Static Pressure Transducer; **RH:** Humidity Measuring Device; **RP:** Rotary Pump; **VP:** Vacuum Pump; I_L: Liquid Injection System

The HPS, on the left, is made of stainless steel and has a length of 1.3 m and a circular inner diameter of 36 mm. The LPS has the same inner diameter except for the widening, which has a diameter of 40 mm. So the ratio of cross sections is 1.23. The local widening is situated 9 cm from the diaphragm, and it has a length of 9 cm. The diaphragm that separates HPS and LPS consists of a polyester membrane (Lexane, thickness 15 μm) clenched in the membrane section MS. The membrane can be weakened by an electrically heated ring–shaped filament (Kanthal), and then is disrupted by the pressure difference between HPS and LPS. The temperature of the filament increases in a period of 50 ms from room temperature up to 600 °C.

3.1. Optical droplet detection

Detection of the small droplets and the subsequent observation of droplet growth is performed by optical means. Small particles scatter incident electromagnetic waves in all directions. The intensity of the scattered light is a function of the properties of the particles such as form, refractive index, and size, and it also depends on the direction of the scattered light and on the direction of the polarization. The light scattering by particles with sizes in the order of the wavelength of the light is described by the Mie–theory (Kerker 1969). According to this theory, the intensity of light scattered at 90° is a typical irregular peak-shaped function of the size parameter α ($\alpha = 2\pi r/\lambda$, r is the droplet radius, λ is the wavelength of the light). When droplets grow during an experiment, this pattern appears as a function of time, and individual peaks can be recognized easily. In this way, droplet size is obtained as a function of time.

Also light attenuation by a cloud of particles is described by the Mie–theory. From the extinction signal, it is possible to determine the concentration of droplets. The extinction of a (laser) beam by a cloud of droplets is described by the law of Lambert–Beer, $I = I_0 \exp(-\beta L)$, where I

is the attenuated intensity of the passed beam, I_0 the reference value, L is the path length through the cloud, and β is the extinction coefficient. The latter is a function of the droplet radius and extinction efficiency, and it is proportional to the droplet concentration. The extinction efficiency, in turn, is a function of reduced droplet radius and refractive index, and it is also described by the Mie–theory. Thus, when the size of a droplet is known (from the scattered light), droplet concentration can be determined from the extinction.

The optical set–up, consisting of a device for 90°–scattering and for extinction measurement, is situated near the endwall of the HPS. Apart from a good optical accessibility through a large window in the endwall, this position has the advantage of a more rapid and deep expansion caused by wave reflection at the wall, and of stagnant droplets.

Droplets are illuminated by an argon–ion laser ($\lambda = 514.2$ nm, 50 mW) which travels through two small windows in the tube wall. To avoid interference from reflections of the laser beam at these two windows, they are not exactly opposite to each other. The windows are situated a little underneath the middle of the tube, so that the angle of incidence of the laser beam is 6°.

Scattered light is focused by a lens on a rectangular stop (11×2 mm). This lens–stop combination limits the solid angle detected by the photomultiplier PM. The θ–angle (angle between scattered light and propagation direction of the laser beam) covers the interval from 88.45° to 90.75°, and the δ–angle (angle between plane of scattering and plane of polarization) ranges from 84° to 96°.

Extinction is measured in the following way: A small part of the transmitted laser beam is focused on a circular stop (ø= 2 mm), in order to minimize the contribution of scattered light to the transmission signal of detector D. Intensity variations of the laser beam are monitored by means of a reference signal.

3.2. Experimental procedure

Before preparation of the gas–vapour mixture, the HPS of the expansion–shock tube and the mixing circuit are evacuated by a turbo–molecular vacuum pump. The vapour component of the gas–vapour mixture is injected in liquid form by a syringe through a septum, into the mixing circuit. When the vapour has evaporated and spread through the HPS, the inert gas (Nitrogen) which acts as a reservoir for the release of latent heat from the growing droplets, is let in. To obtain a homogeneous mixture, it is circulated by a rotary pump RP (Fig.2).The water vapour concentration is measured by a capacitive humidity sensor RH (Humicap) having an accuracy of 2% for vapour pressures down to 0.7 Torr. A piezo resistant pressure transducer is mounted in the tube wall to measure the total pressure in the HPS.

After the membrane was disrupted at $t = 0$, the dynamic pressure is measured by a piezo electric pressure transducer. The temperature of the gas during the experiment is calculated from the isentropic law. The nucleation rate J is calculated from the droplet concentration and the duration of the nucleation pulse Δt; $J = n_p/\Delta t$.

4. Results and discussion

Nucleation rates and droplet growth rates of water were measured for different temperatures and vapour pressures. Fig.3 gives the results of the nucleation experiments, together with data known from literature (Peters and Paikert 1989). The positions of the triangles, circles, and squares indicate the state of the vapour, as it was during the nucleation pulse. The form of the mark gives the order of magnitude of the nucleation rate. The variation in nucleation temperature and pressure is obtained by starting the expansion from different initial vapour pressures. Measured nucleation rates are in the order of $10^{13} \mathrm{m}^{-3}.\mathrm{s}^{-1}$ up to $10^{17} \mathrm{m}^{-3}.\mathrm{s}^{-1}$. In the diagram, also lines of constant nucleation rates according to the Classical Nucleation Theory (Becker and Döring 1935) have been drawn. We observe that most experimental points are in between these lines. Calculations

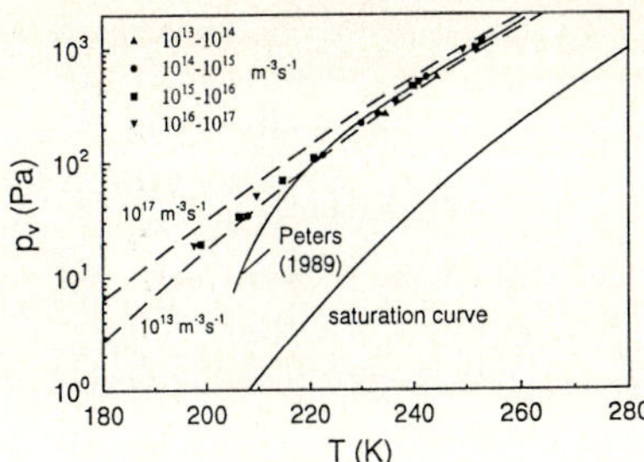

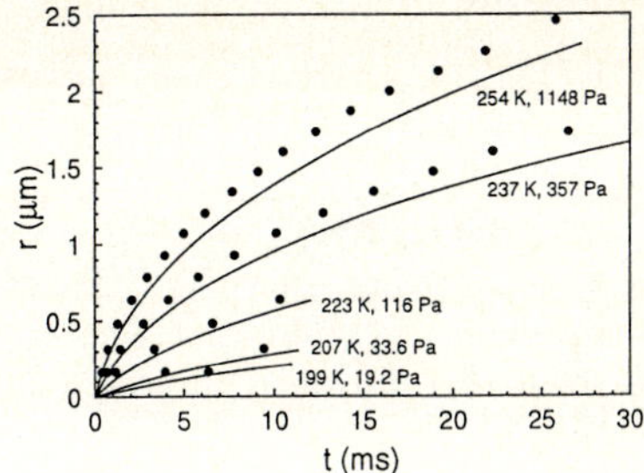

Fig. 3. $p - T$ diagram with nucleation rates of water in nitrogen. The marks indicate the state of the vapour during the nucleation pulse.
—: measurements from literature, $J = 10^{13} - 10^{15}$ $m^{-3}.s^{-1}$ (Peters and Paikert 1989)
- - -: lines of constant nucleation rates according to the Classical Nucleation Theory

Fig. 4. Droplet size of water as a function of time. Various curves correspond to different temperatures and vapour pressures during droplet growth.
● : experiment
—— : theory

show that the experimental results agree with the theory within two orders of magnitude over the whole temperature range.

Growth curves of water droplets for different nucleation temperatures and vapour pressures have been plotted in Fig.4. Circles mark the position of the peaks in the recorded Mie–signal. These peaks proved to be almost as sharp as the theoretically predicted peaks. This indicates that indeed a very homogeneous cloud of droplets is formed by the nucleation pulse technique. The lines represent theoretical curves according to a droplet growth model based on the droplet growth theory of Gyarmathy (1982). Experimental values appear to be larger than the theoretical predictions, by about 10% over the whole range of pressures and temperatures. A decisive explanation has not been found yet.

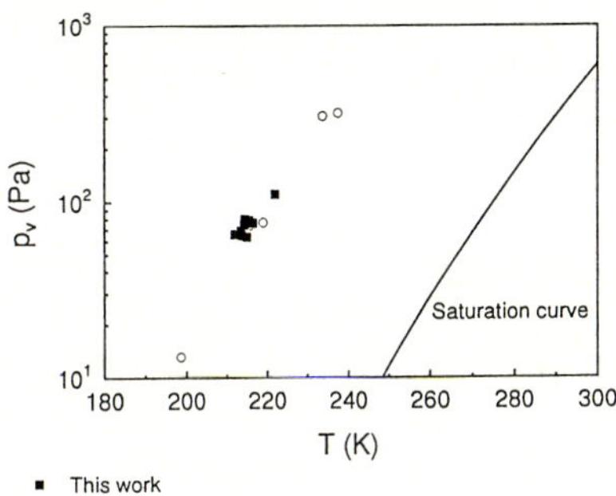

Fig. 5. $p - T$ diagram with vapour state during nucleation of n-nonane in nitrogen. Nucleation rates are of the order of 10^{14} $m^{-3}.s^{-1}$. The results of Wagner and Strey (1984) were obtained in a two-piston cloud chamber

In addition to the water experiments, nucleation rate measurements of n–nonane were performed (Fig.5). The magnitude of supersaturation at which nucleation takes place turns out to be of the order of 10^2. This is in accordance with measurements found in literature (Wagner and Strey 1984).

5. Conclusions

A modified expansion–shock tube for nucleation studies according to the nucleation pulse method
has been presented. Nucleation rates and droplet growth rates can be measured accurately. Nucleation rates of water are in the order of $10^{13} - 10^{17}$ $\mathrm{m}^{-3}.\mathrm{s}^{-1}$ and show good agreement with
Classical Nucleation Theory over the temperature range of 197 K – 260 K. From the sharply
peaked light scattering signal, it is concluded that the nucleation pulse indeed creates a very
mono–dispersed cloud of droplets. Droplet growth rates of water exceed the theoretical prediction by about 10%. Measured nucleation rates of n-nonane are in accordance with data found in
literature.

Acknowledgements

The authors wish to thank Mr. H.J. Jager and Mr. E.J. v. Voorthuisen for their technical support,
and Mr. J.A.A. Smits for his contributions to the experiments.

References

Allard EF, Kassner Jr JL (1965) New cloud-chamber method for the determination of homogeneous nucleation rates. J. Chem. Phys. 42:1401-1405

Becker R., Döring W (1935) Kinetische Behandlung der Keimbildung in übersättigten Dämpfen.
Ann. Phys. 24:719-752

Gyarmathy G (1982) The spherical droplet in gaseous carrier streams: review and synthesis. In:
Hewitt GF, Delhaye JM, Zuber N (eds) Multiphase Science and Technology 1. McGraw-Hill,
London, pp 99-279

Kerker M (1969) The scattering of light and other electromagnetic radiation. Academic Press,
New York

Looijmans KNH, Kriesels PC, Van Dongen MEH (1993) Gasdynamic aspects of a modified
expansion–shock tube for nucleation and condensation studies. Exp. Fluids 15:61-64

Peters F (1983) A new method to measure homogeneous nucleation rates in shock tubes. Exp
Fluids 1: 143-148

Peters F, Paikert B (1989) Nucleation and growth rates of homogeneously condensing water vapor
in argon from shock tube experiments. Exp. Fluids 7:521-530

Wagner PE, Strey R (1984) Measurements of homogeneous nucleation rates for n-nonane vapor
using a two-piston expansion chamber. J. Chem. Phys. 80:5266-5275

Wilson CTR (1897) Condensation of water vapour in the presence of dust-free air and other gases.
Phil. Trans. Roy. Soc; London A 189:265-307

High-Frequency Generation of High-Pressure Pulses Using a Diaphragmless Shock Tube

Koji Teshima

Kyoto University of Education, Fushimi-ku, Kyoto, Japan

Abstract.   We have developed a high-pressure pulse generator at a high repetitive rate by combining a diaphragmless shock tube with shock strengthening by area convergence. The diaphragmless shock tube generates shock waves with Mach numbers of up to 2 traveling in the tube at atmospheric pressure. The shock waves are accelerated by reducing the cross sectional area of the tube. The maximum pressure at the tube end, where the cross sectional area was reduced to 1/64 from the original, exceeded 10MPa and the pulse width was about 25 μ sec. These pressure pulses were generated at a rate of one pulse per several seconds. The facility was used to study biomechanical and biological effects of high-pressure pulses on microorganisms and DNA.

Key words: High pressure pulse generator, Diaphragmless shock tube, Shock acceleration

1 Introduction

Propagation of high-pressure pulses such as shock waves in various media is gaining interest in many fields of application. In media such as porous materials and liquids the pressure pulse must be exerted at atmospheric pressure conditions. Since in some experiments like the study of biomechanical and biological effects of high-pressure pulses on tissues and cells a relatively large number of single pressure pulses would be needed, the pressure pulses must be generated successively within a short time interval. The height of the pressure pulses is required to be several MPa to 10 MPa according to our previous experiments (Ohshima 1992), in which the cell destruction occurred by cavitation effects and the pressure was estimated to be above these values. In addition to these essential capabilities the apparatus has to satisfy the following requirements, if it is used for biological studies; (a) easy handling and operation, (b) portable or easy to move, (c) compatibility with the equipments in the biological experiments, and (d) cleanliness. We have succeeded to develop such a high-pressure pulse generator by modifying a diaphragmless shock tube.

2. Apparatus

In order to generate a single pressure pulse in a tube at atmospheric pressure with a high repetitive rate we have made the following devices: First, we adopted a double-piston type diaphragmless shock tube, as shown in Fig.1. The operation principle is similar to that of the inventors, with the followed improvements: The main piston, which separates the driver gas from the atmospheric air in the driven section, is quickly put in motion by a rapid decrease in the pressure behind it. This is accomplished by discharging the gas, which presses the main piston to the inlet of the low pressure tube, into the atmosphere by a quick motion of the sub-piston. The latter action is produced by exhausting its supply gas using a magnetic valve. In order to achieve a high repetition rate the driver gas was supplied all the time and the driver gas for the sub-piston was discharged when the shock wave was generated. Therefore, the shock tube can be operated by only the on/off operation of the magnetic valve. The operation is controlled with the aid of a microcomputer by setting the number of shots and the time interval.

Shock Waves @ Marseille I
Editors: R. Brun, L. Z. Dumitrescu   © Springer-Verlag Berlin Heidelberg 1995

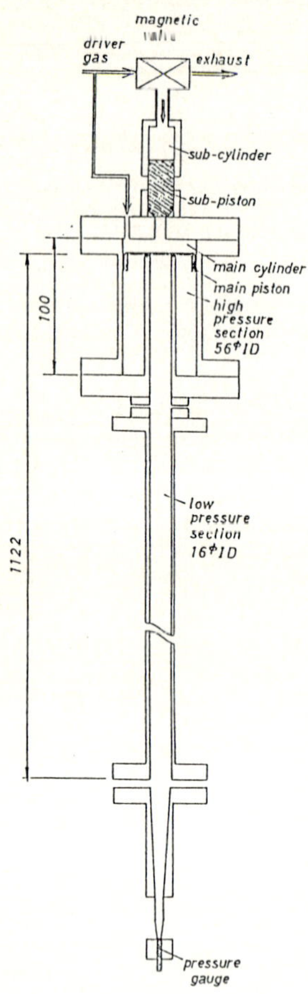

Fig. 1. Double piston-type diaphragmless shock tube

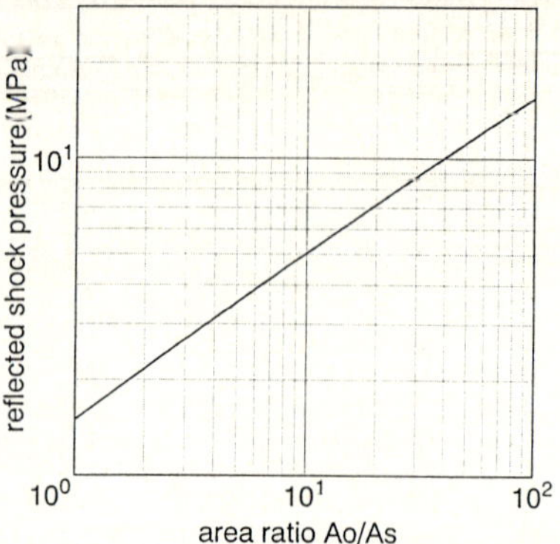

Fig. 2. Attainable pressure at the tube end by area convergence for a shock wave of incident Mach number 2

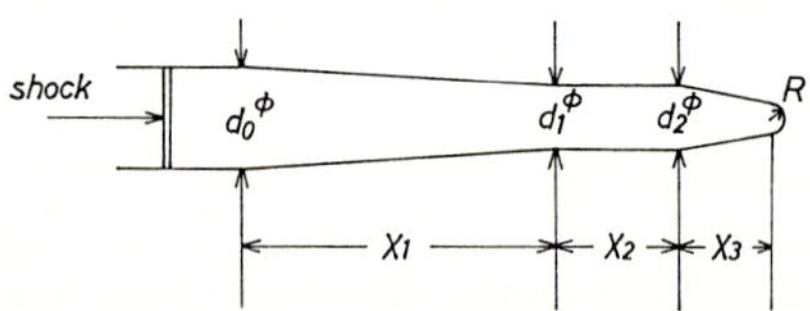

Fig. 3. Shape of area reduction tube and test tube. $d_0 = 16$ mm, $d_1 = d_2 = 6$ mm, $R = 1$ mm, $x_1 = 80$ mm, $x_2 = 17$ mm, $x_3 = 1$ mm

In order to obtain higher shock Mach numbers helium gas was used. The time interval of the operation was then limited by replacement of the driver gas, helium, in the low pressure tube with atmospheric air. This was made by inducing helium with an air flow through a pipe, which was connected to the shock tube far upstream the test section. The flow induced helium with air, which was flowing into the tube from an opening section, described below, and got rid of helium from the low pressure tube. The minimum time interval was determined so as to obtain the same reflected shock pressure as that for a negligible amount of helium. The low pressure tube has a 16 mm inner diameter and a 112 cm length and is made of brass. The inner diameter fits a usual glass test tube. In order to obtain a single pulse the strong wave reflections after the reflection of the incident shock at the tube end must be avoided. For this purpose the low pressure tube was open to the atmosphere just downstream the high-pressure section as in the so-called magic hole type shock tube. Through this opening the air flow, which induces helium, is crossing the low pressure tube. Another opening was located near the tube end just upstream the connection with the test section, where the low pressure tube was separated. This opening was extremely effective to eliminate the subsequent pressure pulses due to wave reflections. The separation distance was adjusted to give minimum effect on the incident shock wave and the maximum effect to the attenuation of the wave reflection. The optimum distance was 20 % of the tube diameter. These openings made it easy to replace the helium gas with air as described above.

For keeping a high repetition rate and for a better durability of the main piston, the practical value of the maximum driver helium pressure was set at 13 atm, which generated a shock wave of Mach number 2 in atmospheric air. The piston borne more than several thousands shots at this pressure. The reflected shock pressure at this shock strength is only 15 atms at the tube end and is not enough for the present purpose. Therefore we have accelerated the shock wave by reducing the cross sectional area of the tube. The strengthening of shock waves by area convergence was analyzed long time ago and it was confirmed that the non-steady one-dimensional theory (Whitham 1958) predicted well the experimental results as far as the area reducing rate was small (Russel 1967). The attainable pressure at the end of a convergent tube was calculated by assuming that the transmitted shock was accelerated according to the non-steady one-dimensional theory as a function of the area ratio A_0/A_s (Fig.2). It suggests that an area reduction of about 1/40 may satisfy the present purpose for an incident shock Mach number of 2.

A small polyethylene test tube, which is used very often in biological experiments, has a shape of a 17 mm -long cylinder of 6 mm inner diameter followed by a converging section of 12 mm length with 2 mm inner dia at the bottom. The test tube was connected to the shock tube end using a uniformly area converging tube from the area of the shock tube to that of the test tube. The apex angle of the convergence was 7 degree and the tube was made in brass. The whole area change downstream the low pressure section is shown in Fig.3. For this tube the area ratio becomes 64 and a pressure of about 12 MPa will be obtained according to the theory.

In order to confirm our measurement system we have made pressure measurements at the bottom of uniform area convergent tubes, with a half apex angle about 3.5 degree and with different diameters at the bottom. These tubes were made in epoxy. The pressure measurements were made with a piezo pressure transducer, a PCB105A23 type. The diameter of the sensor was 2 mm.

3. Experimental results and discussion

The measured pressure histories at the end of the uniformly convergent tubes with $A_0/A_s = 64$ and 7.1 are shown in Fig.4 for an incident shock Mach 2. A sharp pressure rise, to which the pressure transducer presently used would not respond, followed by a rapid decrease due to expansion waves from the region behind the reflected shock, is seen. No other pressure wave was observed after this pressure pulse except small bumps. It can be seen that by reducing the cross sectional area of the tube the shock is strengthened and the pressure pulse is shortened. If we take the pulse width as a full width at half the maximum, it is about 25 μs for the large area ratio tube and about 100 μs for the other. The peak pressure values for these tubes are shown as a function of the incident shock Mach number in Fig.5. For the small area reduction tube the measured pressures are somewhat larger than but not largely different from the theory, whereas for the large area reduction tube it is about 20% less than the theory.

The measured pressure history at the end of the test tube, which is used in the study of the biomechanical effects of shock waves on microorganisms, is shown in Fig.6(a). The peak pressure is less than that for the uniformly convergent tube shown in Fig.3 and the pressure decay becomes slower. The measured values of the peak pressure are also plotted in Fig.5 in comparison with the theory and with those of the uniformly convergent tubes. They are about 60% of the theory and about 80% of that obtained for the uniformly convergent tube with the same area ratio. Since the test tube has a straight part, the transmitted shock accelerated by passing through an area convergent section is decelerated while passing through the straight part and then the shock

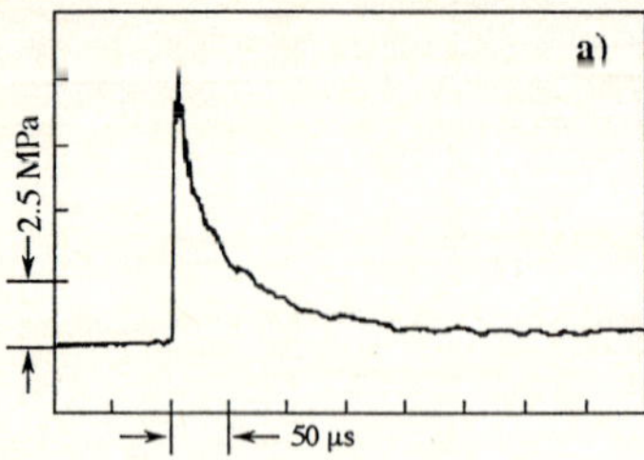

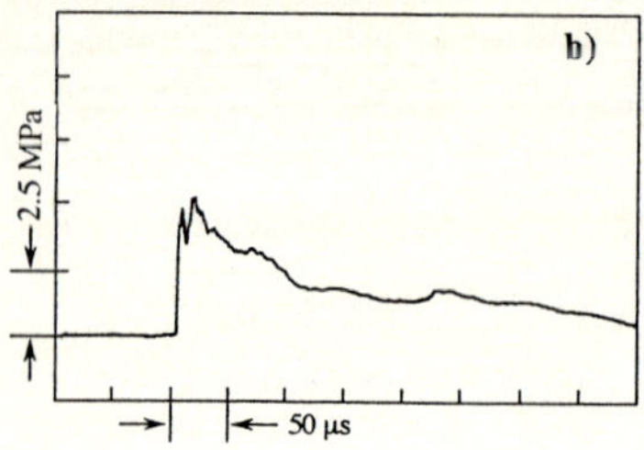

Fig. 4. Measured pressure histories at tube end of the uniform area reduction tube for a shock wave with incident shock Mach number 2. (a) area ratio 64, (b) area ratio 7.1

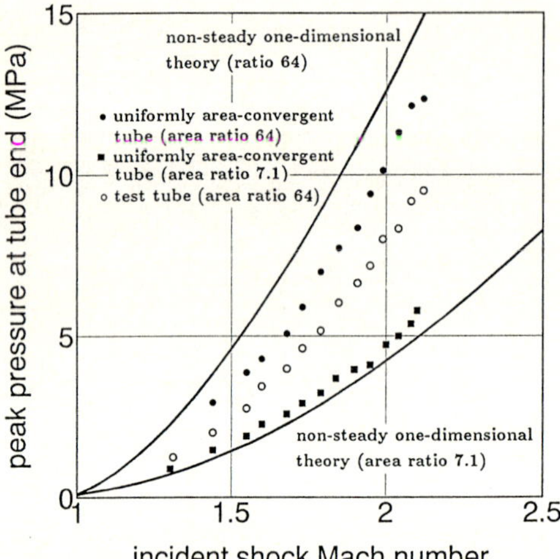

Fig. 5. Peak pressure at the tube end as a function of incident shock Mach number. The solid lines show calculated reflected shock pressure by the non-steady one-dimensional theory (Whitham 1958). • : uniformly area-convergent tube with area ratio 64; o : test tube (area ratio 64); ■ ; uniformly area-convergent tube with area ratio 7.1

strength at the tube end is weakened. The shock wave transmission in the test tube was calculated by using a two-dimensional numerical simulation method by a piece-wise linear method (Saito T et al. 1989). The calculated pressure change at the tube end is compared with the experiments in Fig.6(b). The calculation gives still a larger peak pressure than the experiment, but simulates well the time variation.

In the experiments for the biomechanical effects of shock waves on living cells and microorganisms these are suspended in liquid. A small amount of the solution is laid at the bottom of the tube, where the pressure is expected to be nearly the same as the reflected shock pressure. The remaining volume of the tube is filled with a harmless liquid (e.g. liquid paraffin) and the tube is attached to the shock tube end. The liquid surface is covered by a thin film (polyvinylidene) to avoid splashing the liquid when the shock hits the surface. The pressure at the tube end was measured when the tube was filled with liquid. A typical pressure history is shown in Fig.7. The pressure rise is eased during the transmission in the liquid but the peak pressure becomes larger than in air. The effective pressure width was about 20 μs. The measured values of the peak pressure at the tube end with and without liquid paraffin are plotted as a function of the driver gas pressure of helium in Fig.8.

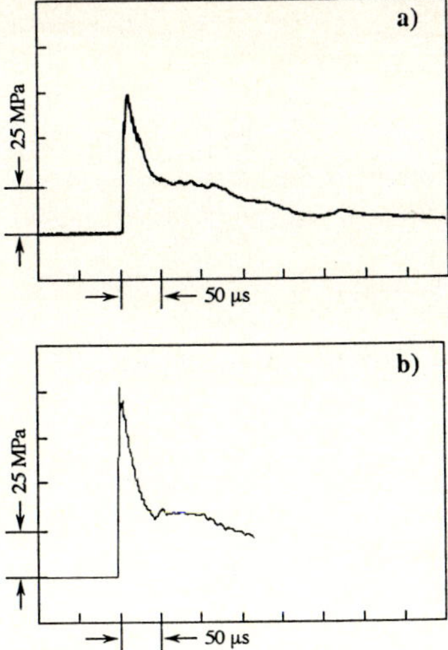

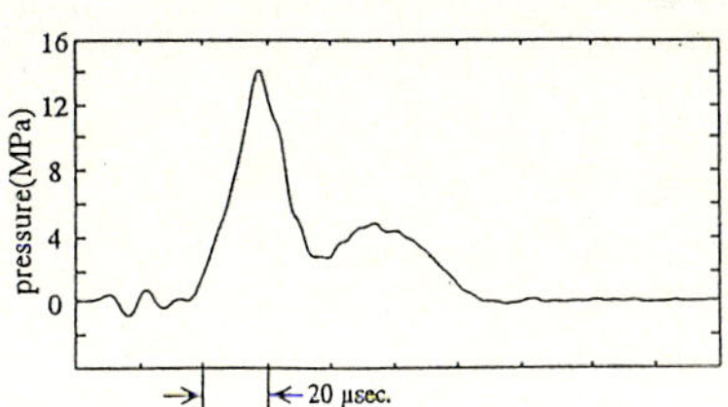

Fig. 6. Comparison of measured pressure history with simulated one for an incident shock Mach number 2 (a) measured pressure history, (b) simulated pressure history

Fig. 7. Measured pressure history at the end of test tube filled with liquid paraffin

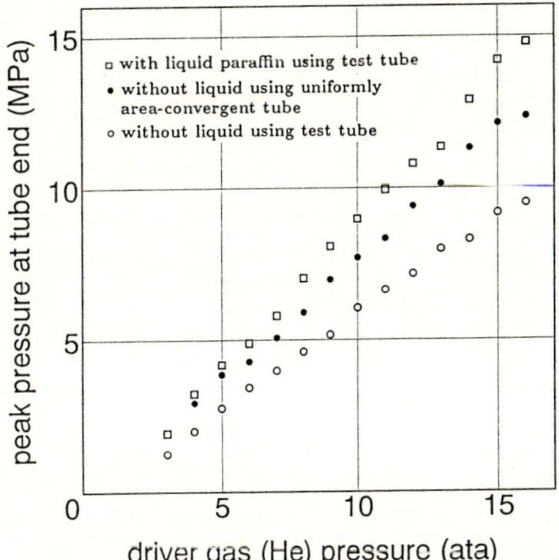

Fig. 8. Peak pressure at tube end with and without filling liquid paraffin as a function of driver helium pressure. □ : with liquid paraffin; • : without liquid using uniformly area-convergent tube (area ratio 64); ○ : without liquid using test tube (Fig.3)

4. Conclusions

High-pressure pulses, not much less than the predicted values, can be achieved with relatively low driver gas pressure. The high pressure was achieved in a restricted portion of the tube and for a short time (about 20 μ sec). These high-pressure pulses can be generated every 4 ~ 5 s. The performance makes it possible to use the facility for biomechanical studies. This facility is small, easy to operate, clean, and less expensive for running, therefore it is suitable to study biomechanical and biological effects on living tissues and cells.

Acknowledgment

The present study was supported by a Grant-in-Aids for the Scientific Research on Priority Area, *Analysis of Shock Wave Phenomena*, from the Ministry of Education, Science and Culture, Japan.

References

Russel DA (1967) Shock-wave strengthening by area convergence J. Fluid Mechs. 27:305

Saito T, Nakatsuji H, Teshima K (1986) Numerical simulation and visualization of freejet flow-fields. Trans. SASS 28:240

Teshima K, Ohshima T, Tanaka S, Nagai T (1993) Biomechanical effects of shock waves microorganisms and λ-phage DNA. These Proceedings

Whitham GB (1958) On the propagation of shock waves through regions of non-uniform area or flow. J. Fluid Mechs; 4:337

A New Diaphragmless Shock Tube Facility for Interface Instability and Mach Reflection Studies

Yong W. Kim

Department of Physics, Lehigh University, Bethlehem, PA 18015, USA

Abstract. A new shock tube facility has been constructed in order to help clarify: (i) the exact nature of Mach reflections in density stratified media and (ii) the growth rate of the interfacial instability under forcing by a shock wave. The shock tube is diaphragmless and has two fast-action valves at each end of the driven section, one to create a well defined interface between two dissimilar gases and the other to produce a shock propagating into the interface. Actuation of the two valves is pneumatically synchronized. The shock tube operates in the Mach number range of 1.01 to 5 with an exceptionally high degree of reproducibility and rapid turn-around time. The design, construction and performance studies of the shock tube are presented.

Key words: Dual driver shock tube, Fast-action valves, Density stratified media, Interfacial instabilities

1. Introduction

Considerable interest and controversy persist in regard to two particular areas of shock wave research. One is the exact nature of Mach reflections in uniform as well as in density-stratified media. (Dewey 1992; Henderson 1991) Another is the matter of the instability of an interface between two gases when it is overrun by a shock wave. This is often referred to as the Richtmyer-Meshkov instability, and the outstanding issue has to do with the order-of-magnitude discrepancy between theory and experiment on the growth rate of the finite amplitude modulation of the interface upon shock passage (Rupert 1992, Remington et al. 1991). Both problems share a common attribute in that the thermodynamic and flow properties across the shock front proper play the central role. On the theoretical side, it is simpler to assume an ideal shock front profile to treat the subsequent development of the instability or reflection process. In reality the shock front or gas interface is not a simple discontinuity. On the experimental side, detailed measurement of the shock structure has remained a difficult unfulfilled task in spite of considerable attention given to the problem some decades earlier.

It appears that measurement of the realistic shock profile in density, temperature and flow velocity over length scales of the order of a micron would contribute toward a resolution of outstanding issues in both areas. Such a determination would also be extended to the interface structure before shock exposure. We have set out to address them by means of the following three experimental implementations: (i) a new rectangular shock tube with two driver sections, each equipped with a fast action valve, in place of a diaphragm; (ii) flow visualization by submicron tracer particles for high resolution measurement of the flow velocity across the shock front; and (iii) pulsed Raman spectroscopic imaging for measurement of the density profile across the shock front as well as the interface according to molecular species. The usual array of diagnostics such as pressure transducers and Schlieren and interferometric imaging is also employed. Gated, intensified 2-D video camera with digital frame-grabber and pulsed lasers such as Q-switched ruby and Q-switched second harmonic Nd:glass lasers are utilized.

Shock Waves @ Marseille I

Editors: R. Brun, L. Z. Dumitrescu

© Springer-Verlag Berlin Heidelberg 1995

2. New experimental facility

In this paper we describe the design and performance of the new diaphragmless shock tube and some preliminary results of the tracer particle visualization of the shock front. The shock tube is a new implementation in a series of the diaphragmless shock tubes we have advanced in the recent years (Kim 1990). It has a wide-area rectangular driven section and two driver sections, one to produce the shock wave and the other to create a two-gas density interface. The two driver sections are mounted at each end of the driven section. The primary driver section has the shape of an annular cylindrical chamber, surrounding the driven section coaxially. The driven section measures 4.45 cm by 8.89 cm in cross-section and 645 cm in maximum length. The inner and outer diameter of the annular primary driver section are 13.97 and 17.78 cm, respectively, and its length is 91.44 cm; the effective gasdynamic length is at least an order of magnitude longer than this value, however. The two sections are brought together evenly on one end so as to permit free access to the joining area between them. Instead of the usual diaphragm, we have placed a novel fast-action bellows valve here. The overall arrangement of the shock tube facility is shown schematically in Fig.1.

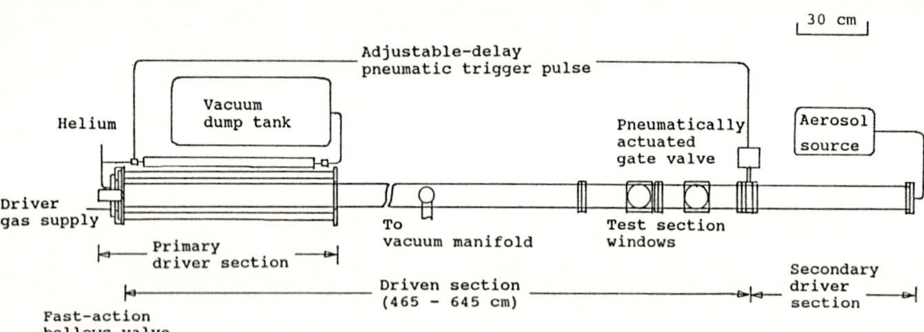

Fig. 1. A schematic diagram of the new diaphragmless shock tube facility. Two fast action valves are shown at the two ends of the common driven section. The primary driver section has an annular cross-section and surrounds the driven section coaxially. The secondary driver section is mounted conventionally at the end of the driven section. Each shock tube run is initiated with actuation of the gate valve

The fast-action valve meets two key objectives: high degree of reproducibility and rapid turn-around time for the experiment. The valve is composed of two metallic sylphon bellows, connected together by a rigid rod but separated into two independent pressure chambers. The bellows assembly is mounted on either side of a cover plate, and the interconnecting rod is shaft-sealed as a motion feedthrough. The sealed end of the bellows on the interior side of the cover plate is fitted with an O-ring and acts as a valve seated within the opening into the driven section. Valve closing is effected pneumatically by pressurizing this interior side bellows with helium. Rapid opening of the valve is accomplished by releasing the helium gas into an evacuated dump tank by free expansion. The exterior side bellows, on the other hand, is used as a spring of controllable strength to assist in rapid valve opening. The exterior bellows is filled also with helium to a preset pressure as a control for the spring constant. The bellows valve assembly as mounted on the primary driver section is shown in Fig.2.

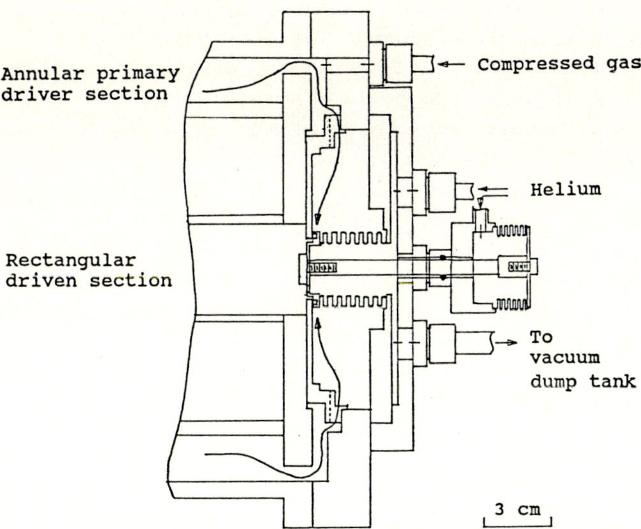

Fig. 2. A detail view of the fast-action bellows valve in place at the joint between the rectangular driven section and the annular driver section surrounding it. The arrowed lines show the path of the free expanding driver gas

The pneumatic actuation of the fast-action bellows valve permits operation of the shock tube for blast simulation because the valve opening and closing can be programmed in temporal sequence and the degree of opening or closing.

The rapid opening action of the bellows valve stems from its intrinsic instability. Once the opening is initiated by sudden release of the helium gas from the interior side bellows, the force due to the pressure difference between the driver and driven sections reverses direction and further accelerates the opening motion. The driver section pressure thus helps shorten the valve opening time; the minimum opening time, however, is determined by the operational condition of the two bellows.

Creating a density discontinuity interface between two gases is far from trivial or simple. There have been a number of methods utilized for this purpose, ranging from use of a thin film membrane to use of a sliding partition (Houas, Chemouni and Touat 1992, Benjamin, Besnard and Haas 1992, Bonazza, Sturtevant 1992). The difficulties with these methods are that the interface density and velocity profile remain complex, ill-defined or unknown over the length scales of importance. The approach we have pursued is a dynamic one in that a second gas is introduced impulsively through the endwall of the shock tube. The actual implementation is in effect to have a second diaphragmless shock tube sharing the same driven section. Firing of the two shock tubes is synchronized by a pneumatic delay line which links actuation of one fast-action valve to that of the other. The time delay between the two actuations is adjustable. Another key element here is the fact that the endwall is made up of two layers of perforated metal foils separated by 1.59 cm. The idea is to breakup the flow of the free-expanding driver gas into small wavelets so as to maintain a planar interface by preventing development of any large eddies. The foils are 0.025 cm thick and have circular apertures of 0.050 cm diameter in a hexagonal grid with 0.100 cm nearest-neighbour separation. The pneumatic valve used for this purpose is of a sliding gate valve design with a piston driver. The gate of the valve is located 0.40 cm from the second layer foil.

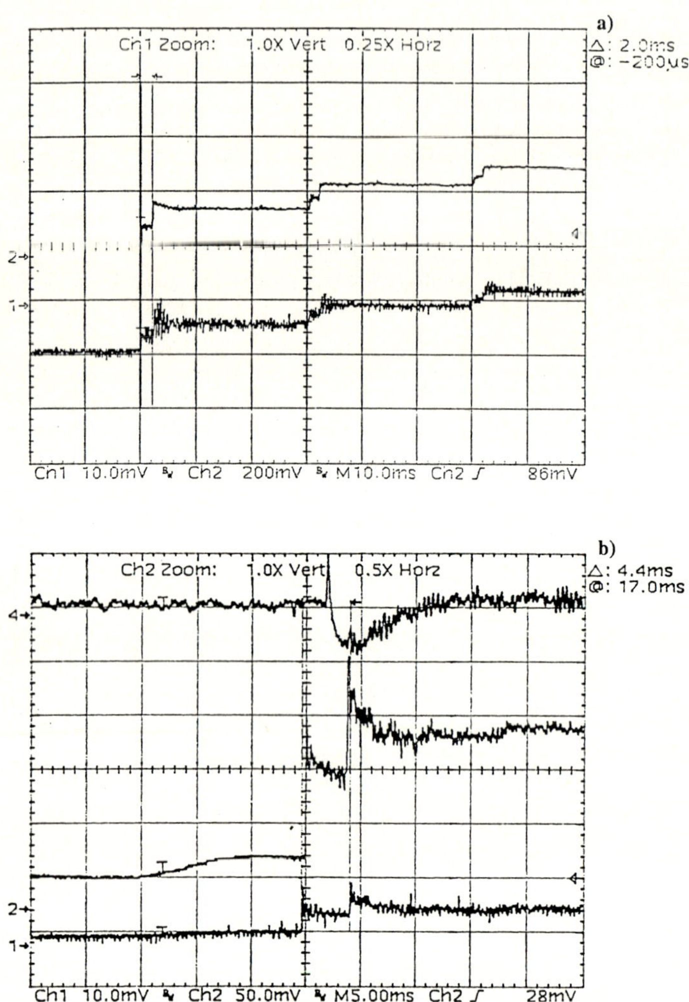

Fig. 3. a) Pressure recordings from two piezoelectric transducers, one at 40.64 cm from the solid endwall (channel 1) and the other at 33.02 cm (channel 2), with only the primary driver section in operation; b) pressure recordings from the two transducers as in a) and light scattering intensity recording (channel 3) from entrained aerosol at 20.32 cm from the perforated metal-foil endwall under dual driver operation

In actual operation the gate valve is set into motion by releasing helium in the driver piston chamber. This helium is directed to another valve actuator mechanism which initiates the bellows valve of the primary driver section. The movement of the helium from the gate valve to the bellows valve is regulated in a controllable manner in order to achieve the time delay between two valve openings.

3. Performance data

The overall performance of the new shock tube has been most satisfactory in regards to all areas of the design objectives. We have achieved a shot to shot reproducibility of better than 0.2% scatter in shock speed using air for both the driver and driven gas over the shock Mach number range of 1.1 to 2.5. The range can be readily extended to less than 1.01 and greater than 5.

In addition to the usual measurements by a number of pressure transducers, we have accomplished full detection and visualization of the shock front by suspension of mineral oil drops in gases. The light scattering intensity shows a translational relaxation of the particles behind the shock front in a size dependent manner. We have developed several methods by which the particle size distribution can be varied. Evidence on hand suggests that the flow field across the shock front can be mapped out in detail by imaging individual particles through reduction of the particle size to about 0.1 microns in radius and using weak shocks.

The shock tube has been successfully operated in the double driver mode as well. The gas interface has been produced between freon-12 and air or sulfur hexafluoride and air. The interaction of the shock wave with the two-gas interface can be effected at an arbitrary location within the driven section in a predictable manner. This permits systematic investigation of the growth rate of the interfacial instabilities over a wide range of interaction time. It is also possible now to create an arbitrary profile of density stratification for shock propagation by selecting the molecular weight and initial pressure of the secondary driver gas.

Fig.3 shows a pair of shock tube runs, one with only the primary driver section operating with a solid endwall (Fig.3a) and the other with both drivers in operation. The two runs are identical in that the primary driver section is filled with air at 3.04 atm and the driven section is filled with air at 300 torr. In the run of Fig.3b the secondary driver was composed of a mixture of freon and mineral oil droplets. The channel 2 trace of Fig.3b shows the arrival of freon before that of the shock. Notice the lowering of shock speed by the presence of freon as well as the significant modifications of the shock profile.

The new shock tube facility requires low maintenance and permits rapid turn-around for experimental runs. The systematic investigation of the interfacial instabilities is currently in progress.

Acknowledgements

The author acknowledges the able assistance of John Gregoris, Bob Wolfinger, Joe Zelinski, Jeff Spirko and Rob Leoni with construction of the shock tube facility and instrumentation.

References

Benjamin RF, Besnard D, Haas JF (1992) Richtmyer-Meshkov instability of shocked gaseous interfaces. In: Takayama K (ed) Shock Waves, Proc. 18th ISSW, Springer-Verlag, p 325

Bonazza R, Sturtevant B (1992) X-ray densitometry of shock-excited Richtmyer-Meshkov instability at an air-xenon interface. In: Takayama K (ed) Shock Waves, Proc. 18th ISSW, Springer-Verlag, p 331

Dewey JM (1992) Mach reflection research - paradox and progress. In: Takayama K (ed) Shock Waves, Proc. 18th ISSW, Springer-Verlag, p 113

Henderson LF, Colella P, Puckett EG (1991) On the refraction of shock waves at a slow-fast gas interface. J. Fluid Mechs. 224:1

Houas L, Chemouni I, Touat A (1992) Experimental investigation of Richtmyer-Meshkov turbulent mixing in shock tube. In: Takayama K (ed) Shock Waves, Proc. 18th ISSW, Springer-Verlag, p 319

Kim YW (1990) Steepening of density gradient in a free expansion. In: Kim YW (ed) Current Topics in Shock Waves, Proc. 17th ISSTW, American Institute of Physics, p 120

Remington BA, Haan SW, Glendinning SG, Kilkenny JD, Munro DH, Wallace RJ (1991) Large growth Rayleigh Taylor experiments using shaped laser pulses. Phys. Rev; Lett; 67:3259

Rupert V (1992) Shock-interface interaction: Current research on the Richmyer-Meshkov problem. In: Takayama K (ed) Shock Waves, Proc. 18th ISSW, Springer-Verlag, p 83; and the references therein

Analysis of Calibration Results in the High-Enthalpy F4 Hot-Shot Wind Tunnel

Ph. Sagnier and G. François
ONERA, Châtillon, France

Abstract. The status of measurements performed in the F4 calibration campaign is presented. Some tunnel hardware modifications which have improved flow uniformity are discussed. A preliminary calibration result analysis, concerning nozzle wall pressure, test section Pitot pressure, hemispherical probe stagnation heat transfer, nozzle exit boundary layer Pitot pressure, in F4 nozzle No.2, has been performed with the help of non-equilibrium flow codes, and is also presented. At low enthalpy, agreement between numerical predictions and experimental data are generally good. However some disagreements occur when enthalpy increases.

Key words: Hypersonic, hot-shot wind tunnel, Non-equilibrium flow

1. Introduction

The ONERA F4 hot-shot wind tunnel was designed to cover part of the HERMES reentry vehicle flight envelope, namely at an enthalpy high enough to produce oxygen dissociation. Although the foreseen performance level has not yet been reached due to technological problems, in fact due to tunnel parts over-heating, F4 still produces hypersonic flows at velocities larger than 4 km/s and at several hundred bars of total pressure with air. F4 is then the most performing hot-shot wind tunnel ever built. Detailed informations about F4 (tunnel sketch on Fig.1) and the attached instrumentation can be found in reports of François et al. (1992), Girard et al. (1990). Before tests on realistic airplane configurations, test section flows have been calibrated mainly by classical techniques such as pressure and heat transfer measurements. Some feasibility tests of advanced optical methods have also been performed but will not be discussed here. Calibration results were compared with numerical simulations performed with tools available at ONERA.

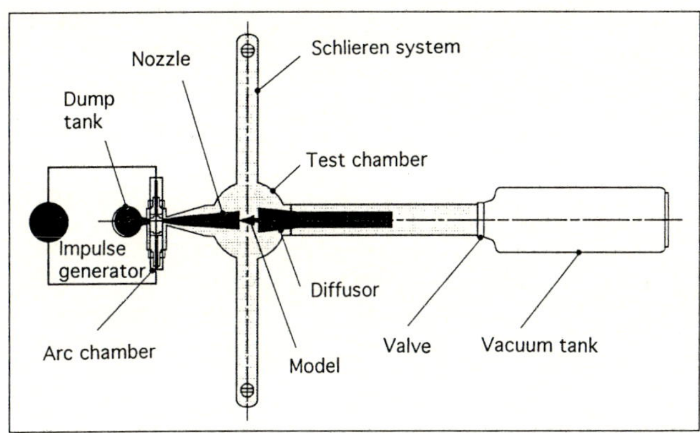

Fig. 1. F4 tunnel and arc chamber

2. Calibration domain

Calibration has been performed, particularly, for 4 conditions (Fig.2) in nozzle No.2, which have been selected by Dassault Aviation for Hermes industrial tests. Some tests have already been performed, like force tests on an Hermes model on a balance with inertia compensation during spring 1993. Fig.2 shows the evolution of total pressure and enthalpy during the runs. Pressure

Shock Waves @ Marseille I
Editors: R. Brun, L. Z. Dumitrescu

and enthalpy decrease once the electric arc is over due to thermal loss in the arc chamber. Furthermore, slope changes on curves represent throat pyrotechnic plug ejection, mostly visible on high-enthalpy tests (indicated by an arrow on one shot of condition 4). Note that controling throat opening by using pyrotechnic plugs has greatly improved test section measurement repeatability. Total conditions are determined as follows: pressure is measured in the arc chamber. With its value and the knowledge of total density, equilibrium calculation (5 species air) gives total enthalpy. When the nozzle throat opens, mass loss in the chamber is also taken into account. In the present calibration analysis, nozzle wall pressure, test section stagnation pressure and heat flux measurements were used.

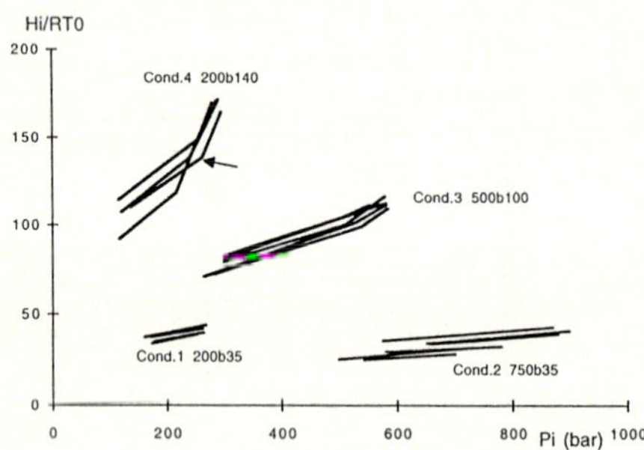

Fig. 2. F4 nozzle No.2 calibration test conditions

3. Numerics used for calibration analysis

To simulate hypersonic nozzle flows, non-equilibrium tools are necessary. Fig.3 shows the different physical nozzle zones, with numerical methods used to compute each one. This methodology was used in the design of the F4 nozzle (Rollin (1990) or in other high-enthalpy hypersonics studies (Sagnier and Marraffa 1991, Devezeaux et al 1992).

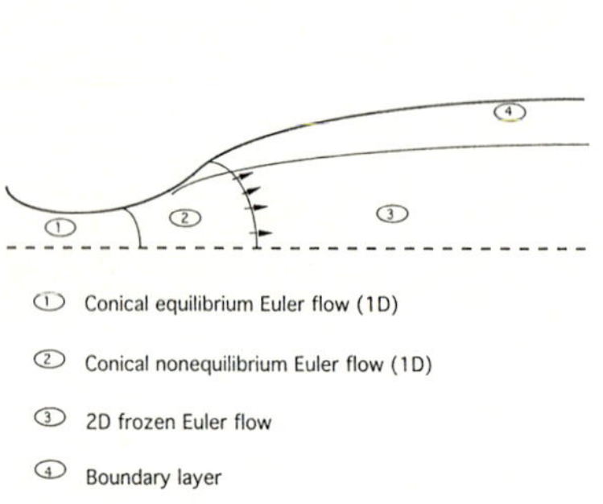

Fig. 3. Calculation zones in high-enthalpy nozzle

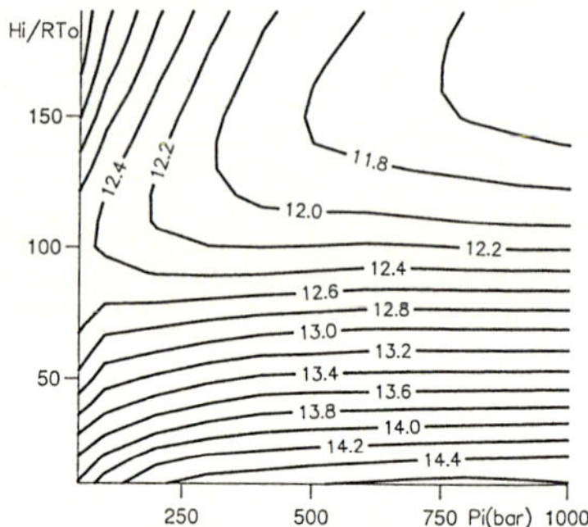

Fig. 4. Chart of nozzle exit frozen Mach number

Quasi-1D non-equilibrium calculations take into account both chemical and vibrational relaxations, while turbulent boundary layer is modelized with Edenfield (1968) correlation. More sophisticated full Navier-Stokes (Flament 1990) or parabolized Navier-Stokes (Hugues, Verant 1991) codes will be also used in the next future when turbulence will be implemented. In F4 nozzle 2, nozzle exit flow charts have been made with equilibrium-non-equilibrium quasi-1D codes coupled with Edenfield boundary layer depending on total pressure and enthalpy. Examples of results are given for frozen Mach number (Fig.4). For these charts, Gardiner's chemical kinetics, Blackman vibrational relaxation times and Treanor-Marrone non preferential vibration-dissociation coupling were used (see Sagnier and Marraffa (1991) for more details). These results and Rollin's ones were used for comparison with calibration data.

4. Analysis of calibration results
4.1. Nozzle static pressure

Static pressure is measured in 5 locations on the nozzle wall. On Fig.5, total pressure-normalized measurements are shown and compared with calculations described in §3, for condition $H_i/RT_0 = 35$, $P_i = 200$ bar. Some tests have been performed with a shield at the nozzle entrance, to get cleaner and quieter flow. The effect of this modification is small on static pressure, while it is more important on test section flow uniformity as shown in next chapter. Back to experimental-theoretical comparisons, 2D calculation (Fig.5) compares well with experiments. 1D calculation results stand logically between 2D wall and axis pressures. But at high-enthalpy, experimental-theoretical comparisons are not so good, as shown on Fig.6, showing condition $H_i/RT_0 = 100$, $P_i = 500$ bar.

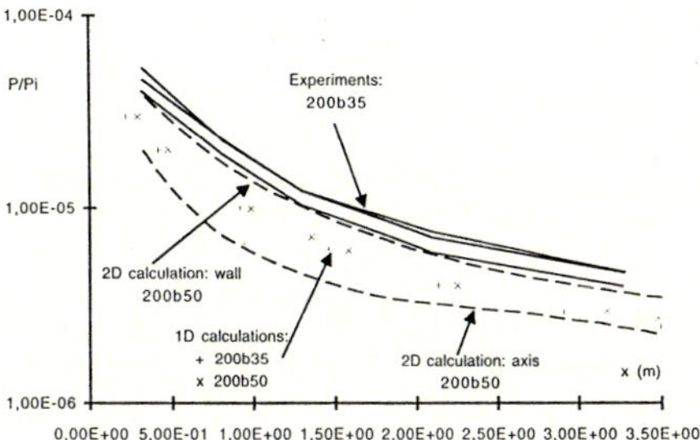

Fig. 5. Experimental/theoretical comparison of nozzle static pressure at $H_i/RT_0 = 35$, $P_i = 200$ bar

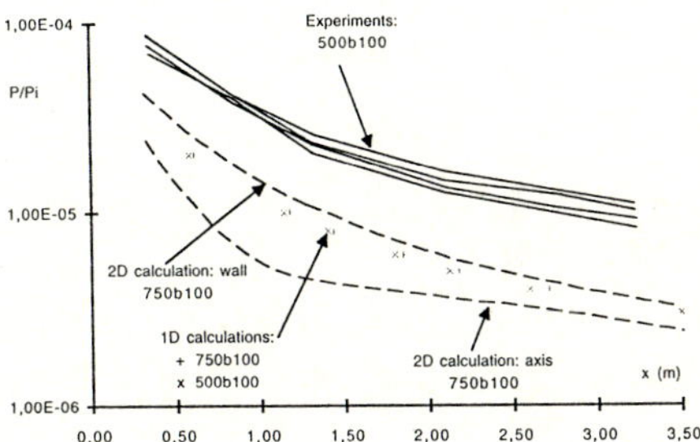

Fig. 6. Experimental/theoretical comparison of nozzle static pressure at $H_i/RT_0 = 100$, $P_i = 500$ bar

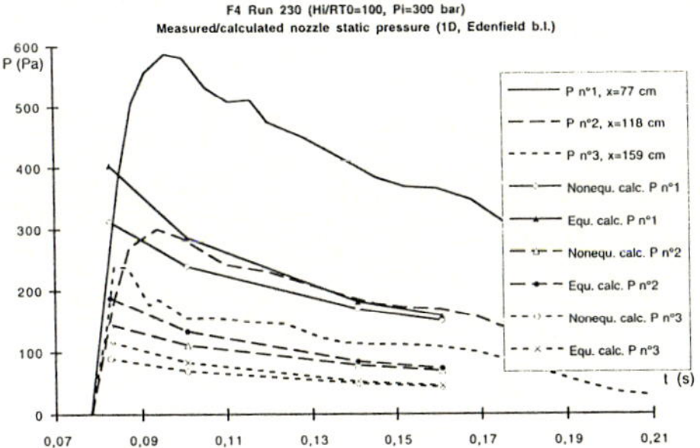

Fig. 7. Experimental/theoretical comparison of nozzle static pressure at $H_i/RT_0 = 100$, $P_i = 300$ bar for nozzle No.1 in function of time. Equilibrium and non equilibrium calculations

This behaviour has not been explained yet since theoretical boundary layer seems to be overestimated at high-enthalpy (see § 4.2) and a thinner boundary layer would yield smaller nozzle static pressure. Furthermore, for other conditions in nozzle No.1, full equilibrium 1D calculations have been performed, showing only slight static pressure increase compared to non-equilibrium (Fig.7).

4.2. Test section stagnation pressure

Stagnation pressure distributions have been measured with the cross-rake, at two axial locations in the test section. First, let us discuss the nozzle entrance shield: its effect is very weak at low enthalpy but dramatic at high-enthalpy as shown on Fig.8, where radial stagnation pressure distributions are plotted, normalized by total pressure, without and with shield, at $P_i = 200$ bar, $H_i/RT_0 = 140$. The test section flow becomes much more uniform when using the shield, indicating the gas is likely more at rest in the arc chamber at low enthalpy. On Fig.8 are also plotted boundary layer thickness (δ) and stagnation pressure ($0.95\rho_\infty V_\infty^2$) given by quasi-1D calculations. Agreement is not good at this high-enthalpy condition while again it is better at lower enthalpy, $H_i/RT_0 = 35$, $P_i = 200$ bar (Fig.9). On Fig.9, stagnation pressure distributions are given for the two rake axial locations.

4.3. Nozzle exit boundary layer

No specific nozzle boundary layer raking has been performed yet but it was possible to use cross rake data to assess nozzle exit boundary layer displacement thickness at low enthalpy condition. The principle of this assessment is shown on Fig.10: since no raking was performed just at the nozzle exit plane, it was checked that the mixing zone was not influencing Pitot probes. Then, seven Pitot probes were used (Fig.9) and, using a perfect gas assumption, the displacement thickness was calculated. Its value (6.2 cm) compares well with the Edenfield value (7 cm). This analysis was not performed at high enthalpy since additional data, such as stagnation enthalpy profiles, would have been necessary.

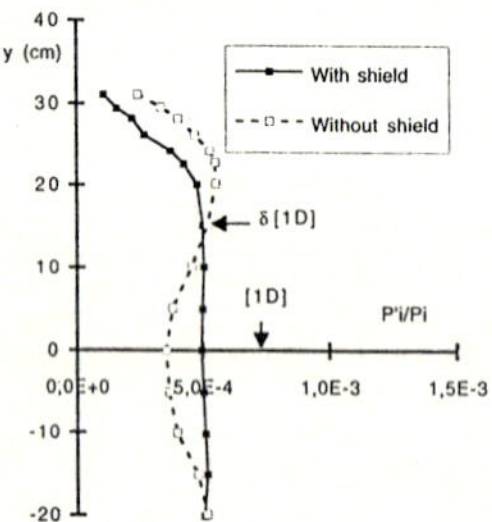

Fig. 8. Shield effect on test section stagnation pressure radial distribution and comparison with 1D calculation results ($H_i/RT_0 = 140$, $P_i = 200$ bar)

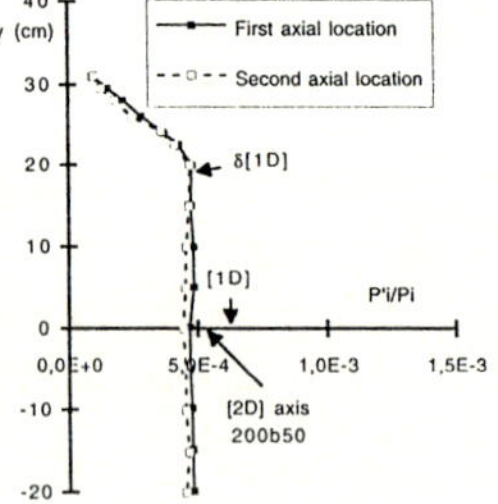

Fig. 9. Test section stagnation pressure radial distribution at 2 axial locations and comparison with 1D and 2D calculation results ($H_i/RT_0 = 35$, $P_i = 200$ bar)

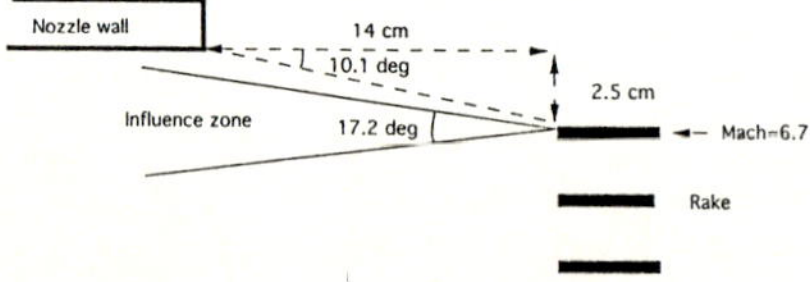

Fig. 10. Experimental / theoretical nozzle boundary layer comparison at low enthalpy ($H_i/RT_0 = 36$, $P_i = 200$ bar). The sketch justifies the use of cross rake measurements for boundary layer analysis

4.4. Total enthalpy determination through stagnation heat flux measurements

Test section total enthalpy is obtained from heat flux probes usung Fay and Riddell's (1958) formulation: from stagnation heat flux and pressure measurements, total enthalpy is calculated. Several heat flux probe types are used, such as constantan and steel thick skin surface temperature or copper calorimetric probes. On the calorimetric type, the spherical volume-averaged heat flux is measured and a stagnation value is derived assuming Lees' (1956) heat transfer distribution. Comparisons between arc chamber enthalpy and total enthalpy given by stagnation flux probes are shown on Fig.11 at low enthalpy ($H_i/RT_0 = 35$, $P_i = 200$ bar) and Fig.12 at high-enthalpy ($H_i/RT_0 = 140$, $P_i = 200$ bar). Arc chamber and test section enthalpies are in agreement at low enthalpy 20 ms after throat opening for all probes but one (Fig.11). At high-enthalpy (Fig.12), scattering between the different probes is larger and probes never match arc chamber enthalpy at the same time. In general, disagreement between arc chamber and test section enthalpies increases with increasing enthalpy. Such differences are rather systematic, excluding some random effects such as solid pollution. Some partial explanation could come from the estimation of the stagnation point velocity gradient used in Fay and Riddell's formula: up to now in F4, the velocity gradient is given by the newtonian assumption, while perfect gas Euler calculations (Barnwell 1971) show significant differences with newtonian values (Fig.13).

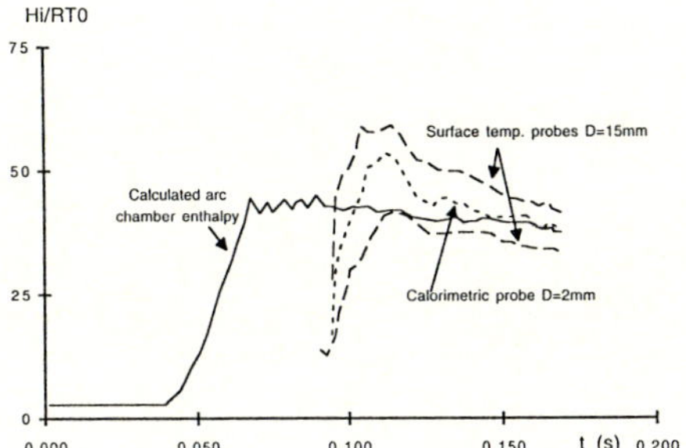

Fig. 11. Comparison of total enthalpies determined in arc chamber and test section in function of time ($H_i/RT_0 = 35$, $P_i = 200$ bar)

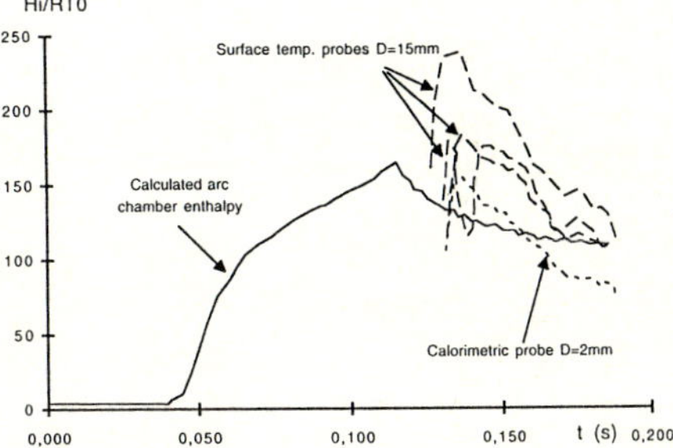

Fig. 12. Comparison of total enthalpies determined in arc chamber and test section in function of time ($H_i/RT_0 = 140$, $P_i = 200$ bar)

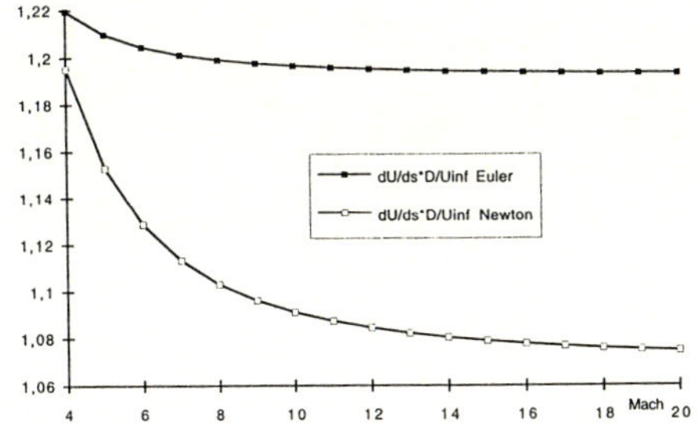

Fig. 13. Comparison of Euler and newtonian normalized stagnation point velocity gradient on a sphere in perfect gas (D =sphere diameter)

Extension of such comparisons to equilibrium and non-equilibrium cases should allow to know more accurately the velocity gradient to use for F4 stagnation heat flux probe data reduction. For example, non-equilibrium Navier-Stokes computations with Flament's code on a spherical nose

Electre model (a highly instrumented Electre model will be used in further F4 test campaign (Muylaert et al 1992)) have shown higher stagnation heat fluxes than those given by Fay and Riddell's formula with newtonian velocity gradient (Sagnier 1993). Furthermore, during the Electre test campaign, direct measurement of free stream velocity will be performed with a pulsed Electron Beam Fluorescence technique which has been successfully tested in the R5Ch ONERA wind tunnel (Mohamed 1992). This technique would allow to deduce total enthalpy from free stream velocity.

6. Conclusion

Preliminary analysis of F4 nozzle No.2 calibration campaign is discussed. Tunnel improvements as pyrotechnic plugs and nozzle entrance shield have greatly improved test section measurement repeatability and flow uniformity. Calibration was performed for the four conditions selected by Dassault Aviation for Hermes industrial tests. These tests have already begun, among them a force measurement campaign on a Hermes model on a balance with inertia compensation during spring 1993. Analysis of calibration measurement results show that for all measured data (nozzle static pressure, test section stagnation pressure and heat flux) agreement with theoretical simulations is good at low enthalpy but gets worse when total enthalpy increases. Possible explanations are still under investigation. Results of a planned test campaign on the Electre model could give additional informations to solve these problems.

References

Barnwell WB, Davis RM (1971) A computer program for calculating inviscid, adiabatic flow about blunt bodies traveling at supersonic and hypersonic speeds at angle of attack. NASA TM X-2334

Devezeaux D, Sagnier P, Gülhan A, Kindler K, Koch U (1992) High enthalpy experiments in LBK and comparison with numerical simulations. Euromech 296, Real Gas Effects in High Enthalpy Flows, Göttingen (RFA)

Edenfield EE (1968) Contoured nozzle design and evaluation for hot shot wind tunnels. AIAA Paper 68-369

Fay JA, Riddell FR (1958) Theory of stagnation point heat transfer in dissociated air. J. Aeron. Sci. 25, 2

Flament C (1990) Ecoulements de fluide visqueux en déséquilibre chimique et vibrationnel: modélisation, applications internes et externes. Thèse de Doctorat, Université de Paris VI

François G et al. (1992) Soufflerie hypersonique à arcs brefs F4. ONERA RT 3/8506 GY

Girard A et al. (1990) Développement de moyens de mesures adaptés aux souffleries hypersoniques - Etat des travaux au 31-12-1989 ONERA RT 24/3409 PN

Hugues E, Verant JL (1991), Non equilibrium parabolized Navier-Stokes code with an implicit finite volume method. AIAA Paper 91-0470

Lees L (1956) Laminar heat transfer over blunt-nosed bodies at hypersonic flight speeds. Jet Propulsion 26, 4

Mohamed AK (1992) Electron beam velocimetry. NATO Advanced Research Workshop-New Trends in Instrumentation for Hypersonic Research, ONERA Le Fauga-Mauzac

Muylaert J et al. (1992) Standard model testing in the european high-enthalpy facility F4 and extrapolation to flight. AIAA 17th Aerospace Ground Testing Conference, Nashville

Rollin G (1990) Soufflerie F4 - Calcul de l'écoulement dans les tuyères No.1 et 2. ONERA RT 1/8507 GY

Sagnier P, Marraffa L (1991) Parametric study of thermal and chemical non-equilibrium nozzle flow. AIAA J. 29, 3

Sagnier P (1993) Chemical kinetics modelling influence in F4 conditions. ONERA RT 47/6121SY

The Taylored Nozzle: A Method for Reducing the Convective Heat Transfer to Nozzle Throats by Gasdynamic Shielding

A. Hertzberg[*], K. Takayama[†], J. Hinkey[*] and S. Itaka[†]

[*]Aerospace and Energetics Research Program, University of Washington, FL-10 Seattle, WA 98195, USA
[†]Shock Wave Research Center, Institute of Fluid Science, Tohoku University, Sendai 980, Japan

Abstract. A novel method to reduce the convective heat transfer rate to the throat walls of nozzles employed in hypersonic research facilities, high performance propulsion devices or high temperature industrial processes is presented. The technique described herein utilizes a light gas film (helium, hydrogen, or perhaps even methane), which is injected tangentially along the wall upstream of the region where the convective heat transfer rate is large. The acoustic impedance of the gas film is matched or tailored ("Taylored") to that of the main gas so as not to interfere significantly with the expansion of the main flow of gas and to minimize mixing. As a result, the light gas film has a significantly lower static and total temperature than the main flow of gas, hence reducing the convective heat transfer rate to the nozzle walls. This reduction in convective heat transfer rate allows the facility, device, etc., to operate at higher total enthalpies for longer durations. The general method is presented, as well as analytical and numerical studies which assess the effectiveness and viability of the technique.

Key words: Nozzle flow, Impedance matching, Convective heat transfer, High enthalpy facilities

1. Introduction

Nozzle throat heating and ablation have long been a problem in hypersonic flow facilities which operate at high total enthalpies. In the past, materials such as stainless steel, and copper and tungsten alloys have been used to mitigate the effects of the extremely large convective heat transfer rates encountered near nozzle throats in short duration (i.e. millisecond) flow simulation facilities. Recent advances in and construction of high enthalpy free-piston shock tunnels and tailored-interface operated hypersonic shock tunnels have pushed this problem to the forefront, since it is the thermal damage to facilities which now defines the limits of their total enthalpy and test time capabilities, as well as contributing to impurities in the test gas (Hornung 1992, Eitelberg et al. 1992). The large convective heat transfer rates due to the high total pressures and temperatures of modern facilities have pushed the conventional method of utilizing materials which have desirable thermal properties (at the expense of mechanical strength) for the nozzle throat section past its practical limits. To offset the effects of high convective heat transfer rates near the nozzle throat and allow the use of conventional high strength materials, a technique has been investigated which utilizes the tangential injection of a low molecular weight gas upstream of the nozzle throat, with the pressure, velocity and acoustic impedance matched to the main flow of gas at the point of injection. The resulting low static and stagnation temperature of the gas film significantly reduces the convective heat transfer rate to the nozzle throat, thereby allowing the use of desirable high strength materials. This technique may also be applicable to long duration flow simulation facilities, high performance propulsion devices, and certain high temperature industrial applications (Mattick et al. 1993).

Shock Waves @ Marseille I
Editors: R. Brun, L. Z. Dumitrescu

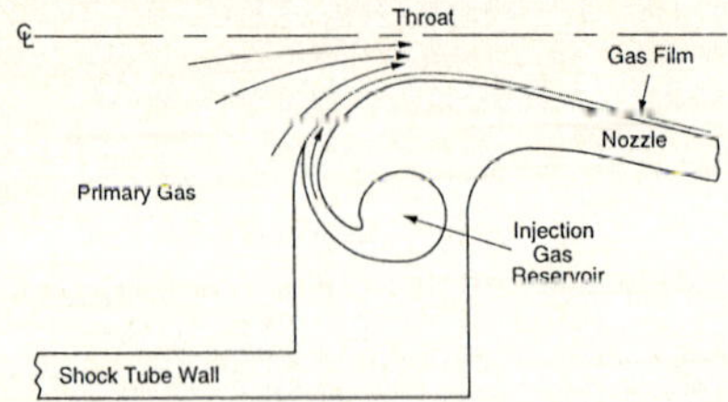

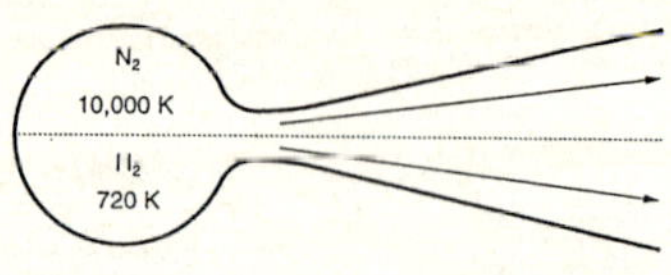

Fig. 1. The general geometry for the tangential injection of a low molecular weight gas upstream of the nozzle throat when using the Taylored nozzle technique

Fig. 2. Illustration of impedance matching using a hypothetical split nozzle and reservoir containing nitrogen gas at 10,000 K on one side and hydrogen gas at 720 K on the other at equal pressure. The gases will expand with equal static pressure and velocity

2. The Taylored Nozzle

The Taylored Nozzle technique has its origins in the tailored-interface hypersonic shock tunnel (Hertzberg et al. 1955, Wittliff et al. 1959). The gases on either side of the interface which separates the test gas from the driver gas are tailored such that the shock wave reflected from the endwall of the shock tube is able to pass through the interface without generating any gas dynamic waves which would disrupt the nozzle supply conditions, thus increasing the effective testing time of the hypersonic shock tunnel. It also turns out that this technique suppresses the Rayleigh-Taylor instability at the interface and hence the name of the Taylored Nozzle. The basic geometry for injection of the low molecular weight gas upstream of the nozzle throat is shown in Fig.1. The point of injection is placed at the endwall of the shock tube on the surface which is the start of the transition from the endwall to the nozzle throat. The pressure and velocity of the injected gas is matched to those of the main gas flow at the point of injection, minimizing mixing between the two gases. The acoustic impedance of the injected gas is also matched to that of the test gas at the point of injection, such that there is minimal effect on the expansion of the test gas to the desired flow conditions, i.e. there is minimal shear between the gas streams as they expand through the nozzle. The matching of the acoustic impedance is accomplished by varying the injected gas temperature and by utilizing a low molecular weight gas such as helium. As a result, the gas film has a significantly lower static and total temperature relative to the main gas, thus significantly reducing the convective heat transfer rate to the nozzle wall. For propulsive or industrial applications, hydrogen or perhaps methane may also be suitable where combustion of the injected gas with the main gas will not occur.

3. Acoustic impedance matching

The acoustic impedance is defined in general as

$$\rho a \tag{1}$$

where ρ is the density and a is the speed of sound. For an ideal gas this can also be written in terms of the specific heat ratio, γ, and static pressure, P, as

$$\gamma \frac{P}{a} \tag{2}$$

If the main gas flow is denoted by $(\ )_1$ and the injected gas by $(\ )_2$ and the acoustic impedances are equated, an expression relating the Mach numbers, M, of the injected gas to the main gas at the point of injection, utilizing the requirement of equal static pressures and particle velocities, is found to be

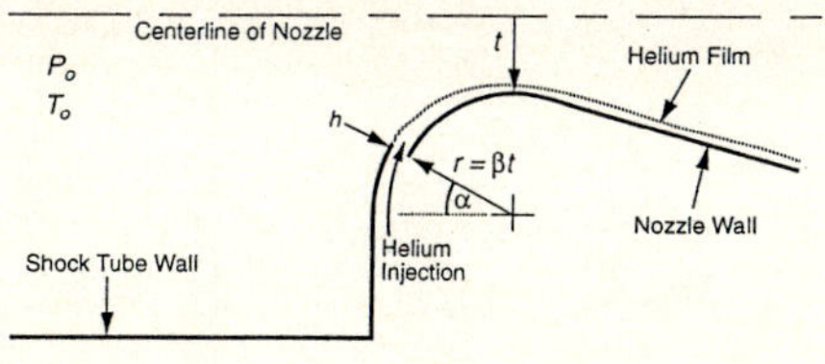

Fig. 3. Generalized two-dimensional planar geometry and notation for the determination of the freestream conditions at the nozzle throat with gas injection upstream. Half-width of nozzle shown

$$\frac{M_2}{M_1} = \frac{\gamma_1}{\gamma_2} \tag{3}$$

with the resulting ratio of the static temperatures, T, given by

$$\frac{T_2}{T_1} = \frac{R_1}{R_2}\frac{\gamma_2}{\gamma_1} \tag{4}$$

where R is the gas constant. It can be calculated from Eq.4 that by using a low molecular weight gas the required injected gas static temperature is much less than that of the main gas, while from Eq.3 the Mach numbers are approximately equal and differ only by the ratio of the specific heat ratios. Since the Mach numbers are nearly equal, the total temperature of the injected gas is significantly less than that of the test gas, while the stagnation pressures are approximately equal.

Impedance matching is illustrated by the split nozzle problem, shown in Fig.2, in which a reservoir and nozzle are split down the middle. If one-half of the nozzle reservoir is nitrogen behaving as an ideal gas at a temperature of 10,000 K, and the other side is hydrogen at the same pressure, the static temperature of the hydrogen need only be 720 K to match the acoustic impedance of the nitrogen. Both gases have equal specific enthalpies (assuming equal specific heat ratios) and will expand at equal pressure and velocity, thereby generating no shear forces between the interface along the center of the nozzle.

4. Heat transfer calculations

4.1. Generalized one-dimensional flow

An estimation of the reduction of the heat transfer rate to the nozzle wall in the vicinity of the throat when utilizing the Taylored Nozzle technique is possible if the freestream conditions of the injected gas are known at the nozzle throat. These conditions may be approximately calculated, without resorting to numerical simulations, by using what are known as the generalized one-dimensional flow relations (Shapiro 1953). These are differential relations which assume the flow to be steady, continuous and quasi-one-dimensional and can be used to model the general effects of such phenomena as area change, heat and work transfer, wall friction, mass addition, internal drag forces, phase change, and changes in molecular weight and ratio of specific heats. These relations may be integrated (usually numerically) to determine qualitatively how a gas flow responds to the above-mentioned phenomena. The generalized one-dimensional flow relations can be combined for two or more adjacent gas streams to form what is known as a "compound-compressible flow" (Bernstein et al. 1967).

A generalized two-dimensional planar geometry for the estimation of the reduction in heat transfer rate is shown in Fig.3. The nozzle consists of a constant radius of curvature section of radius r, which defines the inlet to the nozzle, which has a throat of half-width t. To simplify the calculations the radius of curvature of the inlet to the throat is made a value which is related to t by $r = \beta t$ with β having a constant value. The stagnation temperature and pressure of the main flow are denoted by T_0 and P_0 respectively. The angle of injection is denoted by α and the width of the injection slot is h.

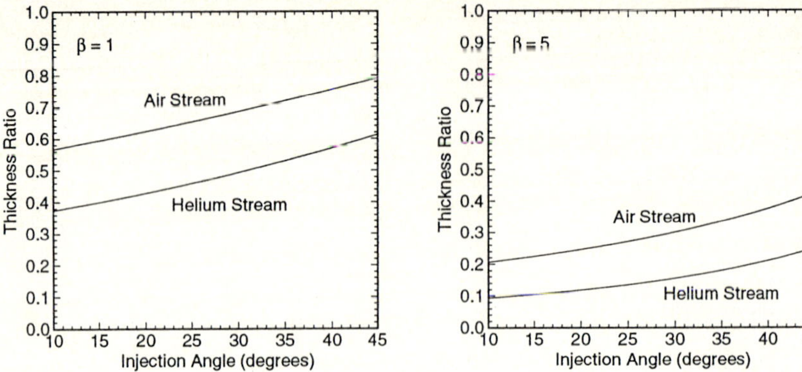

Fig. 4. Ratios of stream thickness at the nozzle throat to initial stream thickness at the point of injection versus angle of injection, α, for β values of 1 and 5. Plots are for helium injection into a test gas (air) having a total temperature of 8000 K. Initial helium stream thickness is 10% of the internal half-width of the nozzle at the point of injection

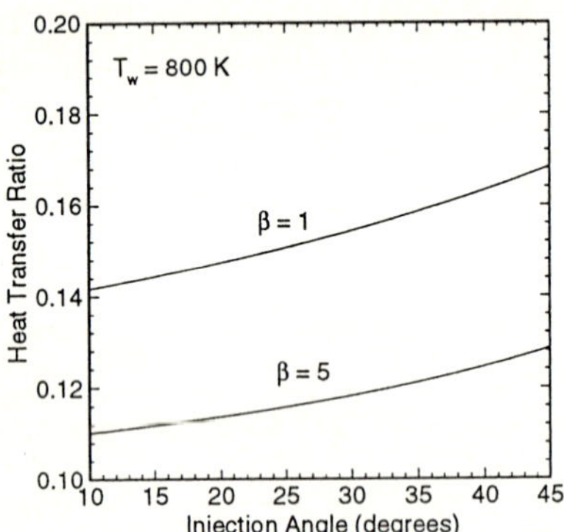

Fig. 5. Ratio of convective heat transfer rate with gas injection to that without injection versus angle of injection; β values of 1 and 5. Helium injection into air test gas with a total temperature of 8000 K. Initial injection thickness is 10% of the internal half-width of the nozzle at the point of injection. Throat wall temperature is assumed to be 800 K

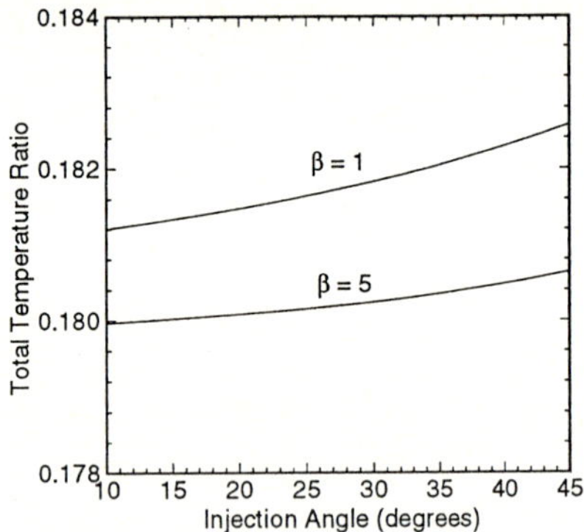

Fig. 6. Ratio of total temperature of injected gas (helium) to the test gas (air) versus angle of injection; β values of 1 and 5. Initial injection thickness is 10% of the internal half-width of the nozzle at the point of injection

The approximate thermodynamic state of the injected gas at the throat is then calculated for various values of injection angle for a fixed value of β. The gases were assumed to be inviscid and only the area (thickness) change effects were modeled. The stream thicknesses, Mach numbers, and static temperatures were calculated at the throat for injection angles from 10 to 45°. Fig. 4 presents plots of the ratios of stream thickness at the throat to initial thickness at the injection point versus angle of injection with β values of 1 and 5 for the injection of helium with a test gas (air) total temperature of 8000 K. The initial stream thickness of the helium injection (h) was assumed to be 10% of the internal half-width of the nozzle at the point of injection. This percentage was not varied since it did not affect the qualitative nature of the calculations.

The helium and air were considered to behave ideally with the specific heat ratio of the air taken as 1.29 to take into account caloric imperfections (vibrational excitation). It can be seen that, for both values of β, as the injection angle increases, the amount by which the injected stream reduces in thickness decreases, which may be desirable, but this leaves more of the nozzle upstream of the injection exposed to the high enthalpy test gas. It is also desirable to inject the gas as far upstream of the throat as possible in order to minimize any disturbances which could degrade the uniformity of the freestream flow farther downstream in the nozzle, but this allows the injected gas film to be thinned considerably (it may decrease in thickness by as much as 90% or more), indicating there may be an optimum injection angle. It was also found that as the injection angle increased, the required total temperature and total pressure of the injected gas increased slightly. An increase in the value of β from 1 to 5 has the effect of increasing the amount of thinning of the injected gas stream. If a greater film thickness at the throat is desired for a given value of β, the initial injection thickness (h) can be increased appropriately.

4.2. Estimation of heat transfer reduction

Now that the conditions of the injected gas at the nozzle throat can be approximately calculated, the heat transfer rate with injection may be qualitatively compared to that with no gas injection by estimating the heat transfer coefficient at the nozzle throat given the freestream conditions (Sibulkin 1954). The results of these estimations are presented in Fig.5, which shows the ratio of the convective heat transfer rate utilizing gas injection to the heat transfer rate with no injection, with a specified wall temperature, T_w, varying injection angles and β having values of 1 and 5. When the wall temperature is at 800 K, the convective heat transfer rate is reduced by over 80%. Although not shown, at the initial wall temperature of 298 K the convective heat transfer rate is reduced by approximately 60%. This reduction in convective heat transfer rate is a direct result of the reduced static and total temperature of the helium at the nozzle throat relative to the main stream of air. A plot of the ratio of the required helium total temperature to the air total temperature is presented in Fig.6 for values of β of 1 and 5. For an air total temperature of 8000 K, the total temperature of the helium is required to be approximately 1500 K, (for a β of 1 or 5) which is a low enough temperature to permit the use of high strength, copper-coated or high temperature alloy throat components.

5. Numerical simulations

Numerical simulations of the injection of the helium gas upstream of the nozzle throat are presently being performed to determine the behavior of the injected gas while it is accelerated in the convergent portion of the nozzle. The stability of the interface between the two gases in the region upstream of the throat, where the flow is subsonic, will be investigated. It will also be of interest to determine what effect the injection of the helium has on the expansion of the flow in the supersonic portion of the nozzle.

A Harten-Yee type second-order accurate upwind TVD finite difference scheme (Harten 1983, Yee 1987) will be used to solve the two-dimensional, planar, two-component Euler and Navier-Stokes equations with both the air and helium assumed to behave as perfect gases. Numerical simulations will be conducted initially for the case of non-viscous flow with viscous calculations following. Since the algorithm is not able to model real gas effects, the numerical simulations will not utilize large total temperatures. However, this will still allow the investigation of the flowfield because real gas effects are not anticipated to have a large effect on the basic nature of the flowfield. Future numerical studies will simulate high stagnation temperature conditions and will include real gas effects once the basic physics of the flow are understood.

The two-dimensional planar nozzle geometries that will be numerically investigated are similar to that shown in Fig.3 with β having a value of 3.33. Simulations of an equivalent nozzle with no injection will be done for comparison. The proposed grid for the simulations are shown in Figs. 7 and 8. Up to six separate grid zones are utilized due to the complicated geometry near the nozzle throat

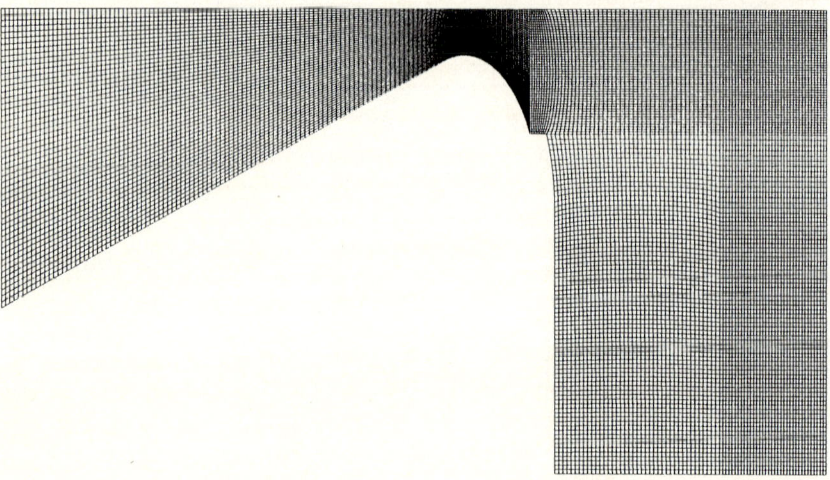

Fig. 7. Two-dimensional planar nozzle, throat, and shock tube endwall grid to be used for the simulation of helium injection upstream of the throat with air as the main gas. Nozzle half-angle of $15°$ and $\beta = 3.33$. Shock tube half-width is $10\,t$. Vertical scale 2 times horizontal scale

Fig. 8. Two-dimensional planar nozzle, throat, and shock tube endwall grid used for the simulation of no gas injection upstream of the throat with air as the main gas. Nozzle half-angle of $15°$ and $\beta = 3.33$. Shock tube half-width is $10\,t$. Vertical scale 2 times horizontal scale

6. Additional considerations

6.1. Endwall/injector design

The tangential injection of the gas upstream of the nozzle throat requires an injection slot or annulus (in the case of axisymmetric nozzles) which may be damaged when the initial shock wave reflects from the shock tube endwall. The thickness of the endwall is small at the injection point and hence is quite weak when subjected to the impulsive loading induced by the shock reflection. To overcome this difficulty the injection slot would be configured such that it is made of a series of individual injection nozzles which would support the endwall at the point of injection. Fig.9 shows one proposed configuration for the injection nozzle geometry which would allow the injection of the gas while simultaneously supporting the endwall.

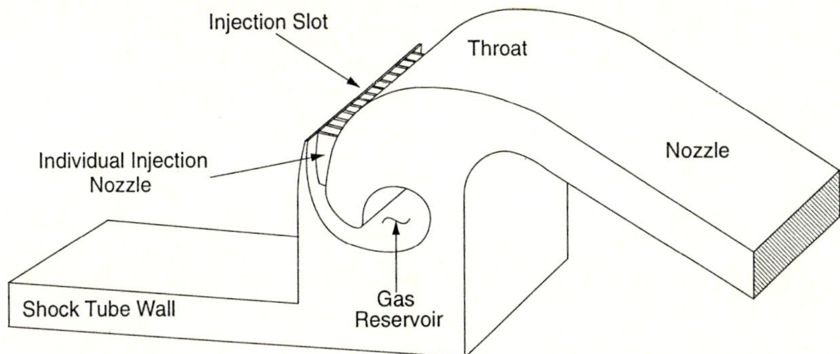

Fig. 9. Configuration of injection slot which would allow support of the endwall at the point of injection. Injection slot is configured into a series of individual nozzles

6.2. Injected gas supply

The gas to be injected upstream of the nozzle throat must be generated separately from the test gas, thus requiring some type of system capable of creating high pressure, moderate temperature gas for a short period of time. Since the injected gas must have a stagnation pressure approximately equal to, but stagnation temperatures much less than the test gas, the supply system must be able to produce a small reservoir of high pressure gas at relatively modest temperatures (approximately 1500 K). For facilities which operate at high stagnation pressures, this may be achieved by utilizing a small secondary shock tube which would be operated simultaneously and would supply the required high stagnation pressure at moderate stagnation temperatures. Secondary shock tubes have been used successfully in flow simulation facilities for similar purposes (Belanger and Hornung 1992). For facilities which operate at lower stagnation pressures some type of small resistance-heated reservoir or pebble-bed heater system may be suitable.

7. Concluding remarks

It has been shown that the Taylored Nozzle technique may be able to significantly reduce the convective heat transfer to throats of convergent-divergent nozzles used in high enthalpy hypersonic research facilities, thereby allowing a corresponding increase in testing time and total enthalpy. This technique may also be applicable to long duration facilities, high- performance propulsion devices and high temperature industrial processes which also utilize convergent-divergent nozzles. The present investigation is preliminary and will be followed by an experimental program coupled with detailed numerical simulations which will verify its effectiveness and suitability for incorporating this technique into existing and future facilities.

References

Belanger J, Hornung HG (1992) A combustion driven shock tunnel to complement the free-piston shock tunnel T5 at GALCIT. AIAA Paper 92-3968

Bernstein A, Heiser WH, Hevenor C (1967) Compound-compressible nozzle flow. Journal of Applied Mechanics, Transactions of the ASME, Sept., pp 548-554

Eitelberg G, McIntyre TJ, Beck WH, Lacey J (1992) The high enthalpy shock tunnel in Gottingen. AIAA Paper 92-3942

Harten A (1983) High resolution scheme for hyperbolic conservation laws. Journal of Computational Physics 49:357-393

Hertzberg A, Smith WE, Glick HS, Squire W (1955) Modifications of the shock tube for the generation of hypersonic flow. C.A.L. Report No. AD-789-A-2, AEDC TN 55-15

Hornung HG (1992) Performance data of the new free-piston shock tunnel at GALCIT. AIAA Paper 92-3943

Mattick AT, Russell DA, Hertzberg A, Knowlen C (1993) Shock-controlled chemical processing. These Proceedings

Shapiro AH (1953) The Dynamics and Thermodynamics of Compressible Fluid Flow - Vol. 1. Wiley New York, pp 219-262

Sibulkin M (1954) Heat transfer to an incompressible turbulent boundary layer and estimation of heat-transfer coefficients at supersonic nozzle throats. Jet Propulsion Laboratory Report No. 20-78

Wittliff CE, Wilson MR, Hertzberg A (1959) The tailored-interface hypersonic shock tunnel. Journal of the Aeronautical Sciences 26(4):219-228

Yee HC (1987) Upwind and symmetric shock-capturing scheme. NASA TM-89464

Dealing with Pressure Oscillations in Stalker Tubes

M.-P. Dumitrescu, R. Brun, M. Billiotte, J.M. Bertoni, Y. Burtschell, A. Canova,
L.Z. Dumitrescu, L. Houas, L. Labracherie and D. Zeitoun
Université de Provence, Lab. IUSTI-CNRS URA 1168, Dept MHEQ, Centre Saint Jérôme, 13397 Marseille, France

Abstract.   During the initial calibration tests performed in the TCM2 free-piston shock tunnel, the helium pressure buildup in the compression tube featured strong oscillations. These oscillations may affect the main diaphragm bursting and the interface stability. It was found that they are caused by acoustic resonance in a cavity present at the downstream end of the compression tube, in front of the main diaphragm. By a novel design of the piston, featuring a hollow front, a smooth pressure buildup has been achieved.

Key words: Free-piston shock tunnels, Pressure oscillations

1. Introduction

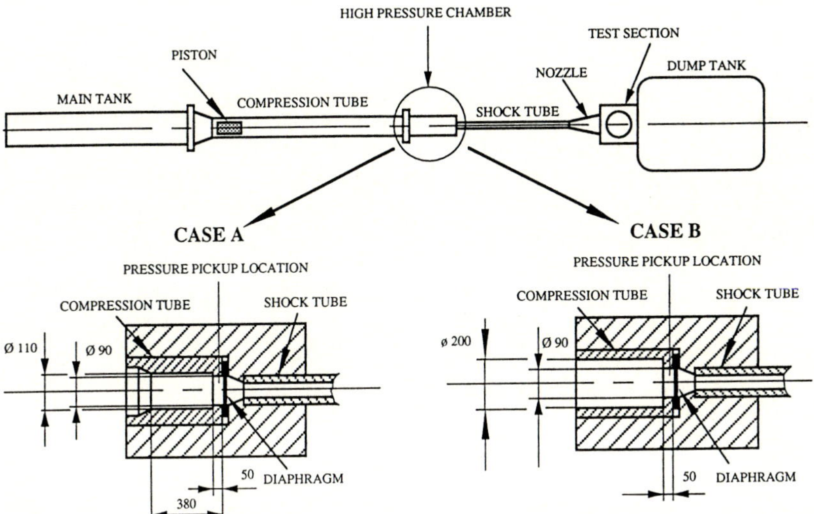

Fig. 1. The TCM2 hypersonic facility

The TCM2 free-piston shock tunnel in Marseille (Fig.1) has undergone several calibration tests. Among these, the driver gas (He) pressure buildup by the piston motion has been investigated using a special plug at the end of the compression tube, in order to prevent the main diaphragm from bursting; a piezoelectric pressure gauge (PCB M109) was placed just in front of the main diaphragm. As shown in Fig.1, the end of the compression tube has two possible configurations; in front of the main diaphragm where, for technical reasons (i.e. making the transition from the compression tube dia. to that of the shock tube), a cavity has to be present, and the two configurations differ in that latter's length. Both of them have been tested, and the pressure recordings obtained are shown in Figs.2 and 3.

Shock Waves @ Marseille I
Editors: R. Brun, L. Z. Dumitrescu

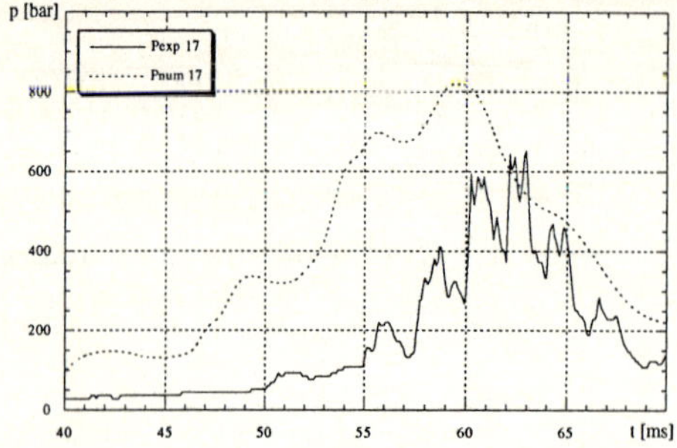
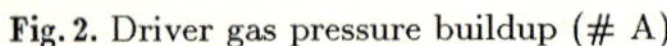

Fig. 2. Driver gas pressure buildup (# A)

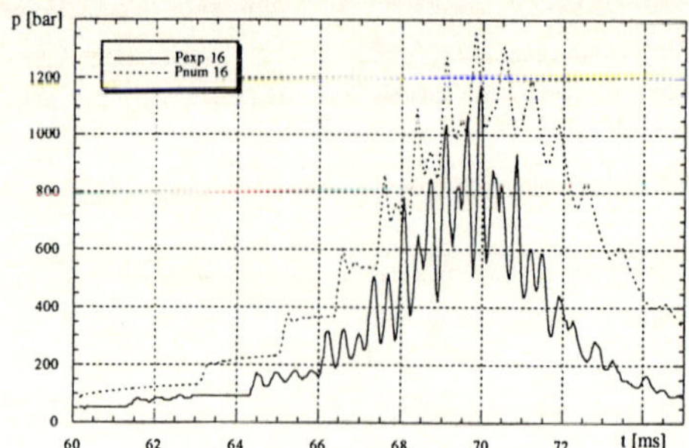

Fig. 3. Driver gas pressure buildup (# B)

2. Discussion

The pressure buildup recorded using configuration # B (Fig.3) reveals the presence of oscillations, of increasing amplitude up to 300 bar and with a frequency ranging from 2.6 to 4 kHz, while the recording made using configuration # A (Fig.2) shows a main frequency of 0.4 to 0.6 kHz with amplitude ranging up to 180 bar and, superimposed on it, a higher frequency of about 3 kHz. As it has been previously reported (Labracherie et al. 1993), the differences between the two pressure recordings stand for evidence that the oscillations are caused by an acoustic resonance which depends strongly on the compression tube's end configuration.

In fact, the piston does not compress smoothly the gas in front of it, but, being very rapidly set in motion at high speed (up to 250 m/sec) it generates a set of compression waves which travel downstream the compression tube, coalescing into a shock wave which subsequently travels forth and back between the tube end and the piston face, and becomes stronger with every reflection (Fig.4). Over these theoretically flat pressure plateaus, oscillations are generated in the cavity at the tube end, at the resonance frequency of the cavity, giving the final aspect of the pressure buildup records. Obviously, in the shorter cavity (configuration # B), the oscillations are at a higher frequency; moreover, one finds that the frequency increases after each shock reflection, since the gas becomes hotter by compression and its sound velocity becomes higher. In configuration # A, as the cavity length is greater, the frequency of the oscillations is lower and the waveforms are less distinguishable.

A numerical simulation of the gas motion in the compression tube was developed, based on a finite-element method (one-dimensional), taking into account the geometry of the compression tube end, in the two configurations (A and B); details are to be found in the cited Reference. Although the actual pressures were overestimated by about 15%, the overall picture of the pressure evolution was fairly well reproduced, and, in Figs.2,3 the computational curves are superimposed over the experimental records. One notices that the coarseness of the grid used had the effect of attenuating somewhat the high-frequency oscillations.

Among the possible consequences of the phenomena discovered, one may expect the main diaphragm to burst prematurely, producing erratic initial conditions for the shock tube flow; also, the contact surface could be disturbed, thus rendering difficult the tailored operation of the shock tube.

To overcome these problems, an efficient shock absorber has to be placed inside the compression tube. It appeared immediately that the absorber could not be placed at the compression tube's end because it would impede the helium flow towards the shock tube, reducing the overall performance of the tunnel. But, if the absorber is placed on the piston's forward face, the incoming shock is sufficiently dissipated.

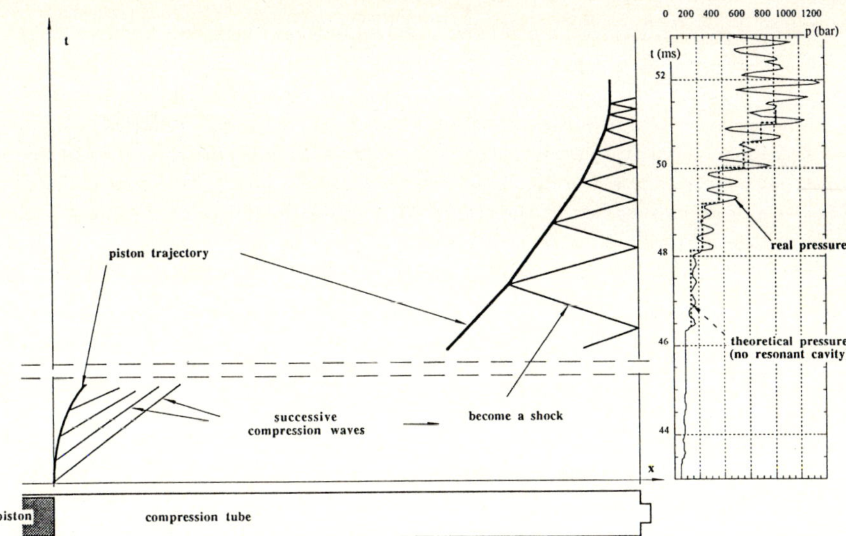

Fig. 4. Schematic of the pressure waves produced in the compression tube by piston motion. Insert: comparison with actual pressure record

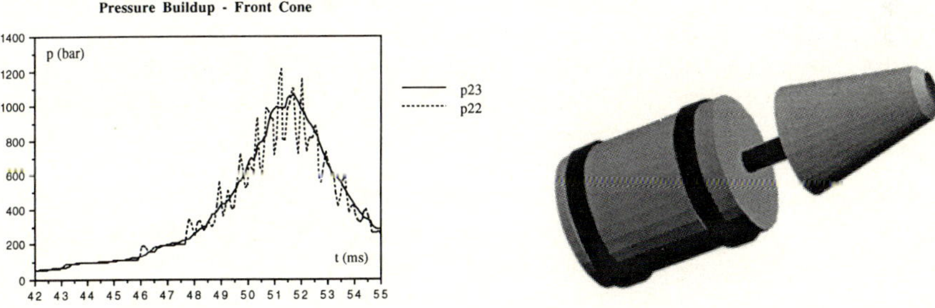

Fig. 5. Suppression of pressure oscillations by the piston equipped with a front cone

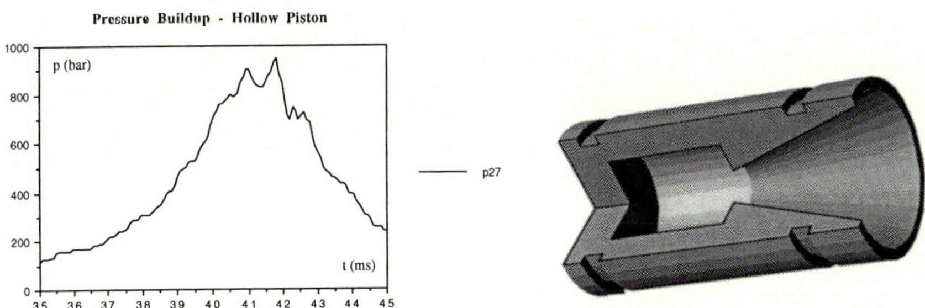

Fig. 6. Final piston configuration for supressing spurious pressure oscillations in the compression tube, and actual pressure record

3. Technical solution and experimental results

A feasible technical solution has been found to be a two-stage device: The first stage consists in a conical "duct" aiming towards sharpening the compression/shock wave, and the second stage is a dissipating cavity. This leads to a piston shaped as in Fig.5: the pressure buildup obained showed a drastic improvement by comparison with the initial one. For practical reasons it appeared that a shorter piston should be easier to use and therefore a second shape has been designed and tested (Fig.6), with satisfactory results. One significant comment on both pressure buildups is that the "twin peaks" which occur very late in the compression process (i.e. at pressures much higher than the design bursting pressure for the main diaphragm) seem to be due to the extremely small distance remaining between the piston and the compression tube end, with the effect that the dissipating cavity is no longer efficient.

4. Concluding remarks

Having found the causes and the remedy for the oscillatory pressure disturbances generated by the piston in he compression tube, it is now possible to have a much improved degree of control of the bursting pressure of the main diaphragm (within one percent), and thus a high degree of shot-to-shot repeatability. The new piston is now routinely used. This has enabled the TCM2 facility to operate in a special, open-end, shock tube configuration, by removing the nozzle throat. In this way, high shock Mach numbers (over 18) have been achieved for the incident shock, with satisfactorily steady flow properties behind it; special experiments are now under way.

Acknowledgement

The first author acknowledges with thanks the many stimulating discussions held with Professor Ray J. Stalker, on the occasion of his visit to Marseille, during which the actual design of the new piston was conceived.

References

Labracherie L, Dumitrescu M-P, Burtschell Y, Houas L (1993) On the compression process in a free-piston shock-tunnel. Shock Waves 3: 19-23

Comparison of the Flow in the High-Enthalpy Shock Tunnel in Göttingen with Numerical Simulations

T.J. McIntyre[*], J.R. Maus[†], M.L. Laster[§] and G. Eitelberg[‡]
[*]Department of Physics and Theoretical Physics, Faculty of Science, ANU, Canberra, Australia
[†]Calspan Corporation/AEDC Operations, Tennessee, USA
[§]AEDC, Arnold Air Force Base, Tennessee, USA
[‡]Institute for Experimental Fluid Mechanics, DLR, Göttingen, Germany

Abstract. Comparisons are presented between numerical simulations of a free-piston driven shock tunnel and experimental data obtained from the new high enthalpy facility in Göttingen, HEG. The models predict with good accuracy the conditions in the driver section of the facility and excellent agreement is obtained for the nozzle reservoir. Agreement between measured Pitot pressures and computed values is only fair.

Key words: Free-piston tunnel, Flow simulation, Comparison theory-experiment

1. Introduction

The high-enthalpy shock tunnel in Göttingen, HEG, described in detail by Eitelberg & al. (1992) is capable of generating operating conditions suitable for simulating the non-equilibrium stage of earth atmospheric re-entry. Currently still in a calibration phase, the facility has already generated free-flight velocities of some 6 km/s with densities up to $5 \times 10^{-3} \mathrm{kg/m^3}$. It is at these conditions that nitrogen dissociation effects are important corresponding also to a region in the re-entry trajectory where heat loads are high. Measurements made in these types of facilities are therefore important in designing re-entry vehicles, in developing the next generation of propulsion units and in the validation of CFD codes.

In designing and operating HEG, a number of computer codes have been used in the prediction of operating conditions. To date, these codes have concentrated on a particular section of the facility, the results of which are passed on to a separate code for predictions further downstream. In the driver section, a code described by McIntyre and Atcitty (1990) based on the method of Hornung (1988) was used to predict the piston trajectory and driver pressure. After diaphragm rupture, predictions of the nozzle reservoir pressure and temperature at the end of the shock tube were performed by an equilibrium code, ESTC, described by McIntosh (1968) while for the nozzle, the one-dimensional non-equilibrium code of Vardavas (1984), STUB was used. Comparisons have also been made with the two-dimensional code of Rein (1989).

While adequate for estimating operating conditions, these codes provide only a time- independent solution as well as requiring inputs that may in themselves be uncertain. A more effective approach is to model the complete shock tunnel with one code. This can be achieved in a number of ways but is here implemented by a Lagrangian formulation of the one-dimensional gasdynamic equations. This allows a prediction of the time history of flow in the facility including motion of the shock wave and contact surface, transient processes and an examination of the test time. Such an approach has also been tried by Jacobs (1993) with good results. This paper presents comparisons between such a code and recent measurements in the HEG tunnel.

2. The experimental facility HEG

HEG is a free-piston driven or Stalker tunnel with high enthalpies generated by a heavy piston which is accelerated by compressed air. The layout for the facility is shown in Fig.1. As with any free-piston driven shock tunnel, the piston is used to compress a driver gas (typically helium) to a high pressure bursting a thick diaphragm. The generated shock wave passes along the shock

Shock Waves @ Marseille I
Editors: R. Brun, L. Z. Dumitrescu
© Springer-Verlag Berlin Heidelberg 1995

tube, heating and compressing the test gas. At the end of the tube, the shock reflects leaving behind a region of high pressure and temperature which is used as a reservoir for a hypersonic nozzle. The current HEG nozzle generates parallel flow with a Mach number in the range 7-10. Dimensions of the facility are given in Table 1.

Table 1. HEG dimensions

	Length	Diameter	Mass
Facility total	60 m		250 000 kg
Compression tube	33 m	0.55 m	
Shock tube	17 m	0.15 m	
Nozzle	3.75 m	0.88 m	
Test section		1.20 m	
Piston			250 - 760 kg

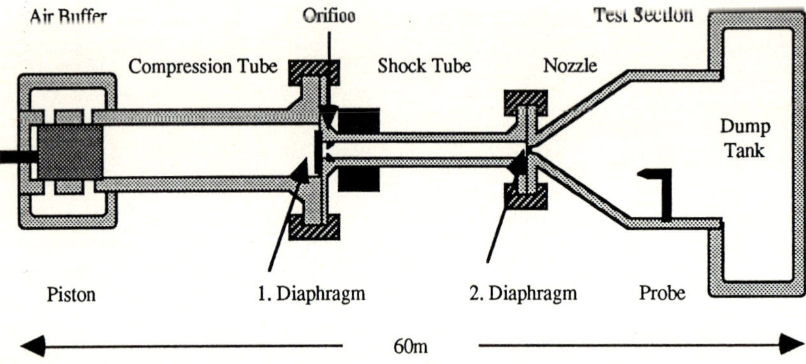

Fig. 1. Schematic diagram of HEG

3. Modelling

The main code used for comparison purposes is a Lagrangian code (hereafter denoted LC) based on a solution scheme for the one-dimensional gas dynamic equations originally developed by Von Neumann and Richtmyer (1950). This code has been used extensively for other types of facilities at the AEDC and has been more recently modified to specifically model free-piston driven tunnels (see Maus et al. 1992).

The code functions by dividing the facility into the regions of the air buffer (modelled as behind the piston), the piston itself, the compression tube, the shock tube and the nozzle. Each of these regions is modelled as a collection of cells or mass points governed by a collective equation of state. Boundary conditions are set at the interfaces to the other regions and the whole facility is treated quasi-one-dimensionally in that area variations are permitted along the tube. The Lagrangian formulation allows the easy tracking of interfaces such as piston location and the contact surface. Furthermore, since mass points become concentrated in regions of high pressure, good resolution is obtained in the regions of interest behind the incident and reflected shocks.

Also used for comparison with experimental measurements of the piston dynamics is a code entitled PaCT described by McIntyre and Atcitty (1990) based on the formulation of Hornung (1988). Here, a set of differential equations for the piston motion and flow of gas out of the compression tube is solved to provide a time history of the piston motion and driver gas pressure. Corrections are made within the code to account for the high pressures encountered in the facility and for heat transfer into the compression tube wall.

4. Results

While a variety of operating conditions were considered when comparing modelling with experiment, only two cases are presented here. These are denoted as HEG Shot # 51, a 50MPa burst condition and Shot # 32, a 200MPa burst condition. The operating parameters for these shots are given in Table 2. These two shots form the extremes of the current operational envelope of HEG. The 50MPa burst condition is the lowest burst pressure and driver conditions can be modelled as practically ideal. For the 200MPa burst condition, this is no longer the case and account must be taken of the higher pressures.

Table 2. Modelled operating conditions

	Shot # 51	Shot # 32
Air buffer pressure (MPa)	5.00	22.6
Compression tube pPressure (kPa)	72.5 He	275 He
Nominal compression ratio	60	60
Diaphragm burst pressure (MPa)	49.3	210
Piston mass (kg)	268	810
Orifice diameter (m)	0.123	0.135
Shock tube pressure(kPa)	23.8 N2	150 N2

Results for the driver section, which consists of the acceleration of the piston by the compressed air, the pressure increase of the driver gas at the end of the compression tube and the diaphragm rupture and subsequent flow of the driver gas into the shock tube, are presented in Fig.2. Two plots are shown for each operating condition. The first plots (Figs.2a and 2b) show the piston trajectory. Axes are labelled so that position is related to the location of the main diaphragm (i.e. all locations in the compression tube are negative) while time is related to the main diaphragm burst time. Shown in each case are the measured HEG values at the various locations together with several modelled curves. For the Lagrangian code, initial predictions of the piston velocity were too high by some 5-10% which was attributed to not taking into account losses at the air buffer - compression tube interface. HEG is designed with a buffer outside of the compression tube and the air must flow through slots in the compression tube wall before accelerating the piston. With a decreased initial air pressure, a piston trajectory matching the HEG data and the PaCT prediction was obtained.

Figs.2c and 2d show the driver pressure comparison. The HEG data for shot # 32 shows high frequency oscillations leading up until shortly after diaphragm rupture. These can be attributed to the motion of waves back and forth between the piston and the end of the compression tube during the compression stage. For shot # 51, these fluctuations are not evident as they have been filtered out during the data reduction stage. Agreement between the Lagrangian code and the HEG data is not so good. The predicted LC pressure reaches a maximum about one millisecond later than the measured value, a fact which cannot be attributed to the inaccuracy in predicting the piston velocity. Here the differences can probably be accounted for by how one defines the diaphragm burst. As is evident in shot # 32, the driver pressure fluctuates significantly as the pressure reaches a maximum. This makes the diaphragm burst pressure very inaccurate. For HEG, the procedure was to smooth the data using Fourier transform methods to remove the high frequency and then define the burst time as when a signal was recorded at the first ionisation gauge immediately downstream of the diaphragm. While this introduces a very small systematic error, it provides a method for defining a diaphragm burst pressure that is comparable between shots. For

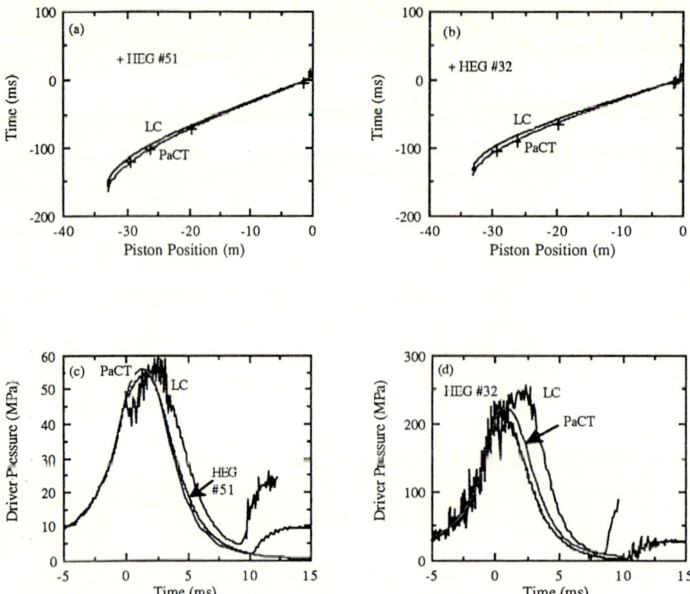

Fig. 2. Measured and modelled parameters for the driver section
(a) Piston trajectory for the 50MPa burst condition, HEG # 51
(b) Piston trajectory for the 200MPa burst condition, HEG # 32
(c) Driver pressure for the 50MPa burst condition, HEG # 51
(d) Driver pressure for the 200MPa burst condition, HEG # 32

the Lagrangian code however, the diaphragm was deemed to have burst when the instantaneous pressure rose above the defined burst pressure. Hence due to the presence of fluctuations in the pressure, the diaphragm has probably 'burst' earlier than in the HEG case.

The predictions of the other code, PaCT, show a better agreement although the nature of the model means that no oscillations are present. For the lower pressure case, agreement is excellent using values for the air sound speed and the air and driver gas polytropic index close to ideal values. For the 200MPa condition, the model tends to overpredict the driver pressure even though the piston trajectory is well modelled (using non-ideal values for the air buffer). This is possibly due to an increase in the polytropic index of the driver gas due to the high pressures. Sychev et al.(1987) have noted that at a pressure of 100MPa with a temperature of 1500K, the polytropic index for helium is about 1.81. Increasing the value used in the code provides better agreement with the measured pressure.

Measurements in the shock tube section consist of monitoring the passage of the shock front at six locations along the shock tube together with pressure measurements at the end of the tube. The results of the comparison with the Lagrangian code are shown in Fig.3. For both the 50MPa and 200MPa burst conditions, excellent agreement is obtained for the shock velocity. Some small offset is evident in each case due again to the definition of the diaphragm burst time however the slopes of the curves are very similar. There is a small attenuation of the shock velocity in each case.

The predicted nozzle reservoir pressure is also in good agreement with the measured value. There is a slight overprediction of the reservoir pressure which can be traced back to the over-prediction of the piston velocity. Correcting this discrepancy leads to almost perfect agreement

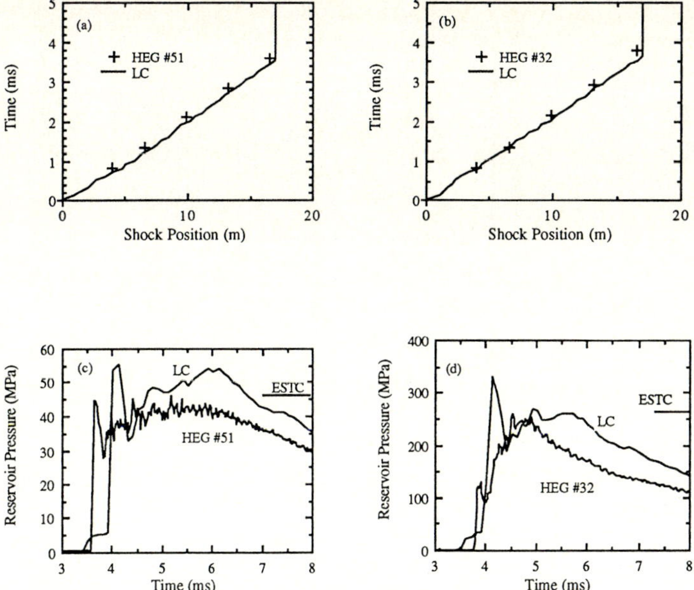

Fig. 3. Measured and modelled parameters in the shock tube
(a) Shock trajectory for the 50 MPa burst condition, HEG # 51
(b) Shock trajectory for the 200 MPa burst condition, HEG # 32
(c) Nozzle reservoir pressure for the 50 MPa burst condition, HEG # 51
(d) Nozzle reservoir pressure for the 200 MPa burst condition, HEG # 32

in the reservoir pressure. Also shown in these figures is the reflected shock pressure as predicted by the ESTC code which seems to be in good agreement with the measured value for shot # 51 but higher than that observed in shot # 32.

Fig.4 shows measured and modelled Pitot pressures. The Pitot pressure has been calculated using the expression $p = 0.92\rho u^2$ where ρ is the calculated free-stream density and u the velocity. There is qualitative agreement between the Lagrangian code and experiment to the extent of the duration of the flow. However in general the modelled pressure is too high especially for the 200 MPa burst condition. Further investigation is required to determine if this results from non-equilibrium chemistry or some incomplete modelling of boundary layer effects. The modelled curves also show several large peaks during the nozzle starting process, not evident in the experimental data. The reason why these are present in the model and not in the experiment is currently unclear although it may be related to the starting time of the Pitot probe. This is somewhat uncertain due to the presence of a cavity in front of the sensor designed to provide heat shielding. Also shown in the figures are the predictions of the one-dimensional non-equilibrium code STUB. Here nozzle reservoir plateau pressures and temperatures were used as input to calculate the free-stream conditions. Excellent agreement is obtained with the measured Pitot pressures in each case. This code, as opposed to the Lagrangian code, takes full account of non-equilibrium chemistry and the HEG nozzle contour which seems to indicate that more complete modelling is required in the Lagrangian code for cases where non-equilibrium chemistry is important.

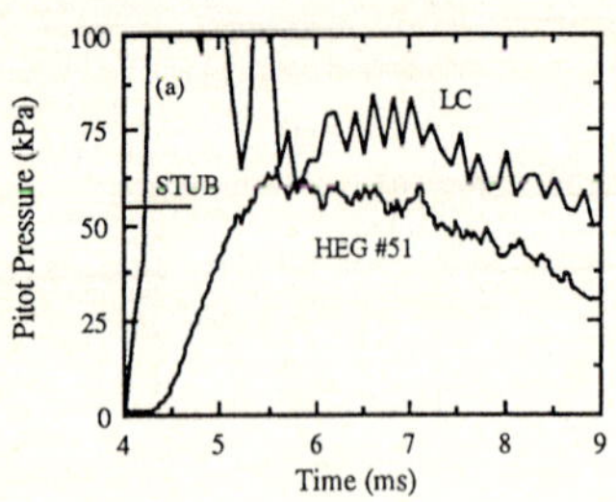
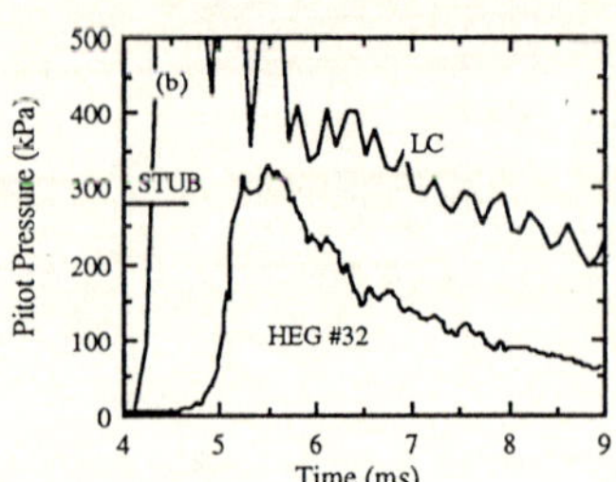

Fig. 4. Measured and modelled parameters in the test section
(a) Pitot pressure for the 50MPa burst condition, HEG # 51
(b) Pitot pressure for the 200MPa burst condition, HEG # 32

5. Conclusions

A Lagrangian code has been used to model the operation of a free-piston driven shock tunnel. The results have been compared with the experimental data from the HEG shock tunnel. Good agreement is obtained for the piston trajectory provided account is taken of losses during the acceleration of the piston. Excellent agreement is obtained for the shock velocity and nozzle reservoir pressures, quantities which have been previously very difficult to predict. Qualitative agreement was achieved for Pitot pressures.

Acknowledgements

The authors would like to thank Mr. J. Lacey, Mr. S. Kortz and Dr. W.H. Beck for their efforts during the experimental measurements. Part of the funding for HEG was provided by ESA through CNES and by the German state of Lower Saxony.

References

Eitelberg G, McIntyre TJ, Beck Wh, Lacey JJ Jr (1992) The high enthalpy shock tunnel in Göttingen. AIAA Paper 92-3942.

Hornung HG (1988) The piston motion in a free-piston driver for shock tubes and tunnels. GALCIT Report # FM88-1

Jacobs PA (1993) Quasi-one-dimensional modelling of free-piston shock tunnels. AIAA Paper 93-0352

McIntosh MK (1968) Computer program for the numerical calculation of frozen and equilibrium conditions in shock tunnels. Dept. of Physics, SGS, ANU, Canberra, ACT

McIntyre TJ, Atcitty C (1990) Pistons and compression tubes - piston motion in the high enthalpy shock tunnel. HEG, DLR IB 222-90 A 20

Maus JR, Laster ML, Hornung HG (1992) The G-Range impulse facility - a high performance free piston shock tunnel. AIAA Paper 92-3946

Rein M (1989) SURF: A program for calculating inviscid supersonic reacting flows in nozzles. GALCIT Report # FM89-1

Sychev VV, Vasserman AA, Kozlov AD, Spiridonov, GA, Tsymarny, VA (1987) Thermodynamic properties of Helium. Hemisphere Publishing Corp. NY,NY

Vardavas IM (1984) Modelling reactive gas flows within shock tunnels. Australian J. Phys., 37, 157-177

Von Neumann J, Richtmyer RD (1950) A method for the numerical calculation of hydrodynamic shocks. J. App. Phys. 21, 232-237

A Numerical and Experimental Study of the Free Piston Shock Tunnel

Katsuhiro Itoh*, Kouichiro Tani*, Hideyuki Tanno*, Masahiro Takahashi*, Hiroshi Miyajima*, Takahisa Asano[†], Akihiro Sasoh[†] and Kazuyoshi Takayama[†]
*National Aerospace Laboratory, Kakuda Research Center, Kakuda, 981-15 Miyagi, Japan
[†]Shock Wave Research Center, Institute of Fluid Science, Tohoku University, Katahira 2-1-1, Aoba, 980 Sendai, Japan

Abstract. In the present study, a CFD-aided tuned operation of a free piston shock tunnel (FPST) was proposed. For that, a numerical method which can accurately predict characteristics of a FPST was developed by using a 4th order pointwise non-oscillatory scheme and models of the flow losses. The tuned operation condition for the FPST of the Shock Wave Research Center(SRC-FPST) at the Institute of Fluid Science, Tohoku University was estimated from the parametric calculations with the present numerical method. The experiments were conducted at the condition, in which the stagnation enthalpy and pressure were 10.6 MJ/kg and 570 atm, respectively. The test flows were observed by holographic interferometry.

Key words: Free-piston shock tunnel, Tuned operation, CFD

1. Introduction

The free piston shock tunnel (FPST) is one of the most useful ground testing facilities which are capable of producing hypervelocity flows around re-entry vehicles and scramjets. The National Aerospace Laboratory is planning to build a FPST having the specifications of stagnation enthalpy 10-30 MJ/kg and stagnation pressure up to 250 MPa.

In order to realize the capability of the FPST, correct operation is essential. Recently, a very effective operation method called "tuned operation" was proposed by Hornung (1990), in which the piston motion is tuned to maintain the driver gas condition constant by having proper speed and stop softly at the compression tube end. Hornung showed with T5, which is the FPST of California Institute of Technology, that well tailored nozzle reservoir pressure could be obtained in a wide enthalpy range by this tuned operation condition (Hornung 1992). However, he estimated the tuned operation condition by assuming that the flow has no losses and the deceleration of the piston is constant after the diaphragm rupture until the piston stops (Hornung 1990). Accordingly, it might be different from the actual tuned operation condition, particularly from the soft landing condition.

Therefore, in the present study, a CFD-aided tuned operation of the FPST is proposed. For that, a numerical method was developed by using a 4th order pointwise non-oscillatory scheme (Itoh et al. 1993), named the KRC scheme, and modeling the flow losses (Itoh 1992, Jacobs 1993, Page et al. 1983) in a suitable form for the quasi one-dimensional governing equations. The reliability of the present method was confirmed by comparing the predicted piston motion with that observed in the compression tube test apparatus of the National Aerospace Laboratory (NAL-CTA) (Tano et al. 1993), in which a transparent tube made of acryl was added at the end of the compression tube. Then, the tuned operation condition for the FPST of the Shock Wave Research Center (SRC-FPST) at the Institute of Fluid Science, Tohoku University was estimated from the parametric calculations with the present numerical method. The experiments were conducted at stagnation enthalpy and pressure of 10.6 MJ/kg and 570 atm, respectively. The test flows were observed by holographic interferometry (Takayama et al. 1983).

Shock Waves @ Marseille I
Editors: R. Brun, L. Z. Dumitrescu

2. Numerical method

2.1. A pointwise non-oscillatory scheme

For the parametric calculation to estimate the tuned operation conditon, a fast numerical scheme with high resolution is required and here a 4th order pointwise non-oscillatory scheme, named the KRC scheme, was developed (Itoh et al. 1993).

The feature of the present scheme is the introduction of a geometrical concept into the construction of a pointwise scheme by using a new flux function which satisfies the following equation with required accuracy:

$$\frac{\partial f(x,t)}{\partial x} = \frac{1}{\Delta x}\left[F\left(x + \frac{\Delta x}{2}, t\right) - F\left(x - \frac{\Delta x}{2}, t\right)\right] \tag{1}$$

For 4th order accuracy, this new flux function is written as follows:

$$F(x,t) = f(x,t) - \frac{\Delta x^2}{24}\frac{\partial^2 f(x,t)}{\partial x^2} \tag{2}$$

Then, the numerical flux can be constructed by approximating $F(x,t)$ at the computational cell boundary, i.e., (1) interpolating for $u(x,t_0)$, (2) smooth solution within cell, and (3) Riemann solution at cell boundary. For more details, see Itoh et al. (1993).

2.2. Loss models

In the present numerical modeling, four typical flow loss types were taken into consideration in suitable form for the quasi one-dimensional governing equations (or details, see Itoh et al. 1993):
1. Discharge loss of the reservoir gas for the piston drive;
2. Heat loss of the driver gas;
3. Discharge loss of the driver gas at the junction of the compression tube, and
4. Wall friction and heat transfer in the shock tube.

3. The SRC-FPST facility

As shown in Fig.1, the SRC-FPST consists of the compression tube of 6 m length and 100 mm ID, the shock tube of 4 m length and 30 mm ID, the air reservoir of 1.5 m length and 250 mm ID, and the nozzle of 0.29 m length, 10 mm throat dia and 100 mm exit dia.

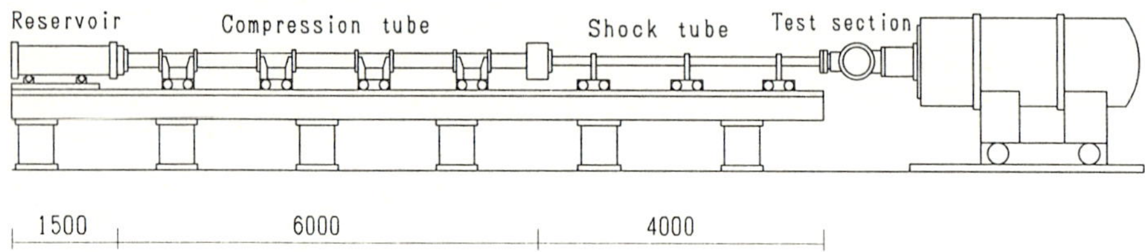

Fig. 1. Schematic diagram of the SRC-FPST

The piston weight was 3.6 kg. The diaphragm was made of SUS304 of 0.6 mm thickness and 500 atm rupture pressure. The test model was a two-dimensional cylinder of 1 5mm diameter. The flow around the test model was observed by double exposure holographic interferometry (Takayama et al. 1983).

4. Tuned operating condition

Assuming that the deceleration of the piston is constant after the diaphragm rupture until the piston stops at the compression tube end and the driver gas pressure is much higher than the pressure behind the piston (Hornung 1990), the tuned operating condition can be written as follows:

- maintaining driver gas condition:

$$\rho_r U_p D^2 = \rho^* a^* d^2 \tag{3}$$

- soft landing:

$$\tfrac{1}{2} W_p U_p^2 = P_r V_r \tag{4}$$

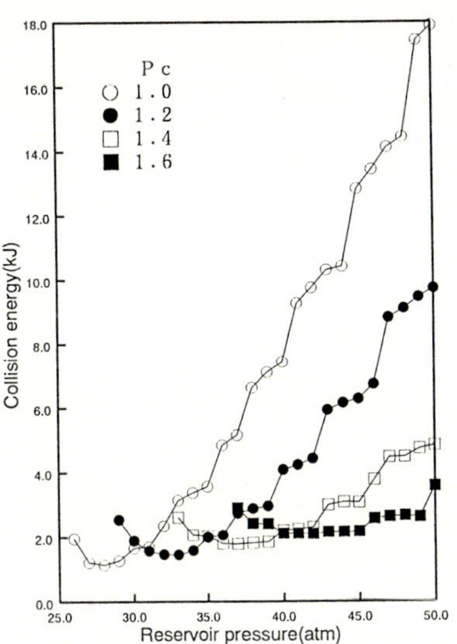

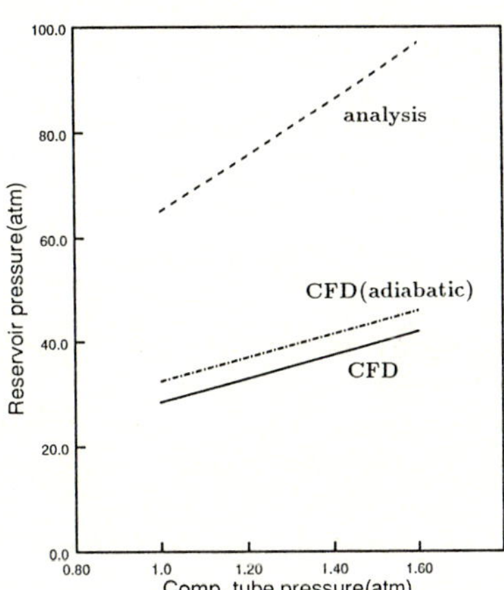

Fig. 2. Collision energy versus reservoir pressure for piston drive

Fig. 3. Soft landing condition

where subscript r indicates the values of the driver gas at diaphragm rupture, asterisc indicates the junction of the compression tube and the shock tube, U_p is the piston speed at diaphragm rupture and W_p is the piston weight. And we define the initial pressure of the compression tube gas and the reservoir pressure for the piston drive as P_C and P_A, and compression ratio as λ. The soft landing condition corresponds to the transition point between two typical piston collision patterns, i.e., collision at the first stroke and collision at the second stroke as the piston rebounds. The tuned operating condition can be estimated by combining the above conditions with the equations of the piston motion and the flow in the compression tube and reservoir, thus it is dominated by the four parameters, i.e., P_C, P_A, W_p and λ or P_r instead. For the analytical estimation of the tuned operating condition shown below, we used Hornung's approximations of the flow equations, i.e., quasi-static adiabatic compression in the compression tube and simple wave theory in the reservoir (Hornung 1988, 1990, 1992).

However, the heat loss in the whole process and the pressure fall some time after the diaphragm rupture are not negligible actually, and both of them decrease the deceleration of the piston. Accordingly, the analytically estimated condition may considerably differ from the actual tuned operating condition, particularly from the soft landing condition. Therefore, we estimated the tuned operating condition from parametric calculations with the numerical method and tried to do this for the SRC-FPST, fixing P_r and W_P as 500 atm and 3.6 kg, respectively.

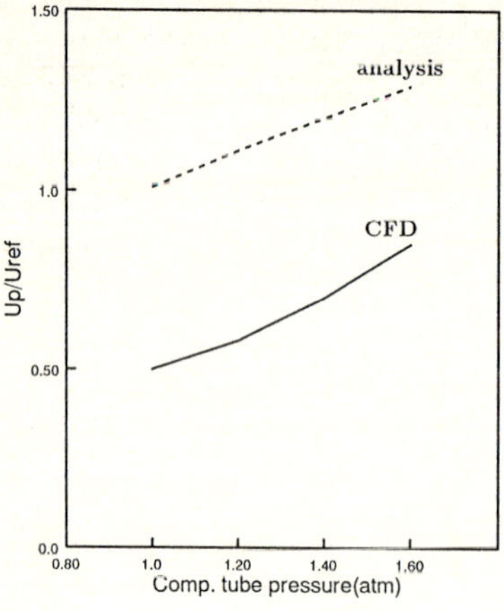

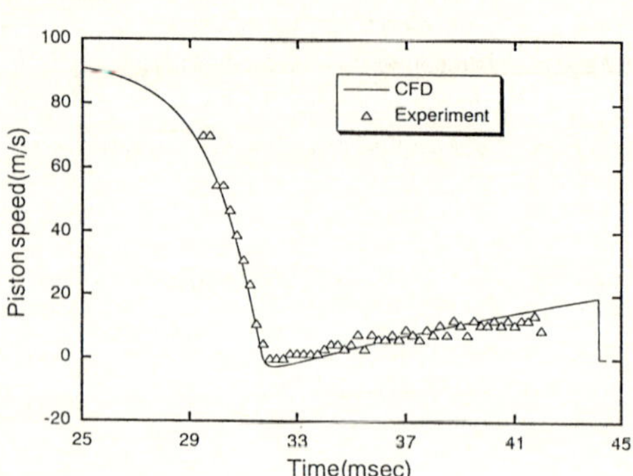

Fig. 4. U_p/U_{ref} at soft landing condition

Fig. 5. Variation of the piston speed

Fig.2 shows the numerical results of the piston collision energy versus P_A for various P_C. As shown in this figure, the collision energy is at first decreasing with decreasing P_A and reaches a minimum, then it starts to increase. The right side of the minimum point is the collision at the first stroke and the left side is that at the rebound. It should be noticed that the minimum energy does not vanish because the compression tube pressure falls below than that in the reservoir at a certain time in the deceleration process and then the piston is re-accelerated. This shows that the pressure fall after diaphragm rupture must be taken into account for the estimation of the soft landing condition.

Next, P_A's at minimum collision energy for each P_C were picked up as the soft landing conditions and ploted in Fig.3, comparing with the analytical one which satisfies Eq.6. As shown in this figure, there is a large discrepancy between the analytical and numerical soft landing conditions.

At the soft landing condition shown in Fig.3, the ratio of the piston speed at diaphragm rupture, U_p, and the reference speed, U_{ref}, which is required to maintain the driver gas condition is ploted in Fig.4. It is seen in this figure that the tuned operation.i.e., $U_p/U_{ref} = 1.0$, can be accomplished when P_C and P_A are 1.8 atm and 46 atm, respectively. However, the maximum P_A of the SRC-FPST is 40 atm, therefore, we set P_A and P_C to 38 atm and 1.4 atm for the experiment with P_r=500 atm, in which the soft landing condition is fully satisfied and the driver gas maintaining condition is 70% satisfied.i.e., $U_p/U_{ref} = 0.7$.

The reliability of the present method for the estimation of the tuned operation condition was confirmed by comparing the piston motion with that observed by the NAL-CTA (Tano et al. 1993), which has a transparent tube made of acryl at the end of the compression tube and is equipped with a high-speed video camera of 4000 frames/sec. Fig.5 shows a comparison of the piston speed variation obtained by the calculation and experiment until the piston stops at the compression tube end. Very good agreement is obtained. For more details, see Tano et al. (1993).

5. Experimental

The experiments were carried out in the conditions established by using the present numerical method, as described in the previous section, i.e. the pressures of diaphragm rupture, in the compression tube, in the reservoir and in the shock tube were 500.0, 1.4, 38.0 and 0.6 atm respectively. The pressures in the shock tube were measured at 29 cm and 229 cm upstream of the endwall with piezo transducers (Kistler 6215 and 6227), and at the endwall with a strain gauge-type transducer. The flow around a two dimensional cylinder was observed by double exposure holographic interferometry (Takayama et al. 1983). The operations were successfully performed at this condition without any damage to the facility. The shock speed measured from the time difference of shock arrival between the endwall and 29 cm upstream was 3030 m/sec. The endwall pressure and enthalpy as the nozzle reservoir condition calculated from this shock speed were 570 atm and 10.6 MJ/kg, respectively.

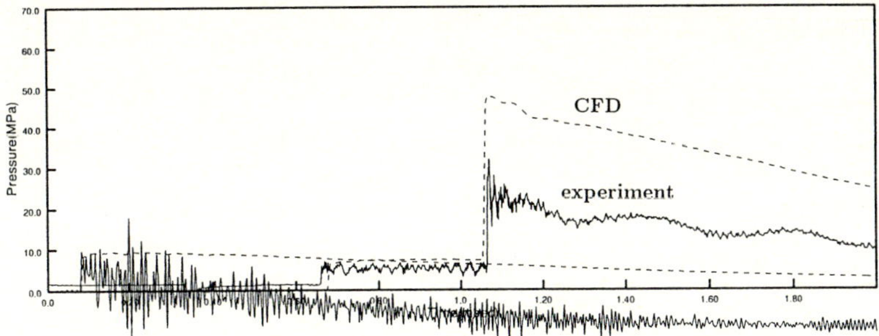

Fig. 6. Pressure at 29 cm and 229 cm upstream of shock tube endwall

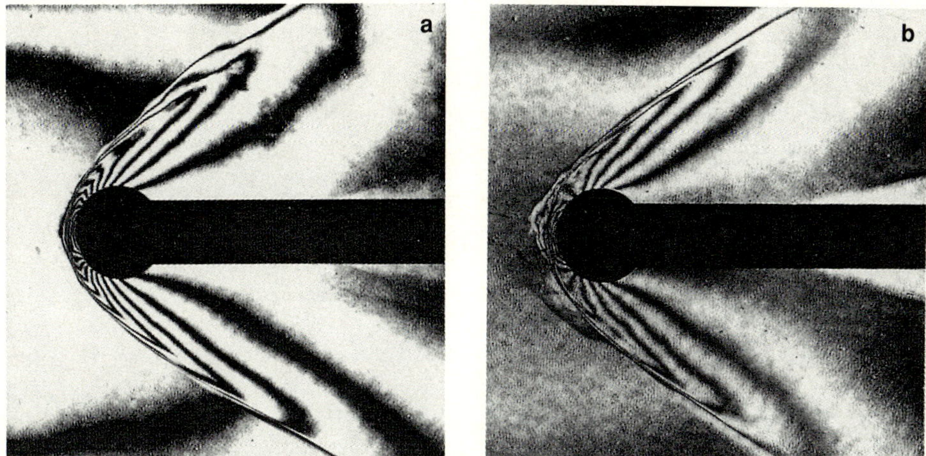

Fig. 7. Interferograms around 2-D cylinder: (a)100μsec after shock arrives at endwall; (b)175μsec after shock arrives at endwall

5.1. Pressure measurements in the shock tube

Fig.6 shows the pressures measured by the piezo transducers at 29 cm and 229 cm upstream of the endwall compared with the numerical results. The timings of the shock arrival of the experiment and calculation agree very well. However, there is a large discrepancy in the pressure level between them. According to the fact that the numerical results agree with the theoretical value calculated from the measured shock speed, this discrepancy seems to be caused by problems with the pressure measurements, such as the effect of high temperature on the piezo transducers. This pressure measurement problem is still being investigated.

5.2. Test flow visualisation

In Figs.7(a), (b), interferograms obtained at 100 and 175 μsec after the shock arrival at the shock tube endwall are shown. The structures of the bow shock, the flow in the shock layer and the jet boundary from the nozzle, which interacts with the bow shock, are clearly observed in these interferograms, (a) just before the flow reaches the steady state, (b), after the steady flow has been already established: the recompression wave can be observed on the supporting rod of the test model. The shock tube condition was slightly under-tailored in this experiment, accordingly, the number of interference fringe of (a) is larger than that of (b), due to the higher pressure in (a). After 150 μsec from (b), the flow patterns, i.e., bow shock angle, etc. begin to change and the contamination of the driver gas seems to begin at this time. Consequently, the test flow duration was only 25% of that expected from the numerical prediction.

6. Conclusion

In the present study, a CFD-aided tuned operation of the FPST was proposed and a numerical method was developed for that. The piston motion after diaphragm rupture predicted by the present method was compared with that observed by the NAL-CTA and the reliability of the present method was confirmed. The present numerical method was applied to estimate the tuned operating condition of the SRC-FPST and compared with that estimated by the simplified analysis. The result showed that the analytical tuned operating condition considerably differs from the numerical one. Experiments were successfully carried out at this numerically estimated tuned operation condition without any damage to the facility. In these experiments, we succeeded in visualizing the detailed structure of the test flow by double exposure holographic interferometry. However, two problems remained, i.e., the high temperature effect to the pressure measurement and the much shorter than expected duration of the test flow.

References

Eckert ERG (1955) Engineering relations for friction and heat transfer to surfaces in high-velocity flow. J. Aeron. Sci. 22:585-587

Hornung HG (1988) The piston motion in a free-piston driver for shock tubes and tunnels. GALCIT report FM88-1

Hornung HG et al. (1990) Role and techniques of ground testing for simulation of flows up to orbital speed. AIAA Paper 90-1377

Hornung HG (1992) Performance data of the new free-piston shock tunnel at GALCIT. AIAA Paper 92-3943

Itoh K (1992) Numerical method for free-piston shock tunnel simulation. Proc.3rd Intl. Workshop on Shock tube Technology, Brisbane, Australia

Itoh K et al. (1993) A pointwise non-oscillatory shock capturing scheme. Proc. 5th Intl. Symp. on CFD, I:370-375

Jacobs PA (1993) Quasi-one-dimensional modeling of free-piston shock tunnels. AIAA Paper 93-0352

Mirels H (1964) Shock tube test time limitation due to turbulent-wall boundary-layer. AIAA J. 2:84-93

Page NW et al. (1983) Pressure losses in free-piston-driven shock tubes. In: Archer RD, Milton BE (eds) Shock Tubes & Waves, Proc. 14th ISSTW, Sydney, pp 119-125

Takayama K et al. (1983) Application of holographic interferometry to shock wave research. Proc.SPIE, p 298

Tanno H et al. (1993) CFD application to optimum design and operation of free-piston shock tubes. Proc. 5th Intl. Symp. on CFD III:211-216

Use of Argon-Helium Driver-Gas Mixtures in the T4 Shock Tunnel

P.A. Jacobs, R.G. Morgan, R.J. Stalker and D.J. Mee
Department of Mechanical Engineering, University of Queensland, St Lucia, 4072, Australia

Abstract. Helium-argon mixtures have been used as the driver gas in the T4 shock tunnel for moderate enthalpy conditions between 3 MJ/kg and 15 MJ/kg. The experimental data showed that constant conditions could be maintained in the shock-reflection/ nozzle-supply region for significantly longer times than for a pure helium driver. Results from quasi-one-dimensional simulations of the facility agreed with the experimental data and provided further detailed information on the performance of the facility.

Key words: Shock tunnel, Free piston driver, Tailored conditions

1. Introduction

The free-piston driver on the T4 shock tunnel facility can use a helium driver gas to generate air flows with total enthalpies in excess of 40 MJ/kg (Stalker and Morgan 1988). However, for many supersonic combustion studies a high pressure level and long, steady test time is desirable for enthalpies in the range 3 MJ/kg to 15 MJ/kg. If a pure helium driver gas is used with a fixed volumetric compression ratio to obtain these (low to moderate) enthalpy conditions, there are significant variations in shock-reflection conditions throughout the test time (Stacey and Simmons 1992). The variations are partly due to undertailoring effects and partly due to driver effects.

The purpose of this paper is to show that constant property conditions can be obtained over a wide range of enthalpies by using a mixture of helium and argon as the driver gas (Wittliff et al. 1959; Jenkins et al. 1991). Increasing the proportion of argon lowers the speed of sound of the driver gas without affecting the ratio of specific heats. Over a wide range of operating conditions, this allows the motion of the piston at the end of the stroke to be controlled so that the driver gas pressure remains essentially constant despite the mass flow into the shock tube. It also allows tailored operation of the shock tube at lower enthalpies.

The following sections describe the T4 facility, the experimental data for some test conditions using pure helium, pure argon and helium-argon driver-gas mixtures and two numerical simulations of the facility at low enthalpy. One condition used a pure helium driver while the other used a pure argon driver. For conciseness, only a compression ratio of $\lambda = 60$ has been considered.

2. The T4 shock tunnel

A schematic diagram of the T4 shock tunnel, together with a wave diagram showing its operating cycle, is shown in Fig.1. The compression tube, which initially contains the driver gas, and the shock tube, which contains the air test gas (state 1), are separated by the primary diaphragm. Attached to the downstream end of the shock tube is a converging-diverging nozzle whose throat is significantly smaller than the inside diameter of the shock tube. The subsonic part of the nozzle effectively closes the the downstream end of the shock tube and forms the shock-reflection (or nozzle-supply) region. The supersonic part of the nozzle empties directly into the test section and dump tank which is evacuated to a low initial pressure. The test gas is retained in the shock tube by a thin secondary diaphragm.

The first stage of operation consists of the launch of the piston and its acceleration along the length of the compression tube. In front of the piston, the driver gas is compressed (approximately

Shock Waves @ Marseille I
Editors: R. Brun, L. Z. Dumitrescu

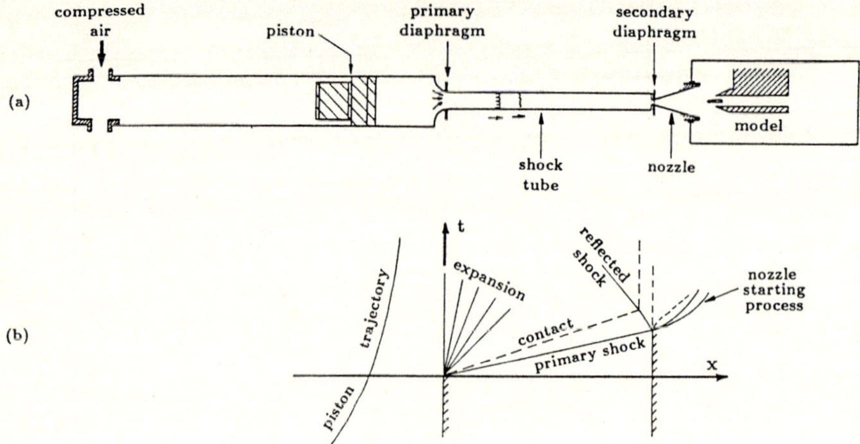

Fig. 1. Diagram of the T4 shock tunnel showing its principal components and a wave diagram

isentropically) and eventually exceeds the rupture pressure of the primary diaphragm (state 4). After diaphragm rupture, the driver gas expands into the shock tube and shock-compresses the test gas before it. The shock wave travels the length of the shock tube, reflects from the nearly-closed end of the tube, and brings the test gas to rest.

Upon shock reflection, the secondary diaphragm ruptures and some of the compressed test gas expands through the nozzle throat into the divergent (supersonic) part of the nozzle. From the point of view of the nozzle, the shock tube is now a reservoir of stagnant, high-temperature, high-pressure test gas. The nozzle expands this stagnant gas to uniform, supersonic conditions in the test section.

3. Numerical simulations

Numerical simulations of two low enthalpy (3 MJ/kg) operating conditions were made using the quasi-one-dimensional code described by Jacobs (1993). The agreement between computed and experimental data for the nozzle-supply pressure is good (Fig.2).

The cause of the pressure decay throughout the test time for the pure helium driver has two components. Fig.2(b) shows that the helium pressure in the driver falls rapidly after diaphragm rupture at $t = 215$ ms. This causes the overall drop in pressure seen in nozzle-supply pressure (P_s) from $t = 220$ ms to $t = 222.5$ ms in Fig.2(a). The second component to the pressure drop is undertailoring which causes a further drop in pressure from $t = 222.5$ ms to $t = 223$ ms in the computed trace. This sudden drop is not seen in the experimental data as the driver-gas/ test-gas interface is not sharp, however, the pressure levels either side of this event agree.

The effects of using an argon driver can be seen on the right column of Fig.2. The nozzle-supply pressure has a lower peak but has a more constant value. The pressure in the driver tube also has a more constant value for 3 ms after diaphragm rupture. In the simulation, this condition is seen to be slightly overtailored with a small shock at $t = 222.3$ ms. There are also some oscillations superimposed on a slight (overall) decay. These oscillations are caused by waves bouncing between the piston and the primary diaphragm station, and being partially transmitted into the shock tube with each reflection at the diaphragm station. With smearing of the driver-gas/ test-gas interface, the pressure jump with overtailoring and the general (but slight) pressure decay might combine to produce the relatively constant supply pressure seen in the experimental data (dashed curve).

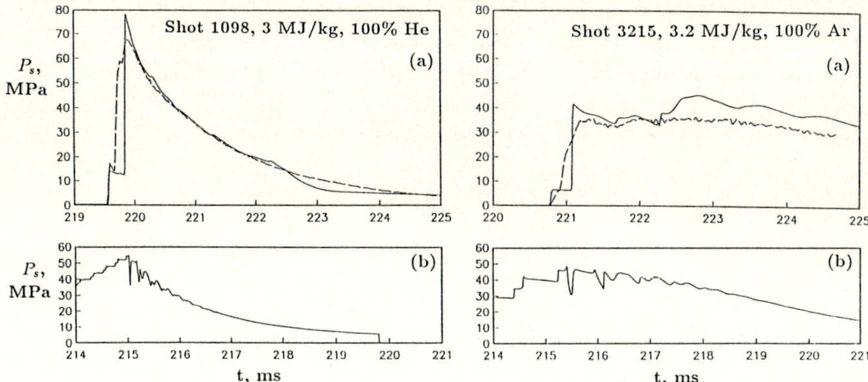

Fig. 2. Pressure histories for the 3 MJ/kg operating condition for a pure helium driver and a pure argon driver: (a) nozzle-supply pressure; (b) pressure in the compression tube, just upstream of the primary diaphragm. The solid line denotes the computed result while the dashed line denotes the experimental data. Time zero corresponds to the release of the piston

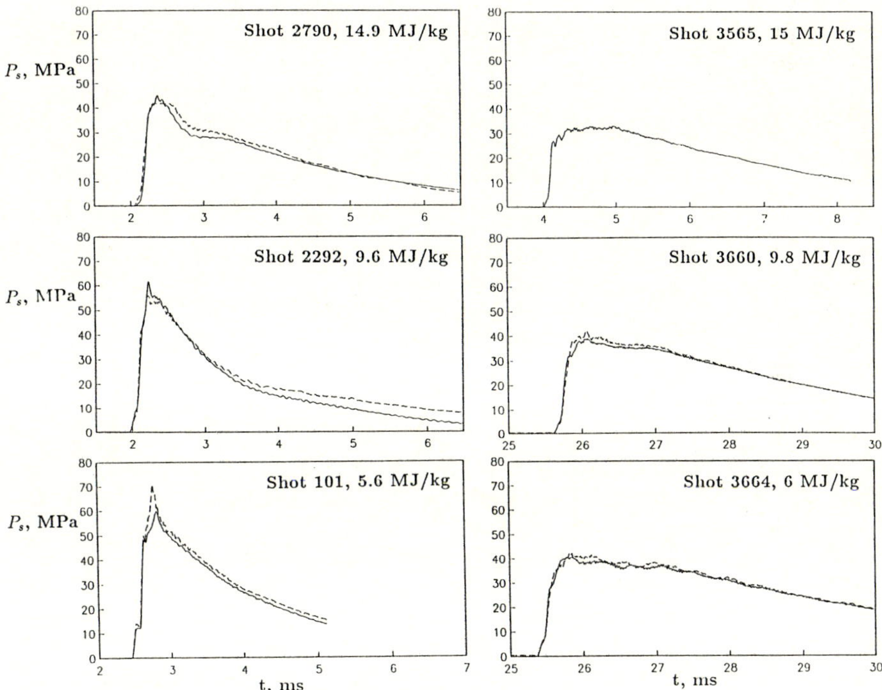

Fig. 3. Histories of the nozzle-supply pressures for a number of enthalpies. The left column shows data for a pure helium driver while the right column shows data for helium-argon mixtures. There were two pressure transducers in the nozzle-supply region so some plots have two traces. The zero on the time scale is arbitrary

Although the argon driver produces a much steadier supply-pressure, there is some concern that, at moderately high enthalpies (say > 12 MJ/kg), it will cause contamination of the test gas at earlier times than the helium driver. To get the same enthalpy as a helium driver, the initial shock-tube filling pressure P_1 has to be lower and the computed wave diagrams (not shown here but similar to those shown by Jacobs, 1993) clearly indicate the shorter length of shock-compressed test gas produced at the end of the shock tube. This implies a shorter time for jetting driver gas to reach the nozzle throat via the mechanism discussed by Stalker and Crane (1978).

4. Experimental data

Fig.3 contrasts pressure histories for tailored/tuned conditions obtained using Helium-Argon driver gas with the undertailored/untuned conditions obtained using a pure helium driver. When compared to low enthalpy operation with a pure helium driver, the drivers with some argon clearly provide test times with more uniform conditions. The test times (based on a constant-property criterion) range from 3 ms at 3 MJ/kg to nearly 1 ms at 15 MJ/kg.

Note that all of the conditions shown in Fig.3. use air as the test gas, have the same (nominal) diaphragm rupture pressure $P_4 = 57$ MPa, and have a volumetric compression ratio $\lambda = 60$. Conditions with rupture pressures ranging from 28 MPa to 72 MPa with both air and nitrogen test gases performed similarly. A nozzle supply pressure ratio of $P_s/P_4 = 0.66$ was representative.

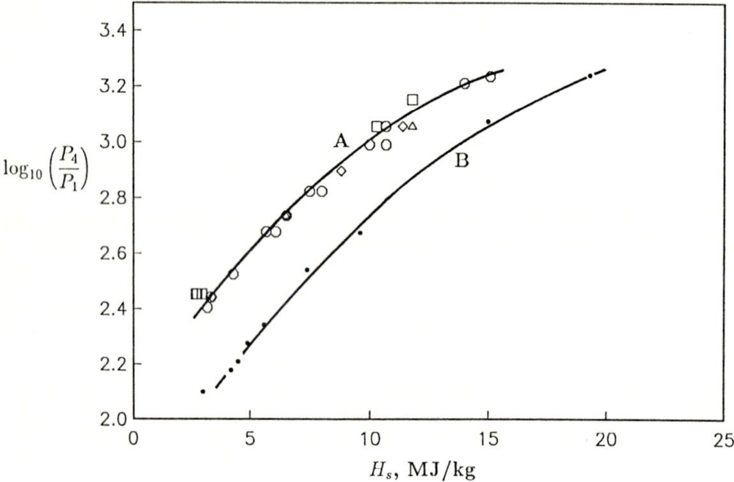

Fig. 4. Diaphragm pressure ratios as a function of total enthalpy for $\lambda = 60$: (A) Open symbols denote tailored/tuned operation with varying argon fractions; (B) Solid symbols denote operation with a pure helium driver

As a guide for the facility user, Figs.4 and 5 show the diaphragm pressure ratio and the argon fractions required to achieve tuned/tailored conditions for a range of enthalpies and pressures. The argon fraction is given as a *volume* fraction. Given a desired nozzle supply pressure, the diaphragm thickness (proportional to rupture pressure P_4) is set. From the enthalpy the initial fill pressure for the shock tube P_1 is obtained from curve A in Fig.4, while the fraction of argon in the driver gas is read from Fig.5. Although not shown, data for low rupture pressures of 15 MPa (equivalent to 1 mm diaphragms) had slightly higher diaphragm pressure ratios than for curve A. The difference may have been caused by viscous effects which were expected to be more significant at the lower operating pressures. Assuming an isentropic compression, the initial driver gas pressure can be inferred from the diaphragm rupture pressure P_4 and the compression ratio λ. (If a pure helium driver is required then curve B in Fig.4 can be used to determine P_1.)

5. Concluding remarks

The use of helium-argon driver gas results in longer periods of constant nozzle-supply pressure and lower peak reflection pressures at the downstream-end of the shock tube (when compared to conditions with a pure helium driver). The average pressure throughout the test time is largely unchanged and is typically 0.66 times the diaphragm rupture pressure for a compression ratio of 60. The helium-argon driver gas also has a lower, but more uniform, shock speed. There is, however, some concern that contamination of the test gas (by the driver gas) will occur at earlier times for moderately high enthalpies.

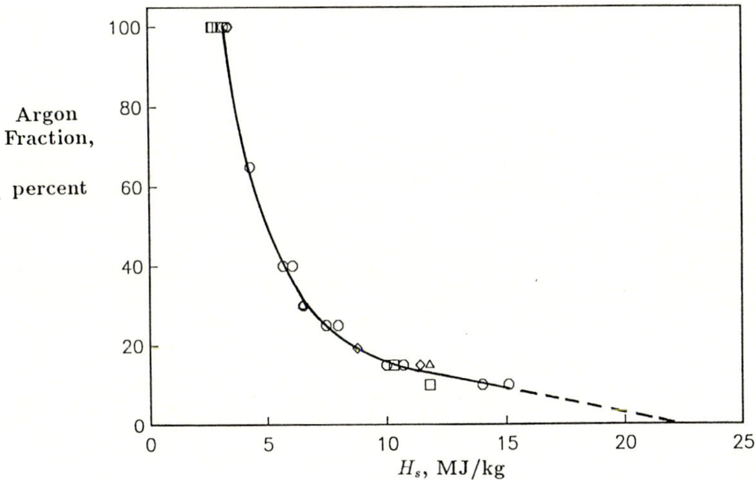

Fig. 5. Argon fractions for tailored/tuned operation with $\lambda = 60$

Acknowledgements

This work was supported the Australian Research Council and by WBM Pty Ltd. The tunnel performance data was accumulated from a number of experiments undertaken by members of the T4 shock tunnel group.

References

Jacobs PA (1993) Quasi-one-dimensional modelling of free-piston shock tunnels. AIAA Paper 93-0352

Jenkins DM, Stalker RJ, Morrison WRB (1991) Performance considerations in the operation of free-piston driven hypersonic test facilities. In: Takayama K (ed) Shock Waves, Proc. 18th Intl. Symposium on Shock Waves, Sendai, Japan pp 597-602

Stacey CHB, Simmons JM (1992) Measurement of shock-wave/boundary-layer interaction in a free-piston shock tunnel. AIAA Journal 30(8):2095

Stalker RJ, Crane KCA (1978) Driver gas contamination in a high enthalpy reflected shock tunnel. AIAA Journal 16(3):277–279

Stalker RJ, Morgan RG (1988) The University of Queensland free piston shock tunnel T4 - initial operation and preliminary calibration. In Fourth National Space Symposium, Adelaide

Wittliff CE, Wilson MR, Hertzberg A (1959) The tailored-interface hypersonic shock tunnel. Journal of the Aerospace Sciences 26(4):219–228

A Velocity Interferometric Study of the Performance of a Gas Gun

T. Matsumura[*], **H. Ohuchi**[†], **N. Narayanswami**[*], **A. Sasoh**[*] and **K. Takayama**[*]

[*]Shock Wave Research Center, Institute of Fluid Science, Tohoku University 2-1-1 Katahira, Aoba, Sendai 980, Japan

[†] SONY Corporation, 6-7-35 Kitashinagawa, Shinagawa, Tokyo 141, Japan

Abstract. As a successful application of the Velocity Interferometry System for Any Reflector (VISAR) to the performance study of a gas gun, in-bore projectile velocity measurements of a 15 mm bore single-stage gas gun with VISAR are presented. For understanding the gas gun operation analytically, a one-dimensional numerical simulation code was developed and the results were compared to that of the experiments. In addition, in order to verify the accuracy of the VISAR measurement, an optical flow visualization of the flow around the launched projectile with holographic interferometry was also carried out.

Key words: Single-stage Gas gun, Velocity measurement, Flow visualization

1. Introduction

Gas guns, such as the two-stage light-gas gun and the single-stage gas gun, are capable of accelerating a projectile up to hypervelocity and are applicable to not only aerospace studies but also to the dynamics of shock in condensed matter generated by hypervelocity impacts. Performance studies of gas guns have been conducted (Charters et al. 1957, Canning et al. 1970, Groth and Gottlieb 1988), however, most of them were semi-empirical. It is still an important research topic to determine the optimal performance of gas guns by using advanced measuring techniques of high accuracy. A VISAR (Velocity Interferometer System for Any Reflector, Barker and Hollenbach 1972) has been installed at the Shock Wave Research Center (SWRC) of the Institute of Fluid Science, Tohoku University. This paper reports the results of an application of the VISAR system to in-bore projectile velocity measurement with a 15 mm bore single-stage gas gun. A series of experiments were conducted for a better understanding of the gas gun performance experimentally; i.e., effect of propellant amount and diaphragm rupturing pressure on the projectile velocity. In order to verify the accuracy of the VISAR measurement, the supersonic flow around the projectile launched into the test section was visualized by double exposure holographic interferometry. The muzzle velocity of the projectile was predicted from the hologram by considering the drag effect, and was compared to that of the measured velocity with VISAR.

2. Experiments

2.1. Single-stage gas gun

The SWRC single-stage gas gun used in the present study consists of a propellant chamber, a launch tube (15 mm bore, 1.1 m long), a flight tube and a test section. Smokeless powder for sports shooting (Nihon Yushi, Co., SS-type) of weight 2 to 10 g is used as the propellant. A high density polyethylene cylinder, 4 g in weight, is used as a projectile. The flight tube forms the initial part of the test section and serves to absorb the muzzle blast and the propellant gases. In order to control the projectile launching velocity, a metal diaphragm of aluminum or steel is set between the launch tube and the propellant chamber. The launch tube and the test section are evacuated to 1 to 10 kPa. For measuring the muzzle velocity of the projectile, pick-up coils are set at the launch tube exit and a small magnet is imbedded in the projectile. The output signal of the pick-up coil is used as a trigger signal for the VISAR operation or for a holographic ruby laser, the light source of the optical flow visualization.

Shock Waves @ Marseille I
Editors: R. Brun, L. Z. Dumitrescu

2.2. VISAR interferometer system

The VISAR interferometer installed in the SWRC is a dual beam type system (ATA Associates, Model 305S). The optical setup for in-bore projectile motion measurement is shown in Fig.1. An argon-ion laser (Spectra Physics, Model 2030) is used as a light source for the VISAR measurement. The interface optical system consists of fiber optic cables and a collimator. In the test section of the gas gun, an expendable mirror of 80×80 mm is set. As shown in Fig.1, a small piece of reflecting tape (Scotchlite tape) is stuck on the projectile front face.

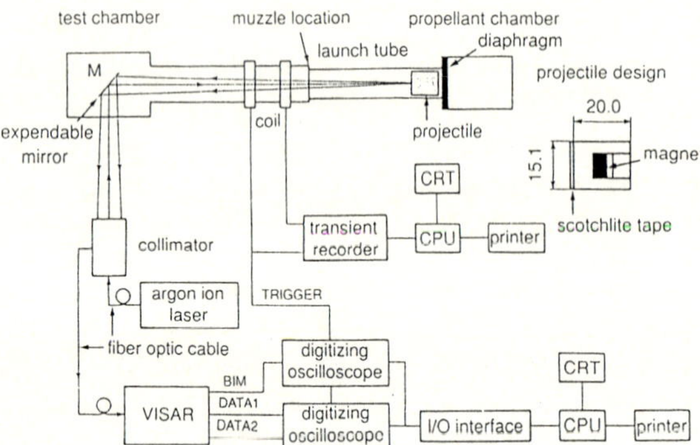

Fig. 1. Optical setup and experimental apparatus

Two sets of 2-channel 250 MHz digitizing oscilloscopes (Hewlett Packard, HP54510A) are used for the data acquisition for the VISAR system, such as output signals of photo multipliers corresponding to beam intensity and $\sin - \cos -$ components of reflected beam from the Scotchlite tape. A 32 bit CPU personal computer (Hewlett Packard, Vectra Q5/20) is used for the data analysis.

2.3. Optical flow visualization

The flowfield around the supersonic projectile shows a bow shock, an expansion fan, a trailing shock and a wake. In order to visualize the flow, various optical flow visualization techniques, such as Schlieren and shadowgraph, for instance, are very useful. In the SWRC, the double exposure holographic interferometry, which has already been successfully used for various shock tube experiments, was applied to the aeroballistic range study with the single-stage gas gun. A holographic ruby laser (Apollo Lasers Inc., Model 22HD, 10 J/pulse) was used as the light source, and the signal from the pick-up coil triggered the laser. By using this image holographic interferometry, a two-dimensional image of the phenomenon is recorded onto the hologram. In the reconstructed hologram, the density change of the flowfield between the two exposure processes is represented by the fringe pattern. This fringe pattern also helps to verify the result of numerical simulation of the supersonic flow around the projectile. From the information of the reconstructed hologram, the muzzle speed of the projectile, V_m, is predicted by the following formula:

$$V_m = V_t \left(2\sqrt{1 + \frac{\rho A L}{m} C_D} - 1 \right) \tag{1}$$

where V_t is the flying speed of the projectile at the test section, C_D is the drag force coefficient of the cylindrical projectile, and L is the distance between the test section and the muzzle. ρ, A and m are the flow density, cross section area and mass of the projectile, respectively.

3. Numerical simulation

In order to understand the physics of the gas gun operation, the various processes within the gun were modeled theoretically. The processes include; (a) propellant combustion, (b) projectile movement and (c) wave dynamics in the gas ahead of the projectile. The wave dynamics (process c) was modeled using a one-dimensional Random Choice (RCM) scheme. The gun barrel (launch tube) geometry, and the geometry of the propellant chamber were accurately specified. For the gas in the launch tube, the governing equations are the one-dimensional nonlinear Euler equations in conservation form, which are written as follows:

$$U_t + F_x = -\overline{A} + H \tag{2}$$

$$U = \begin{bmatrix} \rho \\ \rho u \\ \rho(e + \frac{u^2}{2}) \end{bmatrix}, \quad F = \begin{bmatrix} \rho u \\ \rho u^2 + p \\ u(\rho e + \rho\frac{u^2}{2} + p) \end{bmatrix}, \quad \overline{A} = \frac{1}{A}\frac{dA}{dx} \begin{bmatrix} \rho u \\ \rho u^2 \\ u(\rho e + \rho\frac{u^2}{2} + p) \end{bmatrix}, \quad H = \begin{bmatrix} 0 \\ F_l \\ q \end{bmatrix} \tag{3}$$

where ρ, u, p, γ and e denote the density, velocity, pressure, specific heat ratio and the internal energy per unit mass, respectively; A, F_l and q denote the cross-sectional area, rictional and heat loss factors, respectively. The combustion process of the powder explosives used here is modeled by the lumped parameter method of Groth and Gottlieb (1988), Matsumura et al. (1990), Narayanswami et al. (1993). The burning rate of the powder, V, is assumed to be governed by a pressure exponent law of the form: $V = \beta p^\alpha$ where p is the average pressure inside the propellant chamber, α and β are the burning rate constants of the powder. In general, these two constants are determined from closed bomb tests employing large amounts of the smokeless powder (even up to hundreds of kilograms in certain cases). However, there is little accurate data available for small amounts of powder (< 10 g). In this study therefore, the two coefficients (α and β) were chosen such that they gave a close fit to a reference set of experimental data.

4. Results and discussion

4.1. VISAR measurements

Projectile motion inside the launch tube of the single-stage gas gun was continuously measured with a VISAR interferometer. In the previous study of Matsumura et al. (1990, 1991), the repeatability of this facility was discussed only in terms of the projectile muzzle velocity. By using the VISAR, it is possible to measure the projectile velocity profile along the launch tube continuously and with higher resolution. Fig.2 shows a comparison of the two VISAR measurements of the projectile velocity variations under identical initial conditions. In these shots, an aluminum diaphragm of 1 mm thickness and 6 g and 10 g smokeless powder were used. Very good agreement was obtained for both 6 g and 10 g cases. The maximum difference of velocity in the two cases is less than 4%. The repeatability of the gas gun was thus found to be very good.

Figs.3 and 4 show the projectile velocity profiles for various powder weights. In these two series of experiments, aluminum diaphragms of 1 mm thickness and no-diaphragm, respectively, were used. In the no-diaphragm case, it is found that the initial projectile acceleration is lower and its duration of acceleration is longer than when an aluminum diaphragm was used. From comparison of these two figures, it is seen that the muzzle velocity for 2 g powder weight in the no-diaphragm case attained only 80% of that in the diaphragm-on case. For the case of the powder weight exceeding 6 g, there is no difference between the case without diaphragm and that of the 1 mm aluminum diaphragm.

Fig.5 shows the projectile velocity profiles for various diaphragms. This figure shows the effect of the diaphragm selection, i.e., the rupturing pressure. Three types of diaphragms, (1) aluminum of 1 mm thickness, (2) mild steel of 0.5 mm and (3) mild steel of 1 mm thickness were used: results were compared with the powder weight fixed to 6 g. By increasing the diaphragm rupturing pressure, the initial acceleration of the projectile is increased and the muzzle velocity

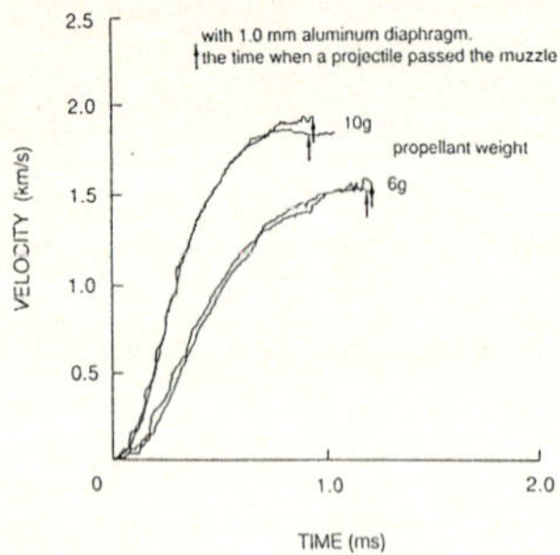

Fig. 2. Comparison of projectile velocities (powder weight: 6 g and 10 g)

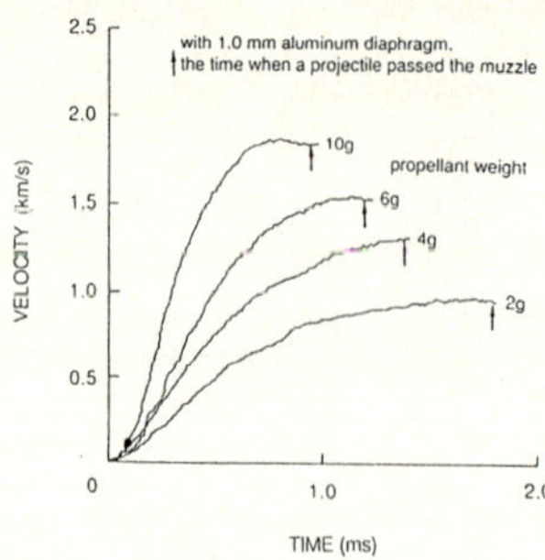

Fig. 3. Projectile velocities for various powder weights (diaphragm: Al 1 mm)

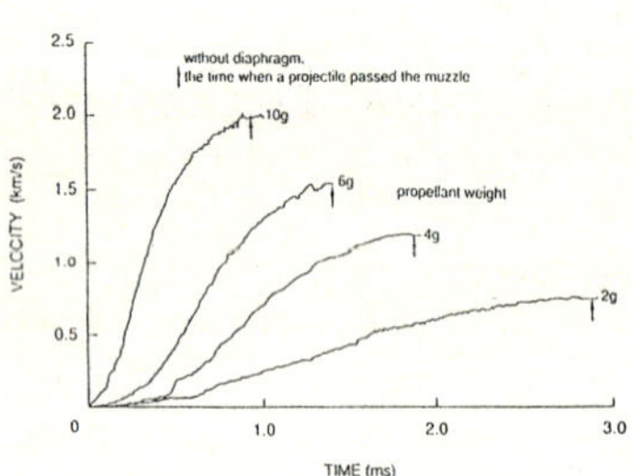

Fig. 4. Projectile velocities for various powder weights (no diaphragm)

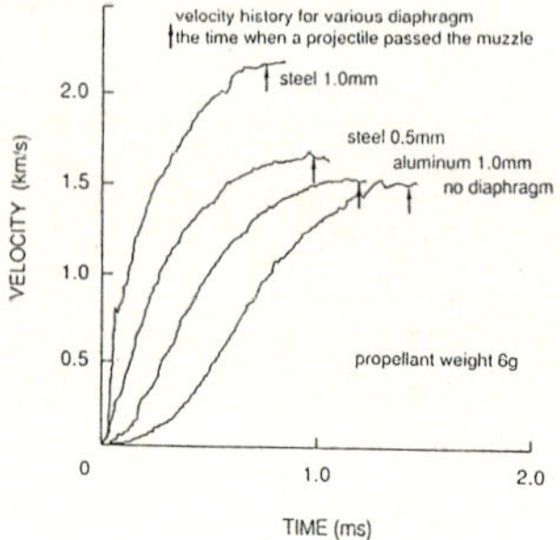

Fig. 5. Comparison of projectile velocities for various diaphragms (powder weight: 6 g)

is also increased. According to the VISAR measurement, the order of the magnitude of the initial acceleration does not change significantly even if the diaphragm rupturing pressure is changed. This implies that there is an optimal selection of diaphragm strength, which, while keeping the initial acceleration low, can increase the muzzle velocity. In particular, in the case of 1 mm steel diaphragm, the muzzle velocity increased up to 2.16 km/s which represents an increase of 144% from the no diaphragm case (1.5 km/s).

4.2. Numerical simulation

The projectile velocity profile along the launch tube obtained by the numerical simulation was compared with the experimental result obtained with the VISAR. Fig.6 shows the no-diaphragm case. The powder weight in this case is 6 g. Except between the time of 1 to 1.5 ms, and the final part, i.e., for time > 1.6 ms, the average shape of the profile is almost the same. In particular, the initial acceleration process is predicted very well. This implies that the burning rate constants values were selected properly. The burning rate constants in this case are $\alpha = 0.761$ and $\beta = 2.0 \times 10^{-5}$, respectively. The 1 mm aluminum diaphragm case is shown in Fig.7. Smokeless powder amount in this case is also 6 g. As in the previous case, the average profile of the present simulation is very similar to the experiment. Burning rate constants in this case are $\alpha = 0.79$ and $\beta = 2.0 \times 10^{-5}$. The value of α is slightly different from that of the no-diaphragm case. This shows that the combustion process of the smokeless powder may be different under the condition with or without using a diaphragm, or under different diaphragm rupturing pressures, even if the powder amount is fixed. As a future work, it must be tested as to whether or not similar results are obtained for other (different) diaphragm cases for a fixed amount of smokeless powder.

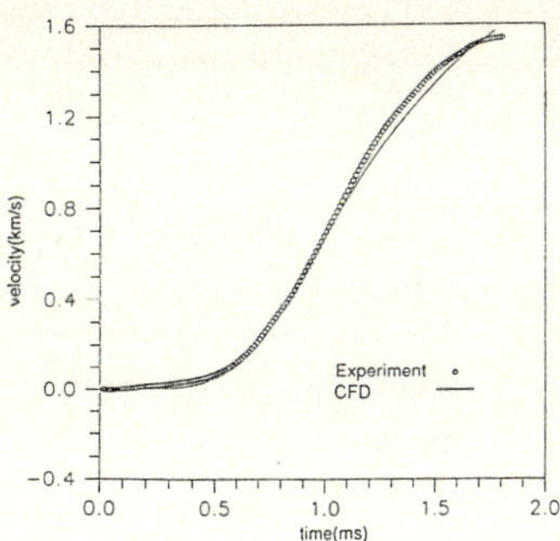

Fig. 6. Comparison of projectile velocity: CFD vs. experiment (powder weight: 6 g; no diaphragm)

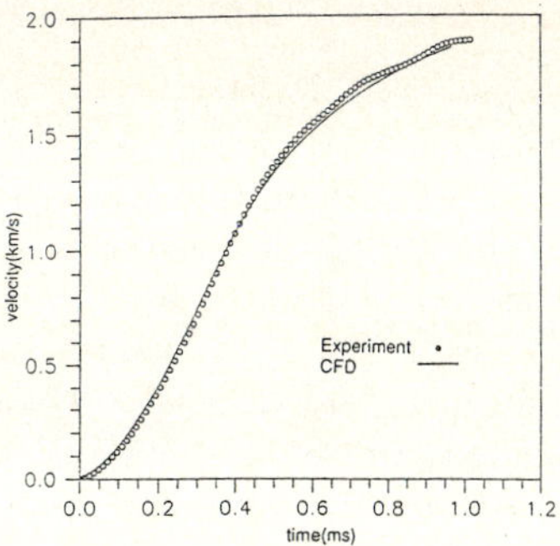

Fig. 7. Comparison of projectile velocity: CFD vs. experiment (powder weight: 6 g; 1 mm-AL diaphragm)

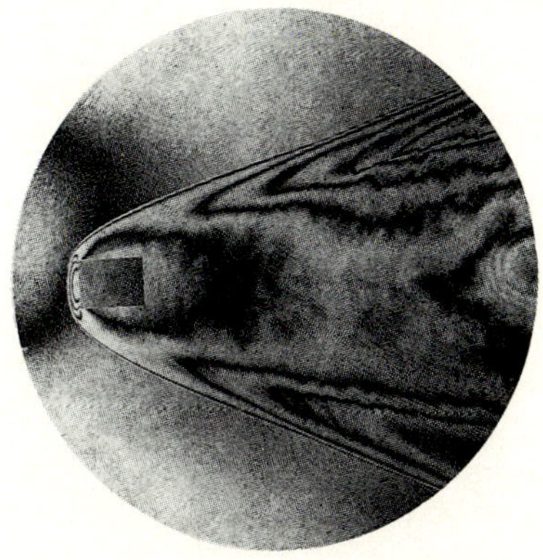

Fig. 8. Reconstructed image hologram (M=4.7)

4.3. Image holographic interferogram

A reconstructed image hologram is shown in Fig.8. Smokeless powder of 6 g, and 1 mm aluminum diaphragms were used. Initial pressure in the test section was 10.12 kPa. A trend of density change is observed in almost all regions behind the bow shock. Symmetric fringes are also observed in the wake region. This suggests that recompression of the flow is occurring in the wake region, and a wake shock is formed. The projectile posture in flight is slightly inclined, however, the flight Mach number of the projectile can be obtained from the stand-off distance of the bow shock. By using Eq.1, the muzzle velocity of the projectile is predicted to be 1.61 km/s. The difference between this predicted value and the other two values of projectile muzzle velocity obtained from (a) the VISAR measurements, and (b) from the numerical simulation, is less than 5% and 10%, respectively. Therefore, the optical flow visualization is very useful not only for quantitative measurements of the flowfield around the flying projectile but also as a verification tool of the projectile muzzle velocity measurement with VISAR interferometer.

5. Summary and future work

In order to analyse the performance of a gas gun, the VISAR interferometer technique was applied to in-bore projectile velocity measurement. In order to understand the gas gun operation, one-dimensional numerical simulations were conducted and the results compared with that of the experiments. In addition, the flow around the flying projectile was visualized with double exposure holographic interferometry, and verifed the accuracy of the VISAR measurements. In the next phase, the following aspects will be given further attention:

(1) Application of the VISAR system to much higher (typically 5 km/s) or lower speed (typically 200 m/s) phenomena.
(2) Improvement of the numerical simulation code, for more accurate predictions of the projectile movement.
(3) Design of an advanced (more efficient) single-stage gas gun facility.

Acknowledgements

The authors express their gratitude to Messrs. O. Onodera, H. Ojima and T. Ogawa of the Shock Wave Research Center of the Institute of Fluid Science, Tohoku University for their assistance in conducting the present experiments. Thanks also to our colleagues Mr. S. Kitashima of Chugoku Kayaku Co., Ltd. for their help on firing the powder explosives and to Mr. N. Okoshi, a Graduate Student of the Faculty of Engineering, Tohoku University, for his helpful cooperation. In conducting the VISAR measurements the kind support of Dr. W. Isbell, president of ATA Associates is acknowledged with thanks.

References

Barker LM, Hollenbach RE (1972) Shock-wave studies of PMMA, Fused silica and Sapphire. J. Appl. Phys. 43, 11:4469-4675

Canning TN, Seiff A, James CS (1970) AGARDograph No.138 on Ballistic-Range Technology. pp.427-465

Groth CPT, Gottlieb JJ (1988) Numerical study of two-stage light-gas hypervelocity projectile aunchers. UTIAS Rep. No.327

Isbell WM (1991) A simplified compact VISAR: concept and construction. Proc. 42nd Aeroballistic Range Association Meeting, Adelaide, South Australia, No.11

Matsumura T, Funabashi S, Saito T, Takayama K (1993) A holographic interferometric study of the axisymmetric supersonic flow around a flying projectile. Rep. Inst. Fluid Sci., Tohoku Univ., Vol.5, pp.89-97

Narayanswami NT, Matsumura T, Sasoh A, Saito T, Takayama K (1993) Theoretical prediction of hypervelocity launching device performance. (Abstract submitted to 5th Int. Symp. on Comp. Fluid Dyn).

Takayama K, Matsumura T, Ohuchi H (1992) Measurements of a projectile motion inside a powder gun with VISAR system. Proc. 43rd Aeroballistic Range Association Meeting, Columbus, OH, Vol.2, No.29

Balances for the Measurement of Multiple Components of Force in Flows of a Millisecond Duration

D.J. Mee, W.J. Daniel, S.L. Tuttle and J.M. Simmons
Department of Mechanical Engineering, The University of Queensland, Australia

Abstract.   Paper reports a new balance for the measurement of three components of force - lift, drag and pitching moment - in impulsively started flows which have a duration of about one millisecond. The basics of the design of the balance are presented and results of tests on a 15° semi-angle cone set at incidence in the T4 shock tunnel are compared with predictions. These results indicate that the prototype balance performs well for a 1.9 kg, 220 mm long model. Also presented are results from initial bench tests of another application of the deconvolution force balance to the measurement of thrust produced by a 2D scramjet nozzle.

Key words: Force measurement, Shock tunnel, Instrumentation

1. Introduction

The measurement of forces on vehicles flying at hypervelocity conditions has been restricted by the short durations for which current experimental facilities can sustain a representative flow. Progress has been made recently in designing balances for use in flows of duration as short as a few milliseconds (Jessen and Gronig 1993; Naumann et al. 1993; Carbonaro 1993). A new technique for measuring a single force component, the deconvolution drag balance, has been developed recently at The University of Queensland (Sanderson and Simmons 1991). This technique involves interpreting transient signals from strain gauges on a long sting connected to the test model to determine the time-history of the drag loading. The deconvolution drag balance has been applied to quite large and complex shapes in flows of duration as short as 1 ms (Porter et al. 1993).

In this paper the deconvolution force balance is extended to the simultaneous measurement of lift, drag and pitching moment. The basic design of this balance is considered and the installation of a prototype balance for measurements with a 15° semi-angle cone is discussed. Results for this model at small angles of incidence in a Mach 5 flow of nitrogen are then presented. A further application of the deconvolution balance to thrust measurement on a two-dimensional scramjet nozzle is also discussed.

2. The three-component balance design

The three-component deconvolution force balance consists of a single, 2 m long sting attached to the test model by four short bars (Fig.1). Each of the short bars is instrumented for measurement of axial strain at its mid-point. A strain gauge bridge is also attached to the sting 200 mm from the model/sting junction. Combinations of the strain signals from the four bars are used to produce two output signals - one responding primarily to a lift input signal and the other responding primarily to a pitching moment input signal. The strain measurement in the sting responds primarily to a drag input. Inevitably there is some coupling amongst these output signals.

The time histories of the three outputs related to lift, drag and moment, $y_L(t)$, $y_D(t)$ and $y_M(t)$, can be related to the time-histories of the lift force, drag force and pitching moment on the model, $u_L(t)$, $u_D(t)$ and $u_M(t)$ via nine impulse response functions. This coupled convolution problem can be written in matrix notation as in Mee et al. (1992)

$$\begin{pmatrix} y_L \\ y_D \\ y_M \end{pmatrix} = \begin{pmatrix} G_{LL} & G_{LD} & G_{LM} \\ G_{DL} & G_{DD} & G_{DM} \\ G_{ML} & G_{MD} & G_{MM} \end{pmatrix} \begin{pmatrix} u_L \\ u_D \\ u_M \end{pmatrix} \tag{1}$$

Shock Waves @ Marseille I
Editors: R. Brun, L. Z. Dumitrescu

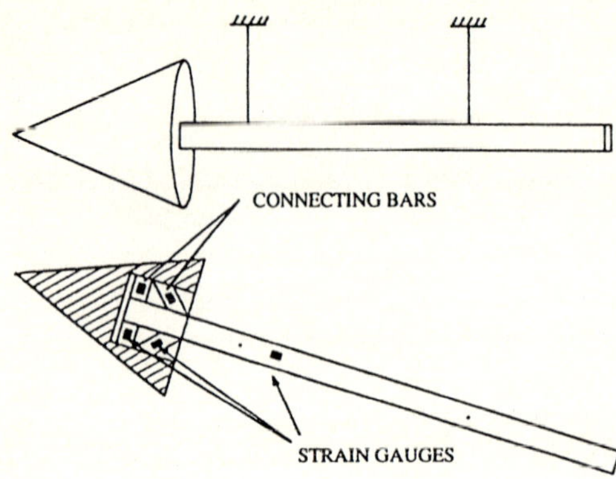

Fig. 1. Schematic of the three-component balance arrangement

where the y vectors are formed from the discretized output signals and the u vectors are formed form the applied load time histories. The square G matrices are formed from the impulse response functions, G_{ij} being the impulse response for the y_i output to a u_j input. If there is no coupling amongst the outputs then the off-diagonal submatrices in the impulse response matrix will be null. Further details on the design of the force balance are given in Mee et al. (1992).

The nine impulse response matrices were found from a series of bench tests in which a weight was attached to the model by a fine wire. The wire was cut, thus producing a step change in the load applied to the model. The output signals resulting from such a load change were recorded and processed to produce the impulse responses. The linearity of the system enables the responses to flow-type loading distributions to be determined by superposition of the results of several tests for single loads applied at various locations on the model.

In experiments in the shock tunnel, each of the y_i outputs was measured and time domain, coupled deconvolution techniques were used to determine the time histories of the lift and drag forces and pitching moment on the model. The experimentally determined impulse response functions were used for this deconvolution.

3. Tests of a prototype balance

The prototype balance was installed in a 220 mm long, 15° semi-angle, aluminium cone as indicated in Fig.2. The cone mass was 1.94 kg. The device was first tested outside the tunnel by cutting wires attaching weights to the model. Results for a single lift load applied to the cone were compared with the results for a distributed loading which gave the same net lift force and centre of pressure. These tests indicated that the balance was not very sensitive to the distribution of load on the model.

Experiments were performed in the T4 free-piston shock tunnel (Stalker and Morgan 1988) to test the performance of the balance. The sting was supported by fine wires (Fig.1), thus allowing it to move freely in the plane of the lift force, and was shielded from the flow so that the only aerodynamic forces were on the surface of the cone. For these tests the cone was set at three angles of attack: 0.0°, 2.5° and 5.0°.

For the tests nitrogen, at the conditions given in Table 1, was used. A typical, filtered trace of Pitot pressure during a shot of the tunnel is given in Fig.3. Note that the zero on this and subsequent time scales is arbitrary. The Pitot trace shows a small initial overshoot which is attributed to the nozzle starting process. The mean level settles to a steady value about 500 μs after the flow starts.

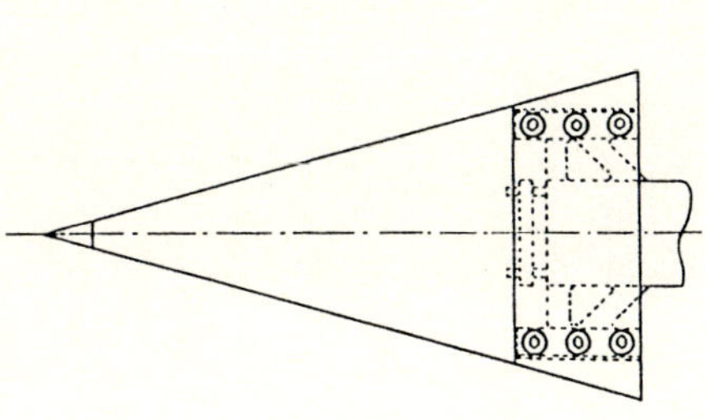

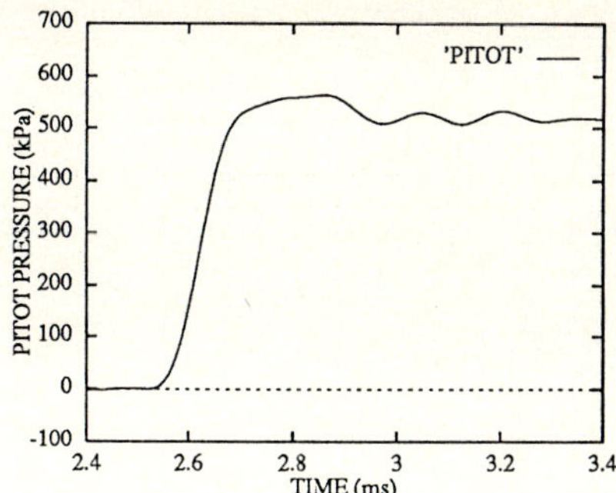

Fig. 2. Details of the connection of the balance to the sting

Fig. 3. Typical Pitot pressure trace for present tests

It takes about 800 μs for plane stress waves initiated by the arrival of the test gas to traverse the 2 m long steel sting, reflect from the free end and return to the strain measurement locations. During this time the flow has traversed approximately 15 model lengths. The deconvolution of the measured signals continues beyond this point but the accuracies of the impulse responses at these larger times deteriorates.

Examples of the raw voltage signals obtained from the five strain gauge bridges are shown in Fig.4. These data are from the test at 2.5°. Results are shown for the strains measured in the two inclined and the two transverse bars joining the model to the sting and for the strain measured in the sting itself. The signal-to-noise ratio is found to be adequate using semiconductor strain gauges. The relatively large oscillations in the signals in the transverse bars are associated with the dynamics of the model/sting configuration.

Table 1. Nominal test conditions - test gas is nitrogen

Supply Enthalpy MJ/kg	Supply pressure MPa	Static temperature K	Static pressure kPa	Static density kg/m^3	Flow speed m/s	Mach number
10.7	34.2	1370	11.4	0.0281	4270	5.8

Deconvolved signals are shown in Fig.5 for the three incidences of the present tests. The deconvolution process leads to amplification of noise on signals and the results shown have been passed through an 8-pole Butterworth filter with a cut-off frequency of 6 kHz. The axial force signals show an initial overshoot during the nozzle starting time, however the normal force signals do not show this overshoot. Both forces show reasonably steady values during the test time of 500 - 700 μs after flow start. The moment signal for the model is defined to be zero when the normal force acts through a point one-third of the cone axial length from the base of the cone - just ahead of the theoretical line of action of the normal force for an inviscid conical flow. The results indicate that the centre of pressure during the test time is about 5 mm (2% of the cone height) further downstream of this point.

The measured results have been compared with the computed data of Jones (1969). There is a small difference in the ratio of specific heats (1.4 in Jones 1969 and 1.32 in the experiments) but a comparison of the results, shown in Table 2, gives an indication of the performance of the balance. There are some uncertainties in the conditions of the flow in the test section which, for example, will lead to an uncertainty in theoretical drag of ±12% (Mee 1993). The absolute values

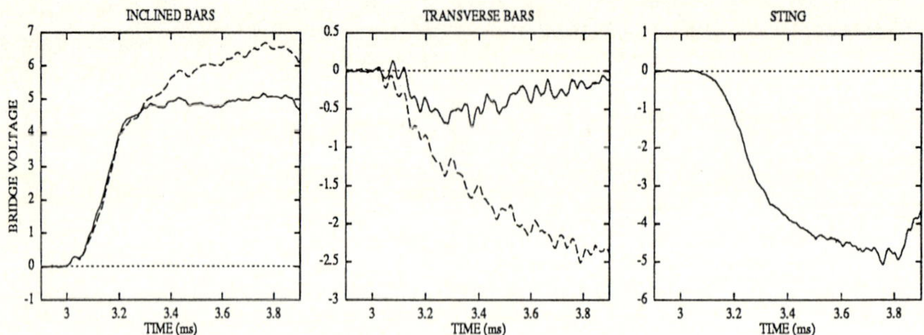

Fig. 4. Raw strain signals from the five bridges for test at 2.5° incidence

for axial and normal forces are about 10% lower in the experiments but the ratios of normal to axial agree well. At this stage the accuracy of the balance has not been quantified.

4. Thrust measurement on a 2D scramjet nozzle

The deconvolution force balance is also being used to measure thrust produced by a symmetrically divergent nozzle with 11° ramp walls. The area ratio is approximately 5.5. Attached to the nozzle are two long stings in which the measurement of tensile strain will be used to find the nozzle's net axial thrust. The system is shown in Fig.6.

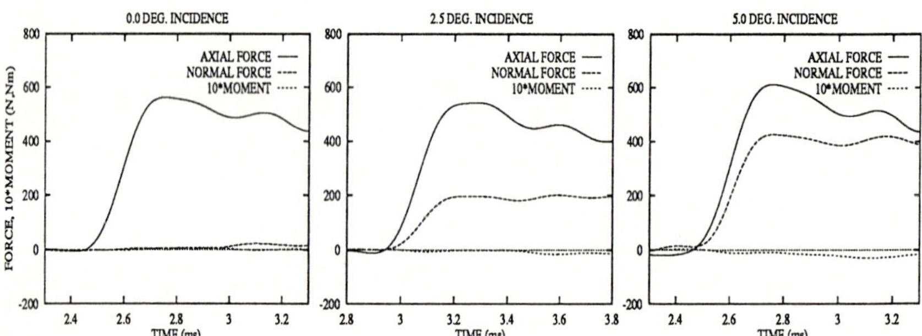

Fig. 5. Deconvolved force and moment signals for 0.0°, 2.5° and 5.0° incidence

Table 2. Comparison of experimental results with computations of Jones (1969)

Test No.	Incidence	Forces-Experiment			Forces-Computation		
	(degs)	Axial, A (N)	Normal, N (N)	N/A	Axial, A (N)	Normal, N (N)	N/A
3745	0.0	510	20	0.04	550	0	0.00
3752	2.5	460	210	0.46	530	220	0.42
3748	5.0	520	420	0.81	590	470	0.80

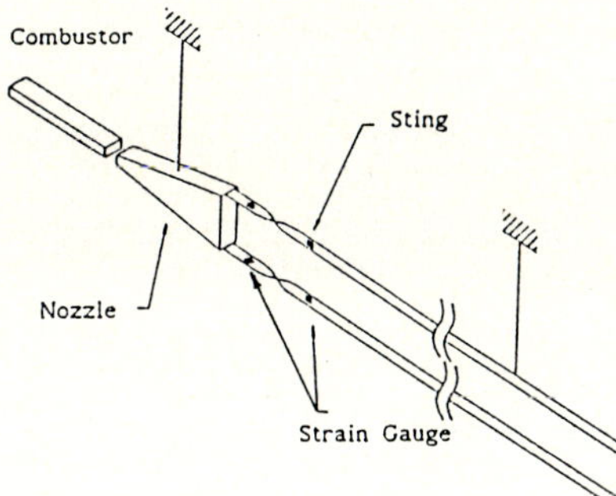

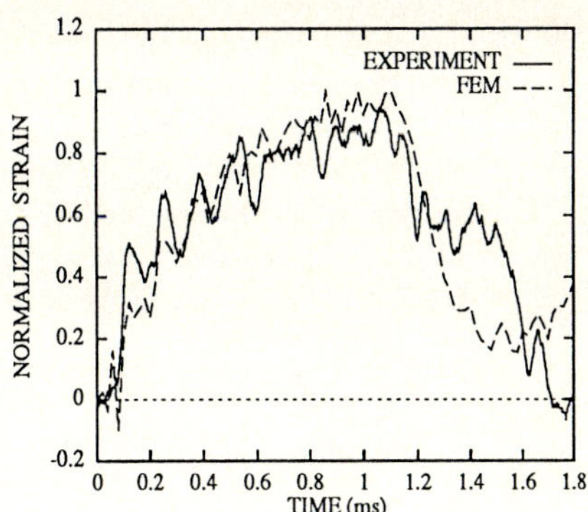

Fig. 6. Schematic of the thrust balance

Fig. 7. Experimental and computed step responses for a point load on the nozzle

The stings are twisted through 90° near to the nozzle to provide rigidity and bending stiffness. Finite element modelling has shown this to be acceptable. There are strain gauges before and after the twist on each sting. The static pressure along the ramp walls and the Pitot pressure across the nozzle's exit plane will be measured in the experiments. Any difference between the thrust indicated by the pressures and that determined with the deconvolution balance will be due to skin friction.

The strains from a finite element simulation for a tunnel-type loading have been deconvolved successfully using an impulse response based on another finite element computation. The model has been manufactured and instrumented and bench testing shows encouraging agreement with the computations (Fig.7). Differences between the step responses after arrival of reflected waves are attributed to different end conditions for the experiment and the computation. Based on these results and previous experience with deconvolution balances it is expected that the present balance will perform satisfactorily in tests in T4.

5. Conclusions

The results of tests using a three-component deconvolution force balance have indicated that the balance performs satisfactorily in a shock tunnel flow of a millisecond duration. The balance performed well on a model that was not restrictively small or light - the model was 220 mm long and had a mass of 1.9 kg. Further testing is required to quantify the accuracy of the balance.

Acknowledgement

The authors are grateful for the support received from the Australian Research Council under grant AE9032029 and the Queen Elizabeth II Fellowship Scheme (for DJ Mee).

References

Carbonaro M (1993) Aerodynamic force measurements in the VKI longshot hypersonic facility. In: Boutier A (ed) New trends in instrumentation for hypersonic research. Kluwer, Dordrecht, pp 317-325

Jessen C, Gronig H (1993) A six component balance for short duration hypersonic facilities. In: Boutier A (ed.) New trends in instrumentation for hypersonic research. Kluwer, Dordrecht, pp 295-305

Jones DJ (1969) Tables of inviscid supersonic flow about circular cones at incidence, $\gamma = 1.4$. AGARDograph 137

Mee DJ, Daniel WJ, Simmons JM (1992) Three-component aerodynamic force measurements in hypervelocity impulse facilities. Proc. 11th Australasian Fluid Mechanics Conference, Hobart, 14-18 Dec. 1992, pp 55-58

Mee DJ (1993) Uncertainty analysis of conditions in the test section of the T4 shock tunnel. Research Report 4/93, Dept. Mech. Eng., The University of Queensland

Naumann KW, Ende H, Mathieu G (1993) Millisecond aerodynamic force measurement technique for high-enthalpy test facilities. In: Boutier A (ed.) New trends in instrumentation for hypersonic research. Kluwer, Dordrecht, pp. 307-316

Porter LM, Mee DJ, Paull A (1993) Drag measurements on blunted cones and a scramjet vehicle in hypervelocity flow. AIAA Paper No. 93-2979

Sanderson SR, Simmons JM (1991) Drag balances for hypervelocity impulse facilities. AIAA J. 29(12):2185-2191

Stalker RJ, Morgan RG (1988) Free piston shock tunnel T4 - initial operation and preliminary calibration. NASA CR-181721

Further Developments of the ISL Millisecond Aerodynamic Force Measurement Technique

K. W. Naumann, H. Ende, G. Mathieu and A. George
French-German Research Institute Saint-Louis ISL, 5, rue du Général Cassagnou, 68301 Saint-Louis, France

Abstract.   The subject of this paper is the development of the ISL millisecond aerodynamic force measuring technique towards operational use with models, which are equipped with laterally blowing jets. The principle of operation consists of a short free-flight of the model during the testing time of the shock tunnel, and simultaneous measurement of model acceleration and Pitot pressure. The models are equipped with a set of small accelerometers, up to three exchangeable nozzles, and two pyrotechnic charges, which supply the side-jets with gas. Paper presents the concept of the models, their design, the gas generation and supply of the side-jets, the installation of accelerometers, and the data evaluation procedure.

Key words: Hypersonic flow, Lateral jets, Force measurement

1. Introduction

The ISL shock tunnel "B" is configured to provide hypervelocity flow at densities typical for the flight of small vehicles in the lower atmosphere, or spaceplanes in the stratosphere. Fig.1 shows a simplified sketch of the facility and the equipment necessary to operate our measuring technique, and Table 1 lists the test flow parameters for the experiments reported here. Notice, that all parameters represent the true flight conditions.

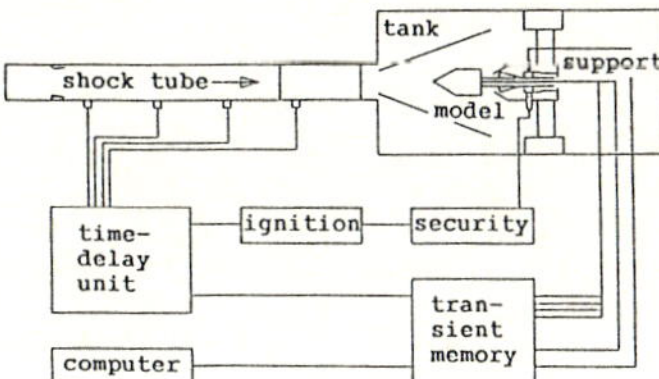

Fig. 1. ISL shock tunnel "B"

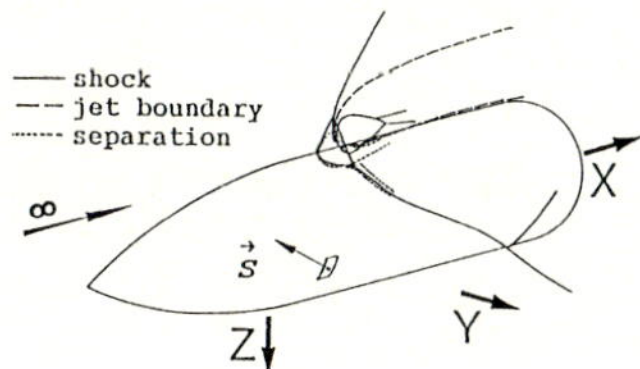

Fig. 2. Side-jet control of supersonic vehicle

The principle of operation (see Naumann et al. 1991) is as follows: Initially the mechanical mounting support fixes the model in the test chamber, at its correct position, with correct attidude. Just before test flow onset, the model is set free for some milliseconds. When the testing time is over, the support closes and fixes the model again, before the afterflow of the driver gas accelerates it too much. Little velocity and displacement allow the use of umbilical wires for data transmission, and control of installations. For the evaluation of the aerodynamic coefficients, for example C_k, we use simultaneously measured acceleration a_k and Pitot pressure $\hat{p}_0$. With model mass m_M , representative model area A_M , and $C_{p_0} = \hat{p}_0/q_\infty = f(Ma_\infty, \gamma)$, we get (Naumann et al. 1993b)

$$C_k = (C_{p_0} a_k m_M) / (A_M \hat{p}_0) \tag{1}$$

This procedure is straightforward (no differentiation), and insensitive against disturbances from test flow start and moderate flow variations, because the output are coefficient histories.

Shock Waves @ Marseille I
Editors: R. Brun, L. Z. Dumitrescu

The investigation on side-jet control (for details see Naumann and Srulijes 1985, Naumann et al. 1993a, 1993c, and the literature cited therein) for these flight conditions is stipulated by the severe aerodynamic heating. This affects particularly thin, delicate and exposed structures like fins, flaps, and actuation devices. In addition side-jet control can provide short response time, and its efficiency increases with flight Mach number. Fig.2 gives an illustration: The expanding jet acts on the oncoming flow like a blunt body, and causes a shock pattern, which propagates around the vehicle. Behind the shocks and in front of the jet the pressure acting on the vehicle surface is augmented; in the jet wake it is lessened. Maximizing efficiency means to maximize the component in the manoever direction of the surface integral of these pressure increments or decrements. The aim of our investigation is to identify and quantify the relevant parameters in order to optimize control efficiency.

Table 1. Test flow parameters

Quantity	Value	Unit
Velocity	1690	m/s
Density	0.48	kg/m^3
Pressure	44.5	kPa
Mach number	4.7	
Reynolds number	45×10^6	1/m
Temperature	310	K
Gas	N_2	
Test section area	200×200	mm

Table 2. Model data

Quantity	Md#1	Md#2	Unit
Model mass	3160	3485	g
Model span	140	190	mm
Model length without stem	140	138	mm
Maximum internal pressure	(50)	100	MPa
Centerline spacing at 3 jets	20	20	mm
Centerline spacing of 2 jets	40	40	mm

2. The side-jet models

The principal demand on our models is, that the gas supply for the side-jets must not disturb the free flight. Hence highly pressurized supply tubes are out of the question, at least for the time of measurement. The jet force has to be of the same order of magnitude as the lift of the model, to provide good resolution. The allowable size, at $M_a \geq 4$, is illustrated by Fig.3. The maximum time available for putting the jets in operation is 13 milliseconds.

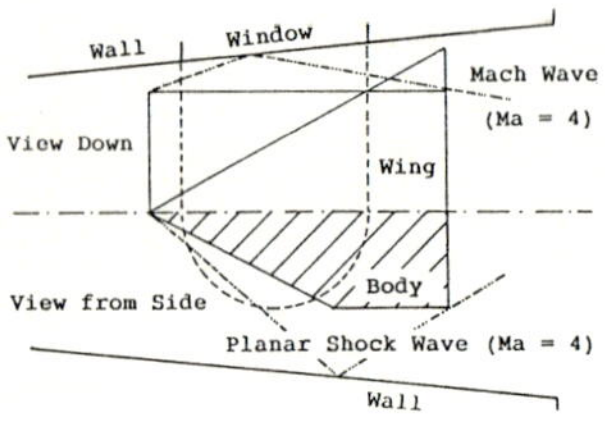

Fig. 3. Allowable size of side-jet models

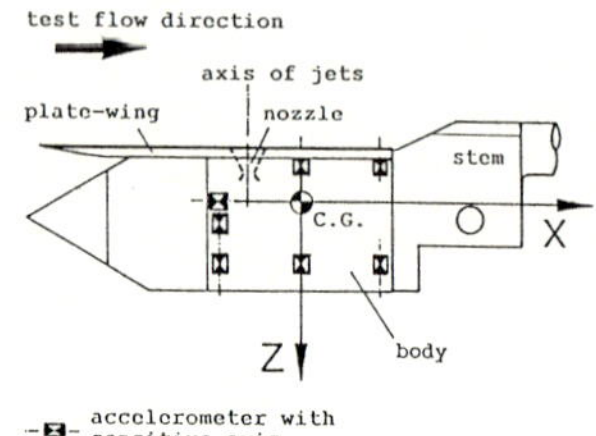

Fig. 4. Principle sketch of Model #1

Fig.4 shows a principle sketch of side-jet Model #1 (Naumann et al. 1993b). The nozzles are located near the C.G., and in the middle of the wing. Two pyrotechnical charges pressurize the plenum chamber. The wires of the transducers run through the support, those for the ignition of the charges (with a curent of about 100 V and 1 A) are placed externally. The model lift and side-jet thrust act in opposite direction in order to limit overall displacement. Fig.5 shows the individual parts of the model, and Table 2 gives some data. This model was successfully used for, to the best of our knowledge, the first measurements of lateral force of side-jet models in a shock tunnel.

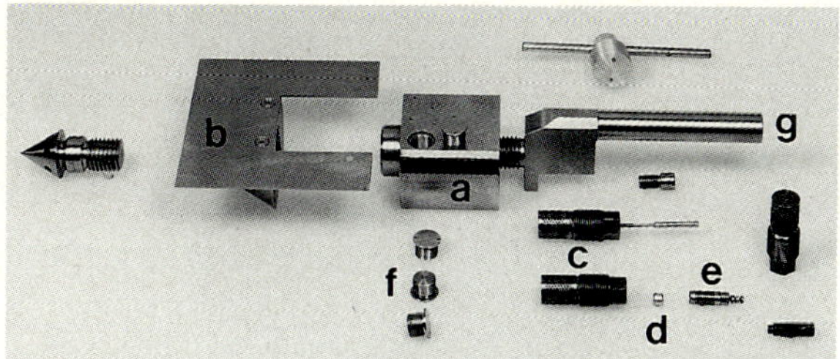

Fig. 5. Parts of Model #1:
(a) body; (b) wing; (c) small cartridge;
(d) fuse; (e) fuse-housing;
(f) nozzles and plugs; (g) stem

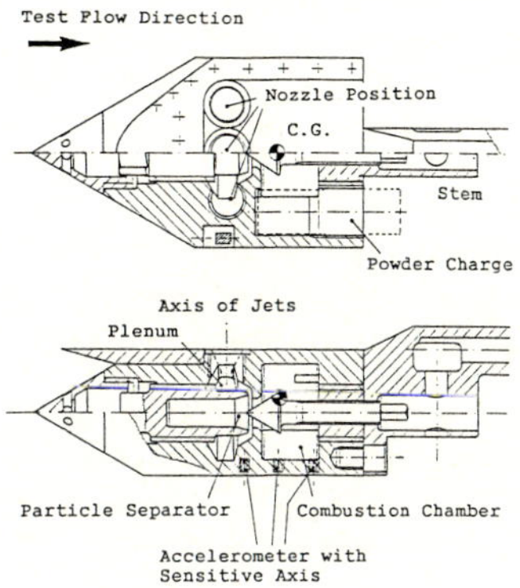

Fig. 6. Sketch of Model #2

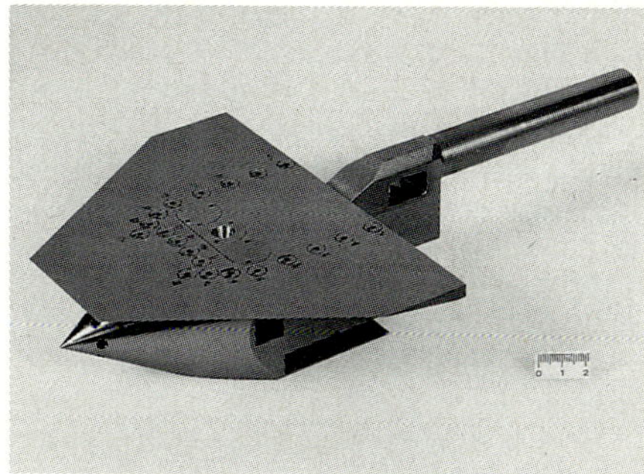

Fig. 7. Model #2

Despite this, it has some shortcomings: (a) Its constructive mechanical strength is not sufficient to allow jet flow examination in the plain laboratory at the later-on necessary plenum pressure of 50 MPa; (b) The rectangular wing produced tip vortices, which impeded flow visualization by differential interferometry; (c) The covered fixation of the wing provided a clean surface, but needed reinforcement; (d) Large gunpowder particles passed the nozzles.

In order to overcome these inconveniencies, Model #2 was built, with some ameliorations (see Fig.6 for a sketch, Figs. 7 and 8 for photographs, and Table 2 for the data): (a) Improved mechanical strength, (b) The swept leading edge of the wing is entirely supersonic; (c) The fixation of the wing is more stable. Moreover, the space in the front part of the body can be used for additional installations. In the case shown here this is a device, which forces the flow to bend sharply outwards, when it passes from the combustion chamber into the plenum chamber. Hence large particles are driven out by centrifugal force and are collected in the central cavity.

From the technical point of view the easiest method of supplying the nozzles is by combustion of a solid propellant. It needs little space, storage is easy, ignition is very rapid, it produces a lot of gas, and it is very suitable to be used in free-flying models. Moreover, it truly represents real vehicles, whose side-jets are mostly operated by solid propellants as well. Nevertheless, in the literature mostly experiments with pure (and cold) gases are reported. Possible reasons are: a) Security regulations; b) The agressive combustion products can cause damage to the test facility.

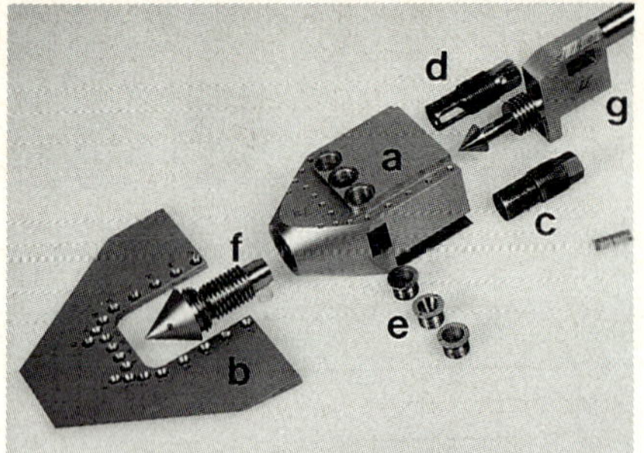

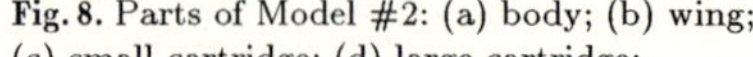

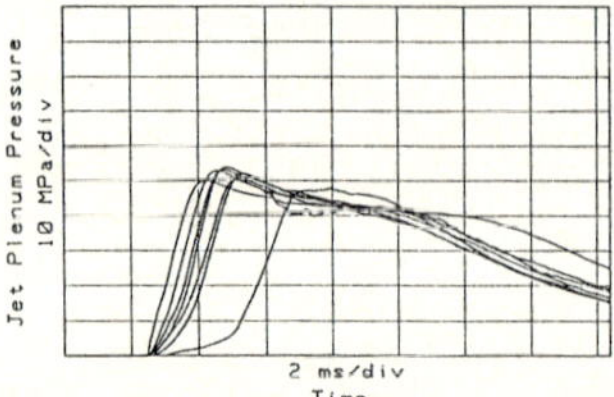

Fig. 8. Parts of Model #2: (a) body; (b) wing; (c) small cartridge; (d) large cartridge; (e) nozzles; (f) particle separator; (g) stem

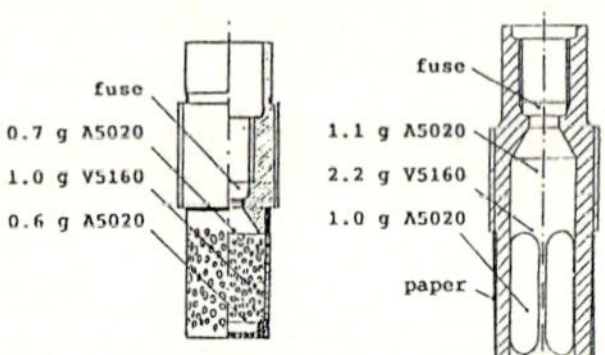

Fig. 9. Pyrotechnic charges

Fig. 10. Plenum pressure histories

For our experiments, we have problems with neither of these considerations. Fig.9 shows a sketch of the two types of pyrotechnical charges, which are presently in use. We compose the charges from different high-explosive gunpowders. The aim is to obtain: (a) A pressure level, which is as constant as possible during the test time; (b) A "moderate" increase of pressure in order to minimize shock loading of the model; (c) A sufficient predictibility of the pressure rise history.

Fig.10 shows a sample of plenum pressure histories from identical charges. As for the combustion of gun powder the results are very well, but it is not possible to predict a pressure history accurately. Fortunately, the proportionality factor between plenum pressure and jet force is constant for the time of measurement of a specific run (Naumann et al. 1993b). Thus the poor reproducibility can be overcome by igniting the powder before test flow onset, in order to evaluate the correct actual relation between plenum pessure and jet thrust.

Despite the relative simplicity of using solid propellant we wish to use pure gases as well, in order to evaluate, whether it is correct to simulate the thrusters by cold gas. The supply of free-flying models with pure cold gases necessitates an on-board storage device, because the filling tube must be depressurized when the model flies free. Fig.11 shows three possible solutions: (a) A reservoir with a fast-acting opening device. To obtain sufficiently hight thrust with a given nozzle throat cross-section area, the fill pressure must be at least the same as with combustion; (b) An on-board Ludwieg tube is able to provide constant flow for some time dependent on its length. The fill pressure must be high as well. The volumetric efficiency in terms of volume of stored gas to overall volume is poor; (c) A shock tube provides constant flow with less fill pressure in its driven part. Disadvantages are its length and the need to change two diaphragms at two different locations prior to each shot. Hence we decided to try solution (a). The necessary device, which opens an area of about 1cm^2 within a few milliseconds, and fits into the limited space inside of our models, had to be developed by ourselves. Model #2 is prepared to accomodate it as well.

The accelerometers are placed in excavations at the exterior of the models (Fig.12 with Model #1). This arrangement yields redundant lift, drag, pitch, roll and yaw accelerations.

Like for the validation experiments (Naumann et al. 1991) the accelerometers are mounted on small damping discs made from resin and micro-balloons. Protection of transducers and wires

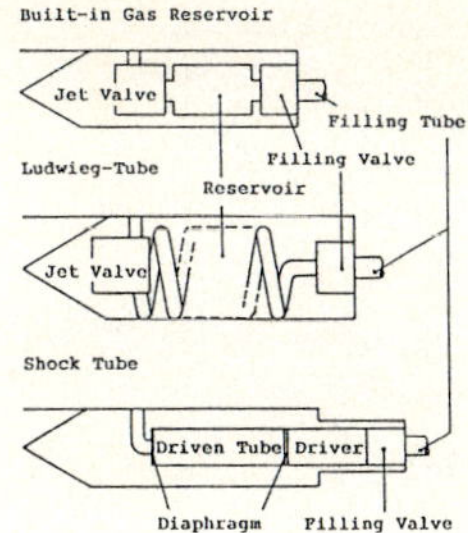

Fig. 11. Solutions for supplying free-flying side-jet models with pure gases

Fig. 12. Installation of accelerometers in Model #1

against the flow is achieved in Model #1 by filling the cavities with plasticine. At Model #2 the accelerometers are mounted in closed square pieces made of the damping material, which are pasted in the cavities. By this method the accelerometers are very well protected, but difficult to dismount. Future arrangements will use individual damping discs and a separate solid shielding.

3. Data processing and evaluation

The output of our experiments are histories of accelerations and pressures. Fig.13 shows, for the time of interest, for an experiment with two jets, and filtered with a low-pass filter using a cut-off frequency of 2 kHz: The lateral acceleration $a_{zt}(t)$ of the C.G. of Model #1, the plenum pressure $p_{0j}(t)$, and Pitot pressure $\hat{p}_0(t)$. Using $\hat{p}_0(t)$ and $p_{0j}(t)$ allows the calculation of the contributions of model lift and side-jet thrust to the total acceleration: $a_{z\infty}(t)$ is calculated from $\hat{p}_0(t)$ and Eq.1, using the previously measured C_L. $a_{zj}(t)$ is calculated from $p_{0j}(t)$ using the actual proportionality factor evaluted before test flow onset. The acceleration $a_{zi}(t)$ due to the force, which results from the interaction of side-jets and oncoming flow is

$$a_{zi}(t) = a_{zt}(t) - (a_{z\infty}(t) + a_{zj}(t)) \tag{2}$$

Without any interaction it would be zero. Fig.14 shows, that after a starting time of almost 1 ms, counted from the first rise of the Pitot pressure signal, $a_{zi} \approx a_{zj}$ for this setup. Fig.15 shows a synopsis of tests with one, two or three nozzles. For more thorough considerations on the interaction effect see Naumann et al. (1993a, 1993c) and the literature cited therein.

The bold line on a_{zi} in Fig.14 shows, that the test flow slug length necessary to establish quasistationary conditions is for this case 12 times the model length. Whether and to what extent a flowfield like this can be quasistationary is not yet clear, because windward supersonic separation zones are known to exhibit unstationary phenomena like oscillations. Hence we think, that the oscillations of the a_{zt} - and a_{zi} -signals filtered with 2 kHz, represent rather oscillations of the flowfield than parasitic effects, and therefore we do not suppress these by excessive filtering.

In all cases we apply numerical filtering only, using an averaging process, whose effect is similar to that of a Bessel filter, but which runs faster and does better. Applying such a filter on a whole history means that insignificant parts of it, like the zero $\hat{p}_0$ before test flow start, or a_{zt} after the model has hit the stop, affect the bounding regions of the useable part of the history. To overcome this, we developed a routine, which allows filtering of selected parts of a digital history. As a consequence we can make better use of the short measuring time, thus increasing the accuracy of the evaluation process. The bold line of a_{zi} in Fig.14 is evaluated by this procedure.

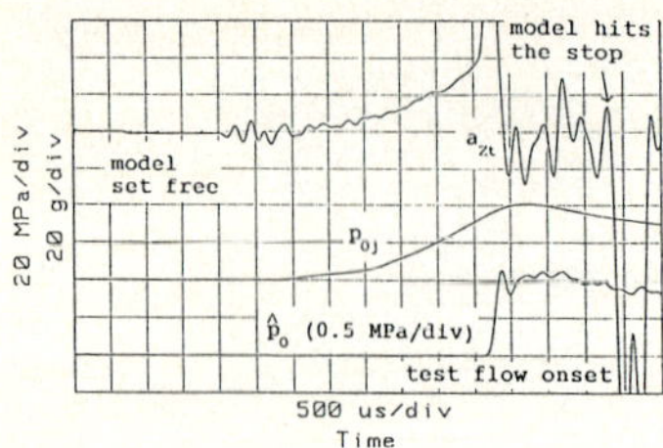

Fig. 13. Measured a_{zt}, p_{0j}, and $\hat{p}_0$ with Model #1 with 2 jets

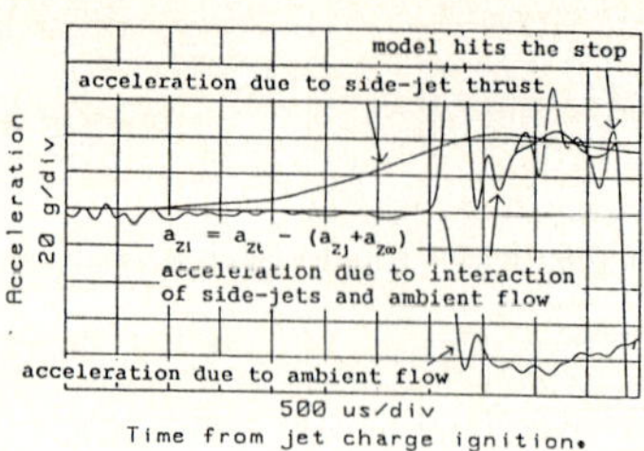

Fig. 14. a_{zi}, a_{zj}, and $a_{z\infty}$ with Model #1 with 2 jets

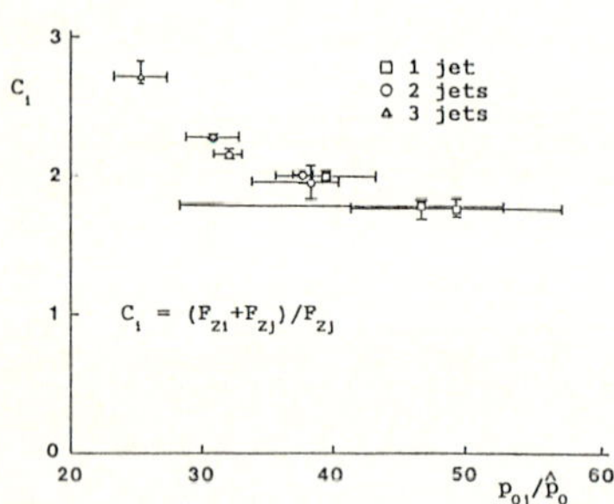

Fig. 15. Experimental C_i

4. Conclusion

In the ISL shock tunnel "B" we carry out experiments on side-jet control efficiency by using our millisecond aerodynamic force measuring technique. The primary developments effected for this are:

- A short-time gas supply by combustion of gun powder, which is installed in the free-flying model and supplies the gas for the side-jets.
- A particle separator, which prevents large unburnt powder grains from passing the nozzles.
- Numerical filtering of selected sequences of a data history.

Our next developments will be:
- A small fast-acting high pressure opening device for on-board stored pure and combustion-generated gases.
- Improved accelerometers, if those which we will test next should display superior performance.

References

Naumann KW, Srulijes J (1985) Die Flugbahnsteuerung mittels seitlich austretender Strahlen. Literaturübersicht. ISL R 117/85

Naumann KW, Ende H, Mathieu G (1991) Technique for aerodynamic force measurement in shock tunnel. Shock Waves 1,3:223-232

Naumann KW, Ende H, Mathieu G, George A (1993a) Experiments on interaction force of jets in hypervelocity cross-flow in a shock tunnel. Paper 21 in: AGARD-CP 534

Naumann KW, Ende H, Mathieu G, George A (1993b) Milisecond aerodynamic force measurement with side-jet model in the ISL shock tunnel. AIAA J. 31, 6:1068-1074

Naumann KW, Ende H, Mathieu G, George A (1993c) Shock tunnel experiments and approximative methods on hypervelocity side-jet control effectiveness. AIAA Paper 93-1929

Six-Component Force Measurement in the Aachen Shock Tunnel

C. Jessen and H. Grönig

Stoßwellenlabor, RWTH Aachen, 52056 Aachen, Germany

Abstract. A six-component strain-gauge balance for short-duration hypersonic facilities is presented. The balance features a new geometry compared to classical strain-gauge balances for model measurements in wind tunnels. The new geometry and the utilization of highly sensitive semiconductor strain-gauges allows for both a short response time, and a high sensitivity. The calibration and data reduction procedures are described in detail. Measurements on a 30° cone at incidence in a $M_\infty = 7.9$ shock tunnel flow are presented. They are compared to inviscid calculations and measurements of other authors, showing that the balance operates reliably.

Key words: Six-component balance, Shock-tunnel, Calibration, Cone measurement

1. Introduction

The state of the art of force measurement in short-duration hypersonic facilities was comprehensively reviewed by Bernstein (1975). Since then different approaches were undertaken to measure forces on a model in high-enthalpy facilities within a testing time of one or several milliseconds.

Naumann and Ende (1989) developed a fast-acting release and clamp mechanism, that allows the model to fly freely within the testing time. The accelerations of the model are measured and the forces are derived from the accelerations and the inertia matrix of the model by applying Newton's law: $\underline{F} = \underline{\underline{m}} \cdot \underline{\ddot{x}}$. Problems arise with this technique when the models are small or the forces low, so that not all components of the accelerations can be measured with sufficient accuracy. Also it is not at all easy to determine the inertia matrix of a complicated model that is equipped with six or more accelerometers. This means that the technique requires an individually equipped model and detailed knowledge about it.

The same holds for another approach by Sanderson et al. (1991). From the stress waves introduced into the model and the sting the instationary forces, responsible for the strains measured on the sting, are evaluated. This requires a complete finite-element model of the sting and the body to be investigated. By means of deconvolution the forces are calculated from the strains.

The strain-gauge balance presented herein seems to be a more general solution to the problem of force measurement within a short testing time. As a sting balance in the classical sense it is virtually independent of the model. Once it is calibrated statically with a high accuracy, it can be used for force measurements on any model having sufficient inner space.

2. A novel design for a strain-gauge balance

Fig.1 shows the finite-element model of the new balance design as it was used for layout purposes. The basic idea of the balance lies in connecting the model to the outer ends of four cross arms. Forces on the model generate deformations in these cross arms measured by strain-gauges. All other parts of the structure between model mounting and cross arms only serve to transmit the forces from the model to the cross arms. The center of the cross is connected to the sting via the inner plug by means of thermal shrinking. The sting is connected to the model mounting in the test section of the tunnel. This compact design leads to a high stiffness of the balance, which manifests itself in a high natural frequency. For the prototype balance discussed in the following, the lowest natural frequency is that of the pitching oscillations, at roughly 2.3 kHz.

Shock Waves @ Marseille I
Editors: R. Brun, L. Z. Dumitrescu

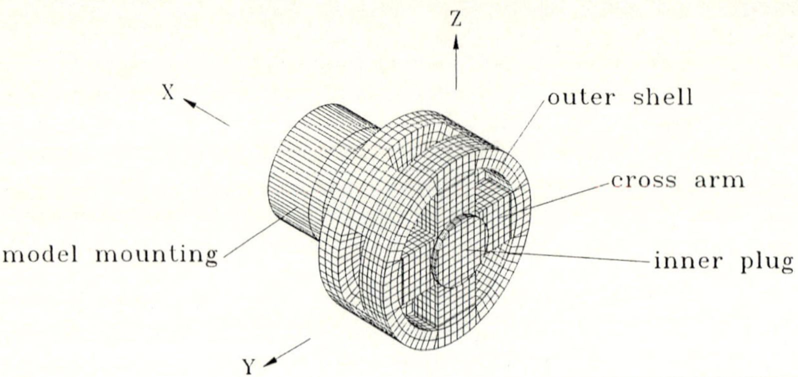

Fig. 1. Finite-element model of the balance

Of course, the natural frequency is considerably lowered by the inertia of the model. To be sure that the response time is still sufficiently small, the natural frequency of the system model-balance should not be lower than 1 kHz (in our application).

From the finite-element calculations it is known, that the axis of the pitching oscillation is located in the cross-section between inner plug and cross. To minimize the response time of the system this cross-section has to be designed to a maximum stiffness on one hand and on the other hand the moment of inertia with respect to this location has to be minimized. This means, that the center of gravity of the balance plus model is best located in that cross-section.

Theoretically a full separation of the six components is possible (Jessen and Grönig 1989), but practically interferences between the components occur. This is primarily due to the fact, that the center of the cross is not fixed symmetrically to the inner plug but only on one side. Therefore an accurate calibration is necessary to separate the six components exactly.

3. Calibration and data reduction

Classical strain-gauge balances yield linear responses with a crosstalk typically below one percent. In this case a mathematical description of the balance behaviour by linear terms is sufficient, but a higher accuracy can be achieved by including square terms.

Unfortunately, a dependence including a square term cannot describe a curve having point symmetry with respect to the origin of coordinates (that is expected due to the physical effects in a balance system). Therefore it seems necessary in the case of comparatively large interferences to use a more sophisticated description of the system taking into account also third order terms (Schnabl 1987). The relation between signal and load is then given by:

$$S_i = R_{0i} + \sum_{j=1}^{6} a_{ij} Z_j + \sum_{j=1}^{6} \sum_{k=j}^{6} b_{ijk} Z_j Z_k + \sum_{j=1}^{6} c_{ij} Z_j^3 \qquad i = 1, 2 \ldots 6 \tag{1}$$

with S_i the signal of the strain-gauge bridge for component i, R_{0i} the signal of bridge i without load, Z_j the total load acting on component j, a_{ij} the calibration factors for linear terms, b_{ijk} the calibration factors for square terms and c_{ij} the calibration factors for cubic terms.

The calibration factors are calculated from the data of a static calibration by means of a least square fit. The S_i are the measured and digitized quantities during an experiment. As the Z_j are the unknowns, the above nonlinear set of equations has to be solved for each set of six values S_i recorded at a certain instant. This is achieved using the Newtonian method of approximation.

The calculated, time-dependent forces and moments are transformed from the balance-fixed into the aerodynamic coordinate system and are non-dimensionalized with the Pitot pressure in the test section – corrected for the static pressure – and characteristic model dimensions.

A procedure is needed to deduce the static load from the instationary measurement. The measured curve is first integrated over the useful testing time. Then we do not simply divide by the testing time to obtain a mean value, but the inclination of the integrated curve is determined by means of a linear regression. It delivers the static portion of the signal directly also in the case where a full oscillation is not completed within the evaluation interval. If this procedure is repeated for increasing time intervals, the history of the stationary coefficients can be monitored.

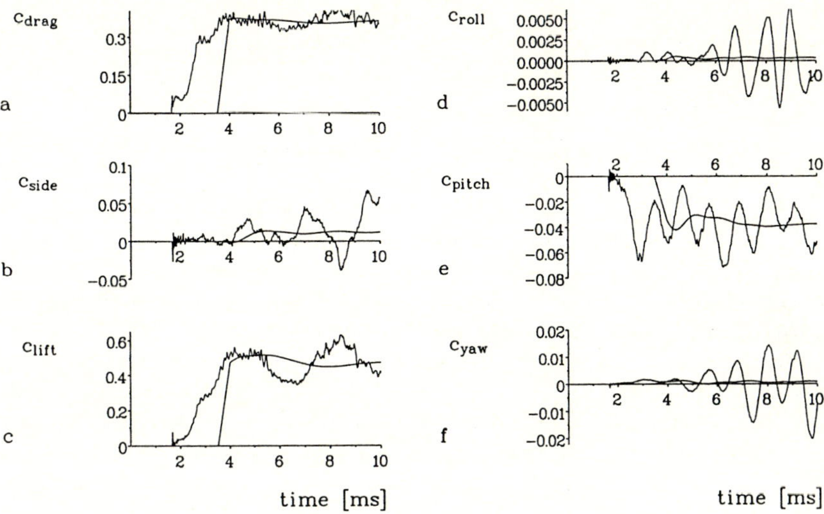

Fig. 2. Coefficients of a 30° cone model at 15° angle of attack in a $M_\infty = 7.9$ flow (The broken lines indicate the beginning of the evaluation interval for the reduced data): (a) Drag; (b) Side Force; (c) Lift; (d) Rolling Moment; (e) Pitching Moment; (f) Yawing Moment

Fig.2 shows an example of the instationary forces together with the reduced data. The first evaluation was carried through between 3.5 ms and 4 ms, so that the first value is given at 4 ms. The broken line at $t = 3.5$ ms indicates the beginning of the evaluation interval. The low-frequency oscillations in the side force and lift plots are generated by the model support inside the test section and not by the balance itself.

4. Measurements on a cone at incidence

A cone of 30° aperture, 200 mm long was investigated with the balance in the Aachen shock tunnel TH 2 at various angles of attack between 0° and 15°. The initial conditions for all tests are $p_4 = 10$ MPa (He) and $p_1 = 0.1$ MPa (air) at ambient temperature. This leads to the following reservoir conditions in front of the nozzle: $p_5 = 6.6$ MPa and $T_5 = 1500$ K. The test gas is expanded in a contoured nozzle to provide a $M_\infty = 7.9$ parallel flow field. The starting process of the nozzle takes about 1.5 ms, the useful testing time under these conditions is about 4 ms.

The data evaluation, as described above, was carried out for each experiment in the time interval identified as "useful testing time" from the Pitot pressure plots. The resulting measuring

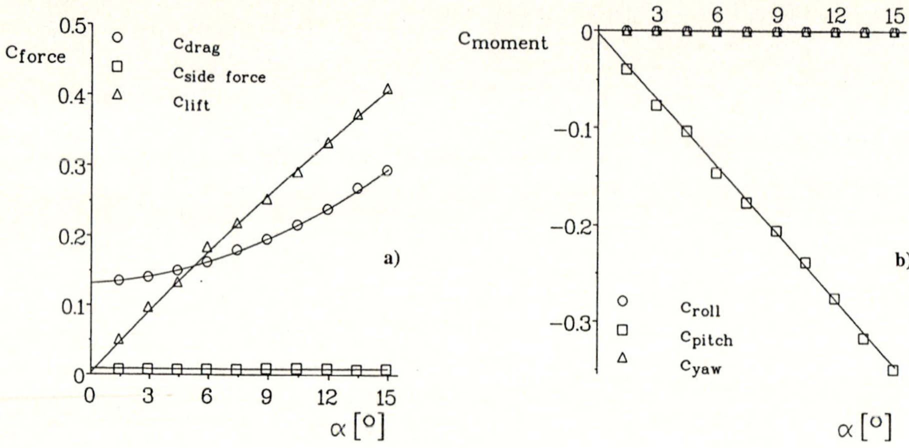

Fig. 3. Coefficients of a 30° cone model at various angles of attack in a $M_\infty = 7.9$ parallel flow field: (a) Force coefficients; (b) Moment coefficients

points are indicated in Fig.3 for all components. They are approximated by second order curves. The base area of the cone was used as reference area for all coefficients. For the moment coefficients the model length served as characteristic length. The moments are given with respect to the center of gravity of the model.

The drag coefficient is obviously well described by a quadratic function, but the lift coefficient increases nearly linearly with angle of attack. The side-force coefficient should be zero in a parallel flow, but the measurements show a coefficient in the order of 0.01. This error has different reasons: as can be seen from Fig.2b, a low-frequency oscillation (≈ 500 Hz) is superimposed over the side force measurement. The data reduction procedure delivers a reasonable mean value of the oscillations, but an error in the order of $\pm 10\%$ has to be expected. Compared with this the error of the static calibration ($\pm 0.5\%$) is negligible. As the unexpected side-force coefficient is nearly constant for all angles of attack investigated, it is assumed that it is due to a misalignment of the sting in the model support of the facility and probably to a not fully symmetrical flow field. A final judgement is only possible after additional tests, where the balance will be turned 180° around its axis. As the loading changes its sign then (from the viewpoint of the balance-fixed coordinate system) a similar indication of the side-force could clearly be attributed to a physical force acting on the model. The moment coefficients in Fig.3b are also approximated by second order curves. While the pitching moment shows a strong dependence on the angle of attack, the rolling and the yawing moment are very close to zero – just as expected.

Taking all sources of error into account, it is conservatively estimated that the reduced curves for drag in Fig.3a and for the rolling moment in Fig.3b are accurate within $\pm 5\%$, while those for side force, lift and the yawing and pitching moment are accurate within $\pm 10\%$.

Jones (1969) gives results of inviscid cone calculations over a wide range of cone angles, angles of attack and Mach numbers for $\gamma = 1.4$. As the contribution of friction is not taken into account, the calculated coefficients are smaller than the ones measured and can be regarded as "lower limit".

Fig.4 shows a comparison between the calculated coefficients for drag and lift and the ones measured by different authors. The reference area for Fig.4 is the planform area of the model. For the lift coefficients the obtained data is obviously slightly too high. This is attributed to the low-frequency oscillation of the model mounting, that leads to a high amplitude inertia force in

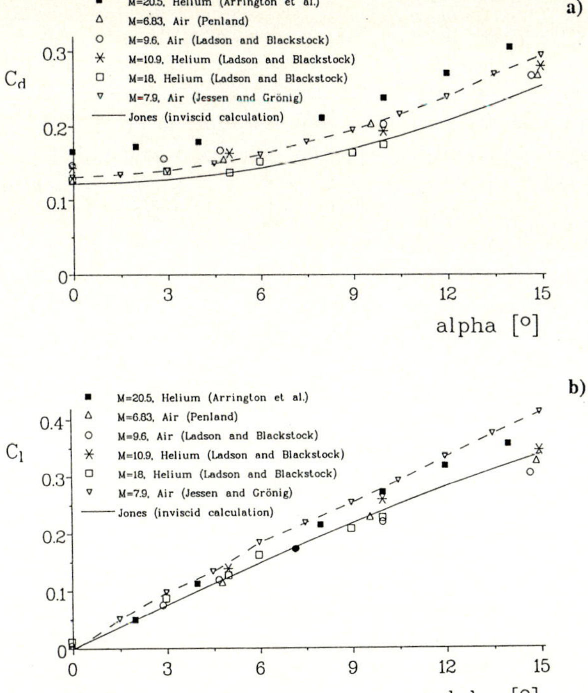

Fig. 4. Force coefficients of a 30° cone: (a) Drag; (b) Lift

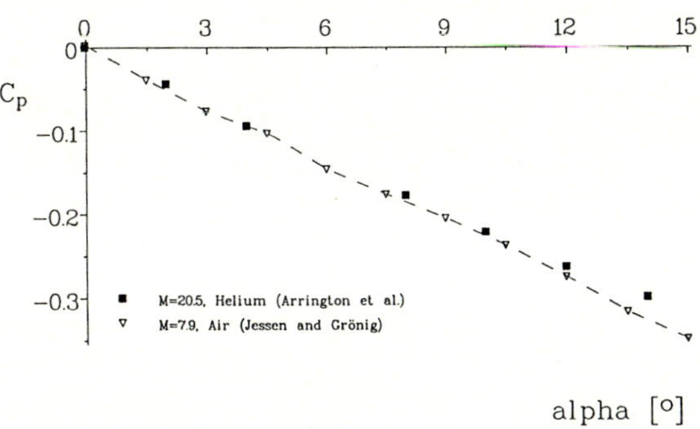

Fig. 5. Pitching coefficient of a 30° cone

the lift component (see Fig.2c). The reduced data at $t = 6$ ms is indicated in Fig.4, if it were taken at $t = 8$ ms (see Fig.2c) a better fit would be obtained. This has not been done, as it is doubtful, whether data obtained after the "useful testing time" – that ends at $t = 6$ ms – should be evaluated.

Fig.5 shows a comparison between measurements of the pitching moment. The moment is given with respect to the tip of the cone. It is noted, that the Mach number varies considerably for the data presented, and that the results were partly obtained in Helium.

5. Conclusions and outlook

A strain-gauge balance with a completely new geometry has been presented. It allows for six-component force measurements on models in the TH 2 shock tunnel flow. The construction of the balance was illustrated and the calibration was explained. A data reduction procedure tailored to the problem makes it possible to extract the stationary force coefficients out of instationary measurements, although sting oscillations of the model support are superimposed. Measurements on a cone compared with inviscid calculations and experimental results of other authors confirm that the balance operates reliably.

With the strain-gauge balance presented, it is possible to conduct force measurements on models without having them equipped with accelerometers. The inertia matrix of the model need not be known, only it must not have too high moments of inertia. This can usually be achieved by proper construction of the model and the use of low density materials like aluminium or magnesium.

Acknowledgements

The authors wish to acknowledge the excellent work of Mr. H. Schobben of the mechanical workshop. Further thanks are due to Dipl.-Ing. M. Lenartz who performed the finite-element calculations. We would also like to thank the DLR Göttingen for their help with the calibration, and MBB Bremen for helping us with the calibration software.

References

Arrington PJ, Joiner RCJr., Henderson AJr. (1964) Longitudinal characteristics of several configurations at hypersonic Mach numbers in conical and contoured nozzles. NASA Technical Note D-2489

Bernstein L (1975) Force measurement in short-duration hypersonic facilities. AGARDograph No. 214

Jessen C, Grönig H (1989) A new principle for a short-duration six-component balance. Exp. in Fluids 8: 231-233

Jones DJ (1969) Tables of inviscid supersonic flow about circular cones at incidence $\gamma = 1.4$. AGARDograph 137, Part 1: 1-16

Ladson CL, Blackstock TA (1962) Air-Helium simulation of the aerodynamic force coefficients of cones at hypersonic speeds. NASA Technical Note D-1473

Naumann KW, Ende H (1989) A novel technique for aerodynamic force measurements in shock tubes. Proc. 13th Intl. Cong. on Instr. in Aerosp. Sim. Fac., Göttingen, Germany, pp 535-544

Penland JA (1961) Aerodynamic force characteristics of a series of lifting cone and cone-cylinder configurations at $M = 6.83$ and angles of attack up to 130°. NASA Technical Note D-840

Sanderson SR, Simmons JM, Tuttle SL (1991) A drag measurement technique for free- piston shock tunnels. 29th Aerospace Sci. Meeting, Reno, Nevada, AIAA Paper 91-0540

Schnabl F (1987) Entwicklung eines Algorithmus zur Auswertung der Eichversuche an 6 - Komponenten - DMS - Waagen. ZFW 11: 342-346

Driver Gas Detection by Quadrupole Mass Spectrometry in Shock Tunnels

J.C. Slade*, K.C. Crane* and R.J. Stalker[†]
*Australian National University, Canberra ACT, Australia
[†]The University of Queensland, Brisbane, Qsld. Australia

Abstract. Experiments are reported in which quadripole mass spectrometry is used to detect the onset of test section flow contamination by helium driver gas in two free-piston shock tunnels, which differ widely in size. The spectrometer flow sampling configuration used in the small tunnel is described, and the results are compared with those for the large tunnel. In both cases, is was found that the measurements agreed satisfactorily with an approximate theory for contamination, based on shock-boundary layer interaction at the reflected shock in the shock tube.

Key words: Shock Tunnels, Contamination, Mass spectrometry

1. Introduction

The last few years have seen a significant increase in the number of high-enthalpy reflected shock tunnels in use throughout the world, thereby stimulating a necessary interest in methods of calibrating the test section flow in such facilities.

One of the most important measurements is the detection of driver gas in the test section flow, since driver gas contamination usually is the critical factor in determining the useful test period. This reduces as the stagnation enthalpy increases, and ultimately becomes so brief that a useful flow cannot be produced. Therefore driver gas contamination does, in effect, limit the stagnation enthalpy produced in a shock tunnel.

This paper presents results of the measurement of time of arrival of helium driver gas contamination in two free piston shock tunnels, using quadrupole mass spectrometry. Measurements were obtained at stagnation enthalpies ranging from 18 MJ.kg^{-1} to 45 MJ.kg^{-1} in both facilities. Since the size of the two facilities varied by a factor of 3.6, this provided a check on the effect of scale on contamination times, and provides a basis for extending the applicability of the results, and the theory which they match, to smaller or to larger facilities.

2. Apparatus

Table 1. Dimensions of shock tunnels tested

	Compression tube		Shock tube		Nozzle	Piston
	Length (m)	bore Dia (mm)	Length (m)	Dia (mm)	Exit Dia (mm)	Mass (kg)
T2	3	76	2	21	112	1.2
T3	6	300	6	76	300	90

The two facilities in which the measurements were made were the T2 and T3 free piston shock tunnels at the Australian National University, Canberra. These are shown diagrammatically in Fig.1, presented according to the relative scale of the two. Table 1 gives the leading dimensions of each facility, and it can be seen that the ratio of shock tube diameters is 3.6. The shock tube diameter has a dominant effect on the time from shock reflection to the passage of driver gas through the nozzle throat; i.e. the "contamination time".

Shock Waves @ Marseille I
Editors: R. Brun, L. Z. Dumitrescu

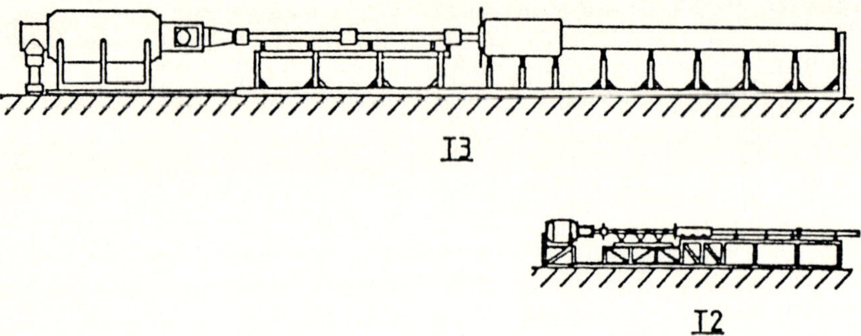

Fig. 1. Shock tunnels

The mass spectrometer arrangement used to detect the arrival of helium in T-2 is briefly described. A more complete description is given in Slade (1970). It was designed simply for the detection of helium, and therefore required less sophistication in construction than the one used for the experiments in T3, where species concentrations also were required. The mass spectrometer arrangement used in T3 is described in Crane and Stalker (1977).

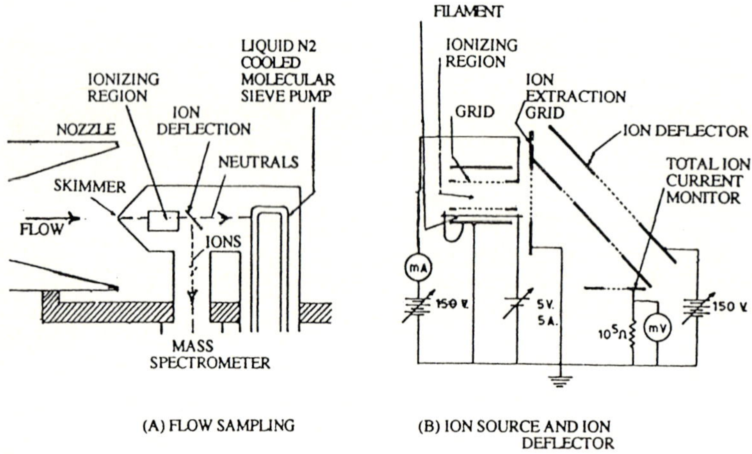

Fig. 2. Flow sampling and ion beam management

As shown schematically in Fig.2, the flow was sampled via a conical skimmer, of 90° included external vertex angle and 70° included internal angle, through an orifice 0.25 mm in diameter. The Knudsen number at the orifice was of the order of unity. The molecules passed from the skimmer orifice to the ioniser, which consisted of a grid at 85 Volts and an earthed tungsten filament electron emitter, surrounded by an earthed metal cylinder. A neutral particle had an estimated 1% probability of being ionised in the ioniser, and this was sufficient to swamp any residual ionisation in the test gas. For the mass spectrometer used in T3, it was necessary to use two skimmers in series, and apply a field between to remove residual ions.

The ions were extracted by the ion extraction grid, and the resultant ion beam was deflected through $90°$ in order to be directed towards the mass spectrometer. Before reaching the mass spectrometer, the beam passed through a grid which acted as a total ion current monitor, to register any shot to shot variation in the ion beam.

The mass spectrometer itself was located on the test section side wall, outside the test section. It was a convential quadrupole mass spectrometer, with four 6.36 mm dia. rods 125 mm long, aligned with axes parallel to the incident ion beam, and positioned in a quadrupole structure with their separation 0.86 times the diameter of the rods. A radiofrequency field was applied to the rods in such a way that the ions experienced a rotating electric field as they passed between the rods, so that essentially only ions of one charge to mass ratio passed through the field region, to be collected by a Cu/Be particle multiplier. The mass spectrometer therefore acted as a mass filter. Though the mass selected could be varied between runs, only one mass could be monitored during any one run of the shock tunnel.

Before a test, the test section and dump tank were evacuated to 10^{-3} torr, and the mass spectrometer cavity to 10^{-5} torr.

3. Measurements

Time resolved records of ion currents for a particular shock tunnel test condition are shown in Fig.3, together with a record of the nozzle reservoir pressure. This shows that nozzle reservoir conditions are constant until well after arrival of the helium which, as shown by the ion current records, occurs 75 μsec after the first detection of the nitrogen test gas. Allowing for an estimated time delay of 15 μsec in detecting helium, and 20 μsec in detecting nitrogen, this gave a period of 70 μsec of uncontaminated test flow.

Since the time for a gas particle to traverse the nozzle is approximately equal to the nozzle starting time for a hypersonic nozzle, this period of uncontaminated test flow may be taken as equal to the contamination time, as defined above.

The sensitivity of helium detection was calculated from the ionisation cross sections as 5% by mol. fraction for an ion current of 10^{-8} amp., which is the minimum current detectable in Fig.3. This does not allow for mass separation in the skimmer flow but as this effect tends to lead to enhanced concentrations at the centre of the beam, the figure of 5% is conservative; i.e. detection levels may be lower.

The contamination times are plotted in Fig.4(b), and contamination times measured in T3 in Fig.4(c), for a fixed driver condition with each facility and a range of shock tube conditions.

4. Theory and comparison with measurements

The mechanism of driver gas contamination was originally proposed by Davies and Wilson (1969), and has been adapted to high enthalpies by Crane and Stalker (1978). It involves interaction of the reflected shock wave in the shock tube with the shock tube wall boundary layer near the downstream end of the shock tube. The pressure rise across the shock forces the boundary layer to separate, causing the shock to take up a "lambda" configuration near the wall. At the centre of the tube, the reflected shock is thus a normal shock, but near the wall, it takes the form of two oblique shocks, which meet the normal shock in a triple point, as shown in Fig.4(a). The boundary layer separated by the first of the oblique shocks reattaches under the influence of the second one, thus forming a "bubble" of boundary layer fluid at the foot of the reflected shock. The mainstream flow therefore takes place in two layers - one just outside the bubble, and the

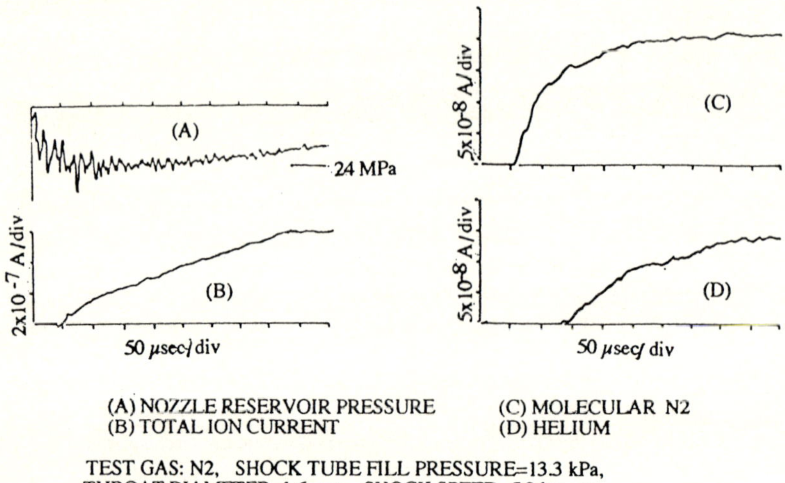

(A) NOZZLE RESERVOIR PRESSURE   (C) MOLECULAR N2
(B) TOTAL ION CURRENT   (D) HELIUM

TEST GAS: N2, SHOCK TUBE FILL PRESSURE=13.3 kPa,
THROAT DIAMETER=1.6 mm, SHOCK SPEED=5.3 km sec

Fig. 3. Typical test results

other outside the triple point, and extending to the centre of the tube. The flow outside the triple point is brought to rest by the normal reflected shock, whereas the flow which passes between the triple point and the bubble experiences the same pressure rise as the normal shock, but taking place through two oblique shocks. It therefore suffers a smaller velocity change than the flow through the normal shock, implying that it is not brought to rest, but continues as an annular jet surrounding the stationary core of gas. When the reflected shock reaches the contact surface, the driver gas enters this jet, and flows to the end of the shock tube where the nozzle throat is located. The driver gas then enters the nozzle, and contaminates the test flow.

This mechanism can be expressed quantitatively as an approximate theory, as outlined by Davies and Wilson (1969) and by Crane and Stalker (1978). In Figs.4(b) and 4(c) it is used to predict contamination times for T2 and for T3 with helium driving air and the appropriate driver condition in each case. It can be seen that the theory agrees closely with the experimental data for T2, but somewhat underpredicts the contamination times at high enthalpy for T3. Given the approximate nature of the theory, this is a satisfactory result.

The theory also predicts that, provided the contamination time does not approach the time for the test gas to drain into the nozzle, then it should be independent of the size of the nozzle throat. The measurements confirm this, showing no greater change than can be attributed to shot-to-shot variation when the throat area is varied by a factor of four.

5. Conclusion

The construction and operation of a sampling system together with a mass spectrometer for detecting the presence of helium driver gas in the test section of a shock tunnel has been briefly described. The detection is 5% by mol. fraction or better. Given that a suitable quadrupole mass spectrometer is usually available from commercial sources, the system is relatively simple to construct.

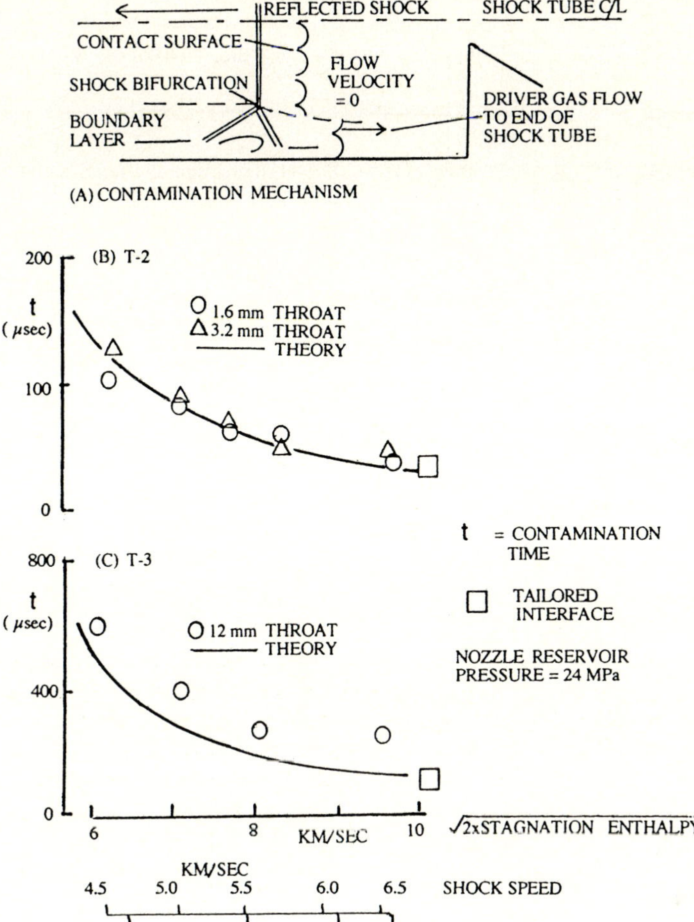

Fig. 4. Helium contamination times

Measurements obtained with this system in the small shock tunnel T2 were compared with measurements taken using a somewhat more sophisticated system in the much larger shock tunnel T3. It was found that the measurements correlated well with an approximate theory for predicting the contamination time, with the results for the larger facility somewhat exceeding the predicted values. It may be concluded that the theory can be used to predict contamination by this mechanism for a large range of shock tunnel sizes, and that it will be conservative for larger shock tunnels.

References

Crane KCA, Stalker RJ (1977) Mass-spectrometric analysis of hypersonic flows. J. Phys. D.: Appl. Phys. 10:679

Crane KCA, Stalker RJ (1978) Driver gas contamination in a high-enthalpy reflected shock tunnel. AIAA Journal 16:277

Davies L, Wilson JL (1969) Influence of reflected shock and boundary layer interactions on shock tube flows. Phys. Fluids Supplement I.I-37

Slade JC (1970) Mass spectrometry in shock tunnels. M.Sc. Thesis, Australian National University, Canberra

Skin Friction Measurements and Reynolds Analogy in a Hypersonic Boundary Layer

G.M. Kelly, J.M. Simmons and A. Paull
Department of Mechanical Engineering, The University of Queensland, Brisbane Q 4072, Australia

Abstract. A new skin friction gauge has been used in measurements of a high enthalpy, Mach 5.5, hypersonic boundary layer on a flat plate in a free-piston shock tunnel. Comparison of skin friction and heat transfer measurements indicates that the Reynolds analogy is valid in hypersonic, laminar boundary layers over the Reynolds number range of 1.0×10^6 to 4.8×10^6 m^{-1}. Onset of transition is detected with the skin friction gauge.

Key words: Hypersonic boundary layer, Skin friction, Shock tunnel, Reynolds analogy

1. Introduction

A skin friction gauge for use in conventional shock tunnels has been reported (Dunn 1981). However, its ability to perform satisfactorily in the very short-duration flows associated with free-piston shock tunnels, typically 1 ms, does not appear to have been demonstrated. This paper reports the development of a transducer having a rise time of about 30 μs and capable of measuring skin friction in the presence of high flow temperatures and pressures. The design is a refinement of the prototype produced by Kelly et al. (1992). Measurements of skin friction and heat transfer rate in a hypersonic, flat-plate boundary layer are presented and used to study the Reynolds analogy and the onset of transition.

2. Gauge design

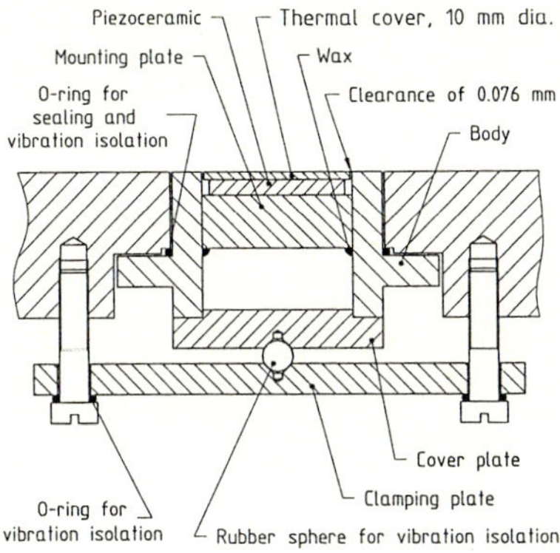

Fig. 1. Skin friction gauge

The gauge is shown schematically in Fig.1. It comprises a thin Invar thermal cover (10 mm diameter and 0.35 mm thick) mounted flush with the surface of the model and bonded to a piezoelectric transducer element (7×7 mm and 1.5 mm thick). The thermal cover is needed to prevent the piezoelectric material from being heated during operation of the shock tunnel to a temperature above its Curie temperature (620 K) at which depolarisation occurs. Because of the

Shock Waves @ Marseille I
Editors: R. Brun, L. Z. Dumitrescu

nominally zero coefficient of thermal expansion of Invar, its use also prevents stressing of the piezoelectric element due to thermal expansion of the cover.

In the earlier design (Kelly et al. 1992), two piezoelectric transducer elements were used, one inverted with respect to the other. In this way the pressure sensitivity of the transducer could then be eliminated by adding the outputs from the two piezoelectric elements. It has now been found that, by precise cutting of the piezoelectric material parallel to the appropriate plane of polarisation, the output from a piezoelectric element can be made insensitive to pressure within the overall accuracy of the transducer. Consequently, only one element is used in the present design. In the experiments reported in this paper, provision was made for the elimination of some pressure sensitivity. Commercial piezoelectric pressure transducers were located close to the skin friction gauges and their outputs were subtracted from those of the skin friction gauges. This precaution proved unnecessary for four of the five gauges which exhibited insignificant pressure sensitivity. The fifth gauge exhibited pressure sensitivity to the extent of 15% of the overall output signal, probably due to inadequate precision in the cutting of the piezoelectric material relative to the polarisation axes or to lack of parallelism in the fastening of the thermal cover to the piezoelectric element. The piezoelectric is PZT-7A. Its rigidity and density are such that, for the configuration in Fig.1, gauge natural frequencies are above 300 kHz. The impulsive nature of the flow in a free-piston shock tunnel results in stress waves in the model, with the skin friction gauge being exposed to a vibration-induced acceleration environment. To counteract this, rubber and felt vibration isolation was incorporated in the design of the transducer and its mounting into the flat-plate model. The technique effectively lowered acceleration-induced output from the transducer to a level that could be handled by digital filtering during signal processing. However, care was required in aligning the soft-mounted transducer with the surface of the model. The shear stress levels on the model were typically only 100-1000 Pa. Charge amplifiers with gains of 0.05 V/pC were located adjacent to the skin friction gauges.

3. Calibration

Because the piezoelectric transducer does not have a static response, dynamic calibration is necessary. Pressure sensitivity was measured in situ with apparatus that enabled fast release of a known pressure to the surface of the model. Attempts to achieve direct calibration for skin friction by applying a known shear force to the center of the thermal cover by way of a thread in tension proved unreliable. At this stage of its development the skin friction gauge must be calibrated indirectly in a known laminar boundary layer on a flat plate. Measurements were made of heat transfer rate close to the site of the gauge and Reynolds analogy was then used to determine the skin friction.

Table 1. Five shock tunnel test conditions

Condition	1	2	3	4	5
Stagnation enthalpy, MJ/kg	4.58	7.27	7.59	9.06	5.25
Stagnation temperature, K	3522	4663	4921	5685	3850
Stagnation pressure, MPa	11.16	9.38	20.0	51.3	10.7
Temperature, K	574.5	1129	1180	1297	695
Pressure, kPa	3.8	4.3	9.17	22.6	3.92
Density, kg/m^3	0.023	0.013	0.027	0.060	0.020
Velocity, m/s	2829	3488	3562	3791	3012
Mach number	5.9	5.3	5.3	5.4	5.8
Unit Reynolds number, m$^{-1} \times 10^{-6}$	2.35	1.04	2.15	4.80	1.88

The Reynolds analogy can be stated as Eq.1 with reasonable accuracy for a hypersonic, laminar boundary layer (Anderson 1989):

$$\frac{C_H}{c_f} = \frac{1}{2}\mathrm{Pr}^{-2/3} \tag{1}$$

where C_H and c_f are the Stanton number and local skin friction coefficient and Pr is the Prandtl number. Using van Driest's determination of skin friction in a compressible, flat-plate, laminar boundary layer (van Driest 1952) and the assumption that the free-stream conditions remain effectively constant, the Reynolds analogy can be reduced to Eq.2 (Kelly 1993):

$$\frac{q_w}{\tau} = \frac{-\mathrm{Pr}^{-2/3}\, c_p\, \left[T_\infty \left(1 + r\frac{\gamma-1}{2}M_\infty^2\right) - T_w\right]}{U_\infty} \tag{2}$$

where T_∞ = free-stream temperature, K τ = skin friction, kPa
       T_w = wall temperature, K c_p = specific heat
       M_∞ = free-stream Mach number r = recovery factor
       q_w = heat transfer rate, kW/m^2 U_∞ = free-stream velocity, m/s

The sensitivities of a set of skin friction gauges, determined from measurements of heat transfer rate at one test condition and Eq.2, were then used for the other test conditions.

4. The experiment

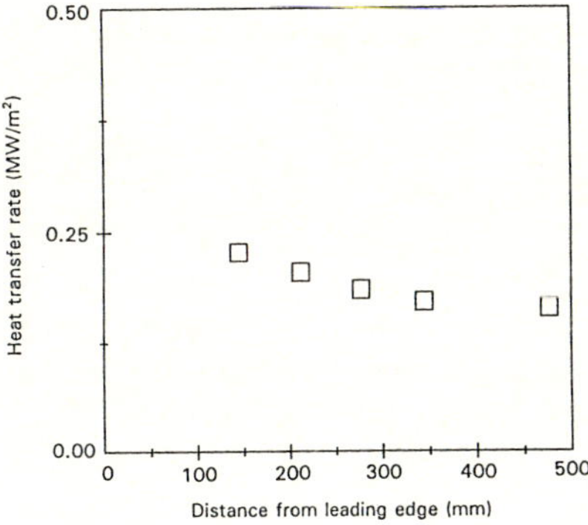

Fig. 2. Heat transfer on plate for calibration of skin friction gauges at Condition 1

The flat plate model used was made of aluminium with a steel leading edge. It was 600 mm in length and 228 mm in width. There were five skin friction gauges in the model, the first at a distance of 145 mm from the leading edge and the others at 65 mm intervals downstream of it. At the site of every skin friction gauge a piezoelectric pressure transducer was positioned 19 mm to one side and a thin film heat transfer gauge 19 mm to the other. Their centres were set at the same distance from the leading edge. The heat transfer gauges were 1.7 mm in diameter and the pressure tappings were 2 mm in diameter. The test conditions used in the T4 free-piston shock tunnel at The University of Queensland are listed in Table 1. Condition 1 is that used for calibration purposes. Fig.2 shows the distribution of heat transfer rate along the plate for the calibration test, approximately 500 μs after the start of the flow. Transition was detected towards the end of the flow by the gauge furthest downstream because, in the mode of operation of the shock tunnel, test section conditions were changing with time. However, at the time chosen for calibration, the boundary layer was laminar.

5. Results

The results presented in this paper are a sample of the full results presented by Kelly (1993). Fig.3a shows the time-histories of heat transfer rate and the corresponding skin friction, 277 mm from the leading edge for Condition 5. If the time-history of heat transfer rate is scaled to a skin friction using Reynolds analogy in the form of Eq.2, the two traces virtually coincide after the unsteady starting flow when the heat transfer rate is high (Fig.3b). This gives some confirmation of the validity of Reynolds analogy in a hypersonic, laminar boundary layer. However, Condition 5 is not greatly different from Condition 1 which was used to calibrate the skin friction gauges against the heat transfer gauges by way of Reynolds analogy. Clearly the boundary layer at this station remains laminar throughout the test flow. A measure of the rise time of the skin friction gauges can be obtained from the starting transients in Fig.3. Kelly (1993) used a small shock tube to show the rise time to be about 30 μs.

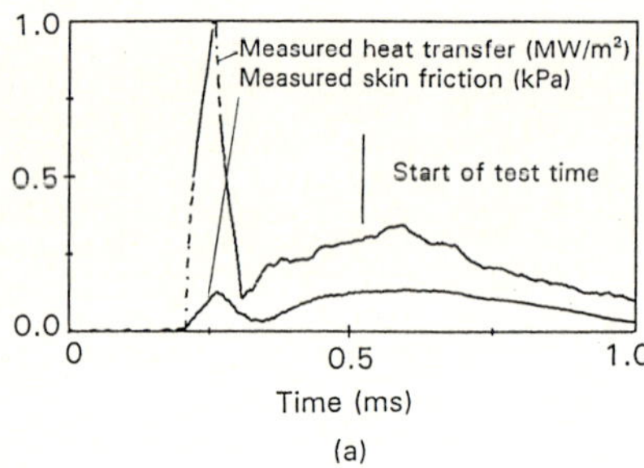

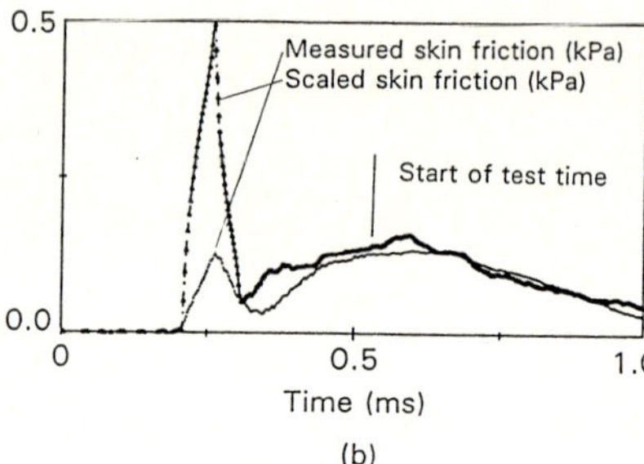

Fig. 3. Time-histories for Condition 5 at 277 mm from leading edge; (a) measured skin friction and heat transfer rate, (b) skin friction measured and scaled from heat transfer rate

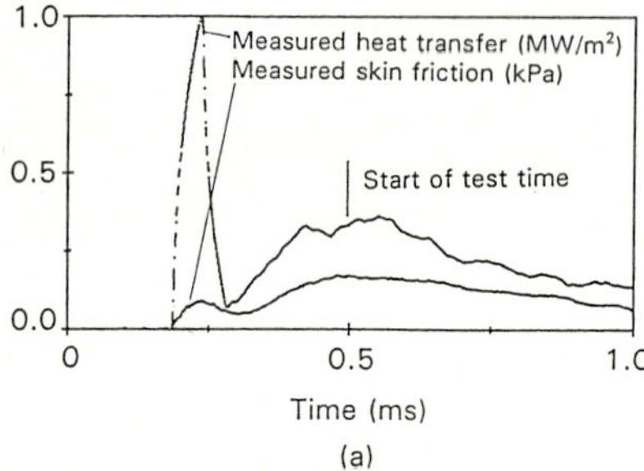

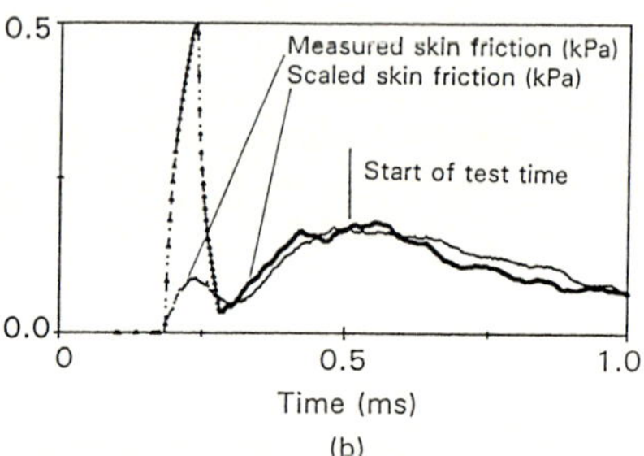

Fig. 4. Time histories for Condition 5 at 212 mm from leading edge; (a) measured skin friction and heat transfer rate, (b) skin friction measured and scaled from heat transfer rate

Again for Condition 5, Fig.4a shows the heat transfer rate and skin friction time-histories, 212 mm from the leading edge. Application of the Reynolds analogy (Fig.4b) shows that after the start of the test time the measured and scaled skin friction traces again agree well.

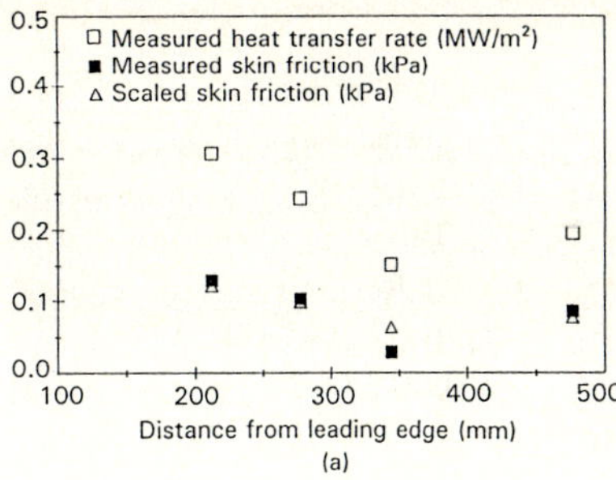
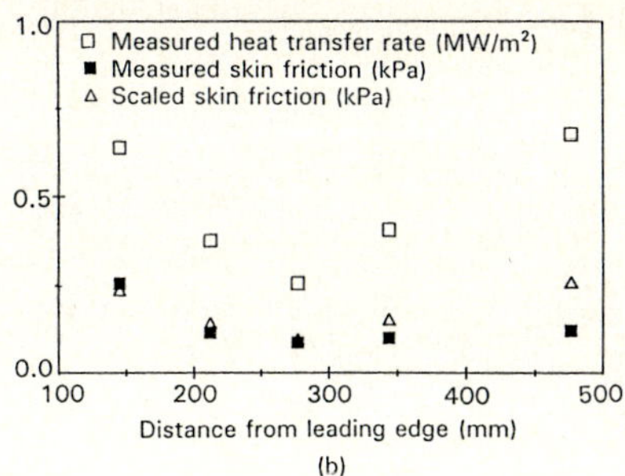

Fig. 5. Skin friction and heat transfer rate along plate; (a) Condition 2, (b) Condition 3

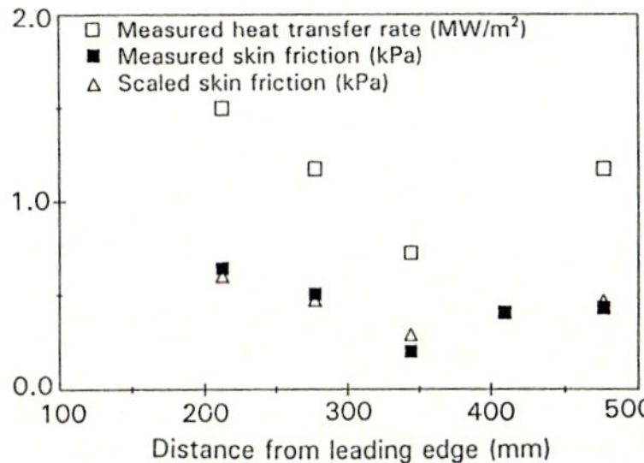

Fig. 6. Skin friction and heat transfer rate along plate
for Condition 4

Fig. 7. Time-histories of measured and scaled skin
friction for Condition 4 at 212 mm from leading edge

Fig.5a shows distributions of skin friction and heat transfer rate along the plate, measured for Condition 2 at the start of the test time. Also shown is heat transfer rate scaled to skin friction using the Reynolds analogy. The increases in skin friction and heat transfer rate indicate that transition occurs between the last two points. This is consistent with a prediction of onset of transition 402 mm from the leading edge (He and Morgan 1989). Fig.5b shows distributions of skin friction and heat transfer rate for Condition 3 which is significantly different from Conditions 1 and 2. Heat transfer rate and skin friction are again consistent with the Reynolds analogy in a laminar boundary layer at the first three points. Transition is predicted by He and Morgan (1989) at 467 mm, a result consistent with measurements at the station furthest downstream.

Fig.6 shows distributions of skin friction and heat transfer rate for Condition 4, a high pressure condition which departs considerably from the calibration Condition 1. The rise in heat transfer rate and skin friction at the station furthest downstream is consistent with prediction of transition at 294 mm. The fact that the Reynolds analogy appears to hold at this station is misleading. Transition is associated with turbulent bursts which may not occur identically over corresponding heat transfer and skin friction gauges because of their separation of 19 mm in the cross-stream direction. This behaviour can be seen in Fig.7. Time-histories of measured skin friction and skin friction scaled from heat transfer rate agree well until transition occurs as the test section conditions change with time. The turbulent fluctuations over the heat transfer and skin friction gauges are clearly different.

6. Conclusions

Although further refinement is needed, the skin friction gauge has been shown to have potential for routine measurement in the short test time of free-piston shock tunnels. The rise time of about 30 μs is sufficiently short for most shock tunnel applications and approaches the rise times needed for expansion tube applications. Calibration of gauges has been achieved in a laminar boundary using measurements of heat transfer rate and Reynolds analogy. A more direct means of calibration is desirable and vibration isolation needs further study.

The gauges have been used to show that Reynolds analogy is applicable to hypersonic, laminar boundary layers over a range of free-stream conditions. They have also been used to detect the onset of transition at locations that are consistent with predictions from the study by He and Morgan (1989).

Acknowledgements

This work was supported by the Australian Research Council under Grant A5852080 and by NASA under Grant NAGW-674. The authors acknowledge the technical contributions of John Brennan and the scholarship support from Zonta International Foundation.

References

Anderson JD,Jr (1989) Hypersonic and High Temperature Gas Dynamics. McGraw-Hill, New York

Dunn MG (1981) Current studies at Calspan utilizing short-duration flow techniques. In: Treanor CE, Hall JG (eds) Proc. 13th Intl. Symp. on Shock Tubes and Waves, Niagara Falls, State University of New York Press, Albany, pp 32-40

He Y, Morgan RG (1989) Transition of compressible high enthalpy boundary layer flow over a flat plate. Proc. 10th Australasian Fluid Mechanics Conf., Univ. Melbourne, 11.17-11.20. Also to appear in Aeron. J.

Kelly GM, Simmons JM, Paull A (1992) Skin-friction gauge for use in hypervelocity impulse facilities. AIAA J. 30:844-845

Kelly GM (1993) A study of Reynolds analogy in a hypersonic boundary layer using a new skin friction gauge. PhD thesis, Univ. of Queensland, Brisbane, Australia

Van Driest ER (1952) Investigation of laminar boundary layer in compressible fluids using the Crocco method. NACA TN 2597

Optical Studies of the Flow Start-up in Convergent-Divergent Nozzles

Klaus-Otto Opalka
U.S. Army Research Laboratory, Aberdeen Proving Ground, Maryland, USA

Abstract.   Shock tube tests were carried out to investigate the influence of the divergent half angle of three planar convergent-divergent nozzles on the incident-shock formation, the flow start-up period, and on the pressure signature downstream from the nozzle. A diaphragm was mounted in the throat of these nozzles and located at the upstream edge of the test section window of high-quality optical glass through which shadowgraphs and Schlieren pictures could be taken. The objectives of this investigation were to obtain optical records of the flow start-up processes in these divergent nozzles immediately after rupturing the diaphragm, and to facilitate comparisons with hydrocode computations. The test set-up and the results of these experiments are presented and compared with the results of one-dimensional hydrocode computations.

Key words: Nozzle, Flow start-up

1. Background

Convergent-divergent nozzles have been extensively investigated, both in theory and experiment, and a summary of these efforts is presented, e. g., by Amann (1968). More recent investigations could not be found in the literature. In all of the known investigations, the diaphragm separating the high-pressure from the low-pressure region was located far upstream from the nozzle under investigation so that a well-formed shock would enter the nozzle from the upstream side. No previous experimental research is known for the flow start-up process in nozzles which have the diaphragm mounted in the throat. Therefore, the present experimental study was initiated.

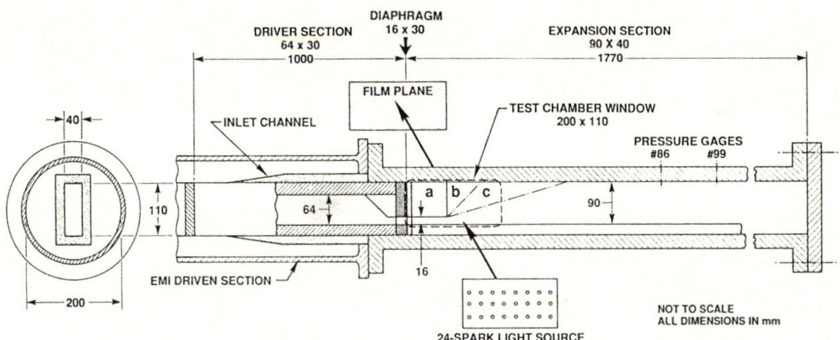

Fig. 1. EMI 200 mm shock tube driver and half-nozzle design

2. Experimental apparatus

The proposed experiments were carried out at the Ernst-Mach-Institut (EMI), Freiburg im Breisgau, Germany (Reichenbach and Opalka 1990). The experimental apparatus at the EMI consists of a 200 mm axisymmetric shock tube which was modified for use in this study, a 24-spark Cranz-Schardin camera with shadowgraph and Schlieren arrangement, and electronic control and

Shock Waves @ Marseille I
Editors: R. Brun, L. Z. Dumitrescu

recording equipment (Fig.1). The diaphragms were composed of two Ultraphan sheets, each 0.1 mm thick, with an ignition wire of 0.08 mm diameter sandwiched inbetween. With this construction, the diaphragm could be burst at an exactly predetermined moment in time by exploding the wire with an electric high-voltage pulse.

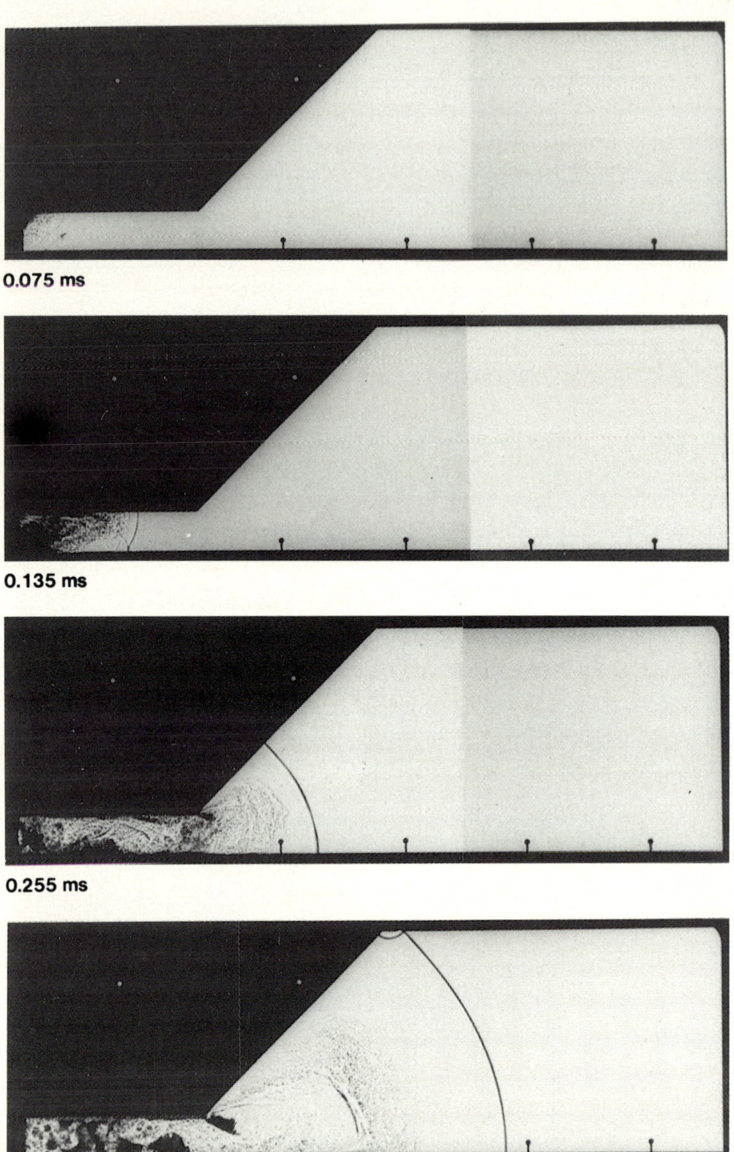

0.075 ms

0.135 ms

0.255 ms

0.385 ms

Fig. 2-1. Early-time shadowgraph sequence for 45° half-nozzle at diaphragm pressure ratio 80

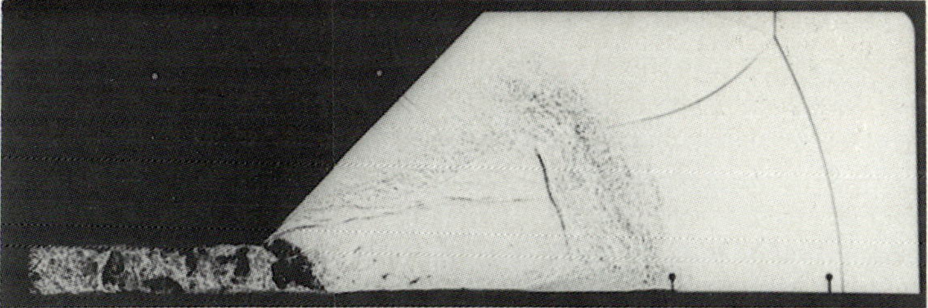

0.503 ms

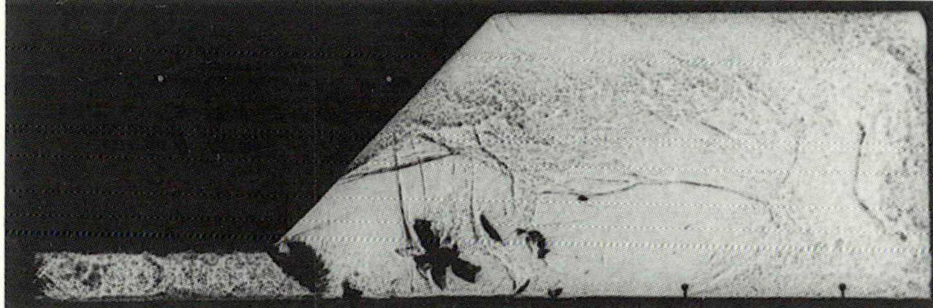

0.803 ms

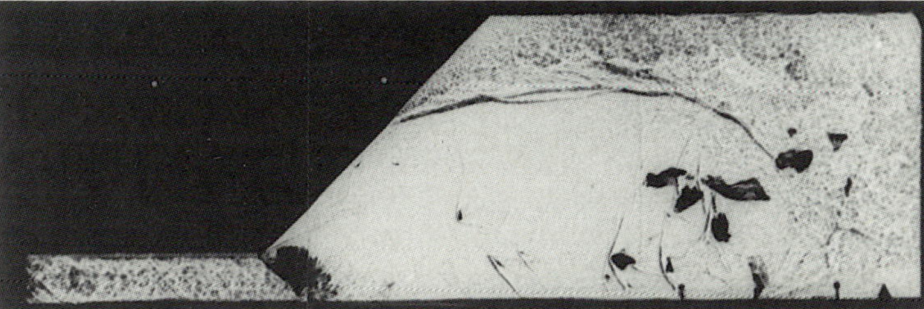

1.103 ms

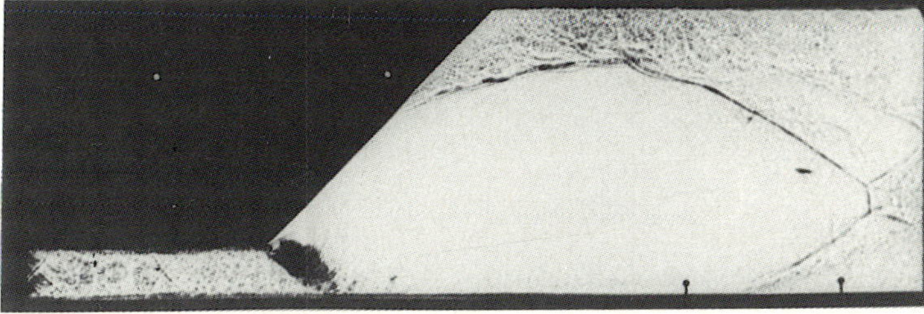

1.463 ms

Fig. 2-2. Late-time shadowgraph sequence for 45° half-nozzle at diaphragm pressure ratio 80

Three planar, divergent half-nozzle models (Fig.1) were built. The convergent section and the throat of the nozzles upstream of the diaphragm are part of the pressure chamber assembly. The planar divergent half-nozzle models are: (a) a throat channel without divergent section, referred to as the 90° divergent half-nozzle; (b) a throat channel and divergent half-nozzle with a 45° angle; and (c) a throat channel and divergent half-nozzle with a 16° angle.

The **initial test conditions** for the three half-nozzle models were diaphragm-pressure ratios of 4, 14, 38, 55, 80, 110, and 188 at room temperature. The models were tested beginning with the lowest specified diaphragm pressure level ($P_{41} = P_4/P_1 = 4$) and commencing with the next highest level ($P_{41} = 14$) up to the highest diaphragm pressure level. The driver operating pressure was 300 kPa (3 bar) above atmosphere for the lowest pressure level ($P_{41} = 4$) and 1300 kPa (13 bar) above atmosphere for all other pressure levels. The driven section was evacuated as necessary until the desired diaphragm-pressure ratio was established.

3. Flow start-up phenomena

After the majority of the optical records had been evaluated, a certain flow start-up pattern became apparent, common to all three nozzles. A sample sequence of shadowgraphs for the 45° divergent half-nozzle at a diaphragm-pressure ratio of 80 (Case N45/080) is shown in Fig.2. It is representative of the flow in the 16° and 90° nozzles. Two photographic test records were combined to show the flow field in and downstream of the half-nozzle at each time step. Flow phenomena like the incident shock, the contact surface, regions of flow separation, and a system of recompression shocks can be seen developing and moving past the observation area.

4. Computational comparison

A systematic comparison of the experimental data of this study with computational results was carried out (Opalka 1991). The experimental test cases were modelled in a quasi one-dimensional (Q1D) context and simulated with the BRL-Q1D code (Opalka and Mark 1986). The computational results for pressure and density were plotted versus distance and time, and the plots were evaluated in the same manner as the shadowgraphs by recording the progressions with time of the incident shock, the contact surface, and various other flow phenomena. The resultant data were entered in $X - t$ diagrams together with the experimental data.

The experimental and computational results for the flow start-up in the 45° half-nozzle at a diaphragm pressure ratio of 80 are compared in Fig.3. The experimental data are plotted as symbols, and the computational data are represented as lines. A careful analysis of the experimental data for the incident shock and the contact surface, reveals that the velocities of the incident shock and of the contact surface initially increase, become constant after a short acceleration phase and then begin to decrease as the flow expands in the divergent nozzle. In the Q1D computational simulation, there is no acceleration phase for the incident shock and contact surface after flow initialization.

5. Experimental results

The **flow start-up period** for the three convergent-divergent half-nozzles is defined as the time from the rupturing of the diaphragm to the establishment of a quasi-steady flow pattern in the nozzle. The data in Fig.4 show that it decreases for all half-nozzles tested with increasing diaphragm pressure ratio and with increasing divergent half-nozzle angle.

The relatively large difference in start-up time for the 90° half-nozzle as compared to the other two half-nozzles can be explained by the fact that the flow start-up period for this nozzle was determined from the data for the throat shocks because the field of view behind the throat exit was too limited to observe the entire start-up process.

The **incident shock formation period** is defined as the time it takes for the incident shock to be well formed. According to the data in Fig.4, the incident shock formation occurs fastest in the 16° half-nozzle and slowest in the 90° half-nozzle. The time difference between the 45° and 90° half-nozzles appears to be approximately 50 μs over the whole range of diaphragm pressure

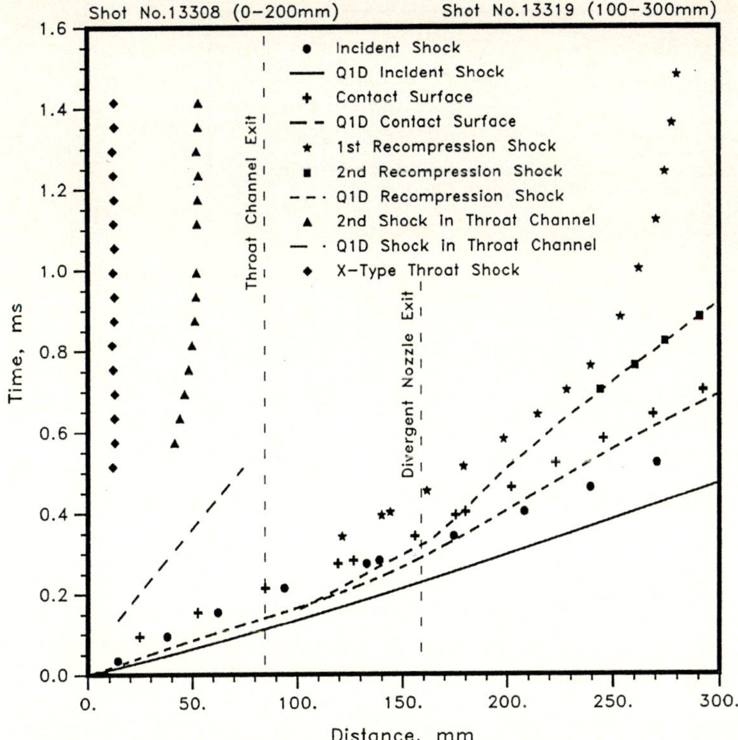

Fig. 3. $X - t$ diagram for 45° half-nozzle at diaphragm pressure ratio 80

ratios. The shock formation distance was found to be constant over the whole pressure range tested for all three nozzle configurations, with statistically significant differences.

6. Conclusions

Three convergent-divergent nozzle configurations were successfully tested at seven diaphragm pressure ratios to study their flow start-up processes. The optical flow records reveal that the flow is very sensitive to local disturbances, e.g., flow asymmetries caused by the breaking of the diaphragm, the passage of the diaphragm fragments through the flow field, and the boundary-layer build-up during the start-up period of the nozzles. From the observations and the analysis of the available data, the following conclusions were drawn:

a) The 90° nozzle effectively forms a divergent nozzle by establishing its particular flow expansion angle in response to the diaphragm pressure ratio. The expansion angles are close to those of the 45° half-nozzle.

b) The shock formation period ranges from 440 ± 50 μs at the low end of the pressure scale to $250\pm$ μs at the full scale, for all half-nozzles. The shock formation distance appears to be constant over the entire pressure range with a mean value of 157 mm (± 16 mm).

c) The start-up period of the 45° half-nozzle ranges from 2.55 ms at the lowest pressure level to 1.12 ms at the highest pressure level. The results on the start-up period of the other half-nozzles are incomplete.

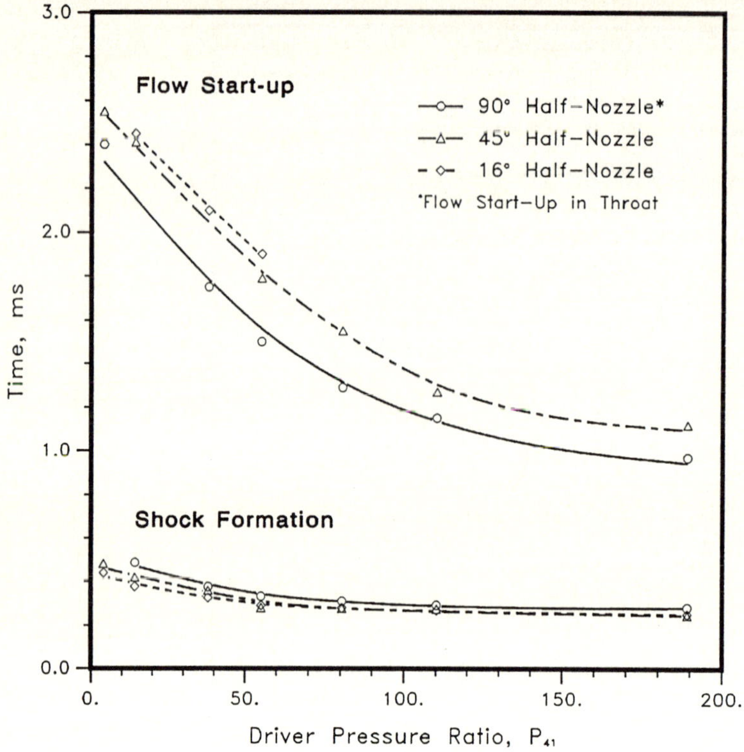

Fig. 4. Incident shock formation and flow start-up periods

d) The BRL-Q1D code proved to be an inappropriate tool for studying the viscous flow phenomena of the start-up process in these nozzles. It remains a useful engineering tool, however, for generating zeroth-order estimates of such flow phenomena and for parametric design studies in shock tubes and tunnels.

References

Amann H-O (1968) Vorgaenge beim Start einer ebenen Reflexions-duese (Flow Start-up Phenomena in a Planar Reflection Nozzle). Bericht Nr. 9/68, Ernst-Mach-Institut, Freiburg/Breisgau, Germany

Opalka KO, Mark A (1986) The BRL-Q1D code: A tool for the numerical simulation of flows in shock tubes with variable cross-sectional areas. BRL-TR-2763, U.S. Army Ballistic Research Laboratory, Aberdeen Proving Ground, Maryland

Opalka KO (1991) Optical studies of the flow start-up processes in four convergent-divergent nozzles. BRL-TR-3215, U.S. Army Ballistic Research Laboratory, Aberdeen Proving Ground, Maryland

Reichenbach H, Opalka KO (1990) An optical study of the flow start-up process in four convergent-divergent nozzles. EMI-Report E 3/90, Fraunhofer Gesellschaft, Ernst-Mach-Institut, Freiburg/Breisgau, Germany.

Two Electric Discharge Methods for Visualizing Three - Dimensional Shock Shapes around Hypersonic Vehicles

Masatomi Nishio

Dept.of Mechanical Engineering, Fukuyama University, Fukuyama 729, Japan

Abstract.   The principles of two electric discharge methods for flow visualization are described. Visualizations of shock shapes around hypersonic vehicles have been performed successfully by using these two electric discharge methods. As an example of the first electric discharge method, the three-dimensional shock shape around a diamond cone in a hypersonic flow has been visualized. Furthermore, the cross-sectional shock shape over a wedge in a hypersonic flow has been visualized by using the second electric discharge method.

Key words: Shock waves, Flow visualisation, Electric Discharge Methods

1. Introduction

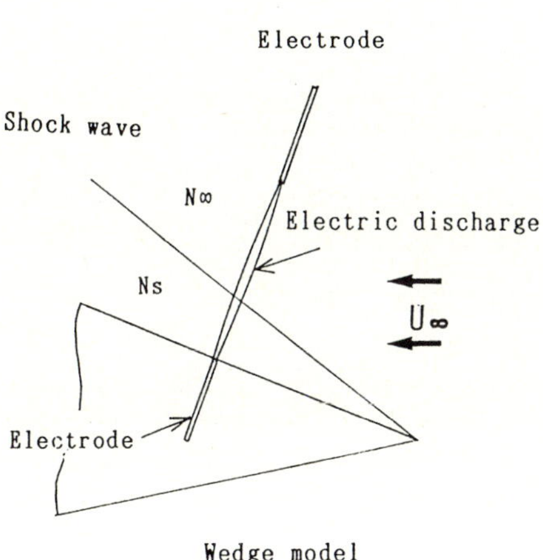

Fig. 1. Illustration of an electrical discharge crossing a shock wave

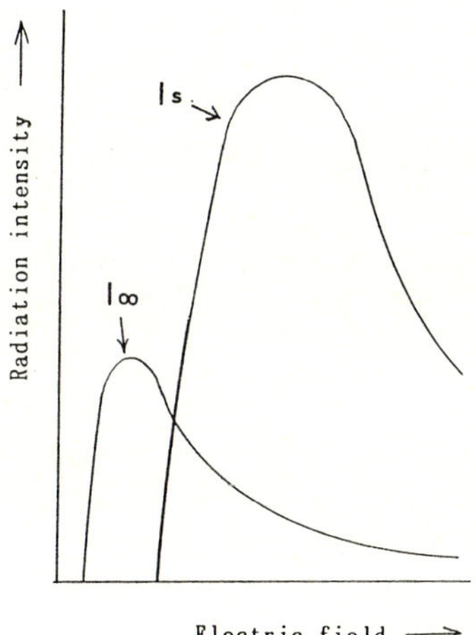

Fig. 2. Radiation intensities I_∞ and I_s from two regions, one in the freestream and the other in the shock layer, vs. the electric field (I_∞: freestream; I_s: shock layer)

The visualization of three-dimensional shock shapes around hypersonic vehicles is very important for understanding the flowfield around the vehicles. However, there are few viable methods for visualizing three-dimensional shock shapes. Optical systems, such as the Schlieren method &c. are not useful for visualizing three-dimensional shock shapes. The electron beam method is available for visualizing three-dimensional shock shapes. However, it is difficult to visualize shock waves whose gas densities are not extremely low.

Recently, two electric discharge methods for visualizing three-dimensional shock shapes around hypersonic vehicles have been developed by M.Nishio (1990, 1992).

The concept of the first electric discharge method is as follows: When an electric discharge is generated across a shock wave, the radiation intensity from the electric discharge column in

Shock Waves @ Marseille I
Editors: R. Brun, L. Z. Dumitrescu

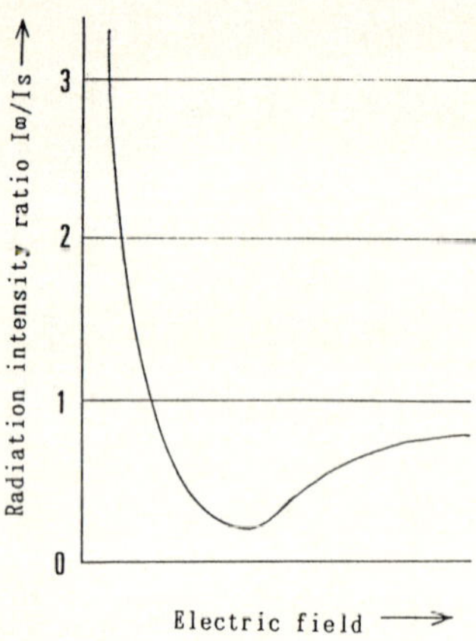

Fig. 3. Radiation intensity ratio I_∞/I_s vs. electric field

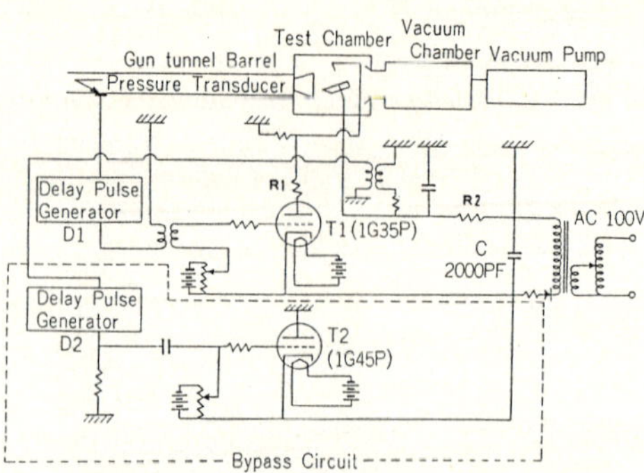

Fig. 4. Electric discharge circuit

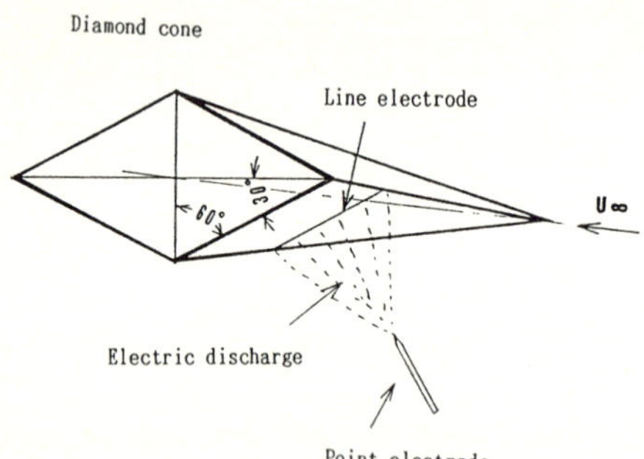

Fig. 5. Arrangement of diamond cone and electrodes

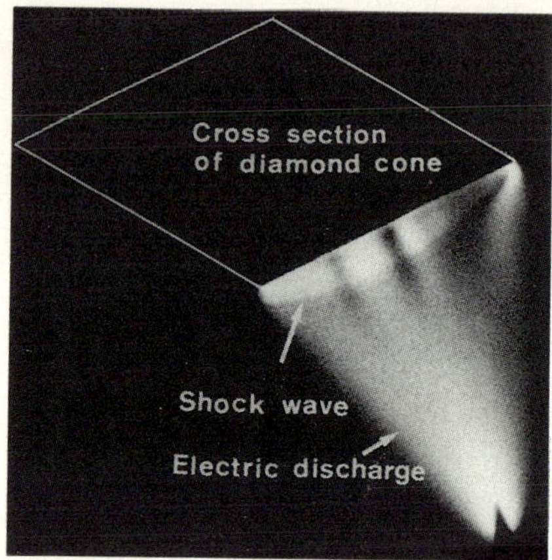

Fig. 6. Visualization of shock wave over diamond cone in hypersonic flow

the shock layer is different from the radiation intensity from the electric discharge column in the freestream. Therefore, three-dimensional shock shapes can be observed by taking a photograph of this electric discharge column in the rear direction of the flow.

The concept of the second electric discharge method is as follows: When an electric discharge is generated across a shock wave, a dark zone at the shock position in the electric discharge can be seen. Three-dimensional shock shapes can be visualized by taking a photograph of the discharge in the rear direction of the flow.

In this paper, three-dimensional shock shapes around hypersonic vehicles are successfully demonstrated by using these two electric discharge methods.

2. The First Electric Discharge Method

The principle of the first electric discharge method is as follows: As illustrated in Fig.1, when an electric discharge is generated across a shock wave, the radiation intensities from the two regions, one in the freestream and the other in the shock layer, are different from each other, because of the gas density difference. The radiation intensity I_∞ from the freestream and the radiation intensity I_s from the shock layer vs. the electric field, vary as shown in Fig.2. Consequently, the radiation intensity ratio I_∞/I_s, namely, the radiation contrast, varies as shown in Fig.3. Judging from this figure, we can select a suitable experimental condition for visualizing shock shapes. Therefore, we can visualize three-dimensional shock shapes by taking a photograph of the electric discharge.

As an example of the first electric discharge method, the shock shape around a diamond cone is visualized. The characteristics of the hypersonic tunnel used in these experiments are as follows: Mach number = 10, freestream density = 0.004 kg/m^3, and duration of freestream = 0.01 sec. The test gas is air. The electric discharge circuit used in these experiments is shown in Fig.4. The diamond cone model and the electrodes are shown in Fig.5. The visualized shock shape is indicated in Fig.6.

3. The Second Electric Discharge Method

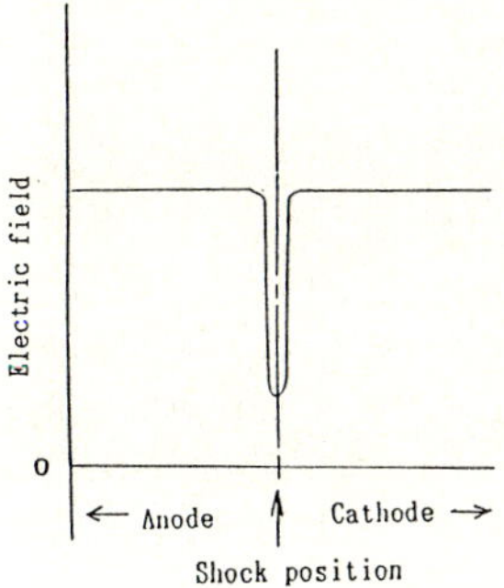

Fig. 7. Electric field near shock wave

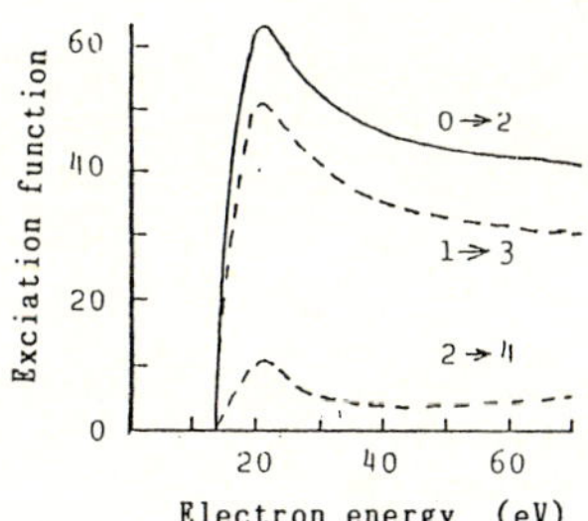

Fig. 8. Excitation function of N_2 vs. electron energy

The principle of the second electric discharge method is as follows: When an electric discharge is generated, as illustrated in Fig.1, we can plot the electric field as shown in Fig.7. As Fig.7 indicates, the electric field at the shock position drops. If the electric field at the shock position is very small, the energy of the electrons traveling from the cathode to the anode becomes very small at that position. Fig.8 indicates the excitation functions of N_2 vs. electron energy. Judging from Fig.8, if the electrons' energy is much smaller than about 14 eV, very few electron excitations would occur. Consequently, it is expected that a dark portion at the intersection of the electric discharge and the shock wave, would appear. Therefore, we can visualize three-dimensional shock shapes by taking a photograph of the electric discharge.

As examples of the second electric discharge method, shock shapes over a wedge are visualized. First, the lateral shock shape is visualized. The arrangement of the model and electrodes is shown in Fig.9. The visualized shock shape is shown in Fig.10. Second, the cross sectional shock shape is visualized. The arrangement of the model and electrodes is shown in Fig.11. The visualized shock shape is indicated in Fig.12.

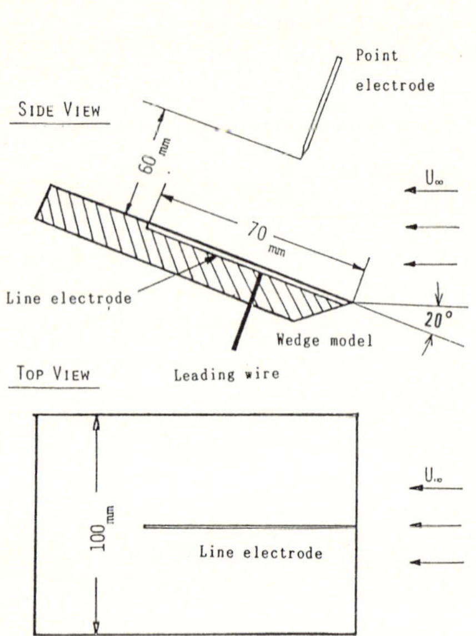

Fig. 9. Arrangement of wedge and electrodes

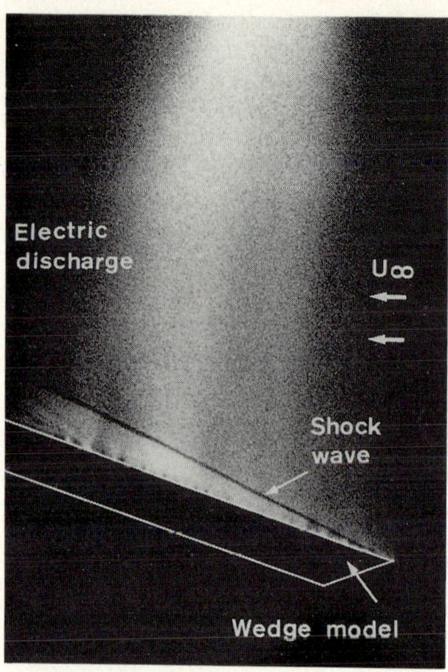

Fig. 10. Lateral shock shape over wedge

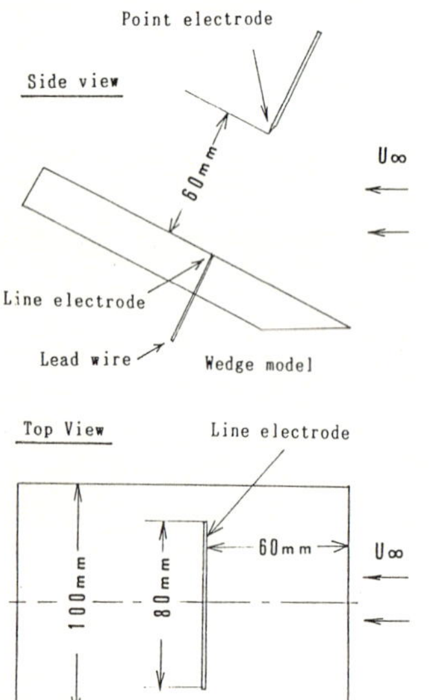

Fig. 11. Arrangement of wedge and electrodes

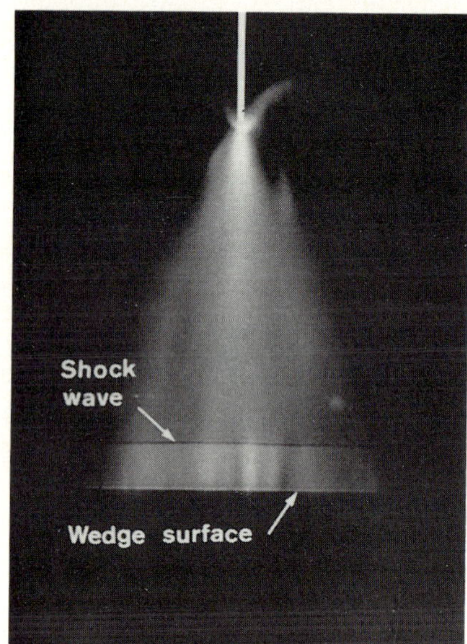

Fig. 12. Cross sectional shock shape over wedge

References

Nishio M (1990) New method for visualizing three-dimensional shock shapes around hypersonic vehicles using an electrical discharge. AIAA Journal 28, 12: 2085-2091

Nishio M (1992) Qualitative model for visualizing shock shapes. AIAA Journal, 30, 9: 2346-2348.

Dual-Laser PLIF Imaging Techniques for Shock Tube Studies of Mixing and Combustion

B.K. McMillin, J.M. Seitzman, J.L. Palmer and R.K. Hanson
Department of Mechanical Engineering, Stanford University, Stanford, CA 94305, USA

Abstract.   Recent advances in the development of planar laser-induced fluorescence diagnostics for imaging of shock-tube-generated combustion flowfields are described, including measurement strategies for (effectively) instantaneous temperature measurements, or alternatively, multiple species imaging. Both techniques are based on the use of two independent, tunable dye laser sources and two time-gated, intensified CCD cameras. In both cases, the lasers illuminate the same planar region but are sequentially pulsed to temporally separate the respective fluorescence signals, and each camera records signal induced by only one of the lasers. Results from nonreacting and reacting jet-in-crossflow experiments are presented, including temperature measurements of fuel-seeded nitric oxide (NO), and simultaneous species imaging of fuel-seeded NO and the combustion-generated hydroxyl radical (OH). The visualizations are used to provide spatially correlated information on the fuel distribution and the location of reaction zones, and the temperature measurements are used to examine the fuel/freestream mixing.

Key words: Planar laser-induced fluorescence, Flow visualization, Temperature measurements

1. Introduction

The use of planar laser-induced fluorescence (PLIF) imaging as a flowfield diagnostic in high enthalpy and high speed combustion flows is becoming increasingly widespread, particularly in shock tube/tunnel facilities where test times are limited (see, for example, Allen et al. 1993, Andresen et al. 1992 and McMillin et al. 1993). PLIF techniques provide a valuable set of diagnostic tools in these types of flows, by combining flow visualization and sensitivity to flowfield properties (species concentration, temperature, pressure and velocity) over a relatively large area, with excellent spatial and temporal resolution. Recent PLIF development work in our lab has been focused on the use of two-laser/two-camera techniques for temporally resolved measurements of temperature or, alternatively, two species' distributions.

In this paper, we briefly describe examples of both types of measurements, in supersonic mixing and combustion flows that were generated within a pressure-driven shock tube. Results presented here include: (1) simultaneous nitric oxide (NO) and hydroxyl radical (OH) visualizations, in which (seeded) NO marks the spatial fuel distribution, and nascent OH marks the reaction zones and entrained combustion gases; and (2) temporally resolved and frame-averaged temperature images of fuel-seeded NO, where the temperature indicates the extent of fuel/freestream mixing. Additional results and details can be found in McMillin (1993) and McMillin et al. (1993).

2. Fluorescence technique

For each fluorescence measurement, one laser is tuned to excite a particular rovibronic transition (originating in the electronic ground state), and the resulting broadband fluorescence from the illuminated plane is collected with an intensified, solid-state camera. In the weak excitation limit, the temporally integrated fluorescence signal, S_f, can be modelled as (Hanson et al. 1990)

$$S_f = C \, B \, E \, \chi_{abs} \, N \, g(\chi_i, N, T) \, f_B(T) \, \Phi(\chi_i, N, T) \tag{1}$$

where C is a constant which depends on the experimental setup; B is the Einstein coefficient for stimulated absorption; E is the laser pulse energy; N is the gas density, g is the overlap integral

Shock Waves @ Marseille I
Editors: R. Brun, L. Z. Dumitrescu

of the absorption and laser lineshapes; T is the gas kinetic/rotational temperature; f_B is the Boltzmann population fraction of the absorbing state; Φ is the fluorescence yield, which depends on the species mole fractions, χ_i, the spontaneous emission (A) and collisional quenching rate coefficients (Q), and is given by $\Phi = A/[A + Q(\chi_i, N, T)]$; and the subscripts '*abs*' and '*i*' refer to the absorbing and *i*th species, respectively.

While the signal generally has a complex dependence on the flowfield properties and composition, with careful transition selection, the signal can be used to (at least qualitatively) visualize the concentration of the absorbing species. In the NO visualizations reported here, for example, the $Q_1 + P_{21}(18.5)$ transition in the $A \leftarrow X(0,0)$ band was excited, providing a fluorescence signal which monotonically decreases with temperature (for $T >\sim 500$ K). The resulting NO fluorescence can therefore be used as a qualitative indication of jet mole fraction, as the cold jet mixes with the hot freestream. For the OH visualizations, the $Q_1(7.5)$ transition in the $A \leftarrow X(1,0)$ band was excited, producing a signal that increases primarily with OH number density and, to a lesser extent, with temperature (due to decreased quenching).

For the temperature measurements, a two-line excitation technique was employed. The essence of the two-line technique is to use the fluorescence ratio, obtained by sequentially exciting two different initial states, to isolate the temperature dependence of the absorbing state populations. By taking the ratio of fluorescence signals, dependence on the number density, absorbing species mole fraction, overlap integral, and collisional quenching is minimized, and consequently, the temperature can be determined, on a pixel-by-pixel basis, from the relative Boltzmann fractions in the ground electronic state. More specifically, the temperature can be determined from the fluorescence ratio, R_{12}, using (see, for example, McMillin et al. 1993)

$$R_{12} = C_{12} \exp\left[- \frac{\Delta\epsilon_{12}}{kT}\right], \tag{2}$$

where $\Delta\epsilon_{12}$ is the rotational energy difference between the initial absorbing states; k is the Boltzmann constant; and C_{12} is a constant that can be determined by independent calibration or from the signal ratio directly (in situ), if the temperature is accurately known at some location within the image. For the temperature measurements described here, the lasers were tuned to the $R_1 + Q_{21}(13.5)$ and $Q_1 + P_{21}(28.5)$ transition pairs in the $A \leftarrow X(0,0)$ band of NO near 226 nm. These transition pairs provided good temperature sensitivity (1531 K) throughout most of the flowfield, where temperatures ranged from 300-1800 K within the region of interest.

3. Experimental details

The experimental facility used in these measurements, shown schematically in Fig.1, is only briefly described here; further details are given in McMillin et al. (1993). The flow studied consists of a sonic, transverse jet in supersonic crossflow, with the injector located downstream of a rearward facing step. The jet stagnation conditions were 3 atm and 300 K, and the jet was composed primarily of hydrogen to simulate a SCRAMJET flowfield. One percent of NO was included in the jet as a fuel marker, and 19% CO was also included to reduce the fluorescence lifetime of NO and thereby minimize the loss in spatial resolution due to motion blurring. The Mach 1.4 freestream conditions were 0.4 atm and 2200 K, with a composition of 21% O_2 in argon in the reacting experiments and 21% CO in argon in the nonreacting experiments. In both cases, the jet-to-crossflow momentum ratio was ~ 1.9, and the jet Reynolds number was $\sim 16,000$, based on the estimated jet/crossflow velocity difference and the observed plume height.

The flows were generated in a pressure-driven shock tube by injecting fuel into the supersonic flow behind an incident shock wave. The test section had a 76×76 mm square cross-section and the fuel was injected through a pulsed solenoid valve with a 2 mm diameter orifice. The valve was triggered so that the jet flow reached steady-state prior to the incident shock arrival, and the

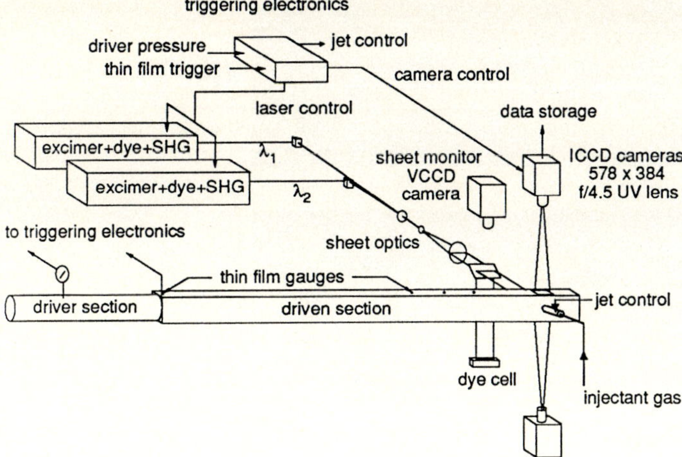

Fig. 1. Schematic of the shock tube and PLIF imaging facility

images were acquired ~ 200 μs after passage of the incident shock, to provide adequate time for the flow to develop.

As illustrated in Fig.1, the frequency-doubled outputs of two XeCl excimer-pumped dye lasers were used as the excitation sources (< 0.5 mJ,$\sim$20 ns, $\sim$0.3 cm^{-1}) in the fluorescence measurements. To maintain temporally separated fluorescence signals, the laser pulses were separated in time by 100 and 250 ns, respectively, for the NO/OH visualizations and NO temperature measurements. The resulting fluorescence decays were imaged onto one of two, intensified, cooled CCD cameras (578×384 array with 23×23 μm pixels) equipped with UV f/4.5 lenses. Images were obtained within two (overlapping) regions, namely $-5 < x/D < 17.5$ and $10 < x/D < 32.5$, with the jet located at $x/D=0$. In all cases, the imaged region was 30×45 mm, but the images have been trimmed to 18×45 mm for presentation. UG-5 and WG-305 Schott glass filters (2 mm thick) were used to block the laser scattering for the NO and OH measurements, respectively. The intensifier gate widths were set to provide $\sim$200 ns and $\sim$60 ns of uniform gating for the NO and OH measurements, respectively.

The raw fluorescence images were corrected for camera dark noise, flat field non-uniform response of the cameras and collection optics, laser energies and spatial distributions, and laser scattering. When necessary during processing, one image was warped using a registration function (including translation, rotation and magnification) to correct for the misalignment of the respective cameras. The processed fluorescence images were software binned 2×2 to improve signal-noise ratio, leading to a nominal resolution of approximately $300 \times 160 \times 160$ μm. However, including the finite resolution of the imaging system and flow motion between laser pulses, the effective spatial resolution was $\sim$0.3-0.5 mm.

4. Results and discussion

An example of simultaneous NO and OH visualizations is shown in Fig.2. As noted above, the NO marks the jet mole fraction, while the OH, formed by the autoignition of jet hydrogen with freestream oxygen, marks the OH number density in the reaction zones and convected combustion gases. In general, the high NO signal in the near field indicates a jet-rich plume, while the lower signal near the wall and in the downstream plume regions indicates a more dilute and, hence,

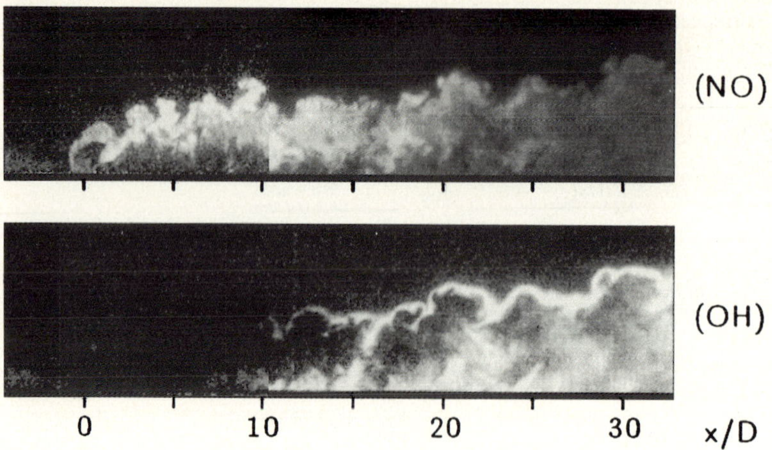

Fig. 2. Temporally resolved, simultaneous NO and OH visualizations. Each image is a composite of two images covering $-5 < x/D < 10$ and $10 < x/D < 32.5$. The images are shown in a logarithmic scale with high signal indicated by pure white

warmer fuel/freestream mixture. The OH is generally found in these hot, lean regions with low NO signal. For example, within the plume OH is usually found in diffuse pockets near the wall for $x/D > 8$, and in a relatively continuous filament along the plume/freestream interface for $x/D > 10$. In the shear layer, the OH signal peaks on the outer edge of the plume, while the NO signal decreases sharply (indicating a sharp increase in temperature and/or a sharp decrease in jet fluid concentration). The large decrease in NO signal coupled with the appearance of strong OH signal suggests that the OH forms on the hot (and lean) side of the jet boundary. Indeed, based on a detection sensitivity analysis of the NO signal, the jet mixture fraction and temperature are estimated to be $\sim$8% and $\sim$2000 K at the location of peak OH signal (at $x/D = 18$). The presence of the OH in the hot and lean shear layer region is consistent with the ignition calculations (McMillin et al. 1993), but differs from a simple turbulent diffusion flame, where the reaction zone is expected closer to the stoichiometric region.

Examples of temporally resolved and frame-averaged NO temperature images from experiments with a nonoxidizing freestream are shown in Fig. 3. In these images, the temperatures are believed to be accurate except within, and immediately downstream of, the barrel shock structure, and within the recirculation zone downstream of the step. In these small regions, the temperatures are not reliable because of fluorescence crosstalk observed between the two cameras. In general, these temperature fields provide an indication of the mixing between the two streams, with the plume temperature increasing as the cool jet mixes with the hot freestream. For example, both images suggest that the mixing within the near wall region is most efficient, with mixture temperatures reaching 1000-1500 K within a few jet diameters. Both images also indicate that the core mixes somewhat less efficiently and that the plume reaches a relatively uniform temperature in about 15 jet diameters. As expected, though, only the temporally resolved image shows the details of the turbulent nature of the plume, including the large vortical structures present along the jet/freestream interface, i.e., the alternating pockets of freestream and jet fluid along the plume boundary.

The mixing can be examined more closely by plotting the frame-averaged, axial temperature distribution of the plume. We typically use the minimum plume temperature to indicate the degree of mixing, since this provides a measure of the maximum jet molar fraction and hence,

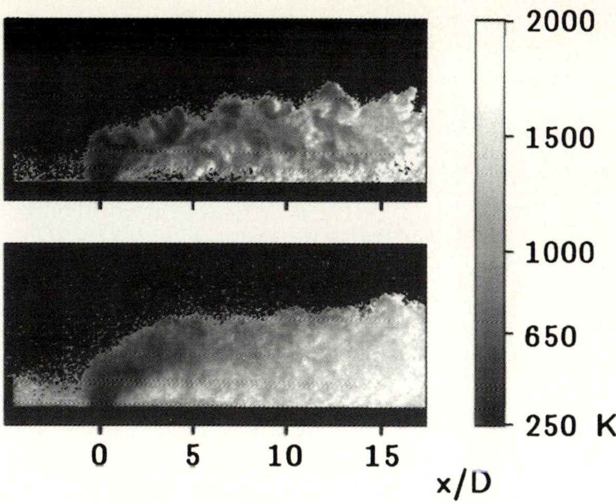

Fig. 3. Temporally resolved and 8-frame-averaged NO temperature images of a jet in a nonoxidizing supersonic crossflow. Each image is trimmed to 18×45 mm and includes about 18 diameters of plume

the least-mixed jet fluid at a given axial location. We normalize the minimum plume temperature and relate it to the jet mole fraction using a simple one-dimensional enthalpy balance given by McMillin (1993):

$$\theta_{min} \equiv \frac{\left(T_{mix,min} - T_\infty\right)}{\left(T_{mix,min} - T_\infty\right) + \frac{C_{p,jet}}{C_{p,\infty}}\left(T_{jet} - T_{mix,min}\right)} \approx \chi_{jet,max} \tag{3}$$

where θ_{min} and $T_{mix,min}$ are the minimum normalized and measured plume temperatures, respectively; C_p is the specific heat; $\chi_{jet,max}$ is the corresponding maximum jet molar fraction; and the subscripts 'jet' and '∞' refer to the initial jet and freestream values, respectively.

The approximate relationship, $\theta_{min} \approx \chi_{jet,max}$, is used here to emphasize that the temperature is not a perfectly conserved scalar in this flowfield; total enthalpy is. Typically, a plot of the minimum temperature distribution is most useful for comparing the mixing observed at different flow conditions or in different geometries. Fig.4 shows an example of a comparison of the distribution from the average temperature field described above and one at the same conditions for a flush wall geometry. In comparing the two cases, the results from the step wall geometry show a lower minimum plume temperature for all downstream locations, which indicates that the mixing rate for that geometry is higher, just as observed in previous cold-flow mixing investigations. Additional results discussed by McMillin (1993) also indicate that, in a given geometry, the jet/freestream mixing is more efficient at lower jet-to-crossflow momentum ratios, which is also consistent with previous investigations.

5. Concluding remarks

The application of two-laser/two-camera PLIF imaging for either NO temperature measurements or simultaneous NO/OH visualizations of shock-tube mixing and combustion flows has been briefly described in this paper. The results illustrate the promise both techniques offer as diagnostic tools for improving our understanding of complex combustion flowfields, by providing

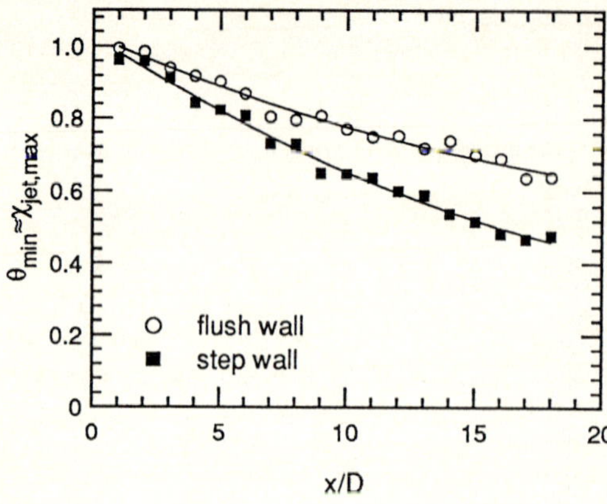

Fig. 4. Comparison of the non-dimensionalized minimum plume temperature as a function of downstream distance for both a flush wall and rearward facing step geometry, at a jet-to-crossflow momentum ratio of ~1.9. The comparison qualitatively indicates the relative rates of mixing for the two configurations

species-specific, spatially and temporally resolved information over a relatively large area. Ongoing research is being directed toward the application of OH temperature measurements and PLIF velocity measurements of NO and OH in high enthalpy and high speed combustion flows.

Acknowledgement

This research was supported by the Air Force Office of Scientific Research, Aerospace Sciences Directorate, with Dr. Julian Tishkoff as Technical Monitor.

References

Allen MG, Cronin JF, Davis SJ, Foutter RR, Parker TE, Reinecke WG, Sonnenfroh DM (1993) PLIF imaging measurements compared to model calculations in high-temperature Mach 3 airflow over a sphere. AIAA 31st Aerospace Sciences Meeting, Reno, NV (January 11-14), AIAA Paper 93-0092

Andresen P, Beck WH, Eitelberg G, Hippler H, McIntyre TJ, Riedl A, Seelemann T, Troe J (1992) A laser induced fluorescence system for the high enthalpy shock tunnel (HEG) in Göttingen. In: Takayama K (ed) Shock Waves, Proceedings of the 18th International Symposium on Shock Waves, Vol. I, Springer-Verlag Berlin Heidelberg, pp. 657-662

Hanson RK, Seitzman JM, Paul PH (1990) Planar laser-fluorescence imaging of combustion gases. Appl. Phys. B 50:441-454

McMillin BK (1993) Instantaneous two-line PLIF temperature imaging of nitric oxide in supersonic mixing and combustion flowfields. Ph.D. thesis, Dept. of Mechanical Engineering, Stanford University, Stanford, CA

McMillin BK, Seitzman JM, Hanson RK (1993) Comparison of NO and OH PLIF temperature measurements in a SCRAMJET model flowfield. AIAA/SAE/ASME/ASEE 29th Joint Propulsion Conference, Monterey, CA (June 28-30), AIAA Paper 93-2035

An Examination of the Aachen Shock Tunnel TH2 Gas Flows Using the HEG PLIF Apparatus

W.H. Beck*, **M. Scheer*** and **M. Vetter**[†]
*Institute for Experimental Fluid Mechanics, DLR, Göttingen, FRG
[†]Shock Wave Laboratory, RWTH Aachen, FRG

Abstract. LIF measurements on TH2 at 3 MJ kg^{-1}/Mach 8.4 conditions are analysed for the freestream and cylinder bow shock gas flows. A freestream NO rotational temperature and concentration were determined. Results show fair agreement with theory, although the poor S/N ratios in the measurements led to large measurement uncertainties. Although a quantitative analysis of the shock-heated gases was not possible at this stage, certain trends and features could be discussed.

Key words: LIF, High enthalpy flow, Shock tunnel, NO

1. Introduction

The influence of real gas effects on the external aerodynamics of space vehicles re-entering a planetary atmosphere is to be examined in the high-enthalpy shock tunnel in Göttingen (HEG) with the aid of (planar) laser induced fluorescence PLIF. A PLIF-apparatus has been built and has been undergoing a series of tests over the past year on other facilities (vacuum wind tunnel V2G at the DLR in Göttingen (Beck et al. 1992), plasma-heated wind tunnel LBK at the DLR in Cologne (Beck et al. 1993) and shock tunnel TH2 at the RWTH in Aachen), in preparation for its implementation at HEG.

This paper will report the results of tests carried out in the shock tunnel TH2 at the RWTH in Aachen. Here measurements were carried out to determine NO rotational temperature and concentration in the test section freestream flow only; in some experiments 2D fluorescence images of the bow shock in front of a cylinder placed transversely in the flow were also recorded. A major purpose of these measurements was to examine the potential difficulties which may arise when applying the PLIF technique to HEG.

2. Experimental setup

The Aachen shock tunnel TH2 has been extensively used in the field of hypersonic research since its re-commissioning in 1987. The tunnel consists of a shock tube, a conical or contoured nozzle and a dump tank with the test section (see Vetter 1993). The shock tube, of 140 mm ID, is divided by a double diaphragm station into a high pressure section (length 6 m) and a low pressure section (15.6 m). Usually helium is used as the driver gas and air as the test (driven) gas. The obtainable Mach numbers in the test section vary between 6 and 12, depending mainly on the area ratio of the nozzle.

For the present series of runs on TH2, He gas driver pressures around 36 MPa led to reservoir conditions with test gas air of $p_0 \approx 17$ MPa and $T_0 \approx 2600$ K (enthalpy $h_0 \approx 3$ MJ kg^{-1}) and the following typical freestream conditions (calculated with non-equilibrium - see Ritzerfeld 1993): $p_\infty \approx 1.3$ kPa, $T_\infty = 180/200/270$ K (frozen/non-equilibrium/equilibrium), $v_\infty \approx 2390$ m.s^{-1}, $M_\infty \approx 8.4$, $[NO]_\infty \approx 1.5 \times 10^{16}/0.5 \times 10^{16}$ cm^{-3} (frozen/non-equilibrium).

For most of the shots, a cylinder with diameter 110 mm and length 300 mm was placed in the test section (with its axis perpendicular to the flow), so that NO fluorescence images of the region between bow shock and cylinder could also be recorded.

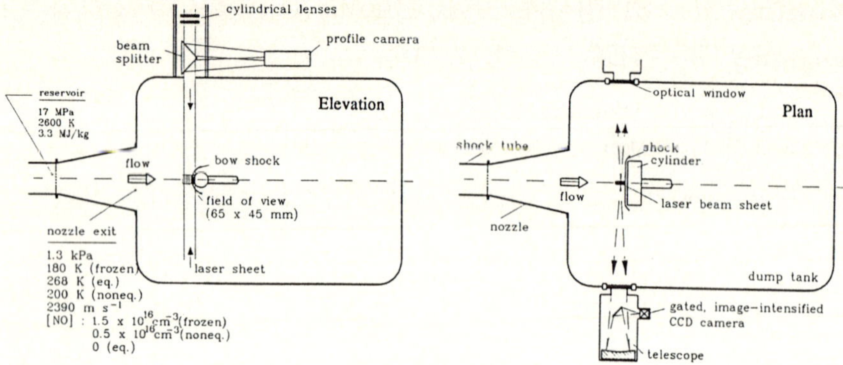

Fig. 1. Block diagram of experimental arrangement for LIF on TH2. Typical nozzle exit (freestream) conditions are also shown

A block diagram of the experimental arrangement for LIF on TH2 is given in Fig.1. The LIF-apparatus uses two excimer lasers (ArF, 193 nm) which are tuned to two NO transitions $(B(7) \leftarrow X(0)$, laser 1 on the $P_2(23.5)$ line and laser 2 on $P_1(27.5))$ and formed to sheets which counter-propagate through the test section to overlap in front of the test model in the flow. The lasers are fired consecutively (1 μs apart) after an appropriate delay, allowing for the flow in the test section to be properly established, and the fluorescence signals are captured in 2D on image intensified, gated CCD cameras via Maksutov-Newton telescopes with a high light gathering power.

Since the laser outputs vary from shot to shot - we have observed beam intensity and profile variations - and since the frequencies may drift, it is necessary to monitor the beams during the test run using a separate profile camera and a frequency monitoring system comprising an NO calibration cell with an OMA (optical multichannel analyser). Further to this, reference measurements using a separate NO test cell placed in the tunnel test section need to be performed. All of these fluorescence and beam images are required to determine NO concentration and temperature in the freestream (in the present case, averaged over a 2D area). The HEG LIF apparatus is described in greater detail elsewhere (Scheer 1994).

3. Results

3.1. Freestream

Of the 8 shots carried out, 4 delivered NO fluorescence images which were usable for an analysis of a rotational temperature in the freestream gas flow. In the last shot carried out (shot 8), the cylinder was removed from the field of view, so that only the freestream fluorescence signals were captured. Fig.2 shows three images for shot 8, all taken during the shot: 1. NO fluorescence from laser 1; 2. the laser beam profiles; 3. the spectrally resolved fluorescence spectrum for both lasers. The signal/noise ratio (S/N) for this shot was poor; it was only discovered after completion of the test campaign that the performance of the optical coatings on the Maksutov-Newton telescopes had degraded in the uv to such an extent that intensity losses of up to a factor of 10 had occurred. In early shots (up to shot 5), there were problems with the synchronisation of timing events in the image capturing system - images were being recorded during a frame transfer of the CCD camera, leading to images with falsified intensities.

In spite of the above-mentioned problems, an attempt was still made to analyse the results. The 2D images were averaged to give a mean value for the fluorescence intensity over the region probed by the laser beams and corrected using the OMA and beam profile measurements. These

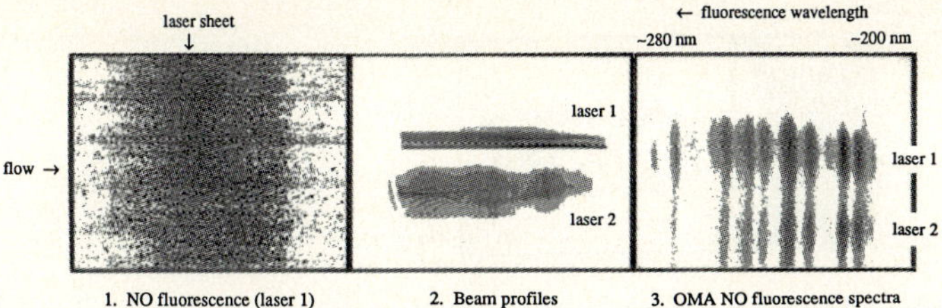

Fig. 2. LIF in TH2 freestream gas flow (shot 8): 1. NO fluorescence image using laser 1; 2. Beam profiles for lasers 1 and 2 (seen end-on); 3. Spectrally resolved fluorescence spectra for lasers 1 and 2, taken during the run using the OMA with the NO calibration cell

average signals, their fluctuations and the derived NO rotational temperatures are summarised in the following Table:

Shot	S/N		Rot. temp.	Comments
number	Laser 1	Laser 2	(K)	
8	25/12	99/48	310	Poor S/N, no image
6	17/7	114/40	220	transfer problems
5	1792/94	3316/222	250	Good S/N, image
4	568/123	505/116	180	transfer problems

These shots were all carried out at nominally the same conditions. An average value for the temperature is 240 ± 60 K, whereby this large standard deviation is due to the problems discussed before. Shot 8 could also be analysed for freestream NO concentration [NO], using calibration measurements carried out in a test cell containing NO placed in the TH2 test section. Appropriate estimations were made for allowance for quenching (Beck et al. 1993). Experimental temperatures and [NO] are compared with theoretical values, calculated for an inviscid, 1D nozzle flow for TH2 (Ritzerfeld 1993), in the following Table; as can be seen, the error in the experimental results, and the small number of results available, do not permit a convincing distinction between calculated equilibrium and non-equilibrium flow to be made at this stage.

	Theory			Experiment		
	Equilibrium	Non-equil.	Frozen			
[NO]					no quench	quench
(cm^{-3})	~ 0	$\sim 0.5 \times 10^{16}$	$\sim 1.5 \times 10^{16}$	line 1	1.8×10^{16}	1.5×10^{16}
				line 2	1.9×10^{16}	1.6×10^{16}
Temp.					310	
(K)	270	200	180		240 (mean)	

3.2 Cylinder bow shock

Fig.3 shows two NO fluorescence images from lasers 1 and 2 and a laser beam profile for shot 6. Whereas image 1 from laser 1 is within the dynamic range of the recording system (235 counts/pixel average), image 2 is saturated. These intensity differences between images 1 and 2 are *not* mainly due to the problems mentioned before. At the high temperatures behind the shock (over 2000 K), the first excited vibrational level of NO is also populated, so that the $D(0) \leftarrow X(1)$ transitions, which have transition moments over two orders of magnitude greater than for $B(7) \leftarrow X(0)$ (Gundlach 1992, and references therein), dominate the spectral range covered by the ArF lasers ($\sim 193 - 194$ nm). There are two near overlaps of $D \leftarrow X$ and $B \leftarrow X$

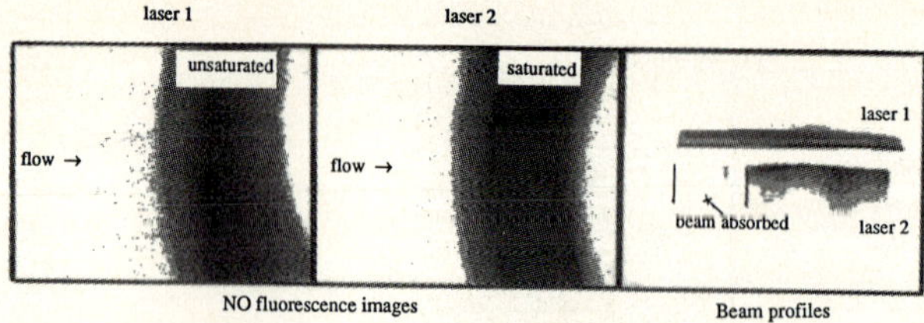

Fig. 3. LIF in TH2 bow shock flow (shot 6). NO fluorescence images and beam profiles for lasers 1 and 2 are shown (note that the grey scales are different). NO image 2 can be seen to be saturated (dark structureless region in the centre). Absorption of profile 2 can be clearly seen

lines - for laser 1, $(D - X)$ $P_2(45.5)$ with $(B - X)$ $P_2(23.5)$, and for laser 2, $(D - X)$ $Q_2(36.5)$ with $(B - X)$ $P_1(27.5)$. However, for laser 1, the transition arises from a very high rotational level $(K = 45.5)$, which is substantially less populated than the level $K = 36.5$ which is accessed by laser 2. For this reason, image 2 has become saturated.

This points to the need for understanding the high temperature spectroscopy of NO to guide one's choice intelligently to suitable transition lines for the shocked gas region. (This study, however, concentrated on the cold freestream conditions, where only $B \leftarrow X$ transitions can be seen.) Work is underway in the testing of a specially constructed heated cell which will permit 2D spatially resolved measurements of NO to be carried out at temperatures up to 1300 K. This cell has two purposes; first, to measure high-temperature NO excitation and fluorescence spectra, and secondly, to be placed in the HEG test section to calibrate the fluorescence detection system for NO concentration measurements.

The beam profile for laser 2 is recorded *after* the beam has passed through the test section. (This is why the beam profiles in Fig.3 are no longer sheets; the lasers have to be focussed slightly to produce the narrowest waist in front of the cylinder, so that they have diverged again before being recorded.) It can be seen in Fig.3 that that part of the laser 2 beam profile for shot 6 which has passed through the shock-heated region in front of the cylinder no longer reaches the profile detection system, i.e. the beam has been absorbed ! This imposes a further restriction on the choice of suitable $D \leftarrow X$ transitions for examining the shocked gas - the strongest lines cannot be used.

A Schlieren picture of the shock was also recorded in a separate experiment (Kleine 1992); it is compared in Fig.4 with an NO fluorescence image taken from shot 6. From these pictures, it can be seen that both techniques agree that the shock stand-off distance $\delta \approx 21$ mm. Using an empirical correlation for δ/R (R = cylinder radius = 55 mm), given by Anderson (1989) for inviscid ideal gas flows, one obtains a slightly larger value of $\delta = 23$ mm. It is not possible to say at this stage whether or not the smaller measured δ is due to real gas effects - at the rather low enthalpies used here (~ 3 MJ kg^{-1}), these effects should not be large.

Whereas the shock front in the Schlieren picture is sharp, the fluorescence intensity rises slowly to its maximum value - this is shown in Fig.4 in the intensity profile along the stagnation line (lower trace). This slow rise can be due to a process of internal relaxation of the NO. The profile in Fig.4 shows that the fluorescence intensity rises over a distance of about 5 mm behind where the shock front is known to be (Schlieren). 1D inviscid calculations for conditions along the stagnation line between shock and cylinder (Ritzerfeld 1993) show that the gas velocity drops from ~ 420 m.s^{-1} at the shock front to ~ 300 m.s^{-1} after 5 mm. Taking an average velocity of 360 m.s^{-1} yields a $1/e$ rise time of about 5 μs. Are there internal relaxation processes which can occur over such a long time?

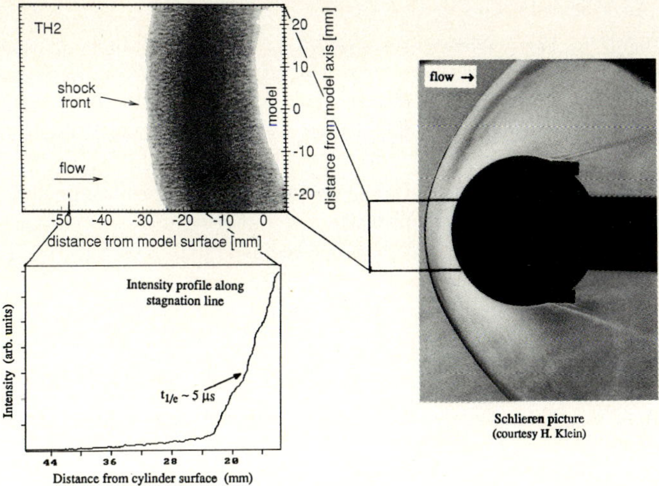

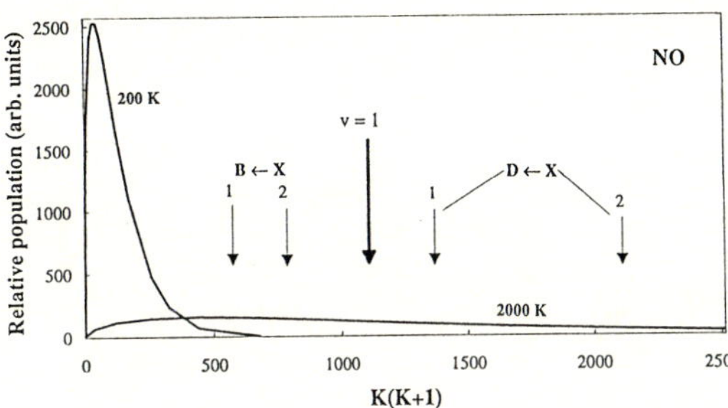

Fig. 4. Schlieren picture of cylinder flow, compared with NO fluorescence image from laser 1 shot 6. An intensity profile along the stagnation line (lower left) shows a fluorescent intensity risetime of $t_{1/e} \sim 5\ \mu s$

Calculations of translational and vibrational temperatures along the stagnation line (Ritzerfeld 1993) show that all $T_{\text{(vib)}}$ lag behind $T_{\text{(trans)}}$, in descending order N_2, NO and O_2, as one would expect from the fundamental vibrational frequencies of these species (2360, 1904 and 1580 cm^{-1} - see Herzberg 1950) - vibrational relaxation times of tens of microseconds are not unusual (Gaydon and Hurle 1963). For fluorescence images which arise from $D(0) - X(1)$ transitions (see image 2 of Fig.3), it is easy to explain such a relaxation distance. However, this distance was also seen in largely $B(7) - X(0)$ transitions (image 1 in Fig.3), where changes in rotational but not vibrational quantum number take place.

Fig. 5. Boltzmann rotational population distributions for NO for two temperatures 200 and 2000 K, plotted against rotational energy ($\propto K(K+1)$, K = rotational quantum number). Also shown are the position of the first excited vibrational energy level $v = 1$, and the locations of the rotational levels from which the transitions for lasers 1 and 2 and for $D \leftarrow X$ and $B \leftarrow X$ transitions occur

Rotational relaxation times are usually much smaller; typical values for N_2 and O_2 from ultrasonic measurements at 1 bar/300 K, extrapolated to 2600 K for TH2 conditions, are around 0.01 μs (Gaydon and Hurle 1963). However, these times may not apply to very high rotational levels where the level spacing grows with quantum number K - no experimental values are known here.

Fig.5 shows a plot of rotational population distribution versus $K(K+1)$ (which is proportional to energy) for NO. Also shown are the positions corresponding to the $B - X$ and the $D - X$

transitions for lasers 1 and 2, as discussed before, and the position of the first vibrationally excited state $v = 1$. When the shock arrives, the population distribution for 200 K must adjust to that for 2000 K; this happens quickly for low K's, but there may be a lag for higher K's. It would be revealing to carry out LIF measurements under TH2 conditions arising from low K states, or, as in HEG where the freestream flow is also hot (~ 700 K), to look at $D - X$ transitions before and after the shock front.

The presence of a shock front leads to a change of properties whose rate depends on a characteristic relaxation time. Although properties such as density relax after only 1 or 2 collisions, leading to sharp images of the shock in interferometric techniques, observation methods such as LIF that probe internal energy modes do not necessarily show sharp shock fronts. There may be a further complication, though; since there is a relaxation region behind the shock, then the population distributions here may not be Boltzmann, so that an internal temperature may not even be defined! This would need to be tested in a facility such as HEG by probing different rovibrational levels in a series of shots with the "same" conditions.

Acknowledgements

The authors would like to thank the following people for their assistance/useful discussions: W. Sader, R. Krek, G. Gundlach (DLR Göttingen); J. Schulte-Rödding, E. Ritzerfeld, H. Kleine (RWTH Aachen). The financial support of the ESA through CNES is also gratefully acknowledged.

References

Anderson JD (1989) Hypersonic and high temperature gas dynamics. McGraw-Hill.

Andresen P, Beck WH, Eitelberg G, Hippler H, McIntyre TJ, Riedl A, Seelemann T, Troe J (1992) A laser induced fluorescence system for the high enthalpy shock tunnel (HEG) Göttingen. In: Takayama K (ed) Proceedings of the 18th International Symposium on Shock Waves, Springer, pp 657-662

Beck WH, Dankert C, Eitelberg G, Gundlach G (1992) Preliminary laser induced fluorescence measurements in several facilities in preparation for application to studies in the high enthalpy shock tunnel Göttingen. Reno, NV, AIAA Paper 92-0143

Beck WH, Koch U, Müller M (1993) Spectroscopic diagnostic techniques for the high enthalpy shock tunnel in Göttingen (HEG): preparatory LIF studies on other facilities. In: Boutier A (ed) New Trends in Instrumentation for Hypersonic Research, Dordrecht, Kluwer, pp 215-224

Gaydon AG, Hurle IR (1963) The shock tube in high-temperature chemical physics. Chapman and Hall, London

Gundlach G (1992) Simulation of laser-induced fluorescence excitation spectra of nitric oxide. DLR internal report DLR-IB 222-92 A 27

Herzberg G (1950) Spectra of diatomic molecules. Van Nostrand Reinhold, NY

Klein H (1992) Private communication

McMillin BK, Lee MP, Hanson RK (1990) Planar laser-induced fluorescence imaging of nitric oxide in shock tube flows with vibrational nonequilibrium. AIAA Paper 90-1519, Seattle WA

Ritzerfeld E (1993) Stationary nonequilibrium flows. Dissertation, RWTH Aachen

Scheer M (1994) Temperature and density measurements on hypersonic wind tunnels using laser-induced fluorescence of NO, including a study of NO spectroscopy in the range 192.8-194.2 nm using a novel heated cell. diplom thesis, DLR/University Göttingen

Vetter M (1993) Hypersonic flows around models in a shock tunnel. Dissertation, RWTH Aachen

Part 4: Numerical Computations

Comparison of Numerical Methods: DSMC Simulations and N-S Predictions on Bluff Bodies

Martin Gilmore

Hypersonic Aerodynamics, DAR, Y25 Bldg, Farnborough, Hampshire, GU14 6TD, UK

Abstract. This paper compares predictions obtained using the Direct Simulation Monte Carlo (DSMC) technique to solve the Boltzmann equation and a Finite Volume technique to solve the Navier-Stokes equations for the flow around a bluff body in a hypersonic regime. Good agreement is obtained between the two approaches for computations of the flowfield around the Marsnet Semi-Hard Lander configuration travelling at Mach 30 at a simulated Earth altitude of 71 km. Predictions of density, temperature and heat transfer are presented.

Key words: Hypersonic flow, Numerical methods, Direct-simulation Monte Carlo, Navier-Stokes

1. Introduction

In the flowfield around bluff vehicles travelling at hypersonic velocities, regions of high and low density may exist (e.g. ahead of a blunt nose and in a planar base region). Due to the differing nature of the two types of flow, different mathematical descriptions must be used in each region to provide a full flowfield simulation. At present no single scheme encompassing both Navier-Stokes (N-S) and Direct Simulation Monte Carlo (DSMC) methodologies is available. The eventual aim of the reported work is to produce a computational scheme capable of utilising both the Navier-Stokes and DSMC methods to provide predictions of the properties in flowfields where both high and low density regions exist.

Before any attempt to link the two methods in a single computational scheme can be made, confidence has to be gained in the ability of the two individual approaches to produce comparable predictions for a common vehicle configuration in a common flow environment. For this work, the flow environment chosen is near the upper range of applicability of the N-S approach and toward the lower end of the computational viability of the DSMC scheme. This region, where the two solvers overlay, can be conveniently thought of as in the 65 - 75 km altitude flow regime in an Earth atmosphere, for the size of vehicle considered.

2. The DMSC and N-S methodologies

The DSMC method is now a well established technique and the reader is referred to the text of Bird (1976) for a detailed discussion. The code used in the present study is a general two-dimensional/ axi-symmetric version of the program reported by Gilmore (1992), Harvey et al. (1989, 1992). Inelastic molecular collisions are modelled using the Variable Hard Sphere (VHS) potential model. Energy is exchanged via a modified Borgnakke-Larsen, variable-ϕ (1974) translational-rotational energy exchange scheme. The standard Gas-Surface Interaction model of Maxwell is used. The Navier-Stokes model is a 3-D, Finite Volume, TVD point implicit time marching code described by Netterfield (1991). The wall slip boundary conditions were treated as being analogous to a fully accommodated diffuse surface interaction model.

The two computational approaches use different molecular potential models. The DSMC scheme uses a VHS potential, while the N-S perfect gas methodology is effectively a hard sphere potential model derived from elementary kinetic theory (see Bird 1976 or Gilmore 1992). The scattering potential could be used in the determination of the viscosity in the DSMC calculation if the simulator gas is in thermodynamic equilibrium. The DSMC scheme treats the gas

Shock Waves @ Marseille I

Editors: R. Brun, L. Z. Dumitrescu

as non-perfect through its microscopic non-equilibrium mechanical simulation of the Boltzmann equation. The DSMC method thus incorporates a reasonable perturbation from the equilibrium state where the concept of Newtonian viscosity is no longer relevant. In the N-S code the viscosity is modelled using a form of Sutherland's Law.

Table 1. Geometry and test conditions

Geometry	60° Sphere/Cone	ρ_∞	$9.15 \times 10^{-5} \text{kg.m}^{-3}$
Base Diameter	2.0 m	Kn_∞	0.00194
Nose Radius	1.25 m	T_∞	139 K
Mass	120 kg	T_W	1500 K
Nose Radius	1.25 m	T_∞	139 K
Mach Number	30	Velocity	7114.5 m/s
Altitude	71 km	$\sigma_{ref}\,(N_2)$	3.681 Å

3. Test case geometry

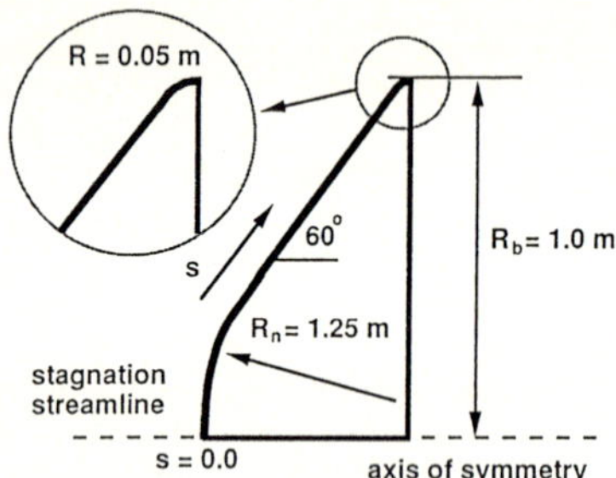

Fig. 1. Mars semi-hard lander entry vehicle geometry

The configuration chosen was the forebody of a Marsnet Semi-Hard Lander (60° sphere cone) entry vehicle shown in Fig.1. The geometry configuration and flow conditions are defined in Table 1. All computations discussed in the text are axi-symmetric. Freestream conditions based on the information from Cupples et al. (1991) were generated, and relate to approximately 71 km altitude in a pure Nitrogen Earth atmosphere.

4. Grid and flowfield definition

In order to determine the sensitivity of the predictions to the defined computational mesh, several grids were generated for each simulation. For the Navier-Stokes computations, two grids were generated, a 60 × 60 and a 90 × 60 grid with 60 cells along the body and with 60 and 90 cells respectively between the body and the free stream in flow. In the Navier-Stokes computations, cells were concentrated near the body surface in the boundary/viscous layer region only, and not in the shock layer. For the DSMC runs, grids of dimension 60 × 70 and 90 × 70 were defined with 70 cells along the body and 60 and 90 cells respectively between the body and the outer computational boundary. These grids have a bi-modal stretching factor thus cells are grouped in the boundary layer, and around the shock. A mesh sensitivity study was also performed examining the cell concentration around the body. As expected the solution was less sensitive to the mesh distribution in this direction.

To determine the effect of the thermal accommodation coefficient, several of the DSMC test cases introduced degrees of specularity, the remainder assumed a fully accommodated (diffuse) model. The accommodation coefficient used in each computation is defined in the figures where appropriate.

5. Results

Fig.2 presents the results of the mesh sensitivity study and shows the density profile taken at the cell centroid position closest to the stagnation streamline for fully diffuse reflection, i.e. at $y/R_B = 10^{-5}$. The cases NS 60, NS 90 and DSMC 60, DSMC 90 refer to the number of cells along the streamline used in the Navier-Stokes and Monte-Carlo computations respectively. The figure shows the predicted shock stand-off distances for all cases while the inset shows a plot of the boundary/viscous layer region around the stagnation point.

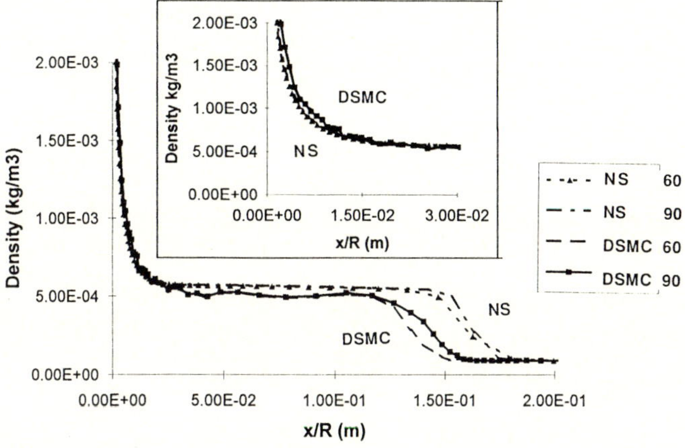

Fig. 2. Comparison of density profiles along the stagnation streamline predicted using the DSMC and N-S codes

The results suggest that a virtually mesh-independent solution has been obtained for each numerical scheme. The difference in predicted density boundary/viscous layer thickness using the two schemes is approximately 5%. The agreement in shock stand-off distance (assumed to be the mid point between the pre and post shock values) is reasonable, and the post shock density plateau values agree to within a few percent. The difference in the shock stand-off prediction could be attributed to the coarser grid definition, around the shock layer, used in the N-S calculations.

The DSMC calculation consistently under-predict the Navier-Stokes value of shock density ratio (approximately 5.75 compared to 5.85). As expected both methods under-predict the Rankine Hugoniot normal shock theory prediction of 5.99. These differences could be due to the different approaches to modelling the descriptive flowfield equations and the different concepts of the viscosity models used.

In the inset plot of the boundary/viscous layer region, the DSMC results show a slightly higher degree of scatter when compared to the N-S cases. This is due to the statistical nature of the DSMC simulation technique. The Navier-Stokes approach is assumed to have reached 'steady state' when all the numerical instabilities have been minimised to within a defined tolerance.

In the boundary/viscous layer region, the local density of the flow is high (of the order (10^{-3}, 10^{-2} kg.m^{-3})) and the applicability of the DSMC code and the form of the Boltzmann equation

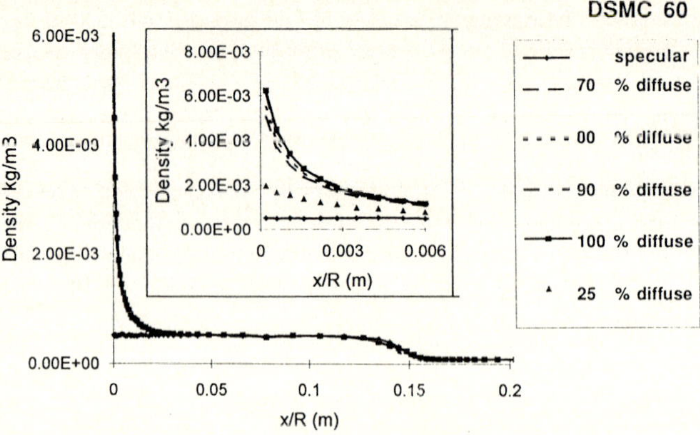

Fig. 3. Comparison of the effects of specular surface reflection assumptions on density

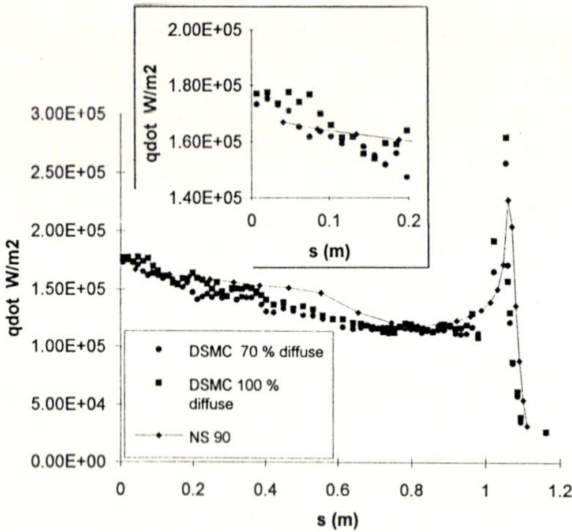

Fig. 4. Heat transfer distributions

used (Bird 1976) may be questioned. The single-particle distribution Boltzmann equation assumes that a gas is dilute and such an assumption in higher density regions (i.e. boundary layer region) may no longer be valid. Fundamental to the DSMC approach with a dilute gas assumption is that binary collisions only are considered. No three body collisions are modelled. Therefore, in a high-density region, there exists the possibility that more high-energy particles will be predicted to be reflected further upstream of the body in the DSMC case than is represented in the N-S case; thus it could be inferred that the predicted boundary/viscous layer would be different for the DSMC predictions.

A degree of specularity may be introduced via the thermal accommodation coefficient. Six cases were chosen varying the thermal accommodation coefficient from fully specular, to 25%, 70%, 80%, 90% and 100% (fully) diffuse. The results are shown in Fig.3. The fully specular case is physically unrealistic but serves to demonstrate the limit of zero accommodation. Reflections are assumed to be perfectly elastic with the molecular velocity component normal to the surface being reversed, while that parallel to the surface, as well as the internal energy, remain unchanged. There is no wall interaction and therefore no formation of a boundary/viscous layer; particles are reflected as if the surface was a mirror. Introducing a degree of specularity should provide a higher percentage of fast moving reflected molecules. It would therefore be expected that the 25% diffuse case would predict a thinner boundary/viscous layer with more slip than the higher percentage diffuse cases. This trend is illustrated in the inset showing the boundary/viscous layer region. Later manipulation of the accommodation coefficient may provide a better DSMC to N-S boundary/viscous layer thickness correlation, with the appropriate velocity and temperature slip conditions incorporated into the N-S code.

Fig.4 presents heat transfer predictions (W/m^2) around the body for the 70% and 100% diffuse reflection, and the NS 90 cases only. Comparing the NS 90 and the 100% diffuse case shows that the agreement at the stagnation point is within 5%. Away from the stagnation point, the agreement is moderate, the DSMC approach under-predicting the Navier-Stokes case by approximately 15% at the sphere cone shoulder. The position of maximum heat transfer ($s \sim 1.05$ m) occurs at the corner radius; the DSMC approach predicts a higher value at this position.

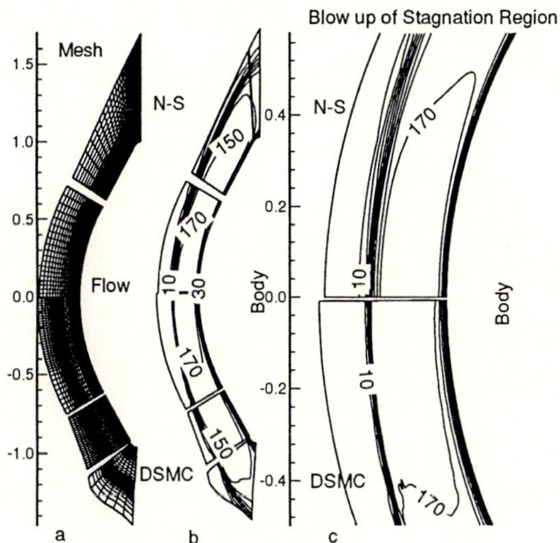

Fig. 5. a) Mesh definition for N-S and DSMC b) Translational temperature contours, Marsnet configuration. Expanded scale: comparison of N-S and DSMC c) Translational temperature contours, Marsnet configuration. Equal scale

Fig.5 shows a direct comparison of the forebody flowfield for the N-S and DSMC methods. The lower half of each figure ($0 < y < -2$) represents the DSMC calculation whilst the upper half of the figure represents the N-S prediction. Fig.5b is a plot of the translational temperature in the whole flow field. Although the two halves of the plot are not identical, the distribution of translational temperature throughout the flowfield for each numerical method is similar. Fig.5c shows the contour profile in the vicinity of the stagnation region. In all cases the temperature contours have been non-dimensionalised with $T_\infty = 139.0$ K.

6. Conclusions

The agreement between the DSMC and N-S predictions is good but it must be noted that like with like has not been strictly compared. The main points of interest to be drawn from this study may be summarised as follows:

1) The results obtained from this comparison exercise provides the confidence in the ability of the two methods to obtain similar results for a common blunt body flow problem.

2) Good agreement has been achieved for the predicted streamline (close to the stagnation streamline) profiles, and shock stand-off distances between the two methods.

3) The DSMC approach predicts a slightly thicker density and temperature boundary/viscous layer than Navier-Stokes method.

4) The DSMC approach predicts a lower post-shock density value than Navier-Stokes approach.

5) The DSMC predicted heat transfer value is slightly lower than that predicted by the Navier-Stokes approach. The introduction of partially specular reflection models reduced the predicted heat transfer rate.

6) The DSMC and N-S solvers predicted similar distributions for the translational temperature throughout the whole of the flowfield.

Acknowledgements

This work was supported by the British Ministry of Defence under the direction of the Defence Research Agency, Farnborough, UK, under contract RAE1a/192.

References

Bird GA (1976) Molecular Gas Dynamics. Oxford Univ. Press, Oxford, UK

Borgnakke C, Larsen PS (1974) Statistical collision models for Monte-Carlo simulations of polyatomic gas. Journal Comp. Phys. 18: 405

Cupples M et al. (1991) Optimisation of aerobrake assisted descent trajectories at Mars. AIAA Paper 91-0057

Gilmore MR (1992) Simulated flows over bluff bodies and wings in the rarefied hypersonic regime. PhD Thesis, University of London

Gilmore MR, Harvey JK (1992) Effects of Mach number, T_{wall}, T_∞ and thermal accommodation coefficient, on the flow around bluff bodies in the rarefied regime. In: Shizgal B (ed) Rarefied Gas Dynamics: 18th Symposium, Vancouver

Harvey JK (1989) Inelastic collision models for Monte-Carlo simulation computations. In: Rarefied Gas Dynamics: Physical Phenomena, AIAA Publication, Vol 117, pp 3-24

Harvey JK ,Celenligil MC, Dominy RG, Gilmore MR (1992) Flat-ended circular cylinder in hypersonic flow. AIAA Journal of Thermophysics and Heat Transfer, 6, 1: 35-43

Netterfield MP (1991) Computations of the aerodynamics of spinning bodies using a point-implicit method. AIAA Paper 91-0339

Application of Multiblock Codes for Computational Aerothermodynamics of Hypersonic Vehicles

Pénélope Leyland, Françoise Perrel and Jan B. Vos
Institut de Machines Hydrauliques et Mécanique des Fluides, Ecole Polytechnique Fédérale de Lausanne, CH-1015 Lausanne, Switzerland

Abstract.   The formation of shock waves over generic and complete hypersonic vehicle shapes is studied using multiblock flow solvers. The influence of the thermochemical modelling and the flight regime on the shock formation is assessed for the flow over the HERMES space shuttle. Particular attention is paid to the multiblock structure and it is verified that shock waves cross block boundaries smoothly.

Key words: Aerothermodynamics, Crossflow shocks, Hypersonics, Multiblock, Navier-Stokes

1. Introduction

The modern design of hypersonic vehicles relies nowadays heavily on numerical simulation. Complete geometry aerothermodynamic calculations are necessary to evaluate the flight feasibilities as well as to determine the heat loads and/or the possibilities of integration of propulsion systems. These simulations cannot be undertaken experimentally for the whole flight regime range, since no facilities exist that can simulate simultaneously all the complex physical phenomena encoutered. For instance, reentry flight aerodynamics must take into account air thermochemistry processes, complex gas-wall interactions, radiative heat transfer or partial ionization. Many of these phenomena are linked to the strong shock waves arising in hypersonic flowfields. For flows over re-entry vehicles different shock structures can be found, from the detached bow shock which surrounds the spacecraft to weaker secondary crossflow shocks. Shock-shock interactions as well as shock / boundary layer interactions can take place and their study is of primary importance in the design of the control surfaces. When the vehicle has an integrated propulsion system additional shock-wave interactions at the forebody and at the intake will be found. Moreover, at the afterbody the coalescing exhaust plume interacts with the structure and with the flowfield passing over the body, leading to very complex flow patterns. The predominant shock structures present within the flowfield around different aerospace configurations, and thus the type of physical phenomena encountered, differ considerably from one another. For example, the separation phase of two - stage to orbit vehicules gives rise to complex multiple transient shocks, whereas during the reentry phase of a space shuttle, the effects of the bow shock (which induces a strong degree of thermochemical nonequilibrium within the shock layer) are the most important.

A family of multiblock solvers is under development at IMHEF/EPFL as part of an international collaboration. It includes inviscid as well as viscous flow solvers. Different thermo-chemical models are implemented: thermally perfect gas and air at chemical equilibrium are considered for both inviscid and viscous flows, whereas chemical and thermo-chemical non-equilibrium air flows are considered for inviscid flows only. In this paper, flowfields over the HERMES space shuttle and associated generic hypersonic geometries are presented for different levels of the thermochemical modelling.

Shock Waves @ Marseille I
Editors: R. Brun, L. Z. Dumitrescu

2. Numerical method and multiblock implementation

Due to the complexity of the flow around a complete hypersonic vehicle, multiblock methods are appropriate to calculate these flows. These methods not only facilitate the generation of meshes, but also permit an efficient use of high performance supercomputers. In principle, they allow the implementation of hybrid discretisations, (structured and unstructured), and hybrid numerical techniques (different schemes in different blocks).

The underlying numerical scheme is based upon a standard finite volume centered scheme with explicit time stepping via a Runge-Kutta multistage scheme, and includes implicit residual smoothing and local time stepping to accelerate convergence to steady state, (Jameson 1985, Leyland et al. 1993, Vos et al. 1992). Second- and fourth-order artificial dissipation terms are used to capture discontinuities and to stabilise the numerical scheme. Directional scaling is employed to improve the artificial dissipation in cells with high aspect ratios (Swanson and Turkel 1992). Regions of discontinuities are found using a normalised second-order pressure difference. The governing conservation equations of species continuity, momentum, total energy, and vibrational energies are coupled directly: i.e. the whole system is solved simultaneously.

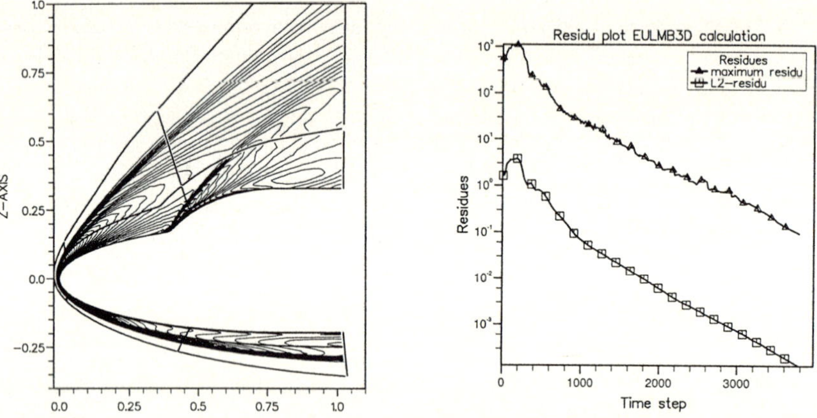

Fig. 1. *Left:* Inviscid flow over a double ellipsoid. Isotherms, non-equilibrium air chemistry for a six-block mesh, $M_\infty = 25, p_\infty = 2.52$ Pa, altitude 75 km. The outer detached bow shock, the inner oxygen dissociation shock, and the canopy shock pass uniformly from one block to another. *Right:* Convergence history for this calculation

The solvers developed at IMHEF/EPFL are built on top of an object-oriented flexible data base system (MEMCOM) including a Dynamical Memory Manager (DMM) which is used to allocate the necessary storage for each block (Merazzi 1988). The multiblock structure is only visible at the level of the database and the outer block loops. Information transfer between different blocks consists of copying variables at the interface of a block into the boundary arrays of the neighbouring block. This is done only once per time step to limit the communication overhead. The boundary arrays are used to fill the ghost cells used to impose the boundary conditions for each flux (convective, viscous and artificial dissipation). In this way, a multiblock flow simulation can be looked upon as a stream of single block calculations. Owing to the C0-continuity of coordinates imposed at block interfaces, shock waves cross block boundaries smoothly. As an illustration, Fig.1 shows the temperature contours for a flow with a free stream Mach number of 25 over a double ellipsoid (generic nose to canopy) at 30° of incidence, assuming non-equilibrium air chemistry. A 6 blocks mesh was used. This figure clearly illustrates the continuity of the detached bow shock and of the canopy shock across the block boundaries.

3. Air thermo-chemistry modelling

For typical flight conditions encountered during a reentry phase, air can be modelled as a mixture of perfect gases, composed of five species, the molecules N_2, O_2, NO and atoms O and N. Hence, ionization effects can be neglected. The chemical reactions are modelled by a seventeen reactions model (Park 1985). The vibrational energy exchange terms for thermal non-equilibrium vibrational energy conservation equations are modelled by the Landau-Teller model using the Millikan-White equation for the vibrational relaxation times (Landau and Teller 1936, Millikan and White 1963). NO is assumed to be in thermal equlibrium, and the loss/gain of the vibrational energy by coupling with molecular dissociation or recombination is neglected. Nonequilibrium modelling is only implemented for inviscid flows.

4. Calculations around the HERMES space shuttle

The flowfield over a HERMES space shuttle with a 12° body flap deflection is calculated at an altitude of 76 km. This corresponds to a freestream Mach number of 25, pressure of 2.14 Pa, and a density of $3.7\ 10^{-5}\ kg/m^3$, with a 30° angle of attack and a zero slip angle. Different air thermo-chemistry models are considered. A monoblock mesh of about half a million points has been used. In the following, equilibrium and non-equilibrium computations are compared for inviscid flows. For air at chemical equilibrium, a comparison of inviscid and viscous computations is also presented.

The main differences between equilibrium and non-equilibrium calculations appear on the leeward side of the body (see Fig.2). For the non-equilibrium computation, the strong dissociation processes taking place behind the detached bow shock lead to a highly dissociated mixture near the stagnation point. Due to the rarefaction along the nose this highly dissociated mixture is frozen within the nose-to-canopy region. The canopy compression increases the temperature which reinduces recombination of N-atoms. On the windward side, the compression of the bodyflap locally increases the temperature, and reinduces nonequilibrium effects. For the chemical equilibrium calculation, the increase in potential energy by the shock wave is completely absorbed by the dissociation processes. Near the body, O_2 is completely dissociated, while N_2 is partly dissociated. The temperatures are thus much lower than in the non-equilibrium case and the canopy shock is weaker. For viscous flows, the heat transfer coefficient (see Fig.3) on the leeward side shows a peak heating spot in the free shear layer reattachment region on the cabin. On the windward side, a very high peak heating rate is present on the compression ramp of the bodyflap. Here, there is a separation shock/induced ramp compression interaction together with shock wave/boundary layer interactions (Leyland 1993, Perrier 1992, Simeonides 1993).

Another important difference between equilibrium and nonequilibrium inviscid flowfields appears in the crossflow shock in the outflow plane between the fuselage and winglet. For non-equilibrium flows this crossflow shock is clearly visible in the Mach number and temperature, (Fig.4 *left*). However, for equilibrium calculations, this shock is smoothed out for these variables, (Fig.4 *right*). The same phenomena can be seen for the viscous calculation at the same Mach number and for a Reynolds number of $2.2\ 10^4$ per meter, (see Fig.3). The smoothing of the crossflow shock is in this case also accentuated by the presence of a growing subsonic boundary layer along the wing.

This phenomenon can be explained as follows: The crossflow shock induces a local increase in temperature. For equilibrium air chemistry, the increase in potential energy is completely absorbed by the chemistry and the crossflow shock is no longer visible in the temperature (or Mach number) contours. However, on closer inspection, it is visible, in the O_2 and O mass fraction contours (see Fig.5). For nonequilibrium flows, the flowfield is partially frozen and the increase in potential energy recombines the frozen N atoms, and although O_2 dissociates, the crossflow shock remains visible. These non-equilibrium effects take place throughout the whole region wing-winglet-fuselage, as illustrated in Figs.2,5 *left*.

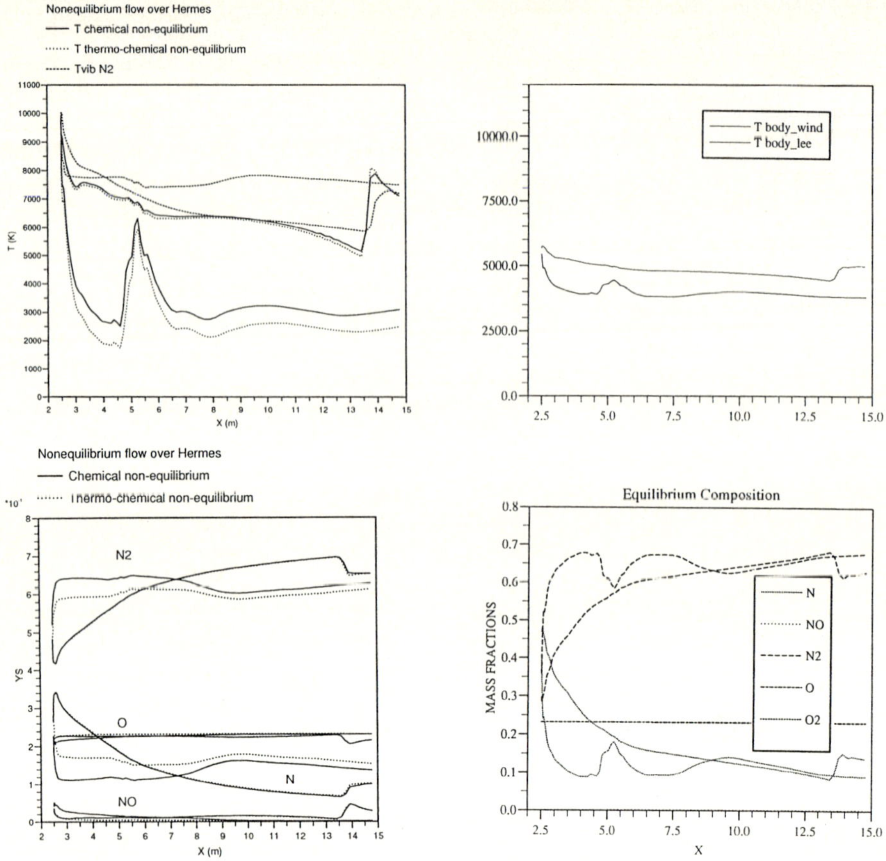

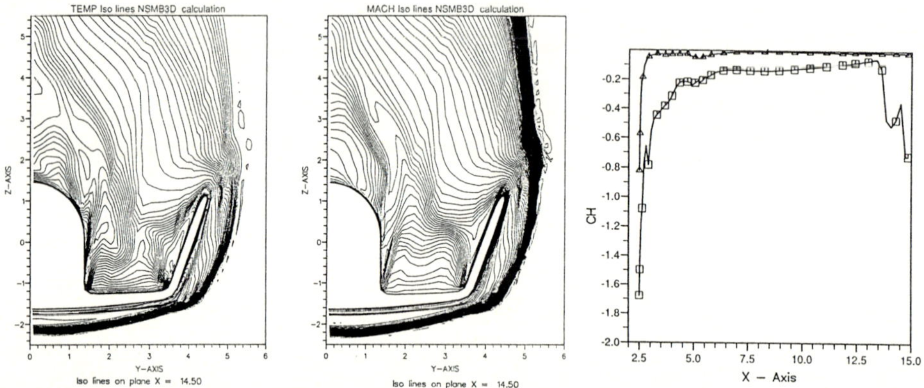

Fig. 2. Comparison of chemical non-equilibrium with thermochemical non-equilibrium *left* and chemical equilibrium *right*; body temperature and mass fractions in the symmetry plane on HERMES; $M_\infty = 25$, $p_\infty = 2.14$ Pa, $\rho_\infty = 3.7 \ 10^{-5}$ kg/m^3, with 30° angle of attack, and bodyflap at 12°.

Fig. 3. Isotherms *left* and Iso Mach contours *centre*; equilibrium chemistry viscous calculation on HERMES. Heat transfer coefficient in the symmetry plane *right*; $M_\infty = 25$, Re/m $= 2.2 \times 10^4$, isothermal wall $T_{wall} = 1500$ K, with a 12° body flap

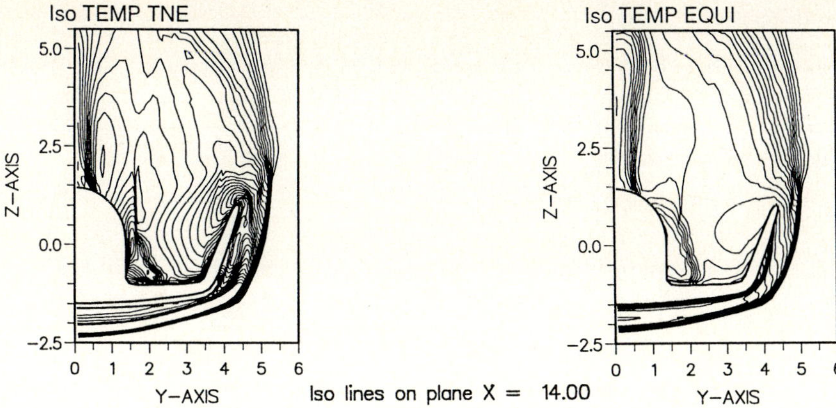

Fig. 4. *Left*: Crossflow shock clearly visible within the isotherms for non-equilibrium flow in the winglet-wing-body section of HERMES at an outflow plane situated at $X = 14$ m. Flight conditions corresponding to 76 km, ($M_\infty = 25$). *Right*: Isotherms for equilibrium flow in the winglet-wing-body section of HERMES at the same outflow plane

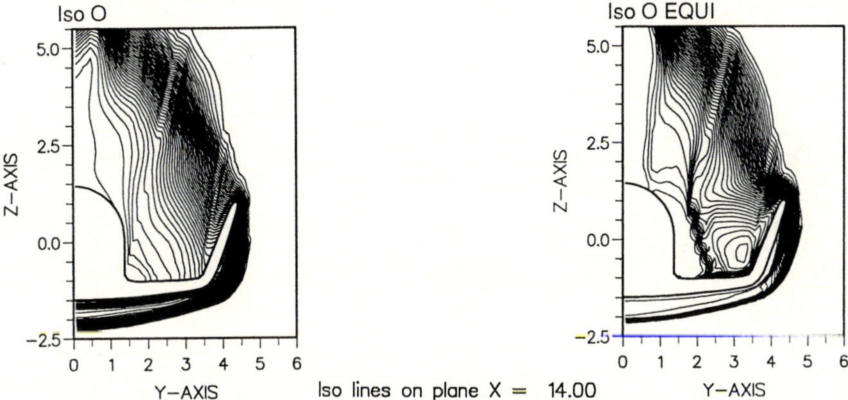

Fig. 5. *Left*: Iso O lines for thermochemical nonequilibrium flow for winglet-body outflow plane, HERMES, flight conditions: 76 km. ($M_\infty = 25$), *Right*: Winglet-body outflow plane crossflow shock for equilibrium flow visible in the nitrogen dissociation region (here Iso O lines)

5. Conclusions

Multiblock methods provide a powerful and efficient tool for simulating flows over complete geometries. The simplicity of the numerical scheme used in this work allows easy extension to different levels of physical modelling. The modelling of the thermo-chemical effects, triggered by the strong detached bow shock, has been shown to have a great influence on the shock structures in the flowfield around the HERMES space shuttle. This applies not only to the bow shock, but is equally valid for secondary crossflow shocks and embedded shocks, as has been shown for the outflow plane of the HERMES space shuttle. For a viscous calculation, the boundary layer also affects the shock structure.

Acknowledgements

The authors would like to acknowledge Dr.H.Wong (ESTEC) for providing the six block double ellipsoid mesh, and Dr. Magnus Bergman (KTH) for the HERMES mesh. Financial support was partially given by the Swiss Commission d'Encouragement de la Recherche Scientifique, (CERS). The first author was on leave from CNRS (France).

References

Jameson A (1985) Numerical solution of the Euler equations for compressible inviscid fluids. In: Angrand F et al. (eds) Numerical Methods for the Euler Equations of Fluid Dynamics, SIAM, Philadelphia

Landau L, Teller E (1936) Zur Theorie der Scalldispersion. S Physik Z.Sowjetunion, 10, 1:34

Leyland P, Perrel F, Vos JB (1993) A family of multiblock codes for computational aerothermo-dynamics: application to complete vehicle hypersonic flows. AIAA Paper 93-3042

Leyland P (1993) Shock wave / boundary layer interactions at hypersonic speeds by means of an implicit Navier-Stokes solver. AIAA Paper 93-2938

Mcrazzi SM (1988) MEM COM: An integrated memory and data management system. MEM-COM User Manual 5.6. SMR TR-5056, SMR Corporation, P.O.Box 41, CH-2500 Bienne

Millikan RC, White DR (1963) Systematics of vibrational relaxation. Journ.Chem Physics, 39,12:3209-3213

Park C (1985) On the convergence of computation of chemical reacting flows. AIAA Paper 85-0427

Perrier P (1992) Aerodynamic and aerothermal challenges for the design of the Hermes spaceplane. AGARD CP - 514, Theoretical and Experimental Methods in Hypersonic Flows, Torino

Simeonides TG (1993) Hypersonic shock wave boundary layer interactions over simplified deflected control surface configurations. AGARD -FDP/ VKI Course, Report R-792, Paper 7, Rhode St.Genèse, Belgium

Swanson RC, Turkel E (1992) On central difference and upwind schemes. Journal Comp. Phys. Vol. 101

Vos JB, Bergman CB, Rizzi A (1992) Hypersonic multi-block flow simulations including non-equilibrium chemistry. 70th Agard FDP Meeting on Theoretical and Experimental Methods in Hypersonic Flows, Torino

Numerical Analysis of Shock/Shock and Shock/Body Interactions for 3D Configurations

G. Hartmann and S. Menne
Deutsche Aerospace AG, Space Group, Dept. RTT314, 81663 München, Germany

Abstract. An overview about a simulation method for inviscid and viscous 3-dimensional hypersonic flow is given. The flow solver is based on a quasi-conservative matrix-splitting method employing finite differencing with Runge-Kutta timestepping and bow shock fitting. Solutions over generic configurations (hyperbola-flare-cylinder, double ellipsoid, blunt deltawing) and real 3-dimensional configurations (Hermes reentry vehicle) for perfect gas, air in chemical equilibrium and in chemical non-equilibrium are presented.

Key words: Inviscid and viscous hypersonic flow, Chemical non-equilibrium, Finite differences

1. Introduction

In recent years a number of hypersonic research programs have been initiated in Europe, USA, and Japan within the development programs of new space transportation systems. Phenomena in supersonic flowfields like imbedded shocks and shock/shock or shock/body interactions are subjects of detailed investigations. The rapid evolution of computer and supercomputer technology and the growing efforts of CFD permit numerical studies and flow simulations of these shock phenomena, both for generic configurations and real three-dimensional configurations like the Hermes space vehicle. An important point in the investigation of those shock phenomena is the influence on the lift, drag and momentum coefficients, as well as the effect on the maneuverability and the trimming of the flight vehicle.

2. Solution method

2.1. Governing equations

The governing equations for viscous flow in chemical non-equilibrium can be written as:

$$\frac{\partial}{\partial t}Q + \sum_{i=1}^{3}\frac{\partial}{\partial x_i}\left(F_{I,x_i} - F_{\nu,x_i}\right) = S \tag{1}$$

where $Q = (\rho, \rho v_{x_1}, \rho v_{x_2}, \rho v_{x_3}, \rho_1, \ldots, \rho_{n_s}, e)^T$ are the conservative variables, F_{I,x_i} the inviscid and F_{ν,x_i} the viscous fluxes and $S = (0,0,0,0,\dot\omega_1,\ldots,\dot\omega_{n_s},0)^T$ are the source terms. The usual notation is adopted here, i.e., ρ denotes the total density, ρ_s are the species densities with $s = 1,\ldots,n_s$.

For chemical equilibrium or for a perfect gas, the source terms are identical to zero, and the species continuity equations are decoupled from the system of equations. In this case the conservative variables are reduced to the vector $Q = (\rho, \rho v_{x_1}, \rho v_{x_2}, \rho v_{x_3}, e)^T$. Further for inviscid flows the viscous fluxes vanish.

2.2. Physical model

2.2.1. Chemical equilibrium.
In chemical equilibrium two independent variables describe the thermodynamic state and therefore two equations are necessary for the four variables p, ρ, T and ε. These equations are the caloric and the thermal equation of state. Connecting these equations gives the desired expression for the pressure as a function of density and energy:

Shock Waves @ Marseille I
Editors: R. Brun, L. Z. Dumitrescu

$$p = \rho RT(\rho, \varepsilon) \, \mathcal{Z}(\rho, \varepsilon) \tag{2}$$

With its defining equation the speed of sound can be expressed by any two independent thermo-dynamic variables, e.g., the density ρ and the internal energy ε, which are governing variables in the fluid-dynamic equations:

$$c^2(\rho, \varepsilon) = \frac{p}{\rho^2} \left(\frac{\partial p}{\partial \varepsilon} \right)_\rho + \left(\frac{\partial p}{\partial \rho} \right)_\varepsilon \tag{3}$$

For a real gas in chemical equilibrium, the implementation of the equation of state in the numerical algorithm is done using $p(\rho, \varepsilon)$, and its derivatives $\left(\frac{\partial p}{\partial \rho} \right)_\varepsilon$ and $\left(\frac{\partial p}{\partial \varepsilon} \right)_\rho$. In order to implement in the numerical algorithm the equation of state for a real gas in chemical equilibrium the function $p(\rho, \varepsilon)$, and its derivatives $\left(\frac{\partial p}{\partial \rho} \right)_\varepsilon$ and $\left(\frac{\partial p}{\partial \varepsilon} \right)_\rho$ are necessary. The state equation for air can be described by Mollier-fits or spline approximations. For arbitrary gas mixtures a special procedure (Hartmann 1993) can be applied generating a spline surface for these function by using the program of Gordon and McBride (1971). For viscous flow simulations the transport coefficients are evaluated by fit or spline functions of the form $\mu(\rho, \varepsilon)$ and $\kappa(\rho, \varepsilon)$.

2.2.2. Chemical non-equilibrium. The total enthalpy H is defined as $H = (e + p)/\rho$ and the internal enthalpy h as $h = H - 0.5v^2$. The internal enthalpy of the mixture is coupled with the species internal entalpies via $h = \sum y_s h_s(T)$ where $y_s = \rho_s/\rho$ denotes the species mass fraction. The internal species enthalpies depend on the mixture temperature only and contain contributions from equilibrium vibrational degrees of freedom and the heat of formation. The assumption of perfect gas for each species leads to an equation of state in the form:

$$p = \bar{R}T \sum_{s=1}^{n_r} \frac{\rho_s}{M_s} = \rho \bar{R} T \mathcal{Z} \tag{4}$$

where $\bar{R}$ is the universal gas constant with $\bar{R} = 8.31441$ J/g-mole-K and M_s is the specific mass of the species s. The density ρ is the total amount of all species densities ρ_s. The reacting air is modeled using the five species (N_2, N, O_2, O, NO) and the 17 nonionizing reaction model of Park (1985). Currently the code uses the assumption of thermal equilibrium of the vibrational degrees of freedom. For the chemical reaction of n_s species in n_r reactions

$$\sum_{s=1}^{n_s} [\alpha^{s,r} \frac{\rho_s}{M_s}] \leftrightarrow \sum_{s=1}^{n_r} [\beta^{s,r} \frac{\rho_s}{M_s}], \quad r = 1 \ldots n_r \tag{5}$$

the source terms are:

$$\dot{\omega}_s = M_s \sum_{r=1}^{n_r} [(\beta^{s,r} - \alpha^{s,r}) J_r], \quad s = 1 \ldots n_s \tag{6}$$

The backward rates $k_b(T)$ are calculated from the temperature fits of the forward rates $k_f(T)$ and the equilibrium constants $K_c(T)$. The point implicit treatment of the chemical source terms (Pfitzner 1990) includes the dependence of the source terms on the species densities only. Normally the temperature dependence can be neglected, but in an equilibrium flow situation it has to be taken into account. For viscous flow the transport terms are added to the governing equations and the transport coefficients have to be determined for the species. A description of the implemented method can be found in Menne et al. (1992).

2.3. Numerical approach

To solve the governing equations for realistic flow cases, curvilinear coordinates are introduced and, after some rearrangements and the introduction of the metric relations, the full conservative form in curvilinear coordinates can be written as:

$$\frac{\partial}{\partial \tau}\left(\frac{1}{J}Q\right) + \sum_{i=1}^{3}\frac{\partial F_{I,\xi_i}}{\partial \xi_i} = \frac{1}{J}S + \sum_{i=1}^{3}\frac{\partial F_{\nu,\xi_i}}{\partial \xi_i} \tag{7}$$

where the generalized fluxes and metric terms are given in Menne et al. (1992). Since a quasi-conservative formulation (Weiland 1986) for Euler flows has been proven to be very accurate, this form is used for the inviscid fluxes. However, the viscous fluxes are considered in full conservative form, because they are much more complicated in a non-conservative formulation than in a fully-conservative formulation. The final form using matrix-splitting is:

$$\frac{\partial Q}{\partial \tau} + \sum_{i=1}^{3}\left(A_{\xi_i}^{+}\frac{\partial Q^{+}}{\partial \xi_i} + A_{\xi_i}^{-}\frac{\partial Q^{-}}{\xi_i}\right) = (S + J\mathcal{R}), \quad A_{\xi_i}^{\pm} = T_{\xi_i}\Lambda_{\xi_i}^{\pm}T_{\xi_i}^{-1} \tag{8}$$

A detailed derivation of the matrix-splitting form of the governing equations is given in Menne et al. (1992), the treatment of the viscous terms summarized in eq. (8) in the single term $\mathcal{R}$ follows closely the formulation given in Menne (1991) and Menne and Weiland (1990). The spatial discretization is done by third-order upwind biased differences for the convective part and by central differencing for the full conservative viscous part. The time integration is performed at the present state by a Runge-Kutta time-stepping scheme.

3. Configurations and results

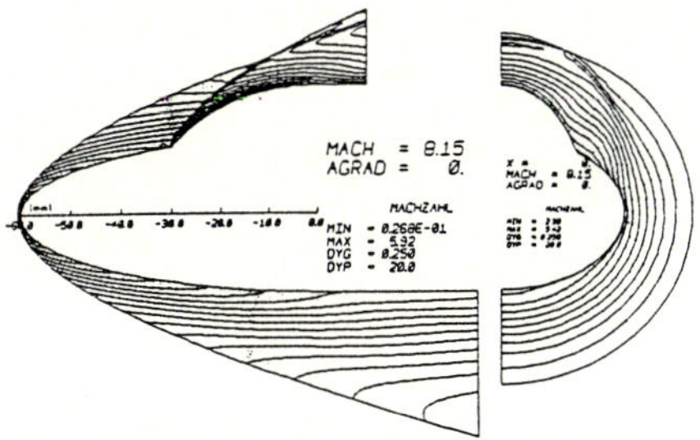

Fig. 1. Double ellipsoid: M_∞=8.15, α=0°, inviscid, perfect gas, plane of symmetry and $X = 0$, Mach number contours, $M_{min,max,\Delta}$=0.0268, 5.92, 0.25

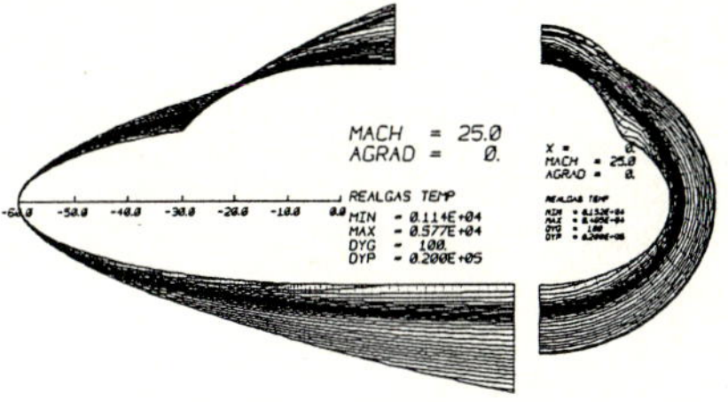

Fig. 2. Double ellipsoid: M_∞=25, α=0°, inviscid, equil. air, plane of symmetry and $X = 0$, temperature contours, $T_{min,max,\Delta}$=1140., 5770., 100.

The double ellipsoid, a generic configuration defined as a Hermes-like configuration for the Antibes workshop, is a good test geometry, because a lot of numerical results from different methods and also experimental data are available. The solutions for the flow cases with an angle of attack of α=0° and freestream Mach numbers of M_∞=8.15 (Fig.1) and 25 (Fig.2) show an interaction of the canopy shock and the bow shock. The bow shock is displaced and a shear layer is produced behind the interaction zone. In the case with M_∞=25 real gas effects (equil. air) are considered and the bow shock is very close to the body. The flowfield between the body and the bow shock contains an area of condensed temperature contours in the transition range between the N_2 and O_2 dissociation.

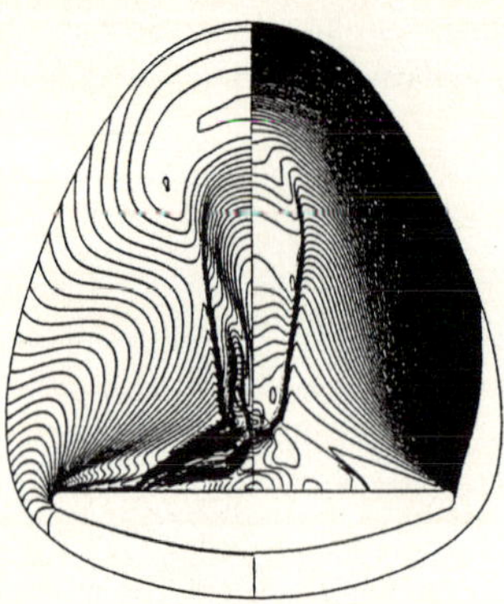

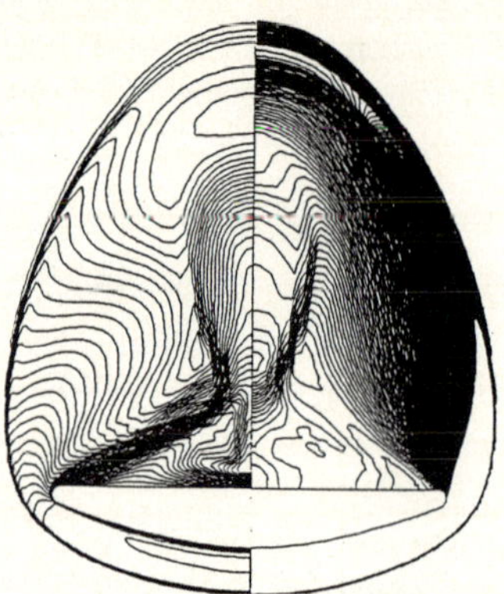

Fig. 3. Blunt delta wing: $M_\infty=8.7$, $\alpha=30°$, inviscid, perfect gas, cross section $X/L = 0.8$, Mach number (left) and C_p (right) contours, $M_{min,max,\Delta}=1.28,9.55,0.2$, $c_{p\,min,max,\Delta}=$-0.02, 0.20, 0.0005

Fig. 4. Blunt delta wing: $M_\infty=8.7$, $\alpha=30°$, viscous, perfect gas, cross section $X/L = 0.8$, Mach number (left) and C_p (right) contours, $M_{min,max,\Delta}=0.$, 8.7, 0.2, $c_{p\,min,max,\Delta}=$-0.02, 0.08, 0.0005

Another configuration of the Antibes workshop is the blunt delta wing. Figs.3 and 4 show the cross section $x/L = 0.8$ of an inviscid and a viscous perfect gas solution and the Mach number contours (left part) and the pressure coefficient C_p (right part) are contrasted. The present viscous delta wing solution was computed by a method described in Menne (1991). The inviscid solution shows a wing shock on the leeward side, which generates a vortex on the upper wing side between the shock and the symmetry line. The shear layer produced by the vortex can be seen in the Mach number contours. In comparison, the flow separates at the wing tip in the viscous solution (Fig.4) and a vortex lies on the upper wing side. In the flowfield above in both solutions the flow direction is turned towards the symmetry line by a deflection shock.

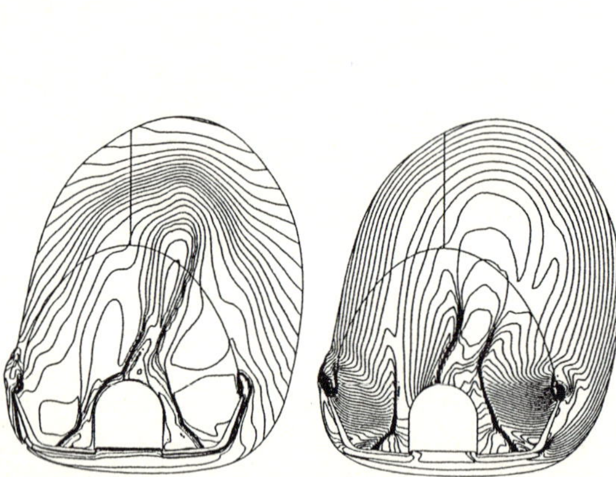

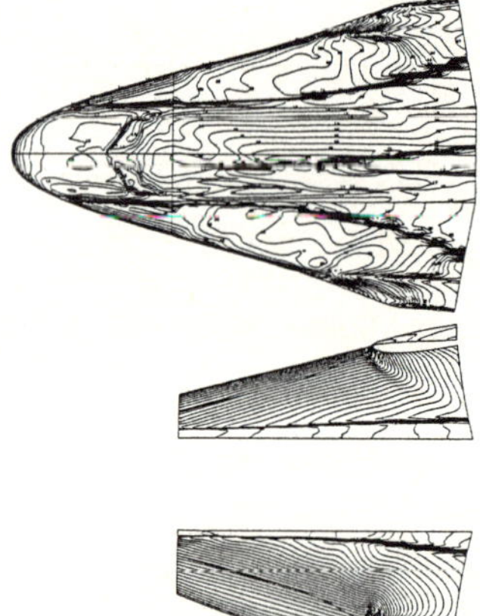

Fig.5. Hermes 1.0: $M_\infty=25$, $\alpha=30°$, $\beta=8°$, inviscid, equil. air, Mach number (left) and $\ln(p/p_\infty)$ (right) contours in cross section, $M_{min,max,\Delta}=1.28,15.1,0.5$, $\ln(p/p_\infty)_{min,max,\Delta}=$-4.49,5.94,0.2

Fig.6. Hermes 1.0: $M_\infty=25$, $\alpha=30°$, $\beta=8°$, inviscid, equil. air, Mach number contours on body and $\ln(p/p_\infty)$ contours in $Z =$const. plane (winglet region)

For the Hermes reentry vehicle the time after the reentry is a critical flight phase. Due to the high Mach numbers in the upper atmosphere a yawing configuration may lead to an interaction between the winglet and the bow shock, inducing high thermal and structural loads at the winglet and the maneuverability may be strongly impaired.

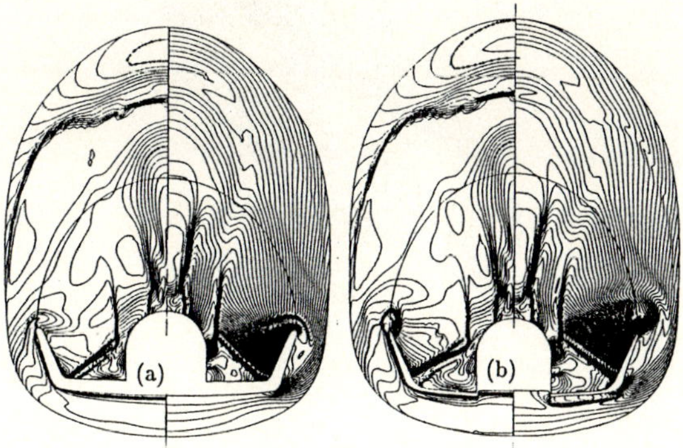
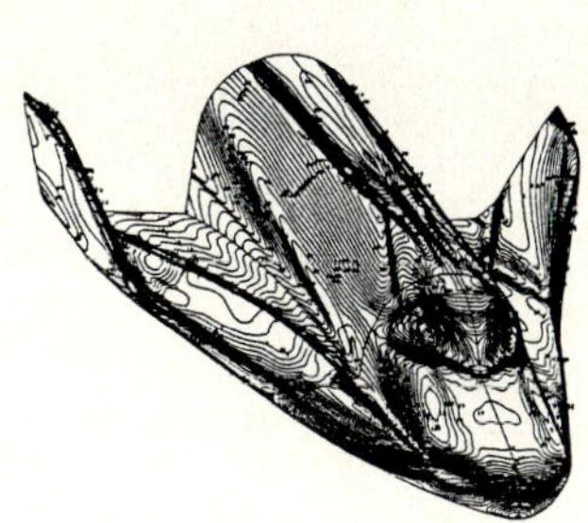

Fig.7. Hermes 1.0: $M_\infty=12$, $\alpha=30°$, $p_\infty=179.9$ Pa, $T_\infty=249$ K, F4 cond., Mach number (left part) and pressure $\ln(p/p_\infty)$ (right part) contours at $x \approx 0.35$ m (a) and $x \approx 0.395$ m (b)

Fig 8. Hermes1.0: $M_\infty=12$, $\alpha=30°$, F4 cond., temperature contours on body, $T_{min,max,\Delta}=190.6$, 3792., 66.

For the flow case with $M_\infty=25$, $\alpha=30°$ and $\beta=5°$, defined as a limit case on the trajectory, no sign of an interaction between the winglet and the bow shock can be seen in the numerical solution. But there is only a small safety range. An increase of the yaw angle up to $\beta=8°$ results in a considerable interaction. The bow shock runs directly to the winglet (Fig.6), displaced only a little by the winglet shock. As usual in shock/shock interactions, a shear layer arises in the flowfield behind, whereas no interaction is present on the opposite side. The cross section plots (Fig.5) display the asymmetrical flowfield containing other imbedded shocks and shear layers. The wing shock interacts with the fuselage shock caused by the change of the flow direction at the fuselage and the shocks on the top of the fuselage, which are generated due to the flow deflection above the orbiter. These shocks are shifted due to the yawing configuration and one of the shocks interacts with the fuselage shock on the right side. In Figs.7 and 8 the result of a numerical non-equilibrium airflow simulation over the Hermes model (1:40) with the F4 conditions (F4 wind tunnel of ONERA, Fauga/Mauzac, France) at a freestream Mach number of $M_\infty=12$ are shown. In Fig.7 two different cross-flow planes in the winglet region are presented, where the Mach number (left part) and the pressure $ln(p/p_\infty)$ (right part) contours are contrasted. The same flow phenomena as for the yawing configuration can be identified: the wing shock in interaction with the fuselage shock, the winglet shock and the shock on the top of the fuselage, which is forced by the flow deflection due to the symmetry condition. In addition a small cross-flow shock can be seen in the fuselage-wing corner and Fig.8, showing the temperature contours on the body, reveals the footprint of the canopy shock.

Finally, in Figs.9 and 10 the results of a parametric study at the generic hyperbola-flare-cylinder configuration are summarized. Fig.9 shows the pressure coefficient distribution for inviscid non-equilibrium air calculations over the axisymmetric geometry, where the flare angle is varied from 40° up to 45° and a viscous non-equilibrium air calculation performed for a flare angle of 45°. Looking at a plane solution (Fig.10) for a flare angle of 50° the viscous flowfield contains a big separation, while the shock in front of the separation and the flare shock are weak compressions forcing only a slight interaction with the bow shock. In the inviscid flowfield a strong detached recompression shock arises and an unsteady separation is produced behind the shock (instant state at 98000 time steps).

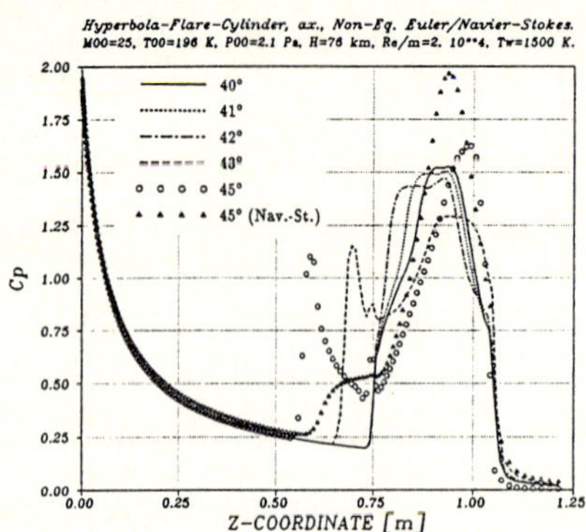

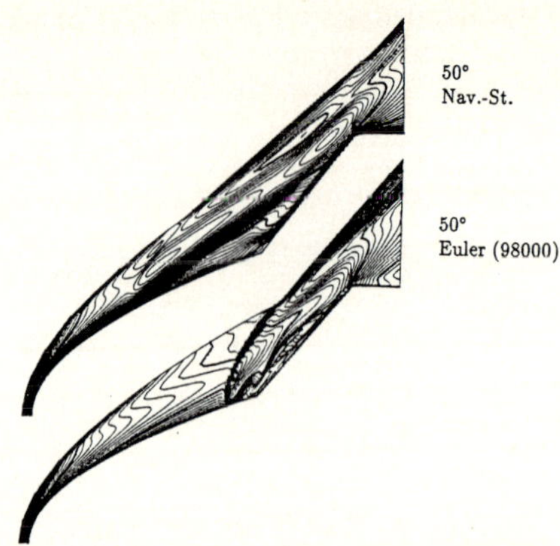

Fig. 9. Hyperbola-flare-cylinder: $M_\infty=25$, $p_\infty=2.106$, $T_\infty=196$, inviscid/viscous, axisymmetric, flare angle $\delta=40$-$45°$, C_p distribution along the body

Fig. 10. Hyperbola-flare-cylinder: $M_\infty=25$, $p_\infty=2.106$, $T_\infty=196$, inviscid/viscous, plane, flare angle $\delta=50°$, temperature contours, $\Delta T=200$ K

4. Conclusions

The results for the 3D generic and real configurations described above emphasize the excellent shock-capturing capability of the present numerical method. Combined with the bow shock-fitting algorithm and the possibility of switching between equilibrium and non-equilibrium flow as well as between viscous and inviscid flow demonstrate that this code is a suitable tool for numerical investigations of shock phenomena and shock/shock and shock/body interactions in hypersonic flow. To study the influence of the shock phenomena on the global coefficients the values of the coefficients can be determined afterwards from the flowfield solution.

References

Gordon S, McBride BJ (1971) Computer program for calculation of complex chemical equilibrium compositions, rocket performance, incident and reflected shocks, and Chapman-Jouguet detonations. NASA SP-273

Hartmann G (1993) Shear layer prediction in nozzles with TEG. DASA Report for ELITE program: Study of Advanced Nozzle Technologies, ESA Contract No. 9974/92/NL/FG

Menne S, Weiland C (1990) Calculation of three-dimensional viscous and inviscid hypersonic flows using split-matrix marching methods. AIAA Paper 90-3070

Menne S (1991) Efficient solution of three-dimensional viscous hypersonic flows. Aerothermodynamics for Space Vehicles, First Europ. Symp., ESA SP-318, pp. 267-272

Menne S, Weiland C, Pfitzner M (1992) Computation of 3-D hypersonic flows in chemical non-equilibrium including transport phenomena. AIAA Paper 92-2876

Park C (1985) On the convergence of computation of chemically reacting flows. AIAA Paper 85-0247

Pfitzner M (1990) Simulations of inviscid equilibrium and non-equilibrium hypersonic flows. Lecture Notes in Physics, Vol. 371, pp. 432-436

Weiland C (1986) A split-matrix method for the integration of the quasi-conservative Euler equations. Notes on Numerical Fluid Mechanics, Vol. 13

Hypersonic Shock-Wave/Boundary Layer Interactions with an Implicit Navier-Stokes Solver

Pénélope Leyland
Institut de Machines Hydrauliques et Mécanique des Fluides, Ecole Polytechnique Fédérale de Lausanne, CH-1015 Lausanne, Switzerland

Abstract. A fully implicit Navier-Stokes solver is used to calculate two-dimensionnal generic hypersonic geometries. Peak heating and shock wave/boundary layer interaction features are illustrated by computations over compression ramps, (representative of flows over control surfaces), and generic nose-to-canopy flows (double ellipse geometry with zero incident flow). Comparisons with experimental data is commented.

Key words: Boundary-layer interactions, Separation, Reattachment, Hypersonics

1. Introduction

Hypersonic vehicles can be decomposed into many generic parts, e.g. an extended sphere for the nose region, an axysymmetric hyperboloid for a symmetry line, a double ellipse geometry for the nose to canopy region, isolated delta wing, afterbody cone flow, deflected ramps for elevon, canopy, rudder design,...and so on (Bertin 1990, Anon. 1992). These specific geometries can be isolated and simulated by high-resolution numerical schemes at a relatively low cost. Such calculations, together with experimental studies, bring deeper insight into particular design needs for each configuration, as well as an evaluation of the validity of more approximate methods such as boundary layer, parabolised Navier-Stokes, and coupled Euler-boundary layer methods.

In this paper, a series of numerical experiments over compression ramps and generic canopies are presented using an implicit full Navier-Stokes solver on unstructured meshes for external flow conditions corresponding to some existing experimental data (Holden 1978, Anon. 1993, Vetter et al. 1993). These flows present strong viscous effects, with multiple shock wave-boundary layer interactions. Real gas effects are also taken into account.

2. Numerical method

The numerical schemes employed are based upon either an upwind approximate Riemann solver or a symmetric TVD scheme for the convective terms on Galerkin finite volumes, with second order extension as with Harten-Yee's modified flux schemes (Harten 1984,Yee 1989). The min-mod limiter is chosen for robustness. A standard finite element approximation is taken for the viscous terms. Time integration to steady state is attained by a pseudo-Newton time stepping procedure, where the linearised Jacobians are inverted by an iterative scheme and a point by point block preconditionned GMRES algorithm for further convergence. An unstructured finite element type grid is used for all calculations, with a structured grid structure within the wall boundary layer. The viscous fluxes are evaluated element by element; a $Q_1 - P_1$ finite element approximation is taken with a four point Gauss quadrature for the near wall elements. This corresponds to a centered approximation in space for the viscous fluxes.

3. Physical modelling

Air is considered to be a thermally perfect mixture of N_2, O_2, NO, N and O. The thermodynamic properties of the individual species are derived by a model proposed in (Park 1989, Anon. 1993). Blottner's model is used for the individual species viscosities, and the thermal conductivities are calculated from Eucken's relation. The mixture transport coefficients are obtained using Wilke's semi-empirical relation. The species diffusion fluxes are approximated using Fick's law for the diffusion velocities (Vincenti and Kruger 1986).

Shock Waves @ Marseille I
Editors: R. Brun, L. Z. Dumitrescu

4. Results

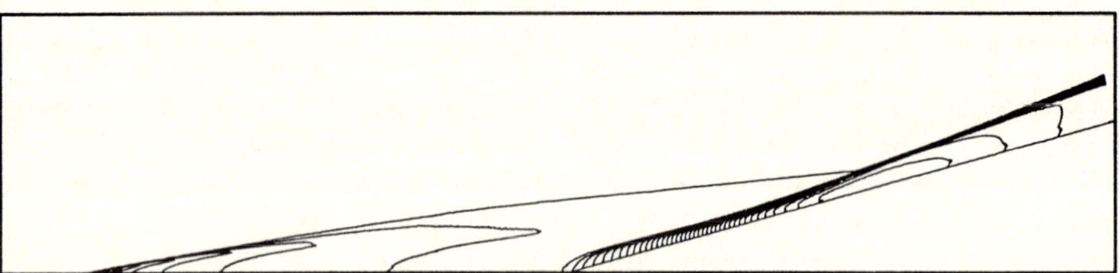

Fig. 1. $M_\infty = 14.1$, Re/m= 2.3×10^5, $p_\infty = 19$ Pa, $T_\infty = 72$ K. Flow over compression ramp. Iso C_p lines showing strong compression

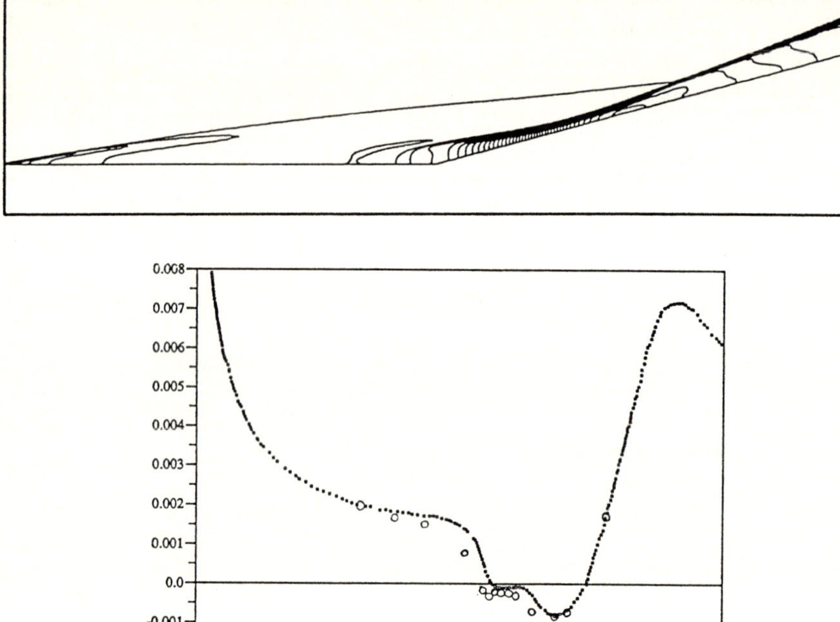

Fig. 2. $M_\infty = 11.98$, Re/m= 5.58×10^5, $p_\infty = 19$ Pa, $T_\infty = 64$ K. Flow over extended compression ramp. Iso Cp lines (top) showing multiple shock-boundary layer interactions, Wall skin friction distribution (bottom left) compares well to experimental data

Laminar viscous flow is simulated over two compression ramps corresponding to well documented test cases. The first one (Fig.1) coresponds to an upstream Mach number of 14.1, a Reynolds per metre of 2.37×10^6, and a pressure of 19 Pa, over a 15° wedge. In this case the leading edge shock impinges on the induced compression fan by the ramp hinge. This interaction induces an expansion fan and an emanating slip line, which is well captured by the numerical scheme. A strong recompression is seen downstream on the ramp, as illustrated by the C_p lines.

The second case (Fig.2) is one of the Antibes Workshop test cases ($M = 11.68$), and presents a large recirculation zone, leading to a secondary separation shock which interacts with the wedge angle shock. Also, there is an interaction with the leading edge shock, producing a complex recompression pattern within the boundary layer. The flow conditions are at the limit of transition. The ramp has been extended for analysis of the pressure gradients and reattachment and the recompression downstream of the multiple interactions. No turbulence modelling is considered here. However, the complex shock/boundary layer interactions are well illustrated by these laminar calculations, as shown in the figures. Experimental results exist for the skin friction coefficient for the second ramp case, and the computational results show a good agreement in this case. The separation region in front of the hinge is well represented. The wall C_p and heat transfer coefficient show clearly the separation, and the significant pressure overhoot before the pressure relaxes to the inviscid pressure level. The computational meshes consisted of structured quadrilateral/structured triangular finite element meshes of the order of 30000 elements. The analysis of such flows is important for estimations of loss (or gain) in control effectiveness in the design of control surfaces, and for the prediction of peak heating in regions of interaction, and downstream of interaction regions.

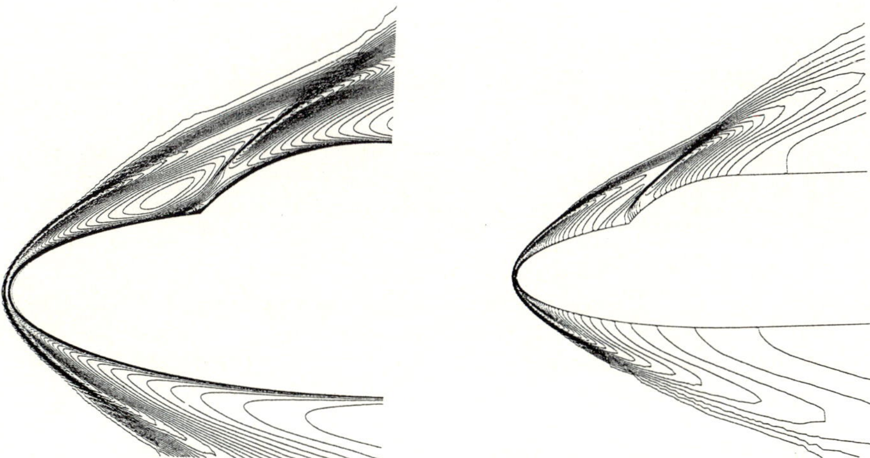

Fig. 3. $M_\infty = 12.7$, Re/m $= 4.3 \times 10^5, p_\infty = 90$ Pa, Iso-density contours (left) and Iso-C_p lines in canopy region (right), showing the highly complex flow structure, and separation bubble during the expansion in front of the canopy

A generic canopy flow is simulated over the leeward side of a double ellipse geometry for high enthalpy experimental flow conditions (Vetter et al. 1993). The double ellipse geometry for zero angle of attack provides an interesting nose-to-car~~ generic shape, and the flowfields around it are highly complex. The free shear layer separates in front of the canopy and reattachs on the cabin, giving rise to peak heating spots within such regions. The detached bow shock impinges on the canopy induced shock and separation shock, as for compression ramps.

A zero incidence Mach 12.7 flow with a Reynolds number per metre of 4.3×10^5, and an upstream pressure of 90 Pa (Figs.3,4) shows a strong viscous effect, with strong dissociation. Cold isothermal wall conditions are assumed ($\approx$ 300 K). A second case for very different conditions,

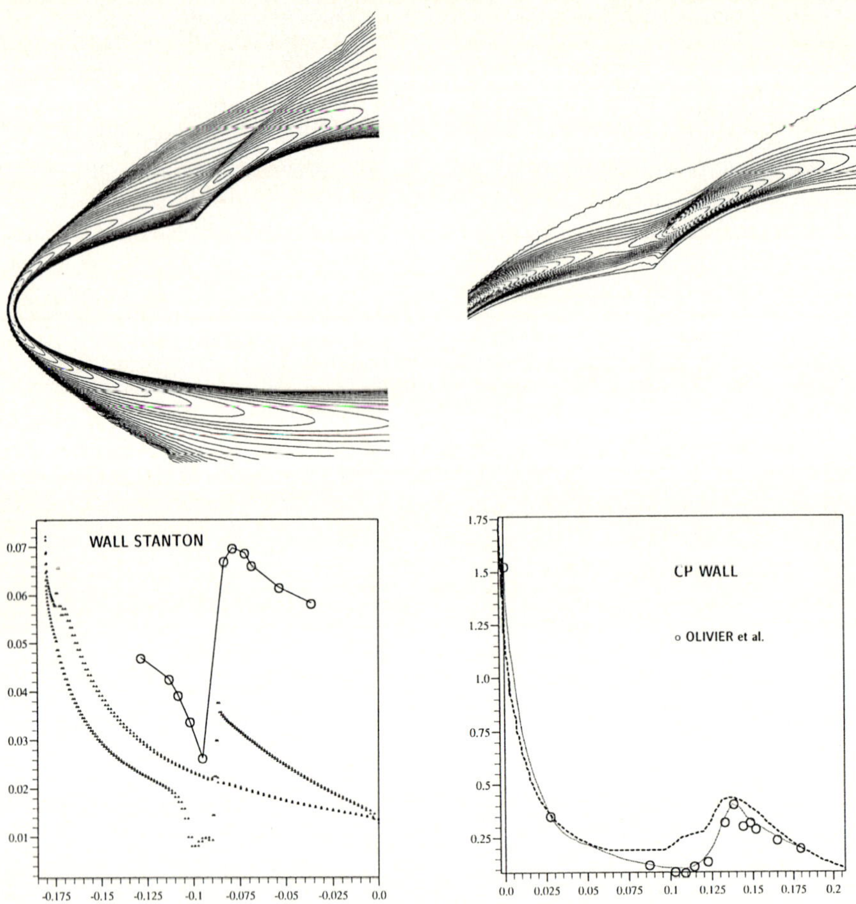

Fig. 4. $M_\infty = 12.7$, Re/m$= 4.3 \times 10^5$, $p_\infty = 90$ Pa. Isotherms (left), and the corresponding Iso-NO lines (right) showing strong dissociation. The wall heat flux and pressure coeffiecient (below) shows a high peak in the reattachment region on the cabin

Mach 6.6, upstream pressure of 5300 Pa and a Reynolds of the order 10^6 (Figs.5,6) leads to more complex shock -boundary layer interactions, due to the higher Reynolds number. The flowfield is similar to the former compression ramp patterns. The severe reservoir conditions imply that, even at such a low Mach, real gas effects can be non-negligeable. The most severe interactions are obtained with the equilibrium gas hypothesis. The shock stand-off distances are smaller, and the interactions closer to the wall in this case. The maximal margins for the various heat loads are thus obtained. Indeed the strong peak in the heat flux within the reattachment region corresponds qualitatively to the experimental data. The computational mesh for the double ellipse consisted of a structured quadrilateral mesh around the geometry, with 110 points in the normal direction, and an unstructured triangular finite element mesh outside this near-wall region.

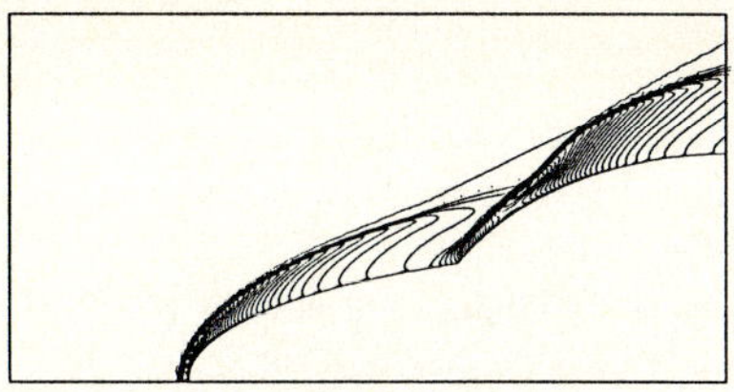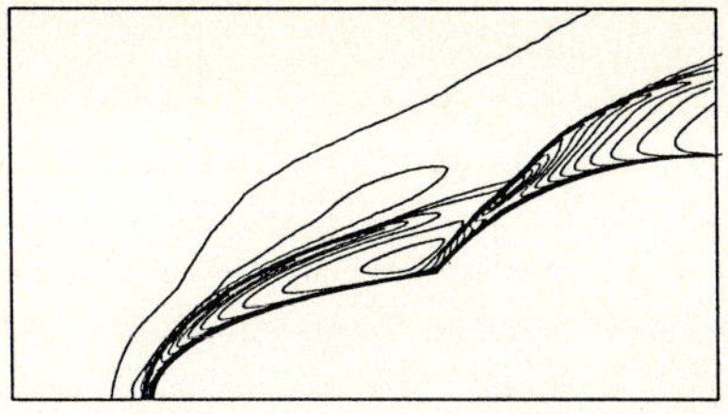

Fig. 5. $M_\infty = 6.6$, Re/m= 2.9×10^6, $p_\infty = 5300$ Pa, $T_\infty = 745$ K (experimental conditions). Flow over double ellipse canopy. Iso-C_p lines (left) and Iso-density lines (right) showing multiple shock -shock and shock-boundary layer interactions for flow considered to be in chemical equilibrium

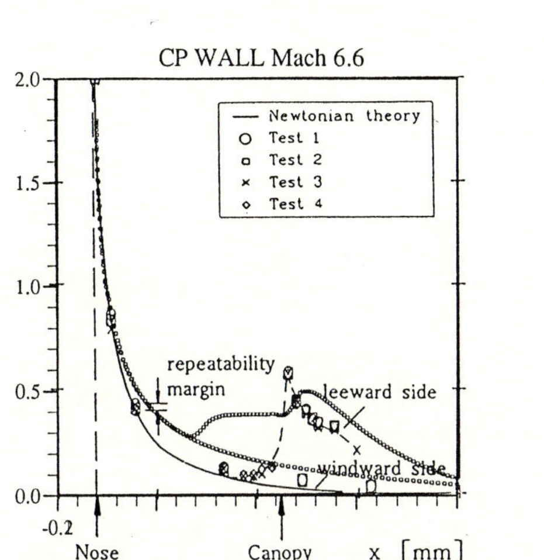

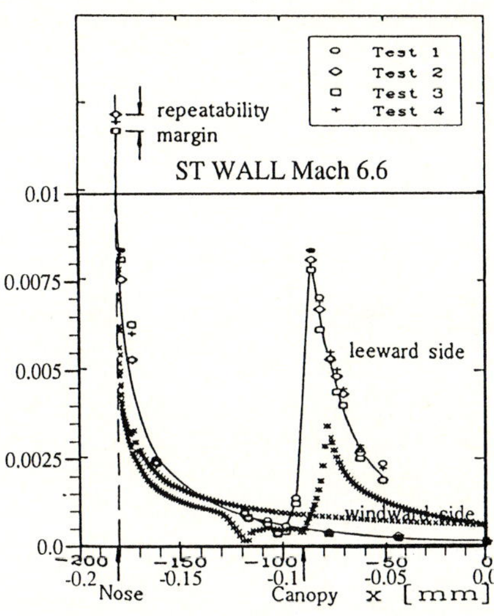

Fig. 6. $M_\infty = 6.6$, Re/m= 2.9×10^6, $p_\infty = 5300$ Pa, $T_\infty = 745$ K (experimental conditions). Flow over double ellipse canopy. Wall pressure coefficient profiles and Stanton number profiles show high peak within the strong recompression region, for equilibrium chemistry. Comparison with experimental results of Vetter et al. (1993) are given

5. Conclusions

Numerical simulation of viscous hypersonic flow over generic geometries illustrate typical aerothermal difficulties in isolated parts of a complete complex flow configuration. Use of the full Navier-Stokes equations and a robust numerical scheme allow precise representations of multiple shock-shock, shock-boundary layer interactions, and an estimation of critical zones for heat loads.

References

Anon. (1992) Proceedings of the Workshop on Hypersonic Flows for Reentry Problems. Springer-Verlag

Anon. (1993) Proceedings of the Second Workshop on Hypersonic Flows for Reentry Problems. Springer-Verlag

Bertin JJ (1990) Aerodynamics for Engineers, John Wiley

Holden MS (1978) A study of flow separation in regions of shock wave - boundary layer interaction in hypersonic flow. AIAA Paper 78-1169

Harten A (1984) SIAM Journal of Numerical Analysis, 54

Park C(1989) A review of reaction rates in high temperature air. AIAA Paper 89-1749

Vetter M, Olivier H, Grönig H. (1993) Flow over double ellipsoid and sphere- experimental results. Proc. Second Workshop on Hypersonic Flows for Reentry Problems, Springer

Vincenti WG, Kruger CH (1986) Introduction to Physical Gas Dynamics. Wiley, New York

Yee H (1989) A class of high resolution explicit and implicit shock capturing methods. NASA Tech. Mem. No. 101088

Visualisation of Shock Waves in Hypersonic CFD Solutions

H.-G. Pagendarm, B. Seitz and S.I. Choudhry
DLR, Deutsche Forschungsanstalt für Luft- und Raumfahrt Göttingen, Germany

Abstract. A new algorithm was added to the visualization system used at DLR's Institute of Theoretical Fluid Mechanics to obtain the shock position from a numerical solution of a flowfield. This assumes that the shock position is given everywhere in the flowfield by the maximal gradient of a quantity like the density along the local flow direction. The local gradient vector is numerically calculated using an integral formulation which is easy to implement and insensitive even to degenerated grid cells which frequently occur in CFD grids. Thresholding techniques allow to distinguish significant portions of the shock patterns from weak numerical artefacts. Complex shock phenomena may be visualized by interactive manipulation of viewing transformations, clipping parameters or threshold values, thus enabling detailed examinations.

Key words: Visualisation, Shock waves, Feature extraction

1. Identification of shock phenomena in numerical data

Today supersonic and hypersonic flows around complex three-dimensional shapes may be simulated numerically using various algorithms from the field of computational fluid dynamics. Such solvers deliver their result at discrete data points. A shock wave within the solution is represented by a change of quantities between several data points. Depending on the resolution of the data and the quality of the numerical solution process these shock patterns are more or less smeared out across a number of data points.

Many recently published results of CFD computations exhibit visualizations which are intended to illustrate the location of shock waves in the flowfield as important elements of the data. Often such visualizations use pseudo colouring or iso-contour plots of some quantity in a series of planes through the flowfield. Once a shock wave cuts through one of these planes selected for visualization, the quantity mapped onto the surface of the plane shows a strong local variation. This can be seen indirectly from a clustering of iso-contours or rapid change of pseudo colours. For simple shock patterns this technique gives a rough impression of the shock position. In three-dimensional complex configurations it may become very difficult to select cutting planes in such a way that a useful visualization of the shock waves will be obtained.

A direct visualization of shock waves as thin layers or surfaces located within the three-dimensional domain of the flow problem becomes highly desirable. In order to obtain such a visualization the location of the shock wave must be known sharply and independent of the grid points of the numerical solution.

2. Numerical results for hypersonic flowfields

The visualization of such shock phenomena will be illustrated using two numerical solutions of hypersonic flows as examples. The first example is a solution for the flow around the front portion of a generic configuration of a SÄNGER type of aircraft. The geometry design of this configuration is based on an interactive grid generation technique (Pagendarm, Laurien and Sobieczky 1988, Sobieczky and Stroeve 1991). The solution was created using an adaptive grid technique by Niederdrenk (1991) and an upwind relaxation method for hypersonic flow simulation by Müller, Niederdrenk and Sobieczky (1990). The second example refers to the configuration for the United

Shock Waves @ Marseille I
Editors: R. Brun, L. Z. Dumitrescu
© Springer-Verlag Berlin Heidelberg 1995

States National Aerospace Plane (NASP) (Kandebo 1990). The NASP configuration is a lifting body with small wings on the rear part of the vehicle. The numerical analysis of the flow around the front part of a generic configuration much similar to some designs suggested for the NASP vehicle exhibits some characteristics which may be found on waveriders as well (Sobieczky and Stroeve 1991). The flow solution was produced using F3D which was originally developed at NASA. The code solves the three-dimensional compressible Euler equations and was adapted to the particular problems of hypersonic waverider flows (Jones, Bauer and Dougherty 1991). The resulting flowfield shows a strong shock underneath the vehicle and a more complex shock pattern above the vehicle and close to the canopy.

3. Shock identification based on density gradients

A new algorithm was developed and implemented into DLR's modular visualization system HIGH-END (Pagendarm 1993 and 1994). This method treats shock visualization in a new fashion by finding the local maxima of the gradient field along the local flow direction everywhere in the field and connecting them to a shock surface. This surface then has to be thresholded in such a way that only parts which are significant for the shock wave are visualized.

A reasonable choice for the position of the shock wave may be given by the position of the steepest gradient of the density field represented at the data points. In one-dimensional problems this position is easily found by calculating the second derivative of ρ with respect to space. The position of the discontinuity is assumed to be at those locations where the second derivative of ρ is zero while the first derivative of ρ is non-zero. In fluid dynamics the quantities to be processed are typically distributed in three-dimensional space. The discrete data points are often given along arbitrary grid lines. The data are defined on computational grids with curved boundaries. All this makes the calculation of derivatives with respect to the physical coordinates time consuming and difficult to implement. Fortunately it turns out to be possible to use an integral expression for the gradient of the desired quantity. This gives a scalar quantity defined everywhere in the field. In order to obtain the gradient vector of the density field from the numerically defined data, the gradient was replaced by a surface integral expression which approaches the gradient for small volumes:

$$\frac{\boldsymbol{u}}{|\boldsymbol{u}|}\nabla\rho = \lim_{V \to 0} \frac{1}{V}\frac{\boldsymbol{u}}{|\boldsymbol{u}|}\left(\oint_A \rho\, dA\right) \tag{1}$$

where ρ is the density and $\boldsymbol{u}$ is the velocity vector.

Now dealing with discrete data a small volume of finite size was evaluated in order to approximate the limiting expression. Staggered grid cells surrounding each node in the numerical grids were defined for that purpose. Besides its simplicity this algorithm has the advantage that there is no possibility for zero divide even in quite curious meshes. This characteristic makes the implementation of boundary treatment extremely simple. The algorithm is insensitive to pathological grid cells where edges have zero length or several corner points collapse. The algorithm does not fail even if cell volumes are zero as long as not all eight neighbouring cells have zero volume.

This algorithm is easily implemented as a computer code. Presently an implementation for structured meshes as presented above is available. For the structured case the points, edges and surfaces are easily annotated by indices, which makes the description simple to read. Nevertheless it is obvious that the same procedure may be implemented for unstructured meshes as well. This simple calculation will be a basic element of the method of visualizing shock waves in flowfields.

The numerical evaluation of Eq.1 allows to compute the local value of the derivative of the density in the local streamwise direction everywhere in the field. A numerical procedure according to the methods explained above was implemented as a module of the visualization system HIGH-END (Pagendarm and Seitz 1993). This scalar quantity is required to be larger than zero for a shock wave to exist, since a compression means an increase of density. Obviously small numerical inaccuracies will also lead to minor positive gradients at various places in the flowfield. These effects which are undesired for shock visualization may be excluded by introducing a threshold value. Values smaller than zero denote expansions.

Along the local flow direction a shock wave is assumed to exist where the second derivative of the density is zero. Therefore as a first step this second derivative of the density is computed over the entire field by applying the derivative module twice. An iso-surface for the value of zero assembles all possible locations for shock waves. Next the result of the first derivative computation needs to be mapped onto this iso-surface in order to be able to distinguish shocks from expansions. This mapping is done by tri-linear interpolation of the discrete values. Afterwards those parts of the iso-surface where the first derivative of the density in local flow direction is smaller than zero are clipped off. This operation is performed for each polygon of the iso-surface by either clipping off a polygon as a whole or by proper subdivision of polygons where necessary. For reasons mentioned above the clipping threshold is always set to a value slightly larger than zero in order to avoid numerical inaccuracies to influence the visualization. As a rule for the numerical value of this threshold cannot be given at this stage, this value is typically adjusted interactively. Fortunately for reasonably good numerical flow solutions, numerical inaccuracies are far below the gradient caused by physical shock phenomena. Therefore the visualization does not depend noticeably on the value chosen for the threshold. The whole procedure is illustrated in the diagram in Fig.1.

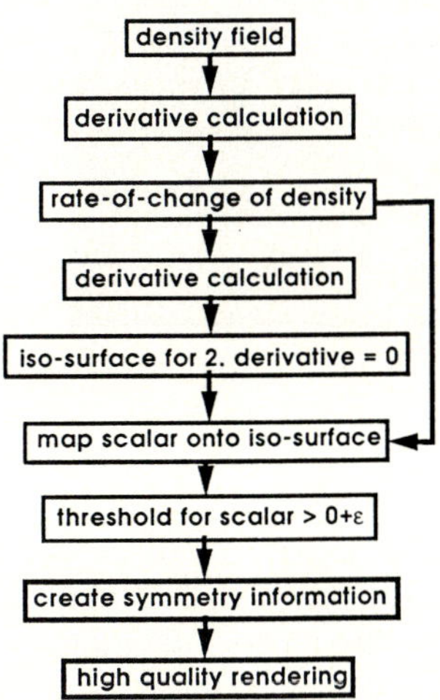

Fig. 1. Simplified scheme for creating high quality shock visualizations

4. Visualization of the shock waves of a hypersonic transport vehicle

It appears that the method may be easily applied to various different fields of scientific data. The essential software components are the calculation of the derivative, the interpolation of the iso-surfaces, and the threshold clipping of surfaces. These algorithms are data transformation algorithms which may be implemented independently of a special visualization system. The first implementation was done for the visualization system HIGHEND which allows easy extension due to its special modular software architecture. Fig.2 shows the location of the shock waves indicated by a transparent surface. Since this image was produced for public relations purposes the geometry of the configuration was extended to both sides of the plane of symmetry and the tail of the aircraft, which was not included in the numerical simulation, is now included.

However, the interactive facilities of the visualization system HIGHEND allow of course to make a detailed analysis of the various shock patterns. Besides the dominant conical nose shock very complex oblique shocks may be visualized close to the strakes of the wings of this configuration.

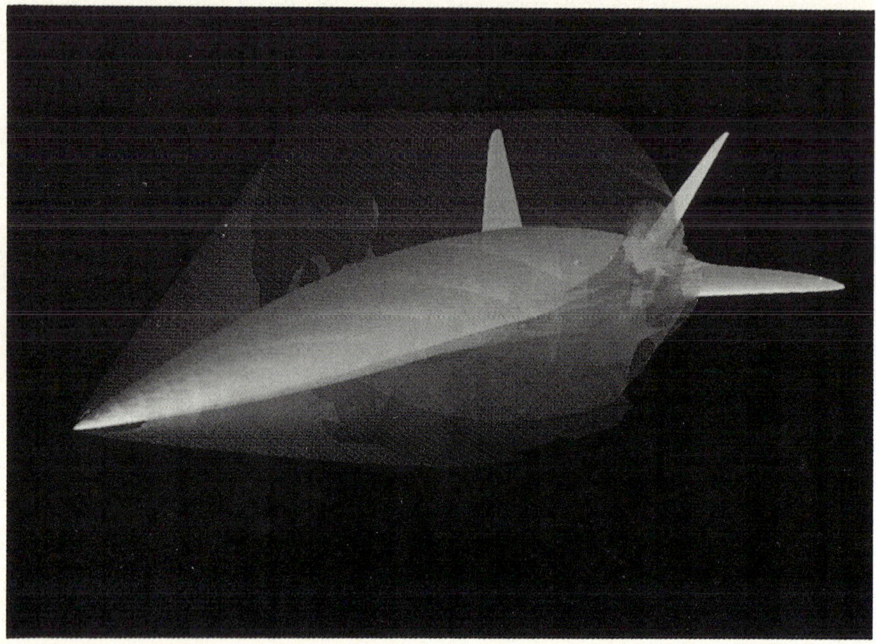

Fig. 2. Shock waves near a hypersonic aircraft

5. Shock visualization using commercial software

DLR's in-house software HIGHEND allows extensions without changing the source code. Since this software may not be available for any purpose, in particular as it was implemented for hardware of SUN Microsystems only up to now, we took a chance to evaluate opportunities to insert this method into commercial software packages. Of course only such systems may be considered which offer some mechanism for extensions by the user. A class of software systems called application builders offers such options. A prominent example of an application builder system is AVS (Coutant and Currington 1991). While the implementation of the derivative calculation within AVS did not pose any severe problem, the implementation of the threshold mechanism turned out to be quite complex. A detailed analysis of the difficulties which occurred when using AVS for shock visualization is given by Pagendarm and Choudhry (1993). However, overall AVS proved to be a useful tool for the visualization of hypersonic flowfields. In spite of the fact that the implementation of such algorithms requires a significant effort the resulting images show a high quality (Fig.3) and the production of animated sequences is well supported.

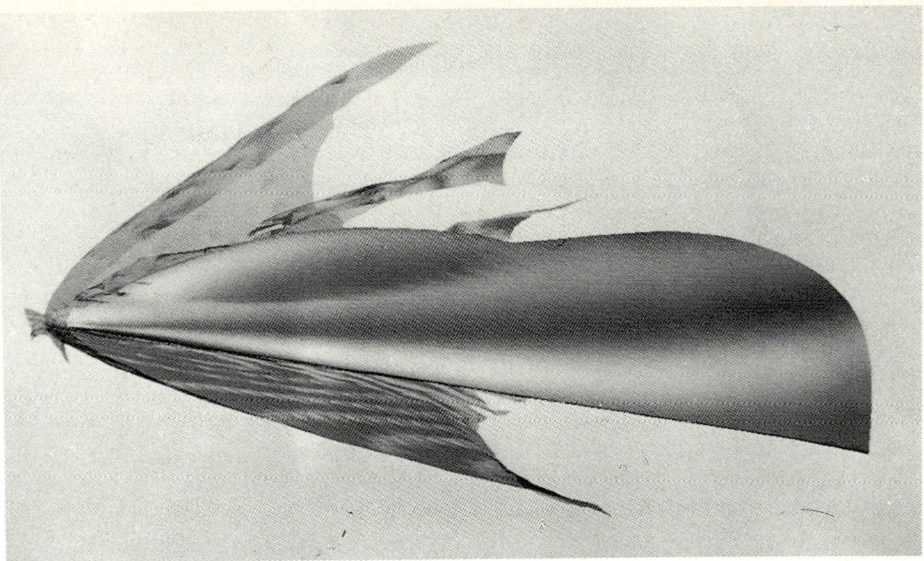

Fig. 3. Shock pattern near the canopy of a configuration designed according to the forebody geometry of a NASP-type vehicle

6. Animated visualization

Since the shock visualization illustrated above turned out to be a useful analysis tool for numerical flow simulation as well as a method which provides images which are easy and intuitively to understand, an animation was produced which allows to demonstrate the various steps involved when processing flow data to finally visualize a shock wave. This animation was entirely produced on a SUN workstation. Single frames were created and stored on the workstation in a new compressed format using SUN's XIL software (Sun Microsystems 1992). This allows to decompress and play back even full resolution images in real time from the disk of the workstation. Applying this technology allows to view and store animated visualization on the workstation. It is also very convenient since the time consuming production of video tapes can be avoided. Only after an animation is completed and all scenes have been produced to a final state it may become necessary to store the animation at a conventional video tape for example for demonstrating the result on a conference. The recording of the video tapes has been performed using a low cost technique described earlier by Choudhry and Pagendarm (1992). The technique has been improved since then by making use of newer hardware and more comfortable software.

Acknowledgements

The authors wish to thank Mrs. M. Hannemann and Mrs. B. Walter who maintain and extend the HIGHEND software system and Mr. R. Zores who helped creating the digital video and image compression software and the video recording software.

References

Choudhry SI, Pagendarm H-G (1992) A low cost approach to animated flow visualization using high quality video. In: Post FH, Hin AJS (eds) Advances in Scientific Visualization, Springer Verlag, Heidelberg

Contant M, Currington I (1991) AVS, a flexible interactive distributed environment for Scientific Visualization applications. Proc. Second Eurographics Workshop on Visualization in Scientific Computing, Delft, The Netherlands

Jones KD, Bauer SXS, Dougherty FC (1991) Hypersonic waverider analysis: a comparison of numerical and experimental results. AIAA Paper 91-1696, AIAA 22nd Fluid Dynamics, Plasma Dynamics & Laser Conference, Honolulu, Hawaii

Kandebo W (1990) Lifting body design is key to single stage to orbit. Aviation Week and Space Technology, Oct. 29

Müller B, Niederdrenk P, Sobieczky H (1990) Simulation of hypersonic waverider flow. First Intl. Hypersonic Waverider Symposium, University of Maryland

Niederdrenk P (1991) Solution - adaptive grid generation by hyperbolic / parabolized hyperbolic P.D.E.s. In: Arcilla AS, Hauser J, Eisemann P, Thompson J (eds) Numerical Grid Generation in Computational Fluid Dynamics and Related Fields, North-Holland, pp 173-184

Pagendarm H-G, Laurien E, Sobieczky H (1988) Interactive geometry definition and grid generation for applied aerodynamics. AIAA 6th Appl. Aerod. Conf., Williamsburg, Virginia

Pagendarm H-G (1993) Scientific visualization in computational fluid dynamics. In: Connors JJ, Hernandez S, Murthy TKS, Power H (eds.) Visualization and Intelligent Design in Engineering and Architecture, Computational Mechanics Publications, Elsevier Science Publishers, Essex, UK, pp 315-330 Baltimore

Pagendarm H-G, Seitz B (1993) An algorithm for detection and visualization of discontinuities in scientific data fields applied to flow data with shock waves. In: Palamidese P (ed.) Scientific Visualization - Advanced Software Techniques, Ellis Horwood Ltd., UK, pp 161-177

Pagendarm H-G, Choudhry SI (1993) Visualization of hypersonic flows - exploring the opportunities to extend AVS. 4th Eurographics Workshop on Visualization in Scientific Computing, Abington, UK (to be published by Springer)

Pagendarm H-G (1994) HIGHEND, A visualization system for 3D data with special support for postprocessing of fluid dynamics data. In: Grave M, LeLous Y, Hewitt WT (eds.) Visualization in Scientific Computing, Springer, see also: Proceedings of the First Eurographics Workshop on Visualization in Scientific Computing, 1990, Clamart, France pp 87-98

Sobieczky H, Stroeve JC (1991) Generic supersonic and hypersonic configurations. AIAA Paper 91-3301

Sun Microsystems (1992) Solaris XIL 1.0 Imaging Library Beta Release Notes, Mountain View

Shock Wave Interactions in Hypersonic Flow

S.L.B. Yeung and I.M. Hall
Department of Engineering, University of Manchester, M13 9PL, England

Abstract. Numerical calculations have been used to examine the differences between two-dimensional and three-dimensional shock-shock interactions. The configurations considered are the intersection of a plane shock wave with the bow shock ahead of a circular cylinder and a sphere. The local flow close to shock intersections has been studied analytically and relations obtained between the shock and streamline curvatures.

Key words: Hypersonic, Shock interaction

1. Introduction

The flowfields around aircraft flying at high supersonic speeds are likely to contain many instances of shock wave interactions. Some of these may have no design significance but others may have critical consequences producing for example local high-temperature regions on the surfaces.

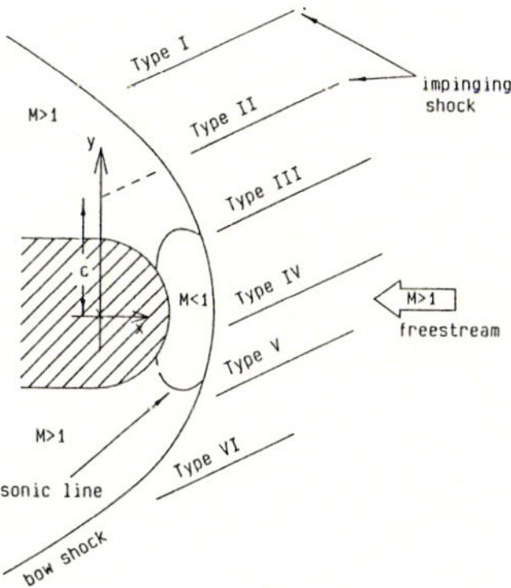

Fig. 1. Schematic diagram for six types of interaction

After an extensive experimental investigation, Edney (1968) classified shock-shock interactions into six types. Where the flow remains supersonic these interaction types are independent of the overall flowfield in which they are produced. For example Types VI and I are simple, sometimes called regular, interactions of weak shock waves of the same and opposing families respectively. If the strengths of one or both of the incoming waves is increased a situation occurs where a regular reflection is not possible and Type V and II interactions occur. Apart from a small local region of subsonic flow generated by the 'Mach stem' the flow is essentially supersonic. However, if one of the shock waves is 'strong' so that the Mach number behind it is subsonic the intersection is no longer 'local'. Such interactions are classified as Type III and IV. The whole range of interaction types can be obtained by the interaction of a plane shock wave generated by a wedge with the bow shock wave on a two-dimensional body such as a circular cylinder as illustrated in Fig.1. As the point of intersection moves from a position well above the cylinder to well below the cylinder the interaction, initially of Type I, changes through Types II, III, IV and V to VI.

Such interactions have been well documented (Klopfer and Yee 1988, Delery 1988) but, since Types III and IV are not 'local' they seemed worthy of further attention. For example, the shock standoff distance is significantly smaller for a sphere than a cylinder and therefore, for

Shock Waves @ Marseille I
Editors: R. Brun, L. Z. Dumitrescu

these types of interaction significant differences can occur between two-dimensional (2D) and three-dimensional (3D) flows. Results of numerical computations are used to comment on the differences.

Fine detail of the flow at a shock wave interaction is difficult to extract from numerical computations and a short foray into old-fashioned analysis is made. Relations between shock shape and stream surface have been obtained.

2. Numerical methods

2.1. Point implicit finite volume method

The disadvantages of early methods with second-order accuracy have been substantially reduced by modern high resolution schemes. In the present work a TVD scheme has been used which is based on the method originally proposed by Harten (1982) and developed by Yee (1987).

In this paper a cell-centred point implicit finite volume scheme is used. The point implicit method is a compromise between fully explicit and fully implicit methods. The implicit method is applied to each cell individually but the scheme is globally explicit. Hence, instead of solving a large system of equations iteratively as would be necessary in a fully implicit method, it is only necessary to solve a 5×5 system (in 3D) by Gaussian elimination. The algorithm used was developed by Gnoffo (1990) and is unconditionally stable according to local linearized analysis.

2.2. Grid generation

The boundaries of the region to be discretized are the body, a cylinder or a sphere, an outer (inflow) boundary upstream of the bow shock wave which is composed of two quadratic curves, and two downstream or outflow boundaries. This region is suitable for simple grid generation methods. However, in order to produce a fine grid around the main features of the flow some distortion of the obvious grid is required. Several methods were tried but the most successful was an elliptic grid generator based on the solution of a Poisson equation with suitable source terms. The source terms depend on the boundary points distribution as proposed by Thompson and Warsi (1982). The boundary points distribution is defined by one-dimensional transformation so that grid clustering at the end point or at an interior point is possible. In 3D calculations, stacking of 2D surfaces was used.

2.3. Boundary conditions

As long as the computational domain is large enough to contain the bow shock and the shock system, the primitive variables at the inflow boundary can be fixed at freestream conditions for both inviscid and viscous calculations. Apart from a small portion of the wall boundary layer, the major part of the flow at the outflow boundary is supersonic. It is assumed that there is no significant upstream effect due to the small region of subsonic flow and zeroth order extrapolation is applied to viscous calculations in exactly the same way as inviscid calculations with supersonic outflow. The no-slip condition is assumed at the body surface where all the velocity components are kept at zero. The pressure is obtained from the zero gradient assumption, $\frac{\partial p}{\partial n} = 0$. A boundary condition on temperature is needed. Fixed wall temperature has been used in the present case.

3. Numerical results and discussion

3.1. Two-dimensional flow

Fig.2a shows the pressure distribution over a cylinder with Mach 8 freestream and shock angle 18.1° at different shock intersection positions. Interaction types vary from Type III to Type V. Fig.2b shows the effects of the impinging shock strength on the pressure distribution. The effect of a very weak impinging shock on the bow shock wave and the downstream flow is considerable. Although the flow deflection across the impinging shock is only 1.25° the bow shock is distorted so that the downstream flow is very close to the maximum deflection angle.

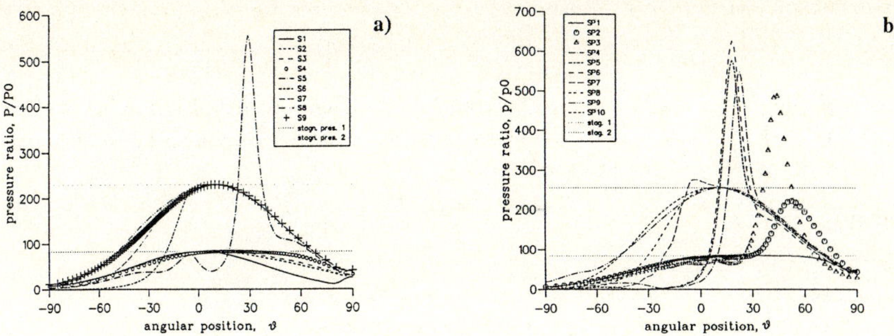

Fig. 2. Pressure distributions (a) with different impinging positions, (b) with various shock strengths

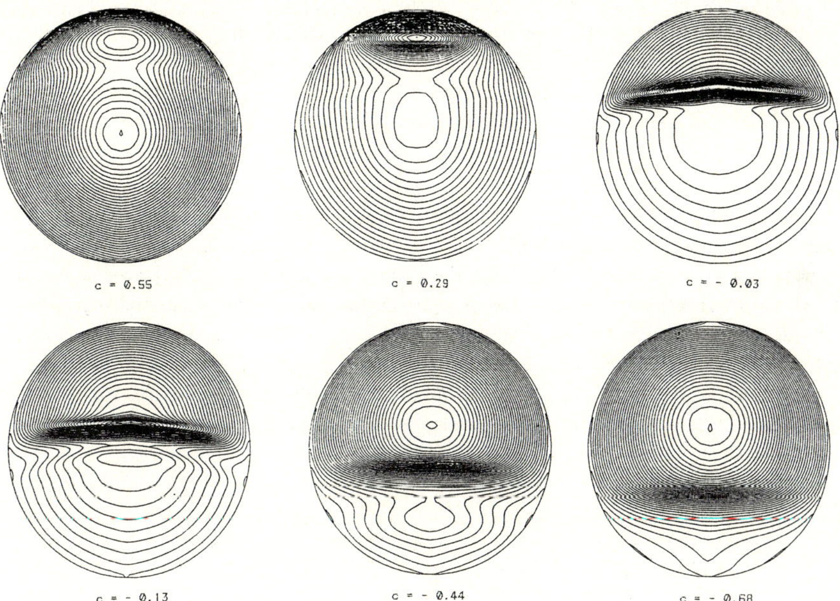

Fig. 3. Surface pressure contours on sphere for six shock interaction cases

3.2. Three-dimensional flow

A series of calculations were carried out for the intersection of a plane oblique shock with the bow shock ahead of a hemispherical body at different locations with freestream Mach number 8.0 and the incident shock angle of 16.0° corresponding to a flow deflection of 10.5°. The grid size is $41 \times 71 \times 21$ for all cases. The position of the impinging shock is described by its virtual intersection, $y = c$, with the y-axis. Results shown are obtained for converged steady state solutions for seven cases where $c = 0.55$, 0.29, -0.03, -0.13, -0.44, -0.68 and -1.04 respectively. The incident shock sweeps from the upper to lower part of the bow shock in order to obtain different intersection points. Interaction types vary from Type III to Type V. Surface pressure contours for the first six cases are shown in Fig.3.

The influence of the impinging shock on a three-dimensional bow shock is relatively less severe than in the corresponding 2D case in terms of the bow shock displacement distance. Centre plane surface pressure distributions, Fig.4, are plotted on the same scale as 2D cases for comparison.

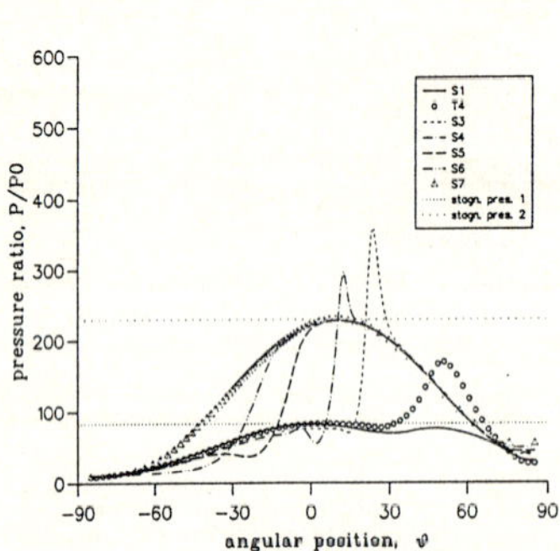

Fig. 4. Centre plane pressure distributions for three-dimensional shock-shock interactions

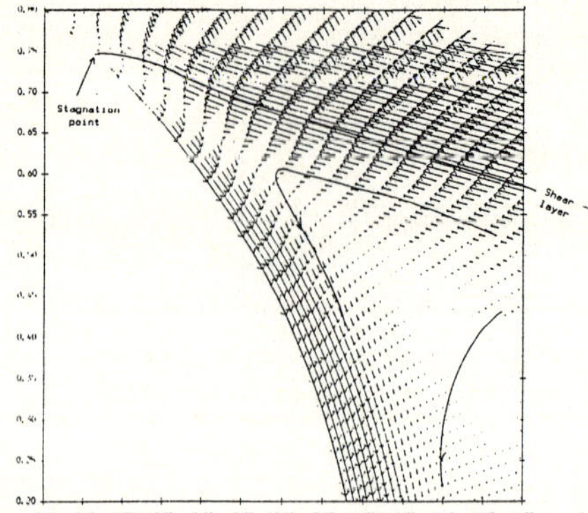

Fig. 5a. Vector plots for Type III interaction, 2D

The pressure peak is lower than the 2D interactions. The relatively coarse grid may account for the smaller pressure jump, but the substantial difference is due to the 3D effect where the jet length is shorter than the 2D cases. Hence, the flow is not allowed to accelerate to higher Mach number before impinging on the sphere surface.

In a typical Type III interaction, the computational results show three stagnation points on the sphere surface, Fig.5a, compared with one in the 2D case, Fig.5b. S_1 is the bow shock stagnation point. S_2 is the shear layer attachment point. The third stagnation point, S_3, is due to the meeting of the flow from S_1 and S_2.

In a Type IV interaction, the main difference between 2D and 3D flows is that the jet flow can expand laterally, the amount of high energy fluid passing over the upper part of the sphere being comparatively less than in the two-dimensional case. Thus, the displacement of the subsonic region is not as significant as appears when the jet impinges on a cylinder. As noticed from the pressure distribution, Type IV interaction is fairly localised. The pressure distribution of the affected region over the sphere at the centre plane is lower than its adjacent planes. This suggests that the flow undergoes two-dimensional expansion and changes are more rapid below the jet impingment point on the centre plane. As seen on the surface contour plot at the jet impingment region, rapid change of the shape can be observed for two Type IV cases. This shows the limited extent of this strong three-dimensional shock-shock interaction.

Table 1. Pressure gradient in streamline direction

M_∞	$\alpha/^\circ$	$\partial p/\partial x$	$\partial^2 p : \partial x^2$
3.50	30.0	-0.3210	-1.1468
5.10	18.0	-0.0437	-0.0968
5.10	33.0	-0.2335	-0.6971
7.10	16.0	-0.0223	-0.0374
7.50	12.0	-0.0118	-0.0210
7.92	30.0	-0.1110	-0.2407
7.95	18.0	-0.0260	-0.0462

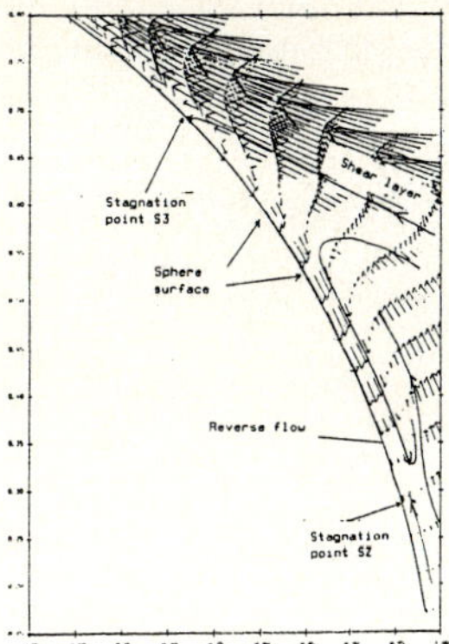

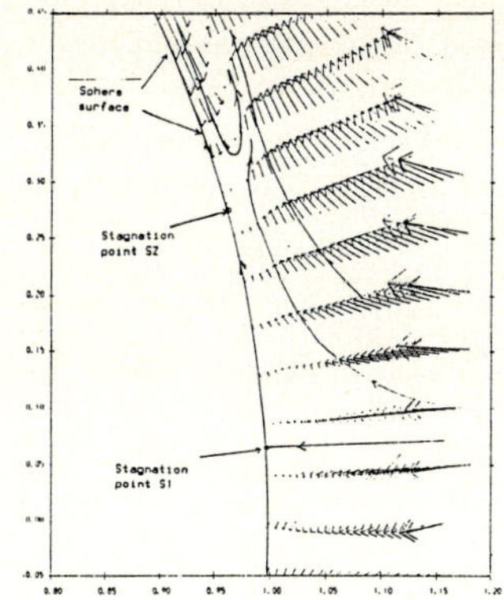

Fig. 5b. Vector plots for Type III interaction, 3D

4. Analytical calculations

Experiments have been carried out on a swept cylinder emerging from a flat plate aligned with a hypersonic freestream (Alziary 1987) and attempts have been made to calculate the flow and to compare theory and experiment. There is, in fact, a shock-shock interaction between a shock from the leading edge of the plate and the bow shock wave on the swept cylinder. However, our interest was whetted by the problem of the shock wave shape at the root. For a highly swept cylinder in a hypersonic stream, the shock wave would be attached in inviscid flow. Hence, the flow behind is known precisely, in principle. Poll (1985) has a criterion for transition based on the velocity derivatives along the leading edge of the cylinder. To find these, the details of the flow behind this attached shock wave were examined.

In 1947, Thomas (1947, 1948) derived a relation between streamline curvature and shock wave curvature in 2D flow. A survey for further work on this can be found in Yeung (1993). An extension to 3D flow is outlined below.

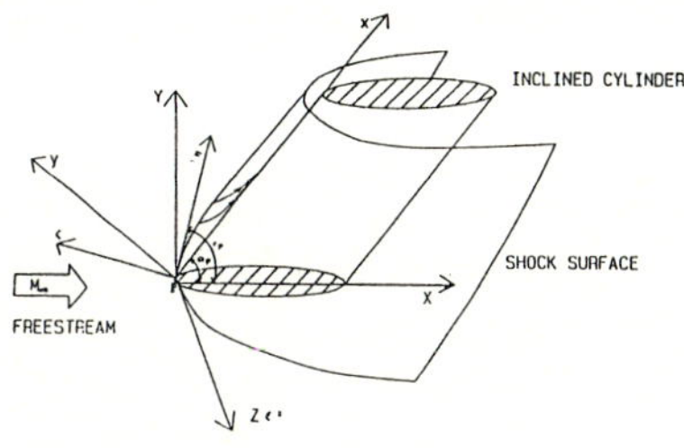

Fig. 6. Shock-aligned and streamline-aligned coordinate systems

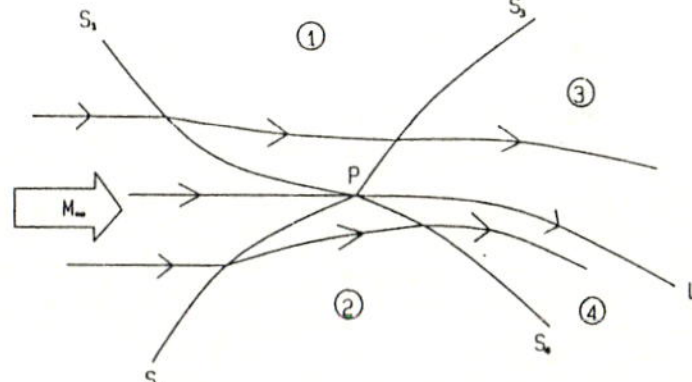

Fig. 7. Schematic for curved shock waves intersection

Two sets of coordinate systems have been used in the calculation, shock aligned and streamline aligned. As shown in Fig.6, the global axis X is positioned in the freestream direction with the Y axis perpendicular to the freestream. X, Y, Z are mutually perpendicular to each other with the origin P lying on the shock surface.

In order to obtain the shock aligned system, the global axes are rotated about the Z axis by the shock angle ϵ_p such that X axis is tangential to the shock surface and Y axis is normal to the shock surface pointing into the freestream. Similar to the shock aligned system, the streamline aligned system is obtained by turning the global axes through δ_p, the deflection angle of the streamline at P, about the Z axis. The orientations of those axes are illustrated as Fig.6; τ_p is the difference between the shock angle and the deflection angle at P.

In the shock-aligned coordinate system the shock surface can be described locally by the polynomial

$$\zeta = \frac{k_1}{2}\eta^2 + \frac{k_2}{2}\xi^2 + \frac{k_3}{3}\eta^3 + k_4\eta\xi^2 + \cdots \tag{1}$$

The first two coefficients k_1 and k_2 are the curvatures of the shock in the $\zeta\eta$ plane and the $\zeta\xi$ plane at P respectively.

The normal to the shock surface is found and the standard Rankine-Hugoniot relations are used to determine conditions downstream of the shock. Velocity components are then transformed to the streamline-oriented system. Collecting together all the information gives 25 equations relating the first derivatives of u, v, w, p and ρ with respect to x, y, z, ξ and η. Hence, for a given body geometry the values of k_1 and k_2 in Eq.1 can be found. After differentiating again and applying two more boundary conditions from the cylinder geometry, k_3 and k_4 can be determined giving more information on the shock shape.

Results have been obtained for seven cases with various combinations of Mach number and angle of inclination of the cylinder. Pressure gradients are tabulated in Table 1.

These analytic results show the existence of a point of inflexion in the spanwise pressure distribution. This point of inflexion may lead to flow instability in the spanwise direction and cause boundary layer transition from laminar to turbulent.

It was realised that a similar analysis could be applied to shock-shock interactions where one or both shock waves is curved as shown in Fig.7. Details are given by Yeung (1993).

References

Alziary T de Roquefort (1987) Leading edge transition in hypersonic flows. Proc. First Europe-US Short Course on Hypersonics, Paris, France

Delery J (1988) Shock/shock and shock-wave/boundary-layer interactions in hypersonic flows. AGARD-FDP-UKI Special Course on Aerothermodynamics of Hypersonic Vehicles

Edney B (1968) Anomalous heat–transfer and pressure distribution on blunt bodies at hypersonic speeds in the presence of an impinging shock. FFA Report 115

Gnoffo PA (1990) An upwind-biased, point-implicit relaxation algorithm for viscous, compressible perfect-gas flows. NASA TR 2953

Harten A (1982) High resolution schemes for hyperbolic conservation laws. J. Comp. Phys. 49:149

Klopfer GH, Yee HC (1988) Viscous hypersonic shock-on-shock interaction on blunt cowl lips. AIAA Paper 88-0233

Poll DIA (1985) Some observations of the transition process on the windward face of a long yawed cylinder. J. Fluid Mech. 150:329

Thomas TY (1947) On curved shock waves. J. Math. Phys. 26:62

Thomas TY (1948) Calculation of curvatures of attached shock waves. J. Math. Phys. 27: 279

Thompson JF, Warsi ZUA (1982) Boundary-fitted coordinate systems for numerical solution of partial differential equations — a review. J. Comp. Phys. 47:1

Yee HC (1987) Upwind and symmetric shock capturing schemes. NASA TM 89464

Yeung SL (1993) Shock wave interactions in hypersonic flow. PhD thesis, University of Manchester

Numerical Simulation of Transient Bluff Body Flows

J. F. Milthorpe
Department of Aerospace and Mechanical Engineering, University College,
University of New South Wales, Australian Defence Force Academy, Campbell, A.C.T. 2600, Australia

Abstract.   An algorithm based on convection of fundamental flow quantities, that is, mass, momentum and energy, has been described previously. Computer codes to implement the algorithm have been developed for two-dimensional and axisymmetric flows. This algorithm has been further developed to improve accuracy and stability of the calculations. In this paper the algorithm is used to calculate the flows about two-dimensional and axisymmetric bluff bodies. Bodies of practical interest in supersonic and hypersonic flow are often sufficiently bluff for a detached shock to develop, and correct prediction of the shock standoff distance and the position of the sonic line is a useful test of numerical methods.

The development with time of these detached shocks should be a further test of the computational method, although the author is not aware of good experimental measurements of standoff development for comparison.

Examples are given of the predicted development of detached shocks following the passage of an incident shock over a bluff body. Results are given for a plate with a flat nose, a body of revolution with a flat nose and a sphere at Mach numbers of 1.177 and 1.69. Comparisons are given with standard experimental results quoted by Liepmann and Roshko (1957) for the steady state standoff.

Key words: Supersonic, Detached shock, Navier-Stokes

1. Introduction

Many problems in supersonic flow involve detached shocks generated by bluff bodies. When a detached shock develops, a significant region of the flow field is subsonic: around the stagnation point the velocity is naturally very small. Some common computational techniques, such as parabolized Navier-Stokes (PNS) rely on convection to make upstream diffusion negligable. The algorithm used in this paper has been described previously, and shown to give good predictions for supersonic flows with attached shocks (Milthorpe 1991a,b, 1992). Prediction of shock shape and stand-off distance for flows around bluff bodies offers a strong test for numerical methods, as experimental data is available for flows around a wide variety of bluff shapes. The development of detached shocks is particularly interesting. This paper reports the prediction of shock development in a variety of flows with detached shocks at low Mach numbers.

2. Conservation algorithm

The algorithm used to calculate the flow values is described fully in Milthorpe (1992). The algorithm is based on conservation of mass, momentum and energy as the simulation proceeds over discrete timesteps. The flow domain is divided into rectangular cells. Each cell contains, at any given time, a certain block of material which possesses a certain momentum and energy. In the course of the next timestep this block, with its associated mass, momentum and energy, is convected a small distance. It may also gain or lose mass, momentum or energy by diffusion and by the action of pressure gradients. The algorithm evaluates the changes in these quantities for the material block. The process is illustrated in Fig.1 for convection with positive u and v velocities. The letters A, B, C, D represent the proportion of the material block located in a particular cell at the end of a timestep dt. These proportions of the quantities in the material block are allocated

Shock Waves @ Marseille I
Editors: R. Brun, L. Z. Dumitrescu

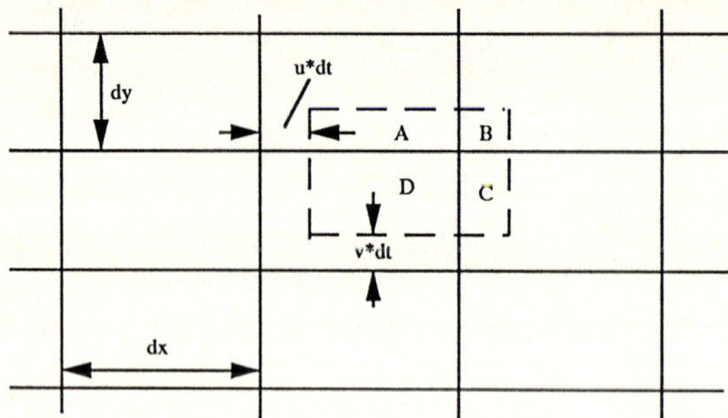

Fig. 1. Convection of material block with velocity u, v during one timestep dt. See text for description of convection algorithm

to their respective cells. When the process has been performed for all cells the new distribution of mass, momentum and energy is available. From this distribution updated values of density, velocity, pressure and temperature can be derived.

The axisymmetric algorithm is derived from the two-dimensional version with a number of changes to suit the geometry. Although the step sizes in the radial and longitudinal directions are constant, the volume of each cell varies with radius, as does the area of the cell surfaces. In addition, the momentum and energy balance in the radial direction are affected by the circumferential pressure term. The computer code has been written with the assumption that the y direction is radial and the x direction longitudinal. Although the examples given have zero radius on the lower edge of the solution region, the program has been written to allow for an offset radius, permitting solution in an annulus.

3. Development of axisymmetric flow over a blunt-ended cylinder

A blunt cylinder is placed in an initially stagnant gas, and a shock is started from the upstream boundary so that the resulting freestream flow is parallel to the axis of the cylinder. After the incident normal shock reaches the blunt end of the cylinder, a detached oblique shock develops which approaches a steady state. This flow has been simulated for inlet Mach numbers of 1.177 and 1.69. Contour plots of density for a Mach number of 1.69 are given in Figs.2 - 4 at a range of times from shortly after the incident shock reaches the cylinder to steady state. The development of the stand-off distance with non-dimensional time, for both Mach numbers, is shown in Fig.8. The non-dimensional time t_{nd} shown is given by

$$t_{nd} = t_{model} \frac{\rho_0 u_0 a_0}{\mu_0}. \tag{1}$$

The time in model units is t_{model} measured from the arrival of the incident shock at the cylinder. The inlet speed of sound is a_0, the inlet density is ρ_0, the inlet speed is u_0 and the inlet viscosity is μ_0. The viscosity varies according to Sutherland's law. Also shown in Fig.8 are the steady-state stand-off distances reported by Liepmann and Roshko (1957). The density gradients visible near the solid boundaries are caused by the cold wall condition, which is used in all the computations reported here.

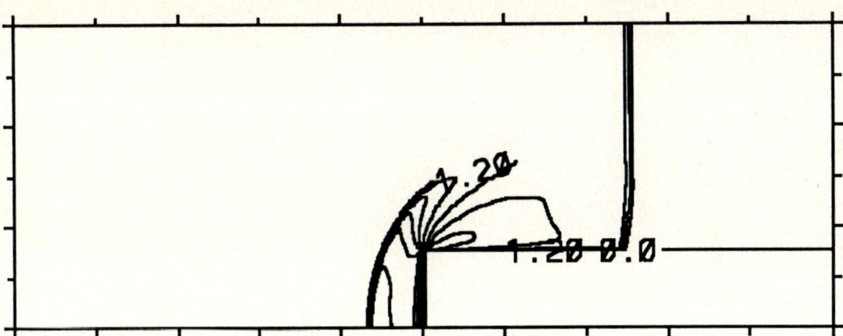

Fig. 2. Density ratio contours for axisymmetric blunt body at Mach 1.69.
Non-dimensional time is 1188. Contours from 0.0 at intervals of 0.3

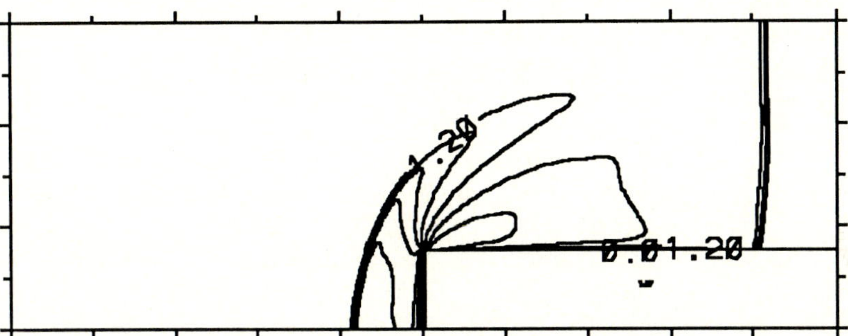

Fig. 3. Density ratio contours for axisymmetric blunt body at Mach 1.69.
Non-dimensional time is 4260. Contours from 0.0 at intervals of 0.3

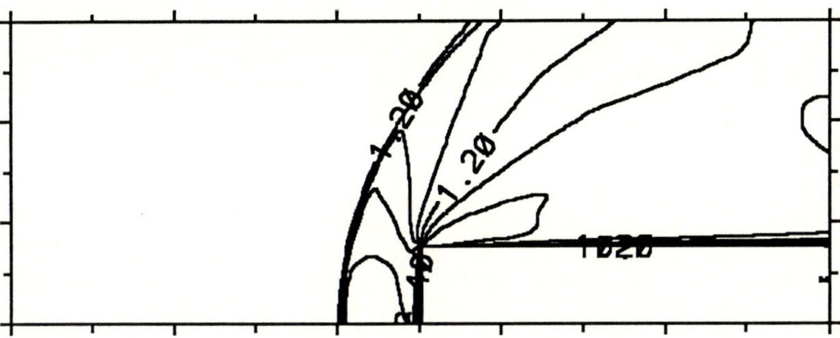

Fig. 4. Density ratio contours for axisymmetric blunt body at Mach 1.69.
Non-dimensional time is 9939 (steady state). Contours from 0.0 at intervals of 0.3

4. Development of flow over a blunt flat plate

The detached flow over a blunt-edged flat plate exhibits much greater stand-off than for an equivalent axisymmetric body. Density ratio contours for such a flow with incident Mach number 1.69 are shown in Fig.5. The flow has developed a steady state. A comparison with the equivalent axisymmetric flow in Fig.4 shows a very different character to the predicted contour shape in

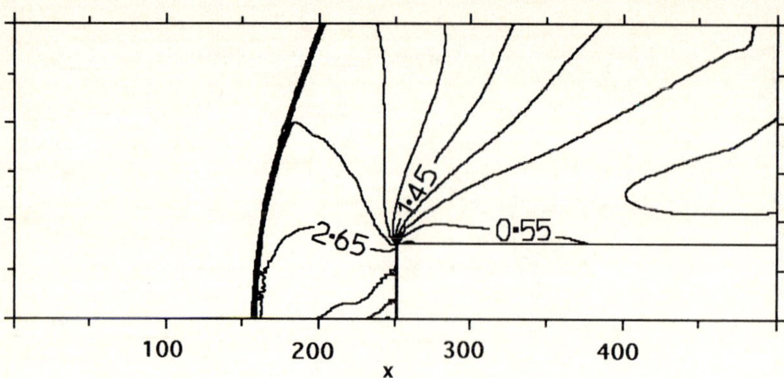

Fig. 5. Density ratio contours for blunt edged flat plate at Mach 1.69
Steady state. Contours from 0.05 at intervals of 0.3

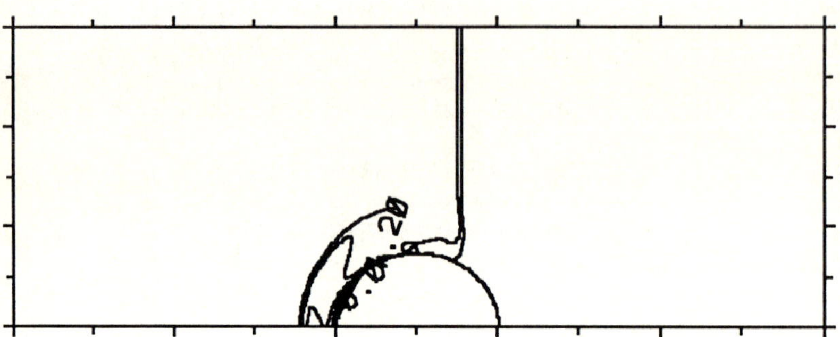

Fig. 6. Density ratio contours for sphere at Mach 1.177.
Non-dimensional time is 659. Contours from 0.0 at intervals of 0.3

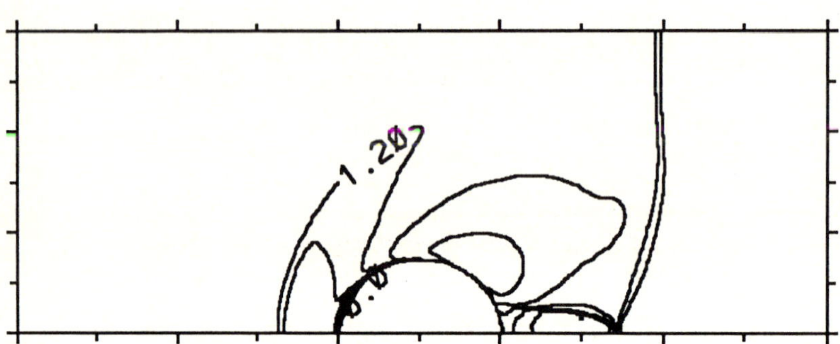

Fig. 7. Density ratio contours for sphere at Mach 1.177.
Non-dimensional time is 1648. Contours from 0.0 at intervals of 0.3

the stagnation region. The axisymmetric flow has a much broader stagnation region, which is
very roughly spherical and has a large region of low density gradients. The flat plate flow has
a stagnation region which is approximtely triangular and has much higher density gradients.
Standoff distances at early times are shown in Fig.8 for Mach 1.69. The early development of the
detached shock is seen to proceed at a very similar rate for all the geometries tested.

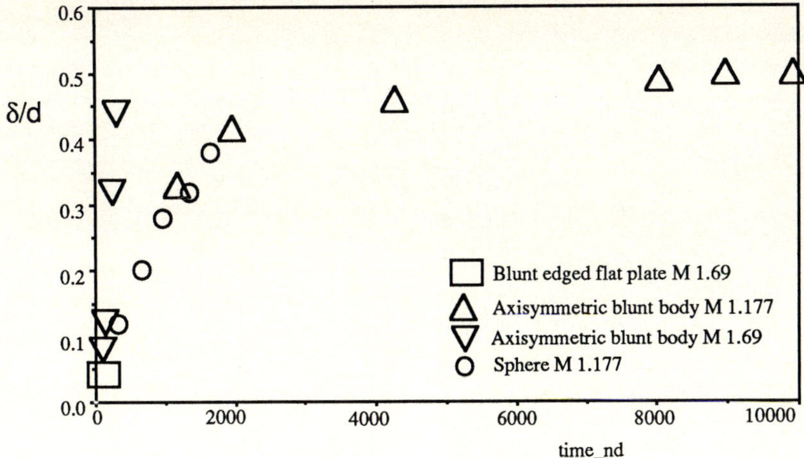

Fig. 8. Development of stand-off distance (δ/d) for detached shocks against non-dimensional time, for various bodies

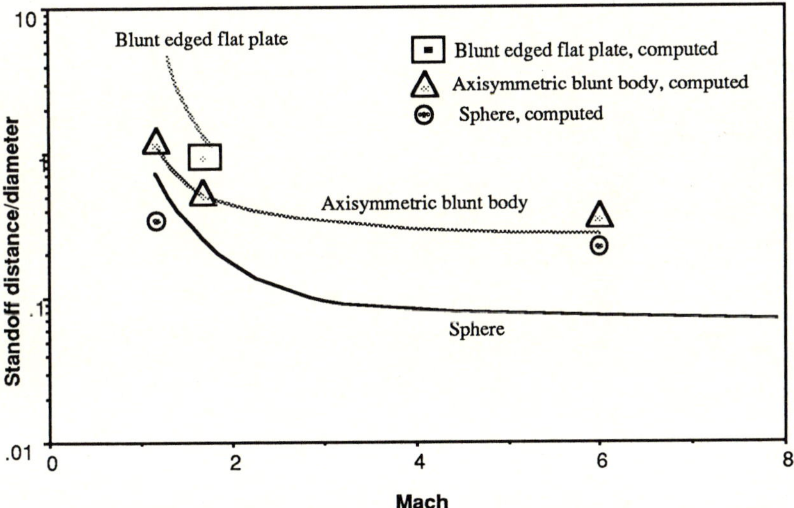

Fig. 9. Comparison of computed and measured shock standoff distances

5. Development of flow over a sphere

The development of the flow over a sphere is shown at two times in Figs.6-7. In the first figure, the incident shock is still passing over the sphere. In the second figure the incident shock is well clear of the sphere and the recompression shock is well developed. The development of stand-off distance with time is shown in Fig.8.

6. Conclusion

The development in time of some supersonic flows with detached shocks has been outlined. The algorithm converges monotonically to the steady state stand-off distances reported by Liepmann and Roshko (1957), which are shown in Fig.9. The method of normalising time does not appear very satisfactory, as the rate of growth of stand-off distance varies considerably with both Mach number and flow geometry.

References

Liepmann, HW, Roshko A (1957) Elements of Gasdynamics. John Wiley and Sons, New York

Milthorpe JF (1991a) Numerical simulation of two-dimensional compressible flows. In: Proc. 7th Intl. Conf. on Numerical Methods in Laminar and Turbulent Flows, Stanford University, Palo Alto, USA

Milthorpe JF (1991b) Numerical simulation of two-dimensional and axisymmetric compressible flows. In: Takayama K (ed) Proc. 18th Intl. Symp. on Shock Waves, Tohoku University, Sendai, Japan, p 1133

Milthorpe JF (1992) Simulation of supersonic and hypersonic flows. Intl. J. Numer. Methods Fluids 14:267-288.

Milthorpe JF (1993) Numerical simulation of two-dimensional compressible flows. Intl. J. Numer. Methods for Heat and Fluid Flow 3:223-231

An Efficient Unstructured Euler Solver for Transient Shocked Flows

A.A. Fursenko*, **D.M. Sharov***, **E.V. Timofeev†** and **P.A. Voinovich***
*Advanced Technology Center, P.O. Box 160, St.Petersburg 198103, Russia
†Ioffe Physical–Technical Institute, RAS, Polytekhnicheskaya 26, St.Petersburg 194021, Russia

Abstract. Paper presents a highly accurate, efficient and flexible approach, combining the advantages of high resolution shock-capturing schemes and adaptive unstructured grids, for the numerical simulation of essentially unsteady 2D shocked gas flows with complex wave patterns. Numerical examples illustrate the accuracy, reliability and efficiency of the method.

Key words: Unsteady shocked gas flows, Godunov-type scheme, Adaptive unstructured grid

1. Introduction

Strongly unsteady shocked gas flows possess some specific features connected mainly with the significant nonlinearity of the mathematical model, and unpredictable, rapidly developing and arbitrary shaped complex wave patterns. As a consequence, the shock-capturing approach represents the most suitable choice for numerical simulation of the above phenomena.

Because the total number of mesh points is typically determined by the desirable resolution at discontinuities rather than by the diversity of the flow domain geometry, dynamic adaptation of the grid to the solution and highly-accurate methods might save considerable computer memory and CPU time.

A range of highly accurate shock-capturing schemes have been introduced during the past decade. Some recent papers are devoted to coupling these schemes with unstructured grids. The latter become popular due to their extreme flexibility to diverse geometries and to their non-fixed data structure, suitable for local rearranging of the grid.

A high-order nonoscillatory Godunov-type scheme and solution-adaptive unstructured grids completed with a time-efficient hierarchical refinement-unrefinement are used in our work. These approaches contribute to considerable improvement in both accuracy and efficiency and result in robust algorithms which require almost no hand-tuning. For lack of space, we present here only a brief description of the proposed technique. Essential details connected mainly with the choice of the appropriate data structure, in view of sparing computer memory, as well as with the methods for improving the accuracy, together with some features of 2D and 3D grid adaptation algorithms and vectorization of the code may be found in Fursenko (1993).

2. Governing equations

The Eulerian gas model is used. The mass, momentum and energy conservation laws underlie the mathematical model. For two dimensions the governing equations are:

$$\int_{\text{Vol}} U\,\mathrm{d}V\ \Big|_{t_0}^{t} + \int_{t_0}^{t} \mathrm{d}t \int_{\sigma} \left(F_x n_x \mathrm{d}s + F_y n_y \mathrm{d}s \right) = 0,$$

$$U = \begin{pmatrix} \rho \\ \rho u_x \\ \rho u_y \\ \rho e \end{pmatrix}, \qquad F_x = \begin{pmatrix} \rho u_x \\ \rho u_x^2 + p \\ \rho u_x u_y \\ (\rho e + p)u_x \end{pmatrix}, \qquad F_y = \begin{pmatrix} \rho u_y \\ \rho u_x u_y \\ \rho u_y^2 + p \\ (\rho e + p)u_y \end{pmatrix},$$

where $\boldsymbol{u} = (u_x, u_y)$ is the velocity vector, p the pressure, ρ the density, $e = \epsilon + 0.5u^2$ the specific total energy. We consider here perfect gases, so the specific internal energy is given by

Shock Waves @ Marseille I
Editors: R. Brun, L. Z. Dumitrescu   © Springer-Verlag Berlin Heidelberg 1995

$\epsilon = C_v T = (1/(\gamma - 1))p/\rho$, where T is the temperature, C_v the specific heat at constant volume, and γ the polytropic index. Vol represents a gas volume bounded by the closed surface σ with the outward normal $\boldsymbol{n} = (n_x, n_y)$.

3. The unstructured finite-volume method

We shall consider here boundary-fitted unstructured grids composed of triangular area elements, which provide more isotropic spatial discretization as compared to traditional rectangular grids (Fursenko et al. 1992). The Delaunay technique is used for grid generation in a given computational domain. Cell-vertex nonoverlapped volumes are established, so that the vertices of the grid triangles correspond to the nodes in which the dependent gasdynamic variables are given. The control volumes are constructed around each node by medians of triangle grid elements. Each edge ij between nodes i and j is associated with the vector area $\boldsymbol{S}_{ij}$.

Numerous schemes belonging to the TVD, TVB, ENO, flux-splitting and high-order Godunov-type classes were employed to compute fluxes through the areas $\boldsymbol{S}_{ij}$. Comprehensive testing has shown that a second-order Godunov-type scheme, proposed by Rodionov (1987) for steady-state supersonic flows and modified for transient flows (see Fursenko et al. 1993a, 1993b) possesses better properties, with regard to the balance between computational efficiency and accuracy for practical applications. It does better also for unstructured grids with arbitrary shaped control volumes.

To enhance the spatial accuracy, a linear reconstruction of primitive gasdynamic variables $(V = (\rho, u_x, u_y, p)^T)$ is used for the whole control volume. The respective gradient is chosen as minimum of average gradient per neighboring triangles $\overline{(\nabla V)}_i$ and doubled gradients $(\nabla V)_e$ in the triangles surrounding node i:

$$\frac{\partial V}{\partial \alpha}_i = \operatorname*{minmod}_e \left[\overline{\frac{\partial V}{\partial \alpha}}_i, \, 2\frac{\partial V}{\partial \alpha}_e \right], \qquad \alpha = x, y; \qquad \overline{(\nabla V)}_i = \frac{1}{\mathrm{Vol}_i} \sum_e (\nabla V)_e \mathrm{Vol}_e.$$

The predictor step employs no Riemann solver, which provides for a high efficiency of the scheme. For the control volume related to nodal point i we obtain the predictor solution $\tilde{U}_i$:

$$\tilde{U}_i \mathrm{Vol}_i = U_i^n \, \mathrm{Vol}_i - \Delta t \sum_j \left\{ F_x \left(V_{ij}^{in} \right) \, n_x S_{ij} + F_y \left(V_{ij}^{in} \right) \, n_y S_{ij} \right\},$$

where V_{ij}^{in} are the primitive gasdynamic variables inside the control volume surface; and

$$V_{ij}^{in} = V_i + 0.5(\nabla V)_i \, \boldsymbol{ij}; \qquad V_{ij}^{out} = V_j + 0.5(\nabla V)_j \, \boldsymbol{ji}.$$

Subscript j denotes grid points surrounding node i; superscript n the time level; Vol_i the value of control volume i; S_{ij} the area of the interface between volumes i and j; $\boldsymbol{n} = (n_x, n_y)$ the unit vector of the outward normal to the surface of control volume i.

Then for each volume i and time level $(n + 1)$ we have the corrector step:

$$U_i^{n+1} \mathrm{Vol}_i = U_i^n \mathrm{Vol}_i - \Delta t \sum_j \left\{ F_x \left(W_{ij} \right) \, n_x S_{ij} + F_y \left(W_{ij} \right) \, n_y S_{ij} \right\},$$

where W_{ij} are primitive gasdynamic variables obtained from the Riemann problem solution for $\tilde{V}_{ij}^{in}, \tilde{V}_{ij}^{out}$ (the Riemann solver by Roe has been used):

$$\tilde{V}_{ij}^{in} = 0.5 \left(\tilde{V}_i + V_i + (\nabla V)_i \, \boldsymbol{ij} \right); \qquad \tilde{V}_{ij}^{out} = 0.5 \left(\tilde{V}_j + V_j + (\nabla V)_j \, \boldsymbol{ji} \right).$$

4. Local grid adaptation

We utilize here a refinement/coarsening criterion based on the values of second-order differences which is consistent with hybrid-type schemes properties:

$$E_i = \max_e \left(\frac{\left| (\nabla U)_e - \overline{(\nabla U)_i} \right|}{|(\nabla U)_e| + \left| \overline{(\nabla U)_i} \right| + \epsilon |U_i| d^{-1}} \right),$$

where e denotes the neighbouring triangles of node i, d the diameter of the cell, and ϵ a filter coefficient.

All nodes are marked according to E_i values. If $E_i > T_r$, all triangles of mutual vertex i should be fractionized (refinement). If $E_i < T_c$, the node i might be removed (derefinement, coarsening). The typical values of the thresholds and of the filter coefficient are: $T_r = 0.02$, $T_c = 0.007$, $\epsilon = 0.05$. As a key variable U, the gas density, is generally used.

For 2D problems the method proposed by Löhner has been taken (Löhner 1987). Unlike Löhner, an original node-based data structure underlies the developed computer code providing excellent efficiency (see Fursenko 1993). The data structure needs 12 words per mesh node to store cross-references from inserted nodes. No memory for storing "parent" and "son" cells is required.

The efficient refinement/coarsening procedure being called at each time step takes about 1 percent of CPU time used by the scheme and saves considerably CPU time and computer memory compared to uniform refinement (by more than one order of magnitude). The vector version of the code has been demonstrated to be approximately 20 times faster as compared to a scalar code for Fujitsu VP-200 implementation.

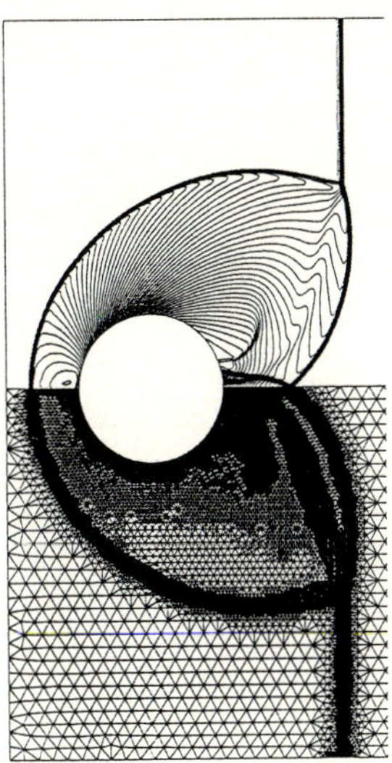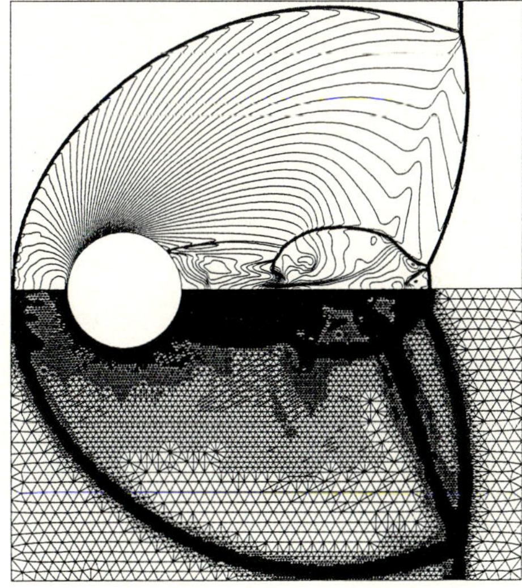

Fig. 1. Shock wave with $M_s = 3$ diffracting over a cylinder in CO_2, $\gamma = 1.289$ (density contour lines and grids)

5. Numerical examples

Several 2D (both plane and axisymmetric) gasdynamic problems have been solved using the above technique. For lack of space, we consider here only some of them (for more examples see Fursenko 1993, Fursenko et al. 1992, 1993a, 1993b).

The classical gasdynamic problem of shock wave diffraction around a circular cylinder involves numerous shocks, contact and slip surfaces and their interactions, it is well studied both experimentally and numerically and may serve as a good test. Comparison of the solution (see Fig.1) with the results obtained earlier using traditional numerical methods indicates a significantly higher resolution of discontinuities. The present results have been obtained using an IBM PC/AT/486 computer and computations have taken about 3 hours.

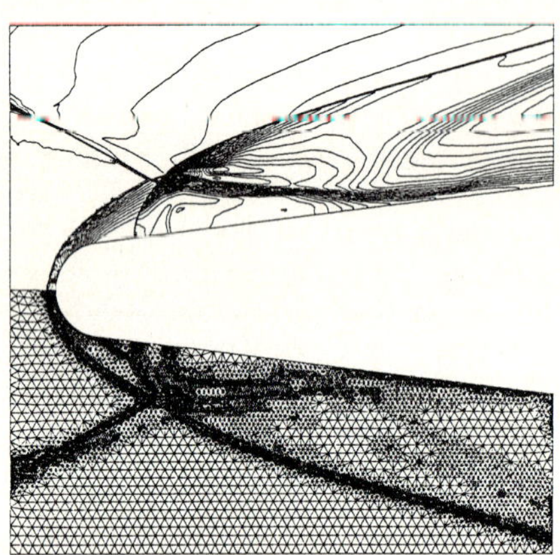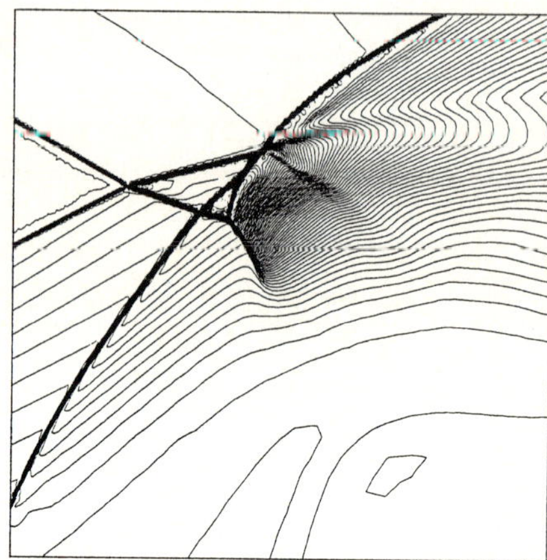

Fig. 2. Converging shock-blunted cone interaction in a steady-state supersonic $M_s = 4.67$ air stream (magnified details of the flow, left - zoom × 10, right - zoom × 100)

The next problem is connected with converging shocks. We have considered supersonic flow in a cylindrical inlet with a central blunt body. The leading edge of the inlet produces a converging oblique shock interacting with the central body and its bow shock. The steady problem has been solved using a time-asymptotic approach and the above explicit scheme. We remark here that the shock collisions of interest occur only in a very small region compared to the spatial scale of the computational domain. Fig.2 presents small details of the flow (density contour lines and respective grid). The left figure comprises 1/100 part of the whole computational domain, while the right figure - 1/100 part of the previous one (near the converging shock - bow shock intersection).

Fig.3 illustrates the shock wave configurations occurring when a strong plane shock interacts with a double wedge. The wedges inclinations considered provide for Mach reflection of both the incident shock wave on the first wedge and the respective Mach stem on the second wedge. The subsequent triple-shocks interaction results in a Mach configuration in accordance with theoretical and experimental data. The computational grid contains only 5000 nodes at the last time moment, however the resolution is sufficient to study the interaction process.

The next problem demonstrates the capability of the method to capture very weak waves. An instant plane explosion near an actual ground surface is considered (Fujii et al. 1990). At the distance of interest - about 2-2.5 km - the overpressure induced by the explosive blast wave

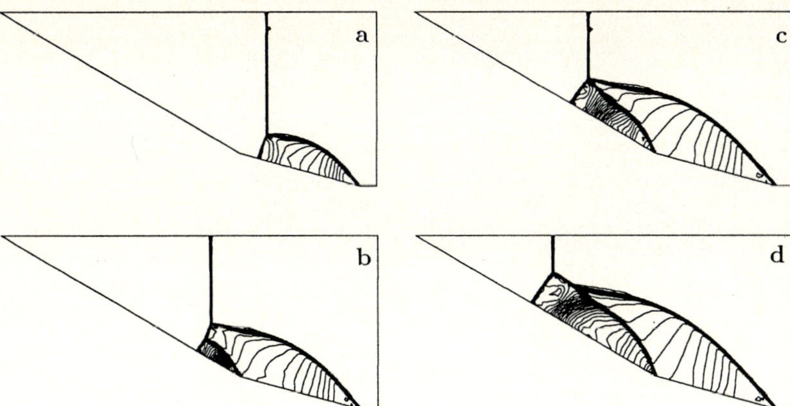

Fig. 3a-d. Flow development for strong shock wave ($M_s \gg 1$) -double wedge ($\alpha_1 = 15°, \alpha_2 = 30°$) interaction. Pressure contour lines

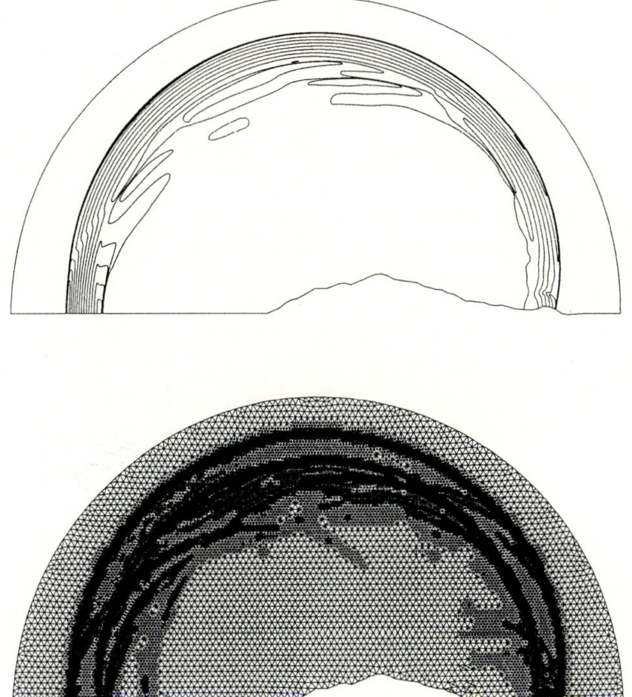

Fig. 4. Blast wave propagation with ground surface effect (pressure contour lines and grid)

is about 2-6% of atmospheric pressure. In spite of the low intensity, Fig.4 clearly shows a high resolution of the leading shock wave and a complex wave structure behind its front.

6. Conclusions

To summarize, the above-presented numerical results clearly indicate the high accuracy, reliability and efficiency of the numerical technique based on unstructured grids. The efficiency of the described technique and respective code allows to use them on vector computers, workstations and even on PCs. The primary attention should be concentrated on the choice of the data structure, as it strongly influences the efficiency, optimization capabilities and sophistication of the resulting code. According to our experience, the unstructured approach is the cheapest compared to the structured one for the class of problems considered here.

Acknowledgments

The authors wish to thank Prof. K. Fujii of the Institute of Space and Astronautical Science, Japan, whose kind support allows for computing on their Fujitsu VP-200 vector supercomputer. We are also thankful to him for bringing to our attention the problem of blast wave - natural landscape interaction. This resarch was partially supported by the Russian Foundation for Fundamental Investigation (Grant 93-02-14797).

References

Fujii K, Shimizu F, Tamura Y, Higashino F, Hinada M, Akiba R (1990) Blast wave simulation with ground surface effect. In: Proc. 17th Intl. Symp. on Space Tech. and Science, Tokyo, Japan, pp 759-762

Fursenko AA (1993) Unstructured Euler solvers and underlying algorithms for strongly unsteady shocked flows. In: Daiguji H (ed) Proc. 5th Intl. Symp. on CFD, Sendai, Japan

Fursenko AA, Sharov DM, Timofeev EV, Voinovich PA (1992) Numerical simulation of shock wave interactions with channel bends and gas nonuniformities. Computers and Fluids 21:377-396

Fursenko AA, Mende NP, Oshima K, Sharov DM, Timofeev EV, Voinovich PA (1993a) Numerical simulation of propagation of shock waves through channel bends. Computational Fluid Dynamics Journal 2:1-36

Fursenko AA, Sharov DM, Timofeev EV, Voinovich PA (1993b) High-resolution schemes and unstructured grids in transient shocked flow simulation. Lecture Notes in Physics, 414:250-254

Löhner R (1987) The efficient simulation of strongly unsteady flows by the finite element method. AIAA Paper 87-0555

Rodionov AV (1987) Improvement of the approximation order in the Godunov scheme. Zh. Vychisl. Mat. Mat. Fiz. 27:1853-1860 (in Russian)

CTH: A Software Family for Multi-Dimensional Shock Physics Analysis

E.S. Hertel,Jr., R.L. Bell, M.G. Elrick, A.V. Farnsworth, G.I. Kerley,
J.M. McGlaun, S.V. Petney, S.A. Silling, P.A. Taylor and L. Yarrington
Sandia National Laboratories, Albuquerque, New Mexico, USA

Abstract. CTH is a family of codes developed at Sandia National Laboratories for modelling complex multi-dimensional, multi-material problems that are characterized by large deformations and/or strong shocks. A two-step, second-order accurate Eulerian solution algorithm is used to solve the mass, momentum, and energy conservation equations. CTH includes models for material strength, fracture, porosity, and high explosive detonation and initiation.

Viscoplastic or rate-dependent models of material strength have been added recently. The formulations of Johnson-Cook, Zerilli-Armstrong, and Steinberg-Guinan-Lund are standard options within CTH. These models rely on the use of an internal state variable (typically the equivalent plastic strain) to account for the history dependence of material response. Comparison with experimental data will be made using a new material strength model. The advancements made in modelling material response have significantly improved the ability of CTH to model complex large-deformation, plastic-flow dominated phenomena.

Key Words: CFD, Shock physics

1. Introduction

The CTH (McGlaun et al. 1990) software family is a complete package for the initialization, integration through time, and visualization of complex phenomena surrounding shock physics. The classes of problems that can be analyzed with CTH include penetration and perforation, compression, high explosive detonation and initiation phenomena, and hypervelocity impact. The software family currently consists of six major components along with several graphics post-processing tools. The major components are CTHGEN which sets up the initial configuration of the problem; CTHREZ which allows the user to rezone a problem and/or combine two or more problems into one; CTH which does the time integration on the conservation equations; CTHED which allows the user to query the problem data base for detailed cell-level information; CTHPLT which allows the user to produce black-and-white or color graphics at a given time of all computational and secondary variables; and HISPLT which allows the user to produce black-and-white or color graphics as a function of time.

CTH uses an Eulerian mesh to solve the conservation equations. Six geometry options are available in CTH: one-dimensional rectangular, cylindrical, and spherical; two-dimensional rectangular, and cylindrical; and three-dimensional rectangular. Up to ten materials and void can occupy a computational cell.

The conservation equations of mass, energy, and momentum are replaced by finite volume approximations. The mesh is generated from three sets of user-specified spatial coordinates $x(i)$, $y(j)$, and $z(k)$ which are logically connected. All quantities are cell centered except the velocities which are face centered. All cell-centered quantities are assumed to be constant across the cell. CTH uses a staggered mesh for the solution of the momentum conservation equation.

Shock Waves @ Marseille I
Editors: R. Brun, L. Z. Dumitrescu   © Springer-Verlag Berlin Heidelberg 1995

2. Solution scheme

Eulerian codes like CTH use a mesh that is fixed in space, and material flows through the mesh in response to boundary and initial conditions. The conservation equations are solved in two steps, a Lagrangian step and a remap step. In the Lagrangian step, the Lagrangian forms of the governing equations are integrated across a timestep. The initial mesh distorts to follow the material motions and there is no mass flux across the cell boundaries. After the Lagrangian step, the remap step is performed. During the remap step, the distorted cells are remapped back to the initial fixed mesh.

The volume, mass, momentum, and energy must be conserved across the Lagrangian step. The mass is conserved trivially because the mesh moves with the material during the Lagrangian step and no mass crosses the cell boundaries. The remaining conservation equations are replaced with explicit finite volume representations of the original integral equations. Although the explicit equations are solvable, the time step must be controlled to prevent information from crossing more than one cell during a single timestep. The timestep is the minimum of a CFL condition and a cell-volume change which limits excessive compression or expansion.

An area of considerable interest is the modelling of material strength. As Eulerian shock physics codes are increasingly used for modelling relatively low-speed impacts, the emphasis has switched from equation-of-state issues to details of material response. A linearly-elastic perfectly-plastic material strength model is available. This model has two yield surface options: a von Mises (constant) yield surface and a pressure-dependent yield surface. Both surfaces limit the second invariant of the stress deviator. The pressure-dependent surface has low strength at low pressure and increasing strength as the pressure increases. Both models have thermal softening and low density degradation corrections available to the user.

In addition to the simple elastic-plastic model, three rate-dependent viscoplastic models are available in CTH: Johnson and Cook (1985), Zerilli and Armstrong (1987), and Steinberg et al. (1980). All of these models utilize complex functional forms of the yield surface which depends on both the local material state and some information about the history or rate-dependent state of the material. The model of Johnson et al. (1990) describing the phenomena of brittle material failure is also available in CTH. This model replaces the normal CTH equation-of-state options with an internal description of the thermal response of the material.

Strength in compression is only one aspect of material response that is important for many of the current applications of CTH. Material fracture can also be critical in making accurate predictions. CTH has historically used a void insertion model to simulate failure. This model monitors the tensile state of a cell and relieves that tension by adding void. The model allows the user to select either pressure or principal stress as the criteria for tension relief. This technique is adequate to predict material failure due to hydrodynamic spall, but does not predict failure due to shear phenomena or large strains. An equivalent plastic strain-based model of Johnson and Cook (1985) has been implemented in CTH. This model predicts failure due to shear deformation. The model is coupled with the void insertion model through the user-specified cut-off tensile pressure or stress. If the user requests the Johnson-Cook fracture model, the user-specified value for the maximum tension that can be supported is degraded by the fraction of failed material. As the failure due to shear deformation increases, the amount of tension that a cell can support decreases until the cell can support no tension. At this point the material has completely failed and will act like a fluid.

A three-term artificial viscosity is used to control the discontinuity associated with shocks and other instabilities. The form used to control shocks is a vector subset of the full viscosity tensor with linear and quadratic terms. The vector includes the diagonal elements xx, yy, and zz. The third viscosity term is linear and controls a singular point in the update of the stress deviators at the axis-of-symmetry for the two-dimensional cylindrical geometry option. Shear viscosity has been found to control non-physical oscillations sometimes seen in normal penetration simulations.

There are three high-explosive detonation models in the production version of CTH. The oldest is the programmed burn model, which is appropriate for simple detonations where the initiation time and location are well known. This model automatically calculates the appropriate amount of energy to be deposited in each cell of high explosive at the correct time. The two additional models are capable of modelling the shock initiation, as well as the detonation of high explosives. They are the Chapman-Jouguet and History Variable Reactive Burn models (Kerley 1992). Both of these models rely on the use of an internal state variable to monitor the reaction parameters of the high explosive initiation.

All of the models discussed above are applied during the Lagrangian step. After this step, the velocities, energies, stress deviators, and any internal state variables must be remapped back to the initial mesh.

The remap step advects the appropriate mass, momentum, energy, and volume from the deformed mesh of the Lagrangian step to the original mesh. The volume flux between the old and new cells is calculated first. An interface tracking algorithm then decides which materials in the old cells are moved with the volume flux. Next, each material's mass and internal energy are moved from the old to the new cells. Finally, the momentum and kinetic energy are moved using the information from the interface tracker.

Operator splitting techniques are used to preform the multi-dimensional remap operation. The resulting one-dimensional convection equations use a second-order accurate conservative scheme developed by van Leer (1977). The scheme used in CTH replaces a uniform distribution in the old cell with a linear distribution. To reduce the asymmetry resulting from the operator splitting, a permutation scheme in direction is applied.

The volume flux is calculated from the geometry of the cell-face motion. Once that is calculated, the volume of materials to be advected must be estimated. The interface reconstruction algorithms are used to estimate the amount of each material to be advected. Two interface reconstruction algorithms are available in CTH. The Simple Line Interface Calculation scheme developed by Noh and Woodward (1976) is available for all geometries. This technique is exact for one-dimensional geometries and effectively first-order for other geometries. A higher resolution (second order) interface reconstruction scheme developed by McGlaun et al. (1990) is also available for two-dimensional geometries.

In most formulations of the momentum advection equation, a staggered mesh is constructed in space and time. The staggered mesh is only used to advance the face-centered (velocity) variables for use by the next Lagrangian step. The staggered mesh requires additional storage and a second and different formulation of the difference equations for momentum advection. A technique developed by Benson (1991) has been recently implemented in CTH as a part of other upgrades. This technique maps the portions of the face-centered information to cell-centered locations and then performs the advection on the cell-centered data. The method requires advection of twice as much information but reduces the complexity of the advection scheme significantly. Fig.1 displays a schematic of the process. The first step is to map the respective halves of the momentum to the adjacent cell-centers. The step leaves two pieces of data at the cell-centers for advection. The second step advects the new cell-centered data using the normal advection algorithm for cell-centered data. After the advection step is complete, the separate halves of the cell-centered momentum are mapped back to the face locations and a face-centered velocity is calculated. This technique has been found to be very effective and guarantees that the accuracy of the advection scheme is maintained and has the additional benefit of also being a monotone scheme.

3. Equations-of-state

Strong shock simulations require sophisticated and accurate models of the thermodynamic behavior of materials. Phase changes, nonlinear behavior, and fracture can be important for accurate predictions. CTH has two major equation-of-state packages available to the user: the Analytic Equation-of-State (ANEOS) package of Thompson and Lauson (1972) and the SNL-SESAME package of Kerley (1991).

The ANEOS package uses a Helmholtz potential to calculate the internal energy as a function of mass density and temperature. The use of a Helmholtz potential assures thermodynamic consistency. The ANEOS package allows for three-phase (liquid-vapor, liquid-solid, and solid-vapor) equations-of-state. In addition to the multi-phase analytic forms, there are three simple analytic expressions within the ANEOS package. Expressions for a linear $U_s - u_p$ (Mie-Gruneisen), Jones-Wilkins-Lee, and ideal gas equations-of-state are subsets of the multi-phase analytic forms. The drawbacks of the ANEOS package are that it can be difficult for the novice to parameterize and is computationally inefficient.

The SNL-SESAME package is based on the SESAME tabular equation-of-state representation. The tabular form allows the equation-of-state to be as sophisticated as the information contained in the table. The tabular representations may contain multiple liquid-vapor, liquid-solid, and solid-solid transitions, whereas ANEOS is restricted to at most three. Table look-up schemes can be made very efficient on vector and parallel computers. Once the tables have been constructed, they are accessible to all users. In addition to the tabular representations, analytic forms for linear $U_s - u_p$ (Mie-Gruneisen), Jones-Wilkins-Lee, and ideal gas equations-of-state are also available. This package also contains two models that can be used to represent porous media.

4. Multi-material cell thermodynamics

Each computational cell can contain up to ten materials and void. The thermodynamic state of the cell must be determined from the volume, energy, and mass of the constituent materials. If only one material occupies the cell, the determination of the thermodynamic state is straight forward. If many materials occupy the cell, the determination of the thermodynamic state is more complicated. Three models are currently available. The first model assumes that all materials in a cell are at the same temperature and pressure. Complex non-linear algebraic equations govern the energy partition for this model. These equations are solved by a multi-variable Newton's iteration. There are several drawbacks associated with this model. It exhibits unrealistic energy flows from hot to cold materials. Also, solid material in a cell containing vapor will not fracture because no tension can be supported. In addition, the iterations can be extremely time consuming.

The second model allows all materials to have different temperatures but identical pressures. In general, this model is superior to the previous model described above but also has deficiencies. In some circumstances, cells containing solid and vapor can unrealistically partition energy between the solid and vapor leading to extreme temperatures in the vapor.

The third model allows all materials to have different temperatures and pressures. Cell quantities for this model are volume weighted averages of the constituent materials. The principal deficiency of this model is that there is no pressure relaxation feature for mixed material cells. In general, this model is the simplest computationally and gives the best results.

5. CTH simulations

The first simulation discussed here is a problem that describes material deformation and failure that is inherently two-dimensional and requires a model that describes rate-dependent constitutive behavior. The initial conditions for this simulation consist of a stationary 6 mm thick tantalum disk and a 3 mm thick tantalum disk that is traveling at 403 m/s. The CTH initial conditions are a simplification of an actual experimental configuration. Both target and projectile are 740 mm in diameter.

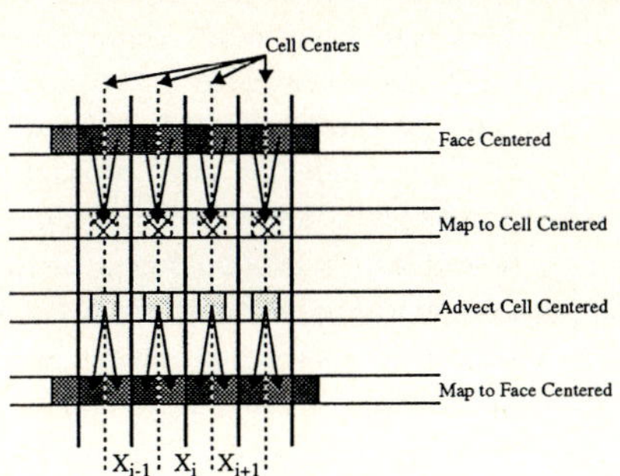

Fig. 1. Cell-centered momentum advection scheme

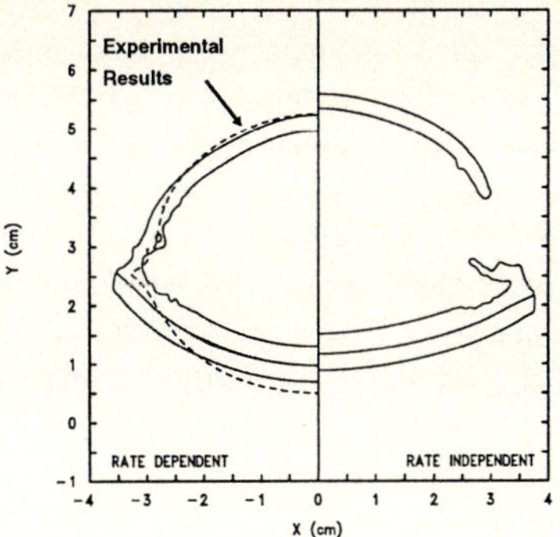

Fig. 2. Strain rate-dependent material deformation

Ideal Gas ($\gamma = 1.4$)

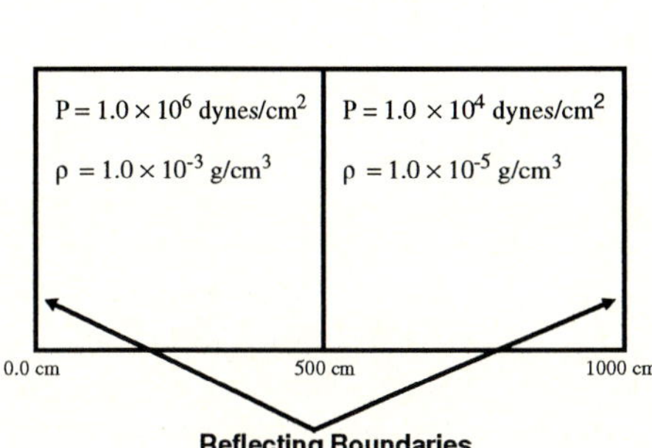

Fig. 3. Sod problem - initial conditions

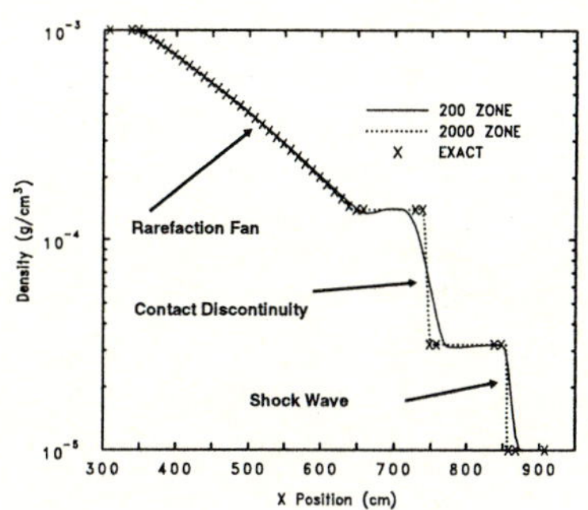

Fig. 4. Sod problem - solutions

Since tantalum is known to possess significant rate-dependent constitutive behavior, the Steinberg-Guinan-Lund model was chosen to represent the material. To assess the effect of rate-dependent behavior, two simulations were completed, one with the full rate dependent treatment and one with the rate term set to zero. Fig.2 displays the results of the two simulations. The right-hand half of Fig.2 displays the terminal state of the target disk 170 ms after impact when the rate dependent term was set to zero. The left-hand side of Fig.2 displays the target disk at the same time when the full rate dependent model was used. A digitized cross section from the experiment is also shown (as a dashed line) on Fig.2. One can readily see the effect of the strain-rate hardening that the Steinberg model predicts. On the right-hand side of Fig.2, the material does not possess adequate strength to resist the shear forces that are deforming the target. The disk thins and finally separates at the margin. When the full rate-dependent model is applied (see Fig.2, left side), the additional strength though strain hardening is sufficient to resist the shear forces at the margin of the disk. In the second case, the agreement with the experimental result is excellent.

The second simulation discussed here is the classic shock tube or Sod (1978) problem. This problem consists of a one-dimensional shock tube in which a higher density and pressure ideal gas is allowed to expand into an ideal gas of lower pressure and density. For this simulation, the initial conditions are shown in Fig.3. This problem is a classical Riemann problem where an exact solution can be derived as long as the generated waves do not interact with the boundaries. In addition, it is a standard test problem for shock physics models because three typical wave structures - a rarefaction fan, a contact discontinuity, and a shock wave occur in the simulation. Fig.4 displays the density field for 200 and 2000 zone CTH simulations as compared to the exact solution. Both CTH simulations are accurate representations of the exact solutions, although the 2000 zone simulation is obviously more accurate. The 200 zone simulation is of principal interest since that represents a typical domain size for two-dimensional simulations and is close to the upper limit for three-dimensional domains. The 200 zone CTH simulation has a small amount of post-contact discontinuity ringing as seen the particle velocity plot. Both the contact discontinuity and the shock are slightly smeared by the diffusion in the advection scheme and the artificial viscosity, respectively. Both of these effects are substantially reduced in the higher resolution simulation.

Both of the simulations shown here demonstrate the effectiveness of the numerical schemes and mechanical models embodied in CTH. It has been shown to possess state-of-the-art numerical treatments and models to represent the mechanical and energetic behavior of materials and can be used to predict a wide variety of shock based phenomena.

Acknowledgment

This work performed at Sandia National Laboratories supported by the U. S. Department of Energy under contract #DE-AC04-76DP000789.

References

Benson DJ (1991) Momentum advection on a staggered mesh. J. Comp. Phys. 100:1

McGlaun JM, Thompson SL, Kmetyk LN, Elrick MG (1990) CTH: A three-dimensional shock wave physics code. Intl. J. Impact Engng. 10:351

Johnson GR, Cook JH (1985) Fracture characteristics of three metals subjected to various strains, strain rates, temperatures, and pressures. Engrg. Frac. Mech. 21:31

Johnson GR, Holmquist TJ, Lankford J, Anderson CE, Walker J (1990) A computational constitutive model and test data for ceramics subjected to large strains, high strain rates, and high pressures. Honeywell report, DE-AC04-87AL-42550

Kerley GI (1991) CTH Reference Manual: The Equation-of-State Package. Sandia National Laboratories report, SAND91-0344

Kerley GI (1992) CTH Equation of State Manual: Porosity and Reactive Burn Models. Sandia Nation Laboratories report, SAND92-0553

van Leer B (1977) Towards the ultimate conservative difference scheme IV. A new approach to numerical convection. J. Comp. Phys. 23:276

Noh WF, Woodward P (1976) SLIC (Simple Line Interface Calculation). Proceedings of the Fifth International Conference on Numerical Methods in Fluid Dynamics, Springer Lecture Notes in Physics 59:330

Sod GA (1978) A survey of several finite difference methods for systems on nonlinear hyperbolic conservation laws. J. Comp. Phys. 27:1

Steinberg DJ, Cochran SG, Guinan MW (1980) A constitutive model for metals applicable at high strain rate. J. Appl. Phys. 51:1498

Thompson SL, Lauson HS (1972) Improvements in the CHARTD radiation-hydrodynamics code III: Revised analytic equations-of-state. Sandia National Laboratories report, SC-RR-71-0714

Zerilli FJ, Armstrong RW (1987) Dislocation-mechanics-based constitutive relations for material dynamics calculations. J. Appl. Phys. 61:1816

Computation of Viscous Shock/Shock Interactions with an Upwind LU Implicit Scheme

D. Darracq* and M. Gazaix[†]
*Société Nationale d'Etudes et de Construction de Moteurs d'Avions (SNECMA), 77550 Moissy Cramayel France
[†]Office National d'Etudes et de Recherches Aérospatiales (ONERA), BP 72 92322 Châtillon Cedex France

Abstract. The development of efficient propulsion systems for the forthcoming hypersonic vehicles requires a detailed understanding of the flowfield in all the components of such systems. In this context, the phenomenon of shock/shock interaction must be studied carefully, since it can increase by a large amount the maximum heat flux on the wall near the inlet cowl lip. To this end, we have undertaken a computational study of type IV interaction, which is very severe in terms of heat flux. The key features of this work are the development of an efficient and robust implicit algorithm to solve the laminar Navier-Stokes equations, upwind discretization with carefully controlled numerical diffusion, and the study of grid convergence and grid adaption.

Key words: Shock interference, Navier-Stokes, Implicit LU scheme, Grid adaption.

1. Introduction

Shock wave interference is a critical problem in the structural design of an hypersonic vehicle. Because of the interaction of the two shocks associated to nosetip and inlet cowl lip, extremely high pressure and heat transfer can locally occur in the stagnation region of the cowl lip surface. Edney (1968) defined a set of six types of shock interference patterns when the bow shock over a cylindrical leading edge is impinged upon by an incident shock. Type IV interaction results in very large heating and pressure loads and therefore, is of greatest interest to researchers.

In this context we have undertaken a detailed computational study of this interaction, performed with the 3D multidomain code FLU3M, developed mainly by ONERA. Initially, the implicit operator was factorized according to the widely used Alternating Direction Implicit (ADI) scheme. Because of the error of factorization of the ADI scheme the CFL number was limited. Since the computations are made on highly stretched meshes, the use of local time steps is mandatory to maintain a reasonable computational cost. However, on the finest mesh used in this study, the ADI scheme was unable to compute a type IV interaction. Because of the very different time steps at each node of the mesh, the pattern of the interaction could not take form and the bow shock was going out of the computational domain.

So, we decided to implement a Lower Upper (LU) decomposition of the implicit operator. This decomposition is robust and well conditioned if the operators are diagonally dominant, and we will show in this paper that it permits to use very high CFL number and global time step. Furthermore, this LU factorization associated with an upwind splitting strongly reduces the computational cost of the implicit operator.

To compute viscous effects, such as wall heat flux with good accuracy, the numerical viscosity must be carefully controlled. With Roe's upwind solver, we show that the entropic correction (leading to numerical diffusion) has to be carefully controlled in the region of the boundary layer.

A grid sensitivity study is presented. With upwind schemes better resolution is achieved when the grid lines are aligned with flow discontinuities. With complex flowfields such as shock/shock interaction this alignment can only be obtained by a technique of grid adaption. We show the improvements of the computed results achieved by an algebraic adaption scheme based on arc equidistribution concept.

Shock Waves @ Marseille I
Editors: R. Brun, L. Z. Dumitrescu

2. Navier-Stokes equations

The two-dimensional Navier-Stokes equations in generalized curvilinear coordinates (ξ, η) read:

$$\partial_t \widehat{Q} + \partial_\xi (\widehat{E} + \widehat{E}_v) + \partial_\eta (\widehat{F} + \widehat{F}_v) = 0. \tag{2.1}$$

where $\widehat{Q}$ is the vector of conservative variables; $\widehat{E}$ and $\widehat{F}$ are the convective flux vectors; and $\widehat{E}_v$ and $\widehat{F}_v$ are the viscous flux vectors.

3. Implicit algorithms

3.1. Linearization

Let us discretize in time the inviscid part of equation (2.1) (the viscous implicit operator is treated in §2.4) :

$$\widehat{Q}^{n+1} - \widehat{Q}^n + \Delta t \left[\partial_\xi \widehat{E}^{n+1} + \partial_\eta \widehat{F}^{n+1} \right] = 0 \tag{3.1}$$

The flux vectors are nonlinear functions of $\widehat{Q}$. Linearizing, we obtain:

$$\begin{cases} \widehat{E}^{n+1} = \widehat{E}^n + \widehat{A}^n \Delta \widehat{Q}^n + O(\Delta t^2) \\ \widehat{F}^{n+1} = \widehat{F}^n + \widehat{B}^n \Delta \widehat{Q}^n + O(\Delta t^2) \end{cases} \tag{3.2}$$

where $\Delta \widehat{Q}^n = \widehat{Q}^{n+1} - \widehat{Q}^n$ and $\widehat{A}$ and $\widehat{B}$ are the flux Jacobians: applying Eq.3.2 to Eq.3.1 produces the *delta form* of the algorithm, where $\widehat{R}$ is the residual:

$$[I + \Delta t \partial_\xi \widehat{A}^n + \Delta t \partial_\eta \widehat{B}^n] \Delta \widehat{Q} = -\Delta t \widehat{R} \tag{3.3}$$

3.2. ADI factorization

The inversion of this implicit operator is too costly in CPU time and memory size. So, an approximate ADI factorization is used:

$$[I + \Delta t \partial_\xi \widehat{A}^n][I + \Delta t \partial_\eta \widehat{B}^n] \Delta \widehat{Q} = -\Delta t \widehat{R} \tag{3.4}$$

A diagonal form proposed by Pulliam and Chaussee (1981) has also been implemented.

   In this work we found that these ADI schemes were not robust enough for a viscous shock/shock interaction computation on the finest mesh used in this study. So, we focused on LU factorization which can lead to significant improvements in robustness and speed of convergence.

3.3. LU factorization

Jameson and Turkel (1981) constructed a scheme involving Lower (L) and Upper (U) triangular matrices:

$$LU.\Delta \widehat{Q} = -\Delta t \, \widehat{R} \tag{3.5}$$

with:

$$\begin{cases} L = I + \Delta t \left(\partial_\xi^- \widehat{A}^+ + \partial_\eta^- \widehat{B}^+ \right) \\ U = I + \Delta t \left(\partial_\xi^+ \widehat{A}^- + \partial_\eta^+ \widehat{B}^- \right) \end{cases} \tag{3.6}$$

Of the many ways of splitting the jacobian matrices, Jameson and Turkel's method augmenting diagonal dominance:

$$\widehat{A}^+ = \frac{1}{2} [\, \widehat{A} + \gamma(\widehat{A}) I]; \qquad \widehat{A}^- = \frac{1}{2} [\, \widehat{A} - \gamma(\widehat{A}) I] \tag{3.7}$$

$$\gamma(\widehat{A}) = \beta \cdot \max(\lambda_\xi) \qquad \beta \geq 1. \tag{3.8}$$

To increase the robustness, Yoon and Jameson (1987) proposed diagonal conditioning:

$$L \; D^{-1} \; U. \; \Delta\widehat{Q} = -\Delta t \widehat{R} \tag{3.9}$$

with:

$$\begin{cases} L = I + \Delta t \left(\partial_\xi^- \widehat{A}^+ + \partial_\eta^- \widehat{B}^+ - \widehat{A}^- - \widehat{B}^- \right) \\[2mm] D = I + \Delta t \left(\widehat{A}^+ - \widehat{A}^- + \widehat{B}^+ - \widehat{B}^- \right) \\[2mm] U = I + \Delta t \left(\partial_\xi^+ \widehat{A}^- + \partial_\eta^+ \widehat{B}^- + \widehat{A}^+ + \widehat{B}^+ \right) \end{cases} \tag{3.10}$$

A great interest of factorising Eq.3.9 and splitting Eq.3.7 is that D is reduced to a scalar form. The implicit boundary procedures are very important for speed of convergence and for accuracy of the solution. So, we introduced the wall boundary conditions in *both* L and U factors.

3.4. Viscous fluxes

The true linearization of the viscous fluxes is substituted by the spectral radius of the jacobian matrix:

$$\nu_M = \nu \cdot \max\left(1; \; \lambda/\mu + 2; \; \gamma/\mathrm{Pr}\right) \tag{3.11}$$

The term $-\Delta t \nu_M \partial_\xi^2 - \Delta t \nu_M \partial_\eta^2$ is added to the left-hand side of Eq.3.4, increasing the diagonal dominance for both ADI and LU factorizations and preserving the scalar structure of the diagonal.

3.5. Diagonal sweeping

By sweeping forward (Lower operator) and backward (Upper operator) in the diagonal direction ($i + j$ constant), the inversion of the two LU factors can be *fully vectorized* avoiding the recursive procedure occuring in the tridiagonal system. Furthermore during the forward and backward inversions all variables needed from the off-diagonals are already updated and can be moved to the right-hand side. Then only scalar diagonal terms remain on the left-hand side and there are *no blocks to invert.*

The LU implicit solver is much more robust (stability ensured at very high CFL number) and faster than the ADI one, even in its diagonal version. CPU time per point and per iteration for the implicit subroutines on a CRAY-YMP is given in the table below for each scheme:

algorithm	CPU time (μs)
ADI full blocks	30
ADI diagonalized	14
LU	10

4. Shock/shock interaction

4.1. Physical description

Type IV interaction occurs when the incident shock strikes the detached shock of the cylinder near the stagnation point. The detached shock is distorted, a supersonic jet develops inside the subsonic pocket, and impinges the wall of the inlet cowl. This phenomenon is responsible for a large increase of the heat flux near the stagnation point. The flow parameters are: $M_\infty = 5.94$, $\mathrm{Re}_D = 186,000$, $T_\infty = 59.5 K$, $T_{\mathrm{wall}} = 408$ K, impinging shock angle $22.75°$.

4.2. Entropic correction effect

With Roe's upwind solver, one must use an entropic correction to avoid expansion shocks. Following Harten (1983), we introduce a parameter δ of entropic correction proportional to the

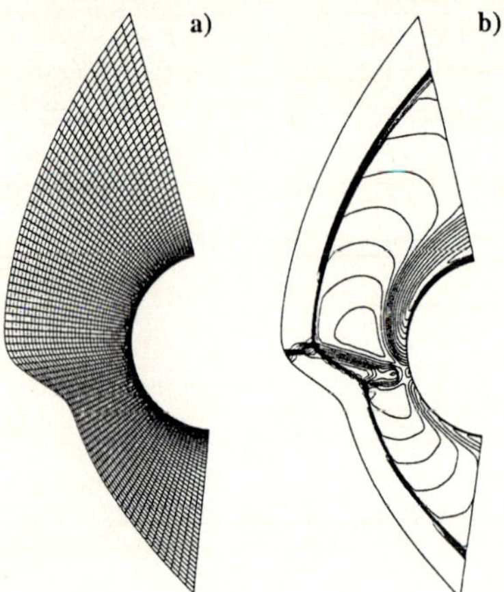

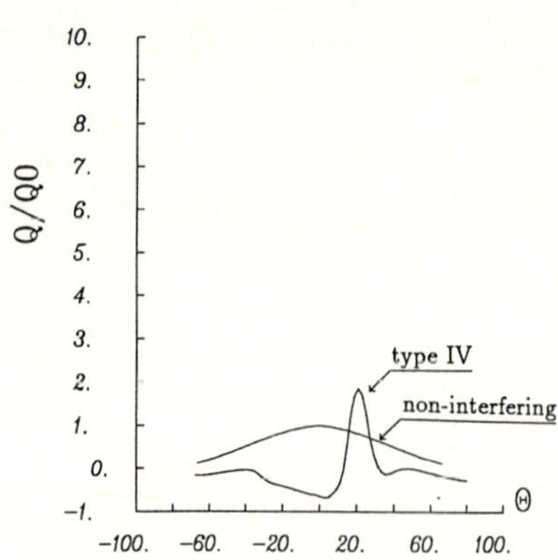

Fig. 1. a: Coarse mesh (78 × 50 points)
b: Iso-Mach lines (increment 0.2)

Fig. 2. Wall heat flux
(normalized by non-interfering stag. flux Q_0)

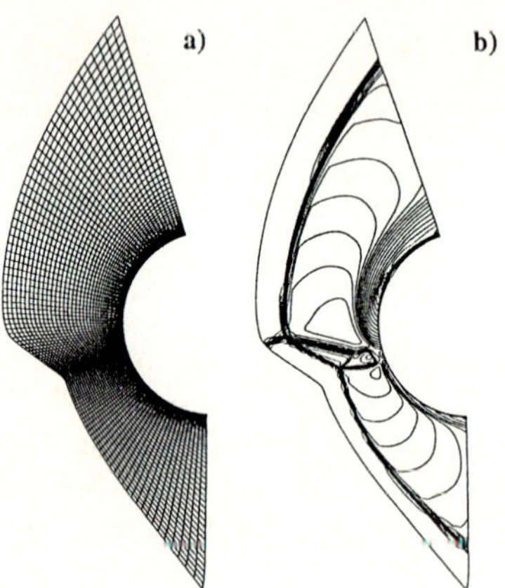

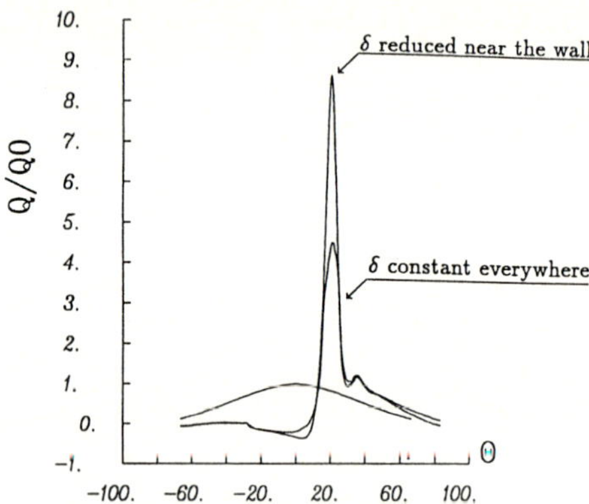

Fig. 3. a: Fine mesh (105 × 80 points)
b: Iso-Mach lines (increment 0.2)

Fig. 4. Wall heat flux rates

maximum of the speed waves. Unfortunately, δ is not independent of the mesh, and must be adjusted. Moreover, the numerical diffusion associated with this correction, must be reduced in the boundary layer. These points will be demonstrated in §4.3 and 4.4.

4.3. Mesh study

To investigate the influence of the grid we performed a mesh refinement study. A relatively coarse mesh, with 78 × 50 points is displayed in Fig.1a. The cell Reynolds number ($\mathrm{Re}_{\mathrm{cell}} = \frac{\mathrm{Re}_D}{D}\Delta\eta_{\mathrm{min}}$, where $\Delta\eta_{\mathrm{min}}$ is the normal spacing at the wall) is 220 and δ is set to 0.1. Iso-Mach lines (Fig.1b) show the essential features of type IV interaction. However shear layers and boundary layer are too thick and the heat transfer rate is underpredicted (Fig.2). To obtain a better estimation of the heat flux, a finer grid (105 × 80) is used (Fig.4). The wall cell Reynolds number is 7.5. The grid refinement in the circumferential direction was found to be very important for capturing correctly the jet and compute accurately the peak heat transfer.

On this grid, in the shock region, δ must not be smaller than 0.4; otherwise, a non-entropic solution is obtained ("carbuncle" phenomenon). The importance of the control of the parameter δ in the boundary layer is shown in Fig.4, where the heat flux computed with $\delta = 0.4$ everywhere, and with δ reduced to 10^{-5} near the wall, is presented. The peak heat flux is roughly doubled with variable δ (note that a similar behaviour has been observed on the coarse mesh).

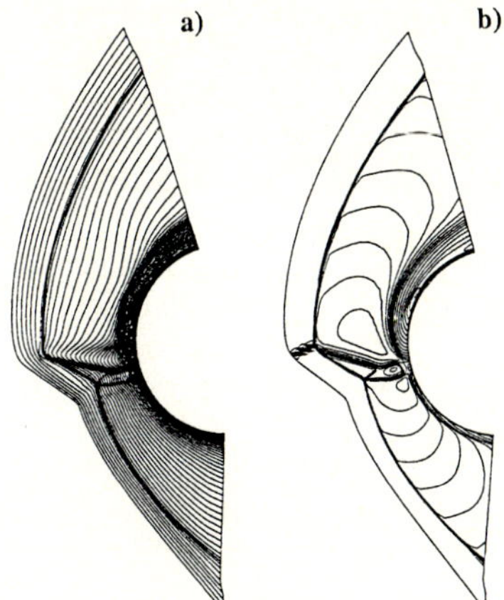

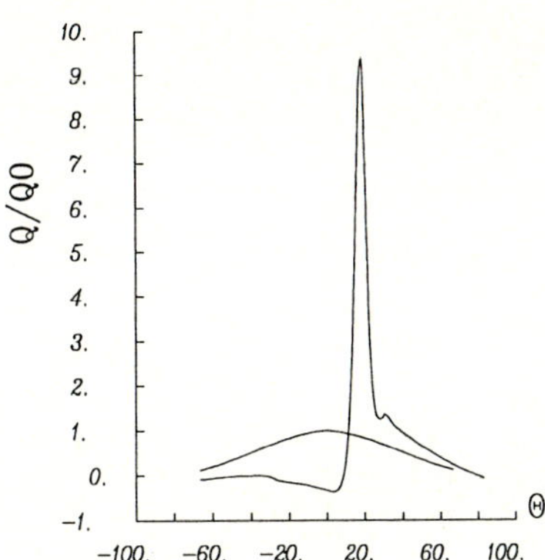

Fig. 5. a: Iso-j lines of adapted mesh
b: Iso-Mach lines (increment 0.2)

Fig. 6. Wall heat flux rates

The ratio of the peak heat transfers evaluated along the η_1-η_2 lines and the η_1-η_3 lines is here equal to 1.01 and to 1.21 on the coarse mesh. This is an indication that grid convergence is achieved.

4.4. Grid adaption

Truncation errors are proportional to grid size and flow gradients. The aim of grid adaption methods is to reduce discretization errors in high gradient regions by grid refinement. Hsu and Lytle's scheme (1989) is based on the equidistribution concept, i.e. redistribution of grid points such that a weight function ω_i is equally distributed over a grid line. In the x-direction:

$$\Delta x_i \omega_i = C \tag{4.1}$$

where Δx_i is the grid size, and C is a constant. For a general coordinate system, we have in the ξ-direction:

$$\Delta S_i \omega_i = \lambda C_i \tag{4.2}$$

ΔS_i is the grid size spacing and ω_i is a function weighted by the flow gradients:

$$\omega_i = 1 + \beta |\partial_\xi u|_i \tag{4.3}$$

β controls the sensitivity to the gradients. C_i is a constant:

$$C_i = \Delta S_i^0 \tag{4.4}$$

with S_i^0 being the arc length of the old grid spacing. To compute λ, the new total arc is required to be equal to the old one (i.e. $\sum_i \Delta S_i^0 = \sum_i \Delta S_i$) and with Eqs.4.2 and 4.4 yields:

$$\lambda = \frac{\sum_i \Delta S_i^0}{\sum_i \frac{C_i}{\omega_i}} \tag{4.5}$$

The fine grid (Fig.3a) is adapted (Fig.5a) on the Mach number of the flow ($\beta = 5$).

The alignment of the grid lines with the strong bow shock allows us to reduce by a factor 2 the entropic parameter in the shock region. The details of the flowfield are now sharply captured (Fig.5b with variable δ). The peak heat flux is increased by 5% with respect to the non-adapted grid.

5. Conclusions

An LU implicit algorithm for the solution of 2-D Navier-Stokes equations has been developed. Viscous shock/shock interaction computations demonstrate the robustness as well as the efficiency of this scheme.

For an accurate heat flux estimation, it is important to reduce both grid size spacing and entropic correction near the wall, i.e., the numerical viscosity has to be low in the region of boundary layer.

Grid adaption increases the stability of the shock location and reduces truncation errors near flow discontinuities. To compute these complex flowfields, this technique is found to be very useful.

We stress that all these points have to be considered together to get an accurate prediction of the peak heat transfer.

References

Coakley TJ (1981) Numerical methods for gas dynamics combining characteristic and conservation concepts. AAIA Paper 81-1257

Edney BE (1968) Anormalous heat transfer and pressure distribution on blunt bodies at hypersonic speeds in presence of an impinging shock. Aeron. Res. Inst. of Sweden, FFA Report 115

Harten A (1983) A high resolution scheme for the computation of weak solutions of hyperbolic conservation laws. J. Comp. Phys. 49: 357-393

Hsu A, Lytle JK (1989) A simple algebraic grid adaptation scheme with applications to two- and three-dimensional flow problems. AIAA Paper 89-1984-CP pp 525-532

Jameson A, Turkel E (1981) Implicit schemes and LU decomposition. Math. of Comp. 37, 156: 385-397

Pulliam TH, Chaussee DS (1981) Diagonal form of an implicit approximate-factorization algorithm. J. Comp. Phys. 39:347-363

Yoon S, Jameson A (1987) Lower-upper symmetric Gauss-Seidel methods for the Euler and Navier-Stokes equations. AAIA Paper 87-600

Numerical Calculations in Support of Complex Shock Interactions

Charles E. Needham, Shuichi Hikida, and Lynn Kennedy
S-CUBED, a Division of Maxwell Laboratories Albuquerque, NM, USA

Abstract.   Results of numerical calculations of complex shock interactions with simple structures are presented. These calculations are the first in a series intended to demonstrate the capabilities of modern high-order codes. The calculational results are compared to experimental data collected at the Ernst Mach Institute in Germany and Waterways Experiment Station in Vicksburg, Mississippi. Specifics of the conditions for the calculations were selected to test the code's capability to properly model shock loading in complex flows in which shocks interact with vortices. The early calculations in the series use non-responding structure elements in two dimensions. Square-wave incident shock waves are used for the first calculation. An ideal gas equation-of-state is used for this case. The second calculation uses a similar structure with an incident two-dimensional, decaying blast wave. The second calculation includes the effects of detonation products as well as real equations-of-state for the air and detonation products. A third calculation is planned for full three-dimensional interactions with the blast wave from a spherical charge. The structure is also non-responding for this case. Shock geometry comparisons are available from the two-dimensional cases. Some cases include laser interferograms, which give measurements of the density distribution in the shocked regions. No pressure data is available from the two-dimensional cases at the present time. Comparisons will be made with the calculated density fields. If data become available, comparisons will be made with pressure waveforms at selected locations as well.

Key words: Computational Fluid Dynamics, Complex shock interactions, Flow visualisation

1. Description of SHARC

SHARC, developed by S-CUBED, is a library of programs and subroutines which make up a general method of solution for compressible, non-conducting hydrodynamic problems in two or three dimensions. The conservation equations of mass, momentum, and energy, with an equation-of-state providing the necessary closure, are solved as a function of time. The method of solution is explicit time-marching and basically Eulerian. SHARC uses a two-phase, second-order solution. The first phase is Lagrangian, followed by a remap phase which restores the original Eulerian grid. SHARC makes use of alternating direction operator splitting and is highly vectorized for CRAY computers.

The SHARC library includes a large number of material equations-of-state. A second-order material interface tracker is used for multi-material problems. The code includes a variety of rezoning techniques, has a shock following subgrid, and activity flags. Solid objects are represented by islands and "shores", or half islands. SHARC will also accept initial conditions from a variety of other one-, two-, and three-dimensional codes.

2. Choice of problems

The first calculation presented here (Fig.1) was a problem with a single material, perfect gas, with constant feed boundary conditions. This represents the most basic type of calculation for a compressible fluid code. This particular calculation was chosen to demonstrate and test the capabilities of state-of-the-art codes for pure hydrodynamics. These capabilities include: shock reflection, vortex formation and growth, complex shock propagation and geometry, and shock

---

Shock Waves @ Marseille I
Editors: R. Brun, L. Z. Dumitrescu

interactions with vortices. The problem consists of a series of regularly spaced baffles in a two-dimensional planar geometry through which a constant pressure shock is passed.

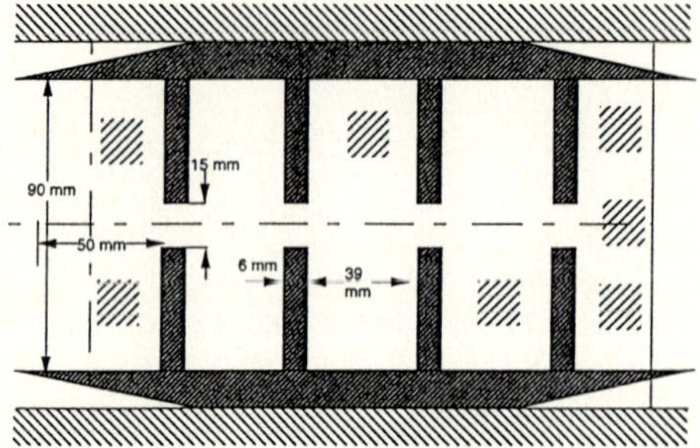

Fig. 1. Initial conditions for square-wave baffle calculation

This problem was chosen, first, because it is well defined, and second, because high quality experimental data exist from the Ernst Mach Institute. The data consist of a time sequence of shadowgraph photos which give precise locations and shock geometries at several times. They also show vortex formation and growth clearly. The complex multiple reflections and demanding resolution provide a challenge for modern high order codes.

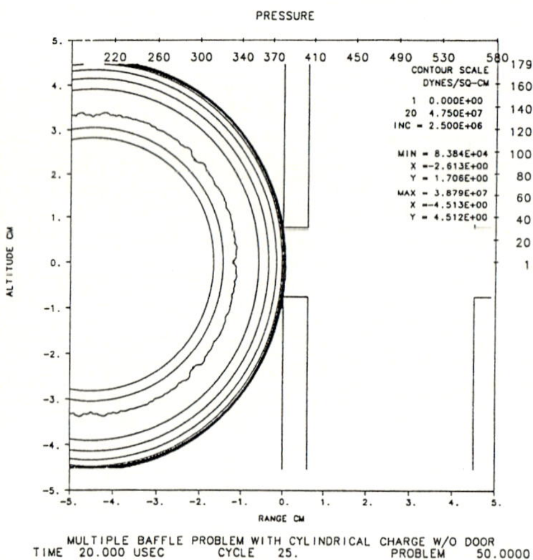

Fig. 2. Initial conditions for cylindrical charge in baffle system

Our calculation used a grid of just over 150,000 zones with dimensions of 200 microns in each direction. The calculation took advantage of symmetry, so only half of the region was calculated. Activity flags were used to activate the grid just prior to shock arrival, thus, only a small fraction of the zones were active initially. Total computer time for the calculation was less than one-half hour.

3. Detonation as a source

The second calculation (Fig.2) uses a cylindrical, condensed high explosive as the shock source. The explosive is placed 4.5 cm in front of the first baffle. Because of the presence of detonation products and the high pressures generated, multiple materials and real gas equations-of-state are required. The size of the charge was small, which makes the duration of the leading shock very short. This short duration and the multiple materials with different equations-of-state present even greater challenges to state-of-the-art codes.

This calculation used a grid of approximately 200,000 zones with dimensions of 250 microns. The calculation again took advantage of symmetry and used activity flags to minimize the number of active zones. Unfortunately, experimental data for this problem are not available.

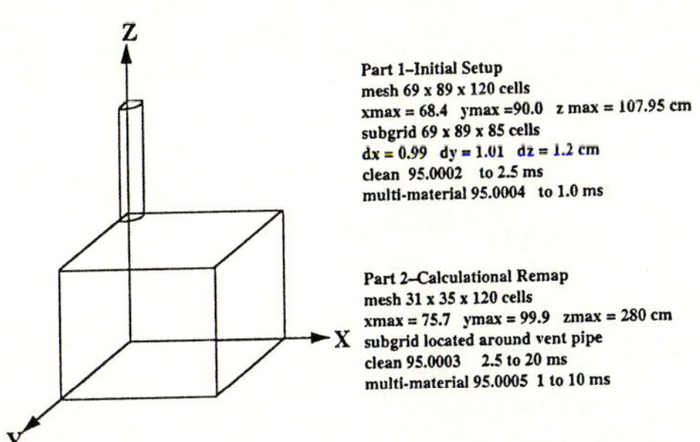

Fig. 3. Initial conditions for cylindrical charge in 3-D box

4. The three-dimensional case

The third calculation (Fig.3) is three-dimensional and consists of a cylindrical charge in a rectangular, non-responding box constructed of a combination of three-inch and six-inc steel plate. Experiments were conducted at Waterways Experiment Station in Vicksburg Mississippi. The initial conditions were supplied by a two-dimensional SHARC calculation of the detonation of the cylindrical charge. The charge was hung from the top and centered in the box. The charge was detonated at the top and burned toward the floor. The three-dimensional calculation began when the detonation was complete and the upwardly-moving shock had nearly reached the top of the box. This calculation requires a multi-material, three-dimensional capability with realistic equations-of-state for the air and detonation products.

Because the charge was centered in the box, we took advantage of four-way symmetry. Only one quarter of the box was calculated. The total number of zones was approximately 130,000. Each zone had dimensions of 1 by 1 by 1.2 cm. The box was indeed non-responding for the charge sizes used in the tests. The explosive was 0.95 pounds of C-4. Pressure measurements were made at several locations within the box; comparisons provide a good check on the code capabilities.

5. Comparisons of calculations and experiments

Three types of photography are commonly used for shock experimental work. Each of these methods measures a different parameter in the flowfield. Laser interferometry provides interference fringes which can be directly related to the density of the gas and can be compared with calculated density contours. Schlieren photographs give an indication of the first derivative of density. Shadowgraph photography is a good indication of the second derivative of the density distribution.

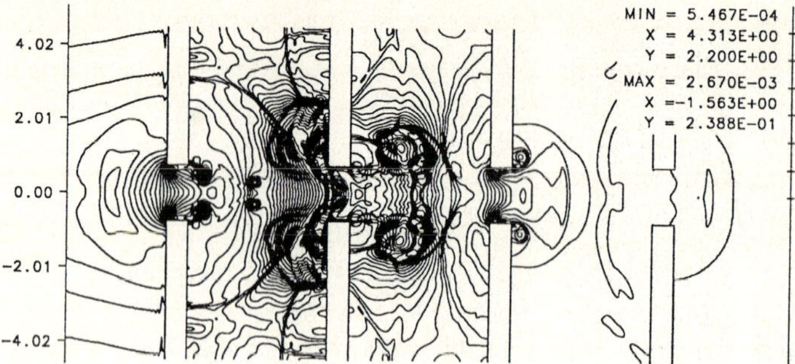

Fig. 4. Density contour for baffle calculation at 395 msec

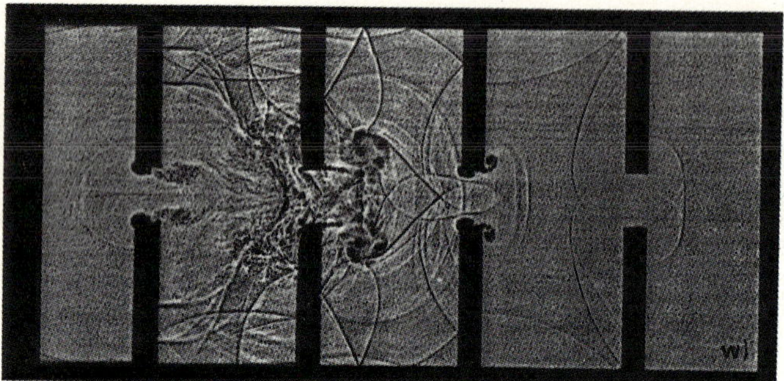

Fig. 5. Shadowgraph at 395 msec

Each of these types of interpretation of the density field shows details of different characteristics. Any of these methods give good representations of the positions of strong shocks. Details of vortices and the interaction of weak shocks are clearly visualized by using shadowgraphs.

The results of the single material, square-wave interaction with a baffle system are compared to the experimental shadowgraph photos at selected times. The first comparisons are with calculated density contour plots. Good-to-excellent agreement is shown between calculation and experiment (Figs.4 and 5) for shock front positions and for vortex position and size. Some differences are seen for interior signals and in regions of vortex interactions. We computed the second derivative of the density and compared these results to the shadow graphs at the same time (Fig.6).

Note that the signal near the second baffle has changed both its position and curvature. This is an example of the variation of flow interpretation that can be caused by the type of measurement used to visualize the flow field.

A similar problem with a different Mach number was measured at the Ernst Mach Institute by using a laser interferogram technique. Another SHARC calculation was made with the modified Mach number for comparison. The comparison of the density contours (Figs.7 and 8) from SHARC results and the interferogram show excellent agreement. Again, this is evidence that different interpretations can be made from the same experiment according to the type of photograph used to observe the experiment.

Fig.9 is a plot of the three-dimensional calculation showing shock convergence in one corner with multiple Mach reflections. There was no interior photography of this experiment. Figs.10

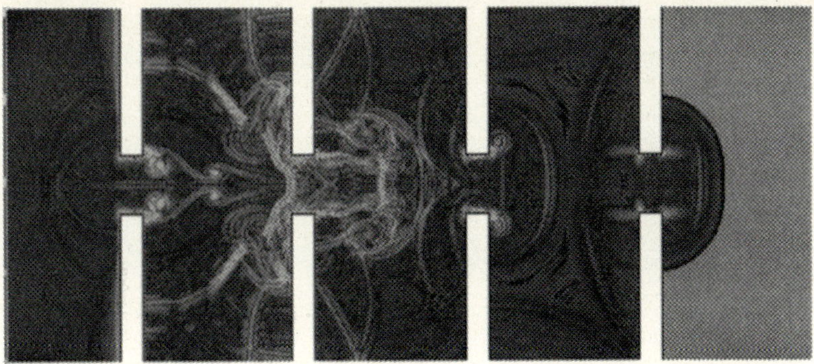

Fig. 6. Calculated second derivative of density contour at 395 μsec

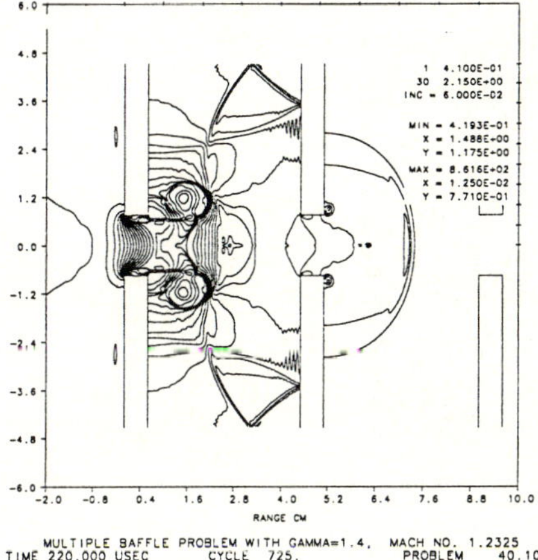

Fig. 7. Calculated density contours at Mach 1.2325

Fig. 8. Color interferogram corresponding to Fig.7

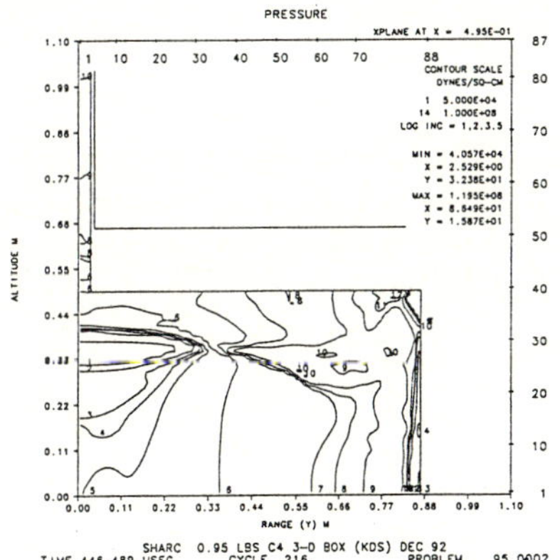

Fig. 9. Pressure contour near corner convergence

and 11 show comparisons of experimental pressure-time histories with calculated time histories on two of the walls of the structure. Note the agreement from shock arrival through the third

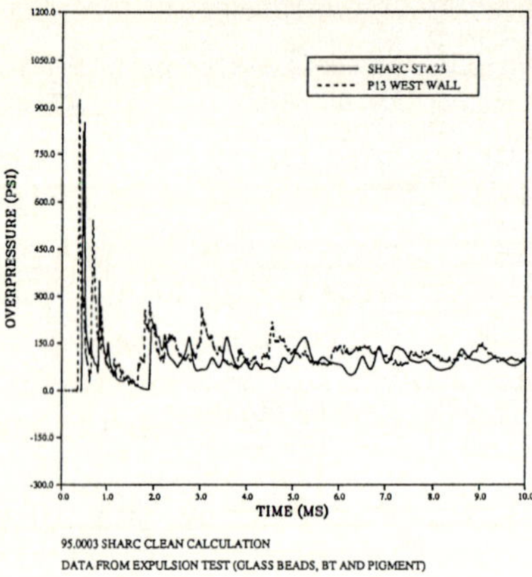

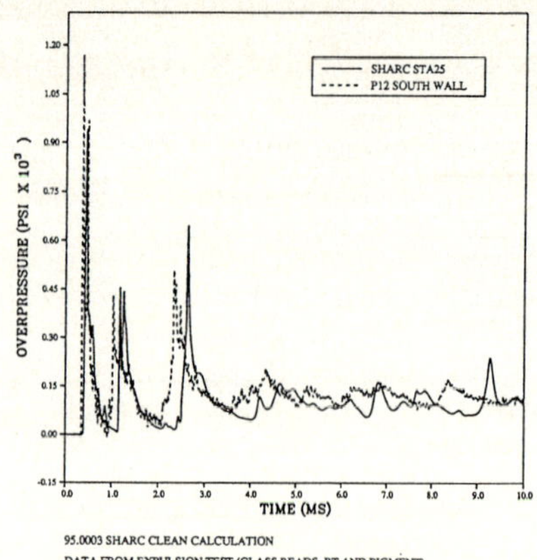

Fig. 10. Overpressure vs. time comparison
of calculated and measured

Fig. 11. Overpressure vs. time, comparison
of calculated and measured

reflection. Timing is slightly delayed in the calculation, probably because of small inaccuracies in
the equation-of-state for detonation products.

6. Conclusions   We have illustrated the differences in flow visualization using two forms of photography by comparison with calculated density contours for interferograms and with computed second derivative of density with shadowgrams. Each emphasizes different characteristics of the flow.

The less than one-half hour running time for the ideal gas, square-wave calculation demonstrates the high efficiency of the SHARC code. The comparisons with experimental data demonstrate its accuracy. In addition, the calculation was set up and run in one morning. Results from our three-dimensional calculation show that no accuracy is lost when going to three dimensions. The resolution required is easily obtained with current codes and equipment.

Acknowledgements

The work reported here was funded by DNA under contract DNA001-92-C-0165 and under a subcontract from APTEK, 3C06-SC03, DNA Prime DNA001-91-C-0075. The calculations were made on the DNA CRAY-YMP

References

Reichenbach H, Kuhl AL (1992) Flow visualisation of shock propagation in baffle systems. Fraunhoffer Gesellschaft, Ernst-Mach-Institut, Freiburg, Germany

Numerical Simulation of Shock Induced Unsteady Aerodynamic Heating with a Highly Dense Mesh System

Kenichi Ohyama[*], Shigeru Aso[*] and Masanori Hayashi[†]
[*]Department of Aeronautics and Astronautics, Kyushu University, Hakozaki, Fukuoka 812, Japan
[†]Nishinippon Institute of technology, Fukuoka 800-03, Japan

Abstract.   The shock reflection processes by a ramp and unsteady aerodynamic heating phenomena are simulated by solving the thin-layer approximated Navier-Stokes equations. The effect of mesh refinement to the calculated results is investigated. Calculations have been conducted for the strong shock wave reflection case with incident shock Mach number $M_s = 3.0$, ramp angle $\theta = 35°$ and Re$= 1.0 \times 10^4$ with 7 mesh systems. Also for moderate incident shock waves calculations have been conducted with $M_s = 2.35$ and 2.75, ramp angle $\theta = 42.5°$ and Re$= 1.0 \times 10^4$ with 5 mesh systems.

Results show that a smaller mesh size is necessary for calculating the precise heat flux distribution and for capturing the fine structure of the shock reflection patterns. Especially, it is revealed that the heat flux distribution is quite sensitive to mesh refinement.

Key words: Aerodynamic heating, Numerical simulation, Mesh refinement

1. Introduction

One of the important problems for designs of high speed winged vehicles is the severe aerodynamic heating and pressure rise caused by the impingements of shock waves on the surfaces (Korkegi 1971, Aso et al. 1989). Especially the unsteady aerodynamic heating caused by shock reflections at a higher shock Mach number has been investigated and the peak heating due to the Mach stem and a second peak heating due to the slip layer have been observed by the present authors (Aso et al. 1990).

In the present study the thin-layer Navier-Stokes equations have been solved in order to investigate unsteady aerodynamic heating phenomena induced by shock impingement on a ramp surface. Especially the effect of mesh refinement to heat flux distributions and the spatial resolution of flow structures are investigated intensively. At high temperatures these aerodynamic heating phenomena might change significantly due to high temperature effects. A study on the effects of high temperature to unsteady aerodynamic heating mechanism is in progress.

2. Numerical methods

Two-dimensional thin-layer Navier-Stokes equations (Baldwin and Lomax 1978) can be written in a general coordinate system as follows:

$$\hat{Q}_\tau + \hat{F}_\xi + \hat{G}_\eta = \text{Re}^{-1}\hat{S}_\eta, \tag{1}$$

where

$$\hat{Q} = J^{-1}\begin{pmatrix} \rho \\ \rho u \\ \rho v \\ e \end{pmatrix}, \quad \hat{F} = J^{-1}\begin{pmatrix} \rho U \\ \rho u U + \xi_x p \\ \rho v U + \xi_y p \\ U(E + p) \end{pmatrix}, \quad \hat{G} = J^{-1}\begin{pmatrix} \rho V \\ \rho u V + \eta_x p \\ \rho v V + \eta_y p \\ V(E + p) \end{pmatrix}, \tag{2}$$

$$\hat{S} = J^{-1}\begin{pmatrix} 0 \\ \mu(\eta_x^2 + \eta_y^2)u_\eta + (\mu/3)\eta_x(\eta_x u_\eta + \eta_y v_\eta) \\ \mu(\eta_x^2 + \eta_y^2)v_\eta + (\mu/3)\eta_y(\eta_x u_\eta + \eta_y v_\eta) \\ \hat{S}_4 \end{pmatrix}, \tag{3}$$

Shock Waves @ Marseille I
Editors: R. Brun, L. Z. Dumitrescu

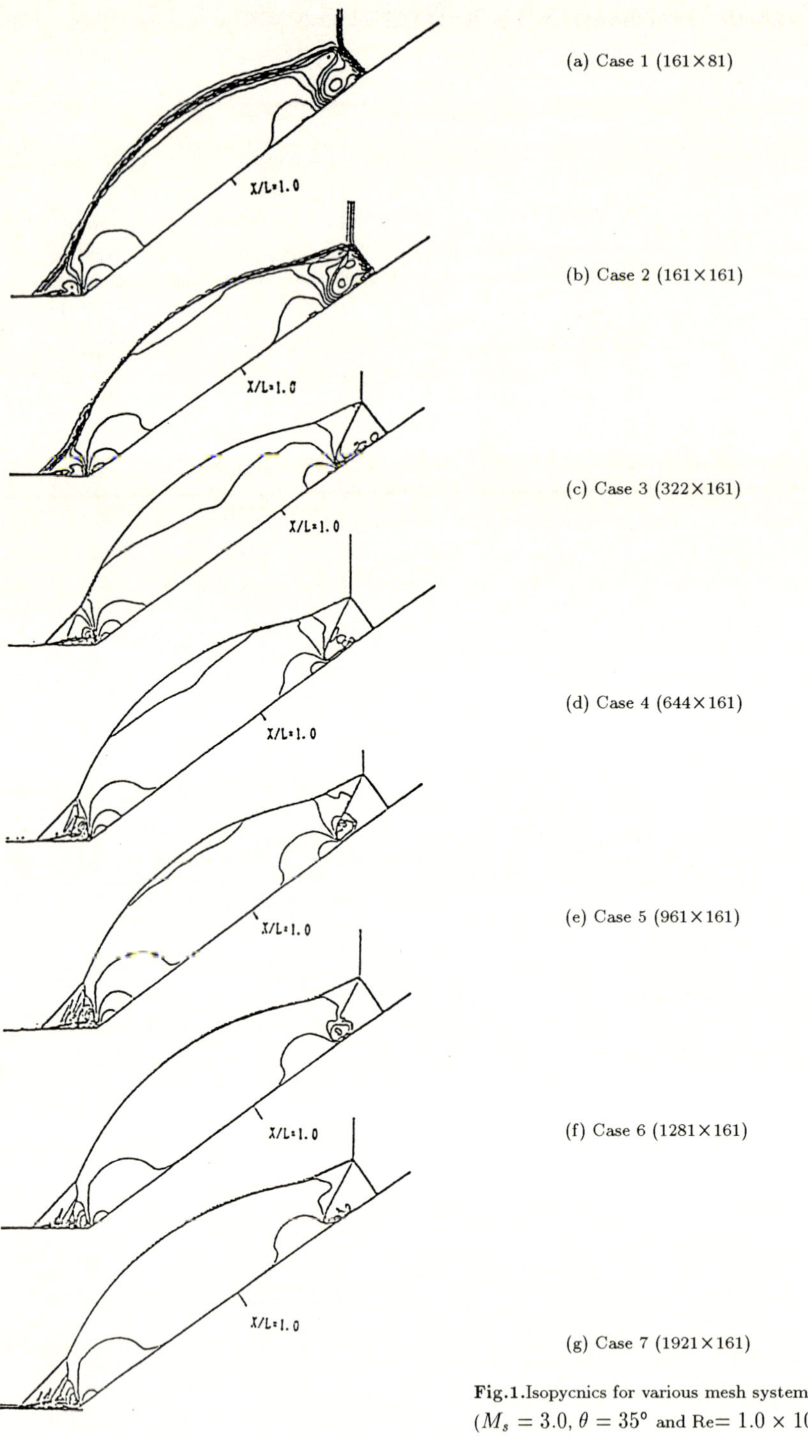

(a) Case 1 (161×81)

(b) Case 2 (161×161)

(c) Case 3 (322×161)

(d) Case 4 (644×161)

(e) Case 5 (961×161)

(f) Case 6 (1281×161)

(g) Case 7 (1921×161)

Fig.1. Isopycnics for various mesh systems
$(M_s = 3.0, \theta = 35° \text{ and Re} = 1.0 \times 10^4)$

$$S_4 = -(\eta_x^2 + \eta_y^2)q_\eta + \mu(\eta_x^2 + \eta_y^2)(u^2 + v^2)_\eta/2 + (\mu/6)[\eta_x^2(u^2)_\eta + \eta_y^2(v^2)_\eta + 2\eta_x\eta_y(uv)_\eta]. \qquad (4)$$

Eq.1 is integrated by a time splitting technique as follows:

$$
\begin{aligned}
L_\xi : & \quad \hat{Q}_\tau + \hat{F}_\xi = 0, \\
L_\eta : & \quad \hat{Q}_\tau + \hat{G}_\eta = 0, \\
L_d : & \quad \hat{Q}_\tau = \mathrm{Re}^{-1}\hat{S}_\eta.
\end{aligned}
\qquad (5)
$$

A Strang type symmetrical operator is used to keep the time accuracy in the computation as follows:

$$\hat{Q}^{n+2} = L_\xi(\Delta t)L_\eta(\Delta t)L_d(\Delta t)L_d(\Delta t)L_\eta(\Delta t)L_\xi(\Delta t)\hat{Q}^n, \qquad (6)$$

or

$$\hat{Q}^{n+2} = L_\xi(\Delta t)\Big[L_\eta(\frac{\Delta t}{m})L_d(\frac{\Delta t}{m})\Big]^m \Big[L_d(\frac{\Delta t}{m})L_\eta(\frac{\Delta t}{m})\Big]^m L_\xi(\Delta t)\hat{Q}^n. \qquad (7)$$

The molecular viscosity is obtained by using Sutherland's law. The molecular thermal conductivity is related to the viscosity through a constant Prandtl number assumption.

For inviscid terms a Harten and Yee TVD scheme (Yee and Harten 1987) is used and for viscous terms a conventional central difference is used. For boundary conditions non-slip conditions are applied for the ramp surface and zero derivatives along freestream are assumed at incoming and downstream boundaries. Also zero physical derivatives normal to incoming flow are imposed for the upper boundary in order to keep the incident shock wave normal. For the energy equation a constant wall temperature condition is assumed.

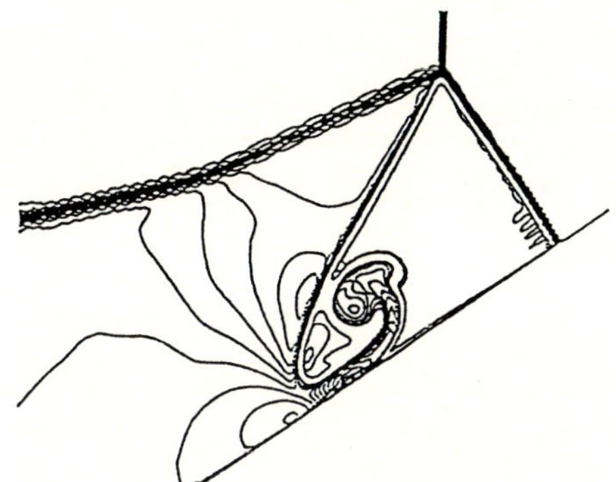

Fig. 2. Close-up of equi-temperature lines near triple point ($M_s = 3.0$, $\theta = 35°$ and Re= 1.0×10^4 with mesh system of Case 5: 961 × 161)

Calculated flow conditions are as follows: incident Mach number, $M_s = 3.0$, ramp angle 35° and Reynolds number, Re= 1.0×10^4. The initial temperature of the gas, T_∞, is kept at 300 K.

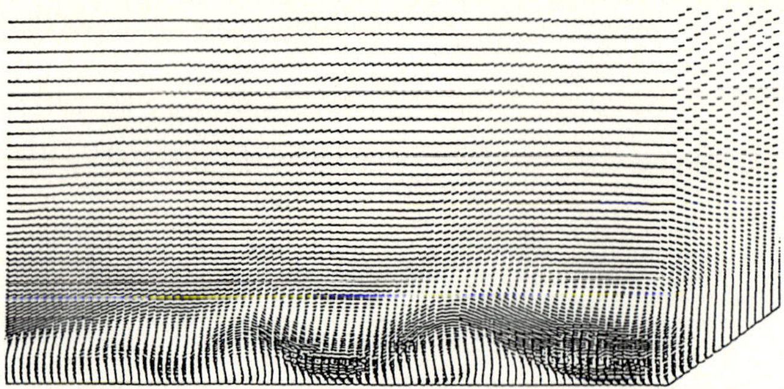

Fig. 3. Close-up of velocity vector diagram near corner ($M_s = 3.0$, $\theta = 35°$ and Re$= 1.0 \times 10^4$ with mesh system of Case 5: 961 × 161)

3. Numerical results and discussion

In the present calculations seven mesh systems with different mesh refinement are used. The same domain is divided by various numbers of grid points such as Case 1 (161 × 81), Case 2 (161 × 161), Case 3 (322 × 161), Case 4 (644 × 161), Case 5 (961 × 161), Case 6 (1281 × 161) and Case 7 (1921 × 161). The first number represents grid points in the freestream direction and the second normal to the freestream direction. Between Case 2 through Case 7, mesh refinements are conducted normal to the freestream direction.

Representative isopycnics at $M_s=3.0$, $\theta = 35°$ and Re$= 1.0 \times 10^4$ for various mesh systems are shown in Fig.1. As shown in the figure the mesh refinement has significant effects for capturing the incident shock wave, Mach stem, reflected shock wave and slip layer as sharp as possible. Especially the detailed flow structure near the triple point and roll-up of slip layer after reaching the ramp surface are captured only by the finest mesh system. The effect of mesh refinement is found to be quite important for capturing the detailed structure of the flow fields.

The close-up of equi-temperature lines near the triple point of Case 5 is shown in Fig.2. The most interesting feature is the strong roll-up of the slip layer. The sharp and straight slip layer generated from the triple point impinges on the wall surface, turns upstream due to the existence of the pressure peak and rolls up. This fine flow structure is not captured for the coarser mesh systems of Cases 1 through 4. The calculated temperature contours show a quite high temperature region in the triangular region bounded by the Mach stem, slip layer and wall. Surprisingly, this fine structure still changes its character slightly for finer mesh systems and a perfectly converged time accurate solution is never obtained in the present calculations. The results suggest that the thin layer approximation of the governing equations couldn't capture the actual physical phenomena where gradients of physical properties parallel to freestream direction are very strong. Further study by full Navier- Stokes equations are needed. A close-up of the velocity vector diagram near the corner of Case 5 is also shown in Fig.3. In the figure several separated vortices are captured and secondary separation bubbles are also observed in the primary separation bubbles. Such a significant fine structure is captured by finer mesh systems. The results suggest that finer mesh systems are necessary to capture detailed vortical structure.

The instantaneous heat flux distributions are shown in Fig.4 for the mesh systems of Cases 1 through 7. The first peak heating counted from the right-hand side increases as mesh points are increased and even at Case 7 the peak value is still higher than that of Case 6. On the other

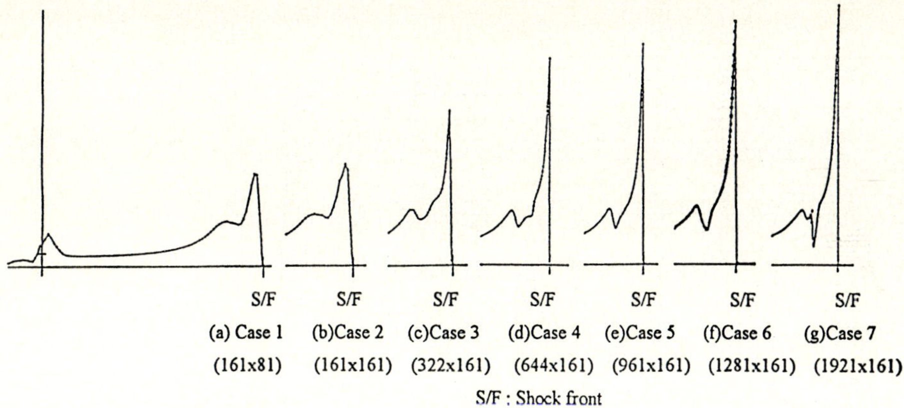

Fig. 4. Heat flux distributions with various mesh systems ($M_s = 3.0$, $\theta = 35°$ and Re$= 1.0 \times 10^4$)

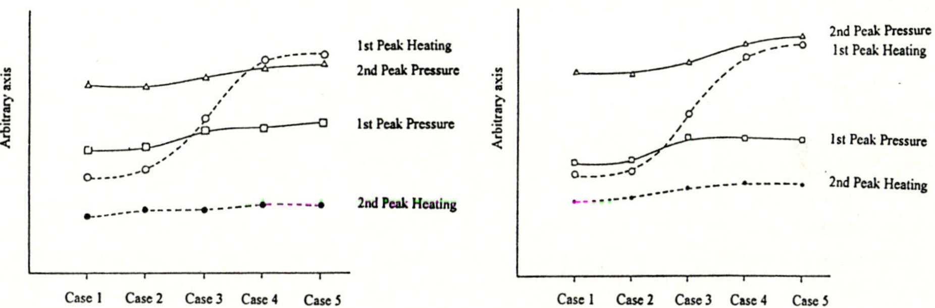

Fig. 5. Change of peak heating and peak pressure due to mesh refinement ($M_s = 2.35$, $\theta = 42.5°$ and Re$= 1.0 \times 10^4$)

Fig. 6. Change of peak heating and peak pressure due to mesh refinement ($M_s = 2.75$, $\theta = 42.5°$ and Re$- 1.0 \times 10^4$)

hand the second peak heating shows an almost converged value for higher mesh systems. Also, the change in first and second peak heating and peak pressure, due to mesh refinement, are shown in Fig.5.

The same calculations have been conducted for $M_s = 2.35$ and 2.75 with ramp angle $\theta = 42.5°$. The change of the first and second peak heating and peak pressure with mesh refinement are shown in Figs.5 and 6. At this lower incident shock Mach number those values can reach almost equilibrium conditions for the finest mesh systems. However, as shown in Fig.7, at a higher incident shock wave Mach number, $M_s = 3.0$, the first peak heating still increases as the number of grid points are increased. The results suggest that the effect of mesh refinement on the prediction of peak heating is quite significant for strong shock waves.

4. Conclusions

The shock reflection processes by a ramp and unsteady aerodynamic heating phenomena are simulated by solving the thin-layer approximated Navier-Stokes equations. The effect of mesh refinement to the calculated results is investigated. The results show that a smaller mesh size is necessary for calculating the precise heat flux distributions and capturing the fine structure of the shock reflection patterns. Especially, the heat flux distribution is quite sensitive to mesh

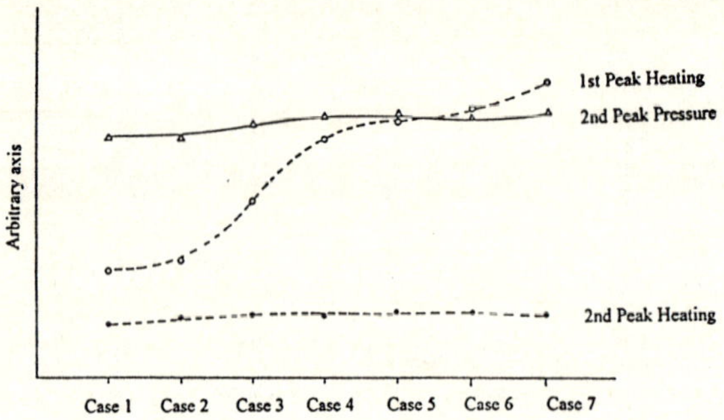

Fig. 7. Change of peak heating and peak pressure due to mesh refinement ($M_s = 3.0$, $\theta = 35°$ and Re= 1.0×10^4)

refinement. Also, the effect of mesh refinement on the prediction of peak heating is quite significant for strong shock waves.

References

Aso S et. al.(1989) In: Grönig H (ed) Proc. 17th Intl. Symp. Shock Waves and Shock Tubes, Aachen, VCH, pp.777-782.
Aso S et. al.(1990) Memo. Fac. of Eng. , Kyushu Univ., 50: 295-308
Baldwin BS, Lomax H (1978) AIAA Paper No.78-257
Korkegi RH (1971) AIAA J., 19: 771-784
Yee HC, Harten A (1987) AIAA J. 25: 266-274

A Comparison Study of Two Finite-Element Schemes for Computation of Shock Waves

Hong Luo[*], Joseph D. Baum[*] and Rainald Löhner[†]
[*]Science Applications International Corporation, 1710 Goodridge Drive, MS 2-3-1 McLean, VA 22102, USA
[†]CMEE, School of Engineering and Applied Science, The George Washington University, Washington, D.C. 20052, USA

Abstract.   This paper describes an extensive comparison study of two high-order-accuracy finite-element schemes: one of upwind type, the other of flux-corrected transport type, for the simulation of shock wave propagation on unstructured meshes. The performance of these two schemes for the solution of the Euler equations on unstructured meshes is compared on the basis of solution accuracy and computational efficiency for some well-known test cases. Advantages and disadvantages of the two schemes are discussed.

Key words: Shock waves, CFD, Upwind, FCT, Unstructured meshes

1. Introduction

The use of unstructured meshes for computational fluid dynamics problems has become widespread due to their ability to discretize arbitrarily complex geometries and due to the ease of mesh adaption in enhancing the solution accuracy and efficiency through the use of adaptive refinement techniques. In recent years, significant progress has been made in developing numerical algorithms for the solution of the compressible Euler and Navier-Stokes equations on unstructured grids. However, little systematic work has been done comparing the ability of different numerical algorithms to predict complex shock structures.

This paper describes an extensive comparison study of two high-order accuracy finite-element schemes for the computation of shock waves on unstructured meshes. The two schemes are widely used in the computational fluid dynamics community due to their ability to capture shocks with high resolution without spurious oscillations. The first scheme belongs to the Godunov-type family. In this scheme, the spatial discretization is accomplished by an edge-based finite-element formulation using Roe's flux-difference splitting. A MUSCL approach is used to achieve higher-order accuracy. Solutions are advanced in time by a multi-stage Runge-Kutta time-stepping scheme. This algorithm has been used previously for computing inviscid flows, including shocks around complex configurations, and has been shown to yield excellent results for steady flows (Luo et al. 1993b). The second algorithm is an edge-based finite-element flux-corrected transport scheme (FEM-FCT) (Löhner et al. 1987, Luo et al. 1993a). The idea behind FCT is to combine a high-order scheme with a low-order scheme in such a way that the high-order scheme is employed in smooth regions of the flow, whereas the low-order scheme is used near discontinuities to yield a monotonic solution. Although FEM-FCT is often criticized as not having a strict mathematical background, our experience has been that it yields excellent resolution for both shock waves and contact discontinuities for the simulation of strongly unsteady compressible flows (Baum and Löhner 1991,1992). The performance of these two schemes for the solution of the Euler equations on unstructured meshes will be compared on the basis of solution accuracy and computational efficiency for some well known test cases. Some conclusions concerning the application of these two methods for the computation of shock waves will be given.

Shock Waves @ Marseille I
Editors: R. Brun, L. Z. Dumitrescu   © Springer-Verlag Berlin Heidelberg 1995

2. Numerical methods

A brief description of the numerical schemes is given in this section. The implementation of both schemes is based on an edge-based data structure, as opposed to a more traditional element-based data structure. The use of this edge-based data structure not only improves the efficiency of the algorithms, but also provides an extremely useful process to construct different numerical schemes (Luo et al. 1993a). Using an edge-based data structure, the right-hand-side is constructed by evaluating fluxes on each edge and then sending their contribution to the nodes.

2.1. Upwind finite-element scheme

The upwind finite-element scheme is constructed using Roe's flux difference splitting, where the flux on each edge is determined by

$$\mathcal{F}_{IJ} = F_I + F_J - \mid A_{IJ} \mid (U_J - U_I) \tag{1}$$

where $\mid A_{IJ} \mid$ denotes the standard Roe matrix. It can be shown that this scheme is equivalent to the first order finite-volume upwind cell-vertex scheme based on a dual mesh. There are many different ways to achieve higher-order accuracy. In the present study, a scheme of higher-order accuracy is achieved by using upwind-biased interpolations of the solution U via the MUSCL approach. With higher-order spatial accuracy, spurious oscillations in the vicinity of shock waves are expected to occur. Some form of limiting is usually required to eliminate these numerical oscillations of the solution and to provide some kind of monotonicity property. Many popular limiters have been tested and implemented in the scheme. Three options exist concerning the choice of interpolation variables: conservative variables, primitive variables, and characteristic variables. Using limiters on characteristic variables seems to give the best results, but sacrifices some computational efficiency.

Our own experience is that the Van Albada limiter based on the conservative variables gives best results for steady problems, and the Minmod limiter based on the characteristic variables produces better solutions for transient problems, as shown in the examples to be presented.

2.2. Finite-element flux-corrected transport scheme

The idea behind FCT is to combine a high-order scheme with a low-order scheme in such a way that the high-order scheme is employed in smooth regions of the flow, whereas the low-order scheme is used near discontinuities in a conservative way, in an attempt to yield a monotonic solution. The implementation of an edge-based FCT scheme is exactly the same as its element-based counterpart (Löhner et al. 1987). However, the use of an edge-based data structure makes the implementation more efficient, which is especially attractive for 3D problems. As the high-order scheme, we employ the edge-based two step Taylor-Galerkin scheme with the consistent mass matrix. The low-order scheme is simply constructed by using the lumped mass-matrix plus mass diffusion. Combined with a modified second-order Lapidus artificial viscosity, the resulting scheme is second-order accurate in space, and fourth-order accurate in phase.

The results available in the literature have shown that FCT could produce results of excellent quality for a single PDE. However, when trying to extend the limiting process to a system of PDEs, the precise details of the limiting mechanism are a little uncertain. Many variations are possible and can be implemented, giving different performance for different problems. The limiter used most frequently for transient problems is the one based on the minimum of density and energy. Unfortunately, the optimal limiter for a specific problem is purely based on empirical experience. Despite this fact, our experience has been that it gives, without much 'user-intervention', excellent resolution for both shocks and contact discontinuities for the simulation of strongly unsteady compressible flows.

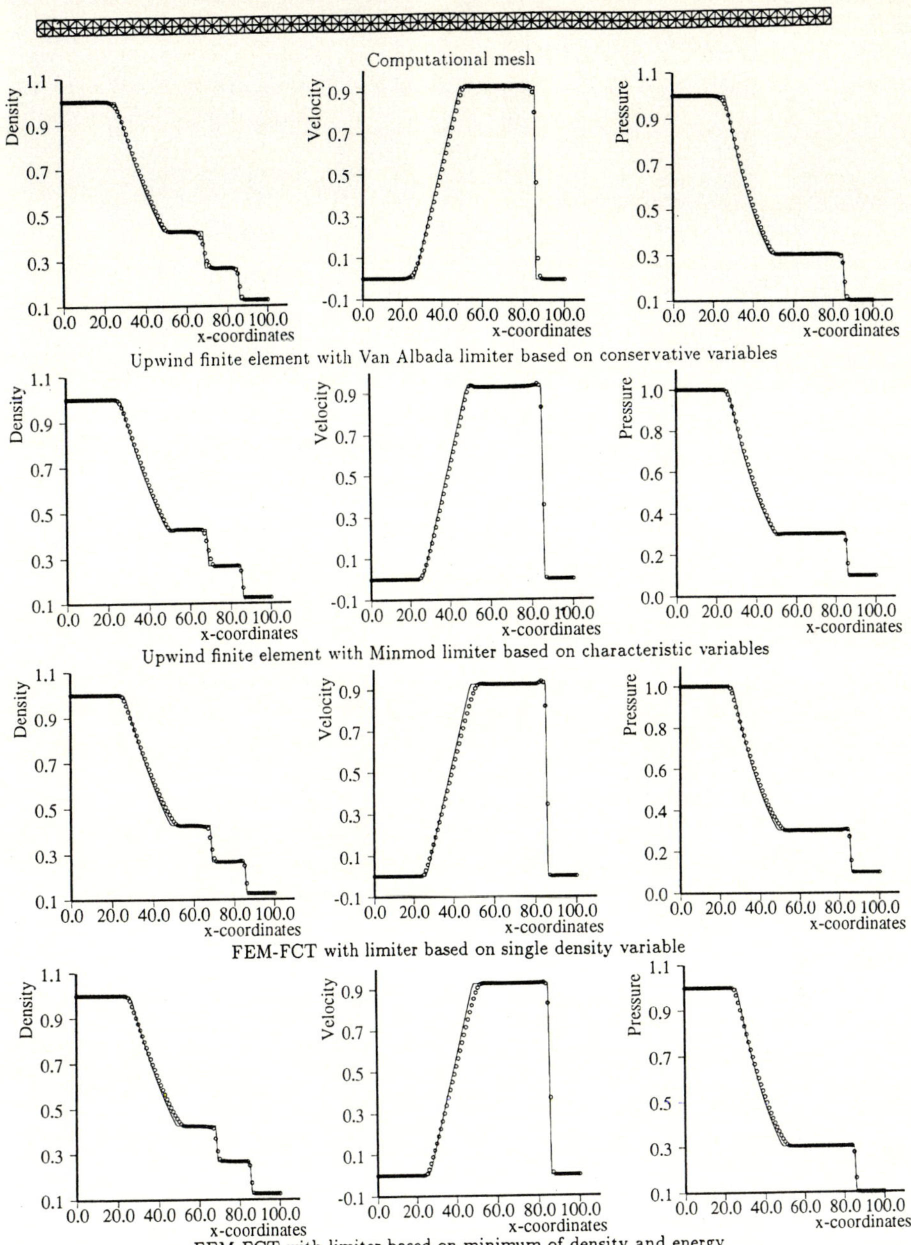

Fig. 1. Computational mesh and results for the shock tube problem

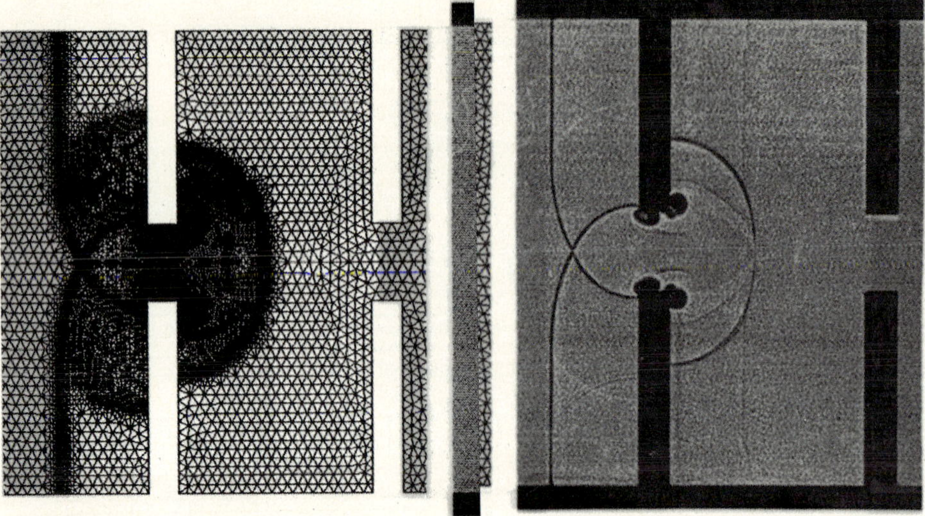

Fig. 2a. Adapted mesh at 60 μs for computing shock diffraction in a baffled tube

Fig. 2b. Shadow cinematograph of shock propagation at 60 μs

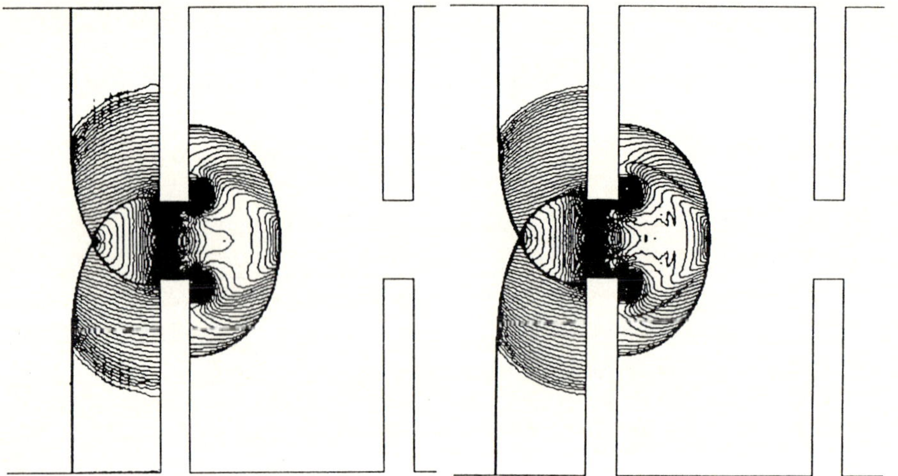

Fig. 2c. Computed density contours at 60 μs using edge-based upwind scheme

Fig. 2d. computed density contours at 60 μs using edge-based FCT scheme

3. Results and discussion

Due to space limitation, results are presented only for two illustrative test cases: a) the classical Sod's shock problem and b) transient shock diffraction (M_s=1.31) within a baffled channel.

Test Case 1. Shock tube problem or Riemann Problem

The shock tube problem constitutes a particularly interesting and difficult test case, since it presents an exact solution to the full system of one-dimensional Euler equations containing simultaneously a shock wave, a contact discontinuity, and an expansion fan. The initial conditions in the present computation are the following:

$$\rho = 1.000, \quad u = 0, \quad p = 1.0, \quad 0.0 \leq x \leq 50.0$$
$$\rho = 0.125, \quad u = 0, \quad p = 0.1, \quad 50. < x \leq 100.$$

Fig.1 shows the computational mesh and the results obtained by two schemes with different limiters. The solid lines in these figures are the exact solutions and the dots represent computed results. This is a 2D simulation of a 1D problem. The mesh consists of 101 points in the X-direction and 3 points in the Y-direction. The accuracy of the computed results is assessed by making comparisons with the exact solution. The results indicate that the solutions obtained by both schemes depend heavily on the limiter used. Both schemes produced fairly good results, with excellent resolution for both shock and contact discontinuity. However, the FCT scheme gives better resolution for the contact discontinuity than the upwind scheme.

Test Case 2. Shock diffraction in a baffled tube
The problem under consideration is shown in Fig.2. A weak shock ($M_s = 1.31$) propagates inside a baffled tube. This problem was selected as a test case because of the available experimental data. Grid adaption is used for this computation. The number of refinement levels allowed in this case was 5 and the grid was modified every 7 timesteps. Density was chosen as the key variable for the error indicator. Fig.2a shows the adapted mesh at 60 μs, while the experimental solution is depicted in Fig.2b. Figs.2c and 2d show the computed density contours obtained by the edge-based upwind finite element scheme with Minmod limiter based on the characteristic variables and the edge-based FEM-FCT with limiter based on the minimum of density and energy, respectively. These represent the best results that could be obtained by these two schemes. The results indicate that both schemes together with the classic h-refinement/coarsening grid adaption scheme can produce excellent results on unstructured meshes for the efficient simulation of strongly unsteady flows, without deteriorating the accuracy of the solutions. Comparison of these results demonstrates that the edge-based FEM-FCT produced better solutions than the edge-based upwind finite element scheme in terms of capturing the weak shock, while running approximately 3 times faster.

4. Conclusions

Based on the results of this study, some general observations can be made:
- The edge-based upwind finite element scheme based on Roe's approximate Riemann solver gives fairly good results for all test cases, although the shocks and contact discontinuities are diffused compared to the results obtained by the FCT scheme. The most significant advantage of this scheme is that it is naturally dissipative, and hence does not require any problem-dependent parameters to adjust.
- The edge-based FEM-FCT scheme gives better resolution of both shocks and contact discontinuities. However, the solutions obtained are dependent on the choice of limiters and the value of the Lapidus dissipation parameter, although the worst choice of those parameters still gives acceptable results.
- The edge-based FEM-FCT scheme requires less computational effort than the edge-based upwind finite element scheme. The present 2D edge-based FEM-FCT code runs at a speed of about 16.4 μ-sec/point/time-step on a single processor Cray 2, about three times faster than the edge-based upwind finite element code. However, the upwind scheme, in conjunction with the implicit residual smoothing, can use a much higher CFL number for the steady computations.

References

Baum JD, Löhner R (1992) Numerical simulation of passive shock deflector using an adaptive finite element scheme on unstructured grids. AIAA Paper 92-0448

Baum JD, Löhner R (1991), Numerical simulation of shock interaction with a modern main battlefield tank. AIAA Paper 91-1666

Löhner R, Morgan K, Peraire J, Vahdati M (1987) Finite element flux-corrected transport (FEM-FCT) for the Euler and Navier-Stokes equations. Intl. J. Num. Meth. Fluids 7:1093-1109.

Luo H, Baum JD, Löhner R, Cabello J (1993a) Adaptive edge-based finite element schemes for the Euler and Navier-Stokes equations on unstructured meshes. AIAA Paper 93-0336

Luo H, Baum JD, Löhner R (1993b) Numerical solution of the Euler equations for complex aerodynamic configurations using an edge-based finite element scheme. AIAA Paper 93-2933

Adaptive Mesh Refinement Computation of Compressible Flow

N. Uchiyama, O. Inoue
Institute of Fluid Science, Tohoku University, Sendai 980, Japan

Abstract. Numerical simulation was performed using an adaptive mesh refinement algorithm combined with a high resolution upwind scheme to study shock wave diffraction phenomena. The particular flow in interest is the shock wave diffraction over a 90 degree sharp corner in which the generation of a vortex at the corner and its associated shock waves are expected. The numerical results of the Euler computation show detailed structures of the flowfield such as growth of Kelvin-Helmholtz instability along the spiraling sheet of the corner vortex and the interaction of the rolled-up vortices with the shock waves, so called "vortex shocks" inherent in the corner vortex.

Key words: Adaptive mesh refinement, Shock wave diffraction, Shock-vortex interaction

1. Introduction

In order to numerically simulate the shock wave propagation and diffraction phenomena by means of finite-difference/volume method, a large number of grid points/cells is required to resolve the fine structure of the flowfield which usually makes the computation very expensive. To overcome this situation, various methods of grid adaptation have been studied by many researchers. Among them, the local adaptive mesh refinement (AMR) algorithm pioneered by Berger et al. (Berger and Colella 1989) has shown successful results in efficiently capturing detailed features of the flowfield in many applications. We have applied this algorithm to our two-dimensional Cartesian Euler code with some simplifications (Uchiyama and Inoue 1992).

In the present study, we apply our code to simulate the flowfield of a planar shock wave with incident shock Mach number of $M_s = 1.5$ diffracting over a 90 degree sharp corner. This flowfield was adopted as a benchmark problem in the poster session held during the 18th International Symposium on Shock Waves at Sendai, Japan. Numerical solutions and experimental pictures presented by twenty-six research groups provided an interesting comparison among them: lots of common features as well as discrepancies appeared (Takayama and Inoue 1991). The current study has been motivated by this comparison; that is we need to further understand the details of this flowfield with the use of AMR computation.

One of the interesting features in this flowfield is the formation of a vortex near the corner edge. Although the Euler equation cannot formally treat the viscous effects, the successful application of Euler computations by Hillier (1991) and Sivier et al. (1991) demonstrate the effectiveness of the Euler equations in simulating the generation of vortices in high speed sharp corner flows.

2. The numerical method

We have applied the AMR algorithm based on the idea of Berger et al. (Berger and Colella 1989) to our two dimensional Euler code in finite-volume formulation. The mesh system of the AMR algorithm consists of hierarchical multilevel submesh panels categorized by their mesh width. The coarsest mesh exists as an underlying base, while the submesh panels with finer mesh width are locally generated by some given criteria in order to dynamically adapt the local region in interest.

The criterion for refinement is given such that any cell in which an associated error indicator ϵ exceeds a prescribed value should be refined. In determining the error indicator, we have adopted

Shock Waves @ Marseille I
Editors: R. Brun, L. Z. Dumitrescu

the absolute values of the local density gradients ρ_x, ρ_y and the second space derivatives of the density ρ_{xx}, ρ_{yy} and also the vorticity ω in the form below:

$$\epsilon = \max \left(\sqrt{|\rho_x|^2 + |\rho_y|^2}, \ \alpha\sqrt{|\rho_{xx}|^2 + |\rho_{yy}|^2}, \ \beta|\omega| \right) \tag{1}$$

Here, α and β are the magnification parameters. In the formulation above, the density gradients work to detect shock waves and contact surfaces and the second space derivatives of the density play a similar role in detecting expansion waves while the vorticty is incorporated to search for the vortex and the shear layers.

In order to refine the mesh in the vicinity of the detected regions, a number of submesh panels with mesh width r times finer than the base mesh width are locally adapted.

Subsequently, an explicit time integration is performed in a nested way such that after the solution on the coarse base mesh is integrated over one coarse time step, the solutions of the submesh panels are integrated r times successively with a time step r times smaller than that of the coarse base mesh, in order to catch up with the same time level. The detailed notes on this process can be found elsewhere (Uchiyama and Inoue 1991).

Having obtained the solutions of both the base mesh and the submeshes at the same time level, a modification step is taken to maintain conservation across different mesh levels. Here, we have followed the method proposed by Berger et al. (Berger and Colella 1989).

The time marching of the whole mesh system is obtained by repeating these procedures. Also by further nesting the algorithm, computation is possible for higher hierarchical mesh systems having refinement levels greater than two.

The definition of the numerical flux is also an important factor in achieving solutions with high resolution. For the present computation, we have adopted the flux-vector splitting (FVS) scheme of Steger and Warming (1981) for its relatively fast and simple character. Also by using MUSCL extrapolation (Van Leer 1977), higher order accuracy in space was achieved. Additional modifications were made in the MUSCL extrapolation to achieve second order accuracy in time with explicit single step integration.

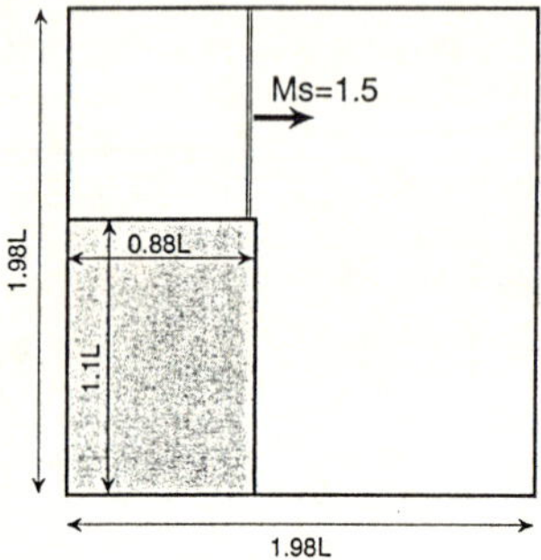

Fig. 1. Schematic of the flowfield

3. Simulation of shock wave diffraction over a 90 degree sharp corner

A planar shock of incident shock Mach number $M_s = 1.5$ diffracts over a 90 degree sharp corner (Fig.1). The ambient medium is considered to be an ideal gas with specific heat ratio $\gamma = 1.4$. The computational domain was set to be a square with its side length $1.98\,L$, where L is a reference scale length and is set to unity in the computation. The coarse base mesh consists of 99 cells in each x and y direction with its width $0.02\,L$. The width and height of the corner block is $0.88\,L$ and $1.1\,L$, respectively. The incident shock wave was set by numerically satisfying the Rankine-Hugoniot condition and was initially located at the corner edge.

We have allowed at most 3 levels of refinement and the refinement ratio between the neigbouring levels was set to $r = 5$. With this mesh system, the finest mesh width on the third mesh level is equivalent to the width of $99 \times 5 \times 5 = 2475$ cells uniformly distributed in the x and y directions of the computational domain. In the compuation, the physical values were non-dimensionalized by the ambient density ρ_∞ and the ambient sound speed a_∞.

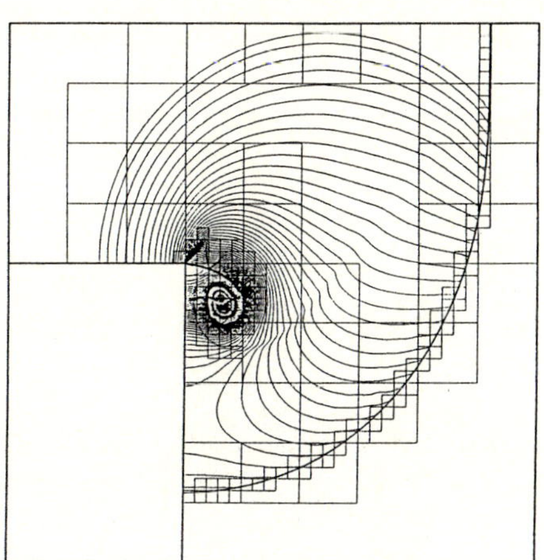

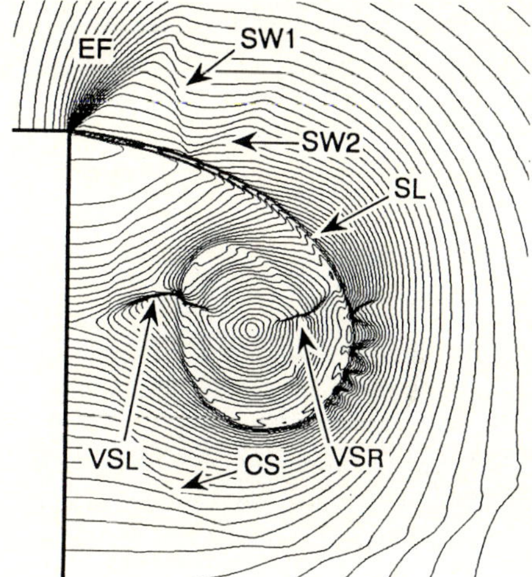

Fig. 2(a). Density contours of the flowfield   Fig. 2(b). Magnified view of the vortex

Fig.2(a) shows the density contours of the entire flowfield at nondimensional time $\tilde{t} = 0.64(L/a_\infty)$. The CPU time required up to this stage took approximately 3 hours on an 8CPU CRAY-YMP. The submesh panels of the second level are dynamically adapted along the diffracted shock wave and to the vicinity of the vortex generated at the corner and also to the outermost wave of the expansion waves. The mesh panels of the third level are adapted more locally to the diffracted shock wave and to the vortex. A magnified view of the vortex (Fig.2(b)) shows the detailed features of the flowfield in this region. A Prandtl-Meyer expansion-fan (EF) at the corner is terminated by a weak shock wave (SW1) on the shear layer (SL) to match the expanded flow to the condition behind the diffracted shock wave. It is to be noted that this shock wave is followed by another weak shock wave (SW2) with its foot located at the same point on the shear layer as the shock wave (SW1). The corresponding experimental picture of Ritzerfeld et al. (Takayama and Inoue 1991) shows a series of shock waves along the shear layer implying that the matching process cannot be accomplished by a single shock wave. A contact surface (CS) below the vortex emanates from a point on the incident shock wave where diffraction starts. Also, two distinct shock waves exist to the left (VSL) and right (VSR) of the vortex core and are located nearly parallel to the radial direction of the vortex. These shock waves are refered to as "vortex shocks" (Hillier 1991) and are often observed in the vortex of a compressible fluid. However, it seems

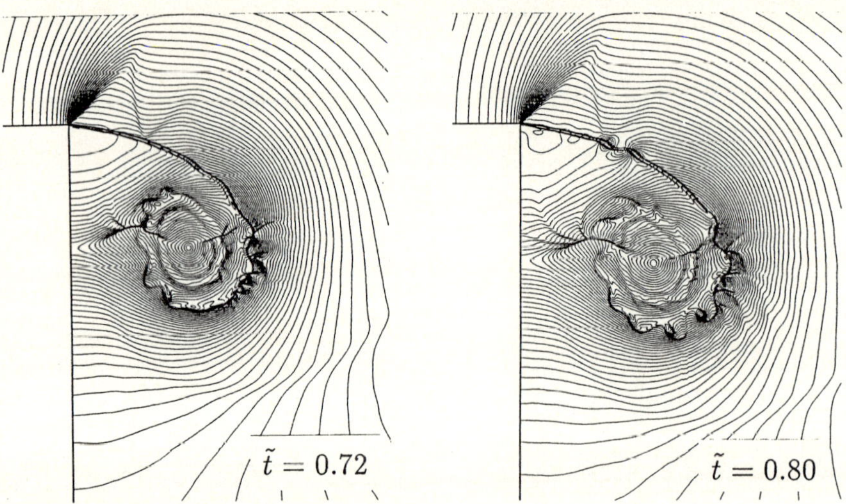

Fig. 3. Evolution of the vortex near the corner (density contours)

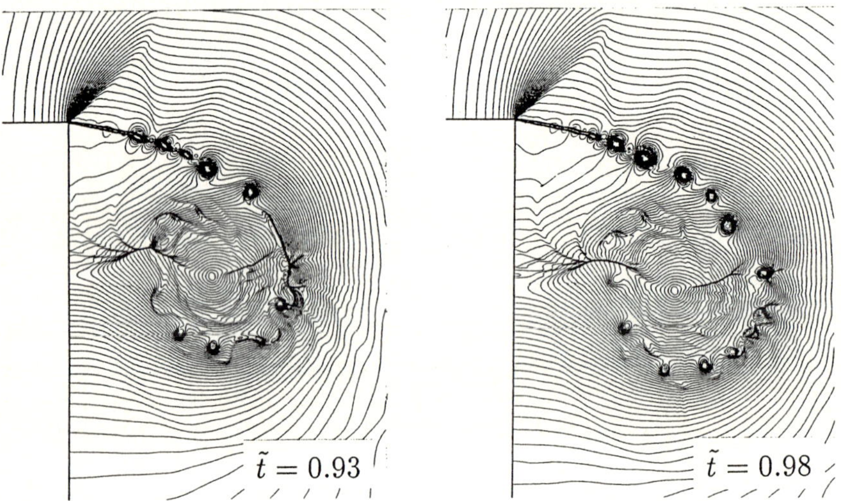

Fig. 3. continued

that either numerical or experimental or even theoretical studies have not been fully addressed
to explain the generation mechanism of these shock waves.

Fig.3 shows the time devolpment of the vortex near the corner edge in density contours. At
time $\tilde{t} = 0.72$ the shear layer forming the vortex is no longer smooth. A wavy pattern emerges
along the shear layer indicating the growth of Kelvin-Helmholtz instability. Also attention should
be paid to the change in the structure of the vortex shocks. In Fig.2(b), both of these shock waves
were observed as single structures. However, as the shear layer becomes unstable, a bifurcation

became apparent on both the left and right vortex shocks. A similar bifuraction of the vortex shock can be observed in the shadowgraph pictures of Schmidt and Duffy (Wang and Widhopf 1990) in the axisymmetric flowfield of a shock wave emission from an open-ended shock tube.

As we advance to $\tilde{t} = 0.80$ the instability of the shear layer and the bifurcation of the left vortex shock become prominent. Also along the shear layer near the corner which was still rather smooth at $\tilde{t} = 0.72$, a roll-up of the vortex sheet emerges near the point where the expansion-terminating shock wave stems. It is likely that the terminating shock wave plays a role in stimulating the shear layer to roll-up; further investigation is required.

At time $\tilde{t} = 0.93$, rolled-up vortices are seen everywhere along the shear layer except for a region near the right vortex shock where the shear layer remains smooth. It is interesting to notice that at this instant the right vortex shock has no apparent bifurcated branch.

At the stage of $\tilde{t} = 0.98$, the vortex roll-ups are observed along the entire part of the shear layer. Also the bifurcation of the left vortex shock is well developed and the reflection of the left-most branch to the wall can be observed.

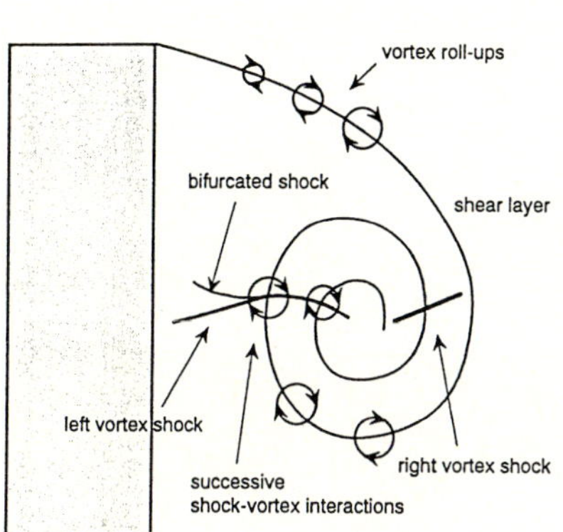

Fig. 4. Schematic of the vortex shock bifurcation

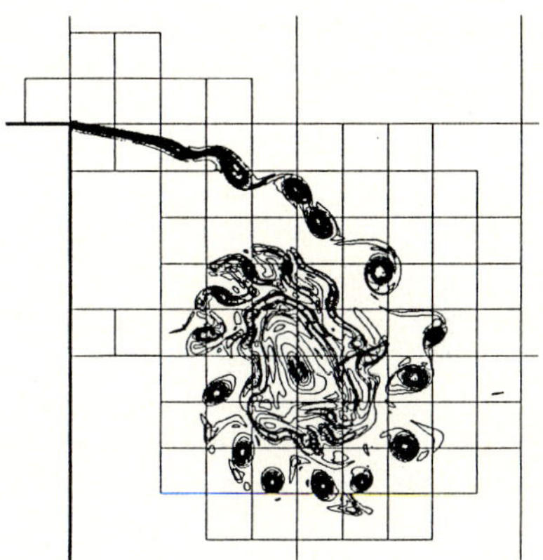

Fig. 5. Vorticity contours near the corner

Fig.4 is a schematic for the explanation of the bifurcation phenomena prominent at the left vortex shock. When the shear layer remains smooth, the left vortex shock looks like a single structure with no bifurcation, compressing the reverse flow between the wall and the vortex core to decelerate. However, as the Kelvin-Helmholtz instability grows along the shear layer, rolled-up vortices having clockwise rotation successively interact with the left vortex shock. It is known from the numerical studies on shock-vortex interaction (Meadows et al. 1989) that a single vortex interaction with a planar shock wave may produce a bifurcated structure of the shock wave. The situation is quite similar in each interaction process of the rolled-up vortices with the vortex shock. Due to the clockwise rotation of the rolled-up vortices, the left portion further compresses the shock while the right portion reacts to expand. Since this interaction occurs successively, the left vortex shock may result in a bifurcated structure with multi-branches. However, the right arm of the left vortex shock extending into the vortex core will experience a second interaction with the rolled-up vortex along the inner path of the spiral. This may act in a way to counterbalance the interaction at the outer path of the spiral. Therefore, we think that the bifurcation of the right arm will not be prominent as the left arm reaches the wall.

Additionally, Fig.5 shows a vorticity contour for $\tilde{t} = 1.00$. The rolled-up vortices are evident along the shear layer. Also it is to be noted that the rolled-up vortices along the inner path of the spiral are skewly deformed.

4. Conclusions

The AMR computation of the Euler equations have been performed to simulate the flowfield of shock wave diffraction over a 90 degree sharp corner. The fine structures of the flowfield were simulated successfully with high resolution, and the Kelvin-Helmholtz instability along the spiraling shear layer of the corner vortex and the bifurcation of the "vortex shocks" were captured. The results suggest that the bifurcation of the vortex shock is a result of the successive interaction between the shock and the rolled-up vortices along the shear layer.

References

Berger MJ, Colella P (1989) Local adaptive mesh refinement for shock hydrodynamics. J. Comp. Phys. 82:64-84

Hillier R (1991) Computation of shock wave diffraction at a ninety degree convex edge. Shock Waves 1:89-98

Meadows K, Kumar A, Hussaini MY (1989) A computational study on the interaction between a vortex and a shock wave. AIAA J. 29:174-179

Sivier S, Baum J, Loth E, Löhner R (1991) Simulations of shock wave generated vorticity. AIAA Paper 91-1668, AIAA 22nd Fluid Dynamics and Plasma Dynamics and Lasers Conference

Steger JL, Warming RF (1981) Flux vector splitting of the inviscid gasdynamic equations with application to finite-difference methods. J. Comp. Phys. 40:263-293

Takayama K, Inoue O (1991) Shock wave diffraction over a 90 degree sharp corner -Posters presented at 18th ISSW. Shock Waves 1:301-312

Uchiyama N, Inoue O (1992) On the performance of adaptive mesh refinement computation. Shock Waves 2:117-120

Van Leer B (1977) Towards the ultimate conservative difference scheme. IV. A new approach to numerical convection. J. Comp. Phys. 23:276-299

Wang JCT, Widhopf GF (1990) Numerical simulation of blast flowfields using a high resolution TVD finite volume scheme. Computers & Fluids 12:103-137

Shock-Capturing Schemes with Entropy Corrections and Dynamically Adaptive Unstructured Meshes

Roland Richter and Pénélope Leyland
Institut de Machines Hydrauliques et Mécanique des Fluides, Ecole Polytechnique Fédérale de Lausanne, CH-1015 Lausanne, Switzerland

Abstract. A dynamical grid-adaptation method coupled with the numerical scheme on unstructured meshes is described. The discontinuity capturing properties of such methods ensure a high precision for the global solution. Entropy corrections for centered space approximation numerical schemes which do not preserve entropy are discussed. In particular, the problem of spurious entropy production by the numerics near solid wall boundaries which present discontinuous contours can be overcome by a localised use of an entropy-preserving Riemann solver.

Key words: Discrete entropy condition, Auto-adaptive mesh, shock capturing

1. Introduction

Discontinuity capturing properties via numerical or artificial dissipation are highly linked to the underlying grid structure, and can be improved by using local refinement on adapted grids. Grid alignment is even necessary when pseudo-one-dimensional solvers are adapted to multidimensional problems, as is the case of many TVD methods. For example, complex flow calculations involving secondary shock formation in shock-shock interactions are particularly difficult to calculate accurately. Such phenomena can be illustrated with greater precision by using mesh adaptivity. The local discretisation error is reduced by refining the underlying mesh in the specific region. The combination of the relative grid dependency of the numerical scheme and auto-adaptive grids using refinement-derefinement algorithms on unstructured meshes provides a robust tool for capturing sharply defined physical phenomena. For instance, shock formations, moving free surfaces, reaction zones or flame propagation in combustion, multiple shock reflections in air engine inlets or turbomachinary cascades, all provide difficult transient or unsteady flows, with not only multiple shock - shock interactions, but also secondary transient shocks and contact discontinuities. The methods described here also give remeshing strategies for unsteady flows, allowing a specific physical phenomena to be "followed" with accuracy, (Richter and Leyland 1993).

Numerical schemes for compressible flows can be decomposed as a flux evaluation term plus some numerical dissipation term. The numerical dissipation can be explicit, as for artificial or matrix dissipation, or internal as, for example, implicit and upwind schemes. The overall scheme should be entropy-preserving for hyperbolic systems, as, for instance, the system of compressible Euler equations used to modelise inviscid flow. However, many popular schemes do not satisfy this property intrinsically, and need to be corrected in order to do so (Khalfallah 1990, Woodward and Colella 1984). Also, shock capturing and contact discontinuity capturing properties of numerical schemes can lead to contradictory constraints. Approximate Riemann solvers and other TVD schemes produce accurate discontinuities at the cost of internal dissipation, whereas symmetric or centered schemes need auxiliary dissipation techniques for stability and discontinuity capturing.

In this paper, entropy corrections are discussed and tested for unsteady transient shock problems. Comparisons with approximate Riemann solvers are given.

Shock Waves @ Marseille I
Editors: R. Brun, L. Z. Dumitrescu

2. Entropy correction schemes

The numerical scheme presented here is based on a standard centered spatial discretisation. Time accuracy is obtained by Runge-Kutta multistage time stepping. The spatial approximation is a Galerkin finite volume method on dual control volumes with additional artificial viscosity terms, necessary for stability requirements and shock capturing properties. The control volumes are taken to be the dual cells of the underlying $P1$ Galerkin finite element approximation. These cells constitute the dual topology of the $P1$ triangular n-simplex space, such that the finite volume formulation over the dual cells is equivalent to a $P1$ Galerkin finite element approximation, (Baba and Tabata 1981, Dervieux 1984). They are constructed by taking the barycentres of the contributing triangles to discretisation node i, as in Fig.1.

The finite volume formulation of the 2D-Euler equations can be considered as the flux balance evaluated by summing all the contributions over the cell interfaces between the node i, and its neighbouring nodes j, ∂C_{ij}:

$$\text{area}C_i \frac{dW_i}{dt} + \int_{\partial C_i} (F(W) \cdot \overrightarrow{n}^i_x + G(W) \cdot \overrightarrow{n}^i_y)d\sigma = 0. \tag{1}$$

where W denotes the state vector of conservative variables, (F, G) denotes the conservative fluxes, $\overrightarrow{n}$ denotes the outward normal to the cell centered around node i. The flux balance is evaluated by summing all the contributions over the cell interfaces between the node i, and its neighbouring nodes j, ∂C_{ij}, (see Fig.1).

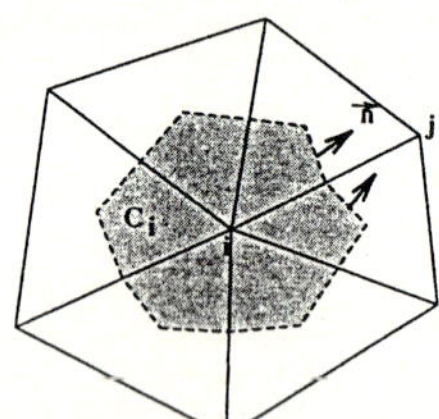

Fig. 1. Construction of a dual cell

The system (1) leads to a set of coupled ordinary differential equations:

$$\text{area}C_i \frac{dW_i}{dt} + R(W_i) = 0$$

where $R(W_i)$ denotes the numerical approximation of the second term in Eq.1. The corresponding numerical flux function of Eq.1 can be written symbolically as

$$\Phi(W_i, W_j) = \frac{F(W_i) + F(W_j)}{2}\eta^x_{ij} + \frac{G(W_i) + G(W_j)}{2}\eta^y_{ij} + \Phi^{\text{diss}}(W_i, W_j, \overrightarrow{\eta}_{ij})$$

$$\equiv \Phi^{\text{exp}} - \kappa\, Q_{ij}(W_j - W_i) - \sum_{l=-m+1}^{m} H_{lij}(W^{n+1} - W^n). \tag{2}$$

where $(\eta^x_{ij}, \eta^y_{ij})$ denote the integrated normals from ∂C_{ij}, $\kappa = \frac{1}{2}(\frac{\Delta x}{\Delta t}, \frac{\Delta y}{\Delta t}) = \frac{1}{2}(\sigma_x, \sigma_y)$. Q and H are matrices defining the scheme (Lerat 1990).

The dissipation flux for the centered scheme contains a second-order term for shock capturing which is improved by a TVD pressure switch, and a fourth-order term for high frequency damping. For details on the construction of these dissipation terms on dual cells, refer to Richter (1993).

The advantages of this scheme are its straightforward implementation and its computational efficiency in terms of CPU time requirements. The quality of the shock capturing shows up in

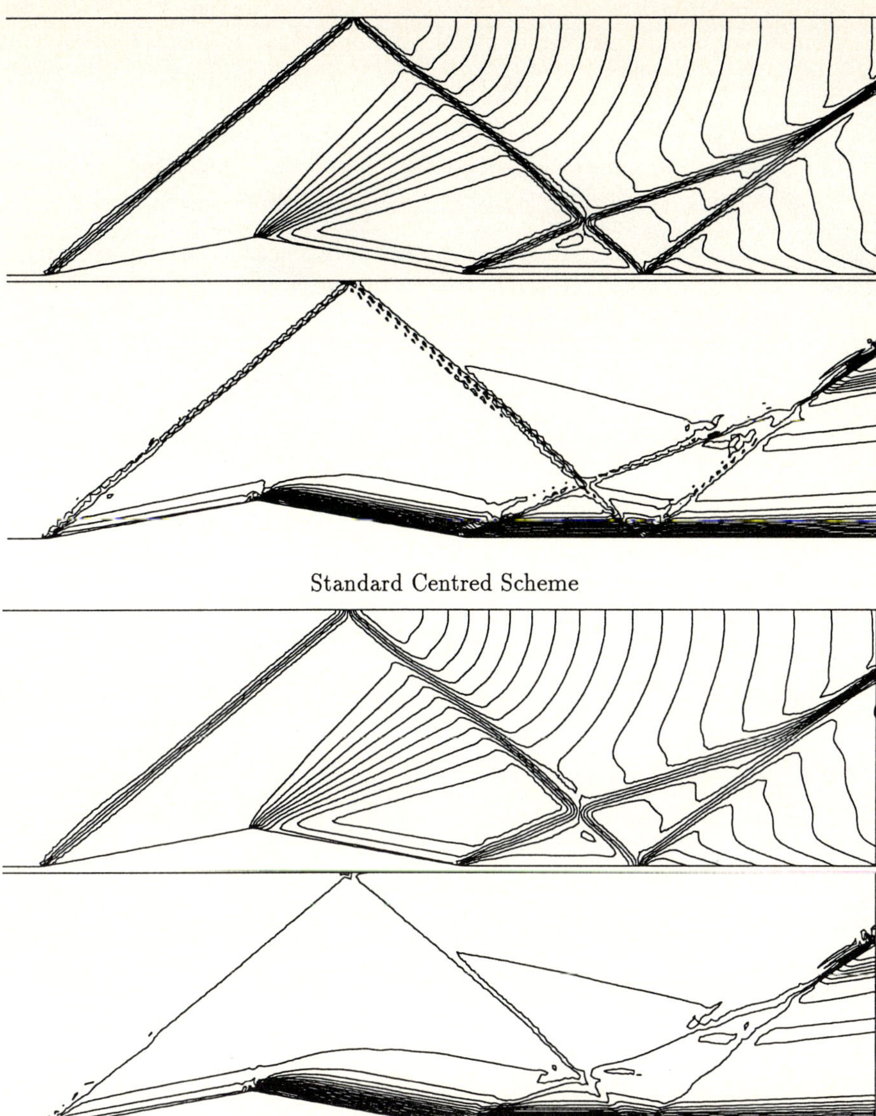

Standard Centred Scheme

Entropy Corrected Centred Scheme

Fig. 2a. Mach 2 flow in a duct, iso- Mach and entropy contours for the standard centered scheme (top) and the corrected scheme (bottom). Oscillations near discontinuities reduced by the entropy correction

extremely reduced over- and under-shoots, which virtually disappear when coupling the method with adaptive meshing, where the mesh points are aligned with the sharp gradients. However, its main drawback is to shows serious entropy oscillations in strong gradient regions and non-physical entropy production at solid wall boundaries.

To eliminate these non-entropic behaviours, an entropy correction similar to that proposed by Tadmor (1987), Khalfallah (1990) and Lerat (1990), has been implemented, which smoothes

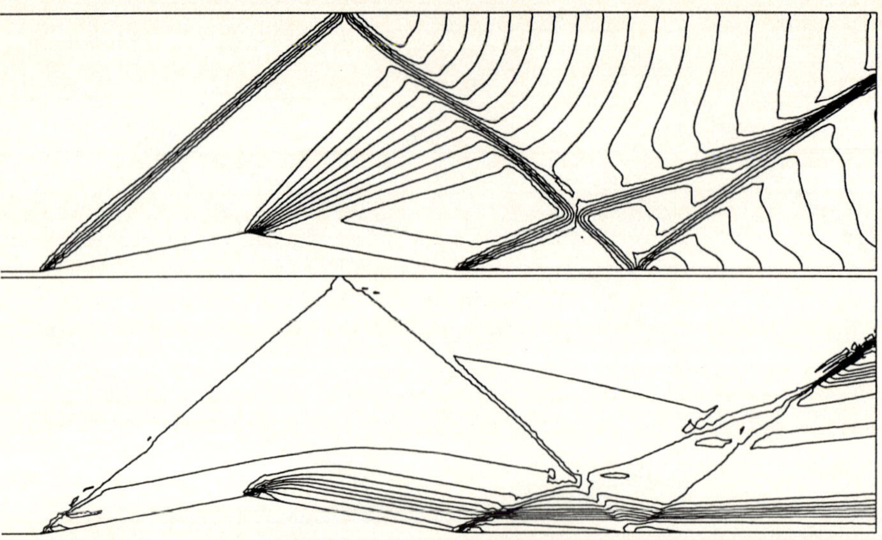

Corrected Scheme plus Riemann Solver at boundaries

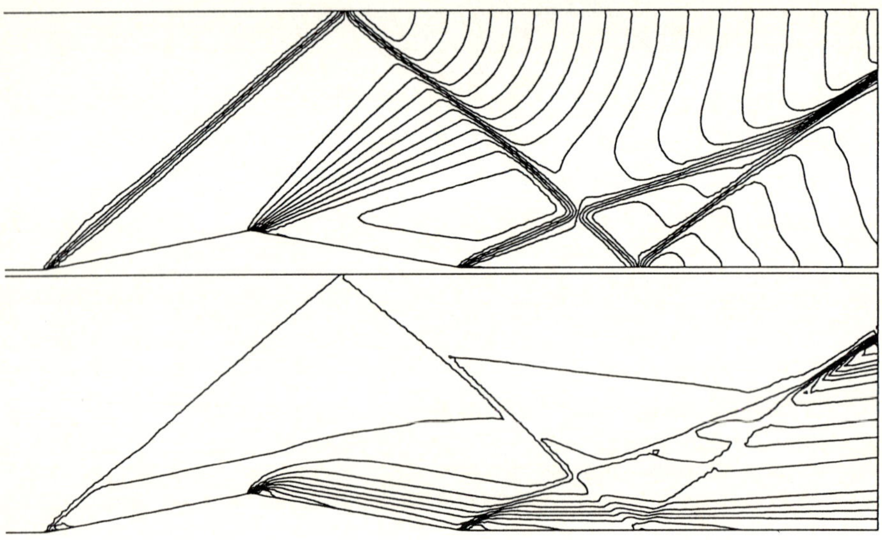

Riemann Solver

Fig. 2b. Mach 2 flow in a duct, non-physical entropy production near boundaries eliminated by use of a centered scheme plus a Riemann solver at boundaries, (top). The comparison with the solution using the Osher scheme globally (bottom) is excellent

the internal field and produces an entropy preserving solution. The explicit part of the numerical flux (Eq.2) is modified by taking into account the dissipation flux matrix:

Notation

$$\delta \mathcal{A} = (\mathcal{A}_j - \mathcal{A}_i), \qquad \mu \mathcal{A} = \tfrac{1}{2}(\mathcal{A}_i + \mathcal{A}_j)$$

$$\widetilde{\Phi}_k^{\mathrm{exp}} = \Phi_k^{\mathrm{exp}} - \tfrac{1}{2}\beta_k \delta W \qquad \text{with} \qquad \beta_k = \begin{cases} 0 & \text{if } \vartheta = 0 \\ \chi \dfrac{\max(q_k^*,\, 0)}{\vartheta_k} & \text{otherwise} \end{cases} \qquad (3)$$

where

$$\begin{cases} q_x^* &= 2\sigma_x \left[\delta\mathcal{F}_1 - (\mu\mathcal{V})^T \,\delta F\right] \\ q_y^* &= 2\sigma_y \left[\delta\mathcal{F}_2 - (\mu\mathcal{V})^T \,\delta G\right] \end{cases} \qquad \text{and } \chi \geq 1.$$

$\mathcal{V}$ represents the entropy variables, and $\mathcal{F}_k$ the entropy fluxes associated with the Euler equations written as a hyperbolic system of conservation laws (Godunov 1961, Lax 1971), and $\vartheta = (\delta\mathcal{V}^T \,\delta W)$. This term acts in fact as an anti-diffusion flux, which can be added to the original second-order difference dissipative flux. However, it will only adjust the dissipation in such a way that the entropy overshoots and undershoots near discontinuities become physical.

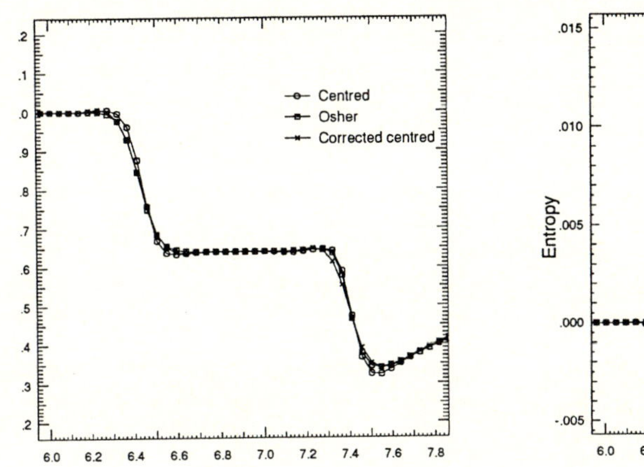
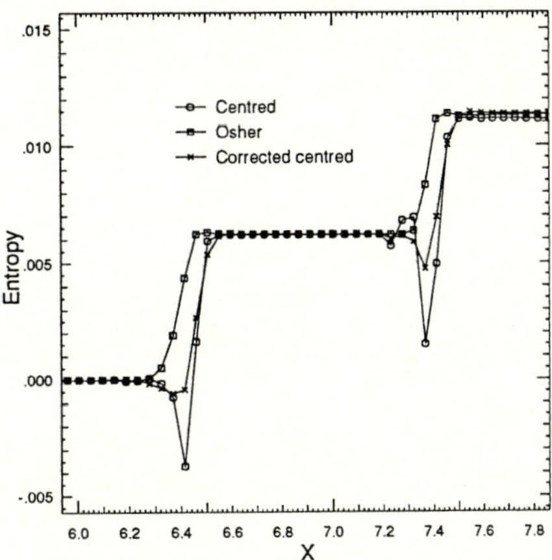

Fig. 3. Supersonic duct flow - $M_\infty = 2$. Cut in a horizontal plane across first reflection. Comparison of Centred, Osher and Corrected schemes

In Fig.2a, a supersonic (Mach 2) flow in a duct with an angular bump is calculated using the centered scheme and the entropy-corrected centered scheme. The oscillations in the vicinity of the discontinuities are eliminated by the correction, even for the weaker reflected shocks. However, the non-physical entropy production at the discontinuous wall boundaries remain. At the points where the wall boundary becomes non-smooth, a localised expansion fan is created. The flow downstream of this expansion fan does not violate the entropy condition, therefore the above entropy correction cannot compensate for the spurious entropy layer coming from the numerics. The way to remedy this is to switch to an exact, or approximate, Riemann solver in the vicinity of such regions. The expansion waves are thus calculated precisely, and indeed, the non-physical entropy production is suppressed, see Fig.2b. An Osher approximate Riemann solver has been used. The adaptation of the standard centered scheme increases of course the CPU cost, but it still results a reduction of more than 50% compared to the calculation performed completely with a non-linear approximate Riemann scheme. By tracing the profiles of the calculated Mach and entropy variables across the first shock reflection in the angular duct, at a height of approximately two-thirds of the way up from the angle, the effect of the numerical dissipation of the scheme shows up (see Fig.3). This violation of the entropy condition at singular points can even lead to completely erroneous solutions: indeed, an unsteady supersonic flow over a forward step generates a strong expansion fan at the corner, which can produce a non-physical entropy layer downstream, and an erroneous solution (see Figs.4).

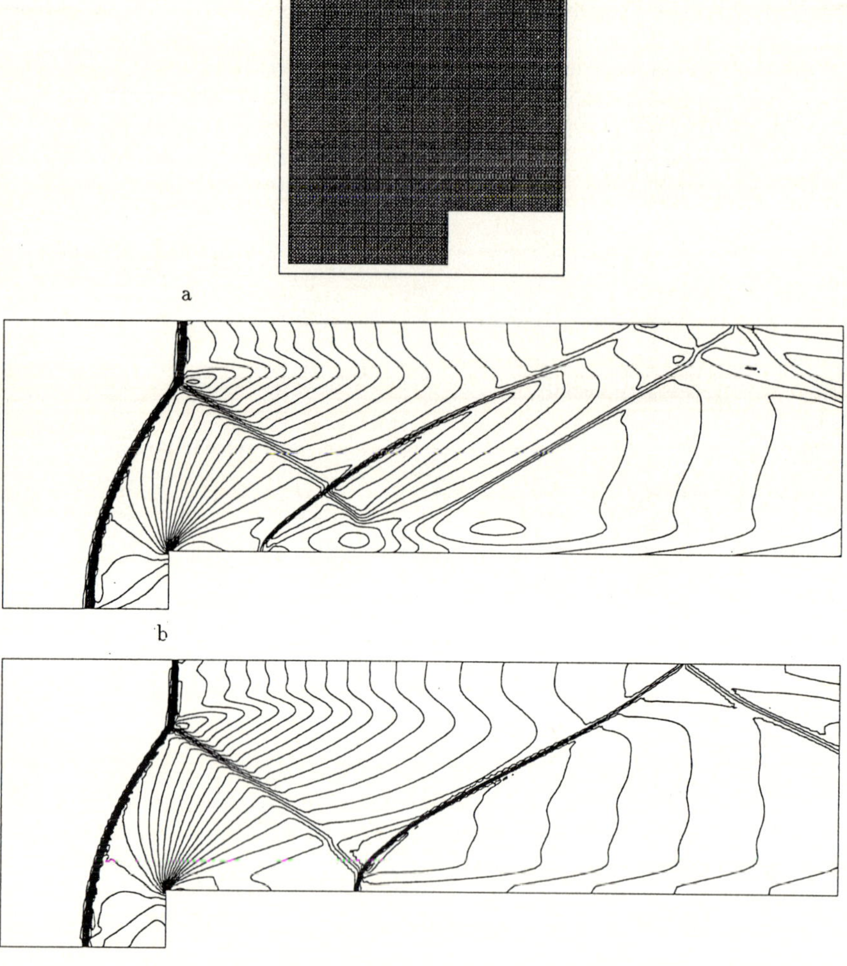

Fig. 4. (a) Partial view of uniform mesh of 16449 nodes
(b) Iso-density lines with centered solver: the solution is false
(c) Iso-density lines with combined centered/Riemann solver

3. Mesh adaptation and an unsteady flow result

The former numerical example was performed on a uniform regular mesh. One of the most important advantages of unstructured grids is the possibility to refine/derefine locally the mesh during the computation. By concentrating mesh points in critical zones and coarsening regions where nodes seem superfluous an optimal relation between precision and calculation cost can

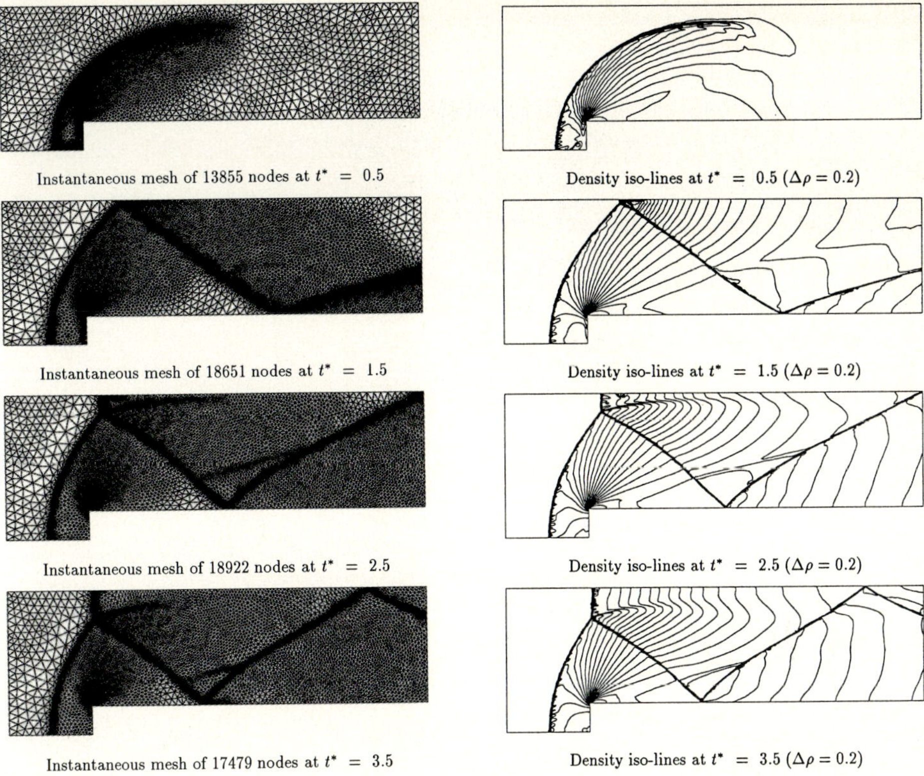

Instantaneous mesh of 13855 nodes at $t^* = 0.5$

Density iso-lines at $t^* = 0.5$ $(\Delta\rho = 0.2)$

Instantaneous mesh of 18651 nodes at $t^* = 1.5$

Density iso-lines at $t^* = 1.5$ $(\Delta\rho = 0.2)$

Instantaneous mesh of 18922 nodes at $t^* = 2.5$

Density iso-lines at $t^* = 2.5$ $(\Delta\rho = 0.2)$

Instantaneous mesh of 17479 nodes at $t^* = 3.5$

Density iso-lines at $t^* = 3.5$ $(\Delta\rho = 0.2)$

Fig. 5. Mesh and solution evolution enabled by dynamic mesh adaptation for a forward step at Mach 3

be achieved. Remeshing techniques are combined with directional scaling, grid alignment to high gradients, and stretching. This further increases precision, and leads to solutions of almost "shock fitted" quality. This dynamically auto-adaptive grid is coupled directly to the above solver.

An example of this procedure is given in Figs.5, where the unsteady flow over a forward-facing step is tracked. The calculation involved 200 intermediate meshes, for an increase in CPU time of about 20%, and a negligible increase in memory requirements. The overall precision of the solutions is remarkable at all stages.

To illustrate both the entropy correction and the coupling with mesh adaptivity, the unsteady Mach 3 flow over a forward step was calculated over an initial non-adapted mesh. If no special boundary conditions are applied, the standard centred scheme does not converge at all to a physically possible solution (Fig.4b). In Fig4c, the computation has been performed using the entropy preserving Riemann Solver at the boundaries and the entropy correction for interior points which assures non-oscillating shock capturing. By dynamically adapting the mesh, this solution is further improved, (Figs.5).

4. Conclusions

Combining entropy corrections for simple solvers gives clearly an interesting alternative to complete Riemann solvers for numerical simulation of many complex unsteady flows. However, non-physical entropy production can destroy the global solution. By locally employing a Riemann solver in such zones an accurate scheme is maintained. Coupling the solvers with dynamical mesh adaptation by enrichment/coarsening on unstructured meshes provides a considerable enhancement on the accuracy.

References

Baba K, Tabata T (1981) On a conservative upwind finite element scheme for convective diffusion equations. RAIRO Numerical analysis, 15: 3-25

Dervieux A (1984) VKI Lecture Series

Godunov S (1961) Une classe intéressante de systèmes quasi-linéaires. Dokl Akad Nauk 139: 521-523

Khalfallah K (1990) Conditions de monotonie et d'entropie et application à une méthode implicite centrée pour les équations d'Euler à grand Mach. Thèse de Doctorat, Université de Paris 6

Lax PD (1971) Shock waves and entropy.In: E. Zarantonello (ed) Contribution to Nonlinear Functional Analysis, Academic Press, p 603-634

Lerat A (1990) Difference methods for hyperbolic problems with emphasis on space-centered approximations. Lecture Series 1990-03, Computational Fluid Dynamics, VKI, Bruxelles

Richter R (1993) Schémas de capture de discontinuités en maillage non-structuré avec adaptation dynamique. PhD Thesis, EPFL

Richter R, Leyland P (1993) Precise pitching airfoil computations by use of dynamic unstructured meshes. AIAA Paper 93-2971

Tadmor E (1987). The numerical viscosity of entropy stable schemes for systems of conservation laws. J. Math. Comp. 49: 217-235

Woodward P, Colella P (1984) The numerical solution of two-dimensional fluid flow with strong shocks. J. Computational Physics 54:115-173

Computing Complex Shocked Flows Through the Euler Equations

A.M. Landsberg, J.P. Boris, T.R. Young and R.J. Scott
Laboratory for Computational Physics and Fluid Dynamics,
Naval Research Laboratory, Washington, D.C. 20375-5344, USA

Abstract. We report on simple, parallel algorithms designed to compute unsteady shocked flows around complex geometries very efficiently. To resolve dynamic shock phenomena without spurious compressions and rarefactions appearing as a result of unphysical regions of local curvature in the numerical representation of the body or boundary geometry, an accurate method to resolve these surfaces is required. The Virtual Cell Embedding (VCE) method is introduced, an efficient new approach for resolving complex geometries on a structured, orthogonal grid with little sacrifice in speed or memory. Unstructured grid methods and adaptive gridding methods for body-fitted structured grids are common alternative techniques to resolve complex geometries. These methods, however, require large amounts of memory and are computationally expensive. In contrast, the VCE method requires very little extra memory and CPU time to define the shape of a body accurately compared to the simple, orthogonal grid solutions computed as if the body were not there. The VCE method has been combined with a new efficient Flux-Corrected Transport (FCT) algorithm in a general simulation program, FAST3D. FAST3D is a three-dimensional unsteady flow solver and runs on vector computers (Convex C210, Cray), modern workstations (IBM RISC 6000, DEC Alpha), and high performance parallel computers (32-node and 512-node Intel iPSC/860) interchangeably. A video has been prepared of several simulations to illustrate this new capability.

Key words: Computational Fluid Dynamics, Flux-Corrected Transport, Complex geometry

1. Introduction

Computational fluid dynamics has advanced to the point where the unsteady flow field about arbitrary complex three-dimensional bodies can be solved given enough memory and CPU time. Unstructured grid methods (e.g. Löhner 1989; Melton et al. 1991), have become popular for resolving complex three-dimensional geometries and the resulting flow solutions, and can produce accurate results. However, these unstructured-mesh flow solvers generally require considerably more memory and CPU time than structured mesh flow solvers and typically do not parallelize as easily. Flow solvers using a block-structured mesh with blocks adapted to the body geometry (e.g. Thompson et al. 1985; Eiseman 1987), generally are more efficient but lack grid flexibility when handling truly complex bodies. Overlapping grid techniques (e.g. Benek et al. 1985, Berger and Oliger 1987) have been used to construct composite grids for very complex geometries but substantial redundant computation occurs and complex interpolation between multiple overlapping grids is necessary. Recently, we have developed an efficient, parallel method to compute unsteady shocked flows around highly complex geometries using a minimal amount of memory and CPU time. The Virtual Cell Embedding (VCE) method is a new capability for resolving complex geometries on a structured, orthogonal grid with little sacrifice in speed or memory.

The basic ideas of this approach and its implementation using an efficient, new, high-resolution Flux-Corrected Transport (FCT) algorithm called LCPFCT (Boris et al. 1993) are described below. The CFD simulation program that resulted, called FAST3D, was designed so that parallel, vector, and workstation implementations can all run interchangeably off the same dump/restart

Shock Waves @ Marseille I
Editors: R. Brun, L. Z. Dumitrescu

files, geometry-defining subroutines, and input data. FAST3D is a three-dimensional unsteady flow solver running on a variety of computers including high performance parallel computers such as the 32-node Intel iPSC/860 and the 512-node Intel Delta. This versatile program has been applied to shocked, supersonic reactive flow through a ram accelerator, to noise radiation from a supersonic jet engine exhaust, to unsteady airflow over a Navy destroyer, and to vortex shedding from a number of complex shapes. The focus of this paper is computing complex shocked flows through the Euler equations.

2. Numerical methods

2.1. The Flux-Corrected Transport algorithm

The unsteady, three-dimensional, gasdynamic Euler equations are a set of partial differential equations in conservation form and are discretized and solved using the Flux-Corrected Transport algorithm in FAST3D. An orthogonal, structured, rectilinear mesh is used but variable spacing between the grid planes, and even motion of the grid planes, is allowed. The Flux-Corrected Transport (FCT) algorithm was originally designed to solve generalized continuity equations for space and time-varying coordinates (Boris and Book 1973, 1976). The most recent version of the algorithm, LCPFCT (Boris et al. 1993), is fourth order accurate in space in smooth regions of the flow and second order accurate in time. The LCPFCT package has the facility built-in to accomodate a wide range of physical boundary conditions at the ends of the integration line segments. This version of FCT uses timestep-split and direction-split techniques to resolve the coupled, 3D continuity equations into a very efficient sequence of 1D integrations. This approach was chosen because an arbitrarily complex geometry divides the computational domain into easily identifiable line segments and because all of the interprocessor communication in the parallel implementation of FAST3D could be collected into a couple of large, efficient, matrix transpose operations executed each timestep.

The LCPFCT algorithm is well-tested, monotone (positivity-preserving), conservative, and relatively high-order. A convected quantity such as density remains positive definite during numerical convection and no new maxima or minima are introduced due to numerical errors. Monotonicity is achieved by introducing a diffusive flux and later correcting the calculated results with an antidiffusive flux modified by a flux limiter. The solution-adaptive, nonlinear flux limiter in FCT acts as a built-in, integrated subgrid turbulence model. This version of the FCT algorithm requires only the minimal five or six words of memory per cell for the gasdynamic variables because of the simplicity of the structured rectilinear mesh. An additional two words per 3D cell are used to carry the boundary condition information for the 1D FCT integrations and to identify cells adjacent to solid bodies. Normally this representation would require that complex bodies be very highly resolved to minimize the "staircasing" effects generated when using an orthogonal, rectilinear mesh. The staircasing effects are created since the body must be aligned to the grid locally in such a simplistic representation, becoming "rough" computationally. To cancel this staircasing effect, the VCE method described below was developed to allow the smooth surfaces of complex bodies to cut cleanly through the rectilinear grid without sacrificing speed or memory significantly.

2.2. The Virtual Cell Embedding method

One method to improve the accuracy around a complex body is to embed smaller cells (subcells) within the cells of the original mesh. Embedding methods for structured, orthogonal grids have been proven effective for resolving the flow field around complex geometries (Krist et al. 1990). Typically these grid embedding methods require extra memory for the embedded cells since these smaller cells are integrated in the flow solution. The computational expense is usually increased further by the need to take proportionately smaller timesteps in the small cells. Even when using embedded fine grids, the staircasing effect still exists; it is simply smaller and more localized.

The VCE method greatly increases the accuracy of the solution around complex geometries by smoothing out the cell corners causing the "staircase" effect usually found with a structured, rectilinear grid. With the VCE method, cells may be fully outside the body, fully inside the body, or partitioned by the body. It is these partitioned, boundary cells that require special treatment. The VCE method computes partial volumes and partial face areas for all of the cells which intersect the body. This information is adequate to compute an average unit vector normal to the surface and to correct fluxes of mass, momentum, and energy through the unobscured portions of the cell faces. The normal vector is used to accurately distinguish flows parallel and perpendicular to the body surface. Since the flow accuracy will ultimately be controlled by the cell size in the flow outside the body, these partial volumes and areas do not have to be computed to very high accuracy.

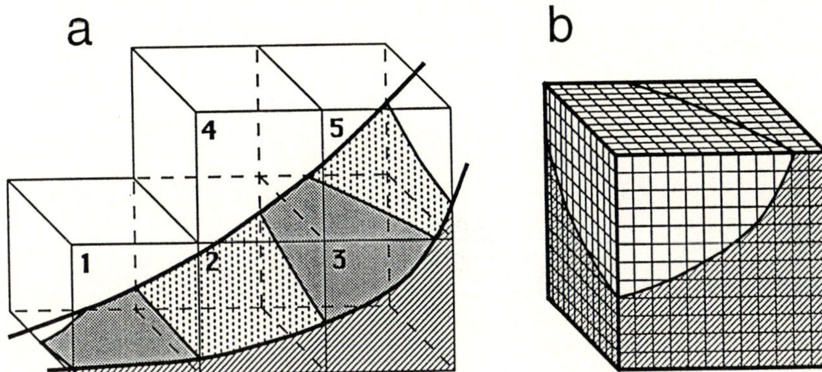

Fig. 1. Illustrating the VCE gridding for a few cells near the surface of a curved three-dimensional body. (a). The boundary cells numbered 1, 2, 3, and 5 are partitioned by the surface. (b) Cell 3 is shown divided into a 12 × 12 × 12 array of subcells. The center of each of these subcells is either inside the body or outside of the body. The partial volume and partial face areas of the cell are determined by simply counting the subcells inside the body and multiplying by the corresponding subcell area or volume

The VCE approach only refines (subdivides) cells next to the body. Therefore CPU time is not sacrificed appreciably since only those cells next to the body require special treatment. Even for very complex geometries there is only about 10% additional memory required and about 5% additional computation. The division into subcells can be made so fine that the body is essentially smooth without staircasing. Fig.1 shows how a mesh of subcells is used to compute the partial cell volumes and face areas. The 12 × 12 × 12 subdivision illustrated generally gives an accuracy near the body surface comparable to the flow solution accuracy in the uniform cells away from the body. The term "virtual" is used since the subcells embedded within a cell are not stored in memory and therefore are not integrated in the flow solution. The partial areas and volumes computed from the subcells are stored in a list at the location indicated by a pointer stored in one of the two extra words of memory associated with each parent cell.

The special treatment required for the partially obstructed cells is a modification to the flux calculation in these cells. To compute the conserved fluxes into and out of each cell, the "correct", partially obstructed face areas and volumes are used for the cells bordering the body. In addition to using these more accurate face areas and volumes, a flux coupling vector between the three integration directions is calculated in boundary cells due to cell blockage from the body. This flux coupling vector is used in the split-direction integration of the flow to correct the apparent fluid compression in each cell caused by non-grid aligned portions of the body. This flux coupling vector, computed only at the surface of the body, explicitly removes the lowest orders of the staircasing effect. Little additional CPU time is needed since the geometric information from the subcells is only computed once and stored and these additional computations are only performed in the cells adjacent to the body.

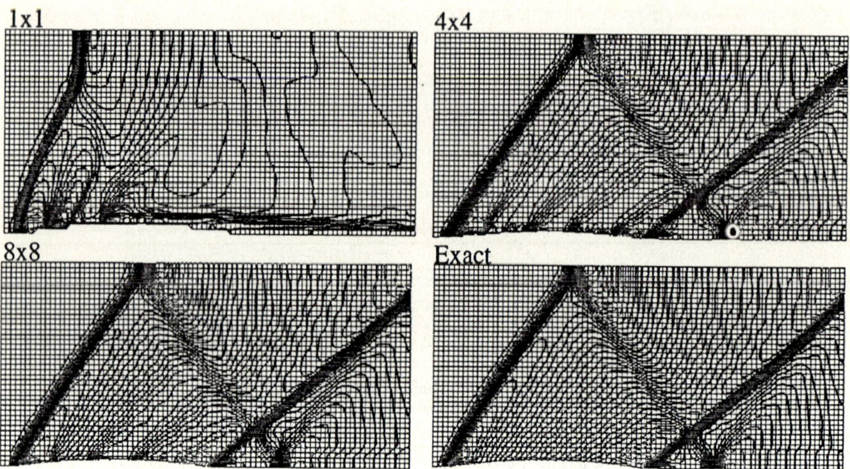

Fig. 2. A 4-degree circular arc airfoil is placed on the lower wall of a 120 x 40 2D mesh of square cells in a Mach 1.4 flow. Contours of pressure are shown for progressively finer subcell refinements. The airfoil is only 1.5 cells high. The degree of VCE subdivision for the boundary cells is indicated in the upper left of each panel. "1 x 1" is the usual staircase grid solution without the VCE algorithm. The panel labelled "Exact" used the analytic shape of the airfoil to compute areas and volumes. Rapid convergence of the VCE procedure is evident. Each calculation required exactly the same computer time and memory

The VCE method is incorporated into FCT through the provisions in the LCPFCT library routines for general one-dimensional geometry (Boris et al. 1993). The model produced accurate results when tested on a series of two- and three-dimensional problems. Fig.2 shows one of these tests for a standard benchmark problem. Other tests included shocks on oblique wedges and flow over spheres and ellipsoids. The FAST3D model is currently in production for ship superstructure airflow design problems and for studying the noise radiated from the oscillating shock structure in a supersonic jet engine exhaust.

3. Summary

The Virtual Cell Embedding (VCE) method has been presented and its implementation with the high-resolution, monotone FCT algorithm LCPFCT in a parallel flow solver called FAST3D has been described. We illustrate this efficient, new, complex geometry CFD capability using a short video showing complex shocked flow through the Euler equations. Two frames from this video are shown in Fig.3.

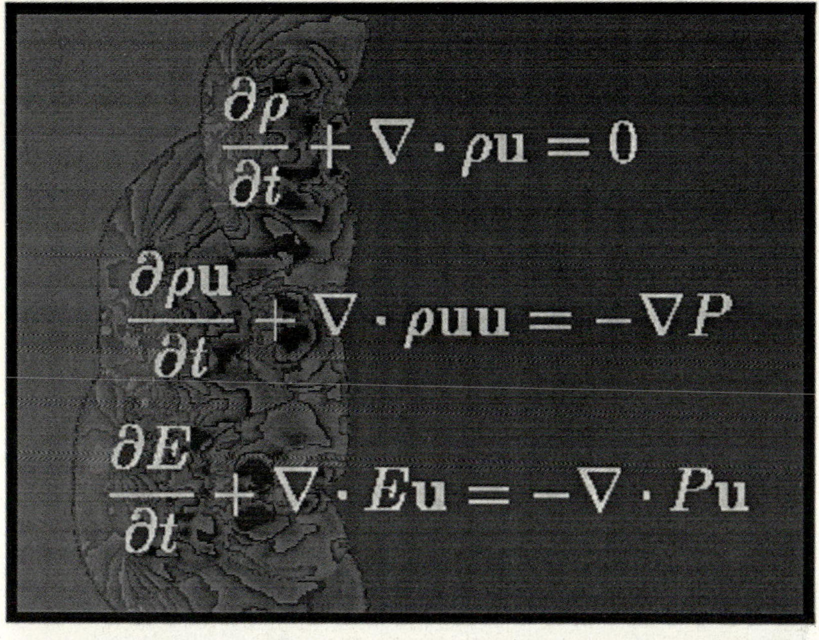

Fig. 3. Two frames from the video "Computing Complex Shocked Flows Through the Euler Equations." Color prints of these figures may be requested of the authors

The Euler equations, in the form of three vector conservation equations, are input to the grid generator to determine the boundary conditions for the flow. A supersonic flow enters the computational domain from the left. Reflected shocks arise from the time partial derivatives in the three vector conservation equations. Behind this complex shock front the fluid is subsonic and displays compressible "turbulence" due to interaction with the divergence terms in the equations. The inhomogeneous source terms on the right hand side of the Euler equations further enhance the complexity of the flow. Behind the reflected shocks, the equations eventually break down and the complex geometry solution is blown away downstream.

Acknowledgements

The authors would like to acknowledge numerous discussion during the course of this work with G. Patnaik, C. Li, E. Oran, D. Fyfe, J. Gardner, and K. Kailasanath. This work was supported by the Office of Naval Research, NAVSEA, and by the Advanced Research Projects Agency.

References

Benek JA, Buning PG, Steger JL (1985) A 3-D Chimera grid embedding technique. AIAA Paper 85-1523

Berger MJ, Oliger J (1987) Adaptive mesh refinement for hyperbolic partial differential equations. J. Comp. Phys. Vol. 53

Boris JP, Book DL (1973) Flux-corrected transport 1. SHASTA, a fluid transport algorithm that works. J. Comp. Phys. 11:38–69

Boris JP, Book DL (1976) Solution of the continuity equation by the method of flux-corrected transport. Chapter 11 in Methods in Computational Physics, Academic Press, New York, 85–129

Boris JP, Landsberg AM, Oran ES, Gardner JH (1993) LCPFCT - a flux-corrected transport algorithm for solving generalized continuity equations. NRL Memorandum Report 6410-93-7192

Eiseman PR (1987) Adaptive grid generation. Computer Methods in Applied Mechanics and Engineering 64:321–376

Krist SL, Thomas WL, Sellers WL, Kjelgaard SA (1990) An embedded grid formulation applied to a delta wing. AIAA Paper 90-0429

Löhner R (1989) Adaptive H-refinement on 3-D unstructured grids for transient problems. AIAA Paper 89-0653

Melton JE, Thomas SD, Cappuchio G (1991) Unstructured Euler flow solutions using hexahedral cell refinement. AIAA Paper 91-0637

Thompson, JF, Warst ZUA, Masten CW (1985) Numerical Grid Generation. North Holland, New York

A Third-Order ENO Scheme on Unstructured Meshes. Application to Shock Wave Calculations

Rémi Abgrall
INRIA, BP 93, 06902 France Cedex

Abstract. We present and illustrate a third-order ENO scheme on unstructured meshes. Some numerical simulations are presented; they show that a higher precision is indeed obtained by this scheme, with a relatively coarse mesh.

Key words: Unstructured mesh, Third-order scheme

1. Introduction

During the past few years, many efforts have been done to develop robust schemes to compute compressible flows. One of the many problems encountered in that work is the treatment of discontinuities, shocks and slip lines. The class of Total Variation Diminishing schemes is one of the answers to that problem. These schemes have been very successful in the treatment of discontinuities due to a clever adjustment of numerical diffusion. They have been widely used with any kind of meshes, structured as well as unstructured. Nevertheless, one of their weaknesses lies in the adjustment of the numerical diffusion: second-order TVD schemes (by far the most used) treat an extrema in the same way as a discontinuity, leading to excessive numerical diffusion. In order to overcome that problem, Harten et al. (1987) have developed the class of so-called essentially non-oscillatory schemes (ENO for short) extended –in their finite volume version– to two-dimensional structured meshes by Casper et al. (1993) for example and by Abgrall (1992b) to unstructured triangular type meshes.The purpose of this paper is to provide a description of the ENO finite volume type scheme on triangular meshes of Abgrall (1992b) (Section 2) and, with the help of a well-known problem, to demonstrate that a higher precision is indeed obtained (Section 4).

2. Description of the numerical scheme

In a domain Ω, we consider a two-dimensional inviscid flow whose equation of state is that of an ideal gas. The ratio of specific heats is set to $\gamma = 1.4$. The flow is described by the density ρ, the x- and y-velocities u and v, the total energy per unit volume E. Its evolution is described by the Euler equations in conservative form with initial and boundary conditions:

$$\frac{\partial W}{\partial t} + \frac{\partial F}{\partial x}(W) + \frac{\partial G}{\partial y}(W) = 0 \tag{1}$$

In this setting, the pressure is given by $p = (\gamma - 1)(E - \rho(u^2 + v^2)/2)$, the vector of conserved quantities is $W = (\rho, \rho u, \rho v, E)^T$. The fluxes are given by very similar expressions, hence only that of F will be written here: $F = (\rho u, \rho u^2 + p, \rho uv, u(E+p))^T$. The Euler equations are discretised in Ω on a triangular mesh $\mathcal{T}$. Around each point M of the mesh, we construct a control volume like in Fezoui et al. (1989): it is obtained by connecting the midpoints of the segments adjacent to M and the barycenter of the triangles M is a vertex of. We denote by $\{C_M\}$ the set of control volumes. In what follows, Δt is the time step, $t_n = n\Delta t$. When one integrates Eq.1 in $C_M \times [t_n, t_{n+1}]$, one gets:

$$\int_{C_M} W(x, t_{n+1})dx - \int_{C_M} W(x, t_n)dx + \int_{t_n}^{t_{n+1}} \int_{\partial C_M} \mathcal{F}_n(W(x, s))dl ds = 0 \tag{2}$$

Shock Waves @ Marseille I
Editors: R. Brun, L. Z. Dumitrescu

where n is the outward unit normal to the boundary ∂C_M of C_M and $\mathcal{F}_n$ is the normal flux in that direction (i.e. the scalar product of (F, G) with n). The vector of conserved variables is approximated around M at time t_n by its cell-average in C_M: $W_M^n \simeq \int_{C_M} W(x, t_n)dx/\text{area}(C_M)$. What remains to do is to approximate the flux term of Eq.2. If one wants that W_M^n approximates the average of $W(x, t_n)$ in C_M with a prescribed order of accuracy (at least formally), one has to evaluate the flux terms with the same accuracy, say order $m+1$. From the constant per cells data W_M^n, one has first to reconstruct W so that the *pointwise* approximation of the reconstruction is of order $m+1$. The procedure to follow is described latter. If one assumes to know the *exact* solution to this generalized Riemann problem (where the initial conditions are given by the reconstruction in each cell), then, the time integration can be accurately approximated with a Runge-Kutta procedure. Here, we have chosen Shu's (1988) third-order TVD Runge Kutta scheme. Then one needs an accurate approximation of flux integrals of the type $\int_{\partial C_M} \mathcal{F}_n dl$ at discrete times $s = s_r$ with the same accuracy. This can be done in three stages:

1. The boundary of C_M is made of few segments $\Gamma_{M,k}$, so that the approximation of the flux on ∂C_M reduces to its approximation on the $\Gamma_{M,k}$'s.

2. This later flux can be approximated by numerical quadrature. Here we choose Gauss' quadrature formulaes, and since we have in mind a third-order scheme, two quadrature points are enough. If the segment $\Gamma_{M,k}$ is mapped onto $[-1, 1]$, the image of the Gaussian points α_1 and α_2 are $\pm 1/\sqrt{3}$. The associated weights are egal to $1/2$ and then (with $l(\Gamma_{M,k})$ the length of $\Gamma_{M,k}$):

$$\int_{\Gamma_{M,k}} \mathcal{F}_n(W(x, s_r))dl \simeq \frac{l(\Gamma_{M,k})}{2} \sum_{1,2} \mathcal{F}_n(W(\alpha_i, s_r))$$

3. In this expression, $W(\alpha_i, s_r)$ represents the *exact* solution that is in general impossible to evaluate. Hence, the flux $\mathcal{F}_n(W(\alpha_i, s_r))$ is approximated by a one-dimensional Riemann solver. Its arguments are not the solution of the previous initial value problem but the values at the α_i's of the reconstruction of the approximation furnished by the previous Runge-Kutta sub-time step.

The boundary conditions of this schemes are implemented as in Fezoui et al. (1989). Now, we describe the reconstruction procedure.

3. Reconstruction procedure

To begin with, let us consider u a real valued function. We follow closely the ideas of Harten et al. (1987) but a completely new reconstruction technique has to be introduced: in Harten et al. (1987), the reconstruction deeply relies on the fact that they were able to design a function – the primitive of the data – whose values at certain nodes are known. For general unstructured meshes, this is in general impossible to do, because it would necessitate that enough recangles of the type $[a, b] \times [c, d]$ be *exactly* covered by a collection of control volumes C_M. We would get a "primitive" of U, and then one could proceed as in Casper and Atkins (1993). The trick is, for a large enough set S of control volumes containing C_M, to consider a polynomial in two variables of total degree m whose average on each element of S is exactly that of the function to reconstruct. In (Abgrall 1992a, 1992b), we show that if S contains $(m+1)(m+2)/2$ cells, it is in general possible to find exactly one such polynomial P. In the same papers, we provide a discussion on the "bad" cases (that never appear in practice), and give an algorithm to compute safely and cheaply P. We recall it here briefly: a natural basis to expand the polynomial is that of the monomials $(x - x_0)^i(y - y_0)^j$, where (x_0, y_0) is any point in the real plane. One can show (Abgrall 1992) that the conditioning number of the linear system one has to solve in this case is

$O(h^{-n})$, where h is the maximum diameter of the cells contained in the stencil S. Clearly, this choice is very bad. A much better choice is the following: since P exists, there are at least three cells, C_{M_1}, C_{M_2}, C_{M_3} of S, and three polynomials, Λ_1, Λ_2, Λ_3 of degree one, the "barycentric coordinates" of C_{M_1}, C_{M_2}, C_{M_3}, such that:

$$\frac{1}{\text{Area}\,(C_{M_i})} \int_{C_{M_i}} \Lambda_j \, dx = \delta_{ij}$$

(δ_{ij} is the Kronecker symbol).

A consequence of the above relations is that

$$\Lambda_1 + \Lambda_2 + \Lambda_3 = 1,$$

so that the set $\left[\Lambda_1^i \, \Lambda_2^i\right]_{0 \leq i+j \leq m}$ is another – legitimate – expansion of P:

$$P = \sum_{l=0}^{m} \sum_{i+j=L} A_{ij} \Lambda_1^i \Lambda_2^j.$$

One can now show (Abgrall 1992), that the conditioning system, or the resulting linear system is now $O(1)$, whatever the diameter h. This idea is borrowed from the finite elements theory. More details can be found in Abgrall (1992). This polynomial has two additional properties:

1. If u admits at least a $m+1$-th bounded derivative on the convex hull K of the union of the elements of S, then, for any $p \leq m$, one has on K, $u^{(p)}(x) = P^{(p)}(x) + O(h^{m+1-p})$ (h is the radius of K).
2. If u admits derivatives up to p_0-th order, $p_0 < m$, and if $u^{(p_0)}$ is discontinuous on a C^1 curve crossing K such that the minimal jump on this curve is $\alpha > 0$ then for $p \geq p_0$, one has $P^{(p)}(x) = O(\alpha \, h^{-(p-p_0)})$.

These two results tell us that if u is smooth enough, then *all* the coefficients of P remains bounded while the highest degree ones blow up when u is not smooth enough. The proof of these results is contained in Abgrall (1992b). This enables to design a hierarchical algorithm in order to choose the stencil that will be used in the reconstruction, following the ideas of Harten et al. (1987). Given a cell C_M, one first consider all the triangles of $\mathcal{T}$ that contain M as vertex. For one of these triangles, one consider the three cells associated to its vertices. In particular, C_M belongs to this set. With these three cells, one can compute a first degree polynomial P_1. We choose the triangle that minimizes, in the L^1 norm, the gradient of the polynomial constructed with the cells associated to its vertices, say T_{min}. This triangle has three sides. Since the mesh is conformal, each of these sides belongs to two triangles, T_{min} and T_1, except perhaps on the boundaries of Ω. The triangle T_1 has three neighbours: T_{min}, and two other ones. The set of the vertices of these four triangles has six elements, except perhaps on the boundaries. They are associated with six cells (among which C_M), and then one can compute a second degree polynomial P_2. One chooses the one whose Hessian has a minimum L^1 norm. In general, the first stage of this method utilizes six triangles, and the second stage uses three stencils (one per edge of T_{min}). In the present method, the reconstruction is first performed on the density, velocity and pressure. Then the vector of conserved variables is obtained. We reduce the degree of the polynomials to 0 for the cells C_M where M is on the boundary. Additional details on these scheme can be found in (Abgrall 1992b).

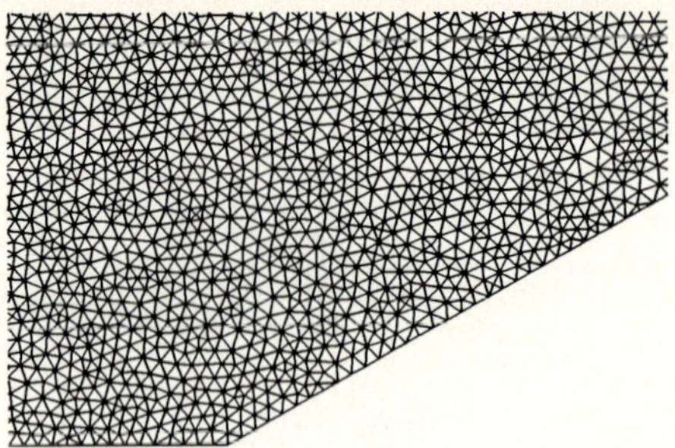

Fig. 1. Portion of the coarse mesh used for the Mach reflection problem

4. Some numerical experiments

Two sets of experiments are presented here, they concern the interaction of a shock on a wedge. In each case, we have used a triangular type mesh like on Fig.1, but we have changed its resolution. In this classical problem, the wedge angle is 30°. A shock at Mach 5.5 is moving in a quiescent fluid (zero velocity, $\rho = 1.4$, $p = 1.4$). One expects a double Mach reflection. The fine mesh has 12814 nodes and 25176 elements, the coarse mesh has 4381 nodes and 8505 elements. Roughly speaking the resolution is twice lower in the coarse mesh compared with the fine one. Since we are mainly interested in showing that an increased order of accuracy does improve the resolution of the solution, we will mainly show details of the flow. Each time, the same contours are used (40 contours at the triple point, Figs.2a-b). They are density plots. It is very clear that the slip line coming from the triple point is much better resolved in the third-order calculation (Fig.2c) than in the second-order one (Fig.2b). The shape of the reflected shock is also cleaner. This implies that the secondary shock on the left is also better resolved despite the fact that our mesh is too coarse to properly capture this weak shock. When we lower the resolution of mesh, these remarks become even more true, see Figs.2c and 2d. For the sake of completeness, we also give global pictures of the density fields in both cases (Figs.2e,f). Besides the differences pointed out before, some wiggles are visible near the corner of the wedge in the third-order solution compared with the second-order one. This may be a problem of our method that has to be explained. From our experience, this type of wiggles appear only in nearly stationary flows.

4. Conclusions

We have presented a third-order ENO finite volume scheme on unstructured meshes and some results of numerical experiments, which clearly indicate that, as we increase the order of accuracy from second-order to third-order, genuine improvements in the solution are observed, especially in the capture of slip lines. This is particularly interesting because in most finite volume methods, discontinuities of this type are excessively smeared. Nevertheless, improvements in the scheme are needed: some small wiggles can be observed in the nearly stationary part of the flow, and the cost of the calculation is still too high: 4 times that of the second-order scheme.

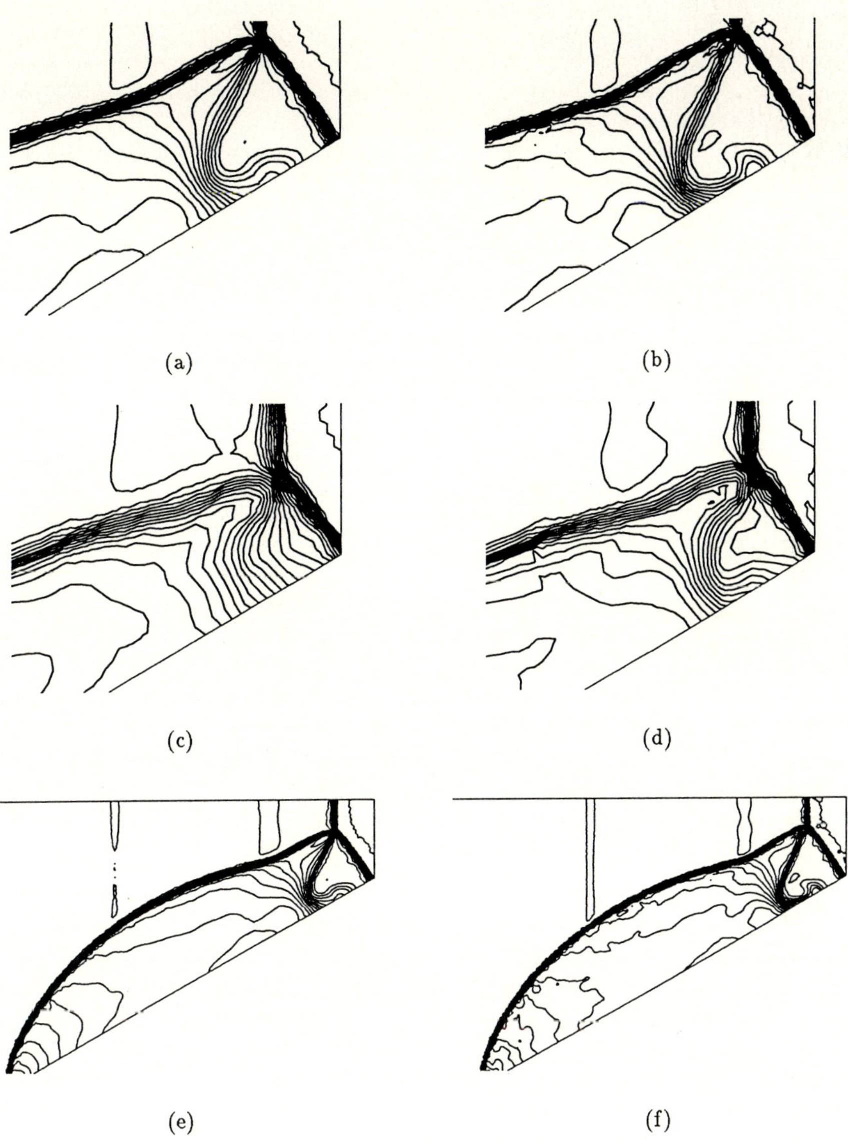

Fig. 2. Comparison of second-order ((a),(c),(e)) and third order ((b), (d), (f)) results for the Mach reflection problem

Acknowledgments

This research has been partly conducted under contract NAS1-19480 at Icase, Nasa Langley Research Center, Hampton Va, USA.

References

Abgrall R (1992a) Design of an essentially non-oscillatory reconstruction procedure on finite-elements type meshes. Icase report 91–84 (December 1991), also INRIA Report No1592 (January 1992), submitted to Math. of Comp.

Abgrall R (1992b) On essentially non-oscillatory schemes on unstructured meshes: analysis and implementation. Icasc Report No 92-74, December, to appear in Journal of Computational Physics

Casper J, Atkins HL(1993) A finite-volume application of high order ENO schemes for two-dimensional hyperbolic systems. Journal of Computational Physics, 106, 1: 62–76

Fezoui L,Stoufflet L (1989) A class of implicit upwind schemes for Euler simulations with unstructured meshes. Journal of Computational Physics 84, 1: 174–206

Harten A, Engquist B, Osher S, Chakravarthy SR (1987) Uniformly high order accurate essentially non-oscillatory schemes III. Journal of Computational Physics 71: 231–303

Shu CW, Osher S (1988) Efficient implementation of essentially non-oscillatory shock-capturing schemes. Journal of Computational Physics, 77: 439–471

Numerical Simulations of Compressible Viscous Flows

C. Mügler*, B. Meltz*, M. Vandenboomgaerde*, S. Gauthier*,
S. Aubert[†], L. Hallo[†], P. Ferrand[†] and M. Buffat[†]
*C.E.A., Centre d'Etudes de Limeil-Valenton, 94195 Villeneuve-Saint-Georges Cedex, France
[†] LMFA, URA 263, Ecole Centrale de Lyon, 36 av. Guy de Collongue, 69131 Ecully Cedex, France

Abstract. Three numerical codes based on high-order finite-volume formulations, with Godunov-type schemes, are described and compared on test cases. Agreements and discrepancies between the results of the three codes are pointed out and analysed.

Key words: Numerical simulation, Finite-volume, Finite-element, Shock tubes, Boundary-layers

1. Introduction

Three numerical codes based on high order finite-volume formulations and Godunov-type schemes are described and compared on some test cases. The first code, CADMEE, will be used to carry out two-dimensional direct numerical simulations of compressible mixing flows, as those occurring in shock tubes. The second code, PROUST, and the third one, NATURng, are designed to treat transonic internal flows in two or three dimensions. These numerical schemes must accurately resolve unsteady waves, like shocks and contact discontinuities, which can interact with each other, or with the walls. All the tools described here are currently well known, but the present work offers coherent solutions in applied CFD. Discrepancies between the results are pointed out and analysed.

2. Governing equations, and description of the three numerical methods

2.1. Governing equations

The motion of a viscous newtonian gas is governed by the Navier-Stokes equations, which express the conservation of mass, momentum and energy:

$$\frac{\partial q}{\partial t} + \nabla \cdot F(q) = \frac{1}{\mathrm{Re}} \nabla \cdot R(q) \tag{1}$$

The state vector q, the convective flux $F(q)$ and the diffusive vector $R(q)$ are defined by:

$$q = \begin{vmatrix} \rho \\ \rho\,\overrightarrow{u} \\ E \end{vmatrix} \qquad F(q) = \begin{vmatrix} \rho\,\overrightarrow{u} \\ \rho\,\overrightarrow{u} \otimes \overrightarrow{u} + p\,\overset{\Rightarrow}{I} \\ (E+p)\,\overrightarrow{u} \end{vmatrix} \qquad R(q) = \begin{vmatrix} 0 \\ \overset{\Rightarrow}{\sigma} \\ \overset{\Rightarrow}{\sigma}\,\overrightarrow{u} - \dfrac{1}{\mathrm{Pr}}\,q \end{vmatrix}$$

$$\qquad\qquad (2) \qquad\qquad\qquad (3) \qquad\qquad\qquad\qquad (4)$$

where ρ is the density, $\overrightarrow{u}$ is the velocity with three components u_1, u_2, u_3, and E the total energy. $\overset{\Rightarrow}{\sigma}$ is the viscous stress tensor, Re denotes the Reynolds number, Pr the Prandtl number and γ is the ratio of specific heats. The pressure p is related to temperature by the perfect gas law, and q is the heat transfert described by Fourier's law.

2.2. Brief description of code CADMEE

This first method is intended to numerically simulate full complex flows in shock tubes. It is derived from the code CAVEAT (Addessio et al. 1990), and solves the 2D unsteady Navier-Stokes equations on quadrangular structured meshes. Mixing is described by a concentration, governed by an advection-diffusion equation. The state variables are cell centered, and a second order spatial differencing technique, coupled with the approximate Riemann solver of Dukowicz,

Shock Waves @ Marseille I
Editors: R. Brun, L. Z. Dumitrescu

is used. The computation is performed in two phases: a Lagrangian phase and a remapping phase in which conservative variables are transfered from the Lagrangian mesh to an arbitrary specified mesh. This approach is the so-called Arbitrary Lagrangian-Eulerian (ALE) formulation. After the Lagrangian step is taken, a new mesh is specified. This rezoning can follow the fluid motion (pure Lagrangian), return the mesh to the same initial configuration after each cycle (pure Eulerian), or recast the mesh in various other ways, for example, to follow features of interest moving through the mesh such as a shock wave. For the remapping phase, two algorithms can be used: a fluxing algorithm in the case of pure Eulerian calculation, or when high mesh refinement is required, an algorithm of high accuracy which makes use of an alternate direction technique. The latter method may be used even though the final mesh is different from the initial one: so, it does not need to be used at each cycle. As a result, less numerical diffusion occurs in comparison with pure Eulerian methods. All diffusive terms can be either handled explicitly or implicitly: the resulting linear system is solved by means of an iterative Orthomin-like algorithm. This method will be referenced as the Godunov approach.

2.3. Brief description of code PROUST

The second method was designed to compute three-dimensional unsteady flows in turbomachiner-ics where interactions of waves and coupling between flow and structure occur. It is based on a MUSCL finite volume formulation over moving structured quadrangular meshes (Aubert et al. 1992). It uses the van Leer flux vector splitting, with Mulder limiter which controls spatial accuracy in the vicinity of discontinuities. Natural boundary conditions, including periodicity, symmetry and slipping walls, are implicitly implemented by means of compatibility relations. The whole scheme is built to ensure coherence in time, with a peculiar attention being paid not to generate numerical phase differences in the solution. Diffusive fluxes are computed using a second order centered difference scheme. An explicit time integration technique, forward Euler, stable under CFL condition is employed. It is subsequently referenced as the van Leer approach.

2.4. Brief description of code NATURng

The goal of this third method is to predict steady two and three-dimensional internal turbulent transonic reactive flows. To take into account complex geometries and to allow mesh refinement, unstructured finite element meshes are used (triangles in 2D and tetrahedrons in 3D). The finite volume integration cells are built around each nodes by mean of the medians. The approximate Riemann solver of Roe with the van Albada limiter is employed, which give a quasi-TVD scheme, (Aubert et al. 1992). The approximation of the time derivative, the diffusive terms and the source terms use a Galerkin finite-element approximation, and the so-called mass lumping technique approximation. A linearized implicit time integration scheme has been introduced to improve the rate of convergence. Boundary conditions are implemented using explicit compatibility relations through a numerical flux. The linear system arising from the linearization are solved by help of CGS method with SSOR preconditioning technique. Mixing is described by concentrations, governed by advection-diffusion equations. In fact, the implicit time integration is performed on the uncoupled equations (mean Navier-Stokes equations / advection diffusion equations), which limits only slightly the numerical stability of a fully implicit scheme. This approach is subsequently referenced as the Roe method.

3. Results

Many practical external or internal transonic flow problems involve complex patterns of waves. They can result from wakes or shocks interacting each other, or from reflection of waves from a wall. The flow can be either steady or unsteady, with static or travelling waves. Such situations appear in turbomachinery where rotor/stator interaction occurs, or in supersonic combustion chambers with interaction of shocks with boundary layers. After all, the numerical simulation of complex flows in shock tubes must be carried out with accurate codes in order to get all the intrinsic evolution of waves. It is the purpose of the two test cases presented below.

3.1. One-dimensional interaction of waves

This test case involves two shock tubes sharing the same low pressure chamber. Initially, the pressure and density ratios are the same across the two diaphragms. Temperature is uniform, and the gas is at rest. Just after the two diaphragms have been broken, pressure waves and contact discontinuities move toward the center of the tube, while rarefaction waves travel outwards. Due to initial conditions, the pattern is symmetric about the center of the tube.

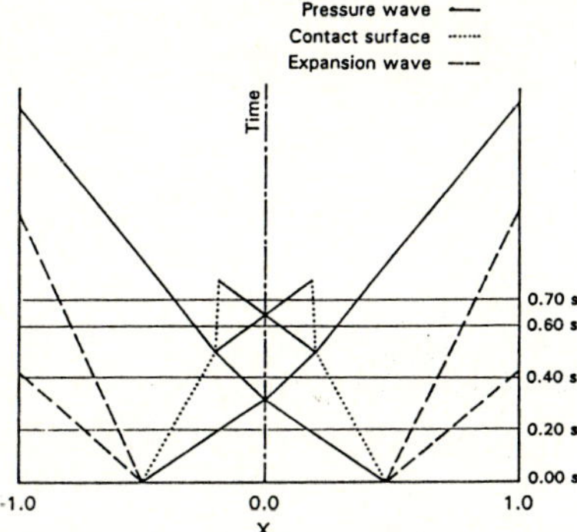

Fig. 1. Theoretical patterns of wave interactions

The initial conditions are a pressure and a density ratio of 4 between the high pressure and low pressure chambers. Temperature is such that pressure to density ratio is 1, and velocity is 0. The tube extends from $-1.$ to $1.$, and is divided into 200 equal cells. The initial diaphragms positions are -0.495 and 0.495. The time step is constant, and fixed at 0.001 s. Under such conditions, the initial pressure wave, which is the quickest, needs 6 time steps to cross one cell. The exact solution of the problem is known, and the theoretical diagram of wave evolutions is presented in Fig.1.

At time around $t = 0.3$ s, there is interaction between the shocks, and two reflected pressure waves moving outwards are generated, while contact discontinuities are moving towards the center of the tube. At $t = 0.5$ s, there is interaction between the reflected shocks and the contact surfaces. A reflected and a transmitted pressure wave appear, with a new contact discontinuity. At time $t = 0.6$ s, the reflected shocks are near $x = 0$, while the contact surfaces are at $x = -0.2$ and $x = 0.2$. The transmitted shocks are moving outwards. The last interaction occurs roughly at time $t = 0.65$ s, between the two reflected pressure waves. Two small shocks moving outwards are generated. They are visible at $t = 0.7$ s in Fig.2, where density profiles are plotted. We notice that the three methods give the right position and strength of each wave, even after several interactions. However, the Roe profiles for contact discontinuities are slightly sharper than those given by van Leer method. Shocks stay resolved out 2-4 cells for the three methods. Contact discontinuities are resolved out 6-8 cells for Roe and van Leer methods, 3 cells for the Godunov method when the high accuracy rezoning is used every 50 cycles (14 remappings), and 5 cells when the fluxing algorithm is used at each cycle (700 remappings). The smoothing of discontinuities does not increase when interactions occur: even though the discontinuities obtained with mesh

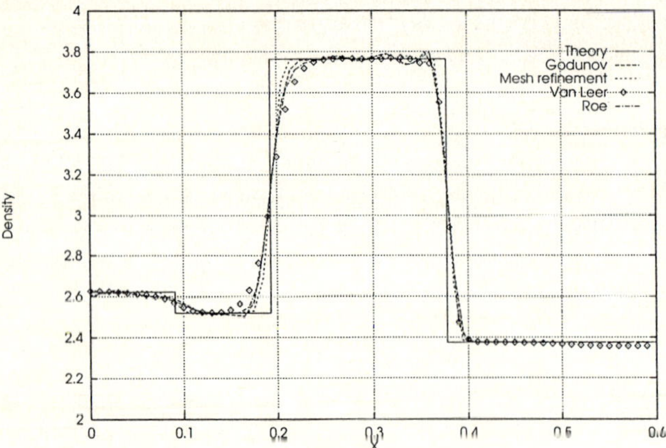

Fig. 2. Density profiles at time $t = 0.7$ s

refinement are sharper (Godunov scheme), the results stay close to each other when the number of interactions increases.

3.2 Supersonic boundary layer

In this condition, we consider a supersonic flow of air over a flat plate, in order to compare the treatment of the viscous terms in the three codes. The external flow, at Mach number 2.3, is prescribed by a pressure P=24586 Pa, a density $\rho = 0.6$ kg/m^3. The dynamic viscosity and the thermal diffusivity are constant. A no slip boundary condition for the velocity, and a fixed temperature (270 K) are imposed on the flat plate. Boundary conditions are imposed upstream of the leading edge. In this interval, symmetrical boundary conditions in the spanwise direction are used for all variables. Similar boundary conditions are used on the top of the computational domain. At the right side of the domain, the outflow boundary conditions are of zero-gradient for the pressure.

The mesh, made of 32 cells in the mainstream x-direction, and 80 cells in the spanwise y-direction, is refined in the x-direction at the leading edge and in the y-direction near the plate so that the smallest cell height is 2.10^{-5} m. A small number of mesh points have been used, however spatial convergence has been reached. Figs.3,4,5 represent the iso-value 99% of the mainstream velocity, which is nothing but the boundary layer thickness. A semi-empirical thickness law (Cousteix 1988), has been superimposed. Fig.6 displays the velocity profiles at $x = 0.7$ m from the leading edge. The units of length and velocity are respectively the boundary layer thickness at $x = 1$ m, calculated from Cousteix (1988), and the mainstream velocity. The shock wave at the leading edge, given by the convective part of the Navier-Stokes equations, is well reproduced by the three numerical codes. Moreover, the thickness of the boundary layer, obtained by CADMEE and NATURng are in good agreement with the theoretical one. However, the boundary layer given by the code PROUST is thicker than that calculated with the Cousteix semi-empirical law. This fact is actually well known: for very low Mach numbers, this numerical scheme is too diffusive.

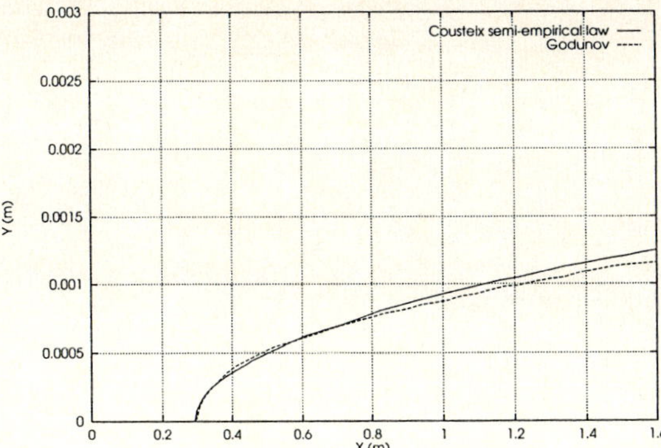

Fig. 3. Iso-value 99% of the mainstream velocity. Code CADMEE

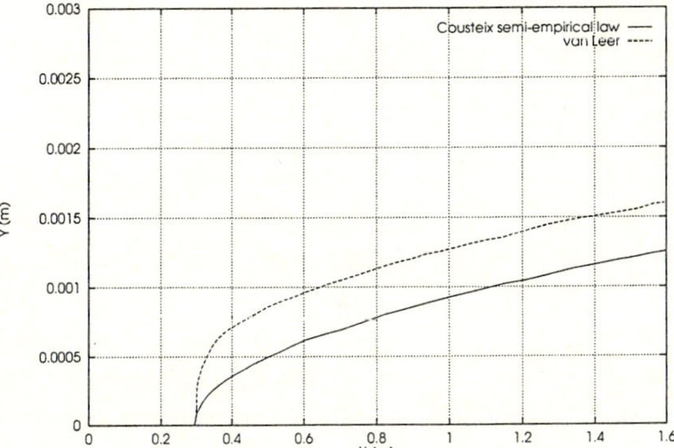

Fig. 4. Iso-value 99% of the mainstream velocity. Code PROUST

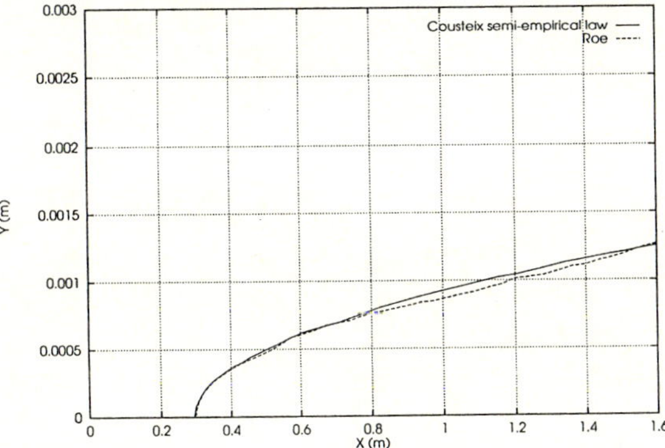

Fig. 5. Iso-value 99% of the mainstream velocity. Code NATURng

4. Concluding remarks

Numerical simulations in inviscid and viscous configurations have been carried out with three different numerical methods. The first calculation is the 1D interaction of waves in a shock tube, as a means to evaluate the spatial accuracy of the convective solvers. The second test case is the

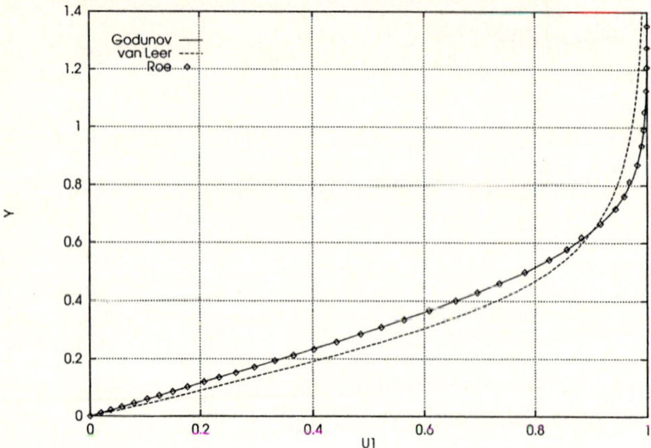

Fig. 6. Mainstream velocity profile at $x = 0.7$ m from the leading edge

simulation of a laminar boundary layer in a supersonic flow. It turns out that the strategy which consists in a Lagrangian phase followed by a remapping phase with a Godunov method, as used in CADMEE, leads to numerical codes with very low numerical dissipation. This is especially true when it is possible to use the high accuracy remapping as in the shock tube test. This conclusion still holds when viscous phenomena occur, and when the fluxing remapping algorithm is used, as in the flat plate flow. The numerical code NATURng gives reasonable results in both test cases proving, in agreement with the literature, that Roe's approximate Riemann solver has a very low numerical dissipation. This is a robust scheme and its numerical dissipation perturbs only slightly the boundary layer velocity profile. Recall that NATURng features an unstructured mesh option making the code well-suited for complex geometries.

The code PROUST gives good results in the shock tube flow. However, as already stated, the van Leer scheme is not adapted to study viscous flows at very low Mach number. The main advantage of the van Leer method is that the fluxes are explicitly known: the Jacobian matrix is known analytically and is simple. As a result, the linearization is exact, making the code well-suited for unsteady flows. To deal with boundary layers, the scheme is generally modified: it becomes close to a centered scheme.

Acknowledgement

The authors acknowledge support from CEA, CNRS, METRAFLU and SNECMA.

References

Addessio FL, Baumgardner JR, Dukowicz JK, Johnson NL, Kashiwa BA, Rauenzahn RM, Zemach C (1990) CAVEAT: a computer code for fluid dynamics problems with large distorsion and internal slip. Los Alamos report, USA

Aubert S, Hallo L, Ferrand P, Buffat M (1992) Numerical behaviour of unsteady waves. Submitted to publication to AIAA Journal

Cousteix J (1988) Couche limite laminaire. Cepadues-Edition

Time-Dependent Simulation of Reflected-Shock/Boundary Layer Interaction in Shock Tubes

G.J. Wilson[*], S.P. Sharma[†] and W.D. Gillespie[‡]
[*]Eloret Institute, 3788 Fabian Way., Palo Alto, CA 94303, USA
[†]Aerothermodynamics Branch, NASA-Ames Research Center, Moffett Field, CA 94035, USA
[‡]Department of Aeronautics and Astronautics, Stanford University, Stanford, CA 94305, USA

Abstract. An initial experimental/numerical investigation has been conducted to gain a better understanding of the multi-dimensional flow phenomena inside pulse facilities and the influence of these phenomena on test conditions and test times. Experimental data from the NASA Ames electric-arc driven shock tube facility (from cold driver shots) is compared to time-dependent axisymmetric numerical simulations of the complete facility. These comparisons help establish the numerical modelling requirements for simulating shock tube flow and help validate the computations. The numerical simulations are used to study the interaction between the reflected shock wave and the side wall boundary layer and the resulting shock bifurcation. Of particular interest is the effect of the bifurcated shock structure on the driver/driven gas interface. The computations demonstrate how this shock structure introduces a mechanism for the driver gas to contaminate the stagnation region thereby reducing the duration of the test time. The simulations incorporate finite-rate chemistry, a moving mesh and laminar viscosity.

Key words: Shock tube, Computation, Shock/boundary layer interaction

1. Introduction

Knowing the length of the test time and the state of the test gas provided by shock tubes or shock tunnels is critical to interpreting data obtained from these facilities. Unfortunately, it is well established that the test time achieved in these facilities is usually significantly less than ideal theory predicts. Many investigations have been carried out to understand and quantify the physical mechanisms which cause shortened test times. These mechanisms include deformation of the contact discontinuity caused by the diaphragm rupture process, mass transfer of the driven gas into the boundary layer, contact discontinuity instabilities, and shock/boundary layer interaction after reflection of the incident shock off the end-wall.

It is not clear which of the mechanisms mentioned above limits test times most and it may be that the dominant mechanism varies with the experimental facility or the run conditions. There is a large amount of evidence, however, that suggests that the reflected-shock/boundary layer interaction is often a major contributor to the contamination process. It is well established that under many conditions the reflected shock will interact with the boundary layer causing it to bifurcate near the wall (see Fig.1). Mark (1958) developed a model to predict the characteristics of the bifurcation and the conditions under which it will occur. He showed that the flow in the energy deficient boundary layer has a stagnation pressure that is less than the stagnation pressure behind the normal reflected shock and is prevented from passing under the reflected shock. Instead, it separates and collects in a bubble of gas next to the wall. A consequence of the shock bifurcation is a jet of gas near the shock tube wall as depicted in Fig.1. The jet is present because the gas which passes through the oblique shocks at the foot of the bifurcated shock retains a higher velocity than the gas which encounters the normal shock. Particularly clear experimental evidence of this phenomenon can be seen in the recent color Schlieren photographs of Kleine et al. (1991) which include features such as the rolling up of the wall jet as it encounters the end-wall. Davies (1966, 1967, 1969) used Mark's model to show that the wall jet provides a

Shock Waves @ Marseille I
Editors: R. Brun, L. Z. Dumitrescu

mechanism for contamination of the stagnation region by propelling cold driver gas toward the end-wall and into the driven gas. He and others such as Bull and Edwards (1968) have done experiments which measured the time of arrival of the cold driver gas through this mechanism.

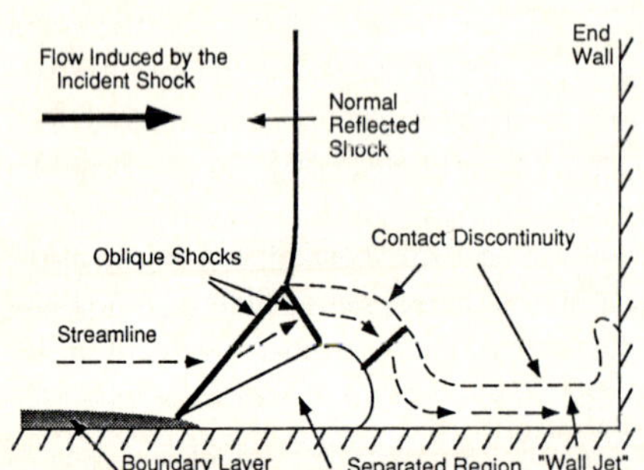

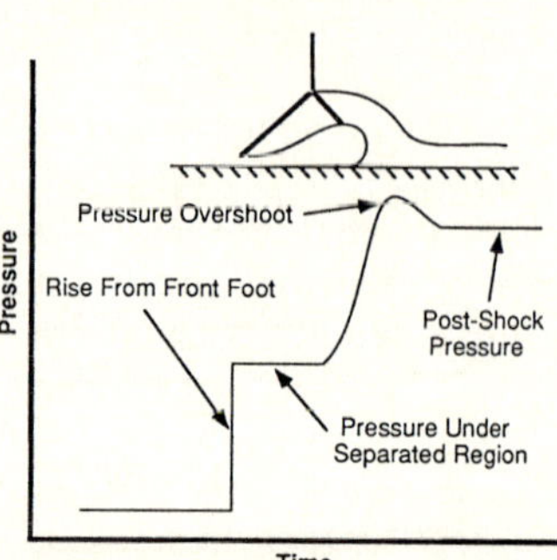

Fig. 1. Schematic diagram of reflected shock/boundary layer interaction

Fig. 2. Schematic diagram of pressure trace under bifurcated shock

The reflected-shock/boundary layer interaction described above can explain much of what is observed experimentally in shock tubes. However, more complicated flow structures such as a pseudo-shock (or shock train) can develop when the shock/boundary layer interaction is strong. Matsuo et al. (1975), Strehlow and Cohen (1959), and Brossard et al. (1985) all show Schlieren photographs of the formation of multiple shocks after the reflection of the incident shock. The effect of these multiple shocks on driver gas contamination has not been studied.

In the absence of optical data, the presence of shock bifurcation can be inferred from side-wall pressure measurements. Fig.2 shows that the passage of the reflected shock is marked by a two-step pressure rise followed by a pressure overshoot as described by Sanderson (1969). It can also be deduced by noting the change of the reflected shock speed compared to inviscid theory (Mark 1958; Strehlow and Cohen 1959; Matsuo et al. 1975).

There have been several recent examples of computations of reflected-shock/boundary layer interaction at the end of a shock tube with the computational domain limited to the stagnation region (e.g. Kleine et al. 1991, Yakano 1991). The present authors are unaware, however, of any computations which have looked at the contamination of the stagnation region with driver gas through the wall jet mechanism proposed by Davies. One way to accomplish this is to begin a simulation of a shock tube at the diaphragm rupture allowing the position of the contact discontinuity and the boundary layer development to be computed. This, in turn, makes it possible for phenomena such as the reflected shock/boundary layer interaction and the reflected-shock/contact discontinuity interaction to be investigated numerically. This approach is adopted herein by computing the time-dependent flow inside the NASA Ames electric-arc driven shock tube (cold driver shots without the arc-driver). The simulations assume that the contact discontinuity is planar at diaphragm rupture and that boundary layer is laminar. Experimental data in the form of wall static pressure traces and heat transfer was gathered to help guide the numerical modelling and validate the simulations. The present work is a continuation of research reported in Wilson et al. (1993) and more details can be found there.

2. Experimental facility

The NASA Ames electric-arc driven shock tube facility has several possible configurations and has a large hypervelocity operating range using its arc-driver (Sharma and Park 1990). However, this work only considers experiments using a cold helium driver with nitrogen in the driven section. These shots were made with a cylindrical driver of .86 m (2.8 ft) in length and 10 cm (3.93 inches) in diameter. The driven section was 4.22 m (13.85 ft) long with the same diameter as the driver. A single self-break diaphragm separated the driver and driven gases.

For the current experiments the instrumentation consisted of pressure transducers flush mounted at three fixed positions on the shock tube walls, at 2, 6, and 24 inches from the end-wall. There were also two pressure transducers flush mounted on the end-wall, one at the center and one 1/4 inch from the tube side wall. In addition, the end-wall was modified so that it could be moved forward and backward at one inch increments. This made it possible to collect data at variable distances from the end-wall (i.e. 1, 2, and 3 inches from the end-wall for the first side wall gauge and, thus, 5, 6, and 7 inches from the end-wall for the second gauge). The pressure transducers were PCB Piezotronics, Inc. Model 113A21 with a circular surface area .218 inches in diameter and a rise time of 1 microsecond. Heat-transfer data has also been collected at the same side wall locations at instrument locations opposite to the pressure transducers.

3. Numerical method and gas model

The gas dynamic equations for the axisymmetric simulations are solved by using an explicit finite-volume form of the Harten-Yee upwind TVD scheme (Yee 1989). The gas model includes the three major species present in the shock tube for the experiments (N_2, N, and He) and accounts for finite-rate chemical processes. A separate equation for vibrational energy is included so that vibrational nonequilibrium effects can be assessed. The present work, however, enforces thermal equilibrium. The numerical method is essentially an extension of the quasi-one-dimensional work in Wilson (1992). The full Navier-Stokes viscous terms are included.

Mesh points are clustered at the contact discontinuity to minimize numerical diffusion and are convected with the gas interface as it travels down the driven tube. This approach has the additional benefit of compressing all of the cells associated with the driven tube into the end-wall region of the shock tube as the driven gas is compressed thereby providing a fine axial mesh during the shock reflection. The grid is also clustered around the incident shock so that the resolution is relatively high where the boundary layer is initially formed behind the shock. Points are also concentrated near the wall to resolve the boundary layer. Because the Euler terms are treated explicitly, the computations are advanced at a CFL number less than 1 based on the inviscid gas dynamics. To avoid the more limiting time step dictated by the viscous terms, the thin-layer viscous terms are treated implicitly (note that all the viscous terms are included explicitly). It is believed that the time accuracy of the solution is not significantly affected by this approach and without it the simulation becomes impractical. The source terms representing the finite-rate chemical kinetics and vibrational relaxation are also treated implicitly. This implicit formulation reduces the formal temporal accuracy to first order.

4. Results and discussion

The results presented in this short paper represent a single test condition with a nominal driver pressure of 4.83 MPa (700 psi) and a driven pressure of 20 torr nitrogen with both sections at ambient temperature (295 K). Three shots are reported with the end-wall position varied for each shot so that there is data 1, 2, and 3 inches from the end-wall (these shall be referred to as the 1 inch position, the 2 inch position, etc.). Because of the self-break single diaphragm, a precise repeatability of shock speed for the three shots was not possible. The shock speed for the 1, 2, and 3 inch positions were 1984 m/sec, 2005 m/sec, and 1953 m/sec, respectively. Fig.3 shows the pressure trace for the 2 inch position. Several of the major events are labeled. These include the passage of the incident and reflected shocks, the arrival of the rarefaction, and the presence of waves reflected off the contact discontinuity. The time axis is adjusted so that zero time corresponds to the shock arrival at the end wall. This allows a meaningful comparison of traces with the end-wall in different positions.

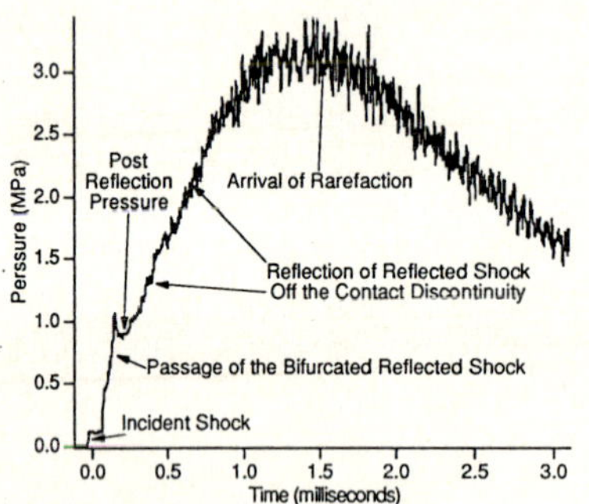

Fig. 3. Experimental and computed pressure traces 2 inches from the driven tube end-wall

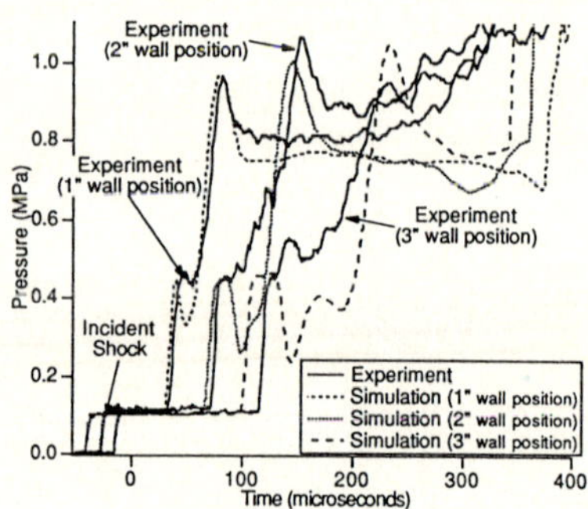

Fig. 4. Experimental and computed pressure traces 1, 2, and 3 inches from the driven tube end-wall

The axisymmetric simulation used an initial driver pressure 11% higher than the experimentally reported value in order for the computed incident shock speed to match the experimental one. The reason for this discrepancy is not known. All other initial conditions matched the experimentally reported ones. A cold wall boundary condition of 295 K was enforced. The computational mesh contained 800 points along the length of the tube (400 each in the driver and driven sections) and 112 points between the tube centerline and an outer wall. The points were exponentially clustered near the wall with a minimum spacing of .150 mm at the wall for the first few meters of the driven tube and was ramped down to a constant .015 mm over the last .75 meters of the tube. This approach eased the computational cost by allowing a larger time step early in the solution. As mentioned before, the solutions assumed an initially planar contact discontinuity and laminar viscosity. Verification that the reflected shock/boundary layer interaction has all of the features depicted in Fig.1 is presented by Wilson et al. (1993).

A composite of experimental and computed pressure traces in the end-wall region for early times after the shock reflection is presented in Fig.4. It is seen that the general features of the experimental and computed pressure traces are quite similar. Evidence of the shock/boundary layer interaction in the pressure traces, as depicted in Fig.2, is clearly seen. The computation and experiment are in good agreement for the trace 1 inch from the end-wall, fair agreement at 2 inch position, and by the 3 inch position it is clear that discrepancies are growing. Further deterioration of the agreement is seen in the traces farther back from the end-wall (not shown here). While it appears that the growth of the bifurcated shock structure is being captured fairly well by the simulations, the computed speed of the bifurcated shock is too fast and the predicted pressures under the separated flow region differ. Even with these differences, the general features of the flow appear accurate enough that a qualitative investigation of the interaction of the reflected shock with the contact discontinuity is deemed worthwhile.

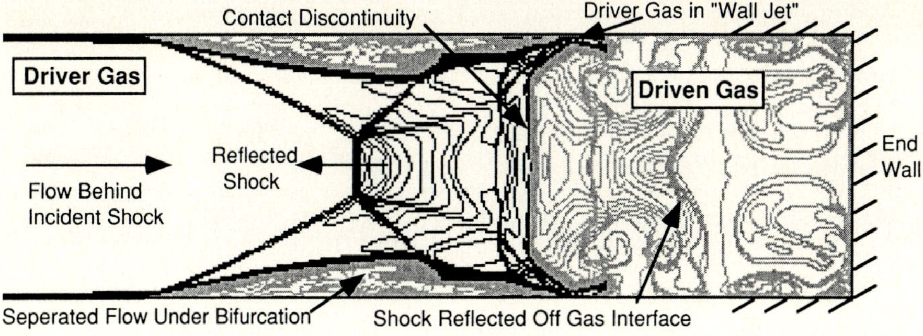

Fig. 5. Computed temperature contours showing reflected shock/contact discontinuity interaction

Fig.5 contains temperature contours after the reflected shock has interacted with the contact discontinuity. The figure shows that the bifurcated shock structure deforms the contact discontinuity near the wall. Hot driven gas in the separated region under the shock bifurcation is carried into the driver gas while the driver gas that passed through the oblique shocks of the bifurcation retains a higher flow velocity and penetrates into the driven gas (i.e. it is part of the wall jet). There is also the additional feature of a shock reflected off the contact discontinuity due to the overtailored conditions. There are many nonuniformities in the stagnation region because of the shock/boundary layer interaction. As waves such as the shock reflected off the contact discontinuity interact with these non-uniformities even more complex features are formed. It is easy to see why pressure trace data can become quite noisy.

There are many possible reasons for the discrepancies seen between the experimental data and the present computations. Experimentally, the point of transition between laminar and turbulent flow is not known. It is believed that the flow near the end-wall is laminar (heat transfer data seems to support this); however, the flow must eventually become turbulent. This may happen in the separated flow region under the bifurcated shock. Data interpretation is further complicated by arbitrary variations in the finer features. For example, the size of the pressure overshoot associated with the shock bifurcation can be notably different on the same shot for two transducers mounted at opposite sides of the tube. Numerically, the authors have found that it very difficult to resolve the shock/boundary layer interaction to the point where solutions become grid independent. Many grid refinement studies have shown that the wall spacing used here (.015 mm) is nearly sufficient;, however, these studies have also shown that the grid spacing along the tube is also important and that the current solutions would be helped by further grid refinement. Additionally, the starting assumption of an sharply defined, planar contact discontinuity causes waves which interact with the interface to be much more sharp (and often of higher magnitude) than those observed experimentally. An initially deformed interface will be tried in the future.

5. Conclusions

Axisymmetric simulations of the NASA Ames electric-arc driven shock tube have been done which include the wall boundary layer and the computations have been compared with experimental data. These simulations have allowed the wall jet created by the reflected shock/boundary layer interaction to be investigated numerically for the first time. These simulations support earlier analytical and experimental work which indicate that this mechanism can contribute to the reduction of the usable test time by allowing the driver gas to contaminate the stagnation region. Before any driver gas contamination, the wall jet creates nonuniformities in the stagnation region which create complex flow patterns, especially when waves due to non-tailored conditions interact with these non-uniformities.

Acknowledgement

Support for G.J. Wilson was provided by a grant from NASA to Eloret Institute (NCC2-420). Computer time was provided by NAS and by the Central Computing Facility at NASA Ames Research Center. The authors also gratefully acknowledge M. A. Sussman's advice on the TVD scheme and J.O. Gilmore's help in data acquisition.

References

Brossard J, Charpentier N, Bazhenova TV, Fokeev VP, Kalachev AA, Kharitonov AI (1985) Experimental study of shock wave reflection in a narrow channel. In: Bershader D, Hanson RK (eds) Proc. 15th Intl. Symp. on Shock Waves and Shock Tubes, pp 163-169

Bull DC, Edwards DH (1968) An investigation of the reflected shock process in a shock tube. AIAA J. 6:1549-1555

Davies L (1966) The interaction of a reflected shock wave with the boundary layer in a shock tube and its influence on the duration of hot flow in the reflected-shock tunnel. Part I. Aeronautical Research Council-CP-880

Davies L (1967) The interaction of a reflected shock wave with the boundary layer in a shock tube and its influence on the duration of hot flow in the reflected-shock tunnel. Part II. Aeronautical Research Council-CP-881

Davies L, Wilson JL (1969) Influence of reflected shock and boundary-ayer interaction on shock-tube flow. Phys. of Fluids, Supplement I, 12:I-37 - I-43

Kleine H, Lyakhov VN, Gvozdeva LG, Grönig H (1991) Bifurcation of a reflected shock wave in a shock tube. In: Takayama K (ed) Proc. 18th Intl. Symp. on Shock Waves, pp 261-266

Mark H (1958) The interaction of a reflected shock wave with the boundary layer in a shock tube. NACA TM 1418

Matsuo K, Kage K, Kawagoe S (1975) The interaction of a reflected shock wave with the contact region in a shock tube. Bull. of the JSME 18.

Sanderson RJ (1969) Interpretation of pressure measurements behind the reflected shock in a rectangular shock tube. AIAA J. 7:1370-1372

Sharma, SP, Park C (1990) Operating characteristics of a 60- and 10 cm electric arc-driven shock tube - Parts 1 and 2: The driver and driven sections. J. Thermophysics and Heat Transfer 4:259-272

Strehlow RA, Cohen A (1959) Limitations of the reflected shock technique for studying fast chemical reactions and its applications to the observation of relaxation in nitrogen and oxygen. J. Chem. Phys. 30:257-265

Wilson, GJ (1992) Time-dependent quasi-one dimensional simulations of high enthalpy pulse facilities. AIAA Paper-92-5096, AIAA 4th Intl Aerospace Planes Conf., Orlando, FL

Wilson, GJ (1993) Time-dependent simulations of reflected-shock/boundary layer interaction. AIAA Paper-93-0480, AIAA 31st Aerospace Sciences Meeting, Reno, NV.

Yakano, Y (1991) Simulations for reflected shock waves in combustible gas in shock tubes. In: Takayama K (ed) Proc. 18th Intl. Symp. on Shock Waves, pp 869-874

Yee, HC (1989) A class of high-resolution explicit and implicit shock-capturing methods. NASA TM 101088

Numerical Simulation of Viscous Flow in a Super-Orbital Expansion Tube

N. Akman and R.G. Morgan
Department of Mechanical Engineering The University of Queensland

Abstract. An axisymetric Gudonov scheme has been implimented to investigate the flow in a free-piston driven superorbital expansion tube. Equilibrium chemistry is assumed , and laminar viscosity is used to compute boundary layer development. The computations start at secondary diaphragm rupture, and follow the secondary shock as it traverses the test gas. An analytical solution is used for the upstream boundary conditions, which arc held constant. The condition of the test gas up to the start of the unsteady expansion has been computed for a shock speed of 10 km/sec. Comparison with experimental results shows good agreement with the macroscopic flow parameters of shock speed and attenuation.

Key words: Expansion tube, Numerical simulation

1. Introduction

A major potential role for impulse facilities lies in simulating the superorbital flow velocities which are associated with entry into planetary atmospheres. A pilot superorbital expansion tube has been commissioned at The University of Queensland to investigate one means of doing this. Whilst the design and development of such facilities is primarily driven by an analytical understanding of the dominant flow mechanisms involved, detailed numerical analysis is required to completely define their performance and limitations. Furthermore, as the velocity range of these facilities increase, the flow density has to be reduced and this in combination with high Mach numbers causes viscous effects to have a dominant influence on the flow. For this reason, a time dependent, axisymmetric, viscous flow simulation of the above superorbital expansion tube is being developed.

The facility uses a compound two stage driver to produce high shock speeds in the acceleration gas. It consists of four tubular sections separated by diaphragms, with an area ratio of 7.2 between the first and second sections. The conditions in the first helium region are chosen to drive an over tailored shock in the second tube. This creates secondary driver gas capable of driving a stronger shock in the tertiary tube than would be possible by directly coupling the first stage to the third without the intermediate section. The use of helium accelerator gas maximizes the energy addition which occurs across the unsteady expansion.

2. Numerical code

The operating principls of the facility are described in (Anon. 1976). The numerical model consists of four constant area circular tubular sections separated by diaphragms. The first two and the last regions contain Helium whereas the third region contains air. The pressure and temperature in the first Helium region are chosen to achieve the same shock speed, as obtained from the experiments, in the second Helium region.

The flow is assumed to be non-viscous in the first two regions and becomes viscous as it enters the third region, where densities are low and viscous effects have a significant influence on the extent and condition of the test flow. No slip and cold wall boundary conditions are prescribed at the the walls of the expansion and acceleration tube sections.

The code uses Godunov's method to calculate the conditions at the cell boundaries. At each time step, the cell properties are updated using the axisymmetric Navier Stokes equations and

Shock Waves @ Marseille I
Editors: R. Brun, L. Z. Dumitrescu

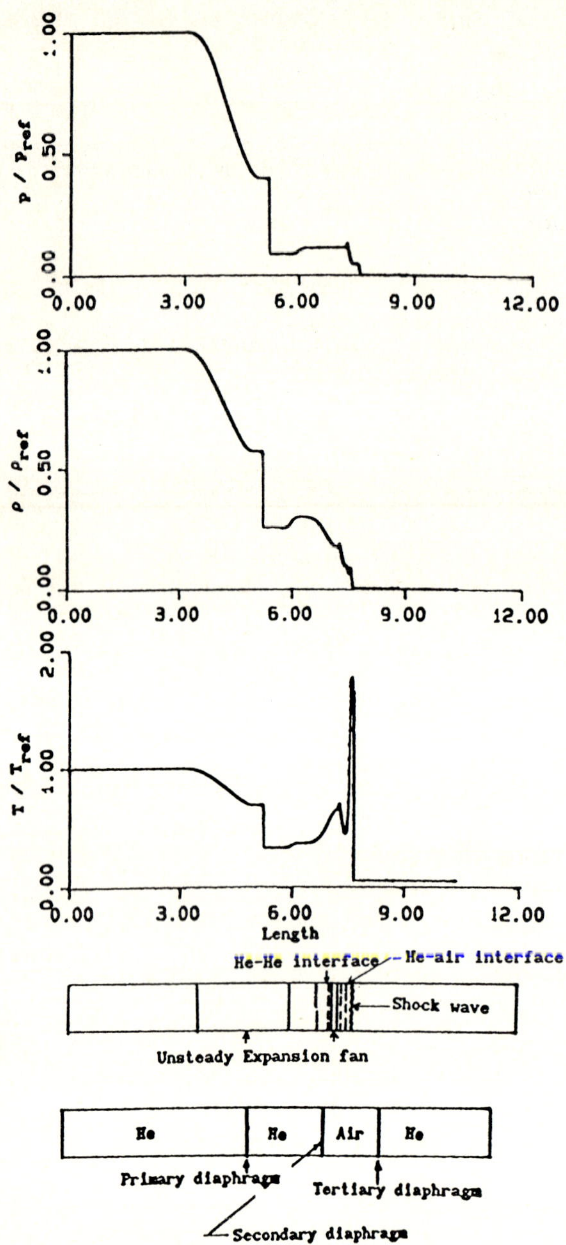

Fig. 1. The centreline axial flow parameter profiles

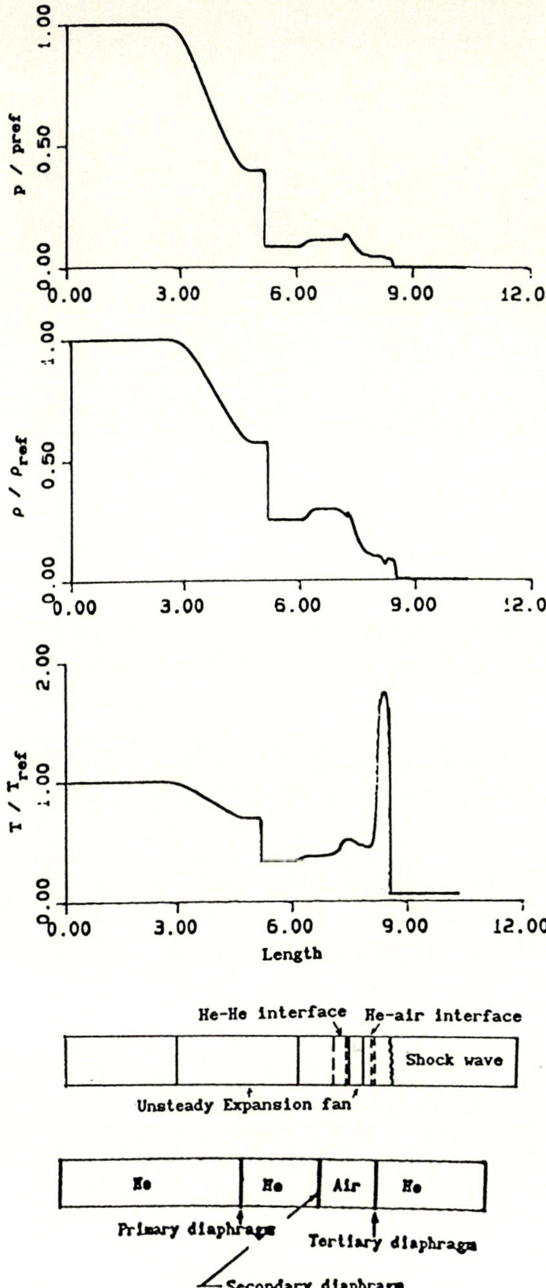

Fig. 2. The centreline axial flow parameter profiles

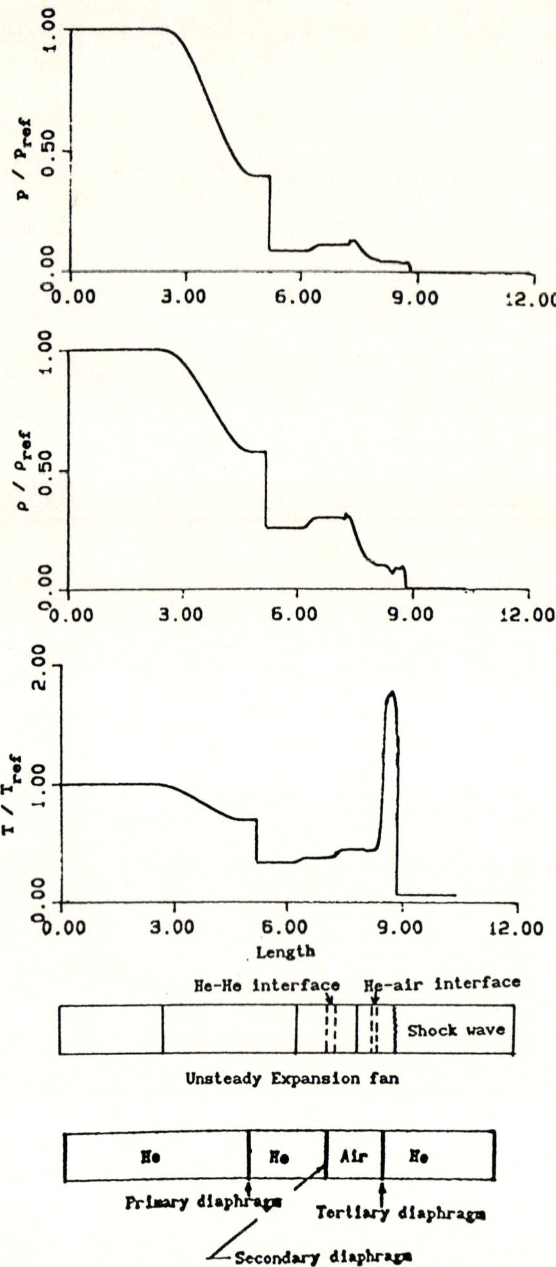

Fig. 3. The centreline axial flow parameter profiles

the concentrations of the species are calculated assuming equilibrium chemistry. The CREK subroutine (Neely and Morgan 1993) is used to calculate species concentrations.The subroutine constructs a Newton-Raphson derivative matrix and solves it by Pivotal Gaussian reduction. An equilibrium solution is obtained by minimizing the Gibb's free energy.

3. Godunov's method

The monotonic first order scheme developed by Godunov (Godunov 1959(?ref.not listed?Ed) for the solution of Lagrangian equations in one dimension was later extended for the Eulerian equations in one- and more dimensions by Godunov, Zabrodin and Prokopov in 1961 (Maurice 1984). In this application, the method is applied to axisymmetric flows with viscosity.

4. Results

The region of interest for the purpose of this study is the propagation of the secondary shock through the test gas and the associated development of high-enthalpy flow. The upstream conditions which create the primary shock wave are not addressed here. A primary driver with cross sectional area equal to that of the driven tubes is postulated, with conditions adjusted to match the measured primary shock speed. The discontinuity in the form of an expansion shock evident in the driver gas, is a problem associated with the Riemann solver at transonic Mach numbers. The downstream boundary conditions of this flow were checked analytically and found to agree with one-dimensional perfect gas theory which applies to the expanding Helium driver gas if the viscous effects are ignored. The numerical anomaly in this region does not influence the conditions of the flow in the test gas. The subsequent development of the test gas is shown in Figs.1 to 3. A clearly defined and expanding core of test gas can be seen between the secondary shock wave and the air-driver gas interface.

In Fig.4 the predicted shock attenuation in this region is compared to the experimentally determined shock locations as indicated by wall-mounted pressure transducers. Excellent agreement is seen. The mechanism for shock attenuation is the growth of the boundary layer on the wall of the shock tube. Because the attenuation agrees well with the experiment it is thought that the code is correctly modelling the boundary layer entrainment of test gas. This is a fundamental step in determining the extent and the condition of the core flow, and provides more precise data than can be obtained by the approximate analytical techniques which are currently being used to categorize the experimental flow parameters.

5. Conclusions

The code has been shown to correctly model the macroscopic features of the flow in the secondary shock region of the superorbital expansion tube. Internal contor plots indicate that a region of uniform core flow exists, but with an axial velocity gradient, which is in agreement with the Mirel's boundary layer analysis. To make full use of the predictive capabilities of the code, it will be extended to cover the acceleration tube section, and to include the effects of tertiary diaphragm inertia. The code has a potential to be an useful tool for both design and operation of expansion tubes.

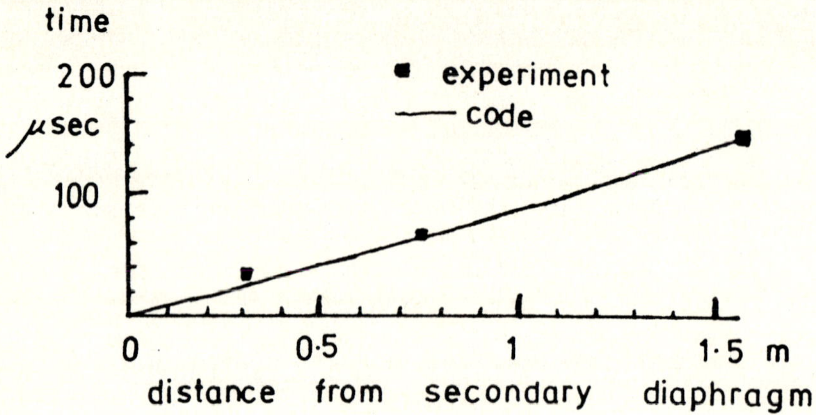

Fig. 4. Secondary shock trajectory

Acknowledgement

This work was supported by the Australian Research Council.

References

Anon. (1976) CREK subroutine, Washington State University
Maurice H (1984) Numerical Metods in Fluid Dynamics, 2nd Rev. ed., Springer-Verlag
Neely AJ, Morgan RG (1993) Hypervelocity aerodynamics in a superorbital expansion tube.
  These Proceedings

Analysis of the Flow Perturbations in a Shock Tube Due to the Curvature of the Diaphragm

V. Daru* and J. P. Damion[†]
*Laboratoire SINUMEF, ENSAM
[†]Laboratoire de Métrologie Dynamique, ENSAM, 151, Boulevard de l'Hôpital, 75013 Paris, France

Abstract.   When using cellophane diaphragms in shock tubes, one can experimentally observe some perturbations superposed to the pressure plateau following the incident and reflected shock. With the help of numerical simulation, based on the solution of the Euler equations, we show that these perturbations are periodic and due to the curvature of the diaphragm separating the high and low pressure chambers. We analyse the development of the waves originating the phenomenon, and we compare the numerical and experimental results which are in excellent agreement.

Key words: Shock tube, Pressure fluctuations, Numerical simulation

1. Introduction

For decades, shock tubes have been used as pressure step generators to characterize pressure transducers on the rise time and natural frequency. To obtain the transfer function of a transducer being given its transient response, it is essential that the incoming step be as pure as possible, and very stable in amplitude. Moreover, if one wishes to characterize the transducer in the low frequency range, the pressure plateau must be as long as possible.

To this end, the Laboratoire de Métrologie Dynamique has built a 23 m long shock tube, having a constant section of 0.2 m. If the stability of the step pressure generated by this tube does not present any problem, the "noise" superposed to the plateau is too important. After making sure that no mechanical phenomenon was involved (dynamic behavior of the transducer or vibrations of the tube), an experimental study has shown that the phenomenon was of pneumatic origin and linked to the operation of the shock tube, and in particular to the curvature of the diaphragm due to the initial pressure difference.

A number of authors have studied the opening process of the diaphragm in a shock tube and the flow perturbations that it causes (Ikui and Kazuyasu 1969, Outa, Tajima and Hayakawa 1973). In our case, this cannot originate the observed perturbations; indeed a measurement of the opening time of the diaphragm (28 μs for a 0.2 m diameter) has shown that it is negligible with respect to the time necessary to the propagation of waves over a significant distance.

2. Numerical simulation

The computations rely on the numerical solution of the Euler equations in an axisymmetrical configuration for a perfect gas. The effects of viscosity are neglected, which is justified by the low pressure ratios and the diameters under consideration. The Euler equations, expressing the conservation of mass, momentum and energy, are written in cylindrical co-ordinates, and a finite volume approach is adopted.

In an uniform mesh defined by constant space increments δr and δz, the system to be solved for each cell Ω_{ij}, is discretized using a Lax-Wendroff type scheme, in a predictor-corrector version involving one predictor for each space direction (Lerat 1982). Here, because of the non-homogeneous term for which we have to know a predicted value, three predictors are computed. It is well known that a centred second-order accurate scheme produces oscillations in the vicinity of discontinuities. It is here crucial to eliminate any spurious oscillation of numerical nature which could hide the

Shock Waves @ Marseille I
Editors: R. Brun, L. Z. Dumitrescu   © Springer-Verlag Berlin Heidelberg 1995

physical phenomena we are studying. We thus have added to the scheme a corrective term aiming to make it Total Variation Diminishing (TVD), similar to the one proposed by Davis (1984), extended independently in each space direction. Although in the case of systems the TVD notion is defined only with respect to the "locally frozen" linear system, because of waves interactions allowing the solution to increase, this correction has given us very good results here.

At the initial state, the pressure discontinuity is supposed to be spherical, and we assign to each cell the mean value of the pressure over the area of the cell. The initial temperature was fixed to 293 K and the initial velocity was zero. We have used a $320(z) \times 60(r)$ cells mesh.

3. Experimental arrangement

The experiments were carried out in closed shock tubes having a constant section whose diameter is 0.10 m, 0.15 m or 0.2 m. A cellophane diaphragm separates the low and high pressure chambers whose lengths are 1 m each. The pressure is measured in the low pressure chamber, on the wall at 0.2 m from the diaphragm position and at the centre of the end wall of the tube, by piezoelectric transducers. The output signal is analogically filtered, then stored on a transients recorder which has a maximum sampling frequency of 2 MHz. The deformation of the diaphragm is measured by a thin stem which crosses the driver chamber and is also used to break the diaphragm. The frequency spectra are computed directly from the experimental results, and after a resampling on the numerical results. The spectral analyses are conducted on the pressure plateau (P_2 on the wall and P_5 at the end) and normalized with respect to the maximum amplitude. For each diameter of the tube the experiments are carried out for low pressure ratios ranging from 1.3 to 2.8, and values of the pitch of the diaphragm from 13 to 47 mm.

4. Results and interpretation

4.1. Comparisons between computation and experiments

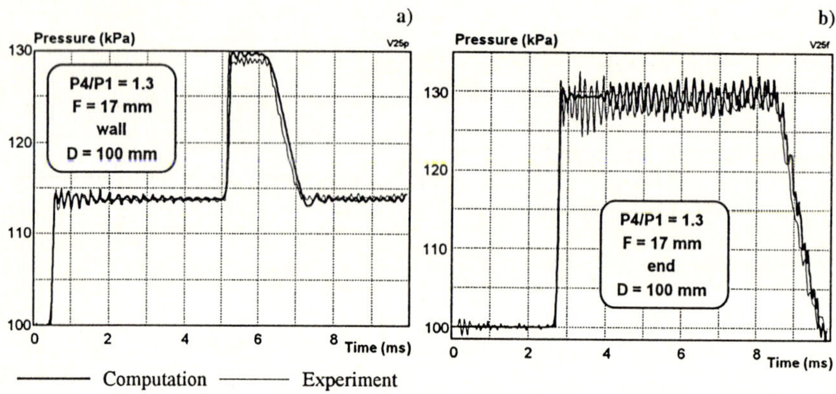

Fig. 1. Comparison between computation and experiment : pressure history, a) wall, b) end

The comparisons are done on the pressure history. Over all the tested cases, we can observe an outstanding agreement between computations and experiments. We can see on Figs.1 and 2 the results obtained on the wall and at the end of the tube, for two tubes of diameters 0.1 m and 0.2 m, the initial pressures being the same (1.3 bar in the high pressure chamber (P_4) and 1 bar in the low pressure chamber (P_1)).

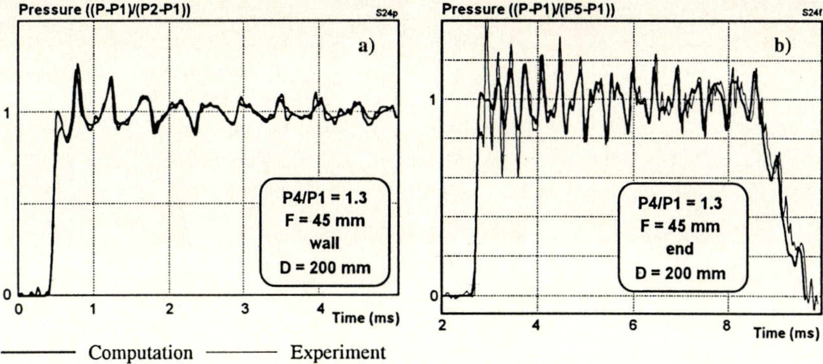

Fig. 2. Comparison between computation and experiment : normalized pressure, a) wall, b) end

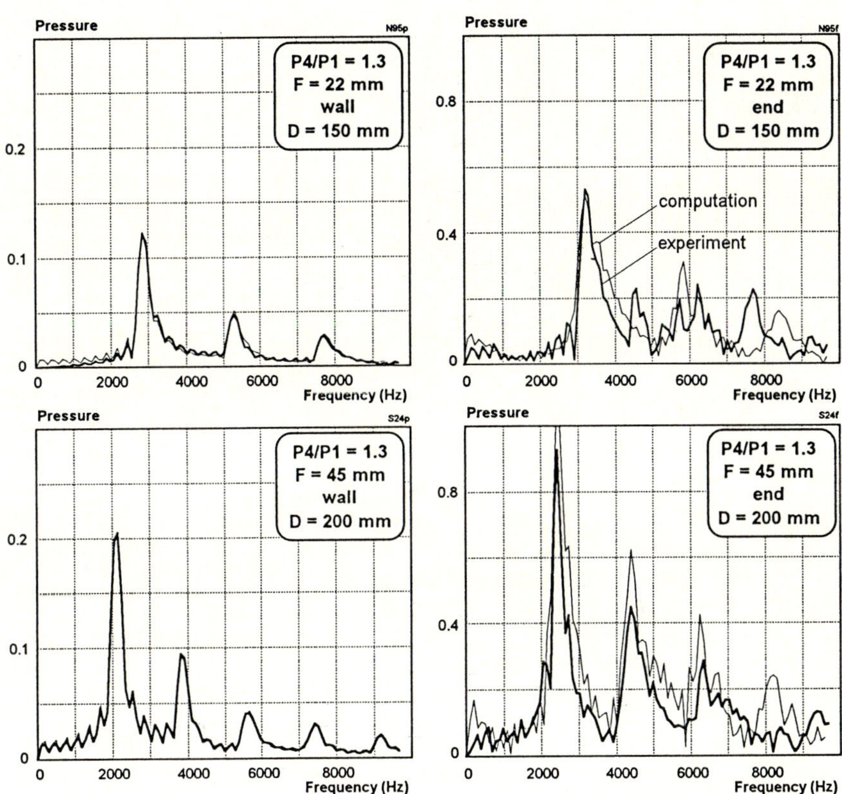

Fig. 3. Frequency spectrum, wall and end, for two diameters of the tube

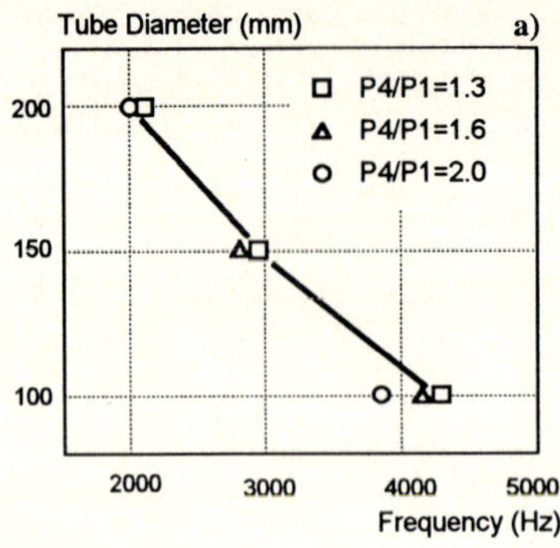

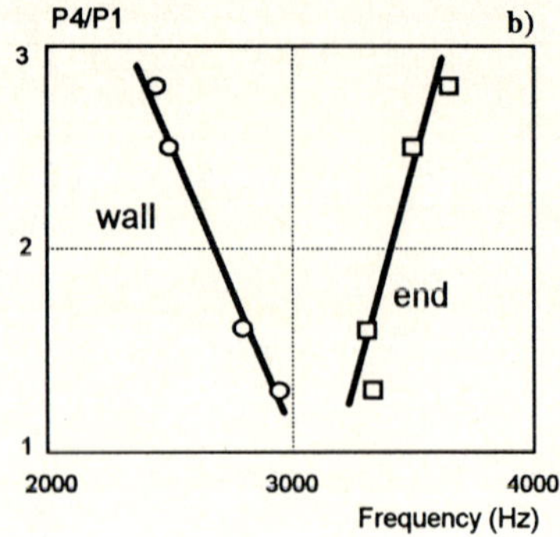

Fig. 4. Main frequency of pressure perturbations, a) vs. tube diameter, b) vs. P_4/P_1; $D = 0.15$ m

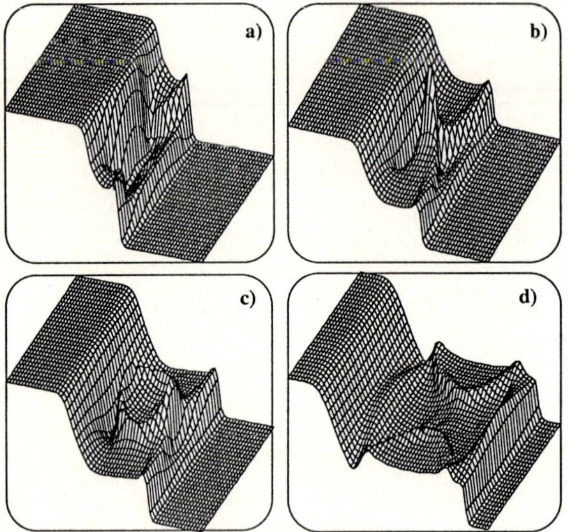

Fig. 5. Pressure field, $D = 0.2$ m, $P_4/P_1 = 2$, $F = 47$ mm; a) $t = 0.2$ ms, 60 iterations, b) $t = 0.3$ ms, 114 it., c) $t = 0.4$ ms, 160 it., d) $t = 0.8$ ms, 280 it.

The perturbations on the pressure after the incident shock (P_2) and the reflected shock (P_5) are of the oscillatory and periodic type. The corresponding spectral analysis (Fig.3) shows very clearly the main frequency of the oscillations, very well represented by the computation. We can make the following observations from these results:
- the main frequency is inversely proportional to the diameter of the tube
- it depends on the initial pressure ratio
- it does not depend on the pitch of the diaphragm F, the amplitude of the oscillations slightly depending on it.

These observations are synthetized on Fig.4, on which are shown the frequency f of the oscillations as a function of the diameter D of the tube for several pressure ratios (Fig.4a), and f as a function of P_4/P_1 (Fig.4b).

Fig. 6. Isobars, $D = 0.2$ m, $P_4/P_1 = 1.3$, $F = 45$ mm; a) $t = 0.2$ ms, 70 it., b) $t = 0.6$ ms, 170 it., c) $t = 2.4$ ms, 700 it.

Fig. 7. Isopycnics, $D = 150$ mm, $P_4/P_1 = 2.8$, $F = 35$ mm, $t = 8.5$ ms, 4000 it.

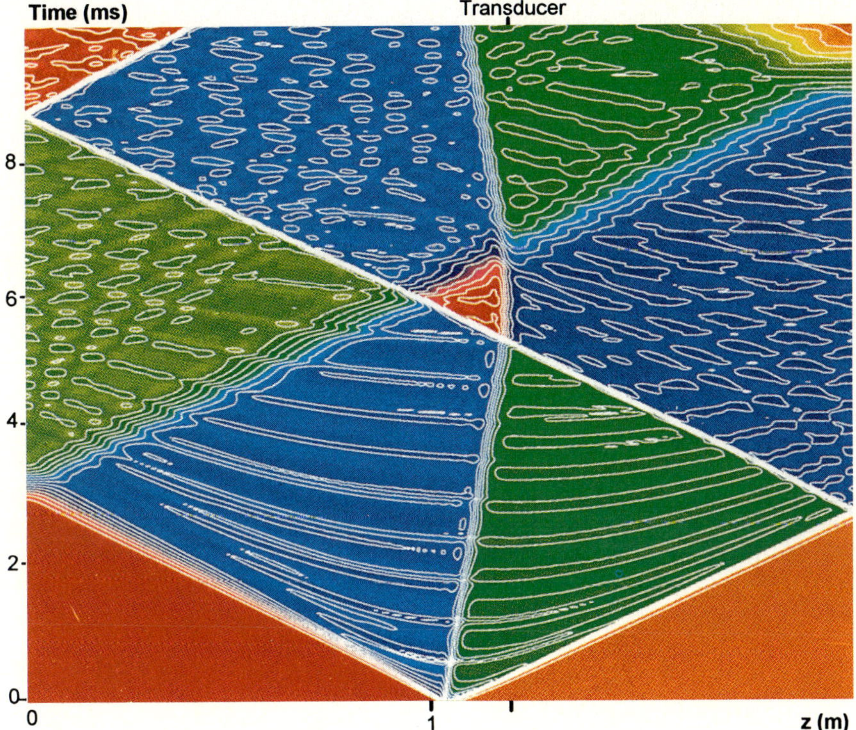

Fig. 8. (z, t) diagram of the density at the wall, $D = 0.2$ m, $P_4/P_1 = 1.3$, $F = 45$ mm

4.2. Interpretation of the phenomenon

By analyzing the numerical results we can explain how the perturbations originate and develop. The pressure field plotted a short time after the breaking of the diaphragm (Figs.5a, 6a) reveal a complex flow pattern, with a Mach reflection due to the oblique configuration of the initial discontinuity on the wall. A transverse shock wave is thus created, which moves toward the axis of the tube, the main shock wave becoming a plane form in a short time as it propagates downstream (Whitham 1974). This transverse wave coalesces at the centre of the tube (Fig.5b), giving rise to a reflection which is a spherical wave expanding toward the wall of the tube (Fig.5c, 6b), where it reflects once again (Fig.5d). This structure reproduces itself when time goes on in a periodic process, moving approximately at the fluid velocity. As it moves, it produces on both sides a "cellular" flow (Fig.6c), bounded at the left by the rarefaction wave and at the right by the shock wave (before they reflect). The periodic oscillations observed on the pressure plateau reflects the fact that the cells move along the transducer. In such complex flows a video film has contributed a lot to the understanding of the phenomenon. After some time, we can notice on Fig.7 the development of a Richtmyer-Meshkov instability due to the crossing of the contact discontinuity by the shock.

A (z,t) diagram of the density (Fig.8) at the wall is also very useful. The perturbations superposed to the uniform states are clearly seen here to be of periodic nature. Moreover, we can observe that the perturbations slow down and that their intensity decreases as they propagate on both sides of the contact discontinuity. It is a complex phenomenon, for which until now, we do not have a clear explanation. It is finally remarkable to observe that the flow keeps a very regular structure while time goes on.

5. Conclusion

In this work we have determined the origin of the spurious oscillations observed on the pressure plateau of a shock tube. The progress which has been made these last ten years in the development of accurate and non oscillatory numerical scheme makes possible the use of computations for such problems, and we have obtained numerical results in outstanding agreement with the experiments. This study shows that using diaphragms having large deformations generates perturbations of the flow with low frequencies in relation to the time of observation. In the case of the characterization of pressure transducers, this restricts greatly the expected performance. The conclusion apply for all the applications where a stable state is needed.

References

Davis SF (1984) TVD finite difference schemes and artificial viscosity. ICASE Report No. 84-20 ICASE NASA Langley Research Center, Hampton VA

Ikui T, Kazuyasu M (1969) Investigations of the aerodynamic characteristics of shock tubes. Bulletin JSME, Vol. 12 No. 52, Part I and II

Lerat A (1981) Sur le calcul des solutions faibles des systèmes hyperboliques de lois de conservation à l'aide de schémas aux différences. Publication ONERA No. 1981-1

Outa E, Tajima K, Hayakawa K (1973) Shock tube flow influenced by diaphragm opening. In: Bershader D, Griffith W (eds) Proc. 9th Intl. Symp. Shock Waves and Shock Tubes, Stanford

Whitham GB (1974) Linear and non linear waves. John Wiley

One-Dimensional Simulation of Free-Piston Shock Tunnel/Expansion Tubes

M. Mitsuda, T. Oda, T. Kurosaka, S. Wakuri and T. Arai

Mechanical Engineering Research Laboratory, Kobe Steel, Ltd., Kobe HYOGO 651-22, Japan

Abstract.   A quasi-one-dimensional simulation for a free-piston shock tunnel/expansion tube was carried out. The governing Euler equations including area change are solved by using the predictor-corrector TVD method. Heat loss, discharge loss, friction loss can be considered. From comparison with data of DLR-HEG, the Stanton number in the compression tube is chosen to be 0.002, the discharge coefficient of the orifice 0.8, the friction coefficient in the shock tube is 0.008. Calculated shock velocities in the shock tube and in the acceleration tube do not agree with experimental data of HEG, TQ. Settling time of nozzle starting predicted by this calculation does not agree with experimental data of T4.

Key words: Free-piston shock tunnel, Expansion tube

1. Introduction

In recent years, hypersonic and high-enthalpy flow research has been performed experimentally by using shock tunnel/expansion tubes with a free-piston driver. The free-piston shock tunnel has better flow quality, longer test time, larger test section than the free-piston expansion tube. Stalker and Hornung (see, e.g. 1969) have worked in the development of free-piston shock tunnels for many years. Recently Hornung (1992) developed the T5 facility at CALTECH, Eitelberg et al. (1992) also developed the HEG facility at DLR. Modification of the G-Range at AEDC is planned by Maus et al. (1992). Paul and Stalker, Neely and Stalker (1991) have developed the free-piston expansion tube TQ at the University of Queensland. Modification of HYPULSE at GASL to free piston driver mode is planned by Tamagno et al. (1990).

Pressure loss and other losses in free-piston shock tunnels have been discussed by Page and Stalker (1983), Hornung (1990), showing their influence on the decrease of shock speed and stagnation pressure. These losses are due to unsteady expansion of compression tube gas during diaphram rupture, interference with the flow at the compression tube end caused by the piston motion, friction and heat losses in the compression tube and shock tube. Maus (1992) reported that the predicted stagnation pressure is 60 % higher than experimental data of T5, but stagnation pressure considering friction (f=0.008) and heat loss (C_h=0.002) agrees with it. Lacey (1990) considered pressure losses at buffer-compression tube interface.

This paper presents the results of a quasi-one-dimensional simulation for free-piston shock tunnels and expansion tubes. The effect of loss factors such as heat loss, friction loss, discharge loss was examined by comparing with the HEG data (Eitelberg 1992). Simulation for the free-piston expansion tube was performed and compared with TQ data (Neely and Stalker 1991). Typical $X-t$ diagrams of the pressure in the compression tube (HEG shot#8), shock tube (HEG shot#13), and expansion tube (TQ) are shown in Figs.1,2,3, respectively.

2. Governing equations and numerical approach

The governing equation is the quasi-one-dimensional Euler equation with area change considering friction loss and heat loss. The perfect gas assumption was adopted. The location of the interface between driver gas and driven gas and/or accelarating gas is monitored at each time step. The gas constant and specific heat ratio are reset at each time step. The governing equation was solved by

Shock Waves @ Marseille I
Editors: R. Brun, L. Z. Dumitrescu

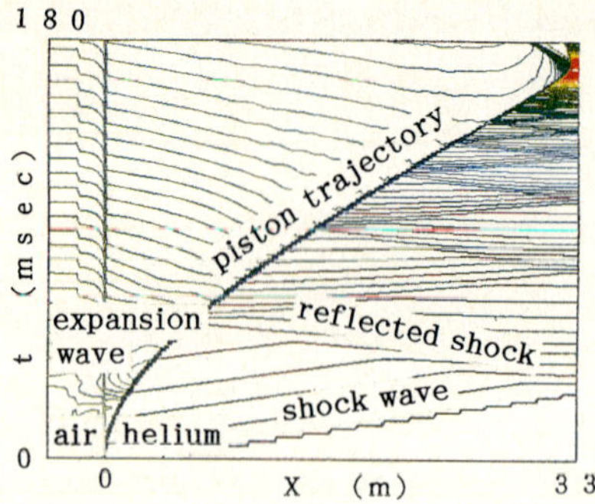

Fig. 1. $X - t$ diagram of pressure in buffer and compression tube: buffer= 44.7 bar, compression tube = 0.639 bar, rupture = 496 bar

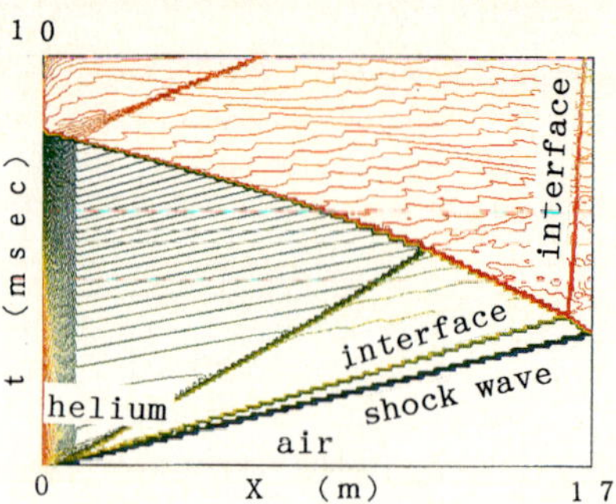

Fig. 2. $X - t$ diagram of pressure in shock tube: buffer = 45.9 bar, compression tube = 0.641 bar, shock tube = 0.33 bar, rupture = 495 bar.

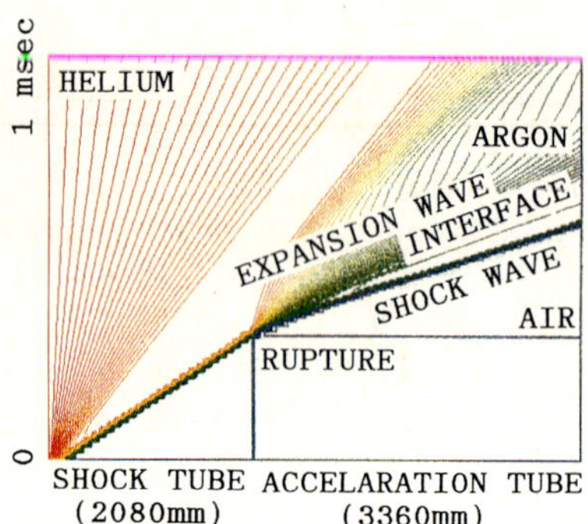

Fig. 3. $X - t$ diagram of pressure in shock tube / acceleration tube: compression tube = 130 KPa, shock tube = 2933 Pa, acceleration tube = 18 Pa

an explicit predictor-corrector TVD scheme (Yee 1989). To evaluate the variables on node (i+1/2), Roe's averaging (1981) was adopted. Courant (CFL) number is unit. The computational domain was divided into three sections, a) buffer and compression tube with free-piston, b) shock tube, c) hypersonic nozzle/expansion tube. The interface boundary condition corresponds to choked flow, though a reflection condition is used when the diaphragm ruptures.

3. Results and discussion

3.1. Compression tube

The buffer volume is 5 m^3, the length and diameter of the compression tube are 33 m and 550 mm, respectively. Operation conditions are the same as at DLR-HEG shot#8. Buffer, compression tube and diaphragm rupture pressures are 44.7 bar, 0.639 bar and 496 bar, respectively. The piston weight is 268 kg. The orifice diameter is 123 mm. Fig.1 shows the $X - t$ diagram for the pressure in the buffer/compression tube. As the piston accelerates, a compression wave travels in helium until diaphram rupture. Expansion waves exists upstream of the piston. Heat loss and discharge loss were considered. Fig.4 shows the pressure at the compression tube end with/without rupture condition without considering heat loss. Assumed discharge coefficient η_d was 1.0, 0.8, 0.6. As η_d decreases peak pressure and holding time increase. Fig.5 shows the effect of heat loss in terms of Stanton number Ch (0, 0.002, 0.004) for η_d=0.8. As Stanton number increases, peak pressure increases, rupture time is earlier. Fig.6 shows the best fitted calculation compared with data (Eiterberg 1992). The discharge coefficient, and the Stanton number are 0.8 and 0.002 in this computation. The predicted pressure (straight line) agrees with experimental data (closed

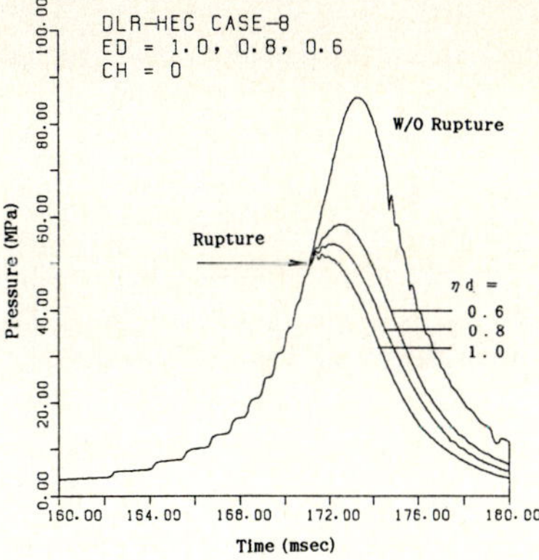

Fig. 4. Pressure at compression tube end: Effect of discharge coefficient η_d=1.0, 0.8, 0.6

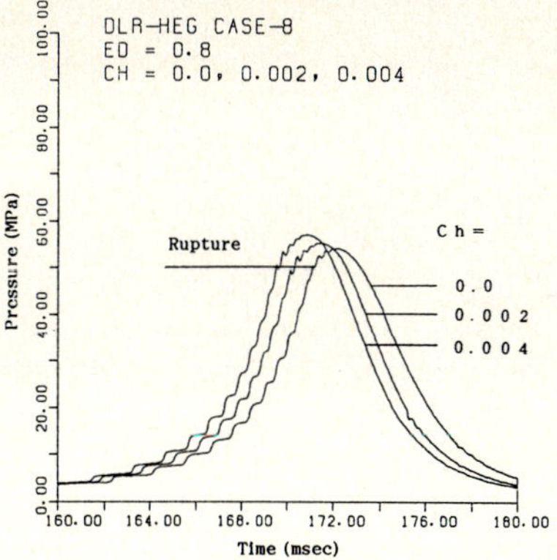

Fig. 5. Pressure at compression tube end: Effect of Stanton number C_h=0, 0.002, 0.004

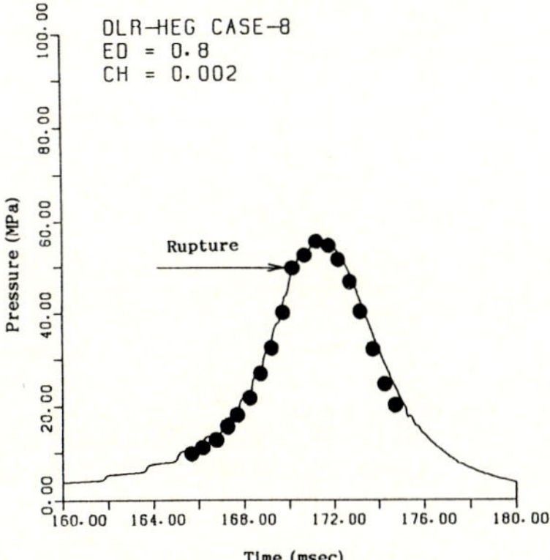

Fig. 6. Pressure at compression tube end: Best fitted result compared with DLR-HEG data, η_d=0.8, C_h=0.002

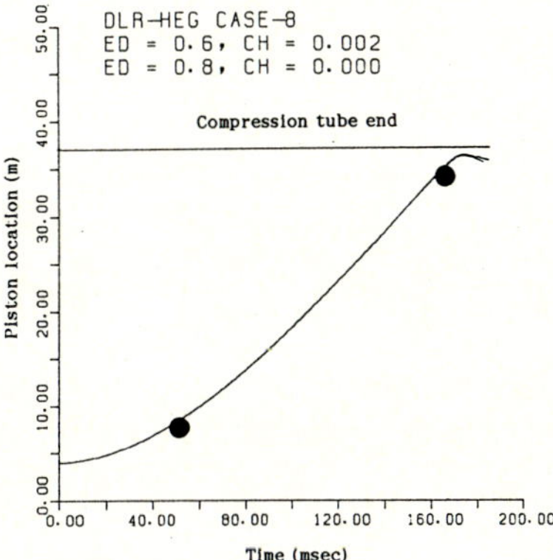

Fig. 7. Piston trajectory compared with DLR-HEG data

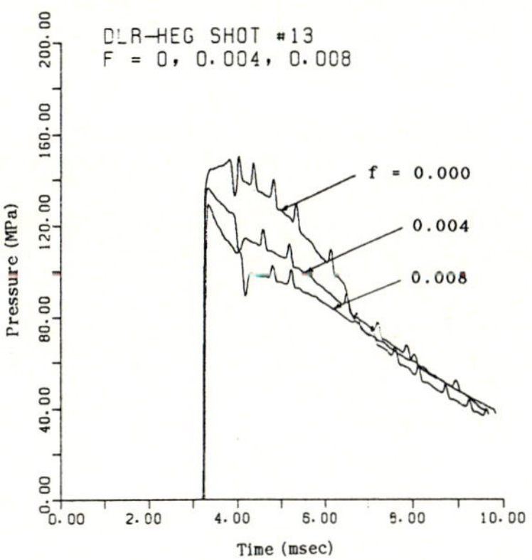

Fig. 8. Stagnation pressure at shock tube end: Effect of friction factor f=0, 0.004, 0.008

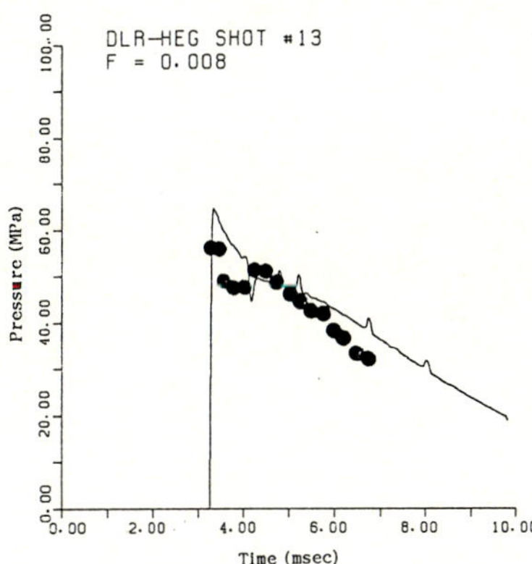

Fig. 9. Stagnation pressure at shock tube end: Best fitted results compared with DLR-HEG data, η_d=0.8, C_h=0.002, f=0.008

symbols). Predicted holding time is 2.2 msec, that of measurement is 2.0 msec. Fig.7 shows the piston trajectory compared with DLR-HEG data, which is not influenced so much by loss parameters.

3.2. Shock tube

The shock tube length/diameter are 17 m/150 mm. Operation conditions are the same as at DLR-HEG shot#13. Buffer, compression tube, shock tube, and diaphragm rupture pressures are 45.9 bar, 0.641 bar, 0.33 bar, and 496 bar, respectively. The piston weight is 269 kg. The orifice diameter is 123 mm. The inlet condition at the orifice, as a function of time, was taken as obtained from compression tube computation. Discharge from the shock tube end was not considered. Fig.2 shows an $X - t$ diagram for the pressure in the shock tube; an overtailored condition is predicted. Fig.8 shows the stagnation pressure at the shock tube end. Assumed friction factor f was 0, 0.004, 0.008. As the friction factor increases, the stagnation pressure behind the reflected shock decreases, and arrival of the incident shock occurs later. Fig.9 shows the predicted stagnation pressure with f=0.008 (straight line) compared with experimental data (closed symbols). Calculated shock speed is 5.1 km/sec, measured one is 4.44 km/sec. The effect of the friction factor on shock speed is negligible.

3.3. Hypersonic nozzle

Length/area ratio of the hypersonic nozzle are 2.187 m and 667, corresponding to T4, University of Queensland (Jacobs 1991). Design Mach number is 8. Assumed stagnation condition is 6010 K, 50 MPa. As inlet condition at the nozzle throat the above stagnation conditions were taken. Inlet conditions are 5008 K, 26.4 MPa, 1419 m/s. Fig.10 shows the Pitot pressure history at the nozzle exit, compared with experimental data. Predicted settling time is 0.8-0.9 msec, on the other hand, that of measurement is 1.2 msec. This difference seems to be caused by viscous effects or multi-dimensonal flow effects.

3.4. Expansion tube

The length/diameter of the shock tube and acceleration tube are 2.08 m/37 mm and 3.36 m/37 mm, respectively. Operation conditions are same as those of Neely and Stalker (1991). Shock tube and accelaration tube pressures are 2933 Pa, 18 Pa. Driver gas condition is constant at 35 MPa, 1800 K. Driver gas, test gas, acceleration gas are helium, argon, and air, respectively. Fig.3 shows an $X - t$ diagram for the pressure in the shock tube and the acceleration tube. No friction loss was considered. Calculated shock velocities in the shock tube and acceleration tube are 6532 m/s and 12640 m/sec, respectively; on the other hand, measured shock velocities are 5346 m/sec, and 9394 m/sec, respectively.

4. Conclusions

A quasi-one-dimensional simulation for free-piston shock tunnel/expansion tubes was performed. It was found that, in comparison with DLR-HEG data, Stanton number and discharge coefficient of compression tube are 0.002, 0.8. Friction factor of shock tube is 0.008. Predicted settling time of the hypersomic nozzle of the T4 facility does not agree with experimental data. Calculation of shock speed of shock tube and acceleration tube did not agree fairly with HEG/TQ data.

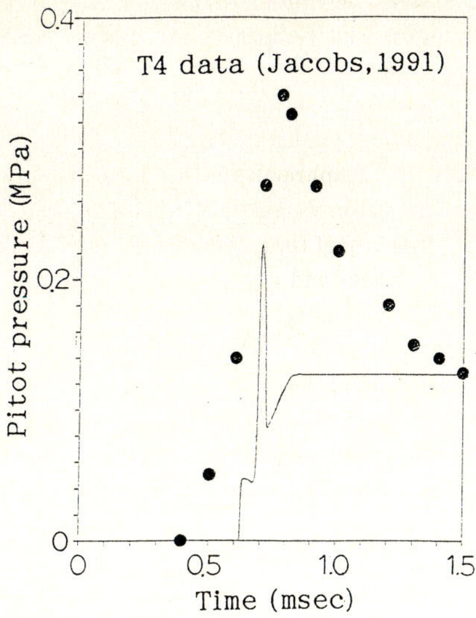

Fig. 10. Predicted Pitot pressure at nozzle exit: comparison with experimental data (T4, Jacobs 1991)

Acknowledgment

We thank specially Mr. John J. Lacey Jr., Dr. Kiyotaka Hayakawa, Dr. Dean F. Long of FluiDýne Engineering Corp., Minneapolis for discussion about the computation of free-piston shock tunnel / expansion tubes.

References

Eitclbcrg G, McIntyrc TJ, Beck WH, Lacey J (1992) The high enthalpy shock tunnel in Gottingen. AIAA Paper 92-3942

Hornung HG, Belanger J (1990) Role and techniques of ground testing for simulation of flows up to orbital speed. AIAA Paper 90-1377

Hornung HG (1992) Performance data of the new free-piston shock tunnel at GALCIT. AIAA Paper 92-3943

Jacobs PA (1991) Transient hypervelocity flow in an axisymmetric nozzle. AIAA Paper 91-0295

Lacey J, Long D (1990) A wave diagram computational method with application to a free-piston shock tube. AIAA Paper 90-1378

Maus J, Laster M, Hornung HG (1992) The G-Range impulse facility a high performance free-piston shock tunnel. AIAA Paper 92-3946

Neely AJ, Stalker RJ (1991) Hypervelocity flows of argon produced in a free piston driven expansion tube. In: Takayama K (ed) Proc. 18th Intl. Symp. on Shock Waves, Sendai, pp 997-1004

Page NW, Stalker RJ (1983) Pressure losses in free-piston driven shock tubes. In: Archer Rd, Milton BE (eds) Proc. XIV Intl. Symp. on Shock Tube and Waves, Sydney, pp 118-125

Paull A, Stalker RJ (1991) Acoustic waves in shock tunnels and expansion tubes. In: Takayama K (ed) Proc. 18th Intl. Symp. on Shock Waves, Sendai, pp 697-704

Roe PL (1981) Approximate Riemann solvers, parameter vectors and difference schemes. J. Comp. Phys. 43:357-372

Stalker RJ, Hornung HG (1969) Two developments with free-piston drivers. In: Glass II (ed) Proc. 7th Intl. Shock Tube Symp., Toronto, pp 242-258

Tamagno J, Bakos R, Pulsonetti M, Erdos J (1990) Hypervelocity real gas capabilities of GASL's expansion tube (HYPULSE) facility. AIAA Paper 90-1390

Yee HC (1989), A class of high-resolution explicit and implicit shock-capturing methods, NASA TM 101088

Author Index (Volumes I - IV)

Contents - Volume II

Part 2: Combustion Kinetics

Part 3: Non-Equilibrium Flow

Part 4: Plasmas, Astrophysics

Contents - Volume III

Part 2: Shocks in Porous Media

Part 3: Shocks in Condensed Matter

Part 4: Industrial Applications and Environment

Part 5: Biological Aspects

Contents - Volume IV

Part 2: Shock Reflection and Diffraction

Part 3: Shock - Interface Interactions

Shock Waves

An International Journal

Editor-in-Chief: I.I. Glass, Downsview, Canada

Managing Editors: R.A. Graham, Albuquerque, USA;
H. Grönig, Aachen, Germany;
A.L. Kuhl, Livermores, USA;
K. Takayama, Sendai, Japan

This journal is aimed at publishing theoretical and experimental results on shock-wave phenomena gases, liquids, solids, and two-phase media from both the fundamental research and the applications points of view. It collates strictly refereed reviews, contributed papers, and short notes in the field, at present spread over numerous publications, in one focused journal.

ISSN 0938-1287

Subscription information 1995:

Vol. 5 (6 issues) DM 368,- (suggested list price)
plus carriage charges:
FRG DM 15,60; other countries DM 39,30

Springer

Springer-Verlag
and the Environment

We at Springer-Verlag firmly believe that an international science publisher has a special obligation to the environment, and our corporate policies consistently reflect this conviction.

We also expect our business partners – paper mills, printers, packaging manufacturers, etc. – to commit themselves to using environmentally friendly materials and production processes.

The paper in this book is made from low- or no-chlorine pulp and is acid free, in conformance with international standards for paper permanency.